Plants That Fight Cancer
Second Edition

Plants That Fight Cancer
Second Edition

Edited by
Spyridon E. Kintzios
Maria G. Barberaki
Evangelia A. Flampouri

CRC Press
Taylor & Francis Group
Boca Raton London New York

CRC Press is an imprint of the
Taylor & Francis Group, an **informa** business

CRC Press
Taylor & Francis Group
6000 Broken Sound Parkway NW, Suite 300
Boca Raton, FL 33487-2742

First issued in paperback 2021

ISBN 13: 978-1-03-209154-9 (pbk)
ISBN 13: 978-1-498-72640-5 (hbk)

Publisher's Note
The publisher has gone to great lengths to ensure the quality of this reprint but points out that some imperfections in the original copies may be apparent.

Library of Congress Cataloging-in-Publication Data

Names: Kintzios, Spyridon E., editor.
Title: Plants that fight cancer / [edited by] Spyridon E. Kintzios, Maria G.
Barberaki, Evangelia A. Flampouri.
Description: Second edition. | Boca Raton, Florida : CRC Press, [2019] |
Includes bibliographical references and index.
Identifiers: LCCN 2019007594| ISBN 9781498726405 (hardback : alk. paper) |
ISBN 9780429056925 (ebook)
Subjects: LCSH: Herbs--Therapeutic use. | Cancer--Treatment. | Medicinal
plants. | Materia medica, Vegetable. | Pharmacognosy.
Classification: LCC RC271.H47 P56 2019 | DDC 616.99/4061--dc23
LC record available at https://lccn.loc.gov/2019007594

Visit the Taylor & Francis Web site at
http://www.taylorandfrancis.com

and the CRC Press Web site at
http://www.crcpress.com

Eleni Grafakou, Vaggos Kintzios, and those who are personally fighting cancer.

Contents

Evangelia Flampouri, Spyridon Kintzios, and Maria Barberaki

Preface

Sixteen years after the publication of the first edition of *Plants that Fight Cancer* remarkable progress has been observed in the research and development of plant-derived anticancer drugs. This is in part reflected on the continuously increasing number of published original research articles and reference books related to the topic of cancer phytotherapy. Indeed, more than 3,000 articles in international, peer-reviewed journals have been published each year of the last six years. More significant, however, is the fact that a series of Phase III clinical trials have been completed during the period, decisively providing evidence of the clinical efficiency of an elite group of plant-derived compounds and extracts. In light of these developments, interest in the investigation of the anticancer properties of yet unexplored plant species has been renewed.

The goal of this book is to provide a review, as thorough and updated as possible, of the state-of-art of cancer treatment and research based on the plant kingdom as the source of both known and novel chemical moieties and mixtures, many of them still under investigation. In accordance with the first edition, we decided not to include plant-based cancer-chemopreventive species and substances, since this is a topic that is regularly and extensively covered in other volumes, contemporary and past. On the contrary, we opted to focus on plant genera and species that are either already used in cancer chemotherapy or have been identified with antitumor and antileukemic properties to a bigger or lesser extent.

This second edition follows the organization of its predecessor with a considerable expansion of content to more than double the volume of information. The first part is dedicated to a review of our current knowledge of cancer, its different types and incidence, the molecular pathways of the disease, and the various treatment approaches, with an emphasis on chemotherapy. The second part is a brief journey in pharmacognosy, with detailed information about each of the 14 different chemical groups of plant secondary metabolites, their use in cancer chemotherapy, and updated information on the biotechnological production of the most representative compounds in clinical practice. The third part of the book comprises six chapters and is dedicated to either plant chemotherapeutical approaches to specific cancer types (e.g. bladder, prostate) or specific groups of plant secondary metabolites with novel and promising properties for cancer treatment (e.g. naphthoquinones, lectins, phenanthridone alkaloids). The fourth part is with no doubt the most investigative, containing analytical information on almost 300 individual plant species with established anticancer properties, either on a clinical or *in vitro* level.

Compared to the first edition, this entire book has been expanded to a two-fold, updated content. The scope of covered genera and species reflects our effort to cover—to the highest possible degree—all plant species with recorded cancer chemotherapeutic potential. Naturally, as with every review and given the dynamic research in this field, it is more than possible that we may have omitted a number of species, especially those reported after the second half of 2017. We would be grateful to readers willing to provide information on said species to be included in the next edition.

The review of algal extracts and metabolites with cytotoxic and antineoplastic activity was omitted from the present edition since we decided to focus exclusively on terrestrial plant species.

The chemical structures of the vast number of different compounds reported for the individual plant species are included in the Appendix, which has also been considerably expanded. In addition, all figures and graphic content of the original version have been entirely redrawn to considerably increased quality.

We would like to express our gratitude to Taylor & Francis Publishers for entrusting us with this challenging project. The personal attention and support provided by Alice Oven, Laura Piedrahita, Randy Brehm, Jennifer Blaise, John Sulzycki, and Jill Jurgensen at Taylor & Francis throughout this demanding, but very rewarding, journey was a decisive element in the timely delivery of the second edition. The completion of this book would not be possible without the valuable assistance of the following students of the Department of Biotechnology at the Agricultural University of Athens: Sophia Nikolaou, Eleni Kapetanou, Antonis Kakalis, George Zisios, Panagiotis-Leonardos Doukas, and Pavlos Boulgaris. We warmly acknowledge their contribution to the compilation and classification of the excessive bibliographical data.

We hope that the book at hand will be a useful guide and stimulate further interest for research in the phytotherapy of cancer among the international scientific and medical community as well as inspire students to follow a career in pharmacognosy.

The Editors

Editors

Professor Spyridon Kintzios is the Rector of the Agricultural University of Athens (AUA) and the founder and Director of the Laboratory of Cell Technology and the AUA Center for Applications of Cell Biological Technologies. He received his PhD in Genetics and Plant Breeding from the Technical University of Munich and has a working experience of over 30 years in biotechnology. He is the author/co-author of 110 peer-reviewed articles in cited international scientific journals, as well as the author or co-author of more than 100 international book chapters and conference presentations. He is also the editor of five international books, a fellow of several scientific societies, and a member of several public, academic, and industry committees in Greece and abroad.

Dr. Maria Barberaki graduated from the Department of Agricultural Biotechnology, Agricultural University of Athens, Greece. Her PhD study was focused on the production of secondary products with medicinal use from *Viscum album* using biotechnological methods. Dr. Barberaki has considerable working experience at international pharmaceutical companies. She was one of the editors of the first edition of *Plants that Fight Cancer*. She currently works in Secondary and Vocational Education.

Dr. Evangelia Flampouri holds a PhD in Agricultural Sciences specializing in mammalian toxicology, a Master of Arts degree in Bioactive Products and Protein Technology, and a Bachelor of Arts degree in Agricultural Biotechnology. She has over ten years of experience in academic and industrial research projects. She has worked as a contracted Associate Professor at the Agricultural University of Athens, teaching Pharmacognosy and Bioactive Products. She is the author/co-author of various peer-reviewed articles in cited international scientific journals.

List of Contributors

Maria Barberaki
Laboratory for Cell Technology
Department of Biotechnology
Agricultural University of Athens
Athens, Greece

Jia-hua Cui
School of Pharmacy
Shanghai Jiao Tong University
Shanghai, China

Evangelia Flampouri
Laboratory for Cell Technology
Department of Biotechnology
Agricultural University of Athens
Athens, Greece

Marios Hadjipavlou
Department of Urology
East Surrey Hospital
Surrey, UK

Azamal Husen
Department of Biology
College of Natural and
    Computational Sciences
University of Gondar
Gondar, Ethiopia

Charlie Khoo
Department of Urology
East Surrey Hospital
Surrey, UK

Spyridon E. Kintzios
Laboratory for Cell Technology
Department of Biotechnology
Agricultural University of Athens
Athens, Greece

Shao-shun Li
School of Pharmacy
Shanghai Jiao Tong University
Shanghai, China

Jerald J. Nair
Research Centre for Plant Growth and
    Development
School of Life Sciences
University of KwaZulu-Natal
KwaZulu-Natal, South Africa

Roman Paduch
Department of Virology and Immunology
Institute of Microbiology and Biotechnology
Maria Curie-Skłodowska University
Lublin, Poland

and

Department of General Ophthalmology
Medical University of Lublin
Lublin, Poland

Yiannis Philippou
Department of Urology
East Surrey Hospital
Surrey, UK

Abhay Rane
Department of Urology
East Surrey Hospital
Surrey, UK

Vasileios Tsekouras
Laboratory for Cell Technology
Department of Biotechnology
Agricultural University of Athens
Athens, Greece

Johannes van Staden
Research Centre for Plant Growth and
    Development
School of Life Sciences
University of KwaZulu-Natal
KwaZulu-Natal, South Africa

CHAPTER 1

Cancer
A Brief Overview of the Disease and Its Treatment

Spyridon E. Kintzios

CONTENTS

In this chapter, our knowledge of cancer incidence is briefly reviewed and a review of major cancer types is presented. A list of most known oncogenes and tumor suppressor genes is given, and representative pathways related with tumorigenesis and/or metastasis are described. The various aspects and strategies of conventional, advanced, and alternative chemotherapy drugs and regimes are presented in detail, including their most common side effects. The process of anticancer drug identification and efficacy evaluation is thoroughly reviewed. Finally, an extensive list of animal and cell culture lines used for primary screening of the *in vitro* antitumor properties of plant-derived and synthetic compounds is provided.

1.1 INCIDENCE AND CAUSES: CANCER IN A NUTSHELL

Everybody thinks it cannot happen to them. And yet, more than eight million people die of cancer every year—almost one-third more than at the time (2004) the first edition of this book was published. As expected, the incidence of the disease is much higher: approximately every second to third citizen of an industrialized country will be stricken with cancer sometime during his life (with almost 25% higher probability in men than in women), and approximately 185 new incidents emerge per 100,000 people annually. In spite of these impressive, if not intimidating data, the global deaths resulting from cancer on a population basis is steadily decreasing, with less than 105 deaths recorded per 100,000 people annually. That said, cancer and the efficacy of its treatment remain an extremely multifaceted story, with aberrant differentiation in incidence and mortality between different affected organs/tissues, countries, and sex. An annual increase of the number of diagnosed cancers should be attributed, at least in part, to the availability of more advanced, accessible, and affordable diagnostic tests. Last, an increasing cancer incidence may represent a kind of trade-off to the continuous extension of life expectancy.

Once considered a mysterious disease, cancer has been eventually revealed to investigators. Disease development begins from a genetic alteration (*mutation*) of a cell within a tissue. This mutation allows the cell to proliferate at a very high rate and to finally form a group of fast reproducing cells with an otherwise normal appearance (*hyperplasia*). Rarely, some hyperplastic cells will mutate again and produce abnormal looking descendants (*dysplasia*). Further mutations of dysplastic cells will eventually lead to the formation of a *tumor*, which can either remain localized at its place of origin, or invade neighboring tissues (*malignant tumor*) and establish new tumors (*metastases*).

Cancer cells have some unique properties that help them compete successfully against normal cells. It can be said that they are masters of survival based on a number of features that allow them to survive and proliferate under limited resources and in adverse microenvironments, conditions that are often artificially created by anticancer therapeutical treatments:

1. As already mentioned, under appropriate conditions cancer cells are capable of dividing almost infinitely. With the exception of skin and hair, normal cells have a limited life span (e.g. human epithelial cells cultured *in vitro* commonly are capable of sustaining division for no more than 50 times—the so-called *Hayflicknumber*). In principle, cancer cells are immortal as far as appropriate conditions are maintained (e.g. *in vitro*). A reason for this is the ability of the cancer cells to overcome or neutralize functional and structural deficits due to age or environmental factors, including radiation and chemotherapy. For example, cancer cells are able to preserve telomere length (which is normally reduced with age) due to particular telomerase activity. They are also able to resist oxidative damage from an excessive concentration of reactive oxygen species (ROS) as a result of the frequently hypoxic tumor microenvironment and/or radiation and chemotherapy. In fact, while an abnormally high ROS production leads to apoptosis of normal cells, it may promote mutagenesis and metastatic potential of cancer cells. In general, avoidance of apoptosis is a key feature of the otherwise genetically defective cancer cells. Another survival strategy is demonstrated by the ability of malignant tissues to stipulate their own angiogenetic processes, thus not only satisfying their requirements in the supply of oxygen and nutrients but also facilitating metastatic invasion in other, distant parts of the body. The so-called angiogenic switch of cancer is partly due to the combined, finely tuned activity of coagulation signaling receptors such as tissue factor (TF), the tissue plasminogen activator (tPA), and urokinase plasminogen activator receptor (uPAR) (also acting as suppressors of angiogenesis inhibitors) and inducers of angiogenesis (e.g. VEGF, FGF2, IL-8, PIGF, FGF-β, PDGF). Cancer cells, especially chemoresistant ones, often opt for glycolytic metabolic

pathways and increased mitochondrial fusion with efficient ATP synthesis and transport. Finally, the surface of the cancer cell is a continuously evolving platform of antigenic presentation, making its recognition by the host's immune system extremely difficult.

2. Normal cells adhere both to one another and to the *extracellular matrix*, an insoluble protein mesh filling the space between them. Cancer cells fail the necessary cell–cell adhesion; in addition, they possess the ability to migrate from the site where they began, invading nearby tissues (e.g. melanoma cells to lungs, colorectal cancer cells to liver, and prostrate cancer cells to bones) and forming masses at distant sites in the body, via the bloodstream. This process is known as *metastasis*. Although metastatic cells are indeed a small percentage of the total of cancer cells (e.g. 10^{-4} or 0.0001), tumors composed of such malignant cells become more and more aggressive over time. Even more problematic is the fact that metastatic tumors are genetically different from the original (primary) ones, therefore increasing the complexity of the therapeutical strategy. It must be emphasized that metastasis is the major contributor (over 90%) to all cancer-associated deaths. It is a very complex, multifaceted process, which in solid tumors is based on the so-called epithelial-to-mesenchymal cell transition (EMT). Under non-pathological conditions, EMT allows the transformation of epithelial cells from highly differentiated, polarized, and organized cells into undifferentiated, mesenchymal-like cells with migratory and invasive properties. A key change in EMT is the reduction of cell-to-cell adhesion by the replacement of E-cadherin by N-cadherin, which leads to the formation of weak cell adhesion between adjacent cells. Also important is the substitution of Type IV collagen by Type I and the abundant synthesis of fibronectin. The metastatic invasive process is further reinforced by the expression, on the cancer cell membrane, of an array of fibrinolytic enzymes but also adhesion molecules (e.g. ICAM-1, VCAM-1, E-selectin) to ensure the establishment of the metastatic cell population at the site of migration.

In a general sense, cancer arises due to specific effects of environmental factors (such as smoking or diet) on a certain genetic background. In the hormonally related cancers like breast and prostate cancer, genetics seem to be a much more powerful factor than lifestyle.

Two gene classes play major roles in triggering cancer. *Proto-oncogenes* encourage such growth, whereas *tumor suppressor* genes inhibit it. The coordinated action of these two gene classes normally prevents cells from uncontrolled proliferation; however, when mutated, oncogenes promote excessive cell division, while inactivated tumor suppressor genes fail to block the division mechanism (Table 1.1). Depending on tissue type some genes, such as *SIRT3*, may act either as oncogenes or tumor suppressors. On a molecular level, control of cell division is maintained by the inhibitory action of various molecules, such as pRB, p15, p16, p21, and p53, on proteins promoting cell division, essentially the complex between cyclins and cyclin-dependent kinases (CDKs). Under normal conditions, deregulation of the cell control mechanism leads to cellular suicide, the so-called *apoptosis* or *programmed cell death*. Cell death may also result from the gradual shortening of *telomeres*, the DNA segments at the ends of chromosomes. However, as already mentioned, most tumor cells manage to preserve telomere length due to the presence of the enzyme telomerase, which is absent in normal cells.

Some oncogenes force cells to overproduce growth factors, such as the *platelet-derived growth factor* and the *transforming growth factor alpha* (sarcomas and gliomas). Alternatively, oncogenes such as the *ras* genes distort parts of the signal cascade within the cell (carcinoma of the colon, pancreas, and lung) or alter the activity of transcription factors in the nucleus. In addition, suppressor factors may be disabled upon infection with viruses (e.g. a human papillomavirus). Development of a tumor requires mutations in a number of these genes, i.e. it is a *step-wise process*. Altered forms of other classes of genes may also participate in the creation of a malignancy, particularly in enabling the emergence of metastatic cancer forms. Although tumorigenesis is a very complex, cancer type-dependent, and fine-tuned process, major biochemical pathways (for example, RAS/RAF/MEK/MAPK and PTEN/PI3K/AKT, see also Section 1.3 of this chapter) have been identified as major pharmacogenomic targets in anticancer research.

More recently, novel genetic components have been identified as key regulators of carcinogenesis and metastasis, acting either as oncogenes or tumor suppressors. The most representative class

Table 1.1   Examples of Genes Related to Cancer Incidence in Humans

Type of Gene	Gene	Encodes for	Cancer Type (In Order of Relevance)
Oncogene	CCND1	Cyclin (cell division regulation)	Multiple myeloma
Oncogene	ErbB3	Human epidermal growth factor receptor 3 (HER-3)	Colorectal
Oncogene	PDGF	Growth factor protein	Glioma
Oncogene	RET	Protein signaling within cells	Thyroid
Oncogene	KRAS	K-Ras protein (cell division regulation)	Lung, ovarian, colon, pancreatic
Oncogene	HPC1	Component of the interferon-regulated 2-5A system	Prostate
Oncogene	N-ras	N-Ras protein (cell division regulation)	Leukemia
Oncogene	C-MYC	bHLH transcription factor	Leukemia, breast, stomach, lung
Oncogene	N-MYC	MYC gene with distinct N-termini	Neuroblastoma, glioblastoma
Oncogene	BCL2	Anti-apoptotic outer mitochondrial membrane protein	Breast, head, neck
Oncogene	FADD	Apoptosis-mediating protein	Breast, ovary, bladder
Oncogene	MDM2	Nuclear-localized E3 ubiquitin ligase	Sarcomas
Oncogene	BCR	GTPase activating protein	Leukemia
Oncogene	BRAF	Nuclear transporter	Lung, colon, multiple myeloma
Tumor suppressor	TP53	p53 (cell division suppression)	Various
Tumor suppressor	CDKN2	p16(INK4a), p14(ARF) (cell division suppression)	Melanoma
Tumor suppressor	CDK4 (RB)	Ser/Thr protein kinase	Retinoblastoma, bone, bladder, small cell lung, breast, skin
Tumor suppressor	BRCA1	DNA-repairing protein	Breast, ovarian, prostate, pancreatic
Tumor suppressor	BRCA2	DNA-repairing protein	Breast, ovarian, prostate, pancreatic, acute myeloid leukemia
Tumor suppressor	APC	APC protein (cell division suppression)	Colon, stomach
Tumor suppressor	MSH2, MSH6, MLH1	DNA-repairing protein	Colon
Tumor suppressor	DPC4 (SMAD4)	Nuclear transporter	Pancreas
Tumor suppressor	VHL	Component of the VCB-CUL2 complex, which regulates the hypoxia-inducible factor 2-alpha (HIF-2α)	Kidney
Tumor suppressor	PIK3R1	Phosphatidylinositol 3-kinase (PI3K) (regulation of cell division and migration)	Uterine, glioblastoma, colon, ovary, breast
Tumor suppressor	PTEN	Cell division suppressor	Breast, lung, bladder, prostate, head, neck
Tumor suppressor	CHEK2	Checkpoint kinase 2 (CHK2)	Breast, ovarian, prostate
Tumor suppressor	PALB2	Transporter of the BRCA2 protein	Breast, ovarian
Tumor suppressor	STK11 (LKB1)	Serine/threonine kinase 11	Digestive tract, breast, other organs

(Continued)

Table 1.1 (Continued)   Examples of Genes Related to Cancer Incidence in Humans

Type of Gene	Gene	Encodes for	Cancer Type (In Order of Relevance)
Tumor suppressor	UVRAG	Beclin1-PI(3)KC3 complex, a promoter of cellular autophagy	Oral
Other	CDH1	E-cadherin	Breast, ovarian, prostate
Other	CTLA4	Immune checkpoint (inhibitory protein to T cells)	Breast, melanoma
Other	PRSS1 (mutation)	Trypsinogen	Pancreas
Other	FH	Components of citric acid metabolism	Kidney
Other	NMP22	Nuclear matrix proteins	Bladder
Other	GSTP1 (hypermethylation)	Glutathione transferase	Prostate
Other	BRIP1	Double-strand break repairing protein	Breast, ovarian
Other	ATM	DNA-repairing protein	Breast
Other	SIRT3	Sirtuin 3, an antioxidant protein in mitochondria	Cervical, bladder, breast, liver, leukemia
Other	FLT1 (VEGFR)	Vascular endothelial growth factor receptor (VEGFR) proteins	Colorectal
Other	Chromosome 3 (deletions)		Lung
Other	Microsatellite instability (MSI)		Colon
Other	11q13 amplicon (long arm of chromosome 11)	Amplified genes CCND1, FADD, UVRAG	Oral, breast, ovary, bladder, pancreas, melanoma, esophagus

of such agents is the so-called *microRNAs (MiRNAs or miRs, e.g. miR-21)*, which are small (~22 bp long) noncoding RNAs able to regulate gene expression at the post-transcriptional level. Another class includes *long noncoding RNAs (lncRNAs)* which are longer than 200 bp but lack an open reading frame. The colon cancer associated transcripts 1 and (*CCAT1* and *CCAT2*) are characteristic examples of lcnRNAs overexpressed in various types of cancer and are directly associated with a high metastatic potential.

The environmental part of the causes of cancer comprises an extremely diverse group of factors that may act as *carcinogens*, either by mutating genes or by promoting abnormal cell proliferation. Most of these agents have been identified through epidemiological studies, although the exact nature of their activity on a biological level remains obscure. These factors include chemical substances (such as tobacco, asbestos, industrial waste, and pesticides), diet (saturated fat, red meat, overweight), ionizing radiation, and pathogens (such as the Epstein–Barr virus, the hepatitis B or C virus, papillomaviruses, and *Helicobacter pylori*). However, in order for environmental factors to have a significant effect, one must be exposed to them for relatively long.

Cancer may also arise, or worsen, as a result of psychological stress. For example, a large-scale study in Israel demonstrated that survival rates declined for patients having lost at least one child in war.

1.2 CLASSIFICATION OF CANCER TYPES

There are several ways to classify cancer. A general classification relates to the tissue type where a tumor emerges. For example, sarcomas are cancers of connective tissues, gliomas are cancers of the nonneuronal brain cells, and carcinomas (the most common cancer forms) originate in epithelial cells.

From a purely epidemiological point of view, the top four cancer diseases in order of decreasing *incidence* are lung, breast, colorectal, and prostate. This classification is useful for revealing priority targets for current anticancer research; at the same time, however, it may be misleading in the assessment *of the degree of impact* of the disease. For example, although lung and breast cancer are ranked first and second, respectively, in terms of incidence and cumulative risk of death, prostate cancer is a far less life-threatening disease. Historical trends also indicate a largely variated degree of the impact of disease on patient survival, either increasing (breast, colorectal, esophageal, stomach) or decreasing (lung). It is rather obvious that anticancer research, including the identification of novel plant-derived compounds, could focus on the treatment of those cancer types that are less susceptible to conventional and/or currently available therapeutic options. Examples of such 'recalcitrant' cancers include both major cancer types (e.g. lung) but also liver, pancreatic, and bladder.

In the following, a classification of major cancer diseases is given according to the currently estimated five-year survival rate of the affected patient.

1.2.1 Cancers with Less Than 20% Five-Year Survival Rate

1. *Pancreatic cancer* is associated with smoking, increasing age, consumption of fats, race, and pancreatic diseases, including diabetes. Diagnosis usually lags behind metastasis. This is due both to the challenging identification of localized tumors within the pancreatic tissue and the very aggressive nature of this particular cancer type. Young (under age 44) female patients may have a slightly better survival prediction (up to 30%, compared to a sex average of 14%). For older patients the outlook is much worse (less than 10%) irrespective of sex. CA 19-9, a blood tumor biomarker, can be used to monitor the patient's response to therapy.

1.2.2 Cancers with Five-Year Survival Rates (At All Stages) Between 20 and 50%

1. *Lung cancer* is associated with exposure to environmental toxins like cigarette smoke and various chemicals and has an incidence higher than 17%. It can be distinguished in two types, *small cell* (rapidly spreading) and *non-small cell* disease, the latter of which represents about 85% of all lung cancers. With a percentage of terminally affected patients higher than 26%, it is one of the less curable cancer diseases. It should be noted, however, that according to more recent studies, younger patients (aged under 44) may have a slightly better survival prediction. On the contrary, patients older than 75 years have on average a very low survival expectancy (less than 10%). On the other hand, people with non-small cell lung cancer (NSCLC) at early stages of the disease have a 45–49% rate of five-year survival.

1.2.3 Cancers with Five-Year Survival Rates Between 50 and 80%

1. *Kidney (renal) cancer* is associated with heredity, sex (males), smoking, and obesity. Wilms tumor refers to renal cancer in children under four years of age. If treated at an early stage, survival rate may exceed 75–80%, a considerable increase over the last years.
2. *Non-Hodgkin's lymphoma* is associated with dysfunctions of the immune system, including many different types of disease and corresponding grades (i.e. speed of progress). Survival rates for NHL have been improved over the last 15 years and can be higher than 90% in children.
3. *Uterine (cervical and endometrial cancer) cancer* is associated with hormonal treatment (such as estrogen replacement therapy), race, sexual activity, and pregnancy history. It can be efficiently predicted by the Pap test (named after its inventor, the Greek-American physician George Papanikolaou). Advanced stage (IV) uterine cancer is associated with much lower survival rates.
4. *Leukemia* is distinguished in acute lymphocytic (common among children), acute myelogenous, and chronic lymphocytic leukemia. All cases are characterized by an abnormal proliferation of immature white blood cells produced in the lymphatic system (mainly in the bone marrow). The disease

is associated with genetic abnormalities, viral infections, and exposure to environmental toxins or radiation. Although the survival rate for acute myelogenous leukemia (AML) has increased over time, it still varies among different subtypes of the disease as well as age (with older patients having a lower survival outlook).

5. *Bladder cancer* is the fourth most common cancer in men and is associated with race (predominantly Caucasian), age, smoking, heredity, and exposure to environmental toxins. In this cancer type, tumors originate from abnormally proliferating cells lining the urinary bladder.

1.2.4 Cancers with Five-Year Survival Rates Higher than 80%

1. *Prostate cancer* is associated with increasing age (>50), obesity, and race (African American). The incidence of the disease is high (about 15%) but its mortality low to very low (especially with advanced age). The disease can be efficiently detected at an early stage by using the prostate-specific antigen (PSA) blood test; however, the reliability of the test has been frequently challenged, particularly in view of its limited usefulness for patients demonstrating a low-to-medium, 'gray zone' PSA concentration (4–10 ng/ml). Recently, prostate cancer screening based on DNA methylation profiling in urine has emerged as a promising alternative to PSA testing.

2. *Colorectal cancer* is associated with heredity, obesity, polyps, and infections of the gastrointestinal tract. Prevention of metastases in the liver is crucial, and treatment at early stages is associated with a high probability of survival. The disease is presumably associated with elevated concentrations of the *cancer embryonic antigen* (CEA), although CEA screening is used mainly for following-up the patients' response to chemotherapy.

3. *Ovarian cancer* is associated with increasing age and heredity, especially as far as mutations in the *BRCA1* or *BRCA2* genes are concerned. It should be mentioned that gene mutations account only for about 5% of all ovarian cancer cases. A record of hormonal replacement therapy as well as a diet rich in fats may play a role. The survival rate for ovarian cancer has been improved over the last 15 years. Advanced stages (III–IV) of the disease are usually associated with much lower survival rates. Ovarian cancer at late stages in the majority of patients can be detected by the elevated concentration of CA-125, a blood tumor biomarker.

4. *Breast cancer* is associated with increasing age, heredity (especially as far as mutations in the *BRCA1* or *BRCA2* genes are concerned), sexual activity, obesity, and pregnancy history. Although the incidence of the disease is high (>24%), survival rates have been remarkably increased, even for advanced stages (III) of the disease. The disease can be efficiently detected at an early stage by self-examination and mammography. In addition, the disease is presumably associated with elevated concentrations of CA15-3. Breast cancer is also classified according to the existence of estrogen receptors (ER-positive), progesterone receptors (PR-negative), human epidermal growth factor 2 receptors (HER2-positive), or none at all (*triple negative*), the last one being the least amenable to targeted treatment.

5. *Skin cancer (basal cell skin cancer, squamous cell skin cancer, melanoma)* is mainly associated with prolonged exposure to the sun, race (fair skin color), age (>50), and suppression of the immune system. Detection and treatment at an early stage is extremely crucial, usually leading to a very high probability of curable disease. Patient survival rates decline considerably for Stage III and IV melanoma, which affects deeper layers of the skin and has a higher metastatic potential.

Most cancers are currently increasing in incidence. However, growth in the major pharmacologically treated cancers, namely breast, colorectal, lung, ovarian, and prostate cancer, is driven by shifting demographics rather than any underlying increase in the risk of developing the disease. Breast cancer is the most prevalent cancer today, followed by cancer of the prostate, colon/rectum, lung, and ovaries respectively. Unsurprisingly, given that cancer is a disease driven by imperfections in DNA replication, the risk of developing most cancers increases with increasing age. For some hormonally driven female cancers, the risk of developing the disease increases rapidly around the time of the menopause.

Diagnosis rates are consequently very high, at over 95% of the prevalent population diagnosed for prostate cancer, and over 99% for breast, colorectal, lung, and ovarian cancers. The stage of the patient's cancer at diagnosis varies highly with each individual cancer with survival times associated with the disease falling rapidly with increasing stage of diagnosis.

1.3 THERAPY

1.3.1 Conventional Cancer Treatments

Conventional cancer treatments include surgery, radiation, and chemotherapy.

Surgery is used for the excision of a tumor. It is the earliest therapy established for cancer and the most widely used. Its disadvantages include the possible (and often unavoidable) damage of healthy tissues or organs (such as lymph nodes) and the inability to remove metastasized cancer cells or tumors not visible to the surgeons. In addition, surgery can activate further proliferation of 'latent' small tumors, the so-called 'pet-cancers'. That said, modern, sophisticated surgical techniques are quite advanced and allow for a far more precise removal of the cancerous tissue region than in the (not distant) past.

Radiation (x-rays, gamma rays) of a cancerous tumor, thus causing cancer cell death or apoptosis, preserves the anatomical structures surrounding the tumor and also destroys non-visible cancer cells. However, it cannot kill metastasized cancer cells. Radiation treatment presents some side effects (such as neurotoxicity in children), but patients usually recover faster than from surgery. Additional side effects include weakening of the immune system and replacement of damaged healthy tissue by connecting tissue.

Surgery and radiation, often combined, are indispensable elements of the first line treatment of cancer. Chemotherapy, which is directly related to the topic of this book, is used complementary (*adjuvant therapy*) to the first two types of treatment, especially in order to prevent and/or treat distant metastatic events.

Chemotherapy is based on the systemic administration of anticancer drugs that travel throughout the body via the blood circulatory system. In essence, chemotherapy aims to wipe out all cancerous colonies within the patient's body, including metastasized cancer cells. As already mentioned, the majority of the most common cancers are not curable with chemotherapy alone while surgery and/or radiation suffice in the treatment of cancers at early stages (e.g. colon cancer up to stage II). Chemotherapy has many side effects, such as nausea, anemia, weakening of the immune system, myelosuppression (reduction of bone marrow activity and reduced production of red and white blood cells), diarrhea, vomiting, and hair loss, though their occurrence depends both on the compound used and the individual patient tolerance. Finally, cancer cells may develop resistance to chemotherapeutic drugs.

Drugs in adjunct therapy do not attack the tumor directly, but instead treat side effects and tolerance problems associated with the use of chemotherapy. For example, anti-emetics such as *lorazepam*, *ondansetron*, or *granisetron* reduce levels of nausea associated with some chemotherapies. This improves compliance rates and enables patients to tolerate higher doses of chemotherapy than would normally be the case. Similarly, some drugs such as *epoetin alpha* target deficiencies in red blood cell counts that often result from the use of chemotherapy and enable normal physical function to be restored to some degree.

Many different compounds are currently used (often in combination). Chemotherapy is the most rapidly developing field of cancer treatment, with new drugs being constantly tested and screened. It is no wonder that almost a thousand new cancer drug trials are currently in process in the USA alone. Moreover, new and recent drugs are not based on historically predominant cytotoxic properties but target cancer-specific pathways. In this way, a more selective and efficient antitumor treatment with less side effects activity may be achieved. Exemplary, main target pathways in cancer development and progression are displayed in Figure 1.1. Cancer-associated pathways are related to the expression of genes listed in Table 1.1.

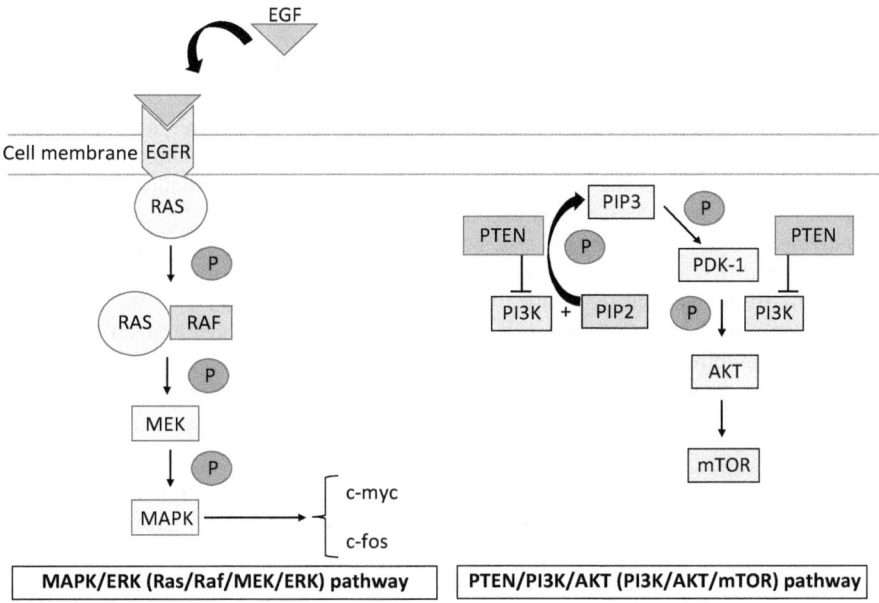

Figure 1.1   Schematic and simplified outline of representative pathways related with tumorigenesis and/or metastasis.

MAPK/ERK (Ras/Raf/MEK/ERK) pathway: The binding of activators to the EFG receptor triggers a cascade of phosphorylation events leading to activation of MAPK and subsequent cell division through c-myc, c-fos activation. EFG = epidermal growth factor, EFGR = epidermal growth factor receptor, MEK = mitogen-activated protein kinase kinase, MAPK (ERK) = mitogen-activated protein kinase (extracellular signal-regulated kinase), RAS = Ras GTPase, RAF = Raf serine/threonine protein kinase, ⓟ = phosphorylation.

PTEN/PI3K/AKT pathway: Loss of PTEN function/PI3K activation results in accumulation of PIP3 and the consequent downstream activation of PDK1 and AKT. AKT in turn suppresses apoptotic processes and promotes cell proliferation, resistance to hypoxia, and migration. AKT = protein kinase B, mTOR = mechanistic target of rapamycin protein kinase, PDK-1, 3-phosphoinositide-dependent protein kinase-1, PI3K = phosphatidylinositol 3 kinase, PIP3 = phosphatidylinositol-3,4,5-triphosphate, PTEN = phosphatase and tensin homologue deleted on chromosome 10.

Conventional chemotherapy is based on *cytotoxic agents* able to kill or impede the growth of tumor cells, at the same time having a relative lower toxicity on normal, usually non-rapidly dividing cells. Mechanisms of cytotoxicity vary widely among different compounds and can be synthetic or plant-derived (the topic of this book). According to a different classification, cytotoxic chemotherapy agents can be discriminated as cell-cycle specific or not, the latter targeting less rapidly growing cancer cells. Quite often, a combination of drugs belonging to different classes is used, provided that every individual drug in the combination is singly effective for a short time and that optimal doses are different among individual drugs. In this way, adjustment of doses is feasible. Cytotoxic chemotherapeutic drugs are classified in ten general groups:

1. *Platinum-based agents* are among the older and most frequently used antineoplastic drugs. They act by crosslinking DNA, thus blocking its repair and replication.
2. *Antimetabolites* act as non-functional analogues of essential metabolites in the cell, thus blocking physiological functions of the tumor.

3. *Alkylating agents* chemically bond with DNA through alkyl groups, thus disrupting gene structure and function, or with proteins, thus inhibiting enzymes.
4. *Topoisomerase inhibitors* inhibit DNA replication in rapidly dividing cells, as is the case with tumors.
5. *Plant alkaloids*, which are nitrogen-containing plant secondary metabolites, also inhibit tumor cell division by blocking microtubule *polymerization*, an essential step for chromosome detachment during mitosis (formation of the mitotic spindle). However, novel plant alkaloids act through other mechanisms as well, which will be analyzed further in this book.
6. *(Plant) taxanes* are diterpenes (hydrocarbon-based secondary metabolites) originally derived from plants of the genus *Taxus*. They inhibit cell division by blocking microtubule *depolymerization* (e.g. they act similarly, but not exactly identically to alkaloids).
7. *Antibiotics* are derived from diverse groups of microorganisms or synthesized and block DNA replication and protein synthesis.
8. *Anthracyclins* are a subgroup of antibiotics, associated with considerable toxic side effects on the heart and bone marrow.

Some representative cytotoxic chemotherapeutic agents along with their most frequent side effects are listed in Table 1.2. The classes of cytotoxic chemotherapy drugs used against major cancer types are listed in Table 1.3.

The success of chemotherapy depends on the treated type of cancer. It can have curative effects on some less common cancers, like Burkitt lymphoma, Wilms tumor, teratomas, and lymphoblastic leukemia. A less satisfactory, though life-prolonging effect is observed on myloblastic leukemia, multiple myeloma, ovarian, prostate, and cervical and breast cancer. Much poorer results must be expected against bronchial, lung, stomach, colorectal, pancreatic, kidney, bladder, brain, glandular, and skin cancer, as well as against bone sarcomas.

Use of pharmacological therapy for cancer varies by both geographic area and tumor type. Lung cancer patients are most likely to be treated with drugs, with around 99% of them being treated with drugs at the first line treatment stage. Prostate cancer patients are least likely to be treated with drugs, with only around 42% of them being treated with drugs at the first line treatment stage.

For those cancers which manifest themselves as a solid tumor mass, the most efficient way to treat them is to surgically resect or remove the tumor mass, since this reduces the tumor's ability both to grow and metastasize to distant sites around the body. If a tumor can be wholly resected, there are theoretically no real advantages in administering drug treatment, since surgery has essentially removed the tumor's ability to grow and spread. For early stage I and II tumors, which are usually golf ball sized and wholly resectable, drug therapy is therefore infrequently used. At stages III and IV, the tumor has usually grown to such a size and/or has spread around the body to such an extent that it is not wholly resectable. For example, rectal tumors at stage III have usually impinged upon the pelvis, which reduces the ability of the surgeon to wholly remove the tumor. In these cases, drug therapy is used either to reduce the size of the tumor before resection, or else 'mop up' stray cancer cells. Drug therapy therefore features prominently for tumors diagnosed at stages III and IV, together with those cancers that have recurred following initial first line treatment and/or metastasized to distant areas around the body.

1.3.2 Advanced Cancer Treatments

The majority of new, advanced, and more sophisticated (compared to cytotoxic chemotherapy) cancer treatments are based on approaches each targeting one or more of the known tumorigenic pathways. This enables a drastic increase in the selective treatment of the disease, although side effects may still present a challenge to routine application of these novel chemotherapy regimens.

Table 1.2 Some of the Cytotoxic Compounds Currently Used in Cancer Chemotherapy

Class	Compound	Cell-Cycle Specific	Common Side Effects at Class Level
Platinum-based	Cisplatin Carboplatin Oxaliplatin	+	Hypersensitivity reactions, nephrotoxicity, neurotoxicity, myelosuppression
Antimetabolites	Azathioprin Capecitabine Cytosine arabinoside 5-fluorouracile (5-FU) Gemcitabine 6-mercaptopurine 6-thioguanine Hydroxyurea Methotrexate Pemetrexed Tegafur/uracil	+	Myelosuppression, diarrhea, stomatitis, cardiotoxicity
Alkylating agents	Bendamustine Busulfan Chlorambucile Cyclophosphamide Dacarbazine Ifosfamide Melphalan hydrochloride Lomustine Thiotepa Mechlorethamine hydrochloride Nitrosureas: Lomustine Carmustine Streptozocin	–	Myelosuppression, vomiting, hepatotoxicity, nephrotoxicity, hypersensitivity reactions
Topoisomerase inhibitors	Amsacrine Irinotecan Topotecan	+	Myelosuppression, hair loss, diarrhea
Plant alkaloids	Etoposide Teniposide Vinblastine Vincristine Vindesine Vinflunine Vinorelbine	+	Neurotoxity
Taxanes	Cabazitaxel Docetaxel Paclitaxel	+	Hypersensitivity reactions, hair loss, diarrhea
Antibiotics	Actinomycin-D Bleomycin Plicamycin	–	Myelosuppression, hair loss, vomiting, hepatotoxicity, skin rash
Anthracyclines	Aclarubicin chlorhydrate Daunorubicin Doxorubicin hydrochloride Epirubicin Idarubicine Mytomycin Mitozantrone Rubidazone		Myelosuppression, vomiting, cardiotoxicity

Table 1.3   Cytotoxic Chemotherapy Drug Groups Used Against Major Cancer Types

	Cancer Type					
Class	Lung	Breast	Colorectal	Prostate	Ovarian	Bladder
Platinum-based	++		+		++	+
Antimetabolites	++	+	++		++	+
Alkylating agents		+			+	
Topoisomerase inhibitors				+		
Plant alkaloids	++	+			+	+
Taxanes		+		++		
Antibiotics						
Anthracyclines						++

++ main class used, + secondary class, often (but not always) used in combination with main class

1.3.2.1 *Protein Kinase Inhibitors and Other Enzymes*

These enzymes target tyrosine and serine/threonine kinases, i.e. protein phosphorylating enzymes either on the cell membrane (receptors such as EGFR or VEGFR, see also Table 1.1) or into the cytoplasm (signal transducers such as RAF and mTOR, see also 3.1). The tyrosine kinase inhibitor imatinib (*Gleevec*) is a historical game-changer in the treatment of chronic myeloid leukemia. Some other enzymes have proteolytic and fibrinolytic properties.

1.3.2.2 *Immunotherapy*

Infectious agents entering the body are encountered by the immune system. They bear distinct molecules called antigens, which are the target of antigen-presenting cells, such as macrophages, that roam the body and fragment antigens into antigenic peptides. These, in turn, are joined to the major histocompatibility complex (MHC) molecules which are displayed on the cell surface. Macrophages bearing different MHC–peptide combinations activate specific T-lymphocytes, which divide and secrete lymphokines. Lymphocines activate B-lymphocytes, which can also recognize free-floating antigens in a molecule-specific manner. Activated B-cells divide and secrete antibodies, which can bind to antigens and neutralize them in various ways.

Lymphocytes are produced in primary lymphoid organs: the thymus (T cells) and the bone marrow (B cells). They are further processed in the secondary lymphoid organs, such as the lymph nodes, spleen, and tonsils, before entering the bloodstream.

In an ideal situation, cancer cells would constitute a target of the patient host immune system. To single out cancer cells, an immunotherapy must be able to distinguish them from normal cells. During the last years, monoclonal antibodies have revealed a large array of antigens that exist on human cancer cells. Many of them are related to abnormal proteins resulting from genetic mutations which turn normal cells into cancer cells. However, cancer cells can elude attack by lymphocytes even if they bear distinctive antigens, due to the absence of proper costimulatory molecules, such as B7, or the employment of immunosuppression mechanisms. The ultimate goal of cancer immunotherapy research is the production of an effective vaccine. This may include whole cancer cells, tumor peptides or DNA molecules, other proteins, or viruses. The idea of a vaccine is an old one indeed: in 1892, William B. Coley at the Memorial Hospital in New York treated cancer patients with killed bacteria in order to elicit a tumor-killing immunoresponse.

The immunotherapy of cancer can be roughly classified in five categories:

1. *Custom-designed monoclonal antibodies (MAbs)* are clonally derived from hybridized B cells (*hybridomas*) genetically engineered to target specific molecular elements of a tumorigenic pathway. They are currently the most frequently used type of advanced anticancer therapy in this category.
2. *Vaccines and cytokines* involve the general stimulation of the immune system and the production of cytokines, such as intereferons, tumor necrosis factor (TNF), interleucins (IL-2, IL-12), and GM-CSF. Such is the case with the use of attenuated tuberculosis (*Mycobacterium bovis*) vaccine, which is used as a first line treatment against bladder cancer. In the few cases where cancer is caused by a known microorganism (e.g. the human papilloma virus) vaccination against the causative agent is a powerful prophylactic measure. Cytokines can also be administered to the patient. However, they represent a less than preferable option due to their considerable toxicity.
3. *Passive* involves the use of 'humanized' mice-derived monoclonal antibodies (MAbs) bearing a toxic agent (such as a radioactive isotope or a chemotherapeutic drug).
4. *Vaccine-like* are made on the basis of human antitumor antibodies.
5. *Adoptive* involves lymphocytes from the patient himself.

A number of targeted therapies based on MAbs have been practically employed in clinical cancer treatment during the last 15 years with varying degrees of success, spectacular in some cases. A representative example is trastumuzab, which targets the human epidermal growth factor receptor 2 (HER2/neu) and is extensively used against breast cancer. MAbs are also used as standard antiangiogenic agents (see below 3.2.3).

Apart from plant-derived compounds, several other agents can stimulate the immune system in a more or less anti-tumor specific manner. Some of the most prominent substances and/or organisms are presented in Table 1.4 (adapted from Koehnlechner 1987).

Table 1.4   Prominent Substances and/or Organisms

Substances	They Activate
Bordetella pertussis Bacillus Calmette–Guerin (BCG) (tuberculosis bacterium a.d. Rind.)[a] **Escherichia coli** Vitamin A	Macrophages
Corynobacterium parvum[b] *C. granulosum* *Bordetella pertussis* **Escherichia coli** Vitamin A[d]	B-lymphocytes
Bordetella pertussis BCG (tuberculosis bacterium a.d. Rind.)[a] **Escherichia coli** Vitamin A[d] Poly-adenosin-poly-urakil Saponine Levamisol[c] Lentinan Diptheriotoxin Thymus factors	T-lymphocytes

[a] In combination with radiotherapy can cause a 40% reduction of leukemia incidence in mice. Has been reported to prolong life expectancy in bladder cancer (systemic use) and leukemia patients who received conventional treatment.
[b] Has been used for the treatment of melanomas, lung, and breast cancer.
[c] A former anti-worm veterinarian drug, levamisol has displayed slight post-operative immunostimulatory and survival-increasing properties in patients suffering from bronchial, lung, and intestinal cancer.
[d] Has been used for the treatment of various skin cancers.

Other compounds include trace elements (selenium, zinc, lithium), haemocyanin, propionibacteria, etc.

1.3.2.3 Angiogenesis Inhibitors

A promising therapeutic strategy focuses on blocking tumor angiogenesis, i.e. the inhibition of the growth of new blood vessels in tumors. Such drugs not only have performed impressively in experimental animal models but also offer an alternative means of tackling multidrug-resistant tumors that have proved intractable to conventional chemotherapy. The link between angiogenesis and tumor progression was first established by Judah Folkman of Boston Children's Hospital. His observations led to the notion of an 'angiogenic switch', a complex process by which a tumor mass expands and overtakes the rate of internal apoptosis by developing blood vessels, thereby changing into an angiogenic phenotype. Drugs that target blood vessel growth should have minimal side effects, even after prolonged treatments. The ready accessibility of the vasculature to drugs and the reliance of potentially hundreds of tumor cells on one capillary add to the benefits of such therapies, which however are limited to the subfraction of tumor capillaries expressing the immature angiogenic phenotype. Another problem is the heterogeneity of the vasculature within tumors. Many approaches for inhibiting angiogenesis are still in development, with several antiangiogenic drugs in clinical trial. At this point it is worth mentioning that the angiogenesis-inhibitor *squalamine* is based on dogfish shark liver. Shark cartilage has been sold as an alternative treatment for cancer since the early 1990s when a book entitled *Sharks Don't Get Cancer* by William Lance was published. It suggested that a protein in shark's cartilage kept the fish from getting cancer by blocking the development of small blood vessels that cancer cells need to survive and grow. The idea spawned a market for shark cartilage supplements that is estimated to be worth $50 million a year. Researchers have since discovered that sharks do get cancer but they have a lower rate of the disease than other fish and humans. Danish researchers tested the treatment on 17 women with advanced breast cancer that had not responded to other treatments. The patients took 24 shark cartilage capsules a day for three months, but the disease still progressed in 15 and one developed cancer of the brain. The Danish results support earlier research that found powdered shark cartilage did not prevent tumor growth in 60 patients with an advanced cancer.

Modern angiogenesis inhibitors are based on MAbs; Bevacizumab (*Avastin*), targeting the varscular epithelial growth factor (VEGF), is prominently used for the treatment of metastatic colorectal and breast cancer.

1.3.2.4 Hormones

Hormones are substances that interfere with other chemotherapeutic agents by regulating the endocrine system. They find specific application against carcinomas of the breast, prostate, and endometrium. This is no wonder, since hormones contribute to the initiation and progress of these, and possibly other, cancer types. Hormones are generally classified into steroidal (estrogens and similar compounds acting intracellularly) and non-steroidal (protein-based, active on membrane receptors).

Breast cancer represents the best established paradigm in hormone-based cancer chemotherapy. At least half of the patients are positive for estrogen-responsive genes and associated receptors in the cancer cells, the activation of said receptors by estradiol (ER) or progesterone (PgR) resulting in upstream positive regulation of receptor-expressing tumors. Antagonistic inhibition of the target receptors would logically lead to reduced tumor growth. This is basically a simplified description of the therapeutic principle, because different receptor subtypes exist with different sensitivities to estrogens and antagonists as well as intra-receptor allosteric relationships. Obviously, the decision to follow a hormone-based therapy will be based on the identification of estrogen-responsive genes in the individual patient. Immunohistochemical tests to determine the expression levels of ER and

PgR are now standard tools of a patient-tailored process known as *predictive, preventive, personalized, and participatory medicine (P4 medicine)*. It is worth mentioning that regular administration of some anti-estrogens, such as tamoxifen, to healthy, ER-positive, high-risk individuals is used for preventing breast cancer. As expected, the response rate to hormone therapy is lower if patients are checked negative for either ER or PgR.

Apart from anti-estrogens, inhibitors of the enzymes responsible for the final stages of conversion of endocrine precursors to estradiol are also used in advanced hormone-based therapy. For example, aromatase inhibitors (AIs) block the p450 aromatase and drastically reduce estradiol concentration in blood.

In the case of prostate cancer, a successfully tested approach is the use of so-called anti-androgens, i.e. agents acting as agonists or antagonists either (i) to the human luteinizing hormone-releasing hormone (LHRH) which is an upstream regulator of testosterone synthesis produced by gonadotropic cells in the anterior pituitary gland or (ii) to the androgen receptor.

Some representative advanced chemotherapy agents along with their most frequent side effects are listed in Table 1.5, while their group-level specific activities against selected cancer types are listed in Table 1.6.

Table 1.5   Representative Advanced Chemotherapy Agents and Their Most Frequent Side Effects

Class	Compound	Target	Common Side Effects At Class Level
Tyrosine kinase inhibitors and other enzymes	Afatinib Erlotinib Gefitinib Imatinib Lapatinib Temsirolimus Vemurafenib	EGFR /HER2 EGFR EGFR PDGFR/ BCR-ABL EGFR /HER2 mTOR BRAF	Myelosuppression, rash, diarrhea
Immunotherapeutic agents *MAbs* *Cytokines* *Immunomodulators*	Bevacizumab Cetuximab Ipilimumab Trastuzumab Interleukin-2 Interferon-α Bacillus Calmette–Guerin (BCG)	VEGFR EGFR CTLA4 HER2 Immune system stimulation Immune system stimulation	Thromboembolism, rash, diarrhea, cardiotoxicity, skin lesions Cardiotoxicity Fatigue, neutropenia Minimal
Angiogenesis inhibitors *MAbs* *Tyrosine kinase inhibitors*	Bevacizumab Axitinib Nintedanib Pazopanib Regorafenib Sunitinib	VEGFR VEGFR VEGFR VEGFR VEGFR VEGFR	Thromboembolism, leukopenia, proteinuria Myelosuppression, thromboembolism, rash, diahhrea
Hormones *Estrogens* *Anti-estrogens* *Progestins* *Aromatase inhibitors* *Anti-androgens* *Adreno/ glucocorticoids* *Luteinising hormone release hormone (LHRH) analogues*	Diethylstilboestrol Tamoxifen Raloxifen Fulvestrant Megestrol acetate Medroxyprogesterone acetate Abiraterone acetate Anastrazole Exemestane Letrozole Bicalutamide Enzalutamide Prednisolone Goserelin Triptorelin	Anti-androgenic ER ER ER PgR, anti-androgenic PgR, anti-androgenic Aromatase AR GCR LHRH receptor	Fluid retension, hypertension, thromboembolism Thromboembolism, hot flashes, endometrial cancers Fluid retention, nausea, hypertension, thromboembolism Osteoporosis, arthritis, fluid retention, hypertension Breast pain, gynecomastia, hot flushes, sexual dysfunction Fluid retention, osteoporosis, hypertension, mood swings Fluid retention, hot flashes, hypertension, mood swings

Table 1.6 Advanced Chemotherapy Drug Classes Used Against Major Cancer Types

Class	Cancer Type									
	Lung	Breast	Colorectal	Chronic Myelogenic Leukemia	Prostate	Renal	Lymphoma	Pancreas	Melanoma	Bladder
Tyrosine kinase inhibitors and other enzymes	+	+		++		+	+	+	++	
Immunotherapeutic agents	+	+	+	++	+	+	++		++	++
Angiogenesis inhibitors	+	+	+			++		+		
Hormones		++			++					

++ main class used, + secondary class, often (but not always) used in combination with main class

1.3.3 Other Advanced Therapies

Advanced cancer therapies also include the use of tissue-specific cytotoxic agents. For example, novel mutagenic cytotoxins (interleukin 13 (IL-13)) have been developed against brain tumors, which do not interact with receptors of the normal tissue but only with brain gliomas.

1.3.4 Alternative Cancer Treatments

These include diverse, mostly controversial methods for treating cancer while avoiding the debilitating effects of conventional methods. The alternative treatment of cancer will probably gain in significance in the future, since it has been estimated that roughly half of all cancer patients currently turn to alternative medicine. The most prominent alternative cancer treatments include:

1. The delivery of *antineoplastons*, peptides considered to inhibit tumor growth and first identified by Stanislaw Burzynski in blood and urine. According to the Food and Drug Administration (FDA) the drug can be applied only in experimental trials monitored by the agency and only on patients who have exhausted conventional therapies. However, the therapy has found a significant amount of political support, while attracting wide publicity.
2. *Hydrazine sulfate*, a compound reversing cachexia of cancer patients, thus improving survival.
3. Various *herbal extracts*, some of which are dealt with in this book.

1.4 FROM SOURCE TO PATIENT: TESTING THE EFFICIENCY OF A CANDIDATE ANTICANCER DRUG

Drug development is a very expensive and risky business. On average, a new drug takes 12 years from discovery to reach the market, costing approximately one billion $US. Considerable efforts have been made by public organizations and private companies to expedite the processes of drug discovery and development, by expanding on promising results from preliminary *in vitro* screening tests. The United States National Cancer Institute (NCI) has set forward exemplary strategies for the discovery and development of novel natural anticancer agents. Over the past 40 years, the NCI has been involved with the preclinical and/or clinical evaluation of the overwhelming majority of compounds under consideration for the treatment of cancer. During this period, more than 400,000 chemicals, both synthetic and natural, have been screened for antitumor activity.

Plant materials under consideration for efficacy testing are usually composed of complex mixtures of different compounds with different solubilities in aqueous culture media. Furthermore, inert additives may also be included. These properties render it necessary to search for appropriate testing conditions. In the past, model systems with either high complexity (animals, organ cultures) or low molecular organization (subcellular fractions, organ and cell homogenates) were used for evaluating the mechanism of action of phytopharmaceuticals. The last decade, however, has seen an enormous trend towards isolated cellular systems, primary cells in cultures and cell lines. In particular, the combination of different *in vitro* assay systems may not only enhance the capacity to screen for active compounds, but may also lead to better conclusions about possible mechanisms and therapeutic effects.

1.4.1 Preclinical Tests

Preclinical tests usually comprise evaluating the cytotoxicity of a candidate antitumor agent *in vitro*, i.e. on cells cultured on a specific nutrient medium under controlled conditions. Certain neoplastic animal cell lines have been repeatedly used for this purpose. Alternatively, animal systems bearing certain types of cancer have been used. For example, materials entering the NCI drug discovery program from 1960 to 1982 were first tested using the L1210 and P388 mouse leukaemia models. Most of the drugs discovered during that period, and currently available for cancer therapy, are effective predominantly against rapidly proliferating tumors, such as leukemias and lymphomas, but, with some notable exceptions such as paclitaxel, show little useful activity against the slow-growing adult solid tumors, such as lung, colon, prostatic, pancreatic, and brain tumors. In the last decade, three-dimensional (3D) cell cultures, such as hepatic spheroid cultures, are becoming increasingly popular intoxicology assays since they resemble the *in vivo* tissue conditions in a more realistic way.

A more efficient, disease-oriented screening strategy should employ multiple disease-specific (e.g. tumor type-specific) models and should permit the detection of either broad-spectrum or disease-specific activity. The use of multiple *in vivo* animal models for such a screen is not practical, given the scope of requirements for adequate screening capacity and specific tumor-type representation. The availability of a wide variety of human tumor cell lines representing many different forms of human cancer, however, offered a suitable basis for development of a disease-oriented *in vitro* primary screen during 1985 to 1990. The screen developed by NCI currently comprises 60 cell lines derived from nine cancer types, organized into subpanels representing leukaemia, lung, colon, central nervous system, melanoma, ovarian, renal, prostate, and breast cancer. A protein-staining procedure using sulforhodamine B (SRB) is used as the method of choice for determining cellular growth and viability in the screen. Other, more sophisticated methods are referred to in the literature. In addition, cell lines used in the *in vitro* screen can be analyzed for their content of molecular targets, such as p-glycoprotein, p53, Ras, and BCL2. Each successful test of a compound in the full screen generates 60 dose-response curves, which are printed in the NCI screening data report as a series of composites comprising the tumor-type subpanels, plus a composite comprising the entire panel. Data for any cell lines failing quality control criteria are eliminated from further analysis and are deleted from the screening report. The *in vitro* human cancer line screen has found widespread application in the classification of compounds according to their chemical structure and/or their mechanism of action. Valuable information can be obtained by determining the degree of similarity of profiles generated on the same or different compounds.

A different approach for *in vitro* screening of cancer therapeutic agents is realized by the use of animal models, mainly mice and rats. There are three basic directions in constructing animal tumor/metastasis models:

- Tumor transplantation into syngeneic (allograft) models (from the same species) or xenograft models (e.g. from human to animal).
- Transgenic mice bearing specific oncogenes or tumor suppressor genes. In the latter case, the contribution of tumor suppressor genes to spontaneous tumor development can be studied by the development of whole-body knockout models where the function of the gene of interest is lost (knockout or null allele). This can be achieved by using site-specific recombinases (SSRs) or more recent targeted genome editing tools like CRISPR/Cas9 and TALENs. It is also possible to over-express oncogenes, for example c-Myc (Eμ-myc mice).
- Chemical induction of tumors on wild or mutated animals, e.g. treating knockout mice with a chemical carcinogen.

In spite of the attractiveness of animal models, in particular their ease of use and accelerated cancer progression (months vs years in humans), tumorigenesis in mice and rats is a less complex and, in many cases, genetically different process than in humans. This in turn makes the interpretation of animal-based chemotherapy screening less reliable than desired.

Some of the most commonly used cell and animal culture lines used for primary screening are listed in Table 1.7.

Table 1.7 Animal and Cell Culture Lines Used for Primary Screening the *In Vitro* Antitumor Properties of Plant-Derived and Synthetic Compounds

Tissue/Organ Specification	Cell/Animal Line	Human Origin	Target Pathway/Effect
Breast	MCF-7 vec	+	p53-dependent pathway, insulin-like growth factor binding proteins (IGFBP) BP-2; BP-4; BP-5, estrogen receptors
	MCF-7 HER2	+	
	MCF-7	+	
	MDA-MB-453	+	Downregulation of Bcl-2 protein expression and upregulation of
	MDA-MB-231	+	Bax expression
			Fibroblast growth factor (FGF) expression
			Mitochondria-related apoptosis, cell cycle arrest
			Cell cycle arrest
			Cell cycle arrest
			c-MYC expression
			EGFR-mediated cell survival
			Cell cycle arrest
			PI3K/Akt/NF-κB pathway
			p53-dependent pathway
			transforming growth factor alpha (TGF alpha) expression
	HER2/neu-overexpressing transgenic mice		Apoptosis
			Apoptosis
	Athymic BALB/c mice transplanted with KPL-1 cells		Inhibition of cell proliferation
			Patient-derived xenografts (PDXs)
	Progestin-dependent tumor in Sprague–Dawley rats		Transfected with the mouse mammary tumor virus long terminal repeat (MMTV-LTR)
	Immunosuppressed NSG mice		regulatory element and overexpressing Her2/neu
	MMTV-ErbB2 mice		
Colorectal	HT-29	+	c-MYC, u-PAR, RAS, p53 expression
	SW480, SW 620	+	Histone hyperacetylation
	Colo 205	+	Apoptosis through β-catenin/TCF
	Colo 320 DM	+	Apoptosis
	HCT-116	+	EGFR and human EGFR-2 pathways
	MG63	+	Cell cycle arrest
	Caco-2	+	TRAIL-induced apoptosis
	LoVo	+	c-MYC, u-PAR, RAS, p53 expression
	SK-CO-1	+	ROS-induced apoptosis
	RKO	+	Apoptosis
			Caspase-induced apoptosis
			Caspase-induced apoptosis
			tNOX
			(tumor-associated NADH oxidase) upregulation
			Caspase-induced apoptosis
			Inhibition of metastatic potential
			Cell cycle arrest
			Cytotoxicity
			Cytotoxicity, c-MYC, u-PAR, RAS, p53 expression
			Cytotoxicity, c-MYC, u-PAR, RAS, p53 expression
			Inhibition of metastatic potential
			Inhibition of metastatic potential
	$Lgr5^{\text{GFP-ires-CreErt2}}$ $APC^{\text{LoxP/LoxP}}$ $Kras^{\text{LoxPstopLoxP-G12D}}$ mice		KRAS and APC knockout models
Gastric	AGS	+	c-MYC expression
Head and neck	KB	+	EGFR and p-Erk1/2 expression, cytotoxicity
	SCC-1, SCC-9	+	Apoptosis
	A431	+	
Acute myeloid leukemia	HL-60	+	c-MYC expression, cell cycle arrest

(Continued)

Table 1.7 (Continued)   Animal and Cell Culture Lines Used for Primary Screening the *In Vitro* Antitumor Properties of Plant-Derived and Synthetic Compounds

Tissue/Organ Specification	Cell/Animal Line	Human Origin	Target Pathway/Effect
Leukemia/ lyphoma	MOLT-4	+	Terminal deoxynucleotidyl transferase (TdT) expression, cytotoxicity
	K 562		
	P388		Tumorigenic in mice, cytotoxicity
	L5178Y		Tumorigenic in mice, cytotoxicity
	Eμ-myc mice		c-MYC overexpression
Liver	Huh7	+	c-MYC expression, apoptosis
	HepG2	+	Inhibition of metastatic potential, IGF expression
	L5		Chemically induced; cytotoxicity
	MDR2KO mice		Reduced beta-catenin synthesis
Lung	SPC-A1	+	P16, P21, P27 expression
	NCI-H1975	+	CCAT1 interference
	NCI-H1975	+	CCAT2 interference
	SpC-Myc mice		c-MYC overexpression
Melanoma	B16		miR-21, inhibition of metastatic potential
	Cdk4 [R24C/R24C] :: *Tyr-NRas* [Q61K]		CDK4, NRAS overexpression

Sophisticated methods for determining cellular growth and viability in primary screens include:

- Suppression of 12-O-tetradecanoylphorbol-13-acetate (TPA)-stimulated ^{32}Pi-incorporation into phospholipids of cultured cells.
- Epstein–Barr virus activation.
- Suppression of the tumor-promoting activity induced by 7,12-dimethylbenz[a]anthracene (DMBA) plus TPA, (calmodulin involved systems).
- Production of the tumor necrosis factor (TNF), possibly through stimulation of the reticuloendothelial system (RES).
- Stimulation of the uptake of tritiated thymidine into murine and human spleen cells.
- Inhibition of RNA, DNA, and protein synthesis in tumoric cells.
- Analysis of endogenous cyclic GMP: cyclic GMP is thought to be involved in lymphocytic cell proliferation and leukemogenesis. In general, the nucleotide is elevated in leukemic vs normal lymphocytes, and changes have been reported to occur during remission and relapse of this disease.
- Determination of DNA damage in Ehrlich ascites tumor cells by the use of an alkaline DNA unwinding method, followed by hydroxylapatite column chromatography of degraded DNA.
- The brine shrimp lethality assay for activity-directed fractionation.
- Suppression of the activities of thymidylate synthetase and thymidine kinase involved in de novo and salvage pathways for pyrimidine nucleotide synthesis.
- Suppression of the induction of the colonic cancer in rats treated with the chemical carcinogens 1,2-dimethylhydrazine (DMH), N -Nitrosodiethylamine, or azoxymethane/dextransulfate sodium (AOM/DSS).
- Inhibition of Epstein–Barr virus early antigen (EBV-EA) activation induced by 12-O-tetradecanoylphorbol-13-acetate (TPA).
- Inhibition of calmodulin-dependent protein kinases (CaM kinase III). These enzymes phosphorylate certain substrates that have been implicated in regulating cellular proliferation, usually via phosphorylation of elongation factor 2. The activity of CaM kinase III is increased in glioma cells following exposure to mitogens and is diminished or absent in nonproliferating glial tissue.
- Inhibition of the promoting effect of 12-O-etradecanoylphorbol-13-acetate on skin tumor formation in mice initiated with 7,12-dimethylbenz-[a]anthracene.
- Inhibition of two-stage initiation/promotion [dimethylbenz[a]anthracene (DMBA)/croton oil] skin carcinogenesis in mice.
- The MTT [3-(4,5-dimethylthiazol-2-yl)-2,5-diphenyl tetrazolium bromide] colorimetric assay.

1.4.2 Clinical Trials

Clinical trials are clinical research studies that involve human subjects. They are used to acquire medical information, namely to evaluate the safety and the effectiveness of new interventions. There are interventional studies and observational studies.

Cancer clinical trials (or cancer clinical studies) are designed to:*

- Treat cancer
- Find and diagnose cancer
- Prevent cancer
- Manage symptoms of cancer and side effects from its treatment

Some examples include:

- Prevention trials designed to keep cancer from developing in people who have not previously had cancer
- Prevention trials designed to prevent a new type of cancer from developing in people who have had cancer
- Early detection trials to find cancer, especially in its early stages
- Treatment trials to test new therapies in people who have cancer
- Quality of life studies to improve comfort and quality of life for people who have cancer
- Studies to evaluate ways of modifying cancer-causing behaviors, such as tobacco use

1.4.2.1 Phases of Clinical Trials

Most clinical research that involves the testing of a new drug progresses in an orderly series of steps.† This allows researchers to ask and answer questions in a way that expands information about the drug and its effects on people. Based on what has been learned in laboratory experiments or previous trials, researchers formulate hypotheses or questions that need to be answered. Then, they carefully design a clinical trial to test the hypothesis and answer the research question. It is customary to separate different kinds of trials into phases that follow one another in an orderly sequence. Generally, a particular cancer clinical trial falls into one of three phases.

1.4.2.1.1 Phase I Trials

These first studies in people evaluate whether a new drug is safe, well-tolerated and in what dosage. A Phase I trial usually enrolls only a small number of patients, as well as about 20 to 80 normal, healthy volunteers. The tests study a drug's safety profile, including the safe dosage range. The studies also determine how a drug is absorbed, distributed, metabolized, and excreted, and the duration of its action. This phase lasts about a year.

1.4.2.1.2 Phase II Trials

A Phase II trial provides preliminary information about how well the new drug works and generates more information about safety and benefit. Each Phase II study usually focuses on a particular type of cancer. Controlled studies of approximately up to 100 volunteer patients assess the drug's effectiveness and take about two years.

---

* https://www.cancer.gov/about-cancer/treatment/clinical-trials/what-are-trials
† http://www.cancerresearchuk.org/about-cancer/find-a-clinical-trial/what-clinical-trials-are/phases-of-clinical-trials#phase1

1.4.2.1.3 Phase III Trials

These trials compare a promising new drug, combination of drugs, or procedure with the current standard. Phase III trials typically involve large numbers of people in doctors' offices, clinics, and cancer centers nationwide. This phase lasts about three years and usually involves 200 to 3,000 patients in clinics and hospitals. Physicians monitor patients closely to determine efficacy and identify adverse reactions.

1.4.2.1.4 Phase IV Trials

Phase IV trials or post-marketing studies are studies done after a drug has been shown to work and it is already available on the market. The main reasons for running Phase IV trials are to find out:[*]

- More about the side effects and safety of the drug (pharmacovigilance)
- What the long term risks and benefits are
- How well the drug works when it's used more widely

1.4.2.2 Clinical Trial Protocols

Clinical trials follow strict scientific, clinical and administrative procedures called clinical trial protocols. These protocols deal with many areas, including the study's design, who can be in the study, and the kind of information people will be given when they are deciding whether to participate. Protocols ensure also the protection of the patient participating the study, the protection of private sensitive data and the ethical conduct of the study. Every trial has a chief investigator, who is usually a doctor. The investigator prepares a study action plan, called a protocol. This plan explains what the trial will do, how, and why. For example, it states:

- The reason for doing the trial
- How many people will be in the study
- Who is eligible to participate in the study
- What study drugs participants will take
- What medical tests they will have and how often
- What information will be gathered

REFERENCES

Alali, F.Q., X.X. Liu, and J.L. McLaughlin 1999. Annonaceous acetogenins: Recent progress. *J. Nat. Prod.* 62:504–540.

Amirouchene-Angelozzi, N., C. Swanton, and A. Bardelli 2017. Tumor evolution as a therapeutic target. *Cancer Discov.* 7:805–817.

Anderson, N.M., P. Mucka, J.G. Kern, and H. Feng 2018. The emerging role and targetability of the TCA cycle in cancer metabolism. *Protein Cell* 9:216–237.

Anonymous 2000. Loss of loved one affects cancer risk and survival. *Healthcentral News.*

Aulas, J.J. 1996. Alternative cancer treatments. *Sci. Am.* 275:162–163.

Bankovic, J. (Ed.) 2016. *Anticancer Drugs: Nature, Synthesis and Cell.* ExLi4EvA, 188 p. ISBN 978-953-51-2814-4.

Barbounaki-Konstantakou, E. 2004. *Chemotherapy.* Beta Medical Arts, Athens, 136 p. ISBN 960-8071-67-4.

[*] http://www.cancerresearchuk.org/about-cancer/find-a-clinical-trial/what-clinical-trials-are/phases-of-clinical-trials#phase1

Bignucolo, A., E. De Mattia, E. Cecchin, R. Roncato, and G. Toffoli. 2017. Pharmacogenomics of targeted agents for personalization of colorectal cancer treatment. *Int. J. Mol. Sci.* 18:1522.

Brower, V. 1999. Tumor-angiogenesis – New drugs on the block. *Nat. Biotechnol.* 17:963–968.

Carnero, A., C. Blanco-Aparicio, O. Renner, W. Link, and J.F. Leal 2008. The PTEN/PI3K/AKT signalling pathway in cancer, therapeutic implications. *Curr. Cancer Drug Targets* 8:187–198.

Chen, Y., H. Xie, Q. Gao, et al. 2017. Colon cancer associated transcripts in human cancers. *Biomed. Pharmacother.* 94:531–540.

Cragg, G.M. 1998. Paclitaxel (Taxol): A success story with valuable lessons for natural product drug discovery and development. *Med. Res. Rev.* 18:315–331.

Cragg, G.M., S.A. Schepartz, M. Suffness, and M.R. Grever 1993. The Taxol supply crisis. New NCI policies for handling the large-scale production of natural product anticancer and anti-HIV agents. *J. Nat. Prod.* 56:1657–1668.

Debinski, W., N.I. Obiri, S.K. Powers, I. Pastan, and R.K. Puri 1995. Human glioma cells overexpress receptors for interleukin 13 and are extremely sensitive to a novel chimeric protein composed of interleukin 13 and pseudomonas exotoxin. *Clin. Cancer Res.* 1:1253–1258.

Dimitriou, D. 2001. Capitalizing on the Industry Challenges in the Pharma & Biotech Sector: A Novel Approach to Drug Development. *DyoDelta Biosciences Newsletter.*

Eferl, R., and E. Casanova (Eds.) 2015. *Mouse Models of Cancer.* Humana Press, NY, 477 p. ISBN 978-1-4939-2296-3.

Flavahan, W.A., E. Gaskell, and B.E. Bernstein 2017. Epigenetic plasticity and the hallmarks of cancer. *Science* 357:eaal2380.

Folkman, J. 1996. Fighting cancer by attacking its blood supply. *Sci. Am.* 275:150–154.

Fukui, M., J. Azuma, and K. Okamura 1990. Induction of callus from mistletoe and interaction with its host cells. *Bull. Kyoto Univ. Forests* 62:261–269.

Fust, K. (Ed.) 2015. *The Gale Encyclopedia of Cancer*, 4th Ed. Gale, Cengage Learning, Farmington Hills, 2083 p. ISBN 978-1-4103-1740-7.

Gebhardt, R. 2000. In vitro screening of plant extracts and phytopharmaceuticals: Novel approaches for the elucidation of active compounds and their mechanisms. *Planta Med.* 66:99–105.

Greenwald, P. 1996. Chemoprevention of cancer. *Sci. Am.* 275:96–99.

Hanna, L., T. Crosby, and F. Macbeth (Eds.) 2015. *Practical Clinical Oncology*, 2nd Ed. Cambridge University Press, Cambridge, UK, 625 p. ISBN 978-1-107-68362-4.

Hayflick, L. 1981. Cell death *in vitro*. In: *Cell Death in Biology and Pathology.* Bowen, I. (Ed). Springer, Berlin, 243–295.

Jakubke, H.O., and H. Jeschkeit 1975. *Biochemie.* F. A. Brockhaus Verlag, Leipzig. ASIN B00A4200UO.

Kaiser, J. 2000. Controversial cancer therapy finds political support. *Science* 287:2139–2141.

Koehnlechner, M. 1987. *Leben ohne Krebs.* Knaur, Munich, 288 p. ISBN 978-3426260272.

Kunnumakkara, A.B. (Ed.) 2015. *Anticancer Properties of Foods and Vegetables.* World Scientific Publishing, Singapore, 405 p. ISBN 978-9814508889.

Lauterbach, K. 2015. *Die Krebs-Industrie: Wie eine Krankheit Deutschland erobert.* Rowohlt, Berlin, 288 p. ISBN 978-3871347986.

Lombardi, L., F. Morelli, S. Cinieri, et al. 2010. Adjuvant colon cancer chemotherapy: Where we are and where we'll go. *Cancer Treat. Rev.* 36, no. 3:S34–S41.

Markopoulos, G.S., E. Roupakia, M. Tokamani, et al. 2017. A step-by-step microRNA guide to cancer development and metastasis. *Cell. Oncol.* 40:303–339.

Marshall, J.L. (Ed.) 2017. *Cancer Therapeutic Targets.* Springer, NY, 1087 p. ISBN 978-1-4419-0716-5.

Meijer, L., S. Guidet, and M. Philippe (Eds.) 1997. *Progress in Cell Cycle Research*, Vol. 3. Plenum Press, NY, 261–269.ISBN 978-1-4615-5371-7.

Moeckel, B., T. Schwarz, J. Eck, M. Langer, H. Zinke, and H. Lentzen 1997. Apoptosis and cytokine release are biological responses mediated by recombinant mistletoe lectin *in vitro. Eur. J. Cancer* 33:35.

Nossal, G.J.V. 1993. Life, death and the immune system. *Sci. Am.* 269:52–62.

Old, L.J. 1996. Immunotherapy for cancer. *Sci. Am.* 275:136–143.

Petit, G.R., F.H. Pierson, and C.L. Herald 1994. *Anticancer Drugs from Animals, Plants and Microorganisms.* Wiley-Blackwell, NY, 49–158. ISBN 978-0471036579

Ramos-García, P., I. Ruiz-Ávila, J.A. Gil-Montoya, et al. 2017. Relevance of chromosomal band 11q13 in oral carcinogenesis: An update of current knowledge. *Oral Oncol.* 72:7–16.

Rodrigo, L. (Ed.) 2016. *Colorectal Cancer: From Pathogenesis to Treatment*, ExLi4EvA, 432 p. ISBN 978-953-51-2545-7.

Schwab, M. (Ed.) 2008. *Encyclopedia of Cancer*, 2nd Ed. Springer, Berlin, 3306 p. ISBN 978-3-540-36847-2.

Sidransky, D. 1996. Advances in cancer detection. *Sci. Am.* 275:104–109.

Steward, W.P., and K. Brown 2013. Cancer chemoprevention: A rapidly evolving field. *Br. J. Cancer* 109:1–7.

Torrens-Mas, M., J. Oliver, P. Roca, and J. Sastre-Serra 2017. SIRT3: Oncogene and tumor suppressor in cancer. *Cancers* 9, pp. 1–10.

Trichopoulos, D., F.P. Li, and D.J. Hunter 1996. What causes cancer? *Sci. Am.* 275:80–87.

Xu, K. (Ed.) 2016. *Tumor Metastasis*, ExLi4EvA, 259 p. ISBN: 953-51-2631-8.

CHAPTER 2

The Plant Kingdom
Nature's Pharmacy for Cancer Treatment

Spyridon E. Kintzios

CONTENTS

Plants have been used throughout history as an indispensable source of natural products for medicine. In this chapter, the components of plant primary and secondary metabolism are presented in detail, with an emphasis on plant-derived compounds with antitumor properties, their chemical structure, properties, isolation, and mode of action, among other information. Fourteen chemical groups are separately described. Finally, a section is dedicated to describing, with several plant species-specific case analyses, the contribution of biotechnology as an alternative method for the production of sufficient plant biomass and/or quantity of target natural products in a relatively short time, at the same time enabling improved control over bioactive compound biosynthesis.

2.1 BRIEF OVERVIEW OF THE GENERAL ORGANIZATION OF THE PLANT CELL

Although plant cells exhibit considerable diversity in their structure and function, their basic morphology is relatively unique (Figure 2.1). A typical plant cell consists of a *cell wall* (primary and secondary) surrounding a *protoplast*, which is delineated by the plasma membrane (or plasmalemma). The *protoplasm* (the protoplast without the plasmalemma) contains bodies bounded by membranes, known as organelles, as well as membrane structures, which do not enclose a body. The *cytoplasm* is the part of the protoplasm including various membrane structures, filaments, various particles, but not organelles. The *cytosol* is the aqueous phase of the cytoplasm, devoid of all particulate material. All membranes (including plasmalemma) chemically consist of a phospholipid bilayer carrying various proteins. Thanks to the existence of these internal compartments, specific functions can be executed in different parts or organelles of the plant cell. For example, the cell membrane permits

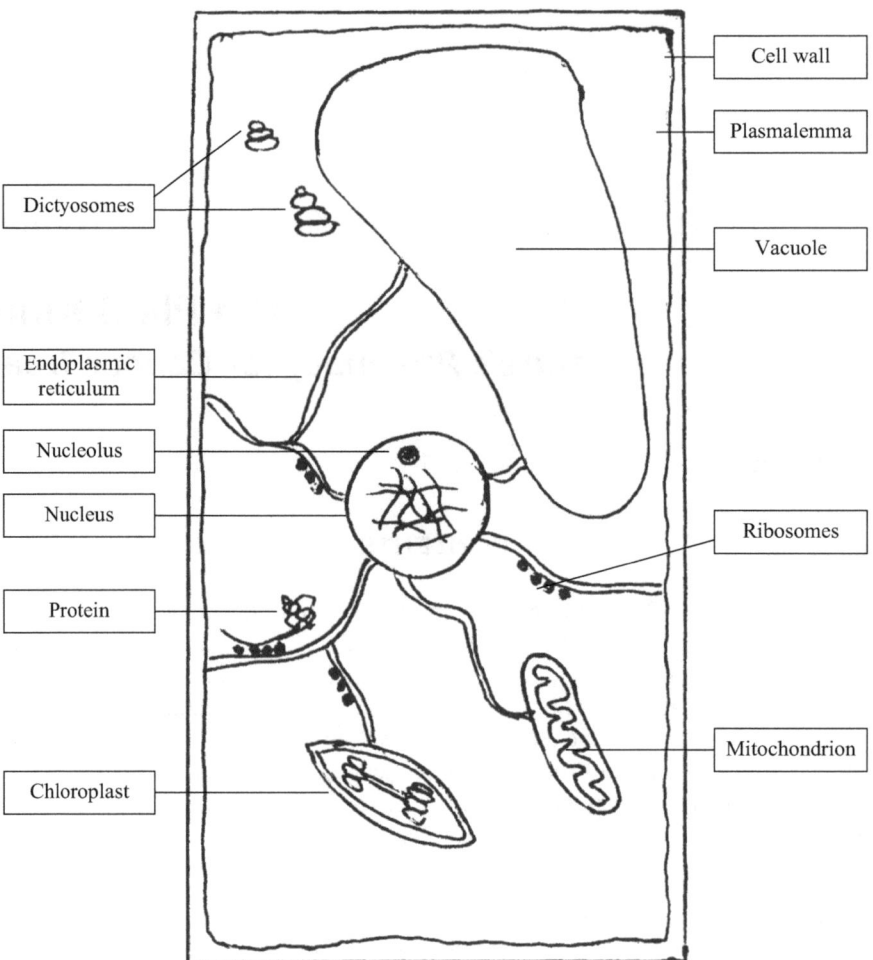

Figure 2.1   General outline of the structure of a plant cell.

the controlled entry and exit of compounds into and out of the cell, while preventing excessive gain or loss of water and metabolic products. The *nucleus* is a large organelle containing chromatin, a complex of DNA and protein. It is the main center for the control of gene expression and replication. Chlorophyll-containing *chloroplasts* are the site for photosynthesis. *Mitochondria* contain enzymes important for the process of oxidative phosphorylation, i.e. the phosphorylation of ADP to ATP with the parallel consumption of oxygen. *Vacuoles* are large organelles (usually only one vacuole is found in mature cells, representing up to 90% of the cell volume). They store water, salts, various organic metabolites, toxic substances or waste products, and water-soluble pigments. Generally, the vacuole content ('the cell sap') is considered to represent, together with the cytosol, the hydrophilic part of the plant cell. *Ribosomes* are small spheroid particles (attached to the cytoplasmic side of the *endoplasmic reticulum*, mitochondria, and chloroplasts), which serve as sites for protein synthesis. *Golgi bodies* (or *dictyosomes*) consist of a stack of about five flattened sacs (cisternae) and are the sites for the synthesis of most of the matrix polysaccharides of cell walls, glycoproteins, and some enzymes. *Microbodies* are small organelles containing various oxidases. Finally, *microtubules* are tubular inclusions within the cytoplasm, consisting of filamentous polymers of the protein tubulin, which can polymerize and depolymerize in a reversible manner. They direct the physical orientation of various components within the cytoplasm.

2.2 THE CHEMICAL CONSTITUENTS OF THE PLANT CELL

Throughout human history, plants have been used as an indispensable source of natural products for medicine. The chemical constituents of the plant cell that exert biological activities on human and animal cells fall into two distinct groups, depending on their relative concentration in the plant body:

2.2.1 Primary Metabolites

By definition, primary metabolism is the total of processes leading to the production of the four major classes of compounds accounting for about 90% of the biological matter and that are, in consequence, required for the growth of plant cells. These are sugars (carbohydrates) (structural and nutritional elements), amino acids (structural elements and enzymes), lipids (constituents of membranes, nutritional elements), and nucleotides (constituents of genes). These compounds occur principally as components of macromolecules, such as cellulose or amylose (from sugars), proteins (from amino acids), and nucleic acids, such as DNA (from nucleotides). Primary metabolites mainly contain carbon, nitrogen, and phosphorous, which are assimilated into the plant cell by three main catabolic pathways: glycolysis, the pentose phosphate pathway, and the tricarboxylic (TCA) cycle, while plants are the most important autotrophic organisms, since carbon assimilation in biological matter is a light-dependent process, known as *photosynthesis*. This is restricted, however, to cells containing chlorophyll and other photosynthetic pigments, e.g. associated with chloroplasts of mesophyll cells.

2.2.2 Secondary Metabolites

Secondary metabolites are compounds belonging to extremely diverse chemical groups, such as organic acids, aromatic compounds, terpenoids, steroids, flavonoids, alkaloids, carbonyles, etc. Their function in plants is usually related to metabolic and/or growth regulation, lignification, coloring of plant parts, and protection against pathogen attack. Even though secondary metabolism generally accounts for less than 10% of the total plant metabolism, its products are the main plant constituents with pharmaceutical properties.

Despite the diversity of secondary metabolites, a few key intermediates in primary metabolism supply the precursors for most secondary products. These are mainly shikimic acid, sugars, acetyl-CoA, mevalonic acid, 1-deoxyxylulose 5-phosphate, nucleotides, and amino acids.

- The shikimic acid (shikimate) pathway leads to the biosynthesis of phenolic compounds, including flavonoids and lignans with anticancer properties, such as podophyllotoxin. It is also related to the synthesis of some amino acids (phenylalanine, tyrosine, tryptophan) and many different types of aromatic compounds.
- Cyanogenic glycosides and glucosinolates are derived from sugars.
- Terpenes and steroids are produced from isoprene units which are derived from acetyl-CoA via the mevalonate and the 1-deoxyxylulose 5-phosphate pathways.
- The nucleotide bases are the precursors to purine and pyrimidine alkaloids.
- The majority of alkaloids and protoalkaloids are synthesized from amino acids.

In addition, many natural products are derived from pathways involving more than one of these intermediates:

- Phenylpropanoids are derived from the amino acid phenylalanine, with acetyl-CoA and sugar units being added later in the biosynthetic pathway.
- The indole and the quinoline alkaloids are derived from the amino acid tryptophan and from monoterpenes.
- The aglycon moieties of cyanogenic glycosides and glucosinolates are derived from amino acids.

Primary and secondary metabolic pathways in plants are summarized in Figure 2.2.

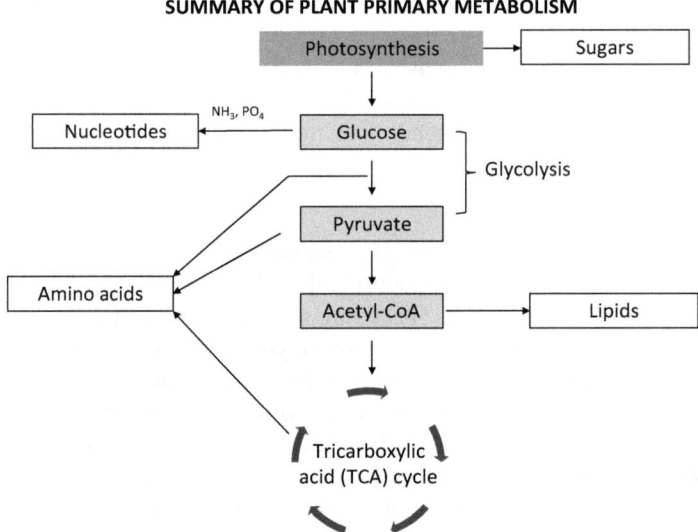

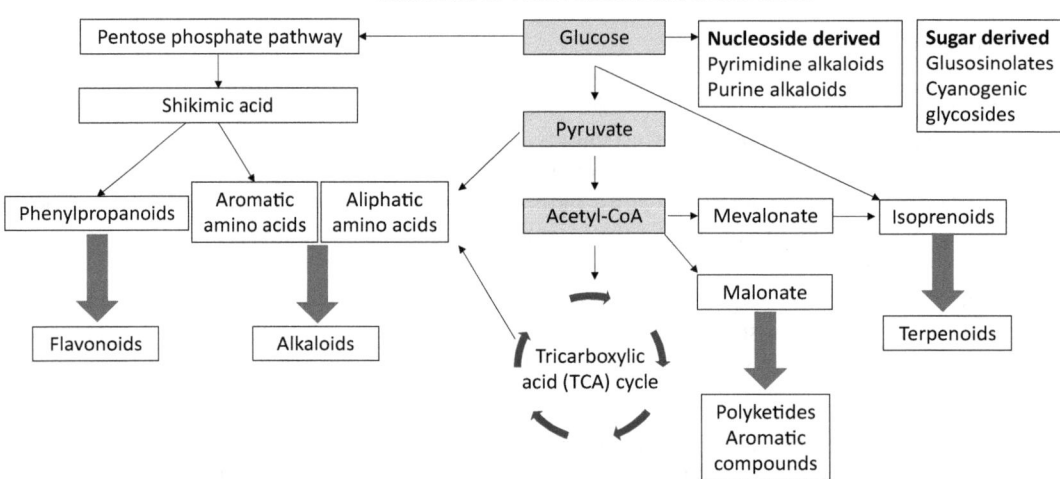

Figure 2.2 Simplified summary of primary and secondary metabolic pathways in plants.

2.3 WHY DO PLANT COMPOUNDS HAVE AN ANTICANCER ACTIVITY?

Some secondary metabolites are considered as metabolic waste products, e.g. alkaloids may function as nitrogen waste products. However, a significant portion of the products derived from secondary pathways serve either as *protective agents against various pathogens* (e.g. insects, fungi, or bacteria) or *growth regulatory molecules* (e.g. hormone-like substances that stimulate or inhibit cell division and morphogenesis). Due to these physiological functions, secondary metabolites are potential anticancer drugs, since either direct cytotoxicity is effected on cancer cells or the course of tumor development is modulated and eventually inhibited. Administration of these compounds at low concentrations may be lethal for microorganisms and small-sized animals, like herbivore insects; however, they affect, more or less specifically, the fastest growing tissues (such as tumors) in a larger organism, as well as humans. Cancer cytotoxic compounds may also act as cancer chemopreventive agents, by inhibiting tumor- and metastasis-specific molecular pathways. A representative list of such targeted properties of plant-derived compounds is presented in Table 2.1.

The anticancer properties of plant secondary metabolites are not always related to direct cytotoxic or otherwise growth-inhibitory effects. For example, terpenoids such as geraniol (a constituent of fragrances from citrus, ginger, citronella, and rose) or betulinic acid (a component of the white birch, *Betula pubescens*, and the ber tree, *Ziziphus mauritiana*) can render tumor cells more sensitive to standard chemotherapeutic drugs, such as 5-Flourouracil. Other modes of action include immunostimulatory effects (a well-established property of mistletoe extracts) and interference with microRNA (miRNA) expression (see also Chapter 1), as documented for phenolic and polyphenolic compounds (e.g. curcumin, garcinol, resveratrol, and epigallocatechin-3-gallate or EGCG), flavonoids (e.g. quercetin), soy isoflavones (e.g. genistein), and carotenoids (e.g. lycopene).

Table 2.1   Examples of Antitumor Properties of Plant-Derived Compounds Determined on the Cellular and Molecular Level

Representative Phytochemical	Target Pathway(s)
Terpenes	
Taxol (paclitaxel)	Inhibition of microtubule depolymerization (cell cycle arrest), inhibition of apoptosis inhibitor protein Bcl-2
Diosgenin	TRAIL-induced apoptosis
Eugenol	ROS-induced apoptosis
Betulinic acid	Melanoma-, mitochondria-specific apoptosis
Alkaloids	
Camptothecin	Inhibition of DNA topoisomerase I
Vincristine	Inhibition of microtubule polymerization (cell cycle arrest)
Protoalkaloids	
Colchicine	Inhibition of microtubule polymerization (cell cycle arrest)
Lignans	
Podophyllotoxin	Inhibition of microtubule polymerization (cell cycle arrest)
Pseudoalkaloids	
Capsaicin	Mitochondria-related apoptosis, cell cycle arrest, apoptosis through β-catenin, tNOX (tumor-associated NADH oxidase) upregulation, caspase-induced apoptosis
Lectins	
Mistletoe lectins (ML)	Immunomodulation, immunostimulation
Flavones	
Luteolin	EGFR-mediated cell survival
Apigenin	p53-dependent pathway, cell cycle arrest
Fisetin	Caspase-induced apoptosis
Flavonoids	
Quercetin	Downregulation of Bcl-2 protein expression and upregulation of Bax expression, cell cycle arrest
Kaempferol	Cell cycle arrest
Anthocyanidins	
Delphinidin	Cytotoxicity, caspase-induced apoptosis
Polyphenols	
Epigallocatechin-3-gallate (EGCG)	Apoptosis, EGFR and human EGFR-2 pathways
Resveratrol	Cell cycle arrest, apoptosis
Organosulfur compounds	
Diallyl disulfide	Histone hyperacetylation, caspase-induced apoptosis

2.4 CHEMICAL GROUPS OF NATURAL PRODUCTS WITH ANTICANCER PROPERTIES

Plant-derived natural products with documented anticancer/antitumor properties can be roughly classified in the following 14 chemical groups:

ALKALOIDS are widely distributed throughout the plant kingdom and constitute a very large group of chemically different compounds with diversified pharmaceutical properties. They are found in more than 300 plant families and include more than 10,000 different compounds.

Many alkaloids are famous for their psychotropic properties, as very potent narcotics and tranquilizers. Examples are *morphine, cocaine, reserpine*, and *nicotine*. Several alkaloids are also very toxic.

Structure and properties: they are principally nitrogen-containing substances with a ring structure (*heterocyclic molecules*) that allows their general classification in one of the groups described in Table 2.2. Most alkaloids with anticancer activity are either indole, pyridine, piperidine, or aminoalkaloids. **Protoalkaloids** (for example, colchicine) are usually derived from L-tyrosine and L-tryptophan and differ from true alkaloids in having their N atom outside the ring structure. On the other hand, **pseudoalkaloids** (for example, capsaicin) are N-containing heterocycles but their nitrogen atom does not originate from amino acids.

Alkaloids are weak bases, capable of forming salts, which are commonly extracted from tissues with an acidic, aqueous solvent. Alternatively, free bases can be extracted with organic solvents.

Distribution: quite abundant in higher plants, less so in gymnosperms, ferns, fungi, and other microorganisms. Particularly rich in alkaloids are plants of the families Apocynaceae, Papaveraceae, Fabaceae, Compositae, Cupressaceae, Leguminosae, Liliaceae, Loranthaceae, Nyssaceae, Piperaceae, Rubiaceae, Rutaceae, Simaroubaceae, Solanaceae, Chenopodiaceae, Lauraceae, Berberidaceae, Menispermaceae, Ranunculaceae, Fumariaceae, Papilionaceae, Loganiaceae, Rubiaceae, Boraginaceae, Convolvulaceae, and Campanulaceae.

Biosynthesis in plant cells: rather complicated, with various amino acids (phenylalanine, tryptophan, ornithine, and lysine) serving as precursor substances.

Basis of anticancer/antitumor activity: alkaloids are mainly cytotoxic against various types of cancer and leukemia. They also demonstrate antiviral properties. More rarely, they demonstrate immuno-modulatory properties.

Some plants containing alkaloids with anticancer properties are indicated in the following table (for more details on each plant, please consult Chapter 9 of this book):

Species	Target Disease or Cell Line (If Known)	Mode of Action (If Known)
Brucea antidysenterica	Antileukemic	Cytotoxic
Calycodendron miluei	Antiviral	
Cassia leptophylla		DNA-damaging (piperidine)
Catharanthus roseus	Hodgkin's disease, leukemia	Tubulin inhibitor
Chamoecyparis sp.	P388	Cytotoxic: inhibition of cyclic GMP formation
Chelidonium majus	Various cancers, lung	Immunomodulator (clinical)
Colchicum antimale	P388, esophageal	Tubulin inhibitor
Ervatamia microphylla	k-ras-NRK (mice) cells	Growth inhibition
Eurycoma lougifolia		Cytotoxic *in vitro*
Fagara macrophylla	P-388	
Nauclia orientalis	Anti-tumor, *in vitro* human bladder carcinoma	
Psychotria sp.		Antiviral
Strylimos usabarensis	Various *in vitro* (liver damage)	Cytotoxic

Table 2.2 General Structural Classification of Alkaloids*

Group Name	Precursor Compound	Base Structure
Pyrrolidine	L-ornithine	
Pyrrolizidine	L-ornithine	
Tropane	L-ornithine	
Piperidine	Lysine	
Indolizidine	Lysine	
Quinolizidine	Lysine	
Indole, terpenoid indole	L-tryptophan	
Quinoline	L-tryptophan	
Isoquinoline	L-tryptophan	

(*Continued*)

Table 2.2 (Continued)   General Structural Classification of Alkaloids*

Group Name	Precursor Compound	Base Structure
Ergot	L-tryptophan	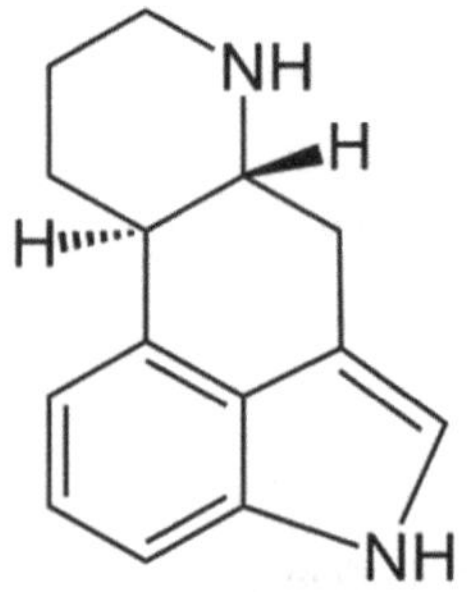
Imidazole	L-histidine	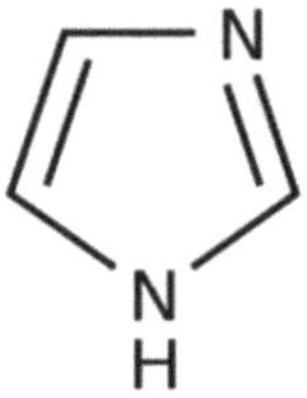
Phenylethylamine (protoalkaloids)	L-tyrosine	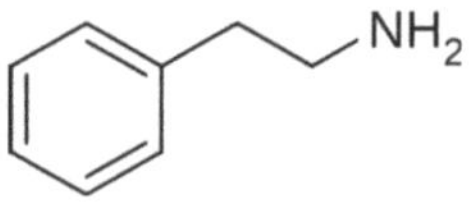
Pyridine	Nicotinic acid	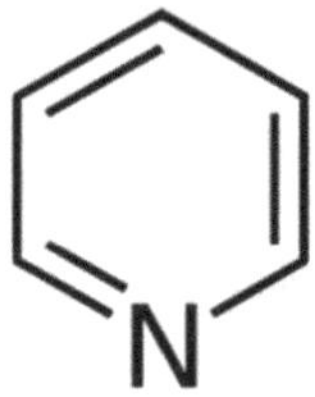
Purine (pseudoalkaloids)	Adenine/Guanine	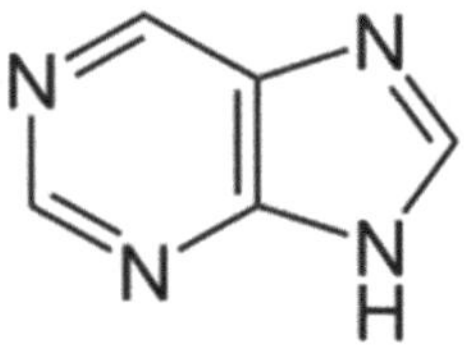
Terpenoid (pseudoalkaloids)	Geraniol	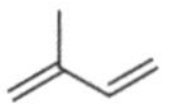
Ephedra (pseudoalkaloids)	Pyruvic acid	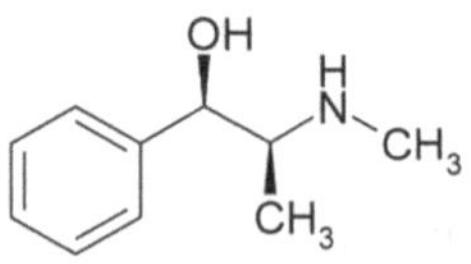

* Some secondary alkaloid groups are not mentioned.

POLYSACCHARIDES and generally carbohydrates represent the main carbon sink in the plant cell. Polysaccharides commonly serve nutritional (e.g. *starch*) and structural (e.g. *cellulose*) functions in plants.

Structure and properties: they are polymers of monosaccharides (and their derivatives) containing ten or more units, usually several thousand. Despite the vast number of possible polysaccharides, only few of the structural possibilities actually exist. Generally, *structural* polysaccharides are straight-chained (not very soluble in water), while *nutritional* (reserve food) polysaccharides tend to be branched, therefore forming viscous hydrophilic colloid systems. *Plant gums* and *mucilages* are hydrophilic heteropolysaccharides (i.e. they contain more than one type of monosaccharide), with the common presence of uronic acid in their molecule.

Depending on their degree of solubility in water, polysaccharides can be extracted from plant tissues either with hot water (pectic substances, nutritional polysaccharides, mucilages, fructans) or alkali solutions (hemicelluloses).

Distribution: they are universally distributed in the plant kingdom. Structural polysaccharides are the main constituents of the plant cell wall (cellulose, hemicelluloses, xylans, pectins, galactans). Nutritional polysaccharides include starch, fructans, mannans, and galactomannans. Mucilages abound in xerophytes and seeds. Polysaccharides also have a key function in the mechanism of biochemical recognition and signal transduction, similar to growth regulators.

Biosynthesis in plant cells: there exists a complex network of inter-related biosynthetic pathways, with various monosaccharides (glucose, fructose, mannose, mannitol, ribose, and erythrose) serving as precursor substances. Phosphorylated intermediates are found in subsequent biosynthetic steps and branching points. The glycolytic, pentose, and UDP-glucose pathways have been defined in detail.

Basis of anticancer/antitumor activity: some polysaccharides are cytotoxic against certain types of cancer, such as mouse skin cancer or tumor lines *in vitro* (e.g. mouse Sarcoma 180). However, most polysaccharides exert their action through stimulation of the immune system (cancer immunotherapy).

Some plants containing polysaccharides with anticancer properties are indicated in the following table (for more details on each plant, please consult Chapter 9 of this book) (Figure 2.3):

Species	Target Disease or Cell Line (If Known)	Mode of Action (If Known)
Angelica acutiloba	Epstein–Barr, skin(mice)	Cytotoxic, immunological
Angelica sinensis	Ehrlich Ascites (mice)	Cytotoxic, immunological
Brucia javanica	Leukemia, lung, colon, CNS, melanoma, brain	
Cassia angustifolia	Solid Sarvoma-180 (mice)	Cytotoxic
Sargassum thunbergii	Ehrlich Ascites (mice)	Immunostimulatory activates the reticuloendothelial system
S. fulvellum	Sarcoma 180 (mice)	Immunomodulatory
Tamarindus indica		Immunomodulatory

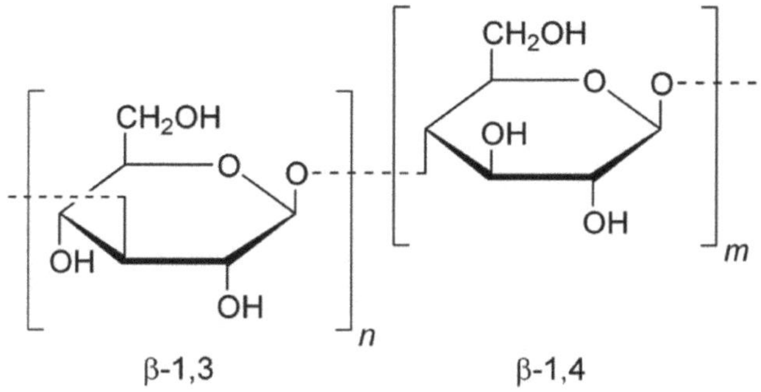

Figure 2.3 β-Glucan, a β-ᴅ-glucose polysaccharide with immunomodulatory properties.

GLYCOSIDES are carbohydrate ethers that are readily hydrolysable in hot water or weak acids. Most frequently, they contain glucose and are named by designating the attached alkyl group first and replacing the -ose ending of the sugar with -oside. The sugar part of a glycoside is known as the glycone and the non-sugar part as the aglycone or genin. Glycosides can be classified according either to their glycone or aglycone part or on the basis of the glycosidic linkage (as N-, O-, C-, or S-glycosides).

Distribution: abundant in plants of the families Apocyanaceae, Leguminosae, Scrophulareaceae, Liliaceae, Dioscoreaceae, Umbelliferae, Rutaceae, Rosaceae, Polygonaceae, Simarubaceae, Myrtaceae, Gentianaceae, and Cruciferae.

Basis of anticancer/antitumor activity: glycosides are mainly cytotoxic against certain types of cancer and also demonstrate antiviral and antileukemic properties.

Some plants containing glycosides with anticancer properties are indicated in the following table (for more details on each plant, please consult Chapter 9 of this book) (Figure 2.4):

Species	Target Disease or Cell Line (If Known)	Mode of Action (If Known)
Phlomis armeniaca	Liver cancer, Dalton's lymphoma (mice) antileukemic	Antiviral, cytotoxic, chemopreventive (human)
Plumeria rubia (iridoids)	P388, KB	Cytotoxic
Wikstroemia indica	Antileukemic	

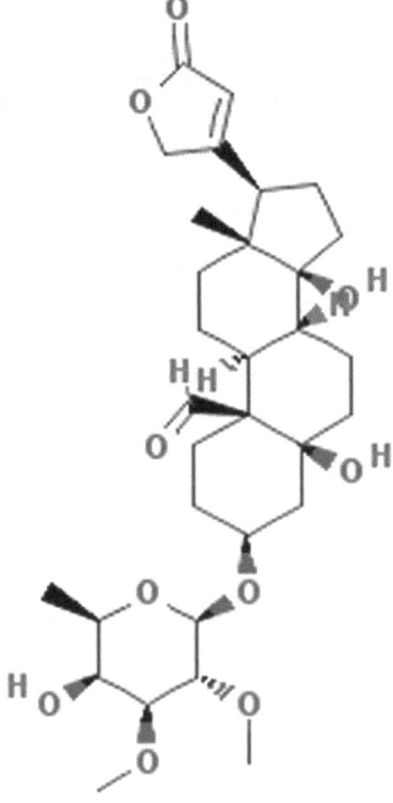

Figure 2.4 Strebloside, a cardiac glycoside from *Streblus asper* with antineoplastic properties.

LIPIDS (saponifiable) include fatty acids (aliphatic carboxylic acids), fatty acid esters, phospholipids, and glycolipids.

Structure and properties: by definition, lipids are soluble only in organic solvents. On heating with alkali, they form water-soluble salts (therefore the designation *saponifiable lipids*). Fatty acids are usually found in their ester form, mostly having an unbranched carbon chain, and differ from one another in chain length and degree of unsaturation.

Distribution: lipids are widely distributed in the plant kingdom. They both serve as nutritional reserves (particularly in seeds) and structural elements (i.e. phospholipids of the cell membrane, fatty acid esters in the epidermis of leaves, stems, fruits, etc.).

Biosynthesis in plant cells: they are derived by condensation of several molecules of acetate (more specifically malonyl-coenzyme A), thus being related to long chain fatty acids.

Basis of anticancer/antitumor activity: saponifiable lipids are cytotoxic against a limited number of cancer types.

Some plants containing lipids with anticancer properties are indicated in the following table (for more details on each plant, please consult Chapter 9 of this book):

Species	Target Disease or Cell Line (If Known)	Mode of Action (If Known)
Nigella sativa	Ehrilch ascites carcinoma	Cytotoxic *in vitro*
Sho-saiko-to, Juzen-taiho-to (extract)	Dalton's lymphoma, sarcoma-180 (clinical)	

FLAVONOIDS are widely distributed colored phenolic derivatives. Related compounds include flavones, flavonols, flavanonols, xanthones, flavanones, chalcones, aurones, anthocyanins, and catechins.

Structure and properties: flavonoids may be described as a series of C_6–C_3–C_6 compounds, i.e. they consist of two C_6 groups (substituted benzene rings) connected by a three-carbon-aliphatic chain. The majority of flavonoids contain a pyran ring linking the three-carbon chain with one of the benzene rings. Different classes within the group are distinguished by additional oxygen-heterocyclic rings and by hydroxyl groups distributed in different patterns. Flavonoids frequently occur as glycosides and are mostly water-soluble or at least sufficiently polar to be well extracted by methanol, ethanol, or acetone; however, they are less polar than carbohydrates and can be separated from them in an aqueous solution.

Distribution: they are widely distributed in the plant kingdom since they include some of the most common pigments, often fluorescent after UV irradiation. They also act as metabolic regulators and protect cells from UV radiation. Finally, flavonoids have a key function in the mechanism of biochemical recognition and signal transduction, similar to growth regulators.

Biosynthesis in plant cells: flavonoids are derived from shikimic acid via the phenylpropanoid pathway. Related compounds are produced through a complex network of reactions: isoflavones, aurones, flavanones, and flavanonols are produced from chalcones; leucoanthocyanidins, flavones and flavonols from flavanonols; and anthocyanidins from leucoanthocyanidins.

Basis of anticancer/antitumor activity: Flavonoids are cytotoxic against cancer cells, mostly *in vitro*. They are commonly valued as cancer chemopreventive agents, less so as chemotherapeutic ones.

Some plants containing flavonoids with anticancer properties are indicated in the following table (for more details on each plant, please consult Chapter 9 of this book) (Figure 2.5):

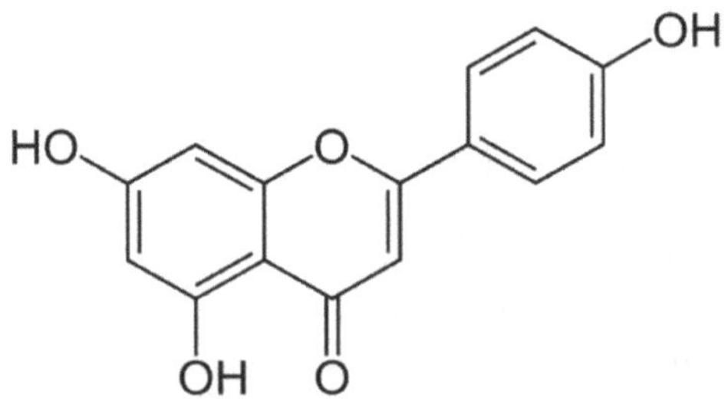

Figure 2.5 Apigenin.

Species	Target Disease or Cell Line (If Known)	Mode of Action (If Known)
Acrougehia porteri	KB	Cytotoxic
Angelica keiskei		Calmodulin inhibitor
Eupatorium altissimum	P-338, KB	Cytotoxic
Gossypium indicum	B16 melanoma	Cytotoxic
Petroselinum crispum	Colon cancer (HT-29)	Cell cycle arrest
Polytrichum obloense	Hela, antileukemic (mice)	
Psorospermum febrifigum	KB	Cytotoxic
Zieridium pseudobtusifolium	KB	Cytotoxic

PROTEINS, like carbohydrates, belong to the most essential constituents of the plant body, since they are the building molecules of structural parts and the enzymes.

Structure and properties: proteins are made up of amino acids, the particular combination of which defines the physical property of the protein. Thus, protein sequences differing in only one amino acid will correspond to entirely different molecules, both structurally (tertiary structure) and functionally. **Peptides** are small proteins, amino acid oligomers with a molecular weight below 6,000. In nature, 24 different amino acids are widely distributed. Sixteen to 20 different amino acids are usually found on hydrolysis of a given protein, all having the L-configuration. Conjugate proteins comprise other substances along with amino acids. Particularly important are glycoproteins, partially composed of carbohydrates. **Lectins** are carbohydrate-binding proteins, macromolecules that are highly specific for sugar moieties and which are critically involved in mechanisms of bio-recognition. Proteins may be soluble in water and dilute salt solutions (albumins), in dilute salt solutions (globulins), in very dilute acids and bases (glutelins), or in ethanolic solutions (prolamines).

Peptides and proteins can be isolated from plant tissues by aqueous extraction or in less polar solvents (depending on the water solubility of a particular protein). Fractionation of the proteins can frequently be achieved by controlling the ionic strength of the medium through the use of salts. However, one must always take precautions against protein denaturation.

Distribution: universal.

Biosynthesis in plant cells: proteins are synthesized in ribosomes from free amino acids under the strict, coordinated control of genomic DNA, mRNA, and tRNA (gene transcription and translation).

Basis of anticancer/antitumor activity: proteins are indirectly cytotoxic against certain cancer types, acting mainly through the inhibition of various enzymes or by inducing apoptotic cell death.

Some plants containing proteins with anticancer properties are indicated in the following table (for more details on each plant, please consult Chapter 9 of this book):

Species	Target Disease or Cell Line (If Known)	Mode of Action (If Known)
Acacia confusa	Sarcoma 180/HeLa cells	Trypsin inhibitor
Ficus cinia		White cell aglutination
Glycyrrhiza uraleusis		SDR-enzymes (antimutagenic)
Momordica charantia	Human leukemia	Inhibits DNA synthesis *in vitro*, immunostimulant
Rugia sp., *R. cordifolia*	P338	
Viscum album	Pancreatic cancer	Immunomodulation

ANNONACEOUS ACETOGENINS are antitumor and pesticidal agents of the Annonaceae family. More than 120 compounds have been identified in organic extracts of the associated plant parts.

Structure and properties: they are a series of C-35/C-37 natural products derived from C-32/C-34 fatty acids that are combined with a 2-propanol unit. They are usually characterized by a long aliphatic chain bearing a terminal methyl-substituted α,β-unsaturated γ-lactone ring with one to three tetrahydrofuran (THF) rings located among the hydrocarbon chain and a number of oxygenated moieties and/or double bonds. Annonaceous acetogenins are classified according to their relative stereostructures across the THF rings.

The Annonaceous acetogenins are readily soluble in most organic solvents. Ethanol extraction of the dried plant material is followed by solvent partitions to concentrate the compounds.

Distribution: exclusively in the Annonaceae family.

Biosynthesis in plant cells: derived from the polyketide pathway, while the tetrahydrofuran and epoxide rings are suggested to arise from isolated double bonds through epoxidation and cyclization.

Basis of anticancer/antitumor activity: Annonaceous acetogenins are cytotoxic against certain cancer species and leukemia. They are the most powerful inhibitors of complex I in mammalian and insect mitochondrial electron transport systems, as well as of NADH oxidase of the plasma membranes of cancer cells. Therefore, they decrease cellular ATP production, causing apoptotic cell death.

Some plants containing Annonaceous acetogenins are indicated in the following table (for more details on each plant, please consult Chapter 9 of this book) (Figure 2.6):

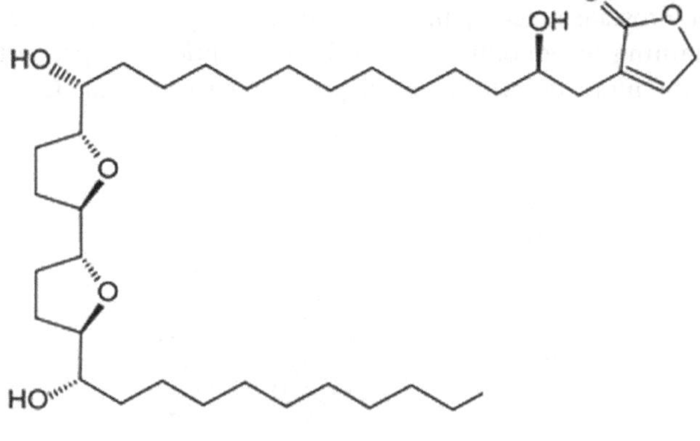

Figure 2.6 Structure of bullatacin, a major annonaceous acetogenin.

(Adapted from Patrikios et al. 2015).

Species	Target Disease or Cell Line (If Known)	Mode of Action (If Known)

Annona muricata,
  A. squamosa
	Prostate adenocarcinoma, human prostate
	  cancer (PC-3), pancreatic carcinoma,
	  human cervical adenocarcinoma (HeLa),
	  human breast adenocarcinoma (MCF-7),
	  human bladder carcinoma (H-460)
		Cytotoxic, mitochondria-related
		  apoptosis

A. bellata	Human solid tumors *in vitro* (colon cancer)	
Gonioothalamus sp.	Breast cancer, *in vitro* various human cancers	Cytotoxic

TERPENOIDS or terpenes are diverse, widely distributed compounds commonly found under groups such as essential oils, sterols, pigments, and alkaloids. They exert significant ecological functions in plants. Mono- and sesquiterpenoids are found as constituents of steam-distillable essential oils. Di- and triterpenoids are found in resins.

Structure and properties: they are built up of isoprene or iso-pentane units linked together in various ways and with different types of ring closures, degrees of unsaturation, and functional groups. Depending on the number of isoprene molecules in their structure, terpenoids are basically classified as monoterpenoids (two), sesquiterpenoids (three), diterpenoids (four), sesterterpenoids (five), triterpenoids (six), and tetraterpenoids (eight). Higher terpenoids (polyterpenoids) contain more than 40 carbon atoms. Sterols share the core structure of lanosterol and other tetracyclic triterponoids, but with only two methyl groups at positions 10 and 13 of their ring system. Steroids occur throughout the plant kingdom as free sterols and their lipid esters.

There exists no general method for isolating terpenoids from plants; however, most of them are non-polar and can be extracted in organic solvents. After saponification in alcoholic alkali and extraction with ether, most terpenoids will accumulate into the ether fraction.

Distribution: wide. Particularly rich in terpenes are plants belonging to the families Annonaceae, Aristolochiaceae, Asteraceae, Celastraceae, Combretaceae, Compositae, Ericaceae, Euphorbiaceae, Flacourtiaceae, Lamiaceae, Leguminosae, Meliaceae, Taxaceae, and Thymelaceae.

Biosynthesis in plant cells: terpenoids are all derived from mevalonic acid or a closely related precursor. The pyrophosphate of alcohol farnesol is a key intermediate in terpenoid biosynthesis, particularly leading to the formation of diterpenoids, triterpenoids, and sterols. Monoterpenoids are derived from geranyl pyrophosphate.

Basis of anticancer/antitumor activity: terpenoids and sterols often possess alkaloidal properties, thus being cytotoxic *in vivo* and *in vitro* against various cancer types, such as human breast cancer, ovarian cancer, prostate cancer, pancreatic cancer, lung cancer, and leukemia.

Some plants containing terpenoids and sterols with anticancer properties are indicated in the following table (for more details on each plant, please consult Chapter 9 of this book) (Figure 2.7):

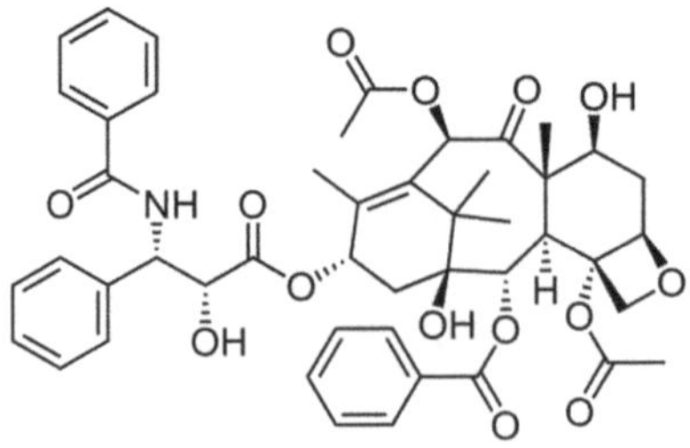

Figure 2.7   Structure of Taxol (paclitaxel), the major anticancer terpenoid from *Taxus* sp.

Species	Target Disease or Cell Line (If Known)	Mode of Action (If Known)
Brucia antidysenterica		Cytotoxic
Caesaria sylvestris		Cytotoxic *in vitro*, apoptotic
Crocus sativus	KB, P388, human prostate, pancreatic, *in vitro*	
Glycyrrhiza sp.	P388	
Melia sp.	Carcinoma, sarcoma, leukemia, AS49, VA13	Apoptotic/inhibits DNA synthesis
Meytemis sp.	Leukemia	
Neurolaena lobata	Human carcinoma *in vitro*	
Polyathia barnesii	Human carcinoma *in vitro*	
Rabdosia trichocarpa	HeLa cells, P388	Cytotoxic
Seseli mairei	KB, P388, L1210	
Stellera chamaejasme	Human leukemia, stem, lung, P388, L1210	Proteinokinase C activator
Taxus sp.	Breast, ovarian cancer	Microtubule inhibitor

NUCLEIC ACIDS, deoxyribonucleic acid (DNA) and ribonucleic acid (RNA), are known as the 'genetic molecules', the building blocks of genes in each cell or virus.

Structure and properties: each nucleic acid contains four different nitrogen bases (purine and pyrimidine bases), phosphate, and either deoxyribose or ribose. DNA contains the bases adenine, quinine, cytosine, thymine, and 5-methylcytosine. The macromolecular structure of DNA is a two-stranded helix with the strands bound together by hydrogen bonds.

Like proteins and polysaccharides, nucleic acids are water-soluble and non-dialyzable. They can be separated from a water extract by denaturing proteins in chloroform-octyl alcohol and then precipitate polysaccharides in a weakly basic solution.

Distribution: universal.

Biosynthesis in plant cells: bases are derived originally from ribose-5-phosphate, purines from inosinic acid, and pyrimidines from uridine-5-phosphate. Nucleic acids are formed after nucleotide transformation and condensation.

Basis of anticancer/antitumor activity: some nucleotides, like cyclopentenyl cytosine (derived from *Viola odorata*), present cytotoxicity against certain cancer species *in vitro*.

LIGNANS are colorless, crystalline solid substances widespread in the plant kingdom (mostly as metabolic intermediaries) and having antioxidant, insecticidal, and medicinal properties.

Structure and properties: they consist of two phenylpropanes joined at their aliphatic chains (dimeric C_6C_3 coupled motifs) and having their aromatic rings oxygenated. Additional ring closures may also be present. Occasionally they are found as glycosides. They are also components of resin phenols.

Lignans may be extracted with acetone or ethanol and are often precipitated as slightly soluble potassium salts by adding concentrated potassium hydroxide to an alcoholic solution.

Distribution: wide.

Biosynthesis in plant cells: lignans are originally derived from shikimic acid via the phenylpropanoid pathway, with p-hydroxycinnamyl alcohol and coniferyl alcohol being key intermediates of their biosynthesis.

Basis of anticancer/antitumor activity: some lignans are cytotoxic against certain cancer types, such as mouse skin cancer, or tumor and leukemic lines *in vitro*.

Some plants containing lignans with anticancer properties are indicated in the following table (for more details on each plant, please consult Chapter 9 of this book) (Figure 2.8):

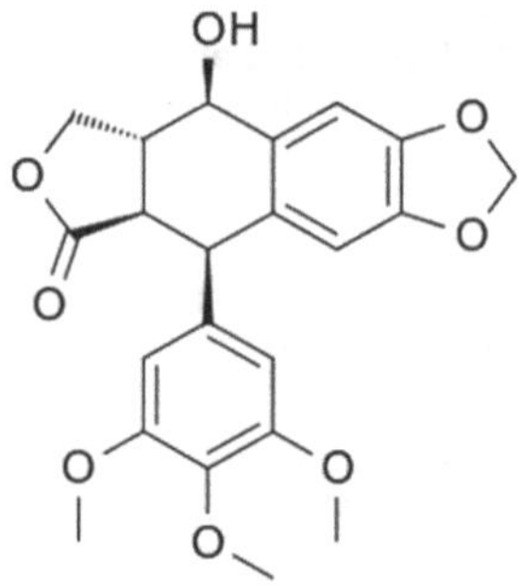

Figure 2.8 Structure of podophyllotoxin, the major anticancer lignin from *Podophyllum* sp.

Species	Target Disease or Cell Line (If Known)	Mode of Action (If Known)
Brussea sp.	KB, P388	
Juniperus virginiana	Liver (mice)	
Magnolia officinalis	Skin (mice)	
Podophyllum sp.	Lung, testicular cancer	Microtubule inhibitor
Plumeria rubia	P388, KB	
Wikstrolemia foetida	P388	Cytotoxic

PHENOLS and derivatives are the main aromatic compounds of plants, whose structural formulas contain at least one benzene ring. They serve as odors, fungicidals, or germination inhibitors. Coumarins are especially common in grasses, orchids, citrus fruits, and legumes.

Structure and properties: simple phenols are colorless solids, which are oxidized by air. Water solubility increases with the number of hydroxyl groups present, but solubility in organic solvents is generally high. Natural aromatic acids are usually characterized by having at least one aliphatic chain attached to the aromatic ring.

Coumarins are lactones of o-hydroxycinnamic acid. Almost all natural coumarins have oxygen (hydroxyl or alkoxy) at C-7. Other positions may also be oxygenated and alkyl side-chains are frequently present. **Furano-** and **pyranocoumarins** have a pyran or furan ring fused with the benzene ring of a coumarin.

Phenolic acids may be extracted from plant tissues in 2% sodium bicarbonate. Upon acidification, acids often precipitate or may be extracted with ether. After removal of carboxylic acids, phenols may be extracted with 5% sodium hydroxide solution. Phenols are usually not steam-distillable, but their esters can be. Coumarins can be purified from a crude extract by treatment with warm dilute alkali which will open the lactone ring and form a water-soluble coumarinate salt. After removal of organic impurities with ether, coumarins can be reconstituted by acidification.

Distribution: wide, abundant in herbs of the families Lamiaceae and Boraginaceae.

Biosynthesis in plant cells: phenolic compounds generally are derived from shikimic acid via the phenylpropanoid pathway.

Basis of anticancer/antitumor activity: phenolic compounds are cytotoxic against certain cancer types *in vitro*. They usually interfere with the integrity of the cell membrane or inhibit various protein kinases. Coumarins, in particular furanocoumarins, are highly toxic.

Some plants containing phenols with anticancer properties are indicated in the following table (for more details on each plant, please consult Chapter 9 of this book) (Figure 2.9):

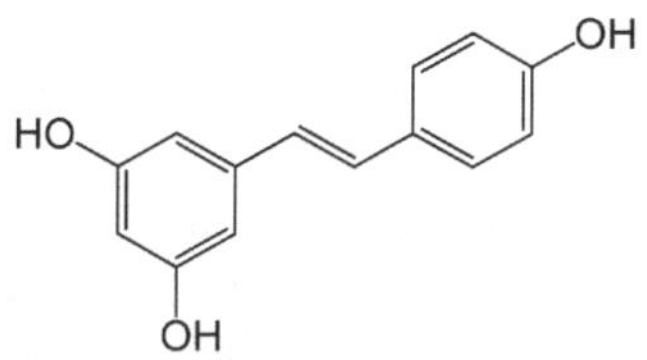

Figure 2.9   Structure of resveratrol, a natural phenol.

Species	Target Disease or Cell Line (If Known)	Mode of Action (If Known)
Angelica gigas, A. decursiva, A. keiskei		Cytotoxic
Gossypium indicum	Murine B16 melanoma, L1210 lymphoma	Cytotoxic
Vitis vinifera	Leukemia (HL-60)	COX-1 inhibition

ALDEHYDES are volatile substances found (along with alcohols, ketones, and esters) in minute amounts and contributing to the formation of odor and flavor of plant parts.

Structure and properties: they are aliphatic, usually unbranched molecules, with up to 12 carbon atoms (C_{12}). They can be extracted from plants by distillation, solvent extraction, or aeration.

Biosynthesis in plant cells: the biosynthesis of aldehydes is related to fatty acids.

Basis of anticancer/antitumor activity: some aldehydes are cytotoxic against certain cancer types *in vitro*, mainly due to inhibition of tyrosinase. Immunomodulatory properties have been also ascribed to this group of secondary metabolites.

Some plants containing aldehydes with anticancer properties are indicated in the following table (for more details on each plant, please consult Chapter 9 of this book) (Figure 2.10):

Species	Target Disease or Cell Line (If Known)	Mode of Action (If Known)
Cinnamonium cassia	Human cancer lines, SW-620 xenograft	Cytotoxic, immunomodulatory
Mondia whitei		Tyrosinase inhibitor
Rhus velgaus		Tyrosinase inhibitor
Sclerocarya caffra		Tyrosinase inhibitor

UNSAPONIFIABLE LIPIDS (in particular **quinones**) are a diverse group of substances generally soluble in organic solvents and not saponified by alkali. They are yellow to red pigments, often constituents of wood tissues, and have toxic and antimicrobial properties.

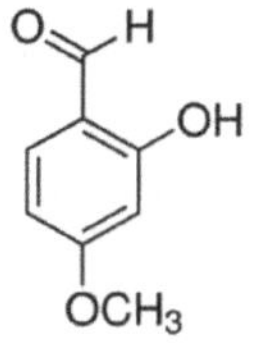

Figure 2.10   Structure of 2-hydroxy-4-methoxybenzaldehyde, a tyrosinase inhibiting aldehyde from *Mondia whitei*.

Structure and properties: naphthoquinones are yellow-red plant pigments, extractable with non-polar solvents, such as benzene. They can be separated from lipids by stem distillation weak alkali treatment. **Anthraquinones** represent the largest group of natural quinones, are usually hydroxylated at C-1 and C-2, and commonly occur as glycosides (water-soluble). Thus, their isolation is carried out according to the degree of glycosidation. Hydrolysis of glycosides (after extraction in water or ethanol) takes place by heating with acetic acid or dilute alcoholic HCl. Phenanthraquinones have a rather more complex structure and can be extracted in methanolic solutions.

Distribution: anthraquinones are particularly found in the plant families Rubiaceae, Rhamnaceae, and Polygonaceae. Phenanthraquinones are rare compounds, having important medicinal properties (e.g. hypericin from *Hypericum perforatum*, tanshinone from *Salvia miltiorrhiza*).

Biosynthesis in plant cells: they are derived by condensation of several molecules of acetate (more specifically malonyl-coenzyme A), thus being related to long chain fatty acids.

Basis of anticancer/antitumor activity: several quinones are cytotoxic against certain cancer types, such as melanoma, or tumor lines *in vitro*.

Some plants containing quinones with anticancer properties are indicated in the following table (for more details on each plant, please consult Chapter 9 of this book) (Figure 2.11):

Species	Target Disease or Cell Line (If Known)	Mode of Action (If Known)
Kigelia pinnata	*in vitro* melanoma, renal cell carcinoma	
Landsbergia quereifolia	P388	
Mallotus japonicus	*in vitro*: human lung carcinoma, B16 melanoma, P388, KB	Cytotoxic
Nigella sativa	MDR human tumor	Cytotoxic *in vitro*
Rubia cordifolia	*in vitro* human cancer lines	
Sargassum tortile	P388	Cytotoxic
Wikstroemia elliptica, W. indica	Erlich ascites carcinoma, MK, P388	

UNIDENTIFIED COMPOUNDS usually refer to complex mixtures or plant extracts, the composition of which has not been elucidated in detail or the bioactive properties of which cannot be assigned to a particular substance only. Ironically, unidentified extracts are usually more potent against various types of cancer than single, well-studied molecules.

Some plants containing unidentified compounds with anticancer properties are indicated in the following table (for more details on each plant, please consult Chapter 9 of this book):

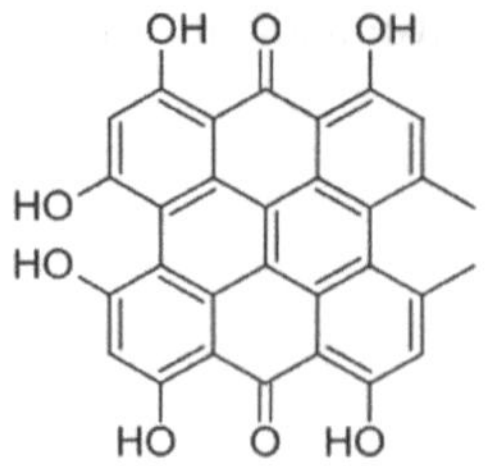

Figure 2.11 Structure of hypericin, an anthraquinone.

Species	Target Disease or Cell Line (If Known)	Mode of Action (If Known)
Chelidonium majus	Esophageal squamous cell carcinoma clinical	Immunostimulant
Menispermum dehiricum	Intestinal metaplasia, atypical hyperplasia of the gastric	Anti-estrogen, LH-RH antagonist (mice)
Palomia sp.	Esophageal squamous cell carcinoma clinical	Immunostimulant
Phyllanthus amarus	Antiviral (HBV)	
Phyllanthus linblica	NK cells	Immunostimulant
Trifolium prateuse		Chemopreventive

2.5 BIOTECHNOLOGY AND THE SUPPLY ISSUE

In spite of the plethora of plant metabolites with tumor cytotoxic or immuno-stimulating properties, plants may not be an ideal source of natural anticancer drugs. There is a number of reasons making the recovery of plant-derived products less attractive than, for example, chemical synthesis of a given compound, should this be possible:

First of all, there is the *supply issue*. In several cases, compounds of interest come from slow-growing plant species (e.g. woody species or perennials like *Podophyllum* sp.), or species that are endangered. In addition, *in vivo* productivity, expressed as the actual content in the compound of interest per unit of plant fresh or dry weight, can be dramatically low; this in turn necessitates the use of an overwhelming amount of plant biomass in order to obtain a satisfactory portion of the natural product. For example, Taxol concentration in needles and dried bark of *Taxus brevifolia* lies in the range of 0.01 to 0.1%. Supply can also be hindered due to inadequate plant production, which, in turn, may be caused by a number of problems related to disease, drought, or socioeconomic factors. The issue of supply can be partly resolved by breeding for high-yielding varieties, introduction of wild species, or derivation (in abundant amounts) of precursors for the hemisynthesis of the desired product.

Secondly, plant-derived pharmaceutical extracts frequently lack the necessary *standardization* that could render them reliable for large-scale, clinical use. This problem is related to a number of factors, such as the dependence of the production on environmental factors and the plant developmental stage (e.g. flowering), as well as the heterogeneity of the extracts, which makes further isolation and purification of the product an indispensable, though costly step. A representative example is mistletoe lectin extract, which presents a remarkable seasonal variation in the levels of ML isolectins (ML I, ML II, and ML III). Furthermore, the specific bioactivity of the extract fluctuates over a prolonged (e.g. two-year) storage time.

Biotechnology could offer an alternative method for the production of sufficient plant biomass and/or quantity of target natural products in a relatively short time. *In vitro* techniques are a major component of plant biotechnology, since they permit to artificially control several of the parameters affecting the growth and metabolism of cultured tissues. Plainly put, plant tissue culture works on the principle of inoculating an explant (that is a piece of plant tissue, such as a leaf or stem segment) from a donor plant on a medium containing nutrients and growth regulators and causing thereof the formation of a more or less dedifferentiated, rapidly growing *callus tissue*. Production of plant-derived anticancer agents could be advantageous over derivation from plants *in vivo* since:

1. By altering the culture parameters, it might be possible to control the quantity, composition, and timing of production of secondary metabolites, both in qualitative and quantitative senses. In this way, problems associated with the standardization of plant extracts could be overcome.
2. By feeding cultures with precursor substances for the biosynthesis of certain metabolites, a higher productivity can be achieved from cultured cells (*in vitro*) than from whole plants.

3. Potentially, entirely novel substances can be synthesized through biotransformation or by taking advantage of somaclonal variation, i.e. a transient or heritable variability of metabolic procedures induced by the procedure of *in vitro* culture.
4. The establishment of a callus culture is the first step required in order to obtain genetically modified cells or plants, e.g. crop plants able to specifically produce a desired product in excessive amount.
5. Protoplasts are plant cells having their cell wall artificially removed. In this way, they can be used in gene transfer experiments and for the creation of hybrid cells, i.e. cells resulting from the direct fusion of two protoplast cells that might have been derived from entirely different species.
6. Plant species that are difficult to propagate could be clonally micropropagated, thus obtaining thousands of seedlings from a very limited mass of donor tissue (essentially from one donor plant only). This can be achieved by plant regeneration via organogenesis (i.e. induction of shoots and roots from callus cultures) or somatic embryogenesis (i.e. the process of embryo formation from somatic (sporophytic) tissues without fertilization).
7. The identification and characterization of genes involved in the biosynthetic pathway of various natural products could allow the cloning of said genes and their heterologous expression in other plant species (e.g. tobacco) or microorganisms (e.g. yeast) that are particularly amenable to scale-up biotechnological processes.

Promising as the perspectives of plant cell culture and genetic engineering may be, established plant-derived commercial anticancer drugs are still produced by isolation from growing plants, eventually drugs are semisynthetically produced from natural precursors also isolated from plant sources *in vivo*. Currently, there are only a few plant-derived natural compounds with antineoplastic properties that are being produced biotechnologically, mostly on the laboratory or pilot-scale level:

Madagascar periwinkle (*Catharanthus roseus*): due to the abundance and wide dispersion of this plant species, the adequate supply of raw material for the extraction of *Catharanthus* alkaloids is not an issue. That said, alkaloids with anticancer properties (more significantly vinblastine and vincristine) represent only a minor fraction of the approximately 130 alkaloids produced by this species and are usually isolated in extremely small quantities (in the order of 10^{-6} per f.w.), following extensive extraction. For this reason, *C. roseus* has been one of the early targets of systematic research for tissue culture application for the purpose of producing secondary metabolites with medicinally important bioactive properties. Progress in this direction has been facilitated by the relative amenability of the species to *in vitro* treatments, including its *Agrobacterium*-mediated transformation. For example, *C.roseus* cell cultures overexpressing the *isopentenyltransferase (Ipt)* gene overproduce cytokinins and can grow without the exogenous supplementation of plant growth regulators. Numerous studies have been conducted on the scale-up indole alkaloid production from cell suspension cultures of *C. roseus*, although success has been mainly reported with the production of the antihypertensive alkaloid ajmalicine, not vinblastine or vincristine. Several factors affecting production have been evaluated, including medium nutrient and growth regulator composition, elicitors (jasmonate, pectinase, UV radiation, and fungal and yeast extracts), immobilization, osmotic stress, and precursor (tryptophan, loganin, secologanin, and tryptamine) feeding. Vinblastine, an antileukemic dimeric indole alkaloid dimmer cannot be directly produced from *C. roseus in vitro*, due to under-expression of the enzyme acetyl-CoA: 17-O-deacetylvindoline 17-O-acetyltransferase (DAT), which catalyzes the formation of vindoline, one of the substrates leading to anhydrovinblastine. Yield values of catharanthine (the second substrate for vinblastine synthesis) up to 17 μg l⁻¹ after fungal induction have been reported. Hairy root cultures have been a rather more efficient approach, with reported catharanthine contents in the range of 0.2 mg/g f.w. Both homologous and heterologous (tobacco, *Cinchona officinalis*, yeast) expression of the vinblastine biosynthetic pathway (including at least 35 intermediates and 32 genes) has been attempted with various degrees of success. Among the cloned genes of interest are the so-called *octadecanoid-responsive Catharanthus AP2-domains (Orca's)*, *strictosidine synthase (STR)* (inducible by yeast extracts), *tryptophan decarboxylase (TDC)*, *cytochrome P450-dependent monooxygenase* *tabersonine-11-hydroxylase (T11H)*, and *DAT* (the last two being regulated by light), as well as cytochrome P450-encoding (CYP) genes. More recently, a yeast strain was developed, by introducing 21 new genes (15 from different plant species and one avian species) and deleting another three, which was able to synthesize strictosidine, a common intermediate of the monoterpene indole alkaloid pathway.

Pacific Yew (*Taxus brevifolia*): since the original isolation of the complex tetracyclic diterpene *Taxol (paclitaxel)* from the bark of *T. brevifolia* and determination of its anticancer properties, the supply of this important molecule has been an issue, not the least due to the slow growth of the source plant species and the quite low Taxol content (0.1–0.6 ‰ d.w.). It is no wonder that naturally derived Taxol enjoys a market price as high as \$600,000 per kilogram. The related *baccatin III* and *10-deacetylbaccatin III (10-DAB)*, used for the semisynthesis of *docetaxel (taxotere)* is more abundant since it can be isolated from the needles of the tree, although it requires extensive purification of the intermediate compounds. On the other hand, the available protocols for the total chemical synthesis of Taxol are quite complex and multi-step, having the disadvantage of low yields and toxic side products. Therefore, biotechnological approaches to increase Taxol availability have been investigated almost from the beginning of the utilization of *Taxus* species as sources of cancer therapeutic compounds. In 1977 NCI awarded contracts for the investigation of plant tissue culture as a source of anticancer drugs, and two of these studied related to Taxol production. Unfortunately, these contracts were terminated in 1980 before any positive results had been obtained. Considerable research effort has once more been focused on the application of this technology to Taxol production. Ironically, *T. brevifolia* has been proven as the least productive source of Taxol and 10-DAB in tissue culture. On the contrary, other Taxus species, such as *T. cuspidata*, *T. x media*, *T. yunnanensis*, *T. globosa*, and *T. chinensis* are more suitable as cell factories for the production of a series of taxanes. Taxol has been also isolated from cell cultures of other species than *Taxus*, including hazel (*Corylus avellana*), ginkgo (*Ginkgo biloba*), and Japanese larch (*Larix leptolepis*) as well as more than 30 endophytic fungal species from Taxol-yielding hosts, including *Fusarium mairei*, *Taxomyces andreanae*, *Nigrospora* sp., and *Phomopsis* sp. The optimization of the *in vitro* biosynthesis of Taxol and other taxanes has been thoroughly investigated in the past 15 years. Although many aspects of the biosynthetic pathway remain to be characterized, an established step is the cyclization of geranyl geranyl diphosphate (GGPP) to taxadiene, catalyzed by plastid-located taxiadiene synthethase (TASY). Various approaches have been tested to increase the *in vitro* yield of taxanes, mainly by using elicitors such as fungal extracts, organic acids (arachidonic acid, methyl jasmonate, 2,3-dihydroxypropyl jasmonate), fungal-derived oligosaccharides (e.g. chitosan), and inorganic compounds (vanadyl sulfate), feeding the cultures with precursors (e.g. L-phenylalaline), heat shock treatment (35–50°C), using specific-type explants (e.g. cambial meristematic cells, CMCs), and modifying culture conditions (e.g. increasing oxygen and ethylene concentration as well as osmotic pressure). The production of up to 1.17% of paclitaxel (Taxol) within five days of elicitation with methyl jasmonate, along with other taxoids, such as *13-acetyl-9-dihydrobaccatin*, *9-dihydrobaccatin III*, and *baccatin VI* was reported during the 1990s. Advanced techniques including co-culture of *Taxus* cells with endophytic fungal cells in bioreactors and use of various elicitors have resulted in paclitaxel yields in the order of 25–67 mg/L. This was further increased to 102 mg/kg fresh weight by using CMCs from *T. cuspidata* as the explant material. Immobilized *T. cuspidata* and *T. baccata* cell cultures have been also developed as a system for the continuous production of baccatin III and paclitaxel at a respectively two- to three-fold increased yield compared to suspension cultures. On the other hand, it has been demonstrated that suspension cultures offer the advantage of active substance release in the liquid medium during the culture process, at rather considerable rates (32% and 69% for paclitaxel and baccatin III, respectively). Homologous expression of key genes involved in Taxol biosynthesis, including TASY and phenylalanine-CoA ligase, has been achieved in *Arabidopsis thaliana*, tomato, *Artemissia annua*, *Nicotiana benthamiana*, and the moss species *Physcomitrella patens*. A trade-off is the growth retardation observed in most transformed species (with the notable exception of *A. annua*). Finally, the heterologous biosynthesis of Taxol in microorganisms is an additional biotechnological approach with promising applications for the scale up production of taxoids. For example, taxadiene has been synthesized from genes encoding enzymes in the Taxol biosynthetic pathway, including GGPP synthase, TASY, and 10-deacetylbaccatin III-10-O-acetyl transferase (TmDBAT) and functionally overexpressed in *Escherichia coli*, leading to an accumulation of 1.3 mg/L of cell culture. Phyton Catalytic (Ithaca, NY) and Samyang Genex (South Korea) were among the pioneer companies to delve into scale-up production of Taxol by means of plant cell culture, while another company, Cytoclonal Pharmaceutics (Dallas, TX), acquired a license from Montana State University to commercialize Taxol production from *T. andreanae*.

American mandrake, mayapple (*Podophyllum hexandrum*, *P. pletatum*): podophyllotoxin (PDT) is a lignan and the major constituent of podophyllin, the resin of the roots and rhizomes, partly of the leaves, of the *Podophyllum* species. PDT has numerous pharmaceutical properties, including antiviral and antitumor ones. PDT is also accumulated in other species of the family Berberidaceae, including *Sinopodophyllum hexandrum* and *Dysosma* sp., as well as members of other families, such as Linanceae and Cupressaceae (at least 35 species in total) and several endophytic fungi, including *Fusarium oxysporum*, *F. solani*, and *Trametes hirsuta*. Due to the enormous demand for PDT, combined with their low seed germination, rhizome development, and plant growth rates, both *Podophyllum* and *Sinopodophyllum* species are considered critically endangered according to the criteria of the International Union for Conservation of Nature (IUCN) (www.iucn.org). Therefore, alternative PDT sources have been investigated. The full chemical synthesis of PDT is not considered economically feasible due to its complex nature. Tissue culture has been applied as the methodological approach of choice, either for enhancing seed germination or for establishing PDT-producing cell factories. Various parameters have been evaluated for optimizing *in vitro* culture conditions, including the effect of light (with prolonged incubation in darkness having a positive effect), cell immobilization, elicitors (e.g. methyl jasmonate, silver, coronalon, ancymidol, coniferin), media composition (in particular, NO_3^-, PO_4^{3-}, Na^+, Fe, and Mn- enriched), and growth regulators (predominantly GA_3). *in vitro* productivity is cell line-dependent, whereas starting materials such as leaves and fungi may be inferior to rhizomes in terms of PTOX accumulation but superior as far as total productivity and cost efficiency are concerned. In addition, the use of hairy root cultures of the aforementioned and other plant species (e.g. *Linum album*) has been successfully tested with three-fold increase in PDT content observed in *P. hexandrum* hairy root lines. Even though PDT accumulation *in vitro* may be 10- to 100-fold lower than *in vivo*, tissue culture allows the continuous and unlimited production of biomass of PDT-producing plant species, while product accumulation can be controlled by regulating culture conditions and/or precursor feeding (e.g. with L-tyrosine). Finally, developments in the molecular dissection of the lignin biosynthetic pathway have created a basis for the perspective heterologous expression of key genes involved in PDT synthesis (a total of 26 genes known so far), including *LuPLR1 (pinoresinol lariciresinol reductase 1)*, *PhCAD (Podophyllum hexandrum cinnamyl alcohol dehydrogenase)*, and *SD (secoisolariciresinol dehudrogenase)*.

Mistletoe (*Viscum album*): Becker and Schwarz (1971) were the first to mention the possible use of mistletoe callus cultures as a source of bioactive products. In 1990, Fukui et al. reported on the induction of callus from leaves of *V. album* var. *lutescens*: they were able to identify in the callus two galactose-binding lectins which were originally observed in mistletoe leaves. Kintzios and Barberaki (2000) succeeded in inducing callus and protoplast cultures from mistletoe leaves and stems in a large number of different growth regulator and media treatments. They have also studied the effects of different plant parts (stems and leaves), harvest times (winter or summer), explant disinfection methods, growth regulators, culture medium composition, and cell wall digestion treatments. They observed a relatively low (8%) somaclonal variation, in the aspect of both the quantitative and the qualitative mistletoe protein production *in vitro* (Kintzios et al. 2000). More recently, the considerably increased cytotoxic properties of somaclonal lectin extracts was demonstrated *in vitro* on PC12 pheochromocytoma and RAW 264.7 macrophage cell cultures (Barberaki et al. 2015). At the same time, considerably lower (essentially zero) toxicity was observed on the non-cancer, immortalized Vero cells. These results suggest that somaclonally variant mistletoe callus cultures could be used as potential cell factories for the production of novel proteins with anticancer activity. Finally, Biosyn (Fellbach, Germany) pioneered the production of mistletoe lectin polypeptides in homologous and heterologous host systems and mistletoe lectin peptides, by cloning different fragments of the ML gene from mistletoe genomic DNA, constructing expression vectors (A- and B-chain coding region), and expressing the single chains separately in *E. coli*. This approach provided a process for producing mistletoe lectins in sufficient quantities, at the same time imitating the diversity in ML-I isoenzymes of the natural mistletoe extract. Recombinant mistletoe lectins (rMLs) depicted promising activity in terms of *in vitro* cytotoxicity against human lymphoblastic leukemia MOLT-4 cells and immune cell activating potency, among other properties.

Cancer tree (*Campotheca acuminata*): this deciduous tree species, native to southern China and Tibet, is rich in the quinoline alkaloid camptothecin, a potent antitumor substance and the source of the semi-synthetic derived prescription anticancer drugs topotecan and irinotecan. Camptothecin is also found in *Nothapodytes foetida* and *Ophiorrhiza* species, including *O. pumila* and *O. rugosa*. Tissue cultures have been established from all species, with shoot culture-regenerants having a higher alkaloid content compared to normal plants (with a maximum content of 0.09% d.w. in the leaves of *O. rugosa,* representing a 450-fold increase). UV-B light and salicylic acid were identified as the best elicitors of camptothecin and 10-hydroxycamptothecin accumulation in cell cultures of *C. acuminata* (with an observed 11-fold and 25-fold increase in content, respectively). Hairy root cultures of *C. acuminata* and *O. pumila* are also able to produce and exude large quantities of camptothecin into the culture medium. Finally, the biotechnological production of camptothecin by endophytic fungi from the bark of *N. foetida* has also been considered.

REFERENCES

Ajikumar, P.K., W.H. Xiao, K.E. Tyo, et al. 2010. Isoprenoid pathway optimization for Taxol precursor overproduction in *Escherichia coli. Science* 330:70–74.

Alamgir, A.N.M. 2017. *Therapeutic Use of Medicinal Plants and their Extracts: Volume 1: Pharmacognosy (Progress in Drug Research).* Springer, Berlin, 546 p. ISBN 978-3319638614.

Anterola, A., E. Shanle, P.F. Perroud, and R. Quatrano. 2009. Production of taxa-4(5),11(12)-diene by transgenic *Physcomitrella patens. Transgenic Res.* 18:655–660.

Aremu, A.O., L. Cheesman, J.F. Finnie, and J. Van Staden. 2011. *Mondia whitei* (Apocynaceae): A review of its biological activities, conservation strategies and economic potential. *S. Afr. J. Bot.* 77:960–971.

Baena Ruiz, R., and P. Salinas Hernández. 2016. Cancer chemoprevention by dietary phytochemicals. *Epidemiol. Evid. Maturitas* 94:13–19.

Barberaki, M., E. Dermitzaki, A.N. Margioris, M. Theodosaki, S. Grafakos, and S., Kintzios. 2015. Protein extracts from somaclonal mistletoe (*Viscum album* L.) callus with increased tumor cytotoxic activity *in vitro. Curr. Bioact. Comp.* 11:104–108.

Barberaki, M., and S. Kintzios. 2002. Accumulation of selected macronutrients in mistletoe tissue cultures: Effect of medium composition and explant source. *Sci. Horticult.* 95:133–150.

Becker, H., and G. Schwarz. 1971. Callus cultures from *Viscum album*. A possible source of raw materials for gaining therapeutic interesting extracts. *Planta Med.* 20:357–362.

Bentebibel, S., E. Moyano, J. Palazón, R.M. Cusidó, M. Bonfill, R. Eibl, and M.T. Piñol. 2005. Effects of immobilization by entrapment in alginate and scale-up on paclitaxel and baccatin III production in cell suspension cultures of *Taxus baccata. Biotechnol. Bioeng.* 89:647–655.

Bestoso, F., L. Ottaggio, A. Armirotti, et al. 2006. In vitro cell cultures obtained from different explants of *Corylus avellana* produce Taxol and taxanes. *BMC Biotechnol.* 6:45.

Beuth, J. 2000. Natural versus recombinant mistletoe Lectin-1. Market trends. In: *Medicinal and Aromatic Plants-Industrial Approaches: The Genus Viscum*. Ed. A. Bussing. Harwood Publishers, Amsterdam, pp. 237–246.

Bhattacharyya, D., S. Hazra, A. Banerjee, R. Datta, D. Kumar, S. Chakrabarti, and S. Chattopadhyay. 2016. Transcriptome-wide identification and characterization of CAD isoforms specific for podophyllotoxin biosynthesis from *Podophyllum hexandrum*. *Plant Mol. Biol.* 92:1–23.

Brown, S., M. Clastre, V. Courdavault, and S.E. O'Connor. 2015. De novo production of the plant-derived alkaloid strictosidine in yeast. *Proc. Natl. Acad. Sci. U.S.A.* 112:3205–3210.

Carnesecchi, S., R. Bras-Gonçalves, A. Bradaia, M. Zeisel, F. Gossé, M.F. Poupon, and F. Raul. 2004. Geraniol, a component of plant essential oils, modulates DNA synthesis and potentiates 5-fluorouracil efficacy on human colon tumor xenografts. *Cancer Lett.* 215:53–59.

Carnesecchi, S., K. Langley, F. Exinger, F. Gosse, and F. Raul. 2002. Geraniol, a component of plant essential oils, sensitizes human colonic cancer cells to 5-fluorouracil treatment. *J. Pharmacol. Exp. Ther.* 301:625–630.

Cha, M., S.H. Shim, S.H. Kim, O.T. Kim, S.W. Lee, S.Y. Kwon, and K.H. Baek. 2012. Production of taxadiene from cultured ginseng roots transformed with taxadiene synthase gene. *BMB Rep.* 45:589–594.

Chen, J., R.J. Henny, P.S. Devanand, and C.T. Chao. 2006. AFLP analysis of Nephthytis (*Syngonium Podophyllum* Schott) selected from somaclonal variants. *Plant Cell Rep.* 24:743–749.

Croteau, R., R.E.B. Ketchum, R.M. Long, R. Kaspera, and M.R. Wildung. 2006. Taxol biosynthesis and molecular genetics. *Phytochem. Rev.* 5:75–97.

Cusido, R.M., M. Onrubia, A.B. Sabater-Jara, et al. 2014. A rational approach to improving the biotechnological production of taxanes in plant cell cultures of *Taxus* spp. *Biotechnol. Adv.* 32:1157–1167.

Davey, M.R., and P. Anthony (Eds.) 2010. *Plant Cell Culture: Essential Methods*. Wiley, New Jersey, 358 p. ISBN 978-0470686485.

Debnath, T., N.C. Deb Nath, E.K. Kim, and K.G. Lee. 2017. Role of phytochemicals in the modulation of miRNA expression in cancer. *Food Funct.* 8:3432–3442.

Evans, W.C. 2009. *Trease and Evans' Pharmacognosy*, 16th ed. Saunders Ltd., Philadelphia, PA, 616 p. ISBN 978-0702029332.

Fu, C., L. Li, W. Wu, M. Li, X. Yu, and L. Yu. 2012. Assessment of genetic and epigenetic variation during long-term *Taxus* cell culture. *Plant Cell Rep.* 31:1321–1331.

Holton, R.A., H.B. Kim, C. Somoza, et al. 1994. First total synthesis of Taxol. 2. Completion of the C and D rings. *J. Am. Chem. Soc.* 116:1599–1600.

Howat, S., B. Park, I.S. Oh, Y.W. Jin, E.K. Lee, and G.J. Loake. 2014. Paclitaxel: Biosynthesis, production and future prospects. *New Biotechnol.* 31:242–245.

Gavamukulya, Y., F. Wamunyokoli, and H.A. El-Shemy. 2017. *Annona muricata*: Is the natural therapy to most disease conditions including cancer growing in our backyard? A systematic review of its research history and future prospects. *Asian Pac. J. Trop. Med.* 10:835–848.

Guo, B., G. Kai, Y. Gong, et al. 2007. Molecular cloning and heterologous expression of a 10-deacetylbaccatin III-10-O-acetyl transferase cDNA from *Taxus × media*. *Mol. Biol. Rep.* 34:89–95.

Gupta, M.L., and A. Dutta. 2011. Stress-mediated adaptive response leading to genetic diversity and instability in metabolite contents of high medicinal value: An overview on *Podophyllum hexandrum*. *Omics* 15:873–882.

Jang, J.Y., J.K. Lee, Y.K. Jeon, and C.W. Kim. 2013. Exosome derived from epigallocatechin gallate treated breast cancer cells suppresses tumor growth by inhibiting tumor-associated macrophage infiltration and M2 polarization. *BMC Cancer* 13:421–421.

Jung, G.R., K.J. Kim, C.H. Choi, T.B. Lee, S.I. Han, H.K. Han, and S.C. Lim. 2007. Effect of betulinic acid on anticancer drug-resistant colon cancer cells. *Basic Clin. Pharmacol. Toxicol.* 101:277–285.

Kajani, A.A., M.R. Mofid, K. Abolfazli, and S.A. Tafreshi. 2010. Encapsulated activated charcoal as a potent agent for improving taxane synthesis and recovery from cultures. *Biotechnol. Appl. Biochem.* 56:71–76.

Kasaei, A., M. Mobini-Dehkordi, F. Mahjoubi, and B. Saffar. 2017. Isolation of Taxol-producing endophytic fungi from Iranian yew through novel molecular approach and their effects on human breast cancer cell line. *Curr. Microbiol.* 74:702–709.

Ketchum, R.E.B., D.M. Gibson, R.B. Croteau, and M.L. Shuler. 1999. The kinetics of taxoid accumulation in cell suspension cultures of *Taxus* following elicitation with methyl jasmonate. *Biotechnol. Bioeng.* 62:97–105.

Kintzios, S., and M. Barberaki 2000. The biotechnology of *Viscum album*: Tissue culture, somatic embryogenesis and protoplast isolation. In: *Medicinal and Aromatic Plants-Industrial Approaches: The Genus Viscum*. Ed. A. Bussing. Harwood Publishers, Amsterdam, pp. 95–100.

Kintzios, S., M. Barberaki, J.B. Drossopoulos, P. Turgelis, J. Konstas, and O. Makri. 2003. Effect of medium composition and explant type on the distribution profiles of selected micronutrients in mistletoe tissue cultures. *J. Plant Nutr.* 26:369–397.

Kintzios, S., M. Barberaki, P. Tourgielis, G. Aivalakis, and A. Volioti. 2002. Preliminary evaluation of soma-clonal variation for the *in vitro* production of new toxic proteins from *Viscum album* L. *J. Herbs Spices Med. Plants* 9:217–221.

Kumar, P., T. Pal, N. Sharma, V. Kumar, H. Sood, and R.S. Chauhan. 2015. Expression analysis of biosynthetic pathway genes vis-à-vis podophyllotoxin content in *Podophyllum hexandrum* Royle. *Protoplasma* 252:1253–1262.

Kumaran, R.S., and B.K. Hur. 2009. Screening of species of the endophytic fungus Phomopsis for the production of the anticancer drug Taxol. *Biotechnol. Appl. Biochem.* 54:21–30.

Kunnumakkara, A.B. 2014. *Anticancer Properties of Fruits and Vegetables: A Scientific Review*. World Scientific Publishing Company, Singapore, 400 p. ISBN 978–9814508889.

Langer, M., H. Zinke, J. Eck, B. Möckel, and H. Lentzen. 1997. Cloning of the active principle of mistletoe: The contributions of mistletoe lectin single chains to biological functions. *Eur. J. Cancer* 33:24.

Lautié, E., M.A. Fliniaux, and M.L. Villarreal. 2010. Updated biotechnological approaches developed for 2,7'-cyclolignan production. *Biotechnol. Appl. Biochem.* 55:139–153.

Lee, C.W., and M.L. Shuler. 2000. The effect of inoculum density and conditioned medium on the production of ajmalicine and catharanthine from immobilized *Catharanthus roseus* cells. *Biotechnol. Bioeng.* 67:61–71.

Li, J.J. 2015. *Top Drugs: History, Pharmacology, Syntheses*. Oxford University Press, Oxford, UK, 224 p. ISBN 978-0199362585.

Li, M., F. Jiang, X. Yu, and Z. Miao. 2015. Engineering isoprenoid biosynthesis in *Artemisia annua* L. for the production of taxadiene: A key intermediate of Taxol. *BioMed. Res. Int.* 2015:504932.

Li, M.F., W. Li, D.L. Yang, L.L. Zhou, T.T. Li, and X.M. Su. 2013. Relationship between podophyllotoxin accumulation and soil nutrients and the influence of Fe^{2+} and Mn^{2+} on podophyllotoxin biosynthesis in *Podophyllum hexandrum* tissue culture. *Plant Physiol. Biochem.* 71:96–102.

Li, S.T., P. Zhang, M. Zhang, et al. 2012. Transcriptional profile of *Taxus chinensis* cells in response to methyl jasmonate. *BMC Genomics* 13:295.

Li, Y.C., W.Y. Tao, and L. Cheng. 2009. Paclitaxel production using co-culture of *Taxus* suspension cells and paclitaxel-producing endophytic fungi in a co-bioreactor. *Appl. Microbiol. Biotechnol.* 83:233–239.

Ma, C., Y. Wang, A. Shen, and W. Cai. 2017. Resveratrol upregulates SOCS1 production by lipopolysaccharide-stimulated RAW264.7 macrophages by inhibiting miR-155. *Int. J. Mol. Med.* 39:231–237.

Ma, J., F. Zeng, C. Ma, et al. 2016. Synergistic reversal effect of epithelial-to-mesenchymal transition by miR-223 inhibitor and genistein in gemcitabine-resistant pancreatic cancer cells. *Am. J. Cancer Res.* 6:1384–1395.

McCreath, S.B., and R. Delgoda 2016. *Pharmacognosy: Fundamentals, Applications and Strategies*. Academic Press, NY, 738 p. ISBN 978-0128021040.

Mishra, R., M.K. Das, S. Singh, R.S. Sharma, S. Sharma, and V. Mishra. 2017. Articulatin-D induces apoptosis via activation of caspase-8 in acute T-cell leukemia cell line. *Mol. Cell. Biochem.* 426:87–99.

Moeckel, B., T. Schwarz, J. Eck, M. Langer, H. Zinke, and H. Lentzen. 1997. Apoptosis and cytokine release are biological responses mediated by recombinant mistletoe lectin *in vitro*. *Eur. J. Cancer* 33:35.

Morris, P., T. Stiefel, W. Voelter, and P. Welters. 2008. Recombinant mistletoe lectins. Patent No CA2320430 C.

Mulsow, K., T. Enzlein, C. Delebinski, S. Jaeger, G. Seifert, and M.F. Melzig. 2016. Impact of mistletoe triterpene acids on the uptake of mistletoe lectin by cultured tumor cells. *PLOS ONE* 11:e0153825.

Nandagopal, K., M. Halder, B. Dash, S. Nayak, and S. Jha. 2017. Biotechnological approaches for production of anti-cancerous compounds resveratrol, podophyllotoxin and zerumbone. *Curr. Med. Chem.* 36:4693–4717.

Nicolaou, K.C., Z. Yang, J.J. Liu, et al. 1994. Total synthesis of Taxol. *Nature* 367:630–634.

Pathania, S., S.M. Ramakrishnan, and G. Bagler 2015. Phytochemica: A platform to explore phytochemicals of medicinal plants. *Database (Oxford)* 2015:bav075.

Patrikios, I., A. Stephanouand, and A. Yiallouris. 2015. Graviola: A systematic review on its anticancer properties. *Am. J. Cancer Prev.* 6:128–131.

Pi, Y., K. Jiang, R. Hou, Y. Gong, J. Lin, X. Sun, and K. Tang. 2010. Examination of camptothecin and 10-hydroxycamptothecin in *Camptotheca acuminata* plant and cell culture, and the affected yields under several cell culture treatments. *Biocell* 34:139–143.

Qian, Z.G., Z.J. Zhao, Y. Xu, X. Qian, and J.J. Zhong. 2005. Highly efficient strategy for enhancing taxoid production by repeated elicitation with a newly synthesized jasmonate in fed-batch cultivation of *Taxus chinensis* cells. *Biotechnol. Bioeng.* 90:516–521.

Rajesh, M., G. Sivanandhan, M. Jeyaraj, R. Chackravarthy, M. Manickavasagam, N. Selvaraj, and A. Ganapathi. 2014. An efficient *in vitro* system for somatic embryogenesis and podophyllotoxin production in *Podophyllum hexandrum* Royle. *Protoplasma* 251:1231–1243.

Ramírez-Estrada, K., T. Altabella, M. Onrubia, et al. 2016. Transcript profiling of jasmonate-elicited *Taxus* cells reveals a β-phenylalanine-CoA ligase. *Plant Biotechnol. J.* 14:85–96.

Ren, Y., W.L. Chen, D.D. Lantvit, et al. 2017. Cardiac glycoside constituents of Streblus asper with potential antineoplastic activity. *J. Nat. Prod.* 80:648–658.

Roja, G. 2008. Micropropagation and production of camptothecin from *in vitro* plants of *Ophiorrhiza rugosa* var. *decumbens*. *Nat. Prod. Res.* 22:1017–1023.

Roja, G., and P.S. Rao. 1998. Anti-cancer compounds from tissue culture of medicinal plants. *J. Herbs Spices Med. Plants* 7:71–102.

Rouck, J.E., B.W. Biggs, A. Kambalyal, W.R. Arnold, M. De Mey, P.K. Ajikumar, and A. Das. 2017. Heterologous expression and characterization of plant taxadiene-5α-hydroxylase (CYP725A4) in *Escherichia coli. Protein Expr. Purif.* 132:60–67.

Roy, S., Y. Yu, S.B. Padhye, F.H. Sarkar, and A.P.N. Majumdar. 2013. Difluorinated-curcumin (CDF) restores PTEN expression in colon cancer cells by downregulating miR-21. *PLOS ONE* 8:e68543.

Ruiz-Sanchez, J., Z.R. Flores-Bustamante, L. Dendooven, E. Favela-Torres, G. Soca-Chafre, J. Galindez-Mayer, and L.B. Flores-Cotera. 2010. A comparative study of Taxol production in liquid and solid-state fermentation with *Nigrospora* sp. a fungus isolated from *Taxus globosa. J. Appl. Microbiol.* 109:2144–2150.

Saito, K., and H. Mizukami 2004. Plant cell cultures as producers of secondary compounds. In: *Plant Biotechnology and Transgenic Plants*. Eds. K.-M. Oksman-Caldentey, and W. Barz. Dekker, New York, pp. 77–109.

Shah, B.A., and S. Avinash 2013. *Textbook of Pharmacognosy and Phytochemistry*, 2nd ed. Elsevier, India. 586 p. ISBN 978-8131234587.

Sharma, A., P. Verma, A. Mathur, and A.K. Mathur. 2018. Genetic engineering approach using early vinca alkaloid biosynthesis genes led to increased tryptamine and terpenoid indole alkaloids biosynthesis in differentiating cultures of *Catharanthus roseus. Protoplasma* 255:425–435.

Sharma, S.H., S. Thulasingam, and S. Nagarajan. 2017. Terpenoids as anti-colon cancer agents – A comprehensive review on its mechanistic perspectives. *Eur. J. Pharmacol.* 795:169–178.

Sirikantaramas, S., T. Asano, H. Sudo, M. Yamazaki, and K. Saito. 2007. Camptothecin: Therapeutic potential and biotechnology. *Curr. Pharm. Biotechnol.* 8:196–202.

Stadler, D., and T. Bach. 2008. Concise stereoselective synthesis of (–)-podophyllotoxin by an intermolecular iron(III)-catalyzed Friedel-Crafts alkylation. *Angew. Chem. Int. Ed. Engl.* 47:7557–7559.

Sun, G., Y. Yang, F. Xie, et al. 2013. Deep sequencing reveals transcriptome re-programming of *Taxus × media* cells to the elicitation with methyl jasmonate. *PLOS ONE* 8:e62865.

Syklowska-Baranek, K., and M. Furmanowa. 2005. Taxane production in suspension culture of *Taxus × media* var. *Hicksii* carried out in flasks and bioreactor. *Biotechnol. Lett.* 27:1301–1304.

Swidergall, M., N.V. Solis, M.S. Lionakis, and S.G. Filler. 2017. EphA2 is an epithelial cell pattern recognition receptor for fungal β-glucans. *Nat. Microbiol.* 3:1074.

Tikhomiroff, C., S. Allais, M. Klvana, S. Hisiger, and M. Jolicoeur. 2002. Continuous selective extraction of secondary metabolites from *Catharanthus roseus* hairy roots with silicon oil in a two-liquid-phase bioreactor. *Biotechnol. Prog.* 18:1003–1009.

van Der Heijden, R., D.I. Jacobs, W. Snoeijer, D. Hallard, and R. Verpoorte. 2004. The *Catharanthus* alkaloids: Pharmacognosy and biotechnology. *Curr. Med. Chem.* 11:607–628.

Verma, P., S.A. Khan, A.K. Mathur, K. Shanker, and R.K. Lal. 2014. Regulation of vincamine biosynthesis and associated growth promoting effects through abiotic elicitation, cyclooxygenase inhibition, and precursor feeding of bioreactor grown *Vinca minor* hairy roots. *Appl. Biochem. Biotechnol.* 173:663–672.

Vongpaseuth, K., and S.C. Roberts. 2007. Advancements in the understanding of *Paclitaxel* metabolism in tissue culture. *Curr. Pharm. Biotechnol.* 8:219–236.

Wiedenfeld, H., M. Furmanowa, E. Roeder, J. Guzewska, and W. Gustowski. 1997. Camptothecin and 10-hydroxy camptothecin in callus and plantlets of *Camptotheca acuminata. Plant Cell Tiss. Organ Cult.* 49:213–218.

Wilson, S.A., and S.C. Roberts. 2012. Recent advances towards development and commercialization of plant cell culture processes for the synthesis of biomolecules. *Plant Biotechnol. J.* 10:249–268.

Xu, J.-P. 2017. *Cancer Inhibitors from Chinese Natural Medicines.* CRC Press, Boca Raton, FL. 772 p. ISBN 9781498787642.

Zhang, C., and P.S. Fevereiro. 2007. The effect of heat shock on paclitaxel production in *Taxus yunnanensis* cell suspension cultures: Role of abscisic acid pretreatment. *Biotechnol. Bioeng.* 96:506–514.

Zhang, P., P.P. Zhou, and L.J. Yu. 2009. An endophytic Taxol-producing fungus from *Taxus* × *media, Aspergillus candidus* MD3. *FEMS Microbiol. Lett.* 293:155–159.

Zhou, M.L., X.M. Zhu, J.R. Shao, Y.M. Wu, and Y.X. Tang. 2012. An protocol for genetic transformation of *Catharanthus roseus* by *Agrobacterium rhizogenes* A4. *Appl. Biochem. Biotechnol.* 166:1674–1684.

CHAPTER 3

Cytotoxic Phenanthridone Alkaloid Constituents of the Amaryllidaceae

Jerald J. Nair and Johannes van Staden

CONTENTS

3.1 INTRODUCTION

The Amaryllidaceae J.St.-Hil. is a large family of bulbous geophytes numbering around 1,000 species in 80 genera (Meerow and Snijman 1998). These monocotyledonous plants have a range covering both tropical and subtropical regions of the globe, but are most speciose in Andean South America, the Mediterranean basin, and South Africa (Meerow and Snijman 1998). Almost one-third of the complement occurs in South Africa where they thrive in the winter rainfall climate of the Western Cape Province (Duncan 2016). Horticulturally, they are reputed for their attractive floral features, epitomized by such popular varieties as 'daffodils' (*Narcissus pseudonarcissus*), 'snowdrops' (*Galanthus nivalis*), and 'snowflakes' (*Leucojum aestivum*) (Meerow and Snijman 1998, Duncan 2016). These now support a vibrant floriculture sector on both sides of the Atlantic, with Britain in particular leading production of daffodil cut flowers with annual sales totaling 30 million dollars (Briggs 2002).

Members of the Amaryllidaceae also have a significant presence in the medicinal practices of indigenous populations around the globe, particularly on the African continent where there is still a strong reliance on traditional medicine (Watt and Breyer-Brandwijk 1962, Hutchings et al. 1996). The use of these plants in the traditional remediation of cancer can be traced back several thousand years, especially in regions noted for their provenance (Graham et al. 2000, Kornienko and Evidente 2008, Caamal-Fuentes et al. 2011, Nair and van Staden 2013). The records of Hippocrates show that the 'oil of narcissus' (*Narcissus poeticus*) was being used in Greek medicine for uterine cancers over 400 years before Christ (Kornienko and Evidente 2008). Furthermore, *Lycoris radiata* and *Zephyranthes rosea* are both documented in the ancient Chinese medicinal literature as a cure for cancer (Hartwell 1967, Graham et al. 2000), whilst *Hymenocallis littoralis* and *Zephyranthes parulla* have for centuries been used as a remedy for cancer in the traditional medicine (TM) of the Inca and Maya people of South and Central America (Hartwell 1967, Caamal-Fuentes et al. 2011). *Amaryllis belladonna*, *Boophone disticha*, and *Crinum delagoense* have been cited by the indigenous Sotho, Xhosa, and Zulu people of southern Africa as remedies for cancer (Nair and van Staden 2013).

The family Amaryllidaceae is also recognizable chemically for its isoquinoline alkaloid constituents (Viladomat et al. 1997, Bastida et al. 2006, Berkov et al. 2012, De Andrade et al. 2012, Jin 2016). The structural diversity showcased by these compounds (1–12) (Figures 3.1 and 3.2) remains unmatched by other alkaloid producing plant families (Jin 2016).They can be divided into several distinct groups based on their relationship to the common amino acid-derived biogenetic precursor norbelladine (1) (Jin 2016). Galantamine (2), lycorine (3), and crinine (4) are representative of the major groups discernible for these alkaloids, whilst homolycorine (5), tazettine (6), and montanine (7) represent the minor groups (Jin 2016). The less conspicuous members of the family include cherylline (8), trisphaeridine (9), and ismine (10) (Jin 2016). Given this structural diversity as well as their challenging molecular architectures, these alkaloids have also served as attractive targets for multi-step orientated synthetic endeavors (Jin 2016).

An impressive feature of these alkaloids is their wide array of biological properties including antiviral, antibacterial, antifungal, antiparasitic, anticancer, antioxidant, anti-inflammatory, and insect antifeedant effects, as well as acetylcholinesterase (AChE), ascorbic acid biosynthesis, and RNA inhibitory activities (Viladomat et al. 1997, Bastida et al. 2006, Berkov et al. 2012, De Andrade et al. 2012, Jin 2016). In terms of commercial success, galantamine (2) stands out amongst

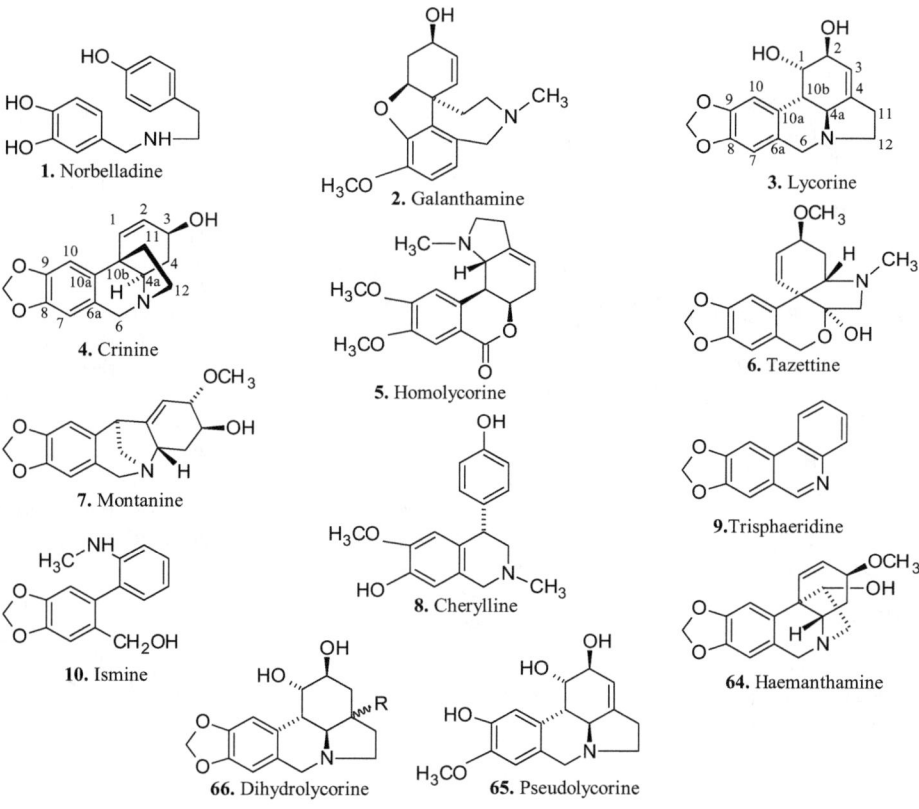

Figure 3.1 Alkaloid representatives of the Amaryllidaceae with significant levels of structural diversity emanating from the common precursor norbelladine.

the Amaryllidaceae as the first FDA approved drug from the family (Heinrich 2010). As a potent and reversible inhibitor of AChE galantamine has gained prominence on the clinical market as an Alzheimer's drug (Heinrich 2010). The pursuit of galantamine as an Alzheimer's drug would not have been possible without indigenous knowledge assimilated from the Caucasus region of Eastern Europe, where its original source *Galanthus woronowii* was shown to be used traditionally for motor-neuron diseases (Heinrich 2010). The success of galantamine as well as the interesting biological activities exhibited by other constituents of the family has rekindled interest in the Amaryllidaceae as a potential source of further chemotherapeutics (Viladomat et al. 1997, Bastida et al. 2006, Berkov et al. 2012, De Andrade et al. 2012, Jin 2016).

Of the various areas of relevance biologically, the Amaryllidaceae alkaloids have been growing most in prominence in the cancer arena (Nair et al. 2016). Although cytotoxic activities in cancer cells have been seen for most group representatives, it is the lycorane, crinane, and phenanthridone alkaloids which hold the most promise (Kornienko and Evidente 2008, Nair et al. 2012a, Nair and van Staden 2014). Of these three groups, the phenanthridones have emerged as the group most likely to spawn an anticancer drug (Kornienko and Evidente 2008, Nair et al. 2016). Exemplified by their chief representatives narciclasine (11) and pancratistatin (12) (Figure 3.2), the phenanthridones have come to the fore as potent and cell line selective antiproliferative agents destined for the clinical arena (Kornienko and Evidente 2008, Nair et al. 2016). This account details the cytotoxic effects of these substances in cancer cells, sheds light on the structural features associated with the anticancer pharmacophore, and affords insight to the molecular mechanisms identified to rationalize the anticancer activity.

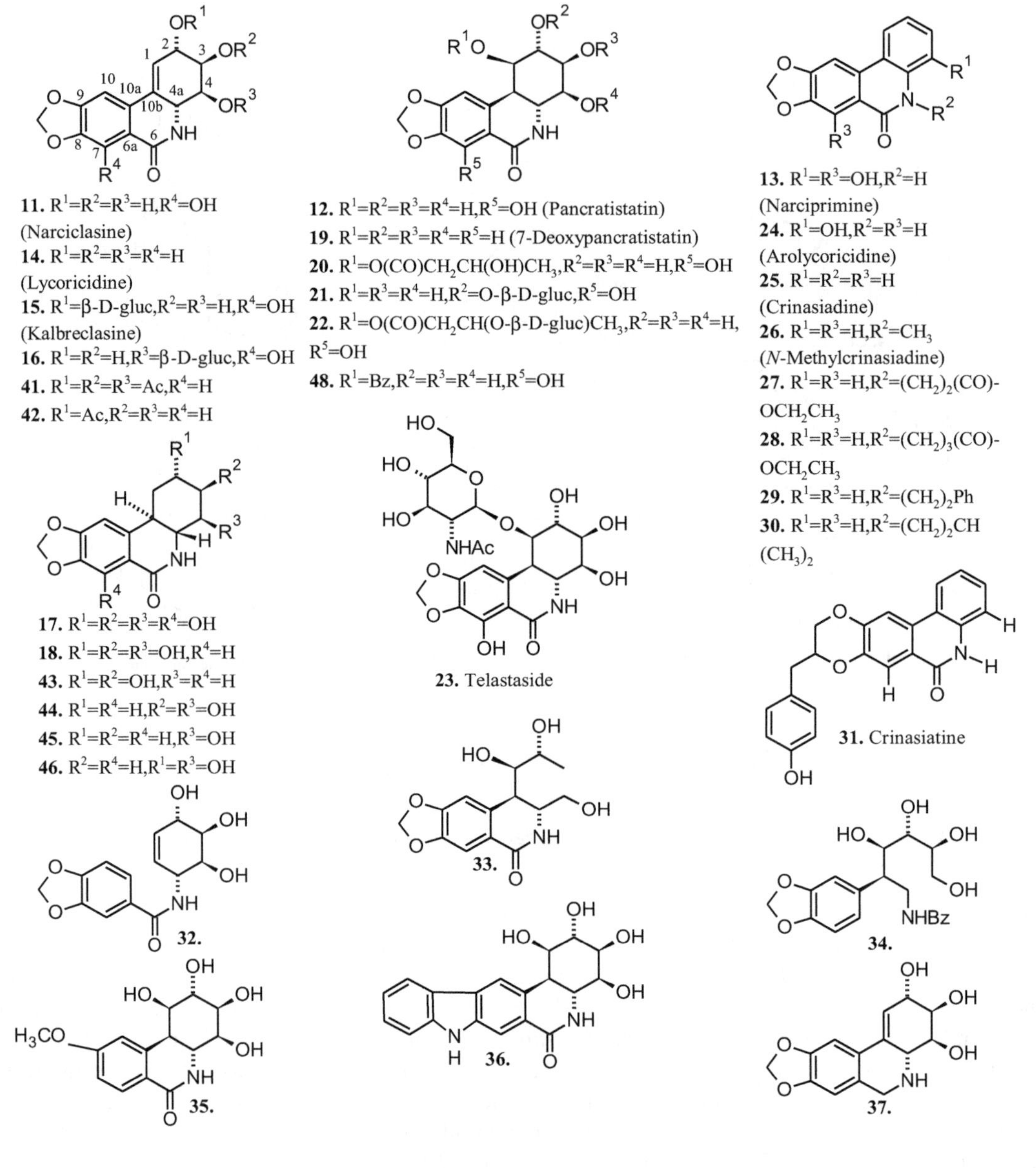

11. R^1=R^2=R^3=H,R^4=OH
(Narciclasine)
14. R^1=R^2=R^3=R^4=H
(Lycoricidine)
15. R^1=β-D-gluc,R^2=R^3=H,R^4=OH
(Kalbreclasine)
16. R^1=R^2=H,R^3=β-D-gluc,R^4=OH
41. R^1=R^2=R^3=Ac,R^4=H
42. R^1=Ac,R^2=R^3=R^4=H

12. R^1=R^2=R^3=R^4=H,R^5=OH (Pancratistatin)
19. R^1=R^2=R^3=R^4=R^5=H (7-Deoxypancratistatin)
20. R^1=O(CO)CH$_2$CH(OH)CH$_3$,R^2=R^3=R^4=H,R^5=OH
21. R^1=R^3=R^4=H,R^2=O-β-D-gluc,R^5=OH
22. R^1=O(CO)CH$_2$CH(O-β-D-gluc)CH$_3$,R^2=R^3=R^4=H,
R^5=OH
48. R^1=Bz,R^2=R^3=R^4=H,R^5=OH

13. R^1=R^3=OH,R^2=H
(Narciprimine)
24. R^1=OH,R^2=R^3=H
(Arolycoricidine)
25. R^1=R^2=R^3=H
(Crinasiadine)
26. R^1=R^3=H,R^2=CH$_3$
(N-Methylcrinasiadine)
27. R^1=R^3=H,R^2=(CH$_2$)$_2$(CO)-
OCH$_2$CH$_3$
28. R^1=R^3=H,R^2=(CH$_2$)$_3$(CO)-
OCH$_2$CH$_3$
29. R^1=R^3=H,R^2=(CH$_2$)$_2$Ph
30. R^1=R^3=H,R^2=(CH$_2$)$_2$CH
(CH$_3$)$_2$

17. R^1=R^2=R^3=R^4=OH
18. R^1=R^2=R^3=OH,R^4=H
43. R^1=R^2=OH,R^3=R^4=H
44. R^1=R^4=H,R^2=R^3=OH
45. R^1=R^2=R^4=H,R^3=OH
46. R^2=R^4=H,R^1=R^3=OH

23. Telastaside

31. Crinasiatine

33.

32.

34.

35.

36.

37.

Figure 3.2 Phenanthridone alkaloids of the Amaryllidaceae which have been examined for cytotoxic effects in various cancer cells.

3.2 ISOLATION OF PHENANTHRIDONES

Given the long history associating the Amaryllidaceae with traditional remedies for cancer, the search and identification of the active constituents had piqued the interest of scientists for some time but was only to see fruition in the latter half of the twentieth century (Nair et al. 2016). Although Amaryllidaceae alkaloids have been detected in roots, stems, and flowers, it is in the bulbs that are found in most abundance, particularly during the flowering stage (Viladomat et al. 1997, Bastida et al. 2006, Berkov et al. 2012, De Andrade et al. 2012, Jin 2016). To date over 20 different natural phenanthridone structures have been described in the literature (Figure 3.2), initially from seven genera of the Amaryllidaceae ratified as *Crinum, Haemanthus, Hymenocallis, Lycoris, Narcissus, Pancratium*, and *Zephyranthes* (Nair et al. 2016). In particular, it is in the genus *Narcissus* that the highest phenanthridone levels have been detected, usually in the region of 30–200 mg/kg of fresh bulbs (Piozzi et al. 1969). They can be divided into three subgroups based on their structural resemblance to either narciclasine (11), pancratistatin (12), or narciprimine (13) (Nair et al. 2016). Narciclasine (11) was the first phenanthridone alkaloid described from the Amaryllidaceae following its identification in *Narcissus pseudonarcissus* (Ceriotti 1967). Although the structure of narciprimine (13) was reported not long after (Piozzi et al. 1968), the appearance of pancratistatin (12) in the literature had to wait almost two decades when Pettit et al. (1984) uncovered its presence in *Zephyranthes grandiflora*. Although the number of natural phenanthridones remains low in comparison to the other groups of Amaryllidaceae alkaloids, given their chemotherapeutic potential there has been significant interest in these compounds from a synthetic standpoint which has bolstered the number of targets now available for biological studies (Kornienko and Evidente 2008, Nair et al. 2016).

3.3 STRUCTURAL FEATURES OF PHENANTHRIDONES

In terms of molecular size phenanthridones are small relative to other alkaloid based anticancer drugs such as camptothecin and vincristine and even smaller than other natural product anticancer agents such as podophyllotoxin and Taxol (Cragg and Newman 2005). Essentially, they comprise a tricyclic skeleton with the central ring (ring-B) distinguished by its heterocyclic character (Jin 2016). Ring-A comprises of a methylenedioxy functionalized aryl system, a β-lactam moiety is characteristic of ring-B, whilst the carbocyclic ring-C is striking for its contiguous hydroxy substitutions (Jin 2016). Narciclasine (11) is distinguished from pancratistatin (12) by the absence of a C-1 hydroxy group, whilst narciprimine (13) is differentiated from these two by possessing an aromatized ring-C (Jin 2016). Five compounds related to narciclasine have been identified including its 7-deoxy analog lycoricidine (14), glucosides (15,16), as well as the dihydro analogs (17,18) (Nair et al. 2016). Similarly, five compounds with close structural proximity to pancratistatin (12) were identified as 7-deoxypancratistatin (19), ester (20), and glucosides (21, 22, 23) (Nair et al. 2016). The narciprimine (13) subgroup was represented by eight natural congeners, which were arolycoricidine (24), crinasiadine (25), *N*-methylcrinasiadine (26), *N*-alkylated derivatives (27, 28, 29, 30), as well as the ring-A 1,4-dioxane analog crinasiatine (31) (Nair et al. 2016).

3.4 SYNTHESIS OF PHENANTHRIDONE ALKALOIDS

Following the discovery of narciclasine (11) and pancratistatin (12) (Figure 3.2) as antiproliferative agents and given the paucity of natural phenanthridones available for biological studies, these compounds have served as attractive targets for which interesting synthetic approaches have been developed (Rinner and Hudlicky 2005, Ghavre et al. 2016). The main challenge of these endeavors has been the densely functionalized ring-C where the installation of the contiguous stereocenters proved to be a significant hurdle to overcome (Rinner and Hudlicky 2005, Ghavre et al. 2016). Nonetheless, nearly 60 total syntheses have to date been published on the key constituents narciclasine (11), pancratistatin (12), lycoricidine (14), *trans*-dihydrolycoricidine (18), and 7-deoxypancratistatin (19) (Rinner and Hudlicky 2005, Ghavre et al. 2016). A further 28 articles have addressed the preparation of unnatural analogs of narciclasine (11) and pancratistatin (12) (Ghavre et al. 2016), whilst five attempts have been made on the synthesis of narciprimine (13) (Nair et al. 2016). Narciprimine is noteworthy in being the first phenanthridone to be synthesized, wherein a photochemical reaction was initially engaged to effect coupling between the two aryl rings (Mondon and Krohn 1970).

Lycoricidine (14) (Figure 3.2) was the first of the polyhydroxylated phenanthridones to be prepared, albeit in racemic form as detailed by Ohta and Kimoto in 1975. The synthesis of racemic pancratistatin (12) was achieved in 1989 (Danishefsky and Lee 1989), whilst its enantiomerically pure (+)-form was prepared six years later (Tian et al. 1995). Interestingly, five of the six total syntheses of narciclasine (11) yielded the natural (+)-enantiomer (Rinner and Hudlicky 2005, Ghavre et al. 2016), with the (−)-enantiomer only secured as the last of these efforts (Matveenko et al. 2008).

3.5 CYTOTOXIC EFFECTS OF PHENANTHRIDONE ALKALOIDS *IN VITRO*

3.5.1 Narciclasine and its Congeners

Motivation for the identification of the antiproliferative principles came early on from studies by Fitzgerald et al. (1958) on the extracts of over 100 species from various genera of the Amaryllidaceae against sarcomas 37 and 180. Here it was evident that extracts of *Narcissus* plants were amongst the most active (Fitzgerald et al. 1958). Inspired by these findings, Ceriotti (1967) subsequently sought to identify the active constituents of *Narcissus pseudonarcissus*, showing convincingly that narciclasine (11) exhibited potent antimitotic activity when administered subcutaneously to mice infected with sarcoma 180 (LD_{50} 5 mg/kg). Its 7-deoxy analog lycoricidine (14) was described not long after from *Lycoris radiata* together with its plant growth inhibitory and cytotoxic effects in the rice seedling and Ehrlich carcinoma tests, respectively (Okamoto et al. 1968). Interestingly these effects were seen to be similar to those manifested by narciclasine (11) which was also found in the plant (Okamoto et al. 1968). Later, Ghosal et al. (1985a, 1986) described the presence of narciclasine (11) in both *Haemanthus kalbreyeri* and *Zephyranthes flava*, whilst its 2-*O*-β-D-glucoside kalbreclasine (15) (Figure 3.2) was identified in the former. Kalbreclasine in this instance was noted for its ability to produce extensive proliferation of splenic lymphocytes in adult mice at a dose of 20 µg or higher (Ghosal et al. 1985a). The mitogenic activation produced by kalbreclasine matched that of the known mitogen concanavalin A (Ghosal et al. 1985a). The 4-*O*-β-D-glucoside (16) of narciclasine following its identification in *Pancratium maritimum* was shown to be cytotoxic to the brine shrimp (*Artemia salina*) (LD_{50} 0.88 µg/mL) (Abou-Donia et al. 1991). Further to this, when assayed on potato discs infected with *Agrobacterium tumefaciens* compound 16 exhibited moderate antitumor activity (53% inhibition of crown gall tumor initiation) (Abou-Donia et al. 1991). In relation to dihydro analogs of narciclasine, Pettit et al. (1990) reported the presence of *trans*-dihydronarciclasine (17) in *Zephyranthes candida* and showed it to be potently active against P388 lymphocytic leukemia cells (ED_{50} 0.0032 µg/mL). Similarly, *trans*-dihydrolycoricidine (18) was reported following a phytochemical study of *Hymenocallis littoralis* and also shown to be active in the P388 system with an ED_{50} of 0.02 µg/mL (Pettit et al. 1990). In terms of potencies of these C-1 deoxyphenanthridones, a comprehensive evaluation was carried out on narciclasine (11), lycoricidine (14), *trans*-dihydronarciclasine (17), and *trans*-dihydrolycoricidine (18) against the NCI 60 cell line panel wherein mean GI_{50} values of 0.016, 0.15, 0.013, and 0.068 µM, respectively, were recorded (Pettit et al. 1993). These findings were central in subsequent efforts to promote narciclasine (11) as a potent antiproliferative agent with significant potential for development into a cancer drug (Pettit et al. 1993).

3.5.2 Pancratistatin and its Congeners

The isolation and characterization of pancratistatin (12) from *Zephyranthes grandiflora* by Pettit et al. in 1984 and its subsequent disclosure as an antineoplastic agent was preceded by much anticipation since the polyhdroxylated phenanthridones up to that stage were only represented by narciclasine (11) and lycoricidine (14). Furthermore, the C-1 hydroxy group in pancratistatin (12) (Figure 3.2) presented a further element of the anticancer pharmacophore to explore since it could potentially serve, as will be seen to be the case later in the text, as a convenient handle for structure–activity relationship study purposes (Nair et al. 2016). In the event, pancratistatin was established as the predominant *in vitro* cytotoxic constituent of *Zephyranthes grandiflora* with an $ED_{50} < 0.01$ µg/mL in the P388 lymphocytic leukemia system (Pettit et al. 1984). After this, Ghosal et al. (1989) reported the presence of 7-deoxypancratistatin (19) and pancratistatin-2-O-β-D-glucoside (21) in *Haemanthus kalbreyeri* and showed that the plant growth regulatory effects of the glucoside (21) were similar to those of kalbreclasine (15). The amino glucoside telastaside (23) was described by Ghosal et al. (1990) from the 'Indian lily moth' *Polytela gloriosa* which is known to consume Amaryllid species such as *Haemanthus kalbreyeri*. In addition to pancratistatin (12), Kojima et al. (1998) found two butanoyl esters (20 and 22) in bulbs of the Mexican plant *Zephyranthes carinata*, all of which were screened against the KB, HeLa, and P388-D1 cancer cell lines. As such, good activities were observed all round and ranged from LD_{50} 41 µM to as low as 0.91 µM with the best activity demonstrated for the hydroxy ester 20 (Kojima et al. 1998). Overall, the P388-D1 cell line proved to be most susceptible to treatment with these three alkaloids (Kojima et al. 1998). Interest in pancratistatin (12) has been sustained owing largely to its potencies against a wide array of cancer cells (Pettit et al. 1993). Encouragingly, a mean GI_{50} value of 0.091µM was determined for pancratistatin in the NCI 60 cell line screen, thus guiding subsequent efforts towards improving its activities in a wide range of cancer cells (Pettit et al. 1993).

3.5.3 Narciprimine and its Congeners

The unsaturated phenanthridone analogs related to narciprimine (13) (Figure 3.2) are significant not only from a cytotoxic perspective, but also from a biosynthetic one since it has been suggested that they may be intermediates of the hydroxylated variants narciclasine (11) and pancratistatin (12) (Piozzi et al. 1968). Accompanying the details of its structure, Piozzi et al. (1968) noted that narciprimine (13), in stark contrast to narciclasine (11), exhibited no antimitotic effect on sarcoma 180 cell growth. As a consequence the ring-C aromatized phenanthridones such as narciprimine (13) have largely been overlooked due to a lack of numerical representation as well as cytotoxic potency. Arolycoricidine (24), which lacks the C-7 phenolic hydroxyl group, was present amongst several alkaloids found in bulbs of *Lycoris sanguine* (Takagi and Yamaki 1974). However, cytotoxicity evaluations almost four decades later showed it to have relatively low *in vitro* activity against L6 (rat myoblast) and KB cells (IC_{50}s > 60 µg/mL) (Kaya et al. 2011). Similarly, Sarikaya et al. (2012) demonstrated that activities for arolycoricidine in MCF7, HeLa, and A431 cells were moderate to mild (28.5–47.9% inhibition at 10 µM). Activities for narciprimine (13) against the same three cells at equimolar concentrations were even lower, ranging from 13.9 to 34.7% inhibition (Sarikaya et al. 2012).

In spite of these findings suggesting that the biaryl phenanthridone analogs may not be promising cytotoxic agents, unrelated studies by Nair et al. (2012b) in fact proved to be quite the opposite. In the five-cell screen involving CEM, K562, MCF7, G361, and HeLa cells, narciprimine was markedly active on CEM cells (IC_{50} 13.3 µM), whilst IC_{50}s > 50 µM were observed for the remainder of the cells (Nair et al. 2012b). Despite this selectivity towards the difficult-to-treat CEM lymphoblastic leukemia cell line, narciprimine was also disappointingly cytotoxic to normal human fibroblast (BJ) cells (IC_{50} 7.9 µM) (Nair et al. 2012b). A series of phenanthridone analogs devoid of phenolic substitution in rings A and C were also identified and shown to exhibit cytotoxic effects against various cancer cells (Ghosal et al. 1985b, Luo et al. 2012). Initially, Ghosal et al. (1985b) identified crinasiadine (25) and crinasiatine (31) in *Crinum asiaticum*, noting at the time that both alkaloids exhibited significant antitumor activity. Luo et al. (2012) more recently isolated several ring-C aromatized, *N*-alkylated derivatives (26, 27, 28, 29, 30) from *Zephyranthes candida* and carried out a screen against five cancer cell lines including, HL60, K562, A549, HepG2, and HT29 cells. Alkaloids 27, 28, and 29 with IC_{50}s that ranged from 0.7 to 39 μM were the most active, whilst N-phenylethylcrinasiadine (29) (Figure 3.2) emerged as the target of choice (IC_{50} 0.70 and 0.81 μM in HL60 and K562 cells, respectively) (Luo et al. 2012). In spite of these good activities, N-phenylethylcrinasiadine was also cytotoxic towards normal human Beas-2B (bronchial epithelial) cells (IC_{50} 7.3 μM) (Luo et al. 2012).

3.6 STRUCTURE–ACTIVITY RELATIONSHIP STUDIES

3.6.1 Truncated Analogs

The shortfall in the availability of natural phenanthridones has by no means impeded their trajectory on the clinical front. The successes realized in target orientated syntheses of the key constituents narciclasine (11), pancratistatin (12), lycoricidine (14), and 7-deoxypancratistatin (19) has, on the contrary, spurred on efforts at expanding the phenanthridone alkaloid repertoire (Rinner and Hudlicky 2005, Ghavre et al. 2016). This then facilitated the consolidation of the structural elements attending the phenanthridone anticancer pharmacophore, which is noteworthy given that the biological target for these molecules is yet to be discovered (Nair et al. 2015). It was apparent from the outset that an intact tricyclic nucleus was central to the performance of these cytotoxic agents. That this was indeed the case can be gauged from the lack of activities observed for several synthetic truncated analogs (Chrétien et al. 1993, McNulty et al. 2008, Manpadi et al. 2009). For example, the B-*seco* derivative (32), which is differentiated from lycoricidine (14) only by the absence of the bond connecting C-10a to C-10b, was inactive against L1210 mouse lymphocytic leukemia cells (Chrétien et al. 1993). By contrast, lycoricidine (14) (Figure 3.2) has been established as a potent antiproliferative agent via the NCI multi-cell screen where it exhibited a mean GI_{50} value of 0.15 μM (Pettit et al. 1993). Furthermore, opening of ring-C proved to be a deleterious operation as in the case of the truncated cyclitol analog (33), which suffered from a complete loss of activity when exposed to both HeLa cervical and MCF/AZ breast adenocarcinomas (Manpadi et al. 2009). Similarly, the B,C-*seco* analog (34), which has all the functional elements of the pharmacophore present but in a conformationally mobile form, was set back markedly in cultures of MCF7 cells where an effective dose (ED_{50}) was observed to be greater than 30 μg/mL (McNulty et al. 2008).

3.6.2 Ring-A Modifications

Studies which explored the effects of the ring-A fragment on cytotoxic activity were understandably drawn to the C-7, C-8, and C-9 positions which hold in place the phenolic hydroxy and methylenedioxy groups. The importance of the C-7 hydroxy group is quite lucid based on the 60 cell line screen carried out at the NCI on the phenanthridones narciclasine (11), lycoricidine (14), *trans*-dihydronarciclasine (17), and *trans*-dihydrolycoricidine (18) (Pettit et al. 1993). These matched pairs could be distinguished by the absence or presence of the C-7 hydroxy group and exhibited mean GI_{50} values of 0.016, 0.15, 0.013, and 0.068 μM, respectively (Pettit et al. 1993). Although 7-deoxypancratistatin (19) has not been evaluated in parallel with pancratistatin (12) (Figure 3.2) in the NCI multi-cell line assay, a screen against P388, NCI-H460, and KM20L2 cells was insightful in accentuating the effect of the C-7 hydroxy group (Rinner et al. 2004a). Here pancratistatin (GI_{50} 0.032, 0.048, and 0.026 μg/mL, respectively) was shown to be far more potent than 7-deoxypancratistatin (GI_{50} 0.44, 0.29, and 0.22 μg/mL, respectively) (Rinner et al. 2004a). Likewise, through various strategic manipulations the 8,9-methylenedioxy moiety was established as an essential feature of the active pharmacophore (Rinner et al. 2004a,b). For example, replacement of this 8,9-dioxy-ring system with a 7-methoxy group as in 35 proved to be highly detrimental in a joint screen against the P388, NCI-H460, and KM20L2 cell lines (Rinner et al. 2004a). Here, a mean GI_{50} value of 3.6 μg/mL for compound 35 was reflective of the severity of this transformation when viewed against 7-deoxypancratistatin (19) (mean GI_{50} 0.32 µg/mL) (Rinner et al. 2004a). Probed further the pharmacophore succumbed to striking losses in activity when the ring-A moiety was modified into a β-carboline unit (36) (Rinner et al. 2004b). To this extent, whereas pancratistatin exhibited a mean growth inhibition index (GI_{50}) of 0.029 µg/mL against P388, BxPC-3, and KM20L2 cells, the β-carboline analog (36) reflected GI_{50} values > 10 µg/mL in each of these cells (Rinner et al. 2004b).

3.6.3 Ring-B Modifications

Given that the 2-piperidone unit making up ring-B is tetra-substituted by way of two further cyclic systems, the resulting compactness makes for little chemical space to probe for SAR purposes. Nonetheless, Pettit et al. (2006) by a series of semi-synthetic manipulations was able to show how important the C-6 oxo group was in the antiproliferative activity of phenanthridone alkaloids. In this regard, as high as a hundred-fold loss in activity could be incurred following its exclusion from ring-B (Pettit et al. 2006). This was ably demonstrated in a six human cell line screen involving BxPC-3, MCF7, SF268, NCI-H460, KM20L2, and Du145 cells where the mean GI_{50} (0.022 µg/mL) determined for lycoricidine (14) was orders of magnitude superior to that of 6-deoxylycoricidine (37) (mean GI_{50} > 10 µg/mL) (Pettit and Meolody 2005, Pettit et al. 2006). However, it was also shown that the loss of activity against the same six cell lines could be significantly recouped by conversion of 6-deoxylycoricidine (37) to its hydrochloride salt (38) (Pettit et al. 2006). The effects of salt formation (38) (Figure 3.3) and β-lactamization (14) are relatively similar since both involve the distribution of charge density away from the nitrogen atom. The difference lies with the fact that the latter is a polarized system wherein the electron density becomes concentrated at the oxygen atom, whilst the former is a charge separated system with individual anionic and cationic species. Furthermore, substitution of the nitrogen with an oxygen atom was an even more deleterious maneuver as shown for the lactones (39, 40) which were debilitated with complete loss of activity compared to the corresponding lactams narciclasine (11) and lycoricidine (14), respectively (Ibn-Ahmed et al. 2004).

3.6.4 Ring-C Modifications

Given the presence of four contiguous substitutable positions in ring-C, this ring has been the focal point of most structure–activity relationship (SAR) studies (Rinner and Hudlicky 2005, Ghavre et al. 2016). Whilst the stereoselective instalment of the hydroxyl groups in ring-C has posed a significant challenge in the synthetic realm, the chemoselective functionalization of these groups has been the desired goal of SAR based studies (Ghavre et al. 2016). In this regard significant strides have been made in this direction from which a more lucid picture of the elements of the pharmacophore pertaining to ring-C has emerged (Ghavre et al. 2016). With few exceptions the innate hydroxy substituents provided the best clues, indicating that small polar groups with hydrogen bond donor–acceptor capability are desirable at the substitutable positions of ring-C (Ingrassia et al. 2008, Kornienko and Evidente 2008). For example, 2,3,4-tri-O-acetyllycoricidine (41) (Figure 3.2) exhibited a mean GI_{50} of 0.42 µg/mL in the seven cells (P388, BxPC-3, MCF7, SF268, NCI-H460, KM20L2, and Du145) examined by Pettit et al. (2006), nearly five times less than that determined for the parent compound lycoricidine (14) (mean GI_{50} of 0.085 µg/mL). In the same instance 2-O-acetyllycoricidine (42) had a mean GI_{50} (0.21 µg /mL) that was 2.5 times less than that of lycoricidine (14) (Pettit et al. 2006). Having established early on that the activity of phenanthridone alkaloids hinged on its hydroxyl substituents, efforts were then redirected to ascertain the degree of this contribution to cytotoxic activity, either individually or in combination with the other groups. From this it emerged that the minimum requirement was dihydroxy substitution either at C-2/C-3 or C-3/C-4 (McNulty et al. 2001, Rinner et al. 2004a). In this regard, 4-deoxydihydrolycoricidine

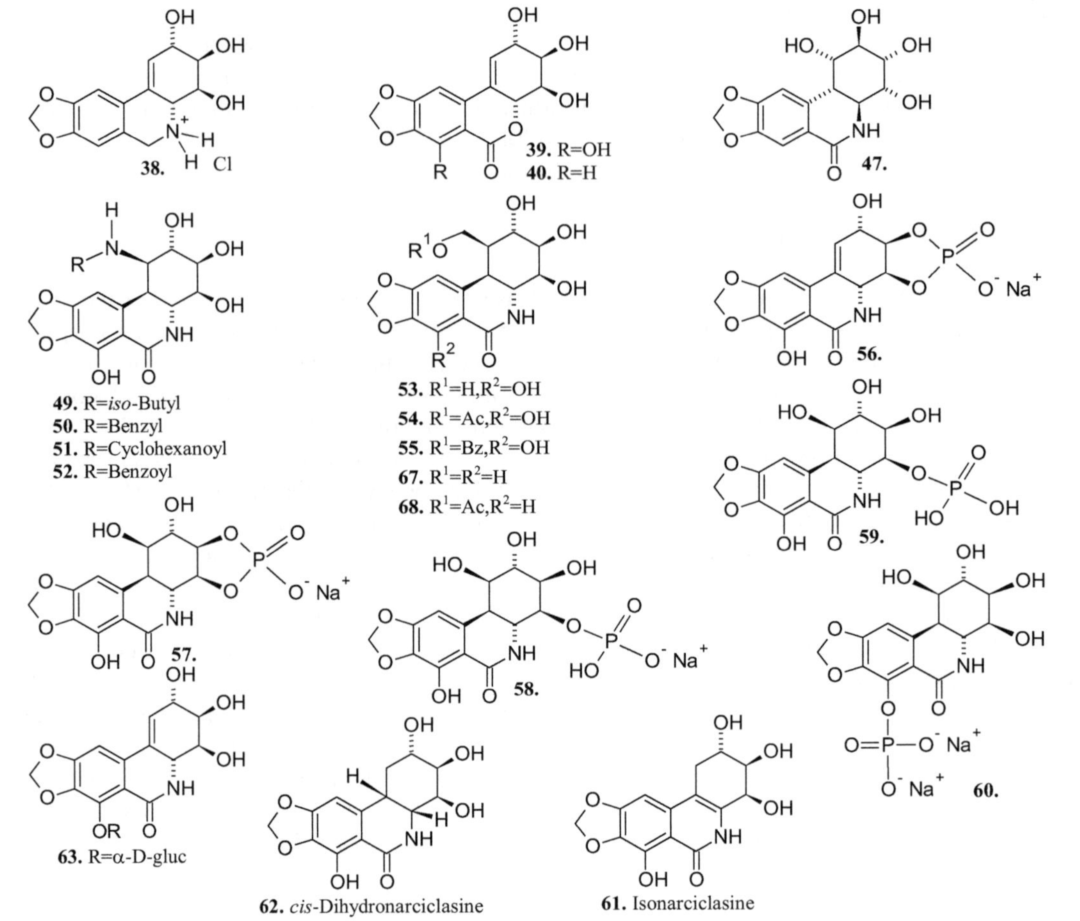

Figure 3.3 Natural and synthetic phenanthridone alkaloids of the Amaryllidaceae which have been screened for cytotoxic effects in various cancer cells.

(43) inhibited proliferation of the non-small cell lung cancer line NCI-H226 (ED_{50} 0.65 µg/mL), P388 (ED_{50} 0.45 µg/mL), as well as two leukemia cell lines: CCRF-CEM (ED_{50} 0.55 µg/mL) and HL-60(TB) (ED_{50} 0.89 µg/mL) (McNulty et al. 2001). Although 2-deoxydihydrolycoricidine (44) was active in the murine P388 lymphocytic leukemia cell system (ED_{50} 1.39 µg/mL), it exhibited weak activities against BxPC-3, MCF7, SF268, NCI-H460, KM20L2, and Du145 cells (ED_{50}s > 10 µg/mL) (Rinner et al. 2004a). 2,3-Dideoxydihydrolycoricidine (45) displayed marginal inhibition of P388 cells (ED_{50} 40.1 µg/mL) (McNulty and Mo 1998). 3-Deoxydihydrolycoricidine (46) was tested via serial dilution (1–10 µM) against MCF7 and Jurkat cells but exhibited no activity even at the highest concentration (McNulty et al. 2005).

The stereochemical arrangement in phenanthridone alkaloids is as expected: one in which each of the four hydroxy groups is in the equatorial plane of the chair-conformed ring-C which is in turn *trans*-fused to ring-B (Jin 2016). In attempts to understand what effects this pattern might have on antiproliferative activity, Hudlicky et al. (2002) synthesized the enantiomer of 7-deoxypancratistatin (19), 7-deoxy-*ent*-pancratistatin (47). Whilst it is reasonable to envisage that the complete reversal of substituent geometry might have a profound effect on cancer cell growth, in reality the depreciation in inhibitory activity was not as pronounced as anticipated. To this extent, 7-deoxypancratistatin (19) (Figure 3.2) was potent in a seven-cell screen where GI_{50} values ranged from 0.22 µg/mL (colon

KM20L2) to 0.47 µg/mL (renal A498) and a mean GI_{50} determined as 0.31 µg/mL (Hudlicky et al. 2002). By contrast, up to a ten-fold loss in activity could be incurred following enantiomerization which was demonstrated for 7-deoxy-*ent*-pancratistatin (47) across the self-same seven cancer cells where GI_{50}s were in the range 2.0–3.4 µg/mL (Hudlicky et al. 2002).

A further aspect addressed in SAR studies of phenanthridone alkaloids is the chemo-regioselective functionalization of the ring-C hydroxyl groups, in itself a challenging feat given the homology of these substituents (Ghavre et al. 2016). In general it was proven that selective or non-selective functionalization of these groups was detrimental towards cytotoxicity (Ghavre et al. 2016). This was apparent from the parallel study of lycoricidine (14), 2-*O*-acetyllycoricidine (42), and 2,3,4-tri-*O*-acetyllycoricidine (41) in the seven cancer cells described above, where mean GI_{50}s were in the order 0.085, 0.21, and 0.42 µg/mL, respectively (Pettit and Melody 2005, Pettit et al. 2006). Whilst chemoselective functionalization of the C-2 allylic hydroxy group in lycoricidine (14) proved to be detrimental, such a maneuver at the C-1 position of pancratistatin (12) was shown to be intriguingly beneficial towards cytotoxic activity (Ghavre et al. 2016). This was initially observed for 1-*O*-benzoylpancratistatin (48), a semisynthetic analog accessed from narciclasine (11) via its C-1/C-2 α-epoxide as key intermediate (Pettit et al. 2001). Remarkably this alkaloid exhibited submicromolar IC_{50}s in four of the seven cell lines screened, low micromolar activity in the remaining three, with the best cytotoxicity index observed for lung-NSC NCI-H460 cells (ED_{50} 0.0001 µg/mL) (Pettit et al. 2001, Collins et al. 2010). More recently several attempts have been made targeting such differentially functionalized C-1 derivatives of pancratistatin (Marion et al. 2010, Vshyvenko et al. 2011, 2012). Marion et al. (2010) synthesized several C-1 nitrogen-substituted analogs of pancratistatin and examined their effects on two tumor cell lines of human origin, A549 and HCT116. In this way four compounds (49–52) (Figure 3.3) were able to significantly improve on the activities of the reference standard narciclasine (IC_{50} 49 and 22 µM, respectively) (Marion et al. 2010). In fact, the 1-benzamide (52) emerged as the best candidate with IC_{50} values of 8.7 and 4.7 µM against the two cells, respectively (Marion et al. 2010). The sustained activities observed for this series of compounds indicate that substituents containing less electronegative elements (nitrogen versus oxygen) are tolerable at C-1. It remains to be seen what effect elements at this position (such as halogens) with greater electronegativity might have on activity. The nitrogen moiety in compound 52 is interesting since, like in pancratistatin, it has the hydrogen atom available for donor-related functions. In addition, its benzoyl group is also nicely positioned for π-stacking with ring-A, as is presumably the case with 1-*O*-benzoylpancratistatin (48). Further study of C-1 functionalization was undertaken by Vshyvenko et al. in 2011 wherein the homologated analogs (53, 54) were prepared and subjected to screening procedures against BxPC-3, DU-145, NCI-H460, and MCF7 cells. In the event mean IC_{50} values of 0.16 and 0.18 µM were determined for the two compounds, respectively, over three times greater than that observed for narciclasine (11) (0.05 µM) (Vshyvenko et al. 2011). This result indicated that there was a significant modulation in activity when the C-1 oxygen substituent becomes shifted away from the parent ring (Vshyvenko et al. 2011). In spite of this the benzoate analog (55) appears to dispel this notion gauging from its low mean IC_{50} value (0.03 µM) in the same four cancer cells (Vshyvenko et al. 2012). The overall potencies of the benzoyl derivatives 1-*O*-benzoylpancratistatin (48), 1-benzamide (52), and analog (55) (Figure 3.3) suggest that expanding the pharmacophore at C-1 via substituents that could electronically interact with ring-A may indeed be a potentiating factor in the antiproliferative effects of phenanthridone alkaloids.

The low aqueous solubility of natural phenanthridones (around 50 µg/mL) has presented a significant hurdle to their progress in clinical evaluations, particularly via intravenous formulations (Pettit et al. 2004). Phosphate esters are known to vastly improve the aqueous solubility of polar substrates, including drugs (Pettit et al. 2003, Pettit et al. 2004, Pettit and Melody 2005). An added advantage of a phosphate prodrug is the ease with which it can be cleaved by cellular phosphatases under relatively mild metabolic conditions to release the parent drug (Pettit et al. 2003, Pettit et al. 2004, Pettit and Melody 2005). With this in mind, Pettit et al. (2004) were able to access a series of phosphate esters from narciclasine (11) and pancratistatin (12) and evaluate their activities in six different cancer cell lines, which were BxPC-3, MCF7, SF268, NCI-H460, KM20L2, and Du145 cells. In this way it was demonstrated that cytotoxicities could be retained following esterification to the respective phosphate derivatives (Pettit et al. 2004). The cyclic phosphate ester (56) prepared from narciclasine exhibited a mean GI_{50} of 0.054 µg/mL, around 13 times less than narciclasine itself (mean GI_{50} of 0.0041 µg/mL) (Pettit et al. 2004). The corresponding 3,4-phosphate ester (57) (Figure 3.3) of pancratistatin was active in each of the six cells (mean GI_{50} 3.2 µg/mL), but notably less so than pancratistatin (mean GI_{50} 0.022 µg/mL) (Pettit et al. 2004). It is not clear why this difference in activity (~145 times) is so disparate from that observed for narciclasine (11) and its corresponding phosphate ester above (Pettit et al. 2004). The uncyclized ester (58) bearing the phosphate group at C-4 had a mean GI_{50} of 0.34 µg/mL, much higher than that of pancratistatin (0.022 µg/mL) but significantly lower than that of the cyclized ester form (57) (3.2 µg/mL) (Pettit et al. 2004). This result also further highlights the importance of the free hydroxy groups in the cytotoxic effects of these alkaloids (Ghavre et al. 2016). Protonation of the sodium salt had almost no effect since the resulting acid (59) exhibited a mean GI_{50} of 0.35 µg/mL, compared to 0.34 µg/mL for the salt (Pettit et al. 2004). The localization of the phosphate ester at C-7 produced a compound (60) that was lower in activity (mean GI_{50} 0.14 µg/mL) compared to pancratistatin (0.022 µg/mL), but better than the case where the phosphate group was present at C-4 (58, 59) (mean GI_{50} 0.34 and 0.35 µg/mL, respectively) (Pettit et al. 2004).

The mean GI_{50} values quantified for narciclasine (11) and pancratistatin (12) (0.016 and 0.091 µM, respectively) in the NCI 60 cell line screen underlines the potency of the former and its potential in the clinical arena (Pettit et al. 1993). Going forward it also highlights the importance of the C-1/C-10b double bond in the development of a phenanthridone anticancer drug target (Ghavre et al. 2016). Although the exact role of the double bond at this stage remains unclear, gauging from the activities (mean GI_{50} values) of *trans*-dihydronarciclasine (17) and *trans*-dihydrolycoricidine (18) in the NCI 60 cell panel it could be said contrariwise that reduction improves activity (Pettit et al. 1993). In this regard there was a slight enhancement in activity comparing narciclasine (11) (0.016 µM) and *trans*-dihydronarciclasine (17) (0.013 µM), but over twice the effect considering lycoricidine (14) (0.15 µM) and *trans*-dihydrolycoricidine (18) (0.068 µM) (Pettit et al. 1993). Isomerization of the double bond so that it becomes localized within the C-4a/C-10b domain is considerably detrimental as isonarciclasine (61) exhibited a mean GI_{50} of 11.8 µM in the NCI evaluation (Pettit et al. 1993). There was a slight modulation in activity following reduction of the double bond in isonarciclasine (61) (Figure 3.3) since the hydrogenated product *cis*-dihydronarciclasine (62) was over three times as active (mean GI_{50} 3.8 µM) (Pettit et al. 1993). This chemical transformation also puts into context the importance of the *trans*-fusion of rings B and C since *trans*-dihydronarciclasine (17) (0.013 µM) was close to 300 times more potent than *cis*-dihydronarciclasine (62) (Pettit et al. 1993) (Table 3.1).

3.7 CYTOTOXIC EFFECTS OF PHENANTHRIDONE ALKALOIDS *IN VIVO*

3.7.1 *In Vivo* Effects of Narciclasine

Accompanying the first description of narciclasine (11) by Ceriotti in 1967 was evidence in support of its *in vivo* cytotoxic activity. Accordingly, oral or intraperitoneal administration of narciclasine produced rapid antimitotic effects in S-180 sarcoma infected mice (Ceriotti 1967). There was a drastic decline in the number of mitoses within 2 h of a single 0.9 mg/kg injection, whilst all observable mitoses were eradicated when the treatment was prolonged for 4 h (Ceriotti 1967). As the dosage was steadily increased the mitoses disappeared more rapidly and the appearance of disrupted cells with globular clumps of chromatin became more frequent (Ceriotti 1967). Morphological analysis

Table 3.1 In vitro Cytotoxic Activities of Phenanthridone Alkaloids in Various Cancer Cells

Alkaloid (No.)	Cell Line Activity[a]									References
	BxPC-3	Du145	H460	KM20L2	MCF7	OVCAR3	P388	SF268	SF295	
Narciclasine (11)	0.026	0.011	0.032	0.021	0.019	0.016	0.0012	0.021	0.012	Pettit et al. (2001), Rinner et al. (2004a)
Pancratistatin (12)	0.028	0.016	0.048	0.026	0.032	0.032	0.032	0.017	0.017	Pettit et al. (2001), Rinner et al. (2004a)
Lycoricidine (14)	0.070	0.051	0.053	0.084	0.046	–	0.019	0.120	–	Pettit and Melody (2005)
trans-Dihydronarciclasine (17)	0.012	0.0066	0.0092	0.015	0.0053	–	0.0024	0.020	–	Pettit and Melody (2005)
trans-Dihydrolycoricidine (18)	0.046	0.040	0.043	0.051	0.034	–	0.029	0.059	–	Pettit and Melody (2005)
7-Deoxypancratistatin (19)	–	–	0.29	0.22	–	0.24	0.44	–	0.29	Hudlicky et al. (2002), Rinner et al. (2004a)
(35)	4.9	2.6	2.8	3.6	4.4	–	4.3	3.3	–	Rinner et al. (2004a)
(36)	> 10	–	–	> 10	> 10	–	18.3	–	–	Rinner et al. (2004b)
6-Deoxylycoricidine (37)	> 10	> 10	> 10	> 10	> 10	–	> 10	> 10	–	Pettit et al. (2006)
(38)	4.4	3.2	3.4	> 10	3.4	–	7.1	2.9	–	Pettit et al. (2006)
2,3,4-(O)-Triacetyllycoricidine (41)	0.2	0.48	1.2	0.18	0.59	–	0.158	0.15	–	Pettit et al. (2006)
2-(O)-Acetyllycoricidine (42)	0.26	0.26	0.32	0.18	0.24	–	< 0.01	0.19	–	Pettit et al. (2006)
(44)	> 10	> 10	> 10	> 10	> 10	–	1.39	> 10	–	Rinner et al. (2004a)
1-(O)-Benzoylpancratistatin (48)	0.0019	0.00021	0.0001	0.00037	0.00031	< 0.001	0.0016	0.00055	0.0013	Pettit et al. (2001), Collins et al. (2010)
(53)	0.22	0.09	0.09	–	0.24	–	–	–	–	Vshyvenko et al. (2011)
(54)	0.07	0.06	0.07	–	0.52	–	–	–	–	Vshyvenko et al. (2011)
(55)	0.01	0.01	0.03	–	0.08	–	–	–	–	Vshyvenko et al. (2012)
(56)	0.069	0.031	0.058	0.06	0.059	–	0.012	0.047	–	Pettit et al. (2004)
(57)	3.3	2.3	3.8	3.7	2.9	–	3.33	2.9	–	Pettit et al. (2004)
(58)	0.18	0.94	0.38	0.24	0.18	–	0.018	0.12	–	Pettit et al. (2004)
(59)	0.36	0.20	0.42	0.33	0.43	–	0.047	0.35	–	Pettit et al. (2004)
(60)	0.20	0.026	0.19	0.17	0.20	–	0.24	0.079	–	Pettit et al. (2004)
(67)	0.77	1.10	0.40	–	0.86	–	–	–	–	Vshyvenko et al. (2011)
(68)	0.34	0.72	0.53	–	1.81	–	–	–	–	Vshyvenko et al. (2011)

[a] Activities given as GI_{50} (μg/mL) except for the P388 cell line and compound 48 where ED_{50} (μg/mL) values are reflected, whilst entries under compounds 53, 54, 55, 67, and 68 are expressed as IC_{50} (μM) units.

indicated that the response of narciclasine was reminiscent of a metaphasic poison with mitoclasic activity, especially at high doses (Ceriotti 1967). An LD_{50} of 5 mg/kg was established via the subcutaneous injection route (Ceriotti 1967). Furusawa et al. (1980) then examined a library of structurally diverse Amaryllidaceae alkaloids *in vitro* against NIH/3T3 mouse embryonic fibroblasts and *in vivo* against Rauscher leukemia in BALB/c mice. Narciclasine with an MTD of 0.005 µg/mL was amongst the most active of the 24 alkaloids *in vitro* (Furusawa et al. 1980). Surprisingly its potent *in vitro* activity could not be replicated *in vivo* where its prolongation of life span index was 31%, compared to other alkaloids in the library whose respective indices exceeded 100% (Furusawa et al. 1980). Further studies were carried out on narciclasine (11) (Figure 3.2) to ascertain if it could minimize the effect of calprotectin-induced toxicity in MM46 mouse mammary carcinoma cells (Mikami et al. 1999). It has been suggested that calprotectin may have a regulatory role in inflammatory processes via its growth-inhibitory and apoptosis-inducing activities towards cells migrating into inflammatory sites (Yui et al. 2003). It was observed that MM46 cells treated with calprotectin underwent an 80% decrease in MTT-reducing activity, which could nevertheless be overcome with narciclasine at doses of 0.001 to 0.01 µg/mL (Mikami et al. 1999). Intraperitoneal administration of narciclasine (1 mg/kg) to male Wistar-Lewis rats produced a marked decrease in swelling judged by the adjuvant-induced arthritis model (Mikami et al. 1999).

A detailed study by Ingrassia et al. (2009) showed that narciclasine was potently active *in vitro* against 22 different cancer cells irrespective of the tumor-type or whether the cells were of human (IC_{50}s 5–99 µM) or rodent (IC_{50}s 28–35 µM) origin. Furthermore, narciclasine proved to be a selective cytotoxic agent as its effects on normal fibroblasts (IC_{50}s > 317 µM) were much lower than on tumor cells (Ingrassia et al. 2009). In addition, its cytotoxicity in human umbilical vein endothelial cells (HUVECs) (IC_{50}s 87–94 µM) was comparable to that in tumor cells, highlighting its potential as an antiangiogenic agent (Ingrassia et al. 2009). In the adjoining *in vivo* study, A549 NSCLC cells were grafted orthotopically into the lungs or brains of immuno-compromised Nu/Nu nude mice to mimic brain metatheses (Ingrassia et al. 2009). In the lung model a single intravenous dose of narciclasine at 1 mg/kg once weekly for five weeks did not increase the survival of the Nu/Nu mice over control animals (Ingrassia et al. 2009). However, narciclasine under the same regimen produced a striking prolongation of lifespan in mice containing A549 brain grafts (Ingrassia et al. 2009). This therefore highlighted the ability of narciclasine to cross the blood–brain barrier with relative ease, an attribute which is highly desirable from a clinical perspective (Ingrassia et al. 2009). Based on these findings for the A549 xenograft model a further two brain cell lines Hs683 and GL19 glioblastoma were explored, but in which narciclasine was shown to have negligible effects (Ingrassia et al. 2009). Nonetheless, its glucoside 7-*O*-β-D-glucosyl-narciclasine (63) (at 1 mg/kg) significantly improved the survival of Hs683 and GL19 grafted mice when administered orally or intravenously five times a week over three weeks (Ingrassia et al. 2009). Further studies of the two glioblastoma cells revealed that intravenous (1 mg/kg/twice per week/five weeks) or oral (1 mg/kg/once per week/five weeks) administration of narciclasine could in fact increase the survival of GL19-infected mice by almost twice that of control animals, which was somewhat attenuated for Hs683-infected mice (LeFranc et al. 2009). Furthermore, increasing the number of doses per week did not increase the survival of Hs683 glioblastoma bearing mice (LeFranc et al. 2009). The results also reflected that the survival rates brought about by narciclasine were similar to those for the chemotherapeutic temozolomide, but at comparatively lower dosage levels (LeFranc et al. 2009). In addition no adverse effects were observed in animals treated with narciclasine (11) at 1, 10, or 25 mg/kg five times per week for three weeks, from which the NOAEL was determined as 1 mg/kg/day (LeFranc et al. 2009).

Van Goietsenoven et al. (2010) then examined the effects of narciclasine on several melanoma cell lines, including the primary human melanoma cell cultures VM-21, VM-47, and VM-48, the human skin melanoma SKMEL-28, as well as the mouse melanoma B16F10. This study was significant since melanomas are known for poor response rates to adjuvant therapies as a consequence of their intrinsic resistance to proapoptotic stimuli (Van Goietsenoven et al. 2010). Encouragingly, narciclasine displayed potent *in vitro* growth-inhibitory activities irrespective of the levels of cellular resistance to proapoptotic stimuli, with IC_{50}s ranging from 0.05 µM in VM-47 cells to 0.005 µM in B16F10 cells (Van Goietsenoven et al. 2010). eEF1A (eukaryotic translation elongation factor 1 alpha) targeting with narciclasine (at 50 µM) resulted in marked actin cytoskeleton disorganization and protein synthesis impairment (Van Goietsenoven et al. 2010). Furthermore, apoptosis induction was only achievable at higher doses of narciclasine (> 200 µM) (Van Goietsenoven et al. 2010). The subsequent *in vivo* study used human melanoma brain metastatic VM-48 cells that were stereotactically implanted into the brains of six-week-old female nude mice (Van Goietsenoven et al. 2010). The results reflected that narciclasine (1 mg/kg/twice per week/three weeks) afforded significant therapeutic benefits in this aggressive melanoma model, similar to the reference drug temozolomide (40 mg/kg/thrice per week/three weeks), with no observable detrimental effects (Van Goietsenoven et al. 2010).

Investigations into the *in vivo* cytotoxicity of narciclasine (11) (Figure 3.2) originate from the initial findings of scientists at the NCI (Ingrassia et al. 2008). Narciclasine was examined at doses ranging from 0.62–12.5 mg/kg in mouse models of B16F10, L1210 (mouse leukemia), P-388, M5076, and Lewis lung carcinoma cells (Ingrassia et al. 2008). P-388-infected animals treated once daily for nine days exhibited T/C indices of 129% (with 1.25 mg/kg/day) and 162% (with 2.5 mg/kg/day) (Ingrassia et al. 2008). M5076-grafted animals on the other hand treated with 3 and 6 mg/kg/day exhibited T/C similar indices of 133 and 160%, respectively (Ingrassia et al. 2008). In spite of these promising outcomes in the mouse models, human xenograft model studies involving HT-29 colon adenocarcinoma, LX-1 lung cancer, and MX-1 breast carcinoma were disappointing (Ingrassia et al. 2008). In addition to narciclasine (11) and 7-*O*-β-D-glucosyl-narciclasine (63), the only other study of double bond containing phenanthridones was that carried out on lycoricidine (14). In this regard, Pettit et al. (1986) identified lycoricidine as an *in vitro* and *in vivo* antineoplastic agent of *Pancratium littorale* with an ED_{50} of 0.02 µg/mL and T/C index of 161% (at 12.5 mg/kg).

3.7.2 *In Vivo* Effects of Pancratistatin

As with the case of narciclasine, the initial description of pancratistatin in 1984 was accompanied by evidence of both its *in vitro* and *in vivo* antiproliferative effects (Pettit et al. 1984). In this investigation, pancratistatin was isolated from bulbs of *Zephyranthes grandiflora* and shown to have potent antiproliferative effects on P388 cells *in vitro* (ED_{50} < 0.01 µg/mL) and *in vivo* (T/C 135–150% at dosages of 0.78–3.1 mg/kg) (Pettit et al. 1984). Further work by these researchers also saw the significant effects of pancratistatin being demonstrated *in vivo* against the M5076 ovary sarcoma cell line (T/C index of 153–184% at 0.38–3.0 mg/kg) (Pettit et al. 1986). Pancratistatin was also subjected to the same mouse xenograft model used for narciclasine (11) (Figure 3.2) by researchers at the NCI, where it exhibited pronounced effects (T/C > 125%) across all six cell lines screened (Ingrassia et al. 2008). Its best activity (T/C 153%) was observed in M5076 tumor-bearing mice at the lowest tested dosage of 0.18 mg/kg (Ingrassia et al. 2008). Good activity was also seen against P-388 leukemia (T/C 139% at 0.75 mg/kg) (Ingrassia et al. 2008). The highest T/C value (203%) was observed in P-388 cells but at the relatively high dose of 12.5 mg/kg (Ingrassia et al. 2008).

In order to improve the notoriously poor aqueous solubility of pancratistatin (~53 mg/mL) Shnyder et al. (2008) synthesized its cyclic 3,4-*O*-phosphate ester (57) (Figure 3.3) and showed that such a modification could drastically improve solubility, in this case to around 20 mg/mL. However, the *in vitro* potency of this phosphate prodrug against human DLD-1 colon adenocarcinoma cells was disappointing, with IC_{50}s of 253 and 19.7 µM after 1 and 96 h exposures, respectively (Shnyder et al. 2008). By contrast the parent pancratistatin (12) was nearly 100 times more potent, suggesting that *in vitro* these cells could be deprived of the requisite endogenous non-specific phosphatases to cleave the ester moiety to afford the active agent pancratistatin (Shnyder et al. 2008). To test that this could be the case, the effect of the phosphate ester (57) was examined *in vivo* on female CD1-Foxnl[nu] immunodeficient nude mice implanted subcutaneously with DLD-1 cells (Shnyder et al. 2008). In this manner it was demonstrated that a single intravenous dose at the MTD of 100 mg/kg significantly increased the mean tumor doubling time (RTV2) from 6.2 days in the controls to 15.1 days in test animals (Shnyder et al. 2008).

Griffin et al. (2011a) examined the *in vitro* and *in vivo* effects of pancratistatin in human colorectal carcinoma cell models. Given the selective cytotoxicity of pancratistatin, this study set out to establish whether it specifically targeted cancer cell mitochondria rather than DNA or its replicative machinery (Griffin et al. 2011a). In this regard pancratistatin decreased MMP and induced apoptotic nuclear morphology in p53-mutant (HT-29) and wild-type p53 (HCT116) colorectal carcinoma cells (IC_{50}s 0.1 μM, respectively), but not in non-cancerous colon fibroblast (CCD-18Co) cells (IC_{50} 10 μM) (Griffin et al. 2011a). Furthermore, it was ineffective against mtDNA-depleted (r0) U87MG glioblastoma cells, confirming that it is a mitochondria-targeting, apoptosis-inducing chemotherapeutic agent (Griffin et al. 2011a). In the subsequent *in vivo* study, it was found that intratumoral administration of pancratistatin (3 mg/kg/twice weekly) for five weeks produced significant reductions in the growth of subcutaneous HT-29 tumors in Nu/Nu mice (Griffin et al. 2011a). The effects of pancratistatin in models of metastatic prostate cancer were also probed using the two cell lines, DU145 and LNCaP (Griffin et al. 2011b). Therein it was shown that in a dose and time dependent manner pancratistatin significantly reduced the viabilities of androgen-responsive (LNCaP) and androgen-refractory (DU145) cells with EC_{50}s of 0.1 μM, respectively (Griffin et al. 2011b). In contrast, normal human fibroblast (NHF) cells remained largely unaffected by such treatments with pancratistatin (Griffin et al. 2011b). A human xenograft model was then set up using DU145 cells implanted subcutaneously in six-week-old male homozygous CD-1 nude mice (Griffin et al. 2011b). Tumors were allowed to grow for four weeks before animals were treated intra-tumorally with vehicle (5 μL DMSO in PBS) or pancratistatin (3 mg/kg) four times per week for three weeks (Griffin et al. 2011b). In this manner it was shown that average tumor volumes were up to 50% less in pancratistatin-treated mice than those exposed to the vehicle (Griffin et al. 2011b) (Table 3.2).

3.8 MECHANISM OF ACTION OF PHENANTHRIDONE ALKALOIDS

3.8.1 Permeability and Solubility

Permeability and solubility are essential criteria in verifying the efficacy of a drug target (Pade and Stavchansky 1998). Maximum absorption occurs when a drug has maximum permeability and maximum concentration (saturation solubility) at the site of absorption (Pade and Stavchansky 1998). Permeability across biological membranes is a key factor in the absorption and distribution of drugs (Pade and Stavchansky 1998). Poor permeability can arise through a number of structural features and membrane-based efflux mechanisms (Pade and Stavchansky 1998). It can lead to poor absorption across the gastrointestinal mucosa or poor distribution throughout the body (Pade and Stavchansky 1998). As such, these factors have received considerable attention in the ongoing pharmacological development of pancratistatin (12) (Beijnen et al. 1995). For example, the solubility of pancratistatin has been improved to around the 1 mg/mL mark through the use of complexing agents such as nicotinamide and cyclodextrin (Beijnen et al. 1995). A further method used to improve the solubility of pancratistatin without compromising its activity, as alluded to above, relied on its functionalization to a phosphate prodrug (Pettit et al. 2003, Pettit et al. 2004, Pettit and Melody 2005). In this way, a solubility value as good as 20 mg/mL was achievable with some of these targets (Pettit et al. 2003, Pettit et al. 2004, Pettit and Melody 2005).

Table 3.2 In Vitro and In Vivo Cytotoxic Effects of Phenanthridone Alkaloids of the Amaryllidaceae Against Human and Animal Cancer Cells

Alkaloid (No.)	Cancer Cell	In Vitro Activity	In Vivo Activity	Animal Model	Admin. Route	Dosage	Interval	Treatment Period	References
Narciclasine (11)	P388	0.042 µM[a]	129%[b]	Mice	ip	1.25 mg/kg	1/day	9 days	Ingrassia et al. (2008)
	M5076	nt	133%[b]	Mice	ip	3 mg/kg	1/day	9 days	Ingrassia et al. (2008)
	S180	nt	active	Nude mice	ip, sc, oa	0.9 mg/kg	Once	4 h	Ceriotti (1967)
	RL	0.005 µg/mL[c]	31%[d]	BALB/c mice	ip	3 mg/kg	3/wk	1 wk	Furusawa et al. (1980)
	A549	0.03 µM[a]	active	Nu/Nu mice	iv	1 mg/kg	1/wk	5 wk	Ingrassia et al. (2009)
	Hs683	0.04 µM[a]	active	Nu/Nu mice	iv	1 mg/kg	1/wk	5 wk	Ingrassia et al. (2009), Le Franc et al. (2009)
	GL19	nt	active	Nu/Nu mice	iv	1 mg/kg	1/wk	5 wk	Ingrassia et al. (2009), Le Franc et al. (2009)
	VM-48	0.04 µM[a]	active	Nude mice	iv, oa	1 mg/kg	2/wk	3 wk	Van Goietsenoven et al. (2010)
Pancratistatin (12)	P388	< 0.01 µg/mL[e]	135–150%[b]	Mice	sc	0.78–3.1 mg/kg	ni	ni	Ingrassia et al. (2008), Pettit et al. (1984)
	M5076	nt	153–184%[b]	Mice	sc	0.38–3.0 mg/kg	ni	ni	Ingrassia et al. (2008), Pettit et al. (1986)
	Du145	0.1 µM[f]	50%[g]	CD-1 nude mice	it	3 mg/kg	4/wk	3 wk	Griffin et al. (2011b)
	HT-29	0.1 µM[a]	Active	Nu/Nu mice	it	3 mg/kg	2/wk	5 wk	Griffin et al. (2011a)
Lycoricidine (14)	P388	0.02 µg/mL[e]	161%[b]	Mice	sc	12.5 mg/kg	ni	ni	Pettit et al. (1986)
Pancratistatin-3,4-O-phosphate (57)	DLD-1	19.7 µM[a]	Active	CD-1-FoxnT^{nu}mice	iv	100 mg/kg	Once	~2 wk	Shnyder et al. (2008)
Narciclasine-7-O-α-D-glucoside (63)	A549	2 µM[a]	Active	Nu/Nu mice	iv, oa	1 mg/kg	5/wk	3 wk	Ingrassia et al. (2009)
	GL19	nt	active	Nu/Nu mice	iv, oa	1 mg/kg	5/wk	3 wk	Ingrassia et al. (2009)

Abbreviations: ip, intraperitoneal; it, intra-tumoral; iv, intravenous; na, not active; ni, not indicated; nt, not tested; oa, oral administration; sc, subcutaneous. Activity: for entries indicated as 'active', no specific cytotoxicity indices were presented in the original work;

[a] IC$_{50}$ value;
[b] T/C index;
[c] maximum toxic dose;
[d] prolongation of mean survival time over control;
[e] ED$_{50}$ value;
[f] EC$_{50}$ value;
[g] reduction in tumor volume.

3.8.2 Efflux Pump Interactions

The P-glycoprotein (P-gp) is an efflux pump which plays a crucial physiological role in protecting cells from toxic xenobiotics and endogenous metabolites (Sharom 2011). It forms a major component of the blood–brain barrier, restricting the uptake of drugs from the small intestine (Sharom 2011). P-gp has also been detected in many human cancers where it is thought to contribute to chemotherapeutic resistance (Sharom 2011). With an excellent Pearson correlation coefficient of 0.447 pancratistatin was identified amongst the top 26 compounds in the NCI library whose cytotoxicity profiles had the highest positive correlations with the ABCG2 transporter (Deeken et al. 2009). ABCG2 has multiple anticancer compounds as its substrates and is known to regulate oral bioavailability and serves a protective role in the blood–brain barrier, the maternal–fetal barrier, and in hematopoietic stem cells (Deeken et al. 2009).

3.8.3 Mitotic Effects

Given that a phenanthridone target could in future appear on the cancer market significant strides have been taken towards understanding the molecular basis to their cytotoxic effects. In spite of such efforts the biochemical target of these antiproliferative agents remains frustratingly aloof. However, it was recognized early on that the cell growth inhibitory effects of phenanthridone alkaloids could be manifesting at some stage of the mitotic cycle. In accompaniment of the first description of narciclasine (11) Ceriotti (1967) observed that its inhibitory effect on sarcoma 180 cells *in vivo* was manifested via an antimitotic effect. Through intraperitoneal (or subcutaneous) injection and oral administration at low dosages (LD_{50} 6 mg/kg) a drastic decline in the number of mitoses was observed after as little as two hours (Ceriotti 1967). From this it was suggested that such characteristics were peculiar of a metaphasic or preprophasic 'poison' with mitoclasic activity at high dosages (Ceriotti 1967). More recently it was revealed that narciclasine induced cytostatic rather than cytotoxic effects in human U373, Hs683, and GL19 glioblastoma multiforme cells (LeFranc et al. 2009). It was demonstrated that narciclasine-induced growth impairment resulted from a marked reduction of mitotic rates (LeFranc et al. 2009).

3.8.4 Effects on Protein Synthesis

Following sustained interest on the interaction of Amaryllidaceae alkaloids with eukaryote ribosomes, Carrasco et al. (1975) showed that narciclasine inhibited protein synthesis in rabbit reticulocytes and yeast cell-free systems by blocking peptide bond formation at the ribosome level, similar to the known inhibitors anisomycin and trichodermin. It was also reported that resistance to narciclasine by the mutant TR_1 strain of *Saccharomyces cerevisiae*, which is anisomycin and trichodermin resistant, is due to an alteration on the peptidyl transferase center of the 60S ribosomal unit (Jimenez et al. 1975a). Interestingly narciclasine (11) and haemanthamine (64), an α-crinane alkaloid of the Amaryllidaceae, shared a common binding site on the peptidyl transferase center of the 60S ribosomal unit that was different from the other Amaryllidaceae alkaloids lycorine (3) and pseudolycorine (65) (Figure 3.1) (Jimenez et al. 1975b). Later on evidence revealed that the alkaloids lycorine (3), haemanthamine (64), pseudolycorine (65), and dihydrolycorine (66) were able to interfere with the binding of [^{3}H]narciclasine to yeast ribosomes suggesting that there was an overlap of their respective binding sites (Baez and Vazquez 1978). At growth inhibitory concentrations narciclasine (11) also blocked protein synthesis in S180 ascites tumors and stabilized HeLa cell polysomes *in vivo*, indicating that it is capable of halting protein synthesis in eukaryote cells by inhibiting peptide bond formation (Jimenez et al. 1976). Studies by Rodriguez-Fonseca et al. (1995) were directed at the ribosomal binding sites of known peptidyl transferase inhibitors including several antibiotics (such as anisomycin and chloramphenicol amongst others) as well as narciclasine (11).

These compounds exhibited varying degrees of specificity for bacterial, archaeal, and eukaryotic ribosomes despite a high level of conservation in the sequence and secondary structure of the peptidyl transferase center of the 23S-like rRNAs (Rodriguez-Fonseca et al. 1995). Binding experiments revealed that these entities were capable of effecting changes to the nucleotides at the peptidyl transferase center, which ranged from one or two changes (in anthelmycin and narciclasine) to eight or nine (virginiamycin M1), from which it was inferred that they were capable of inducing and stabilizing a particular functional conformer of the peptidyl transferase center (Rodriguez-Fonseca et al. 1995).

3.8.5 Topoisomerase Inhibition

Topoisomerases are enzymes which perform regulatory functions in maintaining the topology of DNA, especially its over-winding or under-winding which are commonplace due to its intertwined, double helical structure (Champoux 2001). The clinical significance of topoisomerases is apparent in antibacterial drug therapy, particularly for the broad-spectrum fluoroquinolone antibiotics which act by disrupting bacterial type II topoisomerases (Champoux 2001). Since elevated levels of topoisomerases are detectable in rapidly proliferating cells such as cancer cells, it has been shown that blocking topoisomerase activity with appropriate inhibitor targets is of chemotherapeutic relevance (Champoux 2001). Of the phenanthridones examined, narciprimine (13) and arolycoricidine (24) were shown to inhibit topoisomerase I at concentrations of 0.1–0.25 μg/ml (Sarikaya et al. 2012).

3.8.6 Effects on Calprotectin

Neutrophils are inflammatory cells which accumulate at the site of inflammation during the first phase of inflammation (Yui et al. 2003). They play a significant role in the phagocytosis of microorganisms (Yui et al. 2003). Neutrophils also mediate inflammatory and immunological responses through a variety of protein factors such as proteolytic enzymes and cytokines (Yui et al. 2003). Calprotectin has been identified as a factor in the cytosol of neutrophils, with cytostatic and cytotoxic effects against both normal and tumor cells (Yui et al. 2003). Although the mechanism of action of calprotectin remains to be defined, it has been suggested that it may exert a regulatory role through its growth-inhibitory and apoptosis-inducing activities towards cells migrating into and out of inflammatory sites (Yui et al. 2003). It has also been indicated that calprotectin may cause significant tissue damage in instances where it remains in body fluids for protracted periods (Yui et al. 2003). The identification of targets which could modulate the effects of calprotectin is thus an appealing one and one which could potentially yield rich clinical rewards. In this regard some Amaryllidaceae alkaloids have shown much promise, particularly the phenanthridone narciclasine (11) which inhibited calprotectin-induced cytotoxicity in MM46 cells at markedly low concentrations (IC_{50} 0.001–0.01 μg/mL) (Mikami et al. 1999). Encouragingly, narciclasine (11) was able to achieve such inhibition at concentrations that were up to ten-fold lower than that observed for lycorine (3) (Yui et al. 1998).

3.8.7 Effects on Nitric Oxide

There are compelling pieces of evidence pointing towards the involvement of nitric oxide (NO) in the pathophysiology of cancer (Xu et al. 2002). Increased cellular NO generation could result in increased mutant p53 cellular activity which would contribute to tumor angiogenesis via upregulation of VEGF (vascular endothelial growth factor) (Xu et al. 2002). In addition, NO may modulate tumor DNA repair mechanisms by upregulating tumor suppressor protein p53, poly-ADP ribose polymerase (PARP), and DNA-dependent protein kinase (DNA-PK) (Xu et al. 2002). For these reasons, the discovery of novel cellular NO modulators is of significant interest in the cancer chemotherapy arena. Of the Amaryllidaceae alkaloids examined for NO modulatory effects, narciclasine (11) exhibited potent activity (IC_{50} 0.01 µM) against NO production in LPS-stimulated mouse RAW264 macrophages (Yamazaki and Kawano 2011).

3.8.8 Effects on Tumor Necrosis Factor (TNF)

TNFs are a group of proinflammatory cytokines that have been implicated in tumor regression, septic shock, and cachexia (Hanahan and Weinberg 2000). The involvement of TNFs in programmed cell death (apoptosis) is well established (Hanahan and Weinberg 2000). As part of the apoptosis trigger mechanism, cellular death signals are conveyed by TNF-α binding to its receptor TNF-R1 or via the Fas ligand binding the Fas receptor (Hanahan and Weinberg 2000). Since current TNF-inhibitory therapeutics only involve protein entities such as Adalimumab, Etanercept, and Infliximab, interest in small molecule drug targets such as alkaloids, fatty acids, phenolics, retinoids, sterols, and terpenes has been gaining momentum as cost-effective and viable alternatives (Paul et al. 2006). To this extent, narciclasine (11) exhibited potent TNF-α inhibitory activity in murine RAW264 macrophages stimulated with LPS (IC_{50} 0.02 µM) (Yamazaki and Kawano 2011). Furthermore, it was observed that although TNF-α inhibition by narciclasine was caused by nonselective inhibition of protein synthesis, the pyrrolo-phenanthridine lycorine was able to inhibit TNF-α production at lower concentrations than those required to inhibit protein synthesis in macrophages (Paul et al. 2006).

3.8.9 Apoptosis-Inducing Effects

3.8.9.1 Apoptosis-Inducing Effects of Narciclasine

Of the various biological mechanisms encountered for the cytotoxic effects of Amaryllidaceae alkaloids, their ability to induce apoptosis (or programmed cell death) selectively in cancer cells is arguably the most appealing. The first report on the cytotoxicity of phenanthridone alkaloids is striking in that it also uncovered the antimitotic effect of narciclasine (11), in this case dealing with ascites 180 sarcoma cells where it intervened at either metaphase or preprophase (Ceriotti 1967). This is seminal in the sense that it was the first indication given that Amaryllidaceae alkaloids could be involved in cell cycle arrest in cancer cells (Ceriotti, 1967). The plant growth-inhibitory effects of narciclasine have been known for some time (Bi et al. 1998). However, recent studies have indicated that it could be manifesting such effects via the apoptotic pathway (Lu et al. 2012). As such, it was seen that during narciclasine-induced apoptosis in tobacco Bright Yellow-2 (TBY-2), H_2O_2 together with antioxidant systems such as APX (ascorbate peroxidase) act as signal molecules in regulating mitochondrial activity (Lu et al. 2012). Subsequent mitochondrial dysfunction is manifested via a decrease in MTP (mitochondrial transmembrane potential) (Lu et al. 2012).

Narciclasine (11) had an excellent cytotoxic profile in the NCI 60 cell line screen (mean GI_{50} 0.016 µM) (Pettit et al. 2004, Pettit and Melody 2005). Other studies showed that narciclasine (~1 µM) exhibited selective cytotoxic effects, specifically targeting human breast (MCF7) and prostate carcinoma (PC3) cells but not normal fibroblast cells (Dumont et al. 2007, Ingrassia et al. 2008). These adverse effects were mediated via the apoptotic pathway as indicated by activation of the initiator caspases-8 and -9 of the death receptor pathway (Dumont et al. 2007). In addition, formation of the Fas and death receptor 4 (DR4) death-inducing signal complex was detectable in both MCF7 and PC3 cells (Dumont et al. 2007). Whilst caspase-8 was found to interact with Fas and DR4 receptors in both cell lines, the narciclasine-induced downstream apoptotic pathway in MCF7 cells was distinct from that seen in PC3 cells (Dumont et al. 2007). In the latter caspase-8 activated effector caspases (such as caspase-3) directly in the absence of any further release of mitochondrial proapoptotic effectors (Dumont et al. 2007). The apoptotic process in MCF7 cells by contrast was shown to necessitate an amplification step that is mitochondria-dependent, involving Bid processing, release of cytochrome c, and caspase-9 activation (Dumont et al. 2007).

On the other hand, narciclasine (11) impaired glioblastoma multiforme cell growth by markedly decreasing mitotic rates without inducing apoptosis (LeFranc et al. 2009). It also activated the Rho/Rho kinase/LIM kinase/cofilin signaling pathway by increasing GTPase RhoA activity as well as inducing actin stress fiber formation in a RhoA-dependent manner (LeFranc et al. 2009). Since glioblastoma cells are capable of migrating through the narrow extracellular spaces in brain tissue and traveling relatively long distances, these cancers are a significant challenge for effective surgical management (LeFranc et al. 2009). Consequently, sufferers have a median survival period of only 14 months following the current standard treatment of surgical resection and adjuvant radio- and chemotherapy (LeFranc et al. 2009). Given these dismal prognoses as well as the promising preclinical results acquired for narciclasine, the phenanthridones have been heralded as promising targets in therapeutic approaches towards brain cancer (Van Goietsenoven et al. 2013).

Narciclasine (11) also exhibited good IC_{50} growth-inhibitory values of 30–100 μM against various melanoma cells by targeting the elongation factor eEF1A (Van Goietsenoven et al. 2010). This result is significant since melanomas exhibit intrinsic resistance to pro-apoptotic stimuli and consequently respond poorly to conventional adjuvant therapies (Van Goietsenoven et al. 2010). Found in abundance in eukaryote cells, eEF1A (eukaryotic translation elongation factor 1 alpha) binds to and delivers aa-tRNA (aminoacyl-tRNA) to the vacant A-site of elongating ribosomes (Van Goietsenoven et al. 2010). It also performs several functions in actin cytoskeleton organization, cell migration, cell morphology, protein synthesis, and cell death (Van Goietsenoven et al. 2010). It was shown that eEF1A targeting with narciclasine (at 50 nM) leads to pronounced actin cytoskeleton disorganization and subsequent cytokinesis impairment, as well as protein synthesis impairment via the initiation and elongation steps (Van Goietsenoven et al. 2010). Further study of the apoptosis-inducing ability of narciclasine (11) was carried out in HL60 and HSC2 (human squamous carcinoma) cells against which it had potent cytotoxic effects (IC_{50} 0.018 and 0.05 μM, respectively) (Jitsuno et al. 2011). The apoptotic mode of cell death in this instance was indicated by characteristic morphological changes including, cell shrinkage, chromatin condensation, ladder-like fragmentation pattern for internucleosomal DNA, as well as caspase-3 activation (Jitsuno et al. 2011).

3.8.9.2 Apoptosis-Inducing Effects of Pancratistatin

A significant amount of research has been dedicated towards understanding the apoptotic mode of cell death initiated by the pentahydroxylated phenanthridone pancratistatin (12). In terms of the cellular location of apoptosis induction it was demonstrated that pancratistatin (at 1 μM) specifically targeted the mitochondria of human neuroblastoma (SHSY-5Y) cells, leaving normal human fibroblast (NHF) cells largely unaffected (McLachlan et al. 2005). Mitochondria play a central role in the apoptotic machinery of the cell by triggering the release of caspase activators, caspase-independent death effectors, and cytochrome c at key points which result in the subsequent loss of several essential functions (Elmore 2007). Pancratistatin (at 1 μM) also induced apoptosis selectively in Hs-578-T and MCF7 human breast cancer cells but not in their corresponding non-cancerous counterparts (Hs578Bst and HMEC) (Siedlakowski et al. 2008). As such, pancratistatin increased ROS activity and decreased ATP levels as well as permeabilization of the mitochondrial membrane, indicating activation of the intrinsic apoptotic pathway (Siedlakowski et al. 2008). Exposure to pancratistatin (1 μM) also initiated apoptosis in human melanoma (A375) cells within 72 h, with the mitochondria again indicated as the site of induction (Chatterjee et al. 2011). Furthermore, significant caspase-3 activity was detected in cells treated with pancratistatin (Chatterjee et al. 2011). In addition, it was shown that the estrogen receptor antagonist Tamoxifen sensitized A375 cells to apoptosis induction by pancratistatin (Chatterjee et al. 2011). However, the co-treatment did not affect the viability of non-cancerous NHF cells (Chatterjee et al. 2011).

Pancratistatin (12) reduced cell viability and induced apoptosis in both androgen-responsive LNCaP and androgen-refractory DU145 metastatic prostate cancer cells (Griffin et al. 2011b). Increased ROS activity and collapse of the MMP were the diagnostic markers of pancratistatin-induced apoptosis in both cell lines (Griffin et al. 2011b). This is significant since the majority of metastatic cancers exhibit intrinsic resistant to apoptosis and are thus particularly difficult to manage given that most anticancer drugs function through apoptosis induction (Simpson et al. 2008). Further studies were undertaken to show that pancratistatin indeed targeted the mitochondria of cancer cells rather than DNA or its replicative machinery (Griffin et al. 2011a). As such, it was demonstrated that pancratistatin decreased MMP and induced apoptotic nuclear morphology in both wild-type p53 HCT116 and p53-mutant HT-29 colorectal carcinoma cells but not in normal colon CCD-18Co fibroblasts (Griffin et al. 2011a). Moreover, it had little or no effect on mtDNA-depleted (r0) cancer cells (Griffin et al. 2011a). In addition, pancratistatin-induced cell death did not involve Bax or caspase activation, alteration to β-tubulin polymerization, or double-stranded DNA breaks (Griffin et al. 2011a).

As explained above, a significant improvement in aqueous solubility was achieved by converting pancratistatin into its 3,4-O-phosphate ester (57) (Shnyder et al. 2008). More encouraging was the efficacy of this ester in DLD-1 human colon adenocarcinoma cells where it induced apoptosis via the mitochondrial cascade, impeding the G_2/M phase of the cell cycle (Shnyder et al. 2008). Pancratistatin also caused activation of caspase-3 and exposure of phosphatidylserine on the outer leaflet of the plasma membrane in human lymphoma (Jurkat) cells (Kekre et al. 2005). These events occurred prior to DNA fragmentation and the generation of reactive oxygen species (ROS), indicating that both plasma membrane proteins and caspase-3 were involved in the induction phase of apoptosis (Kekre et al. 2005). This suggested that pancratistatin could directly activate caspase-3 or that it may be specifically targeting some other enzyme on the plasma membrane (Kekre et al. 2005). However, since caspase-3 was not activated in Jurkat cell extracts following incubation with pancratistatin, it is possible that it may be targeting Fas receptors on the plasma membrane which in turn could activate caspase-3 followed by the activation of caspase-8 (Kekre et al. 2005). Pancratistatin was markedly more effective than paclitaxel at inducing apoptosis in Jurkat leukemia cells (80% compared to 25% induction, respectively) (Griffin et al. 2010). *Ex vivo* pancratistatin (12) also induced apoptosis in clinical leukemia isolates but left the non-cancerous peripheral blood mononuclear cells (PBMCs) from healthy individuals intact (Griffin et al. 2010). Hepatoma 5123tc cells stained positively with annexin-V after treatment with pancratistatin indicating the localization of phosphatidylserine on the outer leaflet of the plasma membrane (Pandey et al. 2005). Furthermore, there was no positive indication of annexin-V binding in pancratistatin-treated NHF cells, proving that it did not initiate apoptosis-related changes in the plasma membrane of normal cells (Pandey et al. 2005). In the same instance, the anticancer drugs paclitaxel and VP-16 were shown to induce apoptosis in both cancer and normal NHF cells (Pandey et al. 2005).

Pancratistatin together with two of its ester analogs (20, 22) at concentrations less than 1 μM exhibited strong cytostatic activity in 3Y1 rat embryo fibroblasts (Mutsuga et al. 2002). Cell cycle analysis showed that when cells were arrested at the G_0/G_1 phase via serum deprivation, progression to the S phase was hindered by all three compounds (Mutsuga et al. 2002). Furthermore, cells synchronized at the late G_1/early S phase by hydroxyurea treatment were blocked in progressing through the S phase by compound 20, whilst compound 22 and pancratistatin did not affect cell cycle progress but in fact retarded it (Mutsuga et al. 2002). When the effects of 20 and 22 were evaluated in promyelocytic HL-60RG leukemia cells synchronized at the G_0/G_1 phase, the cells accumulated in the sub G_0/G_1 phase without progress to the S phase, indicative of apoptotic cells (Mutsuga et al. 2002).

The C-1 methyl hydroxy and methyl acetoxy homologated analogs (67, 68) of 7-deoxypancratistatin (19) at 0.5 μM also demonstrated potent apoptotic activities in both Jurkat leukemia and SH-SY5Y neuroblastoma cells (Collins et al. 2010). Encouragingly, compound 68 did not induce apoptosis in non-cancerous NHF and PBMC cells cultured from blood samples of healthy volunteers (Collins et al. 2010). Alkaloid 68 was even more potent in the chemoresistant osteosarcoma (OS) cell lines Saos-2 and U-2 OS (IC_{50}s 0.25 μM, respectively) (Ma et al. 2011a). Both cells succumbed to apoptotic cell death via mitochondrial targeting as indicated by collapsed MMP and increased ROS production (Ma et al. 2011a). It was also revealed that 68 caused the release of endonuclease G (EndoG) and apoptosis inducing factor (AIF) in mitochondria isolated from BxPC-3 cells (Ma et al. 2011a). Alkaloid 68 was shown to be a potent antiproliferative in the pancreatic cancer cells BxPC-3 and PANC-1 at dosages of 0.5–2 μM (Ma et al. 2011b). It induced apoptosis in both cells by targeting the mitochondria where it dissipated MMP, increased the generation of ROS, and caused the release of apoptogenic factors (Ma et al. 2011b). In the same instance normal human fetal fibroblasts were profoundly less sensitive to such treatments (Ma et al. 2011b). To gauge the involvement of caspases in apoptosis induction in BxPC-3 cells, 68 (1 μM) was evaluated in conjunction with the caspase inhibitor benzyloxycarbonyl-Val-Ala-Asp (OMe) fluoromethylketone (Z-VAD-FMK) (Ma et al. 2011b). In this case, the WST-1 colorimetric cell viability assay indicated that Z-VAD-FMK was unable to protect BxPC-3 cells from treatment with 68, indicating a caspase-independent mode of action (Ma et al. 2011b). The acetoxymethyl ester 68 was also effective in human colorectal cancer (CRC) cells, where it induced apoptosis in both p53 positive (HCT116) and p53 negative (HT-29) cells (IC_{50}s 0.25 and 1 μM, respectively) with an efficacy similar to that of pancratistatin (12) (Ma et al. 2012a). The physiological marker effects observed as a consequence included decreased MMP and increased levels of ROS and cytochrome c in the isolated mitochondria of both cell lines (Ma et al. 2012a). These effects could be notably amplified when 68 was administered together with Tamoxifen (Ma et al. 2012a). In addition, minimal cytotoxicity was observed in normal human fetal fibroblast (NFF) and colon fibroblast (CCD-18Co) cells following treatment with 68 (Ma et al. 2012a). The pancratistatin analog 68 (at 1 μM) was also effective against both human breast cancer (MCF7) and neuroblastoma (SH-SY5Y) cells where it induced apoptosis by targeting the mitochondria, with the consequent increase in ROS production and dissipation of MMP (Ma et al. 2012b). Studies in conjunction with Tamoxifen showed that Tamoxifen alone (at 10 μM) induced autophagy (but insignificant cell death) in MCF7 cells, whereas 68 was capable of inducing significant levels of apoptosis with some minor autophagic effects (Ma et al. 2012b). The combined treatment on the other hand produced a drastic elevation in autophagic and apoptotic induction levels (Ma et al. 2012b). However, such treatments had minimal effects on the survival of non-cancerous human fibroblasts (Ma et al. 2012b).

3.8.9.3 Apoptosis-Inducing Effects of Ring-C Unsaturated Analogs

Narciprimine (13) was screened against three human cancers where it was seen via the Calcein AM cell viability assay to be partial towards chronic lymphoblastic leukemia CEM cells (IC_{50} 13.3 μM), compared to breast adenocarcinoma MCF7 and cervical adenocarcinoma HeLa cells (IC_{50}s > 50 μM, respectively) (Nair et al. 2012b). However, the compound could not differentiate between cancerous and non-cancerous cells as normal human fibroblast BJ cells were also considerably affected (IC_{50} 7.9 μM) (Nair et al. 2012b). Flow cytometry was used to quantify the distribution of CEM cells in the various phases of the cell cycle and to determine the subG$_1$ fraction as a marker of the proportion of apoptotic cells (Nair et al. 2012b). Here narciprimine dose-dependently increased the proportion of G$_2$/M phase cells, with concomitant reductions in the proportion of G$_0$/G$_1$ and S cells (Nair et al. 2012b). A slight increase in the subG$_1$ fraction of cells was observed notably after a 20 μM treatment (Nair et al. 2012b). This suggested that narciprimine (13) effectively disturbed cell cycle and induced apoptosis in CEM cells (Nair et al. 2012b). The activities of caspase-3 and -7 (3/7) were then determined in CEM cells exposed to narciprimine for 24 h at 0, 1, 10, and 20 µM, respectively, using the fluorogenic substrate Ac-DEVD-AMC and/or the caspase-3/7 inhibitor Ac-DEVD-CHO (Nair et al. 2012b). From this it was seen that narciprimine induced a 1.8-fold increase in the activity of caspases-3/7 at the highest tested concentration relative to the untreated control (Nair et al. 2012b). Western blot analysis indicated a dose-dependent elevation of tumor suppressor protein p53 levels, whilst expression of the antiapoptotic proteins Mcl-1 and Bcl-2 were unchanged (Nair et al. 2012b). Narciprimine (13) and arolycoricidine (24) as outlined above were assayed for possible adverse effects towards DNA topoisomerases, which are known to be cellular targets for a number of chemotherapeutic drugs (Sarikaya et al. 2012). Both compounds were shown to interfere with human topoisomerase I and II reactions at concentrations ranging from 0.1 to 0.25 µg/ml (Sarikaya et al. 2012). Topoisomerase inhibitors are amongst the most effective apoptosis inducers encountered in clinical practice today (Sordet et al. 2003). Activation of caspases in the cytoplasm by pro-apoptotic elements released from mitochondria is the key signaling pathway leading from topoisomerase-mediated DNA damage to cell death (Sordet et al. 2003). In addition, the death receptor Fas (APO-1/CD95) is also involved with the apoptotic response in some cells (Sordet et al. 2003). These effector pathways are controlled by several upstream-regulated factors that respond to DNA damage induced by topoisomerase inhibitors in apoptotic cells (Sordet et al. 2003). These include the transcription factors NF-κB and p53, the pro-apoptotic c-Abl, Chk2, and SAPK/JNK pathways, as well as the survival PI(3)kinase-AKT-dependent pathway (Oizumi et al. 2002, Sordet et al. 2003). Cellular responses to DNA lesions caused by topoisomerase inhibitors are facilitated by the protein kinases ATM, ATR, and DNA-PK which bind to DNA breaks via their respective sensor molecules (Oizumi et al. 2002, Sordet et al. 2003).

3.8.10 Tumor Invasion and Metastasis

Tumor invasion and metastasis are established amongst the major factors responsible for the high morbidity and mortality statistics in cancer patients (Stock et al. 2013, Wong et al. 2013). It is therefore perplexing that mainstream therapeutic approaches remain centered on the primary tumor as the locus for clinical intervention (Stock et al. 2013, Wong et al. 2013). Furthermore, while several molecular targets such as enzymes, receptors, and signaling pathways have been identified in metastasis drug development, an anti-metastatic drug is yet to see commercial fruition (Stock et al. 2013, Wong et al. 2013). Nonetheless, there have been noteworthy successes in clinical trials with anti-angiogenic agents such as Bevacizumab (Avastin), a recombinant monoclonal antibody which inhibits VEGF-A (vascular endothelial growth factor A) (Los et al. 2007). This protein functions as a chemical signal which stimulates angiogenesis in a variety of diseases, particularly in cancer (Los et al. 2007). Findings with Amaryllidaceae alkaloids are thus poignant in underlining the possibilities which exist in this relatively unexplored area of cancer chemotherapy (Evidente et al. 2009, Evidente and Kornienko 2009). Amongst the alkaloids studied, narciclasine (11) was shown to significantly inhibit the invasion of HeLa cells into collagen type I with only 1–4% invasion observed at non-toxic concentrations (Evidente et al. 2009, Evidente and Kornienko 2009). Since invasive cells secrete metalloproteinases that break down collagen type I fibers, this assay is an excellent model to probe *in vitro* modulation of cell invasion by tissue inhibitors (Evidente et al. 2009, Evidente and Kornienko 2009).

3.9 CONCLUSIONS

Although cancer has been prevalent since the advent of diseases it continues to be a leading cause of morbidity and mortality in society today. With over 200 billion dollars in investment reaching cancer research programs over the past 50 years, significant advances have been made in therapies against this malignancy. Some of the most important anticancer drugs discovered during this period, such as Camptothecin, Podophyllotoxin, Taxol, and Vinblastine are plant-derived substances. Given these successes as well as the rich cultural significance of plants in traditional systems of medicine, they continue to attract the attention of scientists in cancer research across the private and public sectors. Similarly, the discovery of the phenanthridone alkaloids of the Amaryllidaceae as potent and selective antiproliferative agents has sparked considerable interest, particularly in terms of their progress on the clinical stage. Remarkable activities have been demonstrated for the key protagonists narciclasine and pancratistatin against several cancer cells, whilst leaving normal cells largely unaffected. Of the various molecular mechanisms invoked to explain the cytotoxic effects of these compounds, it is their uncanny ability to induce apoptosis selectively in cancer cells that is the most striking. Encouragingly, these properties can be replicated *in vivo* where the early indications are that these compounds are well tolerated in animal models of study. However, the low aqueous solubility of phenanthridones has presented a significant challenge in clinical trials. Nonetheless, this has largely been overcome by innovative semisynthetic derivatization, particularly to phosphate prodrugs, without sacrificing activities. A further challenge presented in the ongoing clinical development of phenanthridones is their limited availability from natural sources. Here, advances in the biotechnological arena have come to the fore in enhancing secondary metabolite production in the Amaryllidaceae, particularly its phenanthridone alkaloid constituents. Even more appealing is the phenomenal successes seen on the synthetic front, where around 60 formal syntheses of narciclasine and pancratistatin have been accomplished over a relatively short period of time. This places them in good stead for future large-scale production.

3.10 ACKNOWLEDGMENTS

The authors are grateful for the generous research funding afforded them by the University of KwaZulu-Natal.

3.11 CONFLICT OF INTEREST

The authors hereby jointly declare no conflict of interest pertaining to any aspect of this manuscript.

3.12 ABBREVIATIONS

Cell lines: A431, human epidermoid carcinoma; A549, human lung carcinoma; B16F10, mouse melanoma; BxPC-3, human pancreatic cancer; DLD-1, human colon adenocarcinoma; Du145, human prostate cancer; GL19, human neuronal glioblastoma; H460, human non-small cell lung carcinoma; HeLa, human cervical adenocarcinoma; HL60, human promyelocytic leukemia; Hs683, human neuronal glioma; HT-29, human colon adenocarcinoma; KB, human oral epidermoid carcinoma; KB-V1, vinblastine-resistant epidermoid carcinoma; KM20L2, human colon cancer; LNCaP, hormone-dependent human prostatic cancer; M5076, murine ovary sarcoma; MCF7, human breast cancer; OVCAR3, human ovarian cancer; P388, murine lymphoid neoplasm; RL, Rauscher leukemia; S180, murine sarcoma; SF268, human glioblastoma; SF295, human glioblastoma; SKMEL-28, human skin melanoma; U373, human glioblastoma astrocytoma; VM-48, human melanoma.

General and specific terms: AIF, apoptosis inducing factor; AKT, protein kinase B; ATM, ataxia telangiectasia mutated; ATR, ataxia telangiectasia and Rad3-related protein; Bax, Bcl-2 associated X-protein; Bcl-2, B-cell lymphoma 2; Bid, BH3-interacting domain death agonist; c-Abl, Abelson murine leukemia viral oncogene homolog 1; DNA-PK, DNA-dependent protein kinase; DR, death receptor; EC, endothelial cell; eEF1A, eukaryotic translation elongation factor 1 alpha; HUVEC, human umbilical vein endothelial cell; JNK, c-Jung N-terminal kinase; LPS, lipopolysaccharide; Mcl-1, induced myeloid leukemia cell differentiation protein; MMP, mitochondrial membrane permeabilization; MND, motor neuron disease; NF-κB, nuclear factor κB; NHF, normal human fibroblast; NO, nitric oxide; NOAEL, no adverse effect level; p53, tumor suppressor protein; PARP, poly(ADP-ribose) polymerase; PBMC, peripheral blood mononuclear cell; PCD, programmed cell death; PI(3)kinase, phosphoinositide 3 kinase; ROS, reactive oxygen species; SAPK, stress-activated protein kinase; TAM, Tamoxifen; T/C, treatment to control index; TNF, tumor necrosis factor; TM, traditional medicine; VEGF, vascular endothelial growth factor.

REFERENCES

Abou-Donia, A.H., A. De Giulio, A. Evidente, et al. 1991. Narciclasine-4-O-beta-D-glucopyranoside, a glucosyloxy amidic phenanthridone derivative from *Pancratium maritimum*. *Phytochemistry* 30:3445–3448.

Baez, A., and D. Vazquez. 1978. Binding of [^{3}H] narciclasine to eukaryotic ribosomes. A study on a structure–activity relationship. *Biochimica et Biophysica Acta* 518:95–103.

Bastida, J., R. Lavilla, and F. Viladomat. 2006. Chemical and biological aspects of *Narcissus* alkaloids. In: *The Alkaloids*, Ed. G.A. Cordell, 87–179. Elsevier, Amsterdam.

Beijnen, J.H., K.P. Flora, G.W. Halbert, et al. 1995. CRC/EORTC/NCI joint formulation working party: Experiences in the formulation of investigational cytotoxic drugs. *British Journal of Cancer* 72:210–218.

Berkov, S., C. Codina, and J. Bastida. 2012. The genus *Galanthus*: A source of bioactive compounds. In: *Phytochemicals – A Global Perspective of their Role in Nutrition and Health*, Ed. R. Venketeshwer, 235–254. InTech Europe, Rijeka.

Bi, Y.-R., K.H. Yung, and Y.S. Wong. 1998. Physiological effects of narciclasine from the mucilage of *Narcissus tazetta* L. bulbs. *Plant Science* 135:103–108.

Briggs, J.B. 2002. Economics of *Narcissus* bulb production. In: *Narcissus and Daffodil – The Genus Narcissus*, Ed. G.R. Hanks, 131–140. Taylor & Francis, London.

Caamal-Fuentes, E., L.W. Torres-Tapia, P. Sima-Polanco, et al. 2011. Screening of plants used in Mayan traditional medicine to treat cancer-like symptoms. *Journal of Ethnopharmacology* 135:719–724.

Carrasco, L., M. Fresno, and D. Vazquez. 1975. Narciclasine: An antitumour alkaloid which blocks peptide bond formation by eukaryotic ribosomes. *FEBS Letters* 52:236–239.

Ceriotti, G. 1967. Narciclasine: An antimitotic substance from *Narcisssus* bulbs. *Nature* 213:595–596.

Champoux, J.J. 2001. DNA topoisomerases: Structure, function and mechanism. *Annual Review of Biochemistry* 70:369–413.

Chatterjee, S.J., J. McNulty, and S. Pandey. 2011. Sensitization of human melanoma cells by tamoxifen to apoptosis induction by pancratistatin, a nongenotoxic natural compound. *Melanoma Research* 21:1–11.

Chrétien, F., S.I. Ahmed, A. Masion, and Y. Chapleur. 1993. Enantiospecific synthesis and biological evaluation of *seco* analogues of antitumor Amaryllidaceae alkaloids. *Tetrahedron* 49:7463–7478.

Collins, J., U. Rinner, M. Moser, et al. 2010. Chemoenzymatic synthesis of Amaryllidaceae constituents and biological evaluation of their C-1 analogues. The next generation synthesis of 7-deoxypancratistatin and *trans*-dihydrolycoricidine. *Journal of Organic Chemistry* 75:3069–3084.

Cragg, G.M., and D.J. Newman. 2005. Plants as a source of anticancer agents. *Journal of Ethnopharmacology* 100:72–79.

Danishefsky, S., and J.Y. Lee. 1989. Total synthesis of (±)-Pancratistatin. *Journal of the American Chemical Society* 111:4829–4837.

De Andrade, J.P., N.B. Pigni, L. Torras-Claveria, et al. 2012. Alkaloids from the *Hippeastrum* genus: Chemistry and biological activity. *Revista Latinoamericana de Quimica* 40:83–98.

Deeken, J.F., R.W. Robey, S. Shukla, et al. 2009. Identification of compounds that correlate with ABCG2 transporter function in the National Cancer Institute anticancer drug screen. *Molecular Pharmacology* 76:946–956.

Dumont, P., L. Ingrassia, S. Rouzeau, et al. 2007. The Amaryllidaceae isocarbostyril narciclasine induces apoptosis by activation of the death receptor and/or mitochondrial pathways in cancer cells but not in normal fibroblasts. *Neoplasia* 9:766–776.

Duncan, G. 2016. *The Amaryllidaceae of Southern Africa.* Kew Publishing, London.

Elmore, S. 2007. Apoptosis: A review of programmed cell death. *Toxicologic Pathology* 35:495–516.

Evidente, A., A.S. Kireev, A.R. Jenkins, et al. 2009. Biological evaluation of structurally diverse Amaryllidaceae alkaloids and their synthetic derivatives: Discovery of novel leads for anticancer drug design. *Planta Medica* 75:501–507.

Evidente, A., and A. Kornienko. 2009. Anticancer evaluation of structurally diverse Amaryllidaceae alkaloids and their synthetic derivatives. *Phytochemistry Reviews* 8:449–459.

Fitzgerald, D.B., J.L. Hartwell, and J. Leiter. 1958. Tumor-damaging activity in plant families showing anti-malarial activity. Amaryllidaceae. *Journal of the National Cancer Institute* 20:763–774.

Furusawa, E., H. Irie, D. Combs, and W.C. Wildman. 1980. Therapeutic activity of pretazettine on Rauscher leukemia: Comparison with the related Amaryllidaceae alkaloids. *Chemotherapy* 26:36–45.

Ghavre, M., J. Froese, M. Pour, and T. Hudlicky. 2016. Synthesis of Amaryllidaceae constituents and unnatural derivatives. *Angewandte Chemie* 55:5642–5691.

Ghosal, S., K. Datta, S.K. Singh, and Y. Kumar. 1990. Telastaside, a stress-related alkaloid-conjugate from *Polytela gloriosa*, an insect feeding on Amaryllidaceae. *Journal of Chemical Research, Synopses* 10:334–335.

Ghosal, S., R. Lochan, A.Y. Kumar, and R.S. Srivastava. 1985a. Chemical constituents of Amaryllidaceae. Part 13. Alkaloids of *Haemanthus kalbreyeri*. *Phytochemistry* 24:1825–1828.

Ghosal, S., K.S. Saini, S. Razdan, and Y. Kumar. 1985b. Chemical constituents of Amaryllidaceae. Part 12. Crinasiatine, a novel alkaloid from *Crinum asiaticum*. *Journal of Chemical Research, Synopses* 3:100–101.

Ghosal, S., S. Singh, Y. Kumar, and R.S. Srivastava. 1989. Chemical constituents of Amaryllidaceae. Part 29. Isocarbostyril alkaloids from *Haemanthus kalbreyeri*. *Phytochemistry* 28:611–613.

Ghosal, S., S.K. Singh, and R.S. Srivastava. 1986. Chemical constituents of Amaryllidaceae. Part 22. Alkaloids of *Zephyranthes flava*. *Phytochemistry* 25:1975–1978.

Graham, J.G., M.L. Quinn, D.S. Fabricant, and N.R. Farnsworth. 2000. Plants used against cancer – An extension of the work of Jonathan Hartwell. *Journal of Ethnopharmacology* 73:347–377.

Griffin, C., C. Hamm, J. McNulty, and S. Pandey. 2010. Pancratistatin induces apoptosis in clinical leukemia samples with minimal effect on non-cancerous peripheral blood mononuclear cells. *Cancer Cell International* 10:6.

Griffin, C., A. Karnik, J. McNulty, and S. Pandey. 2011a. Pancratistatin selectively targets cancer cell mitochondria and reduces growth of human colon tumor xenografts. *Molecular Cancer Therapeutics* 10:57–68.

Griffin, C., J. McNulty, and S. Pandey. 2011b. Pancratistatin induces apoptosis and autophagy in metastatic prostate cancer cells. *International Journal of Oncology* 38:1549–1556.

Hanahan, D., and R.A. Weinberg. 2000. The hallmarks of cancer. *Cell* 100:57–70.

Hartwell, J.L. 1967. Plants used against cancer – A survey. *Lloydia* 30:379–436.

Heinrich, M. 2010. Galanthamine from *Galanthus* and other Amaryllidaceae-chemistry and biology based on traditional use. In: *The Alkaloids*, ed. G.A. Cordell, 157–165., Academic Press, Chennai.

Hudlicky, T., U. Rinner, D. Gonzalez, et al. 2002. Total synthesis and biological evaluation of Amaryllidaceae alkaloids: Narciclasine, *ent*-7-deoxypancratistatin, regioisomer of 7-deoxypancratistatin, 10b-epi-deoxypancratistatin, and truncated derivatives. *Journal of Organic Chemistry* 67:8726–8743.

Hutchings, A., A.H. Scott, G. Lewis, and A.B. Cunningham. 1996. *Zulu Medicinal Plants: An Inventory.* University of Natal Press, Pietermaritzburg.

Ibn-Ahmed, S., M. Khaldi, F. Chrétien, and Y. Chapleur. 2004. A short route to enantiomerically pure benzo-phenanthridinone skeleton: Synthesis of lactone analogues of narciclasine and lycoricidine. *Journal of Organic Chemistry* 69:6722–6731.

Ingrassia, L., F. Lefranc, J. Dewelle, et al. 2009. Structure–activity relationship analysis of novel derivatives of narciclasine (an Amaryllidaceae isocarbostyril derivative) as potential anticancer agents. *Journal of Medicinal Chemistry* 52:1100–1114.

Ingrassia, L., F. Lefranc, V. Mathieu, F. Darro, and R. Kiss. 2008. Amaryllidaceae isocarbostyril alkaloids and their derivatives as promising antitumor agents. *Translational Oncology* 1:1–13.

Jimenez, A., L. Sanchez, and D. Vazquez. 1975a. Location of resistance to the alkaloid narciclasine in the 60S ribosomal subunit. *FEBS Letters* 55:53–56.

Jimenez, A., L. Sanchez, and D. Vazquez. 1975b. Yeast ribosomal sensitivity and resistance to the Amaryllidaceae alkaloids. *FEBS Letters* 60:66–70.

Jimenez, A., A. Santos, G. Alonso, and D. Vazquez. 1976. Inhibitors of protein synthesis in eukaryotic cells. Comparative effects of some Amaryllidaceae alkaloids. *Biochimica et Biophysica Acta* 425:342–348.

Jin, Z. 2016. Amaryllidaceae and *Sceletium* alkaloids. *Natural Product Reports* 33:1318–1343.

Jitsuno, M., A. Yokosuka, K. Hashimoto, et al. 2011. Chemical constituents of *Lycoris albiflora* and their cytotoxic activities. *Natural Product Communications* 6:187–192.

Kaya, G.I., B. Sarıkaya, M.A. Onur, et al. 2011. Antiprotozoal alkaloids from *Galanthus trojanus*. *Phytochemistry Letters* 4:301–305.

Kekre, N., C. Griffin, J. McNulty, and S. Pandey. 2005. Pancratistatin causes early activation of caspase-3 and the flipping of phosphatidyl serine followed by rapid apoptosis specifically in human lymphoma cells. *Cancer Chemotherapy and Pharmacology* 56:29–38.

Kojima, K., M. Mutsuga, M. Inoue, and Y. Ogihara. 1998. Two alkaloids from *Zephyranthes carinata*. *Phytochemistry* 48:1199–1202.

Kornienko, A., and A. Evidente. 2008. Chemistry, biology and medicinal potential of narciclasine and its congeners. *Chemical Reviews* 108:1982–2014.

Lefranc, F., S. Sauvage, G. Van Goietsenoven, et al. 2009. Narciclasine, a plant growth modulator, activates Rho and stress fibers in glioblastoma cells. *Molecular Cancer Therapeutics* 8:1739–1750.

Los, M., J.M.L. Roodhart, and E.E. Voest. 2007. Target practice: Lessons from phase III trials with bevacizumab and vatalanib in the treatment of advanced colorectal cancer. *Oncologist* 12:443–450.

Lu, H., Q. Wan, H. Wang, et al. 2012. Oxidative stress and mitochondrial dysfunctions are early events in narciclasine-induced programmed cell death in tobacco Bright Yellow-2 cells. *Physiologia Plantarum* 144:48–58.

Luo, Z., F. Wang, J. Zhang, et al. 2012. Cytotoxic alkaloids from the whole plants of *Zephyranthes candida*. *Journal of Natural Products* 75:2113–2120.

Ma, D., J. Collins, T. Hudlicky, and S. Pandey. 2012b. Enhancement of apoptotic and autophagic induction by a novel synthetic C-1 analogue of 7-deoxypancratistatin in human breast adenocarcinoma and neuroblastoma cells with tamoxifen. *Journal of Visualized Experiments* 63:3586–3602.

Ma, D., P. Tremblay, K. Mahngar, et al. 2011b. Induction of apoptosis and autophagy in human pancreatic cancer cells by a novel synthetic C-1 analogue of 7-deoxypancratistatin. *American Journal of Biomedical Sciences* 3:278–291.

Ma, D., P. Tremblay, K. Mahngar, et al. 2012a. A novel synthetic C-1 analogue of 7-deoxypancratistatin induces apoptosis in p53 positive and negative human colorectal cancer cells by targeting the mitochondria: Enhancement of activity by tamoxifen. *Investigational New Drugs* 30:1012–1027.

Ma, D., P. Tremblay, K. Mahngar, et al. 2011a. Selective cytotoxicity against human osteosarcoma cells by a novel synthetic C-1 analogue of 7-deoxypancratistatin is potentiated by curcumin. *PLOS ONE* 6:e28780.

Manpadi, M., A.S. Kireev, I.V. Magedov, et al. 2009. Synthesis of structurally simplified analogues of pancratistatin: Truncation of the cyclitol ring. *Journal of Organic Chemistry* 74:7122–7131.

Marion, F., J.P. Annereau, and J. Fahy. 2010. *Nitrogenated Derivatives of Pancratistatin. WO 2010/012714 A1*. World Intellectual Property Organization, Switzerland.

Matveenko, M., M.G. Banwell, and A.C. Willis. 2008. A chemoenzymatic total synthesis of *ent*-narciclasine. *Tetrahedron* 64:4817–4826.

McLachlan, A., N. Kekre, J. McNulty, and S. Pandey. 2005. Pancratistatin: A natural anti-cancer compound that targets mitochondria specifically in cancer cells to induce mitosis. *Apoptosis* 10:619–630.

McNulty, J., V. Larichev, and S. Pandey. 2005. A synthesis of 3-deoxydihydrolycoricidine: Refinement of a structurally minimum pancratistatin pharmacophore. *Bioorganic and Medicinal Chemistry Letters* 15:5315–5318.

McNulty, J., J. Mao, R. Gibe, et al. 2001. Studies directed towards the refinement of the pancratistatin cytotoxic pharmacophore. *Bioorganic and Medicinal Chemistry Letters* 11:169–172.

McNulty, J., and R. Mo. 1998. Diastereoselective intramolecular nitroaldol entry to lycoricidine alkaloids. *Chemical Communications* 29 933–934.

McNulty, J., J.J. Nair, C. Griffin, and S. Pandey. 2008. Synthesis and biological evaluation of fully functionalized *seco*-pancratistatin analogues. *Journal of Natural Products* 71:357–363.

Meerow, A.W., and D.A. Snijman. 1998. Amaryllidaceae. In: *The Families and Genera of Vascular Plants*, Ed. K. Kubitzki, 83–110. Springer, Berlin.

Mikami, M., M. Kitahara, M. Kitano, et al. 1999. Suppressive activity of lycoricidinol (narciclasine) against cytotoxicity of neutrophil-derived calprotectin and its suppressive effect on rat adjuvant arthritis model. *Biological and Pharmaceutical Bulletin* 22:674–678.

Mondon, A., and K. Krohn. 1970. Structure and synthesis of narciprimines. *Tetrahedron Letters* 24:2123–2126.

Mutsuga, M., K. Kojima, M. Yamashita, et al. 2002. Inhibition of cell cycle progression through specific phase by pancratistatin derivatives. *Biological and Pharmaceutical Bulletin* 25:223–228.

Nair, J.J., J. Bastida, F. Viladomat, and J. van Staden. 2012a. Cytotoxic agents of the crinane series of Amaryllidaceae alkaloids. *Natural Product Communications* 7:1677–1688.

Nair, J.J., L. Rárová, M. Strnad, J. Bastida, and J. Van Staden. 2012b. Apoptosis-inducing effects of distichamine and narciprimine, rare alkaloids of the plant family Amaryllidaceae. *Bioorganic and Medicinal Chemistry Letters* 22:6195–6199.

Nair, J.J., L. Rárová, M. Strnad, J. Bastida, and J. van Staden. 2015. Mechanistic insights to the cytotoxicity of Amaryllidaceae alkaloids. *Natural Product Communications* 10:171–182.

Nair, J.J., and J. van Staden. 2013. Pharmacological and toxicological insights to the South African Amaryllidaceae. *Food and Chemical Toxicology* 62:262–275.

Nair, J.J., and J. van Staden. 2014. Cytotoxicity studies of lycorine alkaloids of the Amaryllidaceae. *Natural Product Communications* 9:1193–1210.

Nair, J.J., J. van Staden, and J. Bastida. 2016. Cytotoxic alkaloid constituents of the Amaryllidaceae. In: *Studies in Natural Products Chemistry*, Ed. A.U. Rahman, 107–156. Elsevier, Amsterdam.

Ohta, S., and S. Kimoto. 1975. Total synthesis of (±)-lycoricidine. *Tetrahedron Letters* 27:2279–2282.

Oizumi, S., H. Isobe, S.Ogura, et al. 2002. Topoisomerase inhibitor-induced apoptosis accompanied by down-regulation of Bcl-2 in human lung cancer cells. *Anticancer Research* 22:4029–4037.

Okamoto, T., Y. Torii, and Y. Isogai. 1968. Lycoricidinol and lycoricidine, new plant-growth regulators in the bulbs of Lycoris radiata herb. *Chemical and Pharmaceutical Bulletin* 16:1860–1864.

Pade, V., and S. Stavchansky. 1998. Link between drug absorption, solubility and permeability measurements in Caco-2 cells. *Journal of Pharmaceutical Sciences* 87:1604–1607.

Pandey, S., N. Kekre, J. Naderi, and J. McNulty. 2005. Induction of apoptotic cell death specifically in rat and human cancer cells by pancratistatin. *Artificial Cells, Blood Substitutes, and Biotechnology* 33:1–17.

Paul, A.T., V.M. Gohil, and K.K. Bhutani. 2006. Modulating TNF-α signalling with natural products. *Drug Discovery Today* 11:725–732.

Pettit, G.R., G.M. Cragg, S.B. Singh, et al. 1990. Antineoplastic agents, 162. *Zephyranthes candida*. *Journal of Natural Products* 53:176–178.

Pettit, G.R., S.A. Eastham, N. Melody, et al. 2006. Isolation and structural modification of 7-deoxynarciclasine and 7-deoxy-*trans*-dihydronarciclasine. *Journal of Natural Products* 69:7–13.

Pettit, G.R., V. Gaddamidi, and G.M. Cragg. 1984. Antineoplastic agents. 105. *Zephyranthes grandiflora*. *Journal of Natural Products* 47:1018–1020.

Pettit, G.R., V. Gaddamidi, D.L. Herald, et al. 1986. Antineoplastic agents. 120. *Pancratium littorale*. *Journal of Natural Products* 49:995–1002.

Pettit, G.R., and N. Melody. 2005. Antineoplastic agents. 527. Synthesis of 7-deoxynarcistatin, 7-deoxy-*trans*-dihydronarcistatin, and *trans*-dihydronarcistatin. *Journal of Natural Products* 68:207–211.

Pettit, G.R., N. Melody, and D.L. Herald. 2001. Antineoplastic agents. 450. Synthesis of (+)-pancratistatin from (+)-narciclasine as relay. *Journal of Organic Chemistry* 66:2583–2587.

Pettit, G.R., N. Melody, and D.L. Herald. 2004. Antineoplastic agents. 511. Direct phosphorylation of phenpanstatin and pancratistatin. *Journal of Natural Products* 67:322–327.

Pettit, G.R., N. Melody, M. Simpson, M. Thompson, D.L. Herald, and J.C. Knight. 2003. Antineoplastic agents 500. Narcistatin. *Journal of Natural Products* 66:92–96.

Pettit, G.R., G.R. Pettit III, R.A. Backhaus, et al. 1993. Antineoplastic agents, 256. Cell growth inhibitory isocarbostyrils from *Hymenocallis*. *Journal of Natural Products* 56:1682–1687.

Piozzi, F., C. Fuganti, R. Mondelli, and G. Ceriotti. 1968. Narciclasine and narciprimine. *Tetrahedron* 24:1119–1131.

Piozzi, F., M.L. Marino, C. Fuganti, and A. Di Martino. 1969. Occurrence of non-basic metabolites in Amaryllidaceae. *Phytochemistry* 8:1745–1748.

Rinner, U., H.L. Hillebrenner, D.R. Adams, et al. 2004a. Synthesis and biological activity of some structural modifications of pancratistatin. *Bioorganic and Medicinal Chemistry Letters* 14:2911–2915.

Rinner, U., and T. Hudlicky. 2005. Synthesis of Amaryllidaceae constituents – An update. *Synlett* 3:365–387.

Rinner, U., T. Hudlicky, H. Gordon, and G.R. Pettit. 2004b. A β-carboline-1-one mimic of the anticancer Amaryllidaceae constituent pancratistatin: Synthesis and biological evaluation. *Angewandte Chemie International Edition* 43:5342–5346.

Rodriguez-Fonseca, C., R. Amils, and R.A. Garrett. 1995. Fine structure of the peptidyltransferase centre on 23S-like rRNAs deduced from chemical probing of antibiotic-ribosome complexes. *Journal of Molecular Biology* 247:224–235.

Sarıkaya, B.B., S. Zencir, N.U. Somer, et al. 2012. The effects of arolycoricidine and narciprimine on tumor cell killing and topoisomerase activity. *Records of Natural Products* 6:381–385.

Sharom, F.J. 2011. The P-glycoprotein multidrug transporter. *Essays in Biochemistry* 50:161–178.

Shnyder, S.D., P.A. Cooper, N.J. Millington, et al. 2008. Sodium pancratistatin 3,4-*O*-cyclic phosphate, a water-soluble synthetic derivative of pancratistatin, is highly effective in human colon tumor model. *Journal of Natural Products* 71:321–324.

Siedlakowski, P., A. McLachlan-Burgess, C. Griffin, et al. 2008. Synergy of pancratistatin and tamoxifen on breast cancer cells in inducing apoptosis by targeting mitochondria. *Cancer Biology and Therapy* 7:376–384.

Simpson, C.D., K. Anyiwe, and A.D. Schimmer. 2008. Anoikis resistance and tumor metastasis. *Cancer Letters* 272:177–185.

Sordet, O., Q.A. Khan, K.W. Kohn, and Y. Pommier. 2003. Apoptosis induced by topoisomerase inhibitors. *Current Medicinal Chemistry: Anti-Cancer Agents* 3:271–290.

Stock, A.M., G. Troost, B. Niggemann, et al. 2013. Targets for anti-metastatic drug development. *Current Pharmaceutical Design* 19:5127–5134.

Takagi, S., and M. Yamaki. 1974. On the constituents of the bulbs of Lycoris sanguinea Maxim *Yakugaku Zasshi* 94:617–622.

Tian, X., T. Hudlicky, and K. Koenigsberger. 1995. First total synthesis of (+)-pancratistatin: An unusual set of problems. *Journal of the American Chemical Society* 117:3643–3644.

Van Goietsenoven, G., J. Hutton, J.P. Becker, et al. 2010. Targeting of eEF1A with Amaryllidaceae isocarbostyrils as a strategy to combat melanomas. *FASEB Journal* 24:4575–4584.

Van Goietsenoven, G., V. Mathieu, F. Lefranc, A. Kornienko, A. Evidente, and R. Kiss. 2013. Narciclasine as well as other Amaryllidaceae isocarbostyrils are promising GTP-ase targeting agents against brain cancers. *Medicinal Research Reviews* 33:439–455.

Viladomat, F., J. Bastida, C. Codina, et al. 1997. Alkaloids of the South African Amaryllidaceae. In: *Recent Research Developments in Phytochemistry*, Ed. S.G. Pandali, 131–171. Research Signpost Publishers, Trivandrum.

Vshyvenko, S., J. Scattolon, T. Hudlicky, et al. 2012. Unnatural C-1 homologues of pancratistatin – Synthesis and promising biological activities. *Canadian Journal of Chemistry* 90:932–943.

Vshyvenko, S., J. Scattolon, T. Hudlicky, et al. 2011. Synthesis of C-1 homologues of pancratistatin and their preliminary biological evaluation. *Bioorganic and Medicinal Chemistry Letters* 21:4750–4752.

Watt, J.M., and M.G. Breyer-Brandwijk. 1962. *The Medicinal and Poisonous Plants of Southern and Eastern Africa*. Livingston Ltd, Edinburgh.

Wong, M.S., S.M. Sidik, R. Mahmud, and J. Stanslas. 2013. Molecular targets in the discovery and development of novel antimetastatic agents: Current progress and future prospects. *Clinical and Experimental Pharmacology and Physiology* 40:307–319.

Xu, W., L.Z. Liu, M. Loizidou, M. Ahmed, and I.G. Charles. 2002. The role of nitric oxide in cancer. *Cell Research* 12:311–320.

Yamazaki, Y., and Y. Kawano. 2011. Inhibitory effects of herbal alkaloids on the tumor necrosis factor-α and nitric oxide production in lipopolysaccharide-stimulated RAW264 macrophages. *Chemical and Pharmaceutical Bulletin* 59:388–391.

Yui, S., M. Mikami, M. Kitahara, and M. Yamazaki. 1998. The inhibitory effect of lycorine on tumor cell apoptosis induced by polymorphonuclear leukocyte-derived calprotectin. *Immunopharmacology* 40:151–162.

Yui, S., Y. Nakatani, and M. Mikami. 2003. Calprotectin (S100A8/S100A9), an inflammatory protein complex from neutrophils with a broad apoptosis-inducing activity. *Biological and Pharmaceutical Bulletin* 26:753–760.

CHAPTER 4

Naphthoquinone Constituents of Anticancer Terrestrial Plants

Jia-hua Cui and Shao-shun Li

CONTENT

The naturally occurring naphthoquinones are a class of phenolic compounds with a conjugated diketone moiety on their naphthalene skeletons. As small molecules, they are widely distributed in nature and have a variety of pharmacological effects, such as antibacterial, antifungal, antiviral, antiprotozoal, insecticidal, anti-inflammatory, cytotoxic, and antipyretic activities. Their striking antitumor activity has been the focus of a great deal of research works, and several natural naphthoquinones were recognized as promising lead compounds in the discovery of new anticancer drug candidates. According to the structural similarity of the parent skeletons, these naturally occurring naphthoquinones were categorized into the following three groups shown in Figure 4.1: naphthazarines (1), 1,4-naphthoquinones (2), and 1,2-naphthoquinones (3).

Naphthazarin (Figure 4.1, 1) is a hydroxylated naphthoquinone with the symmetric structural scaffold and also intramolecular hydrogen bonds. The naphthazarin structure-contained natural products were found in a wide range of plant families. The anticancer shikonin (Figure 4.2, 4) and alkannin (5) as a pair of enantiomers, which were extracted and identified from the roots of *Lithospermum erythrorhizon* Sieb. et Zucc. in the Orient and *Arnebia euchroma* (Royle) I. M. Johnst in Europe, belong to the naphthazarin group. 1,4-Naphthoquinones also represent a large class of natural products. Notable examples of naturally occurring 1,4-naphthoquinones include juglone (6, isolated from *Juglans mandshurica* Maxim.), plumbagin (7, obtained from *Plumbago auriculata* Lam.), and α-lapachol (8, obtained from *Tecoma stans* (L.) Juss. ex Kunth). 1,2-Naphthoquinones belonging to the *ortho*-quinones incorporated two adjacent ketone moieties on their naphthalene rings. Saprorthoquinone (9, isolated from *Salvia prionitis* Hance) and its analogs were classified into the 1,2-naphthoquinones group. The following sections will concentrate on naphthoquinone-contained terrestrial plants that had been employed for cancer treatment, the major cytotoxic naphthoquinones from their extracts, the antitumor activities, and the working mechanism of the naphthoquinones. Meanwhile, the related synthetic analogs as promising anticancer agents were also included.

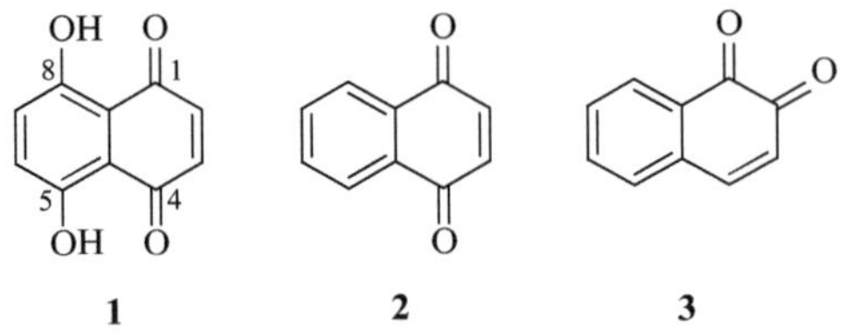

Figure 4.1 Basic structural templates of natural naphthoquinones.

Lithospermum erythrorhizon Sieb. et Zucc.

Family: *Boraginaceae*
Appearance
Perennial herbs
Stem: erect, appressed or spreading; branching distally, 40–90 cm tall, short strigose
Leaves: sessile, ovate-lanceolate to broadly lanceolate, 3–8 × 0.7–1.7 cm
Flower: corolla white, 7-9 mm, sparsely pubescent outside
Roots: dark red, with a copious purple dye
In bloom: June to September
Other anticancer species belonging to the genus *Lithospermum*:
Lithospermum officinale Linn.
Lithospermum hancockianum Oliver (Figure 4.3)

Arnebia euchroma (Royle) I. M. Johnst.

Family: *Boraginaceae*
Appearance
Hairy perennial herbs
Stem: erect, up to 40 cm tall, axillary
Leaves: basal leaves 3.5–12.5 × 5–11 mm, upper ones shorter, lanceolate, strigosely hairy, the hairs
   with bulbous base
Flower: yellow to purple in color, sessile to subsessile, heterostylous calyx 13–17 (–25) mm long, vil-
   lous with yellowish-white to white hairs up to 2 mm long
Roots: thick, exuding a purplish dye
In bloom: June to July (Figure 4.4)

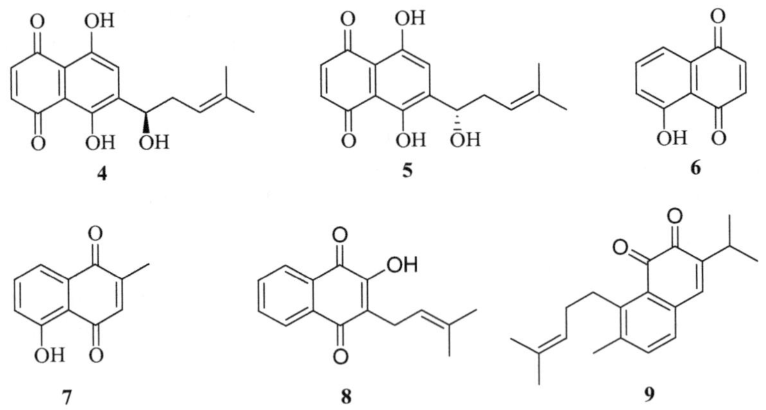

Figure 4.2 Structures of natural anticancer naphthoquinones.

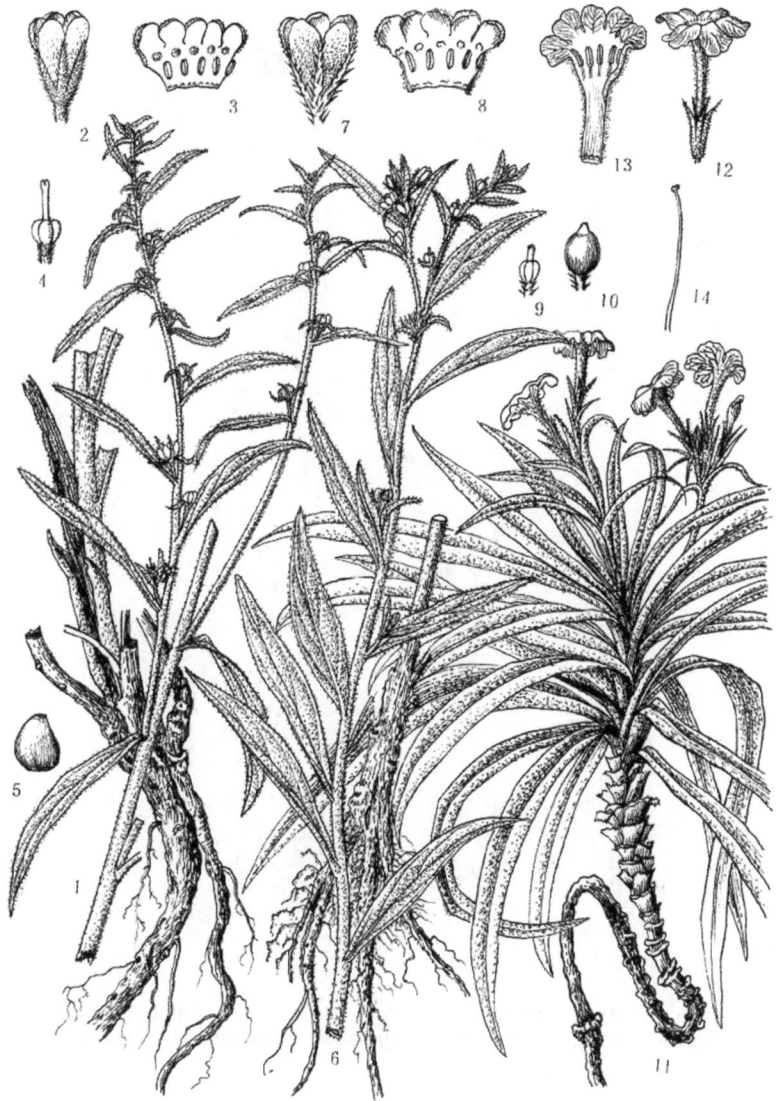

Figure 4.3 The illustration of *Lithospermum erythrorhizon* Sieb. et Zucc. and related anticancer species. 1–5 *Lithospermum officinale* Linn. (1. flowering branch and roots, 2. flower, 3. opened corolla showing stamens, 4. pistil, 5. nutlet) 6–10 *Lithospermum erythrorhizon* Sieb. *et* Zucc. (6. flowering branch and roots, 7. flower, 8. opened corolla showing stamens, 9. pistil, 10. nutlet) 11–14 *Lithospermum hancockianum* Oliver (11. flowering plant, 12. flower, 13. opened corolla showing stamens, 14. style and stigma) (*picture from www.efloras.org*).

The enantiomeric natural naphthoquinone products shikonin (4) and alkannin (5) exhibited striking antitumor activity and also attracted considerable interest from both academia and industry (Papageorgiou et al. 1999). As early as the late 1960s, it was reported that the ethanol extracts of the plants from *Arnebia* genus possessed strong anticancer activity against Walker carcinosarcoma in rats (Bhakuni et al. 1969) and the activity was associated with the naphthoquinone fractions, in which the *S* enantiomer alkannin and its derivatives were the key ingredients (Sankawa et al. 1981, Shukla et al. 1971). In the 1970s, Sankawa et al. discovered that shikonin with the *R* configuration found in the root of *Lithospermum erythrorhizon* Sieb. et Zucc. showed remarkably high antitumor

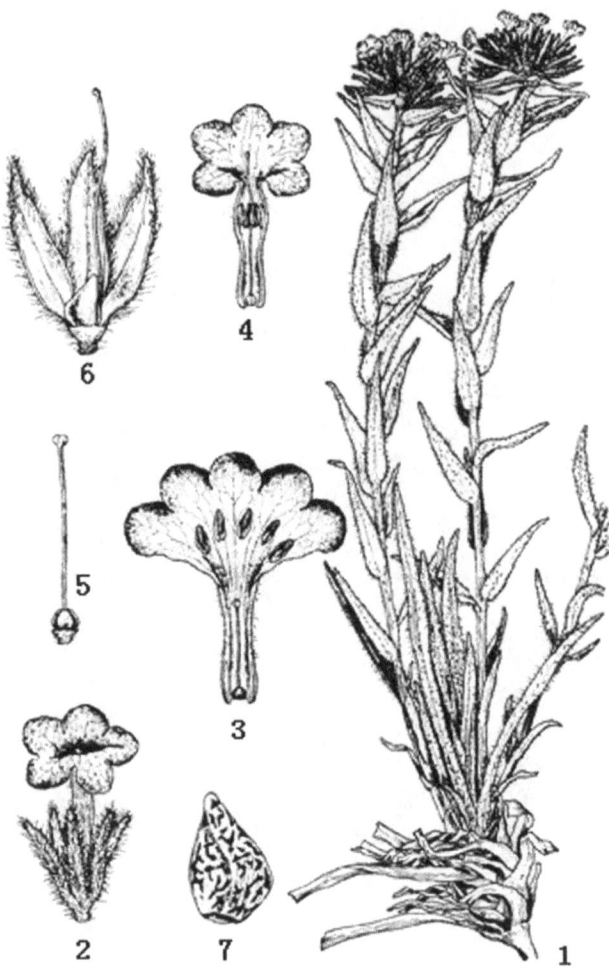

Figure 4.4   The illustration of *Arnebiaeuchroma* (Royle) I. M. Johnst. (1. habit, 2. long styled flower, 3. short styled flower, 4. corolla, 5. carpel, 6. fruiting calyx with one attached nutlet, 7. nutlet) (*picture from www.efloras.org*).

activity against the ascites S-180 cellin ICR mice (Sankawa et al. 1977). The daily intraperitoneal injection (*i.p.*) of shikon in with the dosage of 5–10 mg/kg could significantly increase the survival of the tumor-bearing mice. Subsequently, a great deal of studies were conducted to investigate the anticancer activity of shikonin, alkannin, and related naphthazarin natural products. According to the literature, the two natural naphthoquinones (4 and 5) were active *in vitro* in a dose-dependent manner towards a large variety of human cancer cell lines, including the human leukemia, gastric, colorectal, liver, rectal, prostate, melanoma, and cervical cancers (Andújar et al. 2013, Papageorgiou et al. 1999, Papageorgiou et al. 2006, Tandon, and Kumar 2013). Towards several wild-type and corresponding drug-resistant cell lines, shikonin exhibited similar IC_{50} values (Wu et al. 2013, Xuan and Hu 2009). In addition to their anti-proliferative activities, these two natural products could also inhibit the migration, invasion, and tumorigenic potentials of cancer cells (Andújar et al. 2013).

For more than three decades, a great deal of research works were conducted to investigate the molecular mechanism of shikonin, alkannin, and related natural naphthazarins as potent anticancer agents (Andújar et al. 2013, Chen et al. 2002b, Papageorgiou et al. 1999, Zhang et al. 2017). From the chemical point of view, the molecular mechanism for their anticancer activity

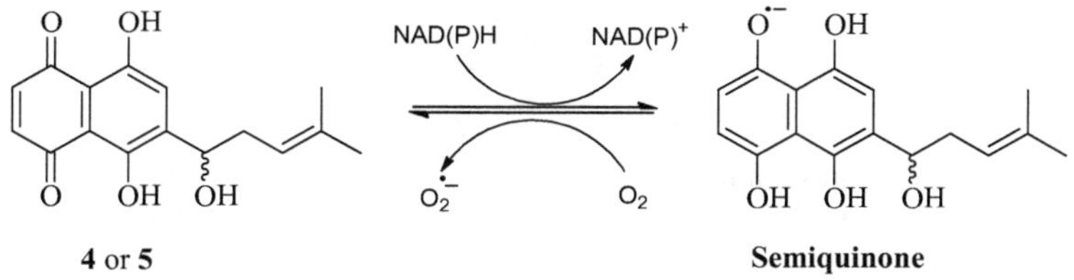

Figure 4.5 The generation of ROS by the shikonin/alkannin redox cycle.

and pervasive toxicity has been demonstrated to be closely related to the generation of reactive oxygen species (ROS, Figure 4.5) and the bioreductive alkylation (Figure 4.6) (Moore 1977, Papageorgiou et al. 1999).

The generated ROS induced the cellular apoptosis, interfered with the MAPKs signal pathway, and also triggered the necroptosis pathway (Figure 4.7) (Zhang et al. 2017). Through the bioreductive alkylation procedures, shikonin and alkannin could inhibit the activity of nuclear factor NF-κB proteins, topoisomerases, protein tyrosine kinases, and telomerase. The alkylation of DNA and key proteins in cellular pathways by these two natural products might also lead to the dysfunction of heterogeneous nuclear ribonucleoprotein A1 (hnRNPA1) (Zhang et al. 2017).

With shikonin and alkannin as the leads, hundreds of their derivatives were designed, synthesized, and also evaluated as excellent antitumor drug candidates. Among these synthetic analogs, several compounds exhibited great potentials as antiproliferative agents in cell-based studies (Kim et al. 2001, Thangapazham et al. 2008, Wang et al. 2009). However, neither shikonin nor its synthetic derivatives have been used as chemotherapeutics in clinics because of their non-specific cytotoxicity *in vivo*. The damage caused by ROS and bioreductive alkylation was nonspecific, affecting a variety of biological macromolecules such as nucleic acids and proteins and resulting in damage not only to tumors but to the normal tissues as well. The previous studies focused on the modifications of the side chain of shikonin (Kim et al. 2001, Thangapazham et al. 2008, Wang et al. 2009), which were insufficient to reduce the generation of ROS and the bioreductive alkylation. Therefore, reasonable modifications to shikonin and alkannin could be vital for obtaining more promising antitumor agents.

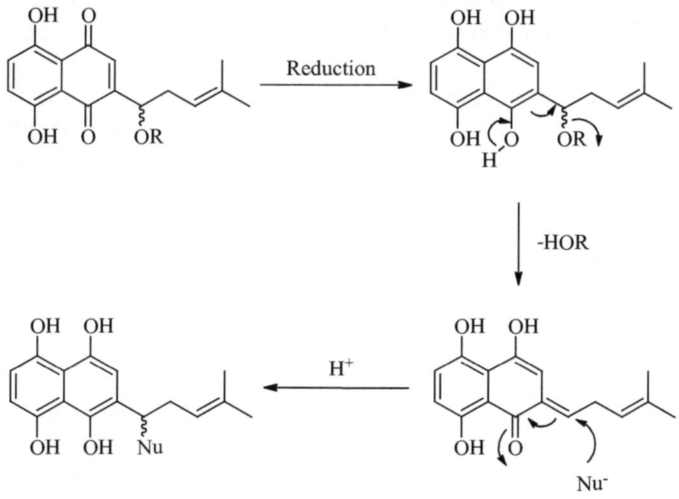

Figure 4.6 The plausible mechanism of the bioreductive alkylation (Moore 1977).

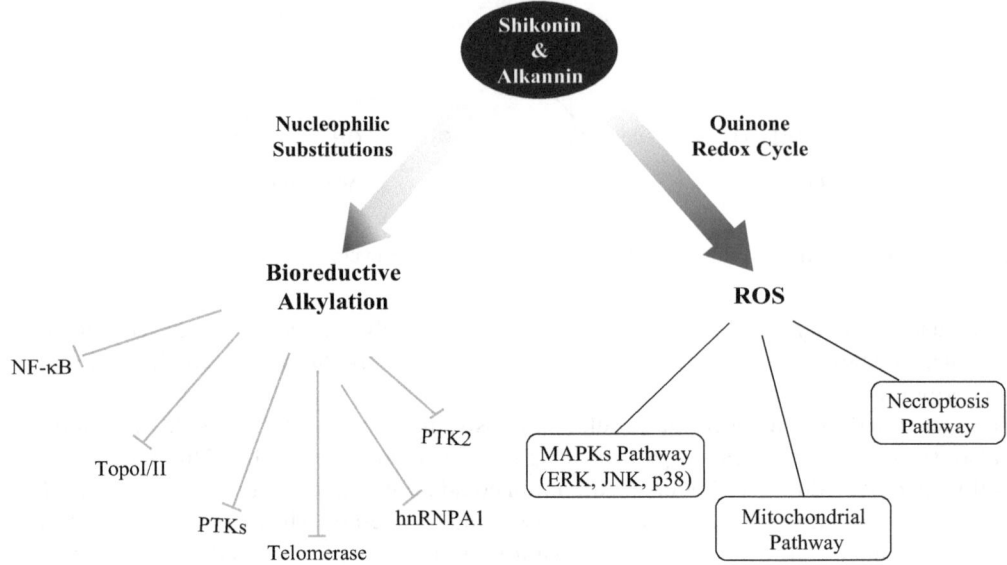

Figure 4.7 Plausible anticancer mechanism of shikonin and alkannin.

In recent studies, the modification of alkannin and shikonin focusing on the naphthoquinone nucleus could obviously minimize their cytotoxicity (Wang et al. 2014a, Wang et al. 2014c, Zhang et al. 2015, Zhou et al. 2010, Zhou et al. 2011). The phenolic hydroxyl groups of the naphthazarin moiety were masked with methyl groups to afford the dimethyl shikonin derivatives (Zhou et al. 2010, Zhou et al. 2011). Further cell-based investigations demonstrated that these dimethylated derivatives were equally effective to shikonin. The selective cytotoxicity toward cancer cell lines was observed together with no toxicity to the normal cell. However, adverse effects such as weight reduction, hypotrichosis, and much bloody ascites were still observable in animal experiments (Figure 4.8).

In further studies, the remaining quinine moiety of these dimethyl shikonin analogs was masked by the acetyl groups or oximes. The obtained derivatives exhibited a dramatic decrease in both ROS generation and bioreductive alkylating capacity (Wang et al. 2014b, Wang et al. 2014c). The dimethylated diacetyl derivatives (Figure 4.9) with poor anticancer activity *in vitro* showed tumor-growth inhibiting effects similar to paclitaxel but without any toxicity *in vivo*. They acted as prodrugs that were activated in plasma.

Further structural optimization afforded the shikonin and alkanninoximes (Figure 4.10), which were more active towards cancer cell lines as compared to the lead naphthoquinones (4 and 5) in cell-based investigation (Wang et al. 2014b, Yang et al. 2016). The results from the *in vivo* efficacy

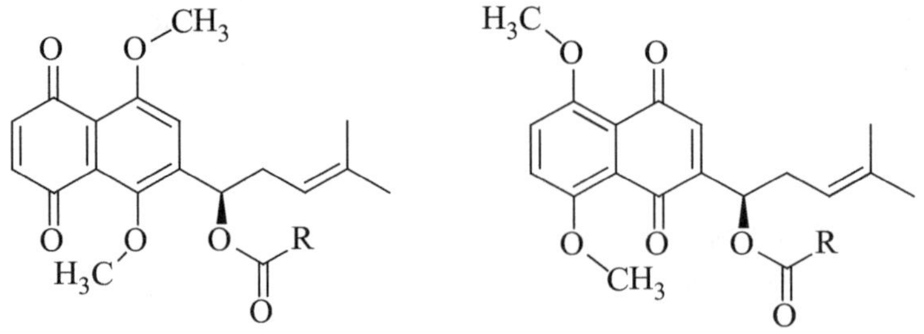

Figure 4.8 Chemical structures of dimethyl shikonin analogues.

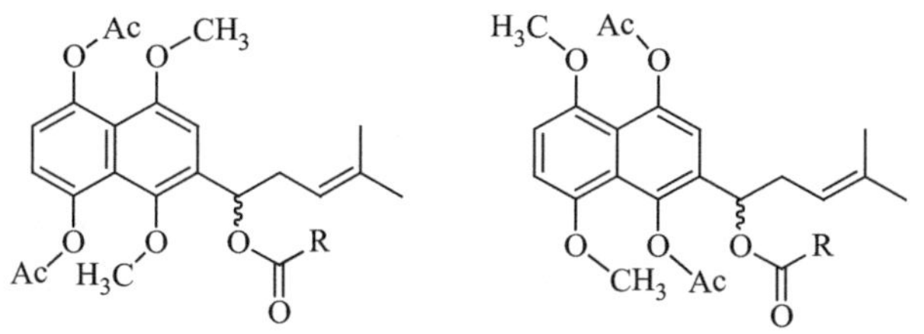

Figure 4.9 Chemical structures of dimethylated-diacetyl-shikonin and alkannin analogues.

evaluation implied that the oximes were as effective as the toxic anti-metabolite 5-Fu towards the growth of HCT-15 xenograft but revealed neither toxicity nor death in mice. Compared with natural shikonin and alkanninas, the leads, these oximes were the most promising compounds to be developed as clinical anticancer drugs.

Juglans mandshurica Maxim.

> Family: *Juglandaceae*
> Appearance
> *Trees or sometimes shrubs*
> *Stem*: erect, to 25 m tall
> *Leaves*: leaves 40–90 cm, petiole 5–23 cm, blade elliptic to long elliptic or ovate-elliptic to long elliptic-lanceolate, 6–17 × 2–7.5 cm, ab axially tomentose or occasionally slightly pubescent
> *Flower*: male spike 9–40 cm, stamens 12–40cm long
> *Fruits*: nuts globose, ovoid, or ellipsoid, 3–7.5 × 3–5 cm; husk densely glandular pubescent, indehiscent; shell thick, rough, with 6–8 prominent ridges and deep pits and depressions
> In bloom: April to May
> Other anticancer species belonging to the genus *Juglans*:
> *Juglans regia* Linn (Figure 4.11)

In the ancient civilizations of the Greeks and Romans, the fungicidal and bactericidal properties of the walnut were employed, and the residents in the American South used the husk of the young nuts for the easy gathering of fish (Soderquist 1973). In traditional Chinese medicine, the use of husk of the young walnut for the treatment of stomach, liver, and lung carcinomas can be traced back many centuries. The *in vitro* experiments indicated that the extracts from the husk had strong proliferative activity towards the human leukemia HL-60, the gastric carcinoma BGC-823, and the cervical carcinoma Hela cells (Wan et al. 2012). In addition, the extracts could also significantly inhibit the growth of both S180 xenografts and H22 ascites tumors (Liu and Ji 2004a). The data from the *in vivo* toxicity evaluation implied that the LD_{50} values for the chloroform fractions and ethyl acetate

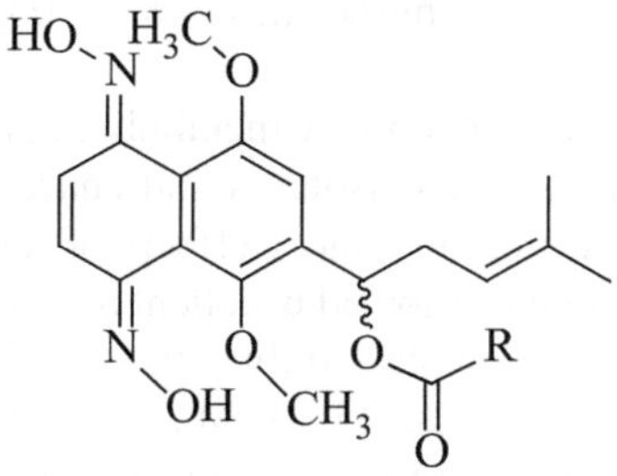

Figure 4.10 Chemical structures of shikonin and alkanninoximes.

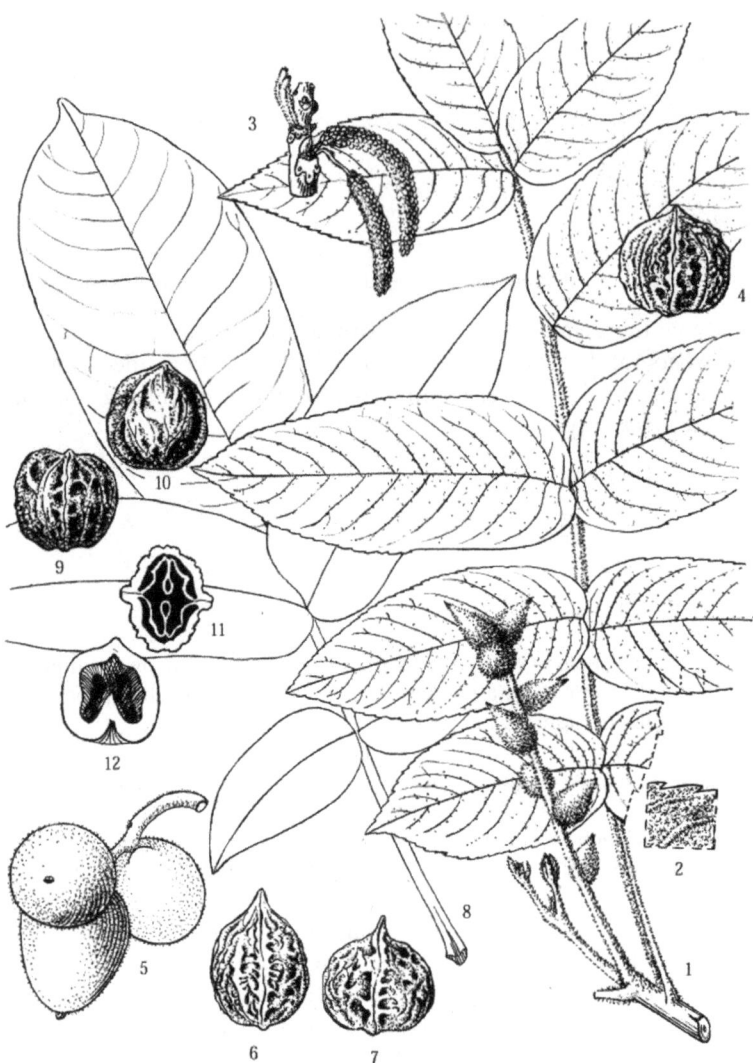

Figure 4.11 The illustration of *Juglans mandshurica* Maxim. and related anticancer species. 1–7 *Juglans mandshurica* Maxim. (1. branchlet with fruiting spike, 2. leaflet ab axial view, 3. male spikes, 4. nuts, 5. fruiting spikes, 6, 7. nuts) 8–12 *Juglans regia* Linn. (8. leaf, 9, 10. nuts, 11. nut cross section, 12. nut longitudinal section) (*Picture from www.efloras.org*).

fractions of the husk extracts were 575.38 mg/kg (*i.p.*) and 1303.59 mg/kg (*i.p.*), respectively (Liu et al. 2004b).

Juglone (Figure 4.2, 6), as the major secondary metabolite in the extracts of bark, root, and husk of *Juglans mandshurica* MAXIM., was first isolated and characterized in the nineteenth century and had been credited with the striking anticancer effects of the extracts. It was recognized as a promising anticancer drug candidate that showed excellent antitumor effects against a large variety of cancer cell lines *in vitro*. Dozens of naphthoquinones as juglone derivatives had been isolated and characterized from *Juglans mandshurica* MAXIM., and a few of them were identified as more active antiproliferative agents as compared to lead (Binder et al. 1989, Chen et al. 2015, Hirakawa et al. 1986, Joe et al. 1996, Zhou et al. 2015). Juglone methyl ether (Figure 4.12, 10), as the derivative of juglone, exhibited much more potent antitumor effects against hepatocarcinoma HepG2 and

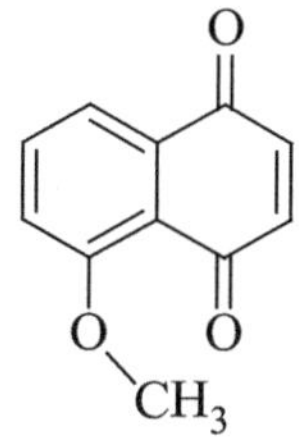

Figure 4.12 The chemical structure of juglone methyl ether.

leukemia LH-60 cell lines than the parent compound (Cui et al. 2015, Montenegro et al. 2010). Towards the growth of breast cancer MDA-MB-435 and the glioblastoma SF-295 cells, compound 10 also exhibited more potent antiproliferative activity than juglone.

The central feature of juglone and its derivatives as quinonoid-based cytotoxins was their ability to generate ROS via the quinine redox cycle, which could then accelerate the intracellular hypoxic conditions (O'Brien 1991, Rossi et al. 1986). Mechanistic investigations indicated that one of the potential pathways through which juglone exerted its anticancer effects was the formation of the juglone semiquinone radical and subsequent production of superoxide anion radical in the mitochondria as well as in the cytoplasm. The semiquinone radical and superoxide anion radical as ROS played important roles in apoptosis induction under both physiologic and pathologic conditions (Simon et al. 2000). The study from Xu et al. (2012) indicated that juglone induced apoptosis of human leukemia HL-60 cells through an ROS-dependent mechanism. The generation of ROS was about two- to eight-fold as compared to the control cells after 24-h treatment with juglone concentrations of 2, 4, and 8 μM, respectively. The glutathione (GSH) depletion was consistent with ROS generation after treatment, and the antioxidant pretreatment would reverse the ROS-induced apoptosis.

Despite excellent anticancer potential, juglone as a naphthoquinone was also reported to exert some toxic effects to normal tissues including acute irritant contact dermatitis (Aithal et al. 2011, Neri et al. 2006). The molecular basis for the initiation of juglone cytotoxicity towards resting or non-dividing cells has also been attributed to the generation of ROS. In future studies, the prodrug approach, which might minimize the pervasive toxicity of juglone, will be a promising strategy in the development of this natural naphthoquinone as a clinical anticancer drug candidate.

Plumbago auriculata Lam.

Family: Plumbaginaceae
Appearance
Evergreen perennial shrubs
Stem: erect, trailing, or climbing, diffusely branched, glabrous or pubescent on youngest shoots
Leaves: sessile, sometimes short-petiolate, blade elliptic, oblanceolate, or spatulate, 2.5–9 × 0.5–2.5 cm, base usually long-attenuate, sometimes auriculate, apex acute or obtuse, mucronate
Flower: corolla pale blue, 37–53 mm, tube 28–40 mm, lobes 10–16 × 6–15 mm
In bloom: all year-round
Other anticancer species belonging to the genus *Plumbago*:
Plumbago zeylanica Linn.
Plumbago indica Linn.
Plumbagella micrantha Spach (Figure 4.13)

Plumbagin (Figure 4.2, 7), a naturally occurring yellow pigment, was found in the root of *Plumbaginaceae* plants (van der Vijver 1972). It has been safely used for centuries in both Ayurveda therapy and traditional Chinese medicine for treating various diseases, such as the infections,

Figure 4.13 The illustration of *Plumbago auriculata* Linn. And related anticancer species. 1–2 *Plumbago zeylanica* Linn. (1. flower branch, 2. calyx) 3–4 *Plumbago indica* Linn. (3. branch, 4. flowering branch) 5–6 *Plumbago auriculata* Lam. (5. flower branch, 6. calyx) 7–8 *Plumbagella micrantha* Spach (7. flowering branch, 8. fruiting calyx) (*Picture from www.efloras.org*).

inflammatory, rheumatosis, and allergic reactions (Krishnaswamy and Purushothaman 1980, Liu et al. 2008, Sung et al. 2012, Wang and Huang 2005). In previous studies, it has been shown to exert antiproliferative activities in cells culture as well as in animal models. Towards the growth of human leukemia Raji cells, lung carcinoma Calu-1 cells, cervical carcinoma HeLa cells, and human transformed epithelial Wish cells, plumbagin exhibited the IC_{50} values of 8.1 ± 3.9, 25.0 ± 8.8, 21.5 ± 2.6, and 21.2 ± 5.0 μM, respectively (Lin et al. 2003). The growth inhibitory effects of this plant-derived naphthoquinone were also observed for human cervical cancer ME-180 cells with the IC_{50} value of 3 μM (Srinivas et al. 2004). It inhibited the invasion of prostate cancer DU145, PC-3, and CWR22rv1 cells (Aziz et al. 2008) and liver cancer HepG2 cells (Shih et al. 2009). The plum bag in treatment at doses of 15 and 20 μM depicted 55 and 70% inhibition of pancreatic PANC1 cells invasion (Hafeez et al. 2014).

In animal experiments, the *i.p.* administration of **7** (2 mg/kg, five injections per week for 11 weeks), beginning three days after the subcutaneous implantation of hormone-resistant prostate cancer DU145 cells (2.5×10^6 cells), delayed tumor growth by three weeks and reduced both tumor weight and volume by 90% (Aziz et al. 2008). The results of *in vivo* efficacy evaluations also indicated that plumbagin (1 mg/kg, *i.p.* for 21 times in 21 days) resulted in a 57% reduction in human ovarian OVCAR-5 xenograft volume (n = 10; 330.0 ± 227.0 mm^3) as compared to the solvent alone group (n = 10; 776.8 ± 155.7 mm^3) (Sinha et al. 2013). Meanwhile, it was also effective in the pancreatic PANC-1 ectopic xenograft model since treatment of the natural 1,4-naphthoquinone (2mg/kg, *i.p.* injection for five days a week) showed significant ($P < 0.01$) inhibition on both tumor growth (50%) and tumor weight (50%), when compared with the control groups (Hafeez et al. 2014). In addition to its antiproliferative activity, plumbagin also exhibited chemopreventive effects against the azoxymethane-induced intestinal carcinogenesis in animals, suggesting its chemopreventive activity (Sugie et al. 1998). It could inhibit both the ultraviolet radiation-induced cutaneous damage and the development of squamous cell carcinomas and was recognized as promising chemopreventive agents for human cancers (Sand et al. 2010).

The mechanistic studies indicated that the cytotoxic effects of plumbagin were probably working through the generation of ROS and subsequent induction of apoptosis (Srinivas et al. 2004). Treatment of cells with this natural 1,4-naphthoquinone caused the loss in mitochondrial membrane potential and further changes in the characteristic of apoptosis, such as phosphatidyl serine translocation, nuclear condensation, and DNA fragmentation. Moreover, plumbagin-induced apoptosis also involved release of mitochondrial cytochrome C and apoptosis-inducing factor (AIF), leading to the activation of caspase-dependent and independent pathways (Srinivas et al. 2004). N-acetyl-l-cysteine (NAC) as a free radical scavenger could reverse the cytotoxic effects of plumbagin (Kawiak et al. 2007). ROS-mediated inhibition of Topo II as an important mechanism also contributed to its anticancer activity (Kawiak et al. 2007).

Tecoma stans (L.) Juss. ex Kunth

Family: *Bignoniaceae*
Appearance
Evergreen perennial shrubs
Stem: up to 2.5 m tall
Leaves: opposite and pinnate, with 3–7 leaflets; leaflets elliptic to elliptic-ovate
Flower: bright yellow and fragrant; calyx tubular-campanulate, about 4.5 mm long; lobes triangular; Corolla tube 32–33 mm long
In bloom: most of the year (Figure 4.14)

α-Lapachol (Figure 4.2, 8) was first isolated from *Tecoma stans* (L.) Juss. ex Kunth (Synonymic name: *Tabebuia avellanedae* Lorentz ex Griseb., *Tabebuia impetiginosa* (Mart. ex DC.) Standl.) in the late nineteenth century and had a well-documented history of anticancer effects (Hussain et al. 2007), including growth inhibition of a variety of cancer cell lines. The administration of α-lapachol produced striking antitumor effects and successfully inhibited the growth of Walker 256 xenografts on mice (Rao et al. 1968). Although it has already been licensed by the Brazil government for general clinical practices (da Silva Júnior et al. 2009b), there was still a debate over the usefulness of α-lapachol as an antitumor agent ascribed to the high doses required for therapeutic efficacy together with the observed side effects (Sunassee et al. 2013b).

α-Lapachone (Figure 4.15, 11) as the analog of naphthoquinone 8 was possibly produced through the intramolecular cyclization mechanism. As shown in Figure 4.15, the protonation of the double bond on the side chain of α-lapachol, the formation of a tertiary carbocation, and further capture of the carbocation by the phenolic hydroxyl group afford the α-lapachone (11). After the Keto-Enol isomerization, the cation ion would be intramolecularly trapped by the oxygen located on C-4 and the β-lapachone (12) as an ortho-quinone was produced (Ravelo et al. 2003).

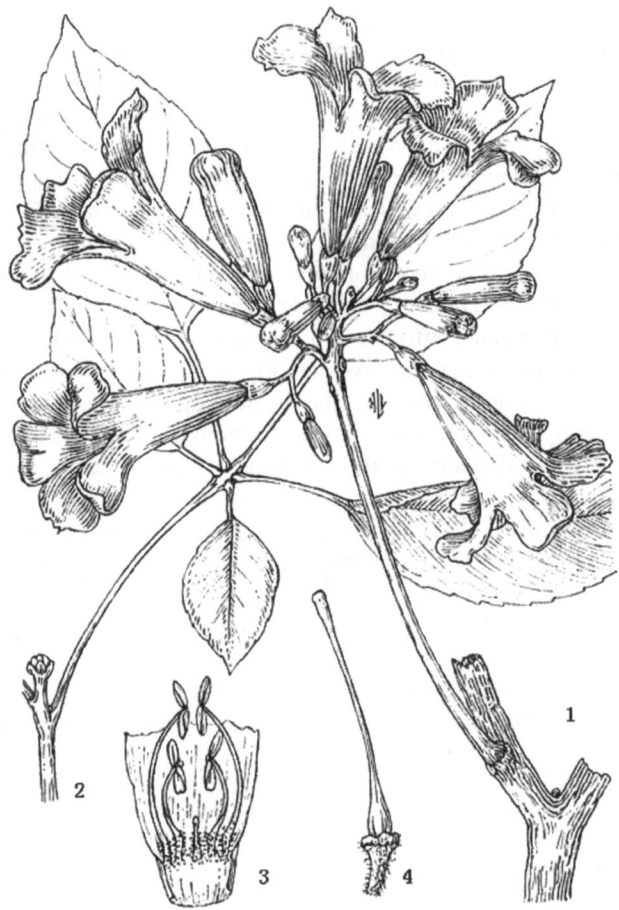

Figure 4.14 The illustration of *Tecoma stans* (L.) Juss. ex Kunth (1. flower branch, 2. leaf, 3. opened corolla showing stamens, 4. gynoecium) (*Picture from www.fcagr.unr.edu.ar*).

α-Lapachone (11) only exhibited weak cytotoxic activity with its IC_{50} values in the range of 3–13 μg/mL against a panel of nine human cancer cell lines and one murine cell line (Peraza-Sánchez et al. 2000). Compared with the lead compound, it showed decreased inhibitory potency against the human leukemia Daudi, K562 and its MDR variant, Lucena-1 cell lines with the IC_{50} values ranging from 40 to 70 μM (Jardim et al. 2015).

The closely related analog β-lapachone (12, also called ARQ 501) was reported to present potent antiproliferative activity against human cancer cell lines originating from leukemia (Planchon et al. 1995), prostate (Choi et al. 2003, Li et al. 1995, Planchon et al. 1995), malignant glioma (Weller et al. 1997), hepatoma (Lai et al. 1998), colon (Woo et al. 2006), breast (Wuerzberger et al. 1998), ovarian (Li et al. 1999a), and pancreatic tumors (Li et al. 1999b, Li et al. 2000),with the IC_{50} values within the 1–10 μM range. In combination with Taxol, β-lapachone was an effective agent against human ovarian and prostate xenografts in mice (Li et al. 1999a). As a promising drug candidate, it has already entered phase II clinical trials in the USA, and the results indicated that this drug candidate had the potential for improved activity and reduced toxicity over other molecular approaches and traditional cancer chemotherapy (NCT00075933).

The exact mechanism of cell death triggered by β-lapachone (12) remains unknown. Apoptosis (Lee et al. 2011), necrosis (Sun et al. 2006), or ROS-mediated autophagy (Park et al. 2011) have

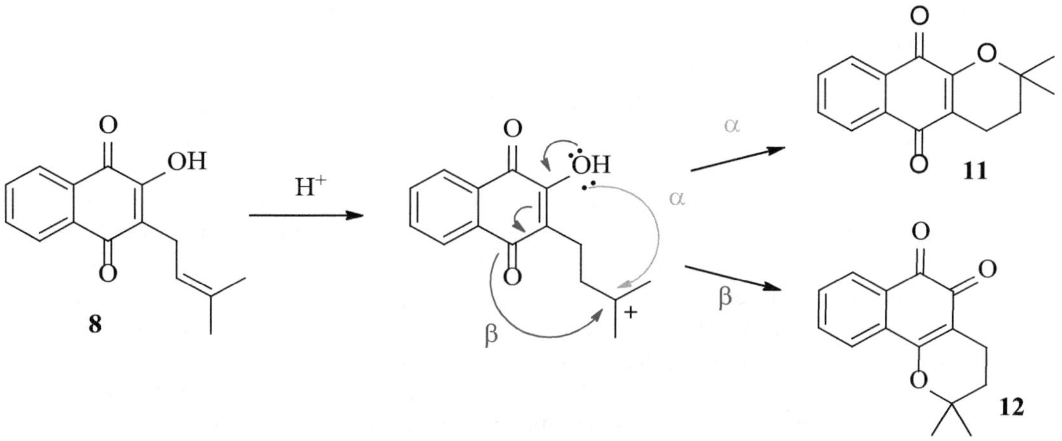

Figure 4.15 The production of α-lapachone (11) and β-lapachone (12) from α-lapachol.

been observed in cells incubated with this quinoid anticancer drug. According to the literature (Bey et al. 2007, Li et al. 2011, Pink et al. 2000a), NAD(P)H quinine oxidoreductase (NQO1) appeared as the most reliable biological molecular target for the natural naphthoquinone 12. It underwent a NQO1-dependent 'futile cycle' (Figure 4.16), since about 60 equivalents of NAD(P)H were consumed in the presence of only one equivalent amount of 12 in the system within five minutes (Pink et al. 2000a, Pink et al. 2000b). As a result, the dramatic elevation of the ROS level together with the released calcium ion from endoplasmic reticulum led to the initiation of the cellular apoptosis, the alteration of cellular metabolism and DNA repair, and further cell death.

The mechanisms of intracellular topoisomerase II inhibition by α- and β-lapachone were also investigated. Krishnan et al. systematically investigated the mechanism of topoisomerase II inhibition by these two naphthoquinones with respect to the steps of the catalytic cycle of the enzyme (Krishnan, Bastow, Krishnan, and Bastow 2000, 2001). β-lapachone inhibited topoisomerase II by inducing religation and dissociation of the enzyme from DNA in the presence of ATP. α-lapachone

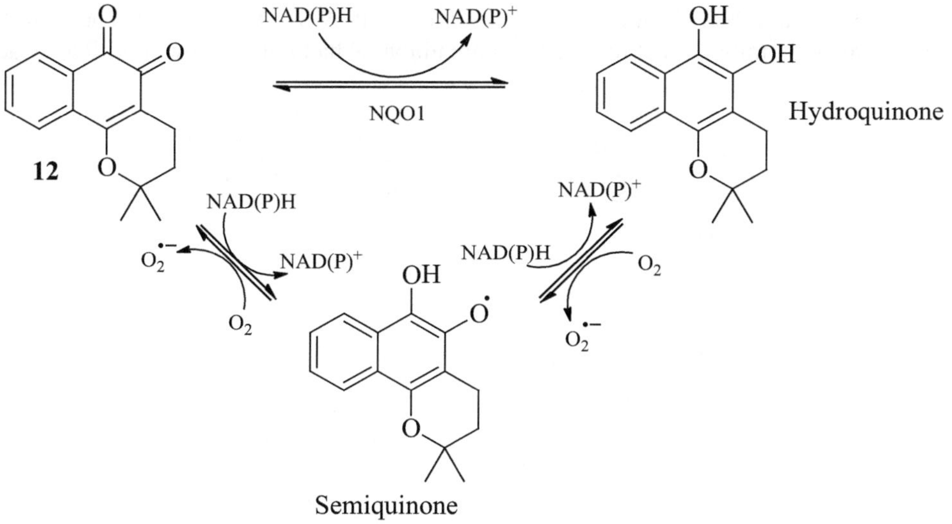

Figure 4.16 NQO1-dependent β-lapachone redox cycle (Pink et al. 2000a).

acted as an irreversible inhibitor of topoisomerase II, which inhibited initial non-covalent binding of topoisomerase II to DNA and induced religation of DNA breaks before dissociating the enzyme from DNA. It was suggested that the oxidation–reduction cycle of these two naphthoquinones might be involved in their topoisomerase II inhibitory activity (Krishnan and Bastow 2001).

In the past two decades, a great deal of studies on the structural optimization of α- and β-lapachone have also been conducted to increase their therapeutic efficacy and reduced the side effects (da Silva Júnior et al. 2010, de Castro et al. 2013, Di Chenna et al. 2001, Sunassee et al. 2013a). Several analogs exhibited increased therapeutic index. Two analogues (Figure 4.17, 13, and 14) that incorporated a benzyl alcohol moiety exhibited enhanced growth inhibition against the esophageal cancer WHCO1 cells (IC_{50} values of 3.0 and 7.3 μM, respectively) as compared to cisplatin (IC_{50} = 16.5 μM) and were non-toxic to the normal fibroblast NIH3T3 cell line (Sunassee et al. 2013a). Two 3-arylamino-nor-β-lapachone derivatives (15 and 16) were more selective than doxorubicin and as active as the precursor β-lapachones (da Silva Júnior et al. 2010). All of these derivatives emerged as promising new lead compounds in anticancer drug development.

Salvia prionitis Hance

> Family: *Lamiaceae*
> Appearance
> *Herbs annual*
> *Stem*: erect, 20–43 cm, densely white hirsute, unbranched or few branched
> *Leaves*: basal, simple or ternate; leaf blades of simple leaves oblong to ovate-lanceolate, 2.5–7.5 × 1.3–4.5 cm, adaxially hirsute, abaxially glabrous, hirsute on veins, base rounded to cordate, margin coarsely crenate
> *Flower*: verticillasters 6–14-flowered, Calyx campanulate, purplish, ca. 4 mm, glandular pilose, throat hirsute annulate, upper lip triangular
> *Fruits*: brown ishnutlets, ellipsoid, ca. 1.3 × 0.7 mm
> In bloom: June to August
> Other anticancer species belonging to the genus *Salvia*:
> *Salvia honania* L. Bailey (Figure 4.18)

More than 40 diterpenoid quinones were isolated from the *Salvia prionitis* Hance, and saprorthoquinone (Figure 4.2, 9) was one of the most important and well-studied compounds within these quinines (Chang et al. 2005, Chen et al. 2002a, Lin et al. 1989, Xu et al. 2006). The results from the *in vitro* evaluation implied that the 1,2-naphthoquinone 9 had potent growth inhibitory activity

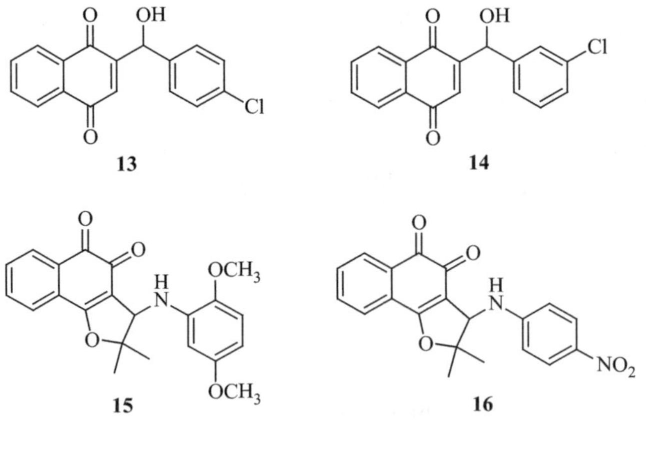

Figure 4.17 The structures of α- and β-lapachone analogues as promising anticancer agents.

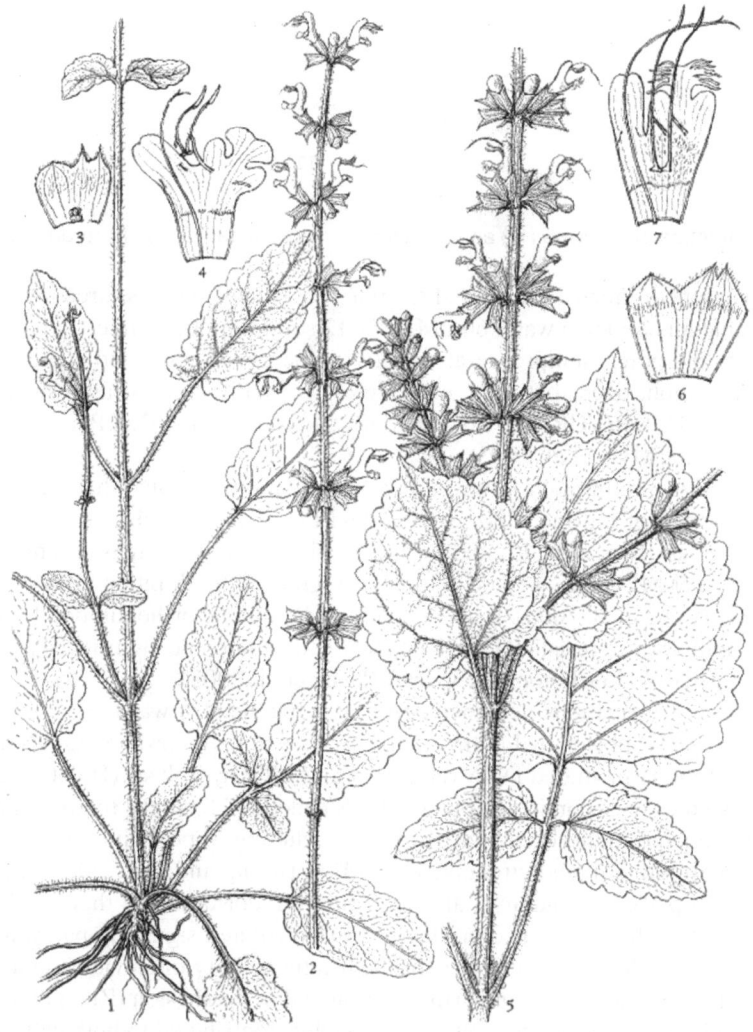

Figure 4.18 The illustration of *Salvia prionitis* Hance and related anticancer species. 1–4 *Salvia prionitis* Hance (1. lower part of the plant, 2. upper portion of the plant showing inflorescence, 3. opened calyx and ovary, 4. opened corolla showing stamens, style, basal hairy annulus, and united lower arms) 5–7 *Salvia honania* L. Bailey (5. upper portion of the plant showing inflorescence, 6. opened calyx showing villous annulus, 7. opened corolla showing stamens, style, basal hairy annulus, united lower arms, and middle lobe of lower lip) (*Picture from www.efloras.org*).

against a wide spectrum of human tumor cells. The initial biological investigations in the early 1990s indicated that saprorthoquinone 9 exhibited cytotoxicity against leukemia P388 cancer cells *in vitro*. Towards the growth inhibition of leukemia P388 and HL-60 cells, lung cancer SPC-A4 cells, and stomach cancer SGC-7901 cells, the orthoquinone showed IC_{50} values of 1.95, 2.36, 2.75, and 1.37 µM, respectively (Jin-Sheng et al. 1999).

In further studies, the systematic structural modification of 9 afforded salvicine (Figure 4.19, 17) as the most pharmacologically active derivative (Deng et al., 2011), which was prepared by the epoxidation of the olefin group on the side chain of the lead 9 and subsequent ring-opening reaction in acidic conditions. Salvicine (17) showed potent growth inhibitory activity against a wide spectrum of human tumor cells *in vitro* (Qing et al. 1999). Compared with etoposide as the positive

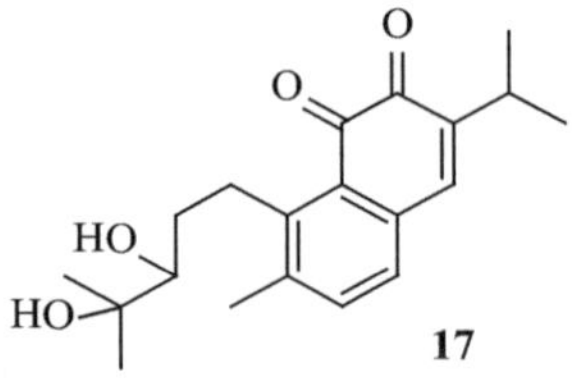

Figure 4.19 The structure of salvicine (17) as a structurally modified derivative of saprorthoquinone.

control, the 1,2-naphthoquinone 17 exhibited similar IC_{50} values towards three leukemia cell lines (the mean IC_{50} value for salvicine was 7.66 μM). For 12 solid tumor cell lines, salvicine (17) exhibited 5.41- and 4.15-fold potent antiproliferative activity than vincristine and etoposide (Qing et al. 1999). It should be emphasized that salvicine 17 effectively kills multidrug-resistant (MDR) cell lines, such as K562/A02, KB/VCR, and MCF-7/ADR, and parental K562, KB, and MCF-7 cell lines to an equivalent degree (Miao et al. 2003).

The results from the preliminary *in vivo* efficacy evaluation indicated that the orthoquinone 17 could inhibit the growth of murine S180 sarcoma and Lewis lung carcinoma in mice (Qing et al. 1999). Salvicine (17) at doses of 6, 12, and 24 mg/kg significantly decreased lung metastatic foci of human breast cancer MDA-MB-435 orthotopic xenograft with inhibition rates of 33.0, 54.0, and 81.6%, respectively (Lang et al. 2005). It was also reported that the orthoquinone 17 had a profound cytotoxic effect on multidrug-resistant (MDR) cancer cells and could significantly reduce the lung metastatic of MDA-MB-435 orthotopic xenograft (Lang et al. 2005).

The mechanistic investigations demonstrated that salvicine (17) was a novel non-intercalative topoisomerase II (Topo II) poison by binding to the ATPase domain, promoting DNA-Topo II binding and inhibiting Topo II-mediated DNA relegation and ATP hydrolysis (Hu et al. 2006, Meng et al. 2001). Further studies have confirmed that the ROS generated from the redox of salvicine played a central role in salvicine-induced cellular response including Topo II inhibition, DNA damage, tumor cell adhesion inhibition, and circumventing MDR (Meng and Ding 2007). Salvicine inactivated $β_1$ integrin that played a fundamental role in tumor metastasis and then inhibited the adhesion of MDA-MB-435 cells to fibronectin via the ROS-mediated signaling pathway. This finding contributed to a better understanding of the antimetastatic activity of salvicine and threw light on the roles of ROS in the regulation of integrin function and cell adhesion (Zhou et al. 2008). In the MDR K562/A02 cells, the generation of ROS by salvicine contributed to both cell killing and Pgp downregulation, thus eliminating the anticancer drug resistance (Cai et al. 2007).

REFERENCES

Aithal, B.K., M.R. Sunil Kumar, B.N. Rao, et al. 2011. Evaluation of pharmacokinetic, biodistribution, pharmacodynamic, and toxicity profile of free juglone and its sterically stabilized liposomes. *Journal of Pharmaceutical Sciences* 100, no. 8:3517–3528.

Andújar, I., J.L. Ríos, R.M. Giner, and M.C. Recio. 2013. Pharmacological properties of shikonin – A review of literature since 2002. *Planta Medica* 79, no. 18:1685–1697.

Aziz, M.H., N.E. Dreckschmidt, and A.K. Verma. 2008. Plumbagin, a medicinal plant-derived naphthoquinone, is a novel inhibitor of the growth and invasion of hormone-refractory prostate cancer. *Cancer Research* 68, no. 21:9024–9032.

Bey, E.A., M.S. Bentle, K.E. Reinicke, et al. 2007. An NQO1-and PARP-1-mediated cell death pathway induced in non-small-cell lung cancer cells by β-lapachone. *Proceedings of the National Academy of Sciences of the United States of America* 104, no. 28:11832–11837.

Bhakuni, D.S., M.L. Dhar, M.M. Dhar, B.N. Dhawan, and B.N. Mehrotra. 1969. Screening of Indian plants for biological activity: Part II. *Indian Journal of Experimental Biology* 7, no. 4:250–262.

Binder, R.G., M.E. Benson, and R.A. Flath. 1989. Eight 1, 4-naphthoquinones from Juglans. *Phytochemistry* 28, no. 10:2799–2801.

Cai, Y., J. Lu, Z. Miao, L. Lin, and J. Ding. 2007. Reactive oxygen species contribute to cell killing and P-glycoprotein downregulation by salvicine in multidrug resistant K562/A02 cells. *Cancer Biology and Therapy* 6, no. 11:1794–1799.

Chang, J., J. Xu, M. Li, M. Zhao, J. Ding, and J.S. Zhang. 2005. Novel cytotoxic seco-abietane rearranged diterpenoids from Salvia prionitis. *Planta Medica* 71, no. 9:861–866.

Chen, G., X.M. Pi, and C.Y. Yu. 2015. A new naphthalenone isolated from the green walnut husks of Juglans mandshurica Maxim. *Natural Product Research* 29, no. 2:174–179.

Chen, X., J. Ding, Y.-M. Ye, and J.-S. Zhang. 2002a. Bioactive Abietane and seco-Abietane diterpenoids from Salvia prionitis. *Journal of Natural Products* 65, no. 7:1016–1020.

Chen, X., L. Yang, J.J. Oppenheim, and M.Z. Howard. 2002b. Cellular pharmacology studies of shikonin derivatives. *Phytotherapy Research* 16, no. 3:199–209.

Choi, Y.-H., H.-S. Kang, and M. Yoo. 2003. Suppression of human prostate cancer cell growth by β-lapachone via down-regulation of pRB phosphorylation and induction of Cdk inhibitor p21 WAF1/CIP1. *BMB Reports* 36, no. 2:223–229.

Cui, J.-H., Q. Cui, Q.-J. Zhang, and S.-S. Li. 2015. An efficient multigram synthesis of juglone methyl ether. *Journal of Chemical Research* 39, no. 9:553–554.

da Silva Júnior, E.N., C.F. de Deus, B.C. Cavalcanti, et al. 2010. 3-Arylamino and 3-alkoxy-nor-β-lapachone derivatives: Synthesis and cytotoxicity against cancer cell lines. *Journal of Medicinal Chemistry* 53, no. 1:504–508.

da Silva Júnior, E.N., M.C.F.R. Pinto, K.C.G. de Moura, et al. 2009b. Hooker's 'lapachol peroxide' revisited. *Tetrahedron Letters* 50, no. 14:1575–1577.

de Castro, S.L., F.S. Emery, and E.N. da Silva Júnior. 2013. Synthesis of quinoidal molecules: Strategies towards bioactive compounds with an emphasis on lapachones. *European Journal of Medicinal Chemistry* 69:678–700.

Deng, F., J.J. Lu, H.Y. Liu, L.P. Lin, J. Ding, and J.S. Zhang. 2011. Synthesis and antitumor activity of novel salvicine analogues. *Chinese Chemical Letters* 22, no. 1:25–28.

Di Chenna, P.H., V. Benedetti-Doctorovich, R.F. Baggio, M.T. Garland, and G. Burton. 2001. Preparation and cytotoxicity toward cancer cells of mono (arylimino) derivatives of β-lapachone. *Journal of Medicinal Chemistry* 44, no. 15:2486–2489.

Hafeez, B.B., N.E. Dreckschmidt, and A.K. Verma. 2014. Plumbagin, a medicinal plant-derived napthoquinone, inhibits the growth of pancreatic cancer cells in *in vitro* and *in vivo* xenograft mouse model systems via targeting NFkB and STAT3 signaling network. *Cancer Research* 70(8 Supplement):LB-424.

Hirakawa, K., E. Ogiue, J. Motoyoshiya, and M. Yajima. 1986. Naphthoquinones from Juglandaceae. *Phytochemistry* 25, no. 6:1494–1495.

Hu, C.X., Z.L. Zuo, B. Xiong, et al. 2006. Salvicine functions as novel topoisomerase II poison by binding to ATP pocket.*Molecular Pharmacology* 70, no. 5:1593–1601.

Hussain, H., K. Krohn, V.U. Ahmad, G.A. Miana, and I.R. Green. 2007. Lapachol: An overview. *Arkivoc* 2:145–171.

Jardim, G.A., T.T. Guimarães, and F.M. do Carmo. 2015. Naphthoquinone-based chalcone hybrids and derivatives: Synthesis and potent activity against cancer cell lines. *Medicinal Chemical Communications* 6, no. 1:120–130.

Jin-Sheng, Z., D., Jian, T., Qin-Mei, et al. 1999. Synthesis and antitumour activity of novel diterpenequinone salvicine and the analogs. *Bioorganic Medicinal Chemistry Letters* 9, no. 18: 2731–2736.

Joe, Y.-K., J.-K. Son, S.-H. Park, I.-J. Lee, and D.-C. Moon. 1996. New naphthalenyl glucosides from the roots of Juglans mandshurica. *Journal of Natural Products* 59, no. 2:159ss–160.

Kawiak, A., J. Piosik, G. Stasilojc, et al. 2007. Induction of apoptosis by plumbagin through reactive oxygen species-mediated inhibition of topoisomerase II. *Toxicology and Applied Pharmacology* 223, no. 3:267–276.

Kim, S.H., I.C. Kang, T.J. Yoon, et al. 2001. Antitumor activities of a newly synthesized shikonin derivative, 2-hyim-DMNQ-S-33. *Cancer Letters* 172, no. 2:171–175.

Krishnan, P., and K.F. Bastow. 2000. Novel mechanisms of DNA topoisomerase II inhibition by pyranonaphthoquinone derivatives – Eleutherin, α lapachone, and β lapachone. *Biochemical Pharmacology* 60, no. 9:1367–1379.

Krishnan, P., and K.F. Bastow. 2001. Novel mechanism of cellular DNA topoisomerase II inhibition by the pyranonaphthoquinone derivatives α-lapachone and β-lapachone. *Cancer Chemotherapy and Pharmacology* 47, no. 3:187–198.

Krishnaswamy, M., and K.K. Purushothaman. 1980. Plumbagin: A study of its anticancer, antibacterial and antifungal properties. *Indian Journal of Experimental Biology* 18, no. 8:876–877.

Lai, C.C., T.J. Liu, L.K. Ho, M.J. Don, and Y.P. Chau. 1998. beta-Lapachone induced cell death in human hepatoma (HepA2) cells. *Histology and Histopathology* 13, no. 1:89–97.

Lang, J.Y., H. Chen, J. Zhou, et al. 2005. Antimetastatic effect of salvicine on human breast cancer MDA-MB-435 orthotopic xenograft is closely related to Rho-dependent pathway. *Clinical Cancer Research* 11, no. 9:3455–3464.

Lee, H., M.T. Park, B.H. Choi, et al. 2011. Endoplasmic reticulum stress-induced JNK activation is a critical event leading to mitochondria-mediated cell death caused by β-lapachone treatment. *PLOS ONE* 6, no. 6:e21533.

Li, C.J., Y.Z. Li, A.V. Pinto, and A.B. Pardee. 1999a. Potent inhibition of tumor survival *in vivo* by β-lapachone plus taxol: Combining drugs imposes different artificial checkpoints. *Proceedings of the National Academy of Sciences of the United States of America* 96, no. 23:13369–13374.

Li, C.J., C. Wang, and A.B. Pardee. 1995. Induction of apoptosis by β-lapachone in human prostate cancer cells. *Cancer Research* 55, no. 17:3712–3715.

Li, L.S., E.A. Bey, Y. Dong, et al. 2011. Modulating endogenous NQO1 levels identifies key regulatory mechanisms of action of β-lapachone for pancreatic cancer therapy. *Clinical Cancer Research* 17, no. 2:275–285.

Li, Y., C.J. Li, D. Yu, and A.B. Pardee. 2000. Potent induction of apoptosis by beta-lapachone in human multiple myeloma cell lines and patient cells. *Molecular Medicine* 6, no. 12:1008–1015.

Li, Y.Z., C.J. Li, A.V.Pinto, and A.B. Pardee. 1999b. Release of mitochondrial cytochrome C in both apoptosis and necrosis induced by beta-lapachone in human carcinoma cells. *Molecular Medicine* 5, no. 4:232–239.

Lin, L.C., L.L. Yang, and C.J. Chou. 2003. Cytotoxic naphthoquinones and plumbagic acid glucosides from Plumbago zeylanica. *Phytochemistry* 62, no. 4:619–622.

Lin, L.-Z., G. Blaskó, and G.A. Cordell. 1989. Diterpenes of Salvia prionitis. *Phytochemistry* 28, no. 1:177–181.

Liu, L., H. Liu, and B. Yang. 2008. Advances in studies on Plumbago zeylanica. *Progress in Modern Biomedicine* 8, no. 3:597–600.

Liu, W., and Y.-B. Ji. 2004a. Study on acute toxicity experiment of mice and anti-tumor function *in vivo* of Chinese traditional medicine Qinglongyi. *Journal of Harbin University of Commerce (Sciences Edition)* 1: 887–890.

Liu, W., W.H. Lin, and Y.B. Ji. 2004b. Study on the acute toxicity experiment of mice and anti-tumor function *in vitro* of the qinglongyi. *China Journal of Chinese Materia Medica* 29, no. 9:887–890.

Meng, L.H., and J. Ding. 2007. Salvicine, a novel topoisomerase II inhibitor, exerts its potent anticancer activity by ROS generation. *Acta Pharmacologica Sinica* 28, no. 9:1460–1465.

Meng, L.H., J.S. Zhang, and J. Ding. 2001. Salvicine, a novel DNA topoisomerase II inhibitor, exerting its effects by trapping enzyme-DNA cleavage complexes. *Biochemical Pharmacology* 62, no. 6:733–741.

Miao, Z.H., T. Tang, Y.X. Zhang, J.S. Zhang, and J. Ding. 2003. Cytotoxicity, apoptosis induction and down-regulation of MDR-1 expression by the anti-topoisomerase II agent, salvicine, in multidrug-resistant tumor cells. *International Journal of Cancer* 106, no. 1:108–115.

Montenegro, R.C., A.J.Araújo, M.T. Molina, et al. 2010. Cytotoxic activity of naphthoquinones with special emphasis on juglone and its 5-O-methyl derivative. *Chemico-Biological Interactions* 184, no. 3:439–448.

Moore, H.W. 1977. Bioactivation as a model for drug design bioreductive alkylation. *Science* 197, no. 4303:527–532.

Neri, I., F. Bianchi, F. Giacomini, and A. Patrizi. 2006. Acute irritant contact dermatitis due to Juglans regia. *Contact Dermatitis* 55, no. 1:62–63.

O'brien, P.J. 1991. Molecular mechanisms of quinone cytotoxicity. *Chemico-Biological Interactions* 80, no. 1:1–41.

Papageorgiou, V.P., A.N. Assimopoulou, E.A. Couladouros, D. Hepworth, and K.C. Nicolaou. 1999. The chemistry and biology of alkannin, shikonin, and related naphthazarin natural products. *Angewandte Chemie* 38, no. 3:270–301.

Papageorgiou, V.P., A.N. Assimopoulou, V. Samanidou, and I. Papadoyannis. 2006. Recent advances in chemistry, biology and biotechnology of alkannins and shikonins. *Current Organic Chemistry* 10, no. 16:2123–2142.

Park, E.J., K.S. Choi, and T.K. Kwon. 2011. β-Lapachone-induced reactive oxygen species (ROS) generation mediates autophagic cell death in glioma U87 MG cells. *Chemico-Biological Interactions* 189, no. 1–2:37–44.

Peraza-Sánchez, S.R., D. Chávez, H.B. Chai, et al. 2000. Cytotoxic constituents of the roots of ekmanianthe l ongiflora. *Journal of Natural Products* 63, no. 4:492–495.

Pink, J.J., S.M. Planchon, C. Tagliarino, M.E. Varnes, D. Siegel, and D.A. Boothman. 2000a. NAD (P) H: Quinone oxidoreductase activity is the principal determinant of β-lapachone cytotoxicity. *Journal of Biological Chemistry* 275, no. 8:5416–5424.

Pink, J.J., S. Wuerzberger-Davis, C. Tagliarino, et al. 2000b. Activation of a cysteine protease in MCF-7 and T47D breast cancer cells during β-lapachone-mediated apoptosis. *Experimental Cell Research* 255, no. 2:144–155.

Planchon, S.M., S. Wuerzberger, B. Frydman, et al. 1995. β-Lapachone-mediated apoptosis in human promyelocytic leukemia (HL-60) and human prostate cancer cells: A p53-independent response. *Cancer Research* 55, no. 17:3706–3711.

Qing, C., J.S. Zhang, and J. Ding. 1999. In vitro cytotoxicity of salvicine, a novel diterpenoid quinone. *Zhongguo Yao Li Xue Bao = Acta Pharmacologica Sinica* 20, no. 4:297–302.

Rao, K., T. McBride, and J. Oleson. 1968. Recognition and evaluation of lapachol as an antitumor agent. *Cancer Research* 28, no. 10:1952–1954.

Ravelo, Á.g., A. Estévez-braun, and E. Pérez-sacau. 2003. The chemistry and biology of lapachol and related natural products α and β-lapachones. *Studies in Natural Products Chemistry*. R. Atta ur (Ed.). Elsevier 29:719–760.

Rossi, L., G.A. Moore, S. Orrenius, and P.J. O'Brien. 1986. Quinone toxicity in hepatocytes without oxidative stress. *Archives of Biochemistry and Biophysics* 251, no. 1:25–35.

Sand, J.M., E.M. Siebers, N.E. Dreckschmidt, and A.K. Verma. 2010. Plumbagin, a medicinal plant-derived naphthoquinone, inhibits ultraviolet radiation-induced cutaneous damage and development of squamous cell carcinomas. *Cancer Research* 70, no. 8:1881.

Sankawa, U., Y. Ebizuka, T. Miyazaki, Y. Isomura, and H. Otsuka. 1977. Antitumor activity of shikonin and its derivatives. *Chemical and Pharmaceutical Bulletin* 25, no. 9:2392–2395.

Sankawa, U., H. Otsuka, Y. Kataoka, Y. Iitaka, A. Hoshi, and K. Kuretani. 1981. Antitumor activity of shikonin, alkannin and their derivatives. II. X-ray analysis of cyclo-alkannin leucoacetate, tautomerism of alkannin and cyclo-alkannin and antitumor activity of alkannin derivatives. *Chemical and Pharmaceutical Bulletin* 29, no. 1:116–122.

Shih, Y.W., Y.C. Lee, P.F. Wu, Y.B. Lee, and T.A. Chiang. 2009. Plumbagin inhibits invasion and migration of liver cancer HepG2 cells by decreasing productions of matrix metalloproteinase-2 and urokinase-plasminogen activator. *Hepatology Research* 39, no. 10:998–1009.

Shukla, Y.N., J.S. Tandon, D.S. Bhakuni, and M.M. Dhar. 1971. Naphthaquinones of Arnebia nobilis. *Phytochemistry* 10, no. 8:1909–1915.

Simon, H.U., A. Haj-Yehia, and F. Levi-Schaffer. 2000. Role of reactive oxygen species (ROS) in apoptosis induction. *Apoptosis* 5, no. 5:415–418.

Sinha, S., K. Pal, A. Elkhanany, et al. 2013. Plumbagin inhibits tumorigenesis and angiogenesis of ovarian cancer cells *in vivo*. *International Journal of Cancer* 132, no. 5:1201–1212.

Soderquist, C.J. 1973. Juglone and allelopathy. *Journal of Chemical Education* 50, no. 11:782–783.

Srinivas, P., G. Gopinath, A. Banerji, A. Dinakar, and G. Srinivas. 2004. Plumbagin induces reactive oxygen species, which mediate apoptosis in human cervical cancer cells. *Molecular Carcinogenesis* 40, no. 4:201–211.

Sugie, S., K. Okamoto, K.M. Rahman, et al. 1998. Inhibitory effects of plumbagin and juglone on azoxymethane-induced intestinal carcinogenesis in rats. *Cancer Letters* 127, no. 1–2:177–183.

Sun, X., Y. Li, W. Li, et al. 2006. Selective induction of necrotic cell death in cancer cells by β-lapachone through activation of DNA damage response pathway. *Cell Cycle* 5, no. 17:2029–2035.

Sunassee, S.N., C.G.L. Veale, N. Shunmoogam-Gounden, et al. 2013b. Cytotoxicity of lapachol, β-lapachone and related synthetic 1,4-naphthoquinones against oesophageal cancer cells. *European Journal of Medicinal Chemistry* 62:98–110.

Sung, B., S. Prasad, S.C. Gupta, S. Patchva, and B.B. Aggarwal. 2012. Regulation of inflammation-mediated chronic diseases by botanicals. In: *Advances in Botanical Research*. L.-F. Shyur, & A. S. Y. Lau (Eds.). Academic Press , Oxford, UK62:57–132.

Tandon, V.K., and S. Kumar. 2013. Recent development on naphthoquinone derivatives and their therapeutic applications as anticancer agents. *Expert Opinion on Therapeutic Patents* 23, no. 9:1087–1108.

Thangapazham, R.L., A.K. Singh, P. Seth, et al. 2008. Shikonin analogue (SA) 93/637 induces apoptosis by activation of caspase-3 in U937 cells. *Frontiers in Bioscience: A Journal and Virtual Library* 13:561–568.

van der Vijver, L.M. 1972. Distribution of plumbagin in the mplumbaginaceae. *Phytochemistry* 11, no. 11:3247–3248.

Wan, Z.-q., W. Li, J.-w. Cui, X.-y. Xu, J.-j. Sun, and J. Liu. 2012. Inhibitory effect of juglone on proliferation of leukemia cells. *Journal of Jilin University (Medicine Edition)* 38, no. 3:486–489.

Wang, R., X. Zhang, H. Song, S. Zhou,and S. Li. 2014a. Synthesis and evaluation of novel alkannin and shikonin oxime derivatives as potent antitumor agents. *Bioorganic and Medicinal Chemistry Letters* 24, no. 17:4304–4307.

Wang, R., X. Zhang, H. Song, S. Zhou, and S. Li. 2014b. Synthesis and evaluation of novel alkannin and shikonin oxime derivatives as potent antitumor agents. *Bioorganic and Medicinal Chemistry Letters* 24, no. 17:4304–4307.

Wang, R.B., W. Zhou, Q.Q. Meng, et al. 2014c. Design, synthesis, and biological evaluation of shikonin and alkannin derivatives as potential anticancer agents via a prodrug approach. *ChemMedChem* 9, no. 12:2798–2808.

Wang, W., M. Dai, C. Zhu, et al. 2009. Synthesis and biological activity of novel shikonin analogues. *Bioorganic and Medicinal Chemistry Letters* 19, no. 3:735–737.

Wang, Y.C., and T.L. Huang. 2005. Screening of anti-Helicobacter pylori herbs deriving from Taiwanese folk medicinal plants. *FEMS Immunology and Medical Microbiology* 43, no. 2:295–300.

Weller, M., S. Winter, C. Schmidt, et al. 1997. Topoisomerase-I inhibitors for human malignant glioma: Differential modulation of p 53, p 21, bax and bcl-2 expression and of CD 95-mediated apoptosis by camptothecin and β-lapachone. *International Journal of Cancer* 73, no. 5:707–714.

Woo, H.J., K.Y. Park, C.H. Rhu, et al. 2006. β-Lapachone, a quinone Isolated from Tabebuia avellanedae, induces apoptosis in HepG2 hepatoma cell line through induction of bax and activation of caspase. *Journal of Medicinal Food* 9, no. 2:161–168.

Wu, H., J. Xie, Q. Pan, B. Wang, D. Hu, and X. Hu. 2013. Anticancer agent shikonin is an incompetent inducer of cancer drug resistance. *PLOS ONE* 8, no. 1:e52706.

Wuerzberger, S.M., J.J. Pink, S.M. Planchon, K.L. Byers, W.G. Bornmann, and D.A. Boothman. 1998. Induction of apoptosis in MCF-7: WS8 breast cancer cells by β-lapachone. *Cancer Research* 58, no. 9:1876–1885.

Xu, H.L., X.F. Yu, S.C. Qu, X.R. Qu, Y.F. Jiang, and da Y. Sui. 2012. Juglone, from Juglans mandshruica Maxim, inhibits growth and induces apoptosis in human leukemia cell HL-60 through a reactive oxygen species-dependent mechanism. *Food and Chemical Toxicology* 50, no. 3–4:590–596.

Xu, J., J. Chang, M. Zhao, and J.S. Zhang. 2006. Abietane diterpenoid dimers from the roots of Salvia prionitis. *Phytochemistry* 67, no. 8:795–799.

Xuan, Y., and X. Hu. 2009. Naturally-occurring shikonin analogues–a class of necroptotic inducers that circumvent cancer drug resistance. *Cancer Letters* 274, no. 2:233–242.

Yang, Y.Y., H.Q. He, J.H. Cui, et al. 2016. Shikonin derivative DMAKO-05 inhibits Akt signal activation and melanoma proliferation. *Chemical Biology and Drug Design* 87, no. 6:895–904.

Zhang, J.S., J. Ding, Q.M. Tang, et al. 1999. Synthesis and antitumour activity of novel diterpenequinone salvicine and the analogs. *Bioorganic and Medicinal Chemistry Letters* 9, no. 18:2731–2736.

Zhang, X., J. Cui, Q. Meng, S. Li, W. Zhou, and S. Xiao. 2018. Advance in anti-tumor mechanisms of shikonin, alkannin and their derivatives. *Mini-Reviews in Medicinal Chemistry* 18, no. 2:164–172.

Zhang, X., J. Cui, W. Zhou, and S. Li. 2015. Design, synthesis and anticancer activity of shikonin and alkannin derivatives with different substituents on the naphthazarin scaffold. *Chemical Research in Chinese Universities* 31, no. 3:394–400.

Zhou, J., Y. Chen, J.Y. Lang, J.J. Lu, and J. Ding. 2008. Salvicine inactivates β 1 integrin and inhibits adhesion of MDA-MB-435 cells to fibronectin via reactive oxygen species signaling. *Molecular Cancer Research* 6, no. 2:194–204.

Zhou, W., Y. Peng, and S.S. Li. 2010. Semi-synthesis and anti-tumor activity of 5, 8-O-dimethyl acylshikonin derivatives. *European Journal of Medicinal Chemistry* 45, no. 12:6005–6011.

Zhou, W., X. Zhang, L. Xiao, J. Ding, Q.H. Liu, and S.S. Li. 2011. Semi-synthesis and antitumor activity of 6-isomers of 5, 8-O-dimethyl acylshikonin derivatives. *European Journal of Medicinal Chemistry* 46, no. 8:3420–3427.

Zhou, Y., B. Yang, Y. Jiang, et al. 2015. Studies on cytotoxic activity against HepG-2 cells of naphthoquinones from green walnut husks of Juglans mandshurica Maxim. *Molecules* 20, no. 9:15572–15588.

CHAPTER 5

Polyphenols and Cancer Immunology

Roman Paduch

CONTENTS

5.1 INTRODUCTION

5.1.1 Tumor Microenvironment

Effective treatment of cancer involves not only efficient elimination of transformed cells but also effective impact on the complex structure of the tumor. The term 'tumor structure', besides tumor cells themselves, also covers the cells that form the cancer microenvironment and the biochemical signaling pathways associated with the direct and indirect interactions of tumor cells with the stroma of a given tissue. The stroma of a pathological mass directly affects the pre-cancerous stages of disease development (initiation and promotion) as well as regulating advanced stages of tumor growth. Because of its important biological function, it is often referred to as the 'reactive stroma' (Gout and Huot 2008). In solid tumors, their microenvironment consists of connective tissue, including extracellular matrix deposits rich in fibrin or collagen type I, fibroblasts, vascular endothelial cells, matrix-associated molecules, as well as the extraordinary variety of cells belonging to the broadly understood immune system. In order to systematize this diverse group, three levels of immunity are distinguished which function in both physiological and pathological conditions. They are the epithelial barrier, cellular immunity, and the presence of antibodies (Gout and Huot 2008, Benencia et al. 2012, Lee and Shin 2014). An important element of communication between tumor and stromal cells is their reciprocal interactions through basement membranes and paracrine interactions based on secreted cytokines, growth factors, and their receptors. A neoplasm and the reactive stroma also actively communicate through microvesicles, alternatively referred to as exosomes. All of these elements allow tumor cells to survive in the primary niche and spread to distant sites, where they form secondary tumors called metastases (Gout and Huot 2008, Peddareddigari et al. 2010, Kopeć-Szlęzak 2012, Hamada et al. 2013). Research on simplified systems, in which only homogeneous cultures of tumor cells are analyzed, does not always reflect the reality and the actual response of tumor cells to treatment. Therefore, to fully characterize a tumor in terms of its local as well as generalized effects on the body, one should not only analyze it at the level of tumor cells themselves, but also investigate their relation to the tissue environment and the host's responses. The dependence of tumor cells on stroma components creates new therapeutic possibilities, encouraging researchers to look for novel treatments supporting the classic healing methods (Szala 2007, Gout and Huot 2008, Peddareddigari et al. 2010). In this chapter, I present the types of immune cells comprising the tumor reactive stroma, briefly characterize the role of these cells in tumor development, and describe the biological activity of selected polyphenols, which regulate the function of these cells thereby affecting the development and progression of the tumor.

The immune cells located in the tumor microenvironment belong to both innate and acquired immunity. Innate immunity involves macrophages, dendritic cells (DCs), natural killer (NK) cells, natural killer T-cells (NKT), as well as granulocytes. On the other hand, acquired immunity of the tumor stroma consists of cancer subpopulations of T and B lymphocytes. Myeloid-derived suppressor cells (MDSCs) are also an extremely important component of the reactive stroma. They are a heterogeneous population comprised of immature granulocytes, dendritic cells, and macrophages (Kopeć-Szlęzak 2012, Hasmim et al. 2015). Depending on the specific activity of immune cells relative to transformed cells, they can be generally divided into those showing antitumor effects and those exhibiting tumor-promoting activity. The former group can include Th1 CD4+ or cytotoxic CD8+ lymphocytes, NKT cells, NK cells, B lymphocytes, dendritic cells, and macrophages M1 (Table 5.1). In turn, cells characteristic of the second group are

Table 5.1 Anti-Tumor Activity of Immune Cells Infiltrating a Tumor and its Microenvironment. This Is Only a Very General and Tentative Division Which May Vary Depending on Many Factors Linked with Tumor and Local Tissue Conditions

The Kind of Immune Cell	Immune Cell Activity
Lymphocytes T CD4+ Th1 (helper)	Tumor cells killing, stimulation of Tc lymphocytes proliferation and activity
Lymphocytes T CD8+ (cytotoxic)	Tumor cell killing
Lymphocytes NKT	Tumor cell killing
Natural Killer (NK) cells	Tumor cell killing
Lymphocytes B	Anti-tumor activity by secreted antibodies
Dendritic cells (DCs)	Presentation of tumor antigens to lymphocytes T
Macrophages type M1	Presentation of tumor antigens to lymphocytes T, killing tumor cells, tumor antigens release

Table 5.2 Pro-Tumor Activity of Immune Cells Infiltrating a Tumor and Forming Its Reactive Stroma and the Broadly Understood Tumor Microenvironment. This Is Only a Suggested Division Which Is Very General and May Vary Depending on Many Factors Linked with Tumor and Local Tissue Conditions

The Kind of Immune Cell	Immune Cell Activity
Lymphocytes Treg CD4+CD25+Foxp3 (regulatory)	Inhibiting of Th1 lymphocytes activity inhibiting of Tc lymphocytes activity inhibiting of NK cells activity inhibiting of T CD8+ lymphocytes activity immunosuppressive activity proangiogenic activity
Lymphocytes Th2 (helper)	Inhibiting of Th1 lymphocytes activity inhibiting of Tc lymphocytes activity inhibiting of NK cells activity inhibiting of T CD8+ lymphocytes activity immunosuppressive activity proangiogenic activity
Lymphocytes Th17 (helper)	Inhibiting of Th1 lymphocytes activity inhibiting of Tc lymphocytes activity inhibiting of NK cells activity inhibiting of T CD8+ lymphocytes activity immunosuppressive activity proangiogenic activity
Myeloid-Derived Suppressor Cells (MDSCs)	Inhibiting of Tc lymphocytes activity inhibiting of Th1 lymphocytes activity inhibiting of DCs activity stimulation of Treg and Th2 lymphocytes proliferation and activity increasing immunosuppressive microenvironment
Macrophages type M2	Inhibiting of T CD8+ lymphocytes activity increasing immunosuppressive microenvironment stimulation of tumor cell proliferation increasing neovascularization of tumor mass promotion of tumor metastasis

regulatory T-cells (Treg) CD4+CD25+Foxp3, Th2 and Th 17 lymphocytes, MDSC cells, and M2-type macrophages (Table 5.2) (Szala 2007, Peddareddigari et al. 2010, Kopeć-Szlęzak 2012, Hasmim et al. 2015). This division is, of course, only very general, contractual, and relative because, depending on the stage of tumor development, its original location, and the specific individual features of the patient's immune system, the activity and function of the immune cells mentioned may strongly diverge from the role attributed to them on the basis of experimental and theoretical studies.

Tumor development is inextricably linked with two processes: local inflammation and immunosuppression. Infiltration of solid tumors by different immune cells induces not only these two processes, but also stimulates both tumor and stroma cells to secrete soluble chemoattractants such as growth factors, cytokines, chemokines, metalloproteinases, cysteine cathepsins, and serine proteases, which accelerate the promotion and progressive development of pathological tissue. In addition, these factors stimulate the process of epithelial–mesenchymal transition (EMT) and angiogenesis, and hence tumor invasion and metastasis (Gout and Huot 2008, Peddareddigari et al. 2010). Knowledge of the biological activity of stromal and tumor-infiltrating cells and understanding of the reciprocal relationships that they form with the transformed and normal cells of the cancer microenvironment may be of use in planning new therapies and improving or supplementing existing ones. Disruption of the mutual relations between these cells may thus represent an interesting alternative to the currently used conventional methods of treatment based mainly on the effects of therapeutic agents on tumor cells alone.

5.2 TUMOR-INFILTRATING CELLS

5.2.1 Macrophages

The macrophages recruited at the initial stages of tumor development are of the M1 (anti-tumor) type. They are part of the immune response associated with Th1 lymphocytes, and produce IL-12, IL-23, IFN-γ, TNF-α, and nitric oxide (NO), which have a potent activity against pathogens and tumor cells. As the disease progresses, however, paracrine pressure of the tumor microenvironment and local hypoxia induce the appearance of M2-type (CD206, pro-tumor) macrophages, also called tumor associated macrophages (TAMs). They constitute a significant portion of the cells forming the reactive stroma of a tumor. Signals that induce M2 polarization of macrophages include IL-4, IL-13, IL-10, as well as macrophage colony-stimulating factor (M-CSF) and toll-like receptor (TLR) ligands. This conversion results in the appearance of tumor-promoting cells, primarily through immunosuppression (production of IL-10 and TGF-β), induction of angiogenesis and lymphangiogenesis (production of IL-8, IL-23, and the platelet derived growth factor—PDGF), stimulation of cell mobility, remodeling of tissue and the extracellular matrix (ECM), as well as metastasis (production of metalloproteinases MMP-2, MMP-9, the migration stimulating factor—MSF, and IL-6). Macrophages are also involved in the release of tumor cells from the primary niche, formation of complexes protecting the migratory tumor cells, and stabilization of those cells at the target site of metastasis. On top of that, they stimulate chemotactic attraction of Treg lymphocytes by secreting the CCL22 chemokine and induce differentiation and activation of these cells primarily via direct cell–cell contact mediated by lymphocyte, or via membrane-bound receptor for TGF-β1. Disturbances in the ratio of effector T-cells to Treg are an important factor in the progression of the disease and potentially portend the patient's short survival. TAMs which show M2 polarization are therefore the major tumor-infiltrating cells responsible, in large part, for the induction of immunosuppression and local inflammation. In most solid tumors, a high level of infiltrating M2 macrophages is associated with poor prognosis for the patient. Hence, any activities aimed at reducing their population, including application of natural plant compounds, are desirable, because they can significantly reduce the development and spread of cancer cells in the body. The therapeutic targets are both the proteins (cytokines, enzymes, or some transcription factors) produced by TAMs as well as TAMs themselves (Szala 2007, Gout and Huot 2008, van der Bij et al. 2008, Peddareddigari et al. 2010, Kopeć-Szlęzak 2012, Feng et al. 2014, Zhao et al. 2014, Hasmim et al. 2015) (Figure 5.1).

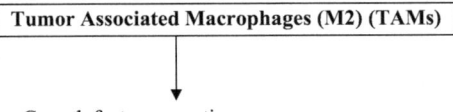

Tumor Associated Macrophages (M2) (TAMs)

- Growth factors secretion

- Immunosuppression

- Inflammation

- Extracellular matrix modulation

- Angiogenesis and lymphangiogenesis stimulation

- Epithelial-Mesenchymal Transition (EMT) induction

- Metastasis stimulation

Figure 5.1   Functions of tumor associated macrophages (TAMs). They express multiple activities enhancing cancer development and metastasis.

5.2.2 Dendritic Cells (DCs)

Dendritic cells belong to the group of immune cells which are most highly specialized in the presentation of antigens. They can be found in peripheral tissues or immunological organs and are considered to provide a link between innate and adaptive immunity. Activated by chemotactic factors (chemokines CXCL12 or CCL20) produced by the tumor microenvironment, DCs migrate to the pathological tissue, contributing to the reactive stroma. They mainly function as antigen presenting cells (APC), presenting antigens to T-cells in the context of MHC I and II. As a consequence, cytotoxic T-cells infiltrate the tumor mass and destroy it as well as stimulating B lymphocytes, which also limit tumor growth by producing antibodies. Cells of the CD11c+CD8a+ DR+ dendritic cell subpopulation directly kill tumor cells and may exert a cytotoxic effect on transformed cells through perforin and granzyme expression. Other types of DCs found in the tumor microenvironment include myeloid dendritic cells (mDCs) CD11c+ with an antitumor activity (which act mainly by inducing cytotoxic T lymphocytes), plasmacytoid dendritic cells (pDCs) CD123+, and so-called immunotolerant dendritic cells (itDCs), which may exert an immunosuppressive activity promoting tumor growth through CD4+ Th2 and Treg cells. Generally, immature DCs (iDCs), which are precursors of mDCs, are also recruited into the tumor. They are inhibited in maturation and stimulated to promote angiogenesis and trans differentiation into endothelial progenitors. Endothelization is characterized by a loss of CD14/45 and appearance of typical markers such as CD31, CD34, von Willebrand Factor (vWF), VE-cadherin, and VEGF receptors. Tumor-Associated Dendritic Cells (TA-DCs) exert an immunosuppressive activity and significantly promote the angiogenic process in the tumor stroma by producing matrix metalloproteinases (MMPs), VEGF, angiogenin, basic Fibroblast Growth Factor (bFGF), and heparanase (Benencia et al. 2012, Francisco et al. 2014, Li et al. 2015, del Cornò et al. 2016).

5.2.3 NK Cells

NK cells are large granular lymphocytes, considered to be major effector cells of innate immunity, which, however, do not exhibit the characteristics of B and T lymphocytes. They show direct cytotoxicity to target cells (tumor and infected cells) without prior sensitization that would require recognition in the antigen/MHC context. NK cells penetrate into the interior of a solid tumor through tumor vasculature. Migration toward tumor is associated with a gradient of tumor-derived chemokine ligands (CXCL9, 10, and 11) and the presence of the CXCR3 chemokine receptor on NK cells. These cells belong to innate immunity, inhibiting tumor growth in its initial phase. They probably act in response to pathogen-associated molecular patterns (PAMPs) and extracellular or cell-associated proteinases. Generally, the main pathway by which NK cells eliminate tumor cells is the release of the content of cytotoxic granules (perforin and granzymes). With tumor growth, however, the immunological balance associated with NK cells is disturbed, mainly due to changes in their surface activating and inhibitory receptors (CD158 and CD16, CD161, and CD69) and cytokine production (IL-2, IL-12, IL-18, IL-21, TNF-α, and IFN-γ). Moreover, in later stages of tumor development, cancer-associated fibroblasts (CAFs) are also involved in reducing NK cell functionality, mainly by affecting their lytic potential. Moreover, DCs can modulate the functions and proliferation of NK cells by coming into direct contact with them, with simultaneous participation of membrane-bound IL-15. NK cell activity can also be suppressed by direct contact, via expression of membrane-bound TGF-β on MDSC cells. Similarly, TGF-β produced by Treg also inhibits the cytolytic functions of NK cells. A similar activity mediated by the hypoxia inducible factor (HIF) is exhibited by the hypoxic tumor microenvironment. Although NK cells belong to innate immunity, the presence of soluble mediators makes these cells actively participate in adaptive immune response. They only mature during an infection or an inflammatory state. By lysing tumor cells, NK cells provide a collection of antigens for mature DCs that present them to T lymphocytes. However, in advanced stages of the disease, NK cells terminate typical interactions between DCs and T lymphocytes by destroying DCs and thereby preventing effective presentation of tumor antigens (Peddareddigari et al. 2010, Kopeć-Szlęzak 2012, Quoc Trung et al. 2013, Lindqvist et al. 2014, Kim et al. 2015, Kim and Lee 2015).

5.2.4 Myeloid-Derived Suppressor Cells (MDSCs)

Myeloid-derived suppressor cells (MDSCs) are a heterogeneous population of immature myeloid cells (granulocytes, DCs, and macrophages) that primarily exhibit immunosuppressive activity. They express surface molecules of CD33, CD15, and CD11b, specific for myeloid cells, but do not have characteristic receptors for monocytes such as CD14 and HLA-DR. Their suppressive activity mainly involves suppression of immune reactions against cancer. MDSCs inhibit anti-tumor activity of T-lymphocytes CD4+ Th1, T-lymphocytes CD8+, and DCs, while simultane-ously stimulating Foxp3-positive Treg and Th2 lymphocytes, which promote tumorigenesis. They also decrease the level of IL-12, thus inducing the immunosuppressive capacity of the tumor microenvironment. The main mechanism of the immunosuppressive action of MDSCs is the secretion of Arginase-1, an enzyme which degrades L-arginine, an amino acid that is essential to the proper functioning of T lymphocytes. Furthermore, they secrete nitric oxide (NO), IL-10, and TGF-β, which all show immunosuppressive activity stimulating macrophage M2 activity, and produce IL-6, which stimulates tumor cell proliferation. The generation of reactive oxygen spe-cies and NO by MDSCs also leads to the formation of peroxynitrite, which nitrosilates receptors of T-cells, exacerbating the suppression effect on the local immune system in the vicinity of the growing tumor. MDSCs also express typical endothelial cell markers such as CD31 and VEGFR2 and can thus supplement the wall of the newly developed tumor vasculature. Lastly, they can dif-ferentiate into fully functional TAM macrophages (Peddareddigari et al. 2010, Kopeć-Szlęzak 2012, Forghani et al. 2014).

5.2.5 T and B Lymphocytes

Tumor-infiltrating lymphocytes (TILs) are a heterogeneous population of cells which infiltrate the structure of solid tumors. TILs include granulocytes, macrophages, and MDSCs originating from myeloid lineage as well as subsets of T, B, and NK lymphocytes. It is believed that the myeloid cells which infiltrate tumors are largely responsible for their progression, while T-cells, especially the CD8+ subset, are associated with tumor regression and a better prognosis for the patient (Yu and Fu 2006, Peddareddigari et al. 2010).

5.2.5.1 T Lymphocytes

The mature population of T-cells includes T αβ (CD4+ or CD8+) lymphocytes as well as T-cells carrying the γδ T-cell receptor (TCR) (CD4- and CD8-). T-cells play an invaluable role in immu-nity by rejecting tumors which are induced by chemical, physical, and biological impacts (LoPresti et al. 2014).

5.2.5.1.1 CD8+ Cytotoxic T-Lymphocytes (CTLs)

Tumors are generally positive for MHC I but do not express MHC II. CD8+ lymphocytes, in turn, are able to recognize antigens presented in the context of MHC class I. Therefore, tumor antigens presented by MHC I are recognized by CTLs, and an antitumor response follows. These cells are generally recognized as critical effectors against tumor cells (Bhattacharyya et al. 2010, Karimi et al. 2015).

5.2.5.1.2 CD4+ Helper T-Lymphocytes

It has been demonstrated that in the absence of CD8+ T lymphocytes, CD4+ T-cells can eliminate tumor cells with sufficient effectiveness in some malignancies. However, because tumor cells only express MHC I, the anti-tumor activity of CD4+ T-lymphocytes is highly limited. This is why they have often been postulated to merely play an auxiliary role in tumor immunity by aiding the activation of CD8+ CTLs. Recent research shows, however, that within the tumor microenvironment, CD4+ T-lymphocytes are often differentiated into regulatory CD4+ T-cell subtype which limits the activity of the CD8+ CTLs effector during the progression of tumor growth. Nevertheless, the presence and activity of CD4+ T-cells is essential for the activation of DCs and conversion of naïve CD8+ lymphocytes into CTLs and their effective response against tumor cells. By contrast, the subpopulation of CD4+ Th17 T-cells secreting the proangiogenic IL-17 exhibits a strong tumor-promoting activity and is associated with a worse prognosis for the patient (Bhattacharyya et al. 2010, LoPresti et al. 2014).

CD4+ T-cells are also divided into Th1, Th2, and Th17 types, depending on the profile of the cytokines they produce. Th1 produce IFN-γ and IL-2 while Th2 secrete mainly IL-4, IL-5, IL-6, IL-10, and IL-13. The action of Th1 is correlated with cellular immunity, including the activity of CTLs and is thus involved in antitumor response. Th2 activity, conversely, is connected with the weakening of the antitumor immune response, promotion of tumor growth, and progression of neoplastic cells. These cells are implicated in humoral immunity. Yet, all this having been said, it is not possible to put these cells into clear-cut classes, because there are reports indicating that Th2 clones are involved in the mechanisms restricting tumor growth, while Th1 phenotype lymphocytes may change their polarization into Th2, Th17, or Tregs and thus promote tumor growth (Yu and Fu 2006, LoPresti et al. 2014).

Th17 lymphocytes are widely known to be implicated in the pathogenesis of cancer and induction of inflammation. They produce IL-17A and IL17F, which stimulate IL-6, Cyclooxygenase-2 (COX-2), VEGF, and NO production and also recruit MDSCs to weaken anti-tumor immunity (Moutia et al. 2016).

Th9 lymphocytes produce IL-9 activating mastocytes and eosinophils, Th22 secrete IL-22 exhibiting antibacterial properties, while T follicular helper cells (Tfh) induce proliferation and differentiation of lymphocytes B in lymph nodes (Kopeć-Szlęzak 2016).

5.2.5.1.3 CD4+ CD25+ FoxP3+ T Lymphocytes (Tregs)

These cells are described in the literature as playing a regulatory function in the overall immune system. In tumor immunity, Tregs are involved in inhibiting the activity of T CD8+ and CD4+ Th1 lymphocytes on the developing tumor as well as in regulating the proliferation of these immune cells. They act by releasing IL-10 and TGF-β, the most powerful immunosuppressive cytokines. This activity, in the tumor context, is directed against CTLs, DCs, and other, like NK, specialized anti-tumor immune cells (Karimi et al. 2015). On the other hand, the inhibition of chronic inflammation may, in some tumors, have a beneficial effect of limiting tumor development and invasion of normal tissues. Therefore, the correct balance of anti-tumor T-cells to Tregs may be important in achieving clinically desirable immune responses to tumor cells (Szala 2007, Peddareddigari et al. 2010, Kopeć-Szlęzak 2012, Karimi et al. 2015).

5.2.5.1.4 γδ T-Lymphocytes

Gamma delta T-cells are an important part of the TILs group. They exhibit antitumor activity, which is associated with the production of cytokines (IFN-γ and TNF-α) as well as stimulation of DCs maturation. Their distinctive feature, which distinguishes them from lymphocytes αβ, is the lack of MHC restriction. In addition, these cells exhibit a strong antitumor activity by exerting a cytotoxic effect. Their activity, however, depends on the type of tumor and location in the tumor. It has been shown that TGF-β produced by TAMs can convert T lymphocytes γδ into cells with Treg properties and, thus, antitumor activity (LoPresti et al. 2014).

5.2.5.2 *B Lymphocytes*

B lymphocytes are the basic component of acquired immunity. They play an important role in tumor immunity, which, however, cannot be clearly defined as pro- or anti-cancer. B lymphocytes reduce the growth of a tumor by the activity of generated antibodies, which also enhances T-cell responses and the secretion of cytokines and chemokines. In this capacity, B-cells can function as local APCs as well as acting via a direct antibody-independent tumor cell killing mechanism. On the other hand, B lymphocytes can act as regulatory B-cells (Bregs) attenuating CD8 + T and CD4 + T-cells and activating tumors, thus promoting tumor development. Tumor-infiltrating B-cells (TIL-Bs) (CD20+) constitute about 40% of the TIL population. However, it is only the coordinated activity of CD8+ and CD20+ TILs that results in a high antitumor response and a better prognosis for the patient (Nelson 2010, Lee-Chang et al. 2013, Shen et al. 2016).

To make a long story short, the role of lymphocytes in the tumor microenvironment is heterogeneous. Their activity and function depend on the subpopulation of these cells as well as on the type of tumor that they infiltrate.

5.2.6 Cancer-Associated Fibroblasts (CAFs)

CAFs are modified and activated fibroblasts constituting important cellular component of the reactive stroma. These cells are contractors and intermediaries in tumor–stroma interactions. They are defined as α-smooth muscle actin (α-SMA) and vimentin positive, spindle-shaped cells expressing fibroblast activation protein (FAP), fibroblast-specific protein (FSP1), platelet-derived growth factor (PDGF-β) receptor, and neuron glial antigen-2 (NG2). In the tumor microenvironment, they organize the stroma by producing matrix proteins (collagens I, III, or IV), proteoglycans, and metalloproteinases (MMPs), as well as chemokines (CXCL1, CXCL2 CCL2, CCL5), angiogenic factors (VEGF, PDGF), and growth factors such as TGF-β, HGF, IGF-1, and EGF. All together, these factors are engaged in enhancement of tumor cell proliferation, angiogenesis, and invasion, ultimately leading to metastasis. Moreover, CAFs recruit tumor MDSCs, decrease antigen presentation, and limit the functionality of CTLs, TIL-Bs, and NK cells. CAFs are thus classified as the main stromal stimulators of tumor growth and the spread of tumor cells in the body (Gout and Huot 2008, Peddareddigari et al. 2010, Ting et al. 2016) (Figure 5.2).

As this brief overview shows, the tumor cellular microenvironment plays a fundamental role in the development of a tumor. Proliferation of tumor cells alone, according to the reductionist

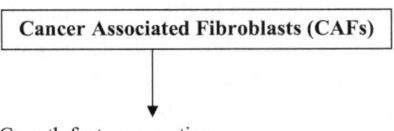

Cancer Associated Fibroblasts (CAFs)

- Growth factors secretion

- Extracellular matrix modulation

- Angiogenesis and lymphangiogenesis stimulation

- Metastasis stimulation

Figure 5.2 Functions of cancer-associated fibroblasts (CAFs). They are the main cellular constituents of the tumor stroma. They express multiple functions facilitating cancer growth and invasion.

model, is completely insufficient for the tumor to grow and for the transformed cells to acquire the metastatic phenotype. Knowledge of the direct and indirect interactions within the highly diverse structure formed by cancer and reactive stroma cells is therefore essential for understanding the biology and immunology of cancer and, ultimately, for the formation of effective therapies. To arrive at a full picture of the processes involved it is necessary to thoroughly analyze the activity of the immune cells forming the tumor microenvironment, because they are not always clearly classified as promoting or inhibiting the process of tumor development. The complicated structure of tumors, both at the cellular and the paracrine level of interactions, makes cancer therapy a difficult problem. The choice of proper therapeutic targets for treatment, which do not necessarily have to be cancer cells themselves, may therefore contribute to a more effective suppression of the development of the disease than is possible with the use of currently applied therapies. An interesting targeted strategy for anticancer activity could thus be focused on normal stromal cells, including the cells of the immune system. It would involve appropriate adjustment of the biologically active substance to the active stroma cells with a view to regulating the processes of cancer promotion and progression. What is important is that the natural substances be carefully selected so that they negatively modulate the activity of immune cells responsible for local immunosuppression, while positively regulating the activity of immunologically competent cells responsible for reducing the spread of cancerous cells and infiltration of the surrounding tissues.

5.3 NATURAL COMPOUNDS IN CANCER THERAPY

Traditional medicine is based on plant formulations, not only herbal ones, which serve as medicinal weaponry against diverse pathogens, undefined ailments, as well as malignancies. Generally, plants are a great source of natural compounds with diverse chemical structures offering pro-health properties. The attraction of phytochemicals, which are organic components of plants, is that, apart from playing physiological functions in these organisms, they can also potentially be used as therapeutic agents in combating cancer in mammalian cells. However, the potential use of phytochemicals in cancer treatment should not be limited only to basic chemoprevention, but they also need to be considered as effective blocking and suppressing agents to be used in supplemental therapies or, after chemical modification, as potential drugs. The task of these compounds and their analogs would be to detoxify the body, modulate carcinogen uptake and metabolism or scavenge reactive oxygen species. In addition, phytochemicals can act at the level of tumor initiation and thus inhibit the promotion and progression of cancer (Spagnuolo et al. 2012). Currently, efforts are being made to develop formulations with a specific level of efficacy and a known purpose and mechanism of action. The problem with natural compounds, in their basic, unmodified form, is that despite having a significant potential therapeutic activity, they are characterized by limited efficacy, complex pharmacodynamics, and poor bioavailability. Hence, there is ongoing work on synthesizing new derivatives combined with new functional groups and small molecules, whose increased activity would, however, still be based on the basic activities of the lead compounds originally isolated from plants. Of course, the introduction of targeted modifications and rational design of new active drugs must be based on solid knowledge about the safety of their application, their new biological properties, and the specific, targeted mechanism of their action (Apaya et al. 2016). Medicinal plants are a source of compounds (micromolecules and macromolecules) with proven anticancer activities. Natural products, however, occur in great diversity, and therefore their biological activity aimed against cancer can vary depending on the component used, its modification, the type of tumor it is used to act on, and even the duration of use. Currently, proper isolation of these components and their subsequent modifications must be based on the knowledge of (1) their basic bioactivity, e.g. a characterization of active groups and the mechanism of action of these compounds, (2) possible modifications allowing to improve their biological activity (increased activity) and performance (improved pharmacological profile), and (3) ways of limiting the induction of potential side effects (decreased generalized toxicity), especially their cytotoxic effect on normal cells. Full knowledge about a substance of natural origin that is potentially useful oncologically should be grounded in research on the structure–activity relationships characterizing the substance, its mechanism of action, the molecular targets of the compound, its metabolism, and the resulting intermediates, as well as its potential for chemical modification. This knowledge is needed to carry out basic research before clinical trials are initiated (Dholwani et al. 2008, Millimouno et al. 2014).

Cancerous mass, besides tumor cells, consists of a heterogeneous population of immune and matrix cells referred to as the active or the reactive stroma. The growth of a tumor, followed by its spread, depends on the dynamic balance between anti- and pro-tumor activity, which is closely linked with the anti- and pro-inflammatory activity of immune cells. Disturbance of these mutual interactions involving the cellular and the paracrine level may be an interesting and, more importantly, an effective way of targeting cancer. Such therapeutic interventions may also lead to sensitization of both differentiated cancer cells and cancerous stem cells (CSCs) to administered drugs. Immunomodulation of the tumor microenvironment could thus be an interesting complementary/supportive strategy to conventional chemotherapy in the treatment of malignant diseases (Pak et al. 2014).

Among natural compounds, polyphenols are most often mentioned as factors capable of modulating the immune inflammatory response of the tumor stroma. This is mainly due to their pleiotropic activity directed at the microenvironment of the tumor, both at the cellular level and at the level of activity of soluble pro-inflammatory mediators (Ghiringhelli et al. 2012, Delmas et al. 2013).

This chapter is focused on the activity of selected polyphenols, including resveratrol, quercetin, apigenin, silibinin, curcumin, epigallocatechin 3-gallate (EGCG), and others, on the immune cells closely associated with cancer development. These polyphenols possess activity targeted not only directly against tumor cells but, in a broader sense, against the tumor immune microenvironment.

5.3.1 Polyphenols

Natural polyphenols or phenolic compounds belong to the group of secondary metabolites produced by plants in response to harmful stimuli and stress signals derived from their environment. Biogenetically, polyphenols come from two pathways: the shikimate and the acetate pathways. Taking into consideration the size of their molecules, polyphenols range from simple molecules (phenolic acids) to polymerized structures (tannins). In nature, they occur in conjugated forms with sugar residues linked to hydroxyl groups, or the sugar unit is directly linked to an aromatic carbon atom. The chemical structure of dietary polyphenols strictly depends on the presence and degree of acylation/glycosylation, conjugation with other phenolics, polymerization (molecular size), and solubility (Bravo 1998). It determines their biological activity as well as how well they are absorbed and metabolized in the mammalian body.

More than 8,000 phenolic compounds have been identified so far. They are often classified according to the chemical structure of the aglycones into phenolic acids, flavonoids, polyphenolic amides, and other polyphenols (Tsao 2010) (Figure 5.3).

It has been suggested that polyphenols exert beneficial healthcare effects when consumed as dietary products. In the context of antitumor therapy, multiple general targets of polyphenol activity have been ascertained. They have been found to interfere with several points on the path leading to cancer development and spread, including mechanisms and pathways of early carcinogenesis, proliferation of tumor cells, the action of apoptosis-related proteins, microenvironmental inflammation, lymphangiogenesis and angiogenesis, resistance to classic therapeutic treatments, as well as metastasis and secondary tumor formation (Asensi et al. 2011).

The anticancer pharmacological properties of polyphenols, due to the amphiphilic character of these compounds, are generally associated with weakening of the protective barrier of tumor cell

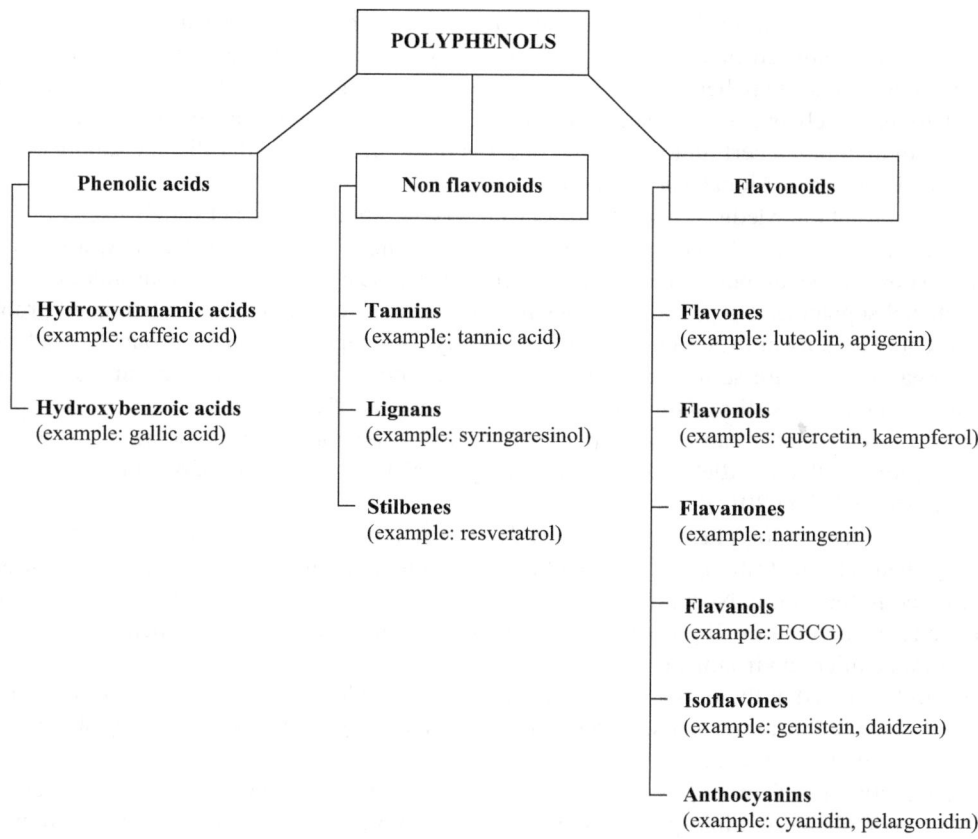

Figure 5.3 Classification of polyphenolic compounds.

membranes. The antitumor effects of polyphenols are connected mainly with their antioxidative abilities but also their cytotoxic, antiproliferative, and immunoregulatory activities (Sajkowska-Kozielewicz et al. 2016).

The antioxidative activity of polyphenols, which involves scavenging of endogenously generated reactive oxygen species by donating an electron or a hydrogen atom, plays a role in the early stages of cancer development or even as a chemopreventive action forestalling the appearance of transformed cells. Polyphenols also exert antioxidative effects by acting as chain-breakers donating an electron to free radicals and thus neutralizing their activity and by inducing antioxidant enzymes, e.g. glutathione peroxidase, catalase, and superoxide dismutase (Tsao 2010).

Beside antioxidant properties, polyphenols also exert pro-oxidant activity. This mechanism may be important in growing cancers and advanced malignancies as it limits the proliferation of tumor cells, inducing their apoptosis and mobilizing immune cells to inhibit cancerogenesis. It has been found that both pro- and antioxidative anticancer activities of polyphenols involve endogenous copper ions Cu (II), possibly bound to chromatin. Mobilization of copper leads to internucleosomal DNA fragmentation, which is an important determinant of induction of apoptotic cell death (Azmi et al. 2006). Polyphenols, which can act as metal chelators in conjunction with co-antioxidants such as vitamins, can also directly reduce Fenton reaction by chelating transition Fe^{2+} ions and preventing the oxidation caused mainly by hydroxyl radicals (Tsao 2010).

Unfortunately, the widely reported anticancer activity of polyphenols is controversial, mainly due to the proportionally low correlations between the results obtained *in vitro* and *in vivo*. This discrepancy may be caused by the fact that peak concentrations of polyphenols in plasma are observed between 1–2 h after dietary intake and are relatively quickly eliminated from the circulation. Moreover, prolonged ingestion does not, generally, produce an increase in the concentration of polyphenols in plasma (Chen et al. 2005). It has also been suggested that some polyphenols may accumulate selectively in certain tissues, especially those in which they can be rapidly metabolized. Therefore, to maintain a certain level of polyphenols in circulation, regular dietary supply of these compounds is required (Manach et al. 2004).

Despite current knowledge of their beneficial effects, polyphenols are not widely used in clinical programs. This can be attributed to several causes, including their limited selectivity, difficulty in estimating doses that can induce therapeutic effects *in vivo*, the need to use supraphysiological doses of polyphenol supplements to achieve the desired biological effect, and the closely related problem of the potential toxicity of these compounds, which may vary depending on the group they belong to and the organism they are acting on. Moreover, the question of what route of administration yields the best healing effect still remains unanswered. Hence, it is often suggested that the therapeutic properties of polyphenols can only be put into service in personalized medicine using selective analogs together with daily diets, which at most may accelerate the activity of parent compounds or approved drugs (Wright 2013).

Widespread use of polyphenols in therapy, due to a number of uncertainties regarding the conditions of their administration, their bioavailability, activity, and metabolism, is still a question to be addressed in the future. Nonetheless, *in vitro* studies point to the efficacy of this group of compounds in reducing tumor cell growth and their potentially therapeutically beneficial effects on the tumor immune microenvironment.

It should be noted at the outset that polyphenols are capable of inhibiting *in vitro* formation of clumps of normal and tumor cells, called spheroids, thus preventing the formation of structures similar to the spatial systems observed *in vivo*.

Under conditions of three-dimensional interactions, normal cells, and cancer cells which form so called tumorspheres, exhibit specific characteristics that often differ from those observed in two-dimensional culture conditions *in vitro* and are similar to those found *in vivo*. It has been shown that, e.g. inhibition of mammosphere formation with physiological polyphenol doses was connected with attenuation of the PI3K/Akt/NF-κB signaling pathway in aggressive tumor cells isolated from the mammary gland. The PI3K/Akt/NF-κB molecular path, among others, features importantly in the regulation of the number of normal and malignant progenitor or stem cells, thus affecting physiological tissue remodeling or, in cancer tissue, proliferation and motility of transformed cells (Montales et al. 2012). It can also have a direct inhibitory effect on the renewal of Cancer Stem Cells (CSCs) and stimulation of cancer growth and spread. These observations indicate that, at a cell-organization level higher than *in vitro* two-dimensional cultures, polyphenols still exhibit anticancer activity. This may suggest that different tissue reactivity *in vivo* may depend not only on physical interactions between cells or the application of high doses of polyphenols, but also on paracrine intercellular interactions and the unusual heterogeneity of both normal and neoplastic tissues. What is common to all these aspects of *in vivo* interactions complexity is that they involve constant penetration and invasion of all tissue structures by soluble mediators and immune system cells.

Tumor development specifically modulates immune system functions. This mainly involves decreasing the reactivity of peripheral blood mononuclear cells (PBMCs), including T-, NK, and B-cells, and inducing apoptosis of selected immune cells. The mechanism of this inhibition may be connected with secretion by tumor cells of a population of membrane vesicles (exosomes), homogenous in size (60–100 nm), which fuse with lymphocytes to limit their reactivity and thus induce a local immunotolerant environment. Polyphenols are capable of reversing this effect by stimulating ubiquitination of exosomal proteins and consequently helping retain the cytolytic activity of NK cells (Zhang et al. 2007). Polyphenols thus have a potent immunomodulatory activity which plays a role in preventing cancer development, among others, by regulating the immune response within the tumor and its microenvironment (niche) (Yi et al. 2017). Based on *in vitro* studies, polyphenols are generally considered to be agonists of selected immune cells. They stimulate cells of the lymphocyte lineage such as NK cells and T- (γδ and αβ) and B-cells. These compounds can also activate lymphocytes in an antigen-independent manner thus potentiating their response to secondary stimuli. What is more, polyphenols have been shown to stimulate lymphocytes (NK and T-cells) to produce interferon-γ (IFN-γ), a cytokine that plays an essential part in immune defense against cancer (Ramstead et al. 2012, Ramstead et al. 2015). Additionally, polyphenols decrease IL-17 production by inhibiting gene expression and thus may be useful in controlling inflammatory diseases, including among others those leading to cancer development (Moutia et al. 2016). These compounds can attenuate the expression of other pro-inflammatory cytokines such as TNF-α or proteases (more precisely metalloproteinases MMP-2 and MMP-9) and in this manner limit the metastatic activity of tumor cells (Nimgulkar et al. 2015). Polyphenols have been reported to influence lysosomal and NO synthase activity and thus stimulate the phagocytic activity of macrophages (mainly M1) against tumor cells and to enhance the migratory capacity of macrophages (Krifa et al. 2013, Nimgulkar et al. 2015). A tumor mass characteristically accumulates MDSCs, which leads to stimulation of the motility of transformed cells, tumor vascularization, and metastasis. Polyphenols improve the antitumor immune response by inducing terminal differentiation of MDSCs and the formation of more mature neutrophil-like cell phenotypes than those observed in typical conditions (Santilli et al. 2013a, Santilli et al. 2013b). Such differentiation and eventual depletion of MDSCs unlocks the functions of T-cells and NK cells inhibited by myeloid cell lines and can thus lead to the activation of immune mechanisms interfering with the development and spread of cancer (Santilli et al. 2013a).

Polyphenols influence differentiation of DCs, expression of their surface markers, the morphology of these cells, and secreted molecules. These compounds have also been reported to reduce the stimulatory action of DCs on the proliferation of naïve allogenic T-cells, induce their ability to generate allogenic IL-10-secreting T-cells, and generally contribute to functional maturation of DCs (Del Cornò et al. 2016). In short, it could be stated that polyphenols modulate immune responses by regulating maturation of DCs and affecting the soluble mediators of these cells (Francisco et al. 2014, Li et al. 2015).

As the discussion above shows, polyphenols have a proven ability to modulate the innate and adaptive immune cells which infiltrate malignancies. Polyphenols can reduce the release of different pro-inflammatory compounds via modulation of signal transduction or through antioxidant effects. The reduction of chronic inflammation and its downstream implications may be considered a main process via which polyphenols affect tumor development and metastasis. Similar effects are obtained when polyphenols interfere with tumor-stimulating immune cells and induce or even restore the activity of cells expressing an antitumor immune response.

In vitro studies conducted to date have demonstrated that plant-derived compounds have a significant impact on inhibiting cancer initiation and development. The mechanisms through which they act, however, are as diverse as the sources from which they are obtained. In the following part of this chapter, I will focus on the interactions of selected polyphenols with cells forming the tumor reactive stroma. Although these substances are most often described in the literature as limiting the growth of tumor cells, their activity on the immune component of the tumor appears to be equally important. In other words, apart from their role as chemopreventive agents, polyphenols could be used as immunomodulators of the tumor-associated immune response and consequently, host-induced anti-cancer response.

5.3.1.1 *Resveratrol*

Resveratrol (3,4′,5-trihydroxy-trans-stilbene) is a naturally occurring polyphenolic phytoalexin primarily isolated from the skin of red grapes, peanuts, and mulberries (Azmi et al. 2005, Shimizu et al. 2006, Cui et al. 2010). This compound is now widely recognized as a chemopreventive agent and suppressor of chronic inflammatory diseases with the ability to inhibit initiation, promotion, and progression of cancer. It also exhibits strong antioxidant and anti-inflammatory properties, being able to inhibit the expression of cyclooxygenase-2 (COX-2), iNOS, and TNF-α (Cui et al. 2010). It has been shown that via this mechanism resveratrol induces apoptosis of leukemic B-cells and macrophage-like cells and inhibits inducible NO synthase (iNOS), suppressing in this way the production of endogenous NO in malignant cells (Quiney et al. 2004). Resveratrol is a compound which acts through various mechanisms to induce apoptosis in tumor cells and infiltrating immune cells. This may, for example, take place by spontaneous oxidative DNA strand breakage or DNA breakage in the presence of transition metal ions, e.g. polyphenol-Cu(II) mediated chemical degradation of DNA (Azmi et al. 2005, Azmi et al. 2006).

Resveratrol also induces apoptosis through a p53-mediated mechanism or activation of the CD95 signaling pathway which enhances CD95L expression or activates caspase-3 activity and decreases the mitochondrial transmembrane potential (Falchetti et al. 2001, Shimizu et al. 2006, Cui et al. 2010). It has also been shown that resveratrol selectively induces programmed death of tumor cells, simultaneously inhibiting this process in healthy cells (Wright 2013). In the presence of resveratrol, the release of pro-inflammatory cytokines and chemokines is reduced, among others through up-regulation of the sirt1 gene encoding the sirtuin 1 protein, also known as NAD-dependent deacetylase sirtuin-1. This process is followed by inhibition of the activity of protein kinase C and NF-κB and nuclear translocation. Sirt1 is a resveratrol receptor involved in peripheral T-cell tolerance. The consequences of sirt1 expression include decreased acetylation of c-Jun, blockage of its translocation into the nucleus, and inhibited activation of T-cells (Rieder et al. 2012, Zou et al. 2013).

All these effects indicate that resveratrol has an immunomodulatory effect on the response of human immune cells, including those cells which infiltrate tumors. This compound in a concentration-dependent manner modulates the development and differentiation of CD4+ and CD8+ T-cells and consequently the production of IFN-γ, IL-2, and IL-4 by these cells. T-cells (Th1 and Th2) treated with resveratrol restrict the production of cytokines in a polyphenol dose-dependent manner. This property can be used in reducing inflammation, including that associated with developing tumors, which is induced by activated T-cells. Additionally, resveratrol is capable of reducing the suppressive activity of Tregs and, possibly, decreasing the number of Th17, thereby limiting tumor development (Zou et al. 2013).

Administration of resveratrol normalizes the CD4+/CD8+ ratio, influences the activity of NK cells and B-cells, and stimulates lymphocyte proliferation, at the same time suppressing the expression of the pro-inflammatory IL-6 cytokine, which may be considered as tumor chemotherapeutic activity (Li et al. 2007). CD8+ T-cells are more sensitive to the modulatory activity of resveratrol than CD4+ T-cells. Resveratrol enhance at lower and suppress at higher concentrations NK cell activity. The mechanism of this action is mainly based on multiple target pathways including AhR, JNK, and ERK. Therefore, the immunomodulatory effect of resveratrol is mediated both by direct influence on cells and on the production of cytokines by T-cells as well as modulation of innate immunity (Falchetti et al. 2001, Li et al. 2014). However, the application of this polyphenol as an anticancer compound requires much caution, because sometimes small differences in the concentration used, the metabolism of the compound, and its availability to tissue may induce different, frequently adverse effects associated with tumor immunity (Li et al. 2014). In the case of resveratrol, although it is sufficiently hydrophobic and is able to transverse cellular membranes, its low bioavailability may be due to rapid biotransformation, before it reaches target tissues, and rapid elimination from plasma (Azmi et al. 2006). One solution to the problem of increased uptake of this compound by tumor cells is to develop modified analogs of this polyphenol, e.g. methoxy derivatives with increased lipophilicity.

The mechanism through which resveratrol can combat cancer is prevention of injury and apoptosis of endothelial cells (ECs) which allows NK cells to reach the pathological mass and kill tumor cells. This may be mediated by inducing the expression of MDSCs, which protect ECs against lysis while decreasing the level of TGF-β and consequently Treg counts. Resveratrol, however, exhibits different, often distinct effects on TGF-β expression and hence on downstream functions of this factor. On the other hand, it has also been shown that resveratrol can act via yet another mechanism impairing the suppressive activity of MDSCs on tumor-infiltrating T lymphocytes (TILs) and thus inducing anti-tumor immunity (Guan et al. 2012).

It has been reported that metastasis of tumor cells involves Fox3+CD4+ Tregs and tBregs expressing TGF-β. Tregs inactivate NK antitumor cells and thus protect migrating tumor cells or clusters. Resveratrol plays an important role in inactivating tBregs and consequently disallows non-Tregs to convert into Fox3+ Tregs. This process, which is necessary to inhibit metastasis, consists in the inactivation of Stat3 phosphorylation and acetylation. An imbalance in the cancer escape-promoting tBregs–Tregs axis activates suppressed CD8+ and NK immune effector cells, thereby limiting the probability or delaying the occurrence of distant metastases (Lee-Chang et al. 2013).

Resveratrol induces S phase cell cycle arrest in malignant B lymphoma cells in a dose- and time-dependent manner (Shimizu et al. 2006). It has also been shown that this polyphenol suppresses the proliferation of leukemia cancer cells by arresting them in the G0/G1-S or G2-M phases of the cell cycle (Ferry-Dumazet et al. 2002, Hu et al. 2012). Leukemia cells subjected to resveratrol treatment have been shown to be more sensitive to the lytic activity of NK cells than native, untreated tumor. It has been established that this effect is conditioned by modulation of the NKG2D receptor/NKG2D ligand expressed by NK cells (Luis Espinoza et al. 2013). There are, however, uncertainties regarding not only the sensitivity of tumor cells and tumor infiltrating cells to resveratrol, which largely depends on the dose of the polyphenol, but also the sensitivity of cancer stem-like cells, which are the main cause of recurrence. It is only when *in vitro* evidence is found that resveratrol exerts activity against cancer stem cells, that scientist will be able to draw any farther-reaching conclusions regarding impact of this compound on initiation and development of cancer.

5.3.1.2 Curcumin

Curcumin [1,7-bis(4-hydroxy-3-methoxyphenyl)-1,6-heptadiene-3,5-dione] is a natural polyphenolic compound derived from the rhizomes of the turmeric (*Curcuma longa* L.) plant. This polyphenol has been shown to possess anti-inflammatory, antioxidant, immunosuppressive, anti-proliferative, cell regulating, and pro-apoptotic activities against tumor cells and the tumor microenvironment *in vitro*. It induces apoptosis of tumor cells via signal transducer inhibition and activation of the STAT3 protein (Fahey et al. 2007). Unfortunately, curcumin is difficult to use as a chemotherapeutic compound due to its lipophilic properties, relative insolubility in water solutions, and low bioavailability when administered orally. On the plus side, its lipid solubility and hydrophobic nature enable it to locate itself in the protein–lipid membrane bilayer and to protect it against lipid peroxyl radicals (lipid peroxidation) and thereby to guard cells against apoptosis or necrosis caused by cell membrane damage (Sebastià et al. 2014). Nevertheless, it is advisable to search for new forms of this polyphenol in order to avoid the need to administer it in the high doses necessary for a significant effective response. One solution that has been proposed is to create a colloid-based preparation of curcumin which acts via the same mechanisms as the native form but is safe and has a better bioavailability than the parental compound (Milano et al. 2013).

The antitumor pleiotropic activity of curcumin has been shown to be mediated through inhibition of the NF-κB and AKT pathways and downstream genes and proteins including: Bcl-2, COX-2, TNF-α, IL-1β, MMPs, and NOS, which are engaged in the proliferation and invasion of tumor cells. This polyphenol also influences direct and paracrine interactions between CAFs and tumor cells. Interference with these interactions by curcumin leads to decreased production of epithelial–mesenchymal transition (EMT) mediators by tumor CAFs and inhibition of EMT in tumor cells (Dudás et al. 2013).

Curcumin has been shown to act as a modulator in leukemia treatment, mainly via strong enhancement of O_2^- generation in leukemic cells. It has been suggested that besides its immuno-potentiating activity, curcumin can also increase the level of free radicals, thus protecting patients against the proliferation of harmful microflora which could induce additional, accompanying infections in patients (Kikuchi et al. 2010). This polyphenol also induces hemoxygenase-1 (HO-1), an enzyme which, apart from having strong antioxidant as well as antiangiogenic properties, limits apoptotic death of cells. Moreover, it has been pointed out that normal cells treated with curcumin may exhibit higher resistance to apoptosis than tumor cells (Andreadi et al. 2006). This may be relevant in the context of possible use of this polyphenol as a supporting agent in clinical therapeutic schemes. Another advantage of curcumin (and resveratrol) for use in the clinical setting is that it possesses radiomodulatory activity, which means it can radioprotect normal cells and radiosensitize tumor ones. This effect has been traced to inhibition of the cell cycle at the G2-phase after radiation (Sebastià et al. 2014).

Curcumin limits the production of inflammatory cytokines by macrophages and enhances T-cell response to IFN-β (Fahey et al. 2007). Moreover, it up-regulates the expression of the co-stimulatory molecule CD86 in DCs and thus improves T-cell activation and conversion to effector T lymphocytes (CTLs), which eradicate tumor cells. Generally, curcumin improves the 'immune phenotype' of DCs and reduces the production of anti-inflammatory cytokines by activated T lymphocytes (Milano et al. 2013).

Curcumin is capable of restoring effector CD4+ and CD8+ T lymphocytes in the circulation of oncological patients, especially in the tumor niche. The depletion of Th1 lymphocytes and increased levels of Th2 cells observed in those patients suggest that curcumin may regulate Th1/Th2 status. This effect is further supported by the suppressive activity of curcumin towards Tregs, which is connected with TGF-β and IL-10 level reduction. The activity of TIL cells, and especially the Th1/Tc1 group, augmented by curcumin treatment plays an important role in antitumor immune response, as it potentiates tumor cell killing (Bhattacharyya et al. 2010).

Curcumin also strongly influences B-cells, down-regulating B-cell-tropic chemokines such as the CXC chemokine receptor (CXC4) and its ligand Stromal cell-Derived Factor-1 (SDF-1). Such activity modulates B-cell malignancies such as chronic lymphocytic or acute lymphoblastic leukemia. Moreover, curcumin modulates the expression of CD20, a specific marker of B-cells, chemokine (C–C motif) ligand 2 (CCL2), which is a chemoattractant for macrophages, cyclophilin D encoded by PPIF, which takes part in mitochondrial membrane permeability transition and thus apoptotic or necrotic death of targeted cells, and proteins from the thyrosine phosphatase family (PTPs), which regulate signal transduction pathways (Skommer et al. 2007). To conclude, curcumin can be taken into consideration as a tumor suppressing factor regulating multiple molecular pathways and cell-mediated immune responses.

5.3.1.3 Quercetin

Quercetin (3,3′,4′,5,7-pentahydroxyflavone) is a dietary flavonoid which can be found in fruits, such as apples, and vegetables, such as onions (Hao et al. 2017).

Quercetin can suppress TNF-α secretion by DCs and thus protect the local microenvironment from chronic inflammation. Moreover, this polyphenol induces the expression of the innate immunity protein called secretory leukocyte protease inhibitor (Slpi), which is secreted by DCs, neutrophils, and macrophages. This protein protects tissues against prolonged inflammatory conditions and favors tissue repair. Therefore, administration of quercetin may also be useful in reducing the development of cancer as an inflammatory disease (De Santis et al. 2016).

Quercetin synergizes with the TNF-related apoptosis-inducing ligand Apo2L/TRAIL to induce cell death in aggressive B-lymphomas. Therefore, use of this polyphenol may bring tangible benefits in the treatment of non-Hodgkin B-cell lymphoma (Jacquemin et al. 2012). Quercetin is also known to inhibit ROS, NO, and inflammatory cytokines by stimulating macrophages and to enhance the release of anti-inflammatory cytokines (Lara-Guzman et al. 2012, Kim et al. 2016). Moreover, quercetin activates the SIRT1 anti-inflammatory protein, inhibiting macrophage polarization and further inflammation (Dong et al. 2014).

Quercetin limits the amount of Mac-3 and CD11b markers, thus reducing the amount of the precursors of macrophages and T-cells, and simultaneously promotes cells which have the CD19 cluster, leading to strong stimulation of B-cell production. Quercetin also stimulates macrophage phagocytosis as well as promoting NK cell activity and thus lysis of transformed cells. Therefore, this polyphenol may be considered as an agent promoting immune response in normal and pathologic tissues (Yu et al. 2010).

5.3.1.4 Green Tea Polyphenols (GTPs)—(–)-Epigallocatechin 3-gallate (EGCG)

Among many catechins present in green tea, EGCG has been reported to express the most potent chemopreventive, photoprotective, and anti-tumor effects modulating the innate and adaptive immune responses (Katiyar et al. 2007). EGCG has been observed to selectively induce time-dependent apoptosis of tumor cells without influencing normal ones. Additionally, it takes part in anti-tumor activity by modulating ROS (Noda et al. 2007). However, there is no agreement in the literature whether this compound expresses pro-oxidant or antioxidant properties (Nakazato et al. 2005). Moreover, its usefulness, like that of other polyphenols, is limited due to the low bioavailability and low stability it shows in aqueous and alkaline solutions (Osanai et al. 2007). Therefore, topical treatment with this polyphenol may exert a stronger effect on tissues than when it is administered orally.

The pro-apoptotic mechanism of EGCG is connected with its potent inhibitory action on the Bcl-2 protein. Moreover, this polyphenol downregulates the expression of antiapoptotic proteins (MCL-1, XIAP) and Bax movement from the cytosol to mitochondria activating the caspase-3 and -9 cascade, blocking VEGF receptors (R1 and R2), and downregulating the CD31 antigen, MMP-2, and MMP-9 and thus sensitizing tumor cells to apoptosis and decreasing metastatic spread of these cells (Mantena et al. 2005, Katiyar et al. 2007, Shanafelt et al. 2009). These mechanisms are also shown to be correlated with decreased cellular proliferation resulting from inhibition of the G0-G1 cell cycle phase. Because EGCG possesses hydroxyl functional groups in its structure, it may be implicated in ROS activity.

It has been shown that EGCG influences cellular immune response by affecting the widely understood population of leukocytes. This compound stimulates macrophages by increasing expression of CD3 molecules on T-cells and CD11b+/CD18 molecules on macrophages, leading to potentiated macrophage lytic effects and phagocytosis. It has also been demonstrated that EGCG can reverse the immunosuppressive activity of MDSCs and ultimately reduce tumor development. This effect is connected with reduction of the levels of TGF-β and PGE2 mediators and adhesion receptors, as well as enhanced arginase production and Treg cell recruitment (Orentas 2013) Moreover, EGCG promotes the B-cell population by increasing the expression of the CD19 marker on these cells. By taking part in all the processes mentioned, EGCG stimulates the immune response, including anti-tumor immune reactions mediated by cytotoxic T-cells (CD8+) and B-cells (Huang et al. 2013). Increased numbers of these cells augment the pool of tumor infiltrating lymphocytes (TILs) which, by acting in the tumor microenvironment, potentiate immunosurveillance and may reduce the tumor mass and its invasiveness (Mantena et al. 2005, Katiyar et al. 2007). This is additionally supported by the fact that EGCG down-regulates IL-10 levels and induces IL-12 production by macrophages, thus promoting antitumor M1 polarization of these cells. In turn, IL-12 strongly influences T-cells, augmenting the Th1-type immune response and secretion of IFN-γ, which expresses anti-proliferative activity leading to growth inhibition and apoptosis induction in tumor cells.

5.3.1.5 Apigenin

Apigenin (4′,5,7-tyrihydroxyflavone) is a cell-permeable polyphenol present in dietary plants, such as parsley, fruits, and herbs (Kang et al. 2009, Erdogan et al. 2016). This flavonoid has been shown to inhibit cancer cell growth by inducing apoptosis via both intrinsic and extrinsic pathways and arresting cell cycle mainly in the G0/G1 phase via regulation/induction of p21 and p27 inhibitors of cyclin-dependent kinases. Apigenin has been observed to upregulate the levels of caspase-8 and -3, Apaf-1, Bcl-2, and TNF-α, leading to apoptotic death of tumor cells. Moreover, by decreasing MMP-2 and MMP-9 levels, apigenin strongly influences the migratory capacity of tumor cells (Erdogan et al. 2016). A relationship has also been suggested between apigenin and macrophage M1/M2 polarization balance, which is achieved through modulating the activation of peroxisome proliferator-activated receptor γ (PPARγ) (Feng et al. 2016). In addition, apigenin can block COX-2 expression without influencing its activity in tumor cells, which is important in inflammation, auto-immunity, and the functioning of immune cells. Decreased autoimmune activity of Th1, Th17, and IL-6 has been observed after treatment with apigenin.

The chemopreventive activity of apigenin is based on modulation of many molecular pathways. This polyphenol has been reported to inhibit the NF-κB pathway in T-cells and macrophages. Moreover, it has been shown to significantly limit autoantigen-presenting functions of Antigen Presenting Cells (APCs), which are necessary for Th and B-cells activation. Apigenin suppresses the activation of NF-κB through the Pl3K-Akt pathway and, as a result, down-regulates NF-κB-regulated anti-apoptotic pathways (Kang et al. 2009). Moreover, apigenin inhibits tumor growth and Cancer Stem Cell (CSC) development and activity through many pathways, which suggests its potential usefulness in cancer treatment (Erdogan et al. 2016).

5.3.1.6 Silibinin

Silibinin is a bioactive flavonolignan, a constituent of silymarin extracted from blessed milk thistle (*Silybum marianum*). It is widely used to treat many ailments and has been reported to possess pleiotropic antitumor activities in various cancer cells. This polyphenol down-regulates MMP expression and activity, extracellular matrix composition (fibronectin-induced motility), and EMT transition, thereby reducing tumor cell invasion, modulating inflammatory pathways, and activating extrinsic and intrinsic apoptotic pathways in tumor cells (Lin et al. 2012, Deep et al. 2014). Silibinin also limits the invasiveness of tumor cells induced by cancer-associated fibroblasts (CAFs), which are an important part of cancer stroma. CAFs are irreversibly and constitutively activated and are considered to be markers of poor prognosis for the patient. Silibinin acts on those cells by limiting their secretion of fibronectin, vimentin, and other matrix proteins. Moreover, CAFs treated with this compound have been found to exhibit decreased expression of TGFβR2 and other typical molecules such as α-Smooth Muscle Actin (α-SMA). In consequence, invasiveness and viability of tumor cells are reduced. Silibinin has also been found to inhibit transformation of naïve fibroblasts into cells expressing the CAF phenotype. In consequence it reduces recruitment of pro-tumor immune cells and restricts the formation of new vessels, decreasing the probability of tumor cells spreading to other organs (Ting et al. 2016). Silibinin also restricts tumor volume and the number of MDSCs (CD11b+Gr-1+) which infiltrate the tumor mass and circulate in blood. This polyphenol is considered to be an anti-inflammatory compound which, by reducing inflammation, also affects the accumulation of MDSCs in tumors. This process may be connected with decreased expression of CCR2 on MDSC cells and decreased immunosuppressive activity. What is more, silibinin increases the number of T-cells which infiltrate the tumor microenvironment. This is the main influence of this polyphenol on immune response during tumor development because an increase in TILs is closely followed by a reduction in MDSCs. Silibinin has also been reported to exert antitumor activity by guiding polarization of tumor microenvironmental macrophages into the M1 phenotype (Forghani et al. 2014).

5.3.1.7 Other Polyphenols

The most abundant polyphenolic compound found in medicinal plants (*Ilex paraguariensis, Bacharis genistelloides, Pimpinella anisum, Achyrochine satureioides, Camellia sinensis, Melissa officinalis,* and *Cymbopogon citratusis*) is chlorogenic acid (CGA), which belongs to the group of (–)-quinic acid esters of hydroxycinnamic acids (caffeic acid, in this case). It has been found that chlorogenic acid possesses anxiolytic properties connected with its antioxidant and possibly anti-inflammatory actions. It protects blood granulocytes from oxidative stress and thus facilitates their lytic functions during infiltration of tumors (Bouayed et al. 2007, Marques and Farah 2009).

Another natural polyphenol which can directly or indirectly influence tumor microenvironment functions is ellagic acid along with its derivatives found in high amounts in *Epilobium hirsitum L. (EH), Terminalia ferdinandiana,* and *Reaumuria vermiculata.* It has been found that this polyphenol induces apoptosis and decreases MMP release in the tumor mass. It has been observed to exert toxic effects on B-lymphocytes isolated from chronic lymphocytic leukemia (CCL) patients, at the same time having limited influence on normal B-cells. Therefore, this compound may be considered as a potential anticancer molecule whose mode of action is targeted mainly at mitochondrial pathways (Salimi et al. 2015, Karakurt et al. 2016).

Kaempferol is another polyphenol compound found in many edible plants, such as broccoli, cabbage, beans, tomato, strawberries, and grapes and in common medicinal plants such as *Ginkgo biloba, Equisetum* spp., *Moringa oleifera,* and propolis. It functions as an anti-oxidant, anti-inflammatory, and thus an anti-cancer agent. It inhibits tumor growth by suppressing VEGF production and ultimately by reducing the occurrence of neoangiogenesis and lymphangiogenesis (Calderón-Montaño et al. 2011, Lin et al. 2015).

Luteolin (3′,4′,5,7-tetrahydroxyflavone), found in *Cuminum cyminum* L., *Bacopa monnieri* (L.) Pennell, or *Achillea millefolium* L. has also been reported to express anti-cancer effects (Srinivasa et al. 2004, Lee et al. 2017). It has been suggested that this flavonoid can suppress TAM Receptor Tyrosine Kinases (RTKs), which are the molecular target of its anti-tumor activity. TAM RTKs are so important because they transduce signals connected with growth, proliferation, and general survival of cells, including tumor ones. Therefore, the inhibitory activity of luteolin on these molecules may be considered as a typical anti-cancer action. (Lee et al. 2017). On the other hand, it has been shown that flavonoids such as luteolin or rutin can facilitate tumor growth by strongly influencing innate and adaptive immunity. In particular, they have been found to reduce the migration of anti-tumor immune cells into tumors and consequently increase the occurrence of metastases. On this basis, it could be concluded that it is newly recruited macrophages but not residential ones that are responsible for the anti-metastatic effect of those polyphenols (van der Bij et al. 2008).

Finally, propolis, with its two important immunomodulatory polyphenolic compounds, caffeic acid (CAPE) and artepilin C, should be mentioned. It has been found that propolis exerts its effects both *in vitro* and *in vivo*. It enhances the cytotoxic action of M1 macrophages against tumor cells, decreases the level of immunostimulatory cytokines, reduces the population of cancer stem cells (CSCs), diminishes the angiogenic potential of tumors, reduces signal transduction in tumor cells, and induces their apoptotic death. CAPE has been found to inhibit the function of T lymphocytes and maturation of DCs. It also suppresses the invasive capacity of tumor cells by inhibiting the expression of MMPs (MMP-2 and MMP-9), hypoxia-inducing factor-1α (HIF-1α), and VEGF. The strong influences propolis has on the tumor microenvironment may be viewed as potentially supporting the 'classic' anti-tumor actions (Chan et al. 2013).

This study shows that anti-tumor effects of polyphenols should be analyzed with special care, taking into consideration not only the general activity of those compounds but also the kind and stage of process that they influence in the tumor microenvironment. This is especially important because often minor changes in the concentration of polyphenols or duration of their action may produce an effect that is contrary to theoretically anticipated or experimentally demonstrated results that are likewise affected by the time (stage of tumor development) of application of polyphenol-based treatments.

One example that illustrates this claim well is the activity of chrysin (5,7-di-OH flavone). This is a natural flavonoid which has been identified as a regulator of peroxisome proliferator-activated receptor γ (PPARγ). Chrysin significantly limits the number of infiltrating macrophages and induces the M2 phenotype, which on the one hand expresses anti-inflammatory properties but on the other may promote tumor growth, as is the case with TAMs, which are mainly of the M2 phenotype (Feng et al. 2016). This polyphenol may thus regulate the M1/M2 status not necessarily to the benefit of the patient and may limit the anti-tumor immunity or, under certain circumstances, even lead to the development of the disease. A similar effect has been observed for 5,7,3′,4′,5′-pentamethoxyflavone (PMFA) which facilitates phenotype shift from M1 to M2 by regulating STAT1/STAT6 signaling. On the other hand, chrysin and PMFA reduce the levels of pro-inflammatory cytokines (IL-6, TNF-α, IL-1β) and in this way can limit the development of cancer, which is described as a typical inflammatory disease (Feng et al. 2014, Feng et al. 2016). Moreover, a synthetic analog of chrysin, 8-bromo-7-methoxychrisin (BrMC), has been found to reverse M2 macrophage polarization through inhibition of the NF-κB factor, suppression of cytokine secretion by these cells, and suppression of the CD163 marker, characteristic of cells originating from the monocyte/macrophage lineage (Sun et al. 2017). It has also been found that chrysin may enhance the cytotoxic and anti-tumor activity of NK cells (Lin et al. 2012).

The example of chrysin shows that, depending on the perspective from which the action of polyphenols is interpreted, sometimes different conclusions can be drawn. Hence, it is necessary to thoroughly examine the activity of individual flavonoids and conduct a multidirectional analysis of the obtained results.

Flavonoids are polyphenolic compounds with potent biological activity. They have anti-inflammatory properties, being able to inhibit the level of prostaglandin E2 (PGE2), IL-6, COX-2, or NO in the tumor stroma. These effects have been shown for isorhamnetin, kaempferol, quercetin, genistein, daidzein, amentoflavone, and isoliquiritigenin (ISL) and other similar compounds. Flavonoids fight inflammation by reducing M2 macrophage polarization and thus limiting interactions of IL-6 with PGE2 or limiting IL-6-induced COX-2 and Arg-1 expression in macrophages. Moreover, they can down-regulate NF-κB activity to inhibit iNOS expression and, ultimately, NO production. Limiting of the amount of this molecule affects, in turn, the functioning of the macrophages. In consequence, the activity of flavonoids leading to the limitation of M2 transformation may prevent tumorigenesis (Hämäläinen et al. 2007, Guruvayoorappan and Kuttan 2008, Zhao et al. 2014, Wang et al. 2015).

Flavonoids such as naringenin or myricetin can also exert anti-tumor effects by potentiating the activity of NK cells. This activity is dose-dependent and involves induction of NKG2D ligands expressed on NK cells. This expression stimulates anti-tumor immunity manifesting *inter alia* in enhanced sensitivity of tumor cells to the cytotoxic action of the activated NK cells (Lindqvist et al. 2014, Kim and Lee 2015).

5.4 CONCLUSIONS

The complex structure of solid tumors prompts scientists to look for new, unconventional therapeutic approaches showing a better efficiency or to seek methods of treatment that could support classic cancer therapies in order to improve clinical outcomes. In recent years, the attention of researchers has been drawn to tumor–stromal interactions, as more and more studies have shown that cancer cells may not be the sole therapeutic target and that stromal cells, which regulate the immune anti-cancer response, should be an equally important focus of treatment interventions. Unchanged intracellular signaling of stromal cells generally implies that the patient's immune system, both innate and adaptive, is effectively combating transformed cells. Normal signaling, in other words, is a sign of a good response to clinical anti-cancer therapy. On the other hand, if the cancer microenvironment transforms the response of infiltrating immune cells to a therapy-resistant pro-tumor response, then the fight against cancer becomes much more difficult. This is why researchers strive to find novel anti-tumor strategies which could be used, often in combination with conventional therapies, to improve clinical outcomes. One of the popular research trends regarding cancer therapy is the study of natural substances, including polyphenols, which can fine-tune the immune system to combat cancer. One of the central issues in this trend is the links among polyphenols, inflammation, and the immune cells which lead to cancer progression or its suppression. These relationships involve various types of immune cells such as macrophages, MDSCs, T and B lymphocytes, NK cells, DCs, granulocytes, and CAFs and their mediators, including cytokines, growth factors, prostaglandins, and leukotrienes. Discovery of the exact mechanisms governing the regulatory effects of polyphenols on the immune cells infiltrating tumors, targeted for anti-cancer activity, may be an important advance in the treatment of this disease.

As natural secondary metabolites, polyphenols target various different mechanisms and molecular pathways connected with tumorigenesis, starting from tumor cell proliferation and ending with metastasis and therapy resistance. However, these compounds do not always act as absolute anti-cancer agents. Depending on many factors, they may promote immune response focused solely on regulating certain processes taking place in the cancer microenvironment. Promise of improvement of the bioavailability of polyphenols and the mechanism of their action is seen in rational modification of their chemical structure and synthesizing of new derivatives with better biological properties, enhanced bioactivity, and improved delivery systems. Unfortunately, the promising results obtained in *in vitro* studies are often not confirmed or substantiated in *in vivo* assays. This may be linked with uncorrelated concentrations of compounds used in the two types of models, conditions of administration, experimental methodology, or simply significant differences in the complexity of the two systems. For the time being, polyphenols remain promising agents that can be applied in personalized therapies of oncological patients as well as in supplementary therapeutic procedures supporting regular therapies or simply used as chemopreventive ingredients to supplement the patient's diet.

REFERENCES

Andreadi, C.K., L.M. Howells, P.A. Atherfold, and M.M. Manson. 2006. Involvement of Nrf2, p38, B-Raf, and nuclear factor-κB, but not phosphatidylinositol 3-kinase, in induction of hemeoxygenase-1 by dietary polyphenols. *Mol. Pharmacol.* 69:1033–1040.

Apaya, M.K., M.T. Chang, and L.F. Shyur. 2016. Phytomedicine polypharmacology: Cancer therapy through modulating the tumor microenvironment and oxylipin dynamics. *Pharmacol. Ther.* 162:58–68.

Asensi, M., A. Ortega, S. Mena, F. Feddi, and J.M. Estrela. 2011. Natural polyphenols in cancer therapy. *Crit. Rev. Clin. Lab. Sci.* 48:197–216.

Azmi, A.S., S.H. Bhat, and S.M. Hadi. 2005. Resveratrol-Cu(II) induced DNA breakage in human peripheral lymphocytes: Implications for anticancer properties. *FEBS Lett.* 579:3131–3135.

Azmi, A.S., S.H. Bhat, S. Hanif, and S.M. Hadi. 2006. Plant polyphenols mobilize endogenous copper in human peripheral lymphocytes leading to oxidative DNA breakage: A putative mechanism for anticancer properties. *FEBS Lett.* 580:533–538.

Benencia, F., L. Sprague, J. McGinty, M. Pate, and M. Muccioli. 2012. Dendritic cells the tumor microenvironment and the challenges for an effective antitumor vaccination. *J. Biomed. Biotechnol.* 2012:Article ID 425476, 15 p.

Bhattacharyya, S.D., D. Md Sakib Hossain, S. Mohanty, et al. 2010. Curcumin reverses T cell-mediated adaptive immune dysfunctions in tumor-bearing hosts. *Cell. Mol. Immunol.* 7:306–315.

Bouayed, J., H. Rammal, A. Dicko, Ch. Younos, and R. Soulimani. 2007. Chlorogenic acid, a polyphenol from *Prunusdomestica* (Mirabelle), with coupled anxiolytic and antioxidant effects. *J. Neurol. Sci.* 262:77–84.

Bravo, L. 1998. Polyphenols: Chemistry, dietary sources, metabolism, and nutritional significance. *Nutr. Rev.* 56:317–333.

Calderón-Montaño, J.M., E. Burgos-Morón, C. Pérez-Guerrero, and M. López-Lázaro. 2011. A review on the dietary flavonoid kaempferol. *Mini Rev. Med. Chem.* 11:298–344.

Chan, G.Ch.-F., K.W. Cheung, and D.M.-Y. Sze. 2013. The immunomodulatory and anticancer properties of propolis. *Clin. Rev. Allerg. Immunol.* 44:262–273.

Chen, C.M., S.C. Li, Y.L. Lin, C.Y. Hsu, M.J. Shieh, and J.F. Liu. 2005. Consumption of purple sweet potato leaves modulates human immune response: T-lymphocyte functions, lytic activity of natural killer cell and antibody production. *World J. Gastroenterol.* 11:5777–5781.

Cui, X., Y. Jin, A.B. Hofseth, et al. 2010. Resveratrol suppresses colitis and colon cancer associated with colitis. *Cancer Prev. Res.* 3:549–559.

De Santis, S., D. Kunde, G. Serino, et al. 2016. Secretory leukoprotease inhibitor is required for efficient quercetin-mediated suppression of TNFα secretion. *Oncotarget* 7:75800–75809.

Deep, G., R. Kumar, A.K. Jain, Ch. Agarwal, and R. Agarwal. 2014. Silibinin inhibits fibronectin induced motility, invasiveness and survival in human prostate carcinoma PC3 cells via targeting integrin signaling. *Mutat. Res.* 768:35–46.

Del Cornò, M., B. Scazzocchio, R. Masella, and S. Gessani. 2016. Regulation of dendritic cell function by dietary polyphenols. *Crit. Rev. Food Sci. Nutr.* 56:737–747.

Delmas, D., C. Rébé, A. Hichami, and F. Ghiringhelli. 2013. Polyphenols as immunomodulators to fight cancers. In: *Polyphenols: Chemistry, Dietary Sources and Health Benefits.* Jian Sun, K. Nagendra Prasad, Amin Ismail, Bao Yang, Xiangrong You and Li Li (Eds.). Nova Science Publishers, Inc, Hauppauge, NY, U.S.A. 539–568.

Dholwani, K.K., A.K. Saluja, A.R. Gupta, and D.R. Shah. 2008. A review on plant-derived natural products and their analogs with anti-tumor activity. *Indian J. Pharmacol.* 40:49–58.

Dong, J., X. Zhang, L. Zhang, et al. 2014. Quercetin reduces obesity-associated ATM infiltration and inflammation in mice: A mechanism including AMPKα1/SIRT1. *J. Lipid Res.* 55:363–374.

Dudás, J., A. Fullár, A. Romani, et al. 2013. Curcumin targets fibroblast-tumor cell interactions in oral squamous cell carcinoma. *Exp. Cell Res.* 319:800–809.

Erdogan, S., O. Doganlar, Z.B. Doganlar, et al. 2016. The flavonoid apigenin reduces prostate cancer CD44+ stem cell survival and migration through PI3K/Akt/NF-κB signaling. *Life Sci.* 162:77–86.

Fahey, A.J., R.A. Robins, and C.S. Constantinescu. 2007. Curcumin modulation of IFN-β and IL-12 signalling and cytokine induction in human T cells. *J. Cell. Mol. Med.* 11:1129–1137.

Falchetti, R., M.P. Fuggetta, G. Lanzilli, M. Tricarico, and G. Ravagnan. 2001. Effects of resveratrol on human immune cell function. *Life Sci.* 70:81–96.

Feng, L., P. Song, H. Zhou, et al. 2014. Pentamethoxyflavanone regulates macrophage polarization and ameliorates sepsis in mice. *Biochem. Pharmacol.* 89:109–118.

Feng, X., D. Weng, F. Zhou, et al. 2016. Activation of PPARγ by a natural flavonoid modulator, apigenin ameliorates obesity-related inflammation via regulation of macrophage polarization. EBio Medicine 9:61–76.

Ferry-Dumazet, H., O. Garnier, M. Mamani-Matsuda, et al. 2002. Resveratrol inhibits the growth and induces the apoptosis of both normal and leukemic hematopoietic cells. *Carcinogenesis* 23:1327–1333.

Forghani, P., M.R. Khorramizadeh, and E.K. Waller. 2014. Silibinin inhibits accumulation of myeloid-derived suppressor cells and tumor growth of murine breast cancer. *Cancer Med.* 3:215–224.

Francisco, V., G. Costa, B.M. Neves, M.T. Cruz, and M.T. Batista. 2014. Anti-inflammatory activity of polyphenols on dendritic cells. In: *Polyphenols in Human Health and Disease.* R. R. Watson, V. Preedy, and S. Zibadi (Eds.). Elsevier Inc., Oxford, United Kingdom. 373–392.

Ghiringhelli, F., C. Rebe, A. Hichami, and D. Delmas. 2012. Immunomodulation and anti-inflammatory roles of polyphenols as anticancer agents. *Anticancer Agents Med. Chem.* 12:852–873.

Gout, S., and J. Huot. 2008. Role of cancer microenvironment in metastasis: Focus on colon cancer. *Cancer Microenviron.* 1:69–83.

Guan, H., N.P. Singh, U.P. Singh, P.S. Nagarkatti, and M. Nagarkatti. 2012. Resveratrol prevents endothelial cells injury in high-dose interleukin-2 therapy against melanoma. *PLOS ONE* 7:e35650, 12 p.

Guruvayoorappan, C., and G. Kuttan. 2008. Amentoflavone stimulates apoptosis in B16F-10 melanoma cells by regulating bcl-2, p53 as well as caspase-3 genes and regulates the nitric oxide as well as proinflammatory cytokine production in B16F-10 melanoma cells, tumor associated macrophages and peritoneal macrophages. *J. Exp. Ther. Oncol.* 7:207–218.

Hamada, S., A. Masamune, and T. Shimosegawa. 2013. Novel therapeutic strategies targeting tumor-stromal interactions in pancreatic cancer. *Front. Physiol.* 4:Article 331, 7 p.

Hämäläinen, M., R. Nieminen, P. Vuorela, M. Heinonen, and E. Moilanen. 2007. Anti-inflammatory effects of flavonoids: Genistein, kaempferol, quercetin, and daidzein inhibit STAT-1 and NF-kappaB activations, whereas flavone, isorhamnetin, naringenin, and pelargonidin inhibit only NF-kappaB activation along with their inhibitory effect on iNOS expression and NO production in activated macrophages. *Mediators Inflamm.* 2007:Article 45673, 10 p.

Hao, J., B. Guo, S. Yu, et al. 2017. Encapsulation of the flavonoid quercetin with chitosan-coated nanoliposomes. *LWT Food Sci. Technol.* 85:37–44.

Hasmim, M., Y. Messai, L. Ziani, et al. 2015. Critical role of tumor microenvironment in shaping NK cell functions: Implication of hypoxic stress. *Front. Immunol.* 6:Article 482, 9 p.

Hu, L., D. Cao, Y. Li, Y. He, and K. Guo. 2012. Resveratrol sensitized leukemia stem cell-like KG-1a cells to cytokine-induced killer cells-mediated cytolysis through NKG2D ligands and TRAIL receptors. *Cancer Biol. Ther.* 13:516–526.

Huang, A.C., H.Y. Cheng, T.S. Lin, et al. 2013. Epigallocatechin gallate (EGCG), influences a murine WEHI-3 leukemia model *in vivo* through enhancing phagocytosis of macrophages and populations of T- and B-cells. *In Vivo* 27:627–634.

Jacquemin, G., V. Granci, A.S. Gallouet, et al. 2012. Quercetin-mediated Mcl-1 and survivin downregulation restores TRAIL-induced apoptosis in non-Hodgkin's lymphoma B cells. *Haematologica* 97:38–46.

Kang, H.K., D. Ecklund, M. Liu, and S.K. Datta. 2009. Apigenin, a non-mutagenic dietary flavonoid, suppresses lupus by inhibiting autoantigen presentation for expansion of autoreactive Th1 and Th17 cells. *Arthritis Res. Ther.* 11:R59, 13 p.

Karakurt, S., A. Semiz, G. Celik, A.M. Gencler-Ozkan, A. Sen, and O. Adali. 2016. Contribution of ellagic acid on the antioxidant potential of medicinal plant *Epilobiumhirsutum*. *Nutr. Cancer* 68:173–183.

Karimi, S., S. Chattopadhyay, and N.G. Chakraborty. 2015. Manipulation of regulatory T cells and antigen-specific cytotoxic T lymphocyte-based tumour immunotherapy. *Immunology* 144:186–196.

Katiyar, S., C.A. Elmets, and S.K. Katiyar. 2007. Green tea and skin cancer: Photoimmunology, angiogenesis and DNA repair. *J. Nutr. Biochem.* 18:287–296.

Kikuchi, H., F. Kuribayashi, N. Kiwaki, and T. Nakayama. 2010. Curcumin dramatically enhances retinoic acid-induced superoxide generating activity via accumulation of p47-phox and p67-phox proteins in U937 cells. *Biochem. Biophys. Res. Commun.* 395:61–65.

Kim, C.S., H.S. Choi, Y. Joe, H.T. Chung, and R. Yu. 2016. Induction of heme oxygenase-1 with dietary quercetin reduces obesity-induced hepatic inflammation through macrophage phenotype switching. *Nutr. Res. Pract.* 10:623–628.

Kim, J.H., and J.K. Lee. 2015. Naringenin enhances NK cell lysis activity by increasing the expression of NKG2D ligands on Burkitt's lymphoma cells. *Arch. Pharm. Res.* 38:2042–2048.

Kim, Y.S., T.J. Sayers, N.H. Colburn, J.A. Milner, and H.A. Young. 2015. Impact of dietary components on NK and Treg cell function for cancer prevention. *Mol. Carcinog.* 54:669–678.

Kopeć-Szlęzak, J. 2012. Rola komórek układu odpornościowego w mikrośrodowisku nowotworów. *Postępy Nauk Med.* s3:15–21.

Kopeć-Szlęzak, J. 2016. Nowe subpopulacje limfocytów T pomocniczych CD4+. *Postępynauk Med* s2:126–131.

Krifa, M., I. Bouhlel, L. Ghedira-Chekir, and K. Ghedira. 2013. Immunomodulatory and cellular antioxidant activities of an aqueous extract of *Limoniastrumguyonianum* gall. *J. Ethnopharmacol.* 146:243–249.

Lara-Guzman, O.J., J.H. Tabares-Guevara, Y.M. Leon-Varela, et al. 2012. Proatherogenicmacrophage activities are targeted by the flavonoid quercetin. *J. Pharmacol. Exp. Ther.* 343:296–306.

Lee, J.B., and Y.O. Shin. 2014. Oligonol supplementation affects leukocyte and immune cell counts after heat loading in humans. *Nutrients* 6:2466–2477.

Lee, Y.J., T. Lim, M.S. Han, et al. 2017. Anticancer effect of luteolin is mediated by downregulation of TAM receptor tyrosine kinases, but not interleukin-8, in non-small cell lung cancer cells. *Oncol. Rep.* 37:1219–1226.

Lee-Chang, C., M. Bodogai, A. Martin-Montalvo, et al. 2013. Inhibition of breast cancer metastasis by resveratrol-mediated inactivation of tumor-evoked regulatory B cells. *J. Immunol.* 191:4141–4151.

Li, J., J. Li, and F. Zhang. 2015. The immunoregulatory effects of Chinese herbal medicine on the maturation and function of dendritic cells. *J. Ethnopharmacol.* 171:184–195.

Li, Q., T. Huyan, L.J. Ye, J. Li, J.L. Shi, and Q.S. Huang. 2014. Concentration-dependent biphasic effects of resveratrol on human natural killer cells *in vitro*. *J. Agric. Food Chem.* 62:10928–10935.

Li, T., G.X. Fan, W. Wang, T. Li, and Y.K. Yuan. 2007. Resveratrol induces apoptosis, influences IL-6 and exerts immunomodulatory effect on mouse lymphocytic leukemia both *in vitro* and *in vivo*. *Int. Immunopharmacol.* 7:1221–1231.

Lin, C.C., C.S. Yu, J.S. Yang, et al. 2012. Chrysin, a natural and biologically active flavonoid, influences a murine leukemia model *in vivo* through enhancing populations of T-and B-cells, and promoting macrophage phagocytosis and NK cell cytotoxicity. *In Vivo* 26:665–670.

Lin, C.M., Y.H. Chen, H.P. Ma, et al. 2012. Silibinin inhibits the invasion of IL-6-stimulated colon cancer cells via selective JNK/AP-1/MMP-2 modulation *in vitro*. *J. Agric. Food Chem.* 60:12451–12457.

Lin, F., X. Luo, A. Tsun, Z. Li, D. Li, and B. Li. 2015. Kaempferol enhances the suppressive function of Treg cells by inhibiting FOXP3 phosphorylation. *Int. Immunopharmacol.* 28:859–865.

Lindqvist, C., M. Bobrowska-Hägerstrand, L. Mrówczyńska, C. Engblom, and H. Hägerstrand. 2014. Potentiation of natural killer cell activity with myricetin. *Anticancer Res.* 34:3975–3979.

LoPresti, E., F. Dieli, and S. Meraviglia. 2014. Tumor-infiltrating γδ T lymphocytes: Pathogenic role, clinical significance, and differential programming in the tumor microenvironment. *Front. Immunol.* 5:Article 607, 8 p.

Luis Espinoza, J.L., A. Takami, L.Q. Trung, and S. Nakao. 2013. Ataxia-telangiectasia mutated kinase-mediated upregulation of NKG2D ligands on leukemia cells by resveratrol results in enhanced natural killer cell susceptibility. *Cancer Sci.* 104:657–662.

Manach, C., A. Scalbert, Ch. Morand, C. Rémésy, and L. Jiménez. 2004. Polyphenols: food sources and bioavailability. *Am. J. Clin. Nutr.* 79:727–747.

Mantena, S.K., S.M. Meeran, C.A. Elmets, and S.K. Katiyar. 2005. Orally administered green tea polyphenols prevent ultraviolet radiation-induced skin cancer in mice through activation of cytotoxic T cells and inhibition of angiogenesis in tumors. *J. Nutr.* 135:2871–2877.

Marques, V., and A. Farah. 2009. Chlorogenic acids and related compounds in medicinal plants and infusions. *Food Chem.* 113:1370–1376.

Milano, F., L. Mari, W. van de Luijtgaarden, K. Parikh, S. Calpe, and K.K. Krishnadath. 2013. Nano-curcumin inhibits proliferation of esophageal adenocarcinoma cells and enhances the T cell mediated immune response. *Front. Oncol.* 3:Article 137, 11 p.

Millimouno, F.M., J. Dong, L. Yang, J. Li, and X. Li. 2014. Targeting apoptosis pathways in cancer and perspectives with natural compounds from mother nature. *Cancer Prev. Res.* 7:1081–1107.

Montales, M.T.E., O.M. Rahal, J. Kang, et al. 2012. Repression of mammosphere formation of human breast cancer cells by soy isoflavonegenistein and blueberry polyphenolic acids suggests diet-mediated targeting of cancer stem-like/progenitor cells. *Carcinogenesis* 33:652–660.

Moutia, M., F. Seghrouchni, O. Abouelazz, et al. 2016. *Allium sativum* L. regulates *in vitro* IL-17 gene expression in human peripheral blood mononuclear cells. *BMC Complement. Altern. Med.* 16:377, 10 p.

Nakazato, T., K. Ito, Y. Ikeda, and M. Kizaki. 2005. Green tea component, catechin, induces apoptosis of human malignant B cells via production of reactive oxygen species. *Clin. Cancer Res.* 11:6040–6049.

Nelson, B.H. 2010. CD20+ B cells: The other tumor-infiltrating lymphocytes. *J. Immunol.* 185:4977–4982.

Nimgulkar, Ch., S. Ghosh, A.B. Sankar, et al. 2015. Combination of spices and herbal extract restores macrophage foam cell migration and abrogates the athero-inflammatory signalling cascade of atherogenesis. *Vasc. Pharmacol.* 72:53–63.

Noda, C., J. He, T. Takano, et al. 2007. Induction of apoptosis by epigallocatechin-3-gallate in human lymphoblastoid B cells. *Biochem. Biophys. Res. Commun.* 362:951–957.

Orentas, R.J. 2013. Reading the tea leaves of tumor-mediated immunosuppression. *Clin. Cancer Res.* 19:955–957.

Osanai, K., K.R. Landis-Piwowar, Q.P. Dou, and T.H. Chan. 2007. A para-amino substituent on the D-ring of green tea polyphenol epigallocatechin-3-gallate as a novel proteasome inhibitor and cancer cell apoptosis inducer. *Bioorg. Med. Chem.* 15:5076–5082.

Pak, F., M. Barati, M. Shokrolahi, and P. Kokhaei. 2014. Tumor immunology and tumor escape mechanisms from immune response. *Koomesh* 15:412–430.

Peddareddigari, V.G., D. Wang, and R.N. DuBois. 2010. The tumor microenvironment in colorectal cancinogenesis. *Cancer Microenviron.* 3:149–166.

Quiney, C., D. Dauzonne, C. Kern, et al. 2004. Flavones and polyphenols inhibit the NO pathway during apoptosis of leukemia B-cells. *Leuk. Res.* 28:851–861.

Quoc Trung, L., J.L. Espinoza, A. Takami, and S. Nakao. 2013. Resveratrol induces cell cycle arrest and apoptosis in malignant NK cells via JAK2/STAT3 pathway inhibition. *PLOS ONE* 8:e55183, 11 p.

Ramstead, A.G., I.A. Schepetkin, M.T. Quinn, and M.A. Jutila. 2012. Oenothein B, a cyclic dimeric ellagitannin isolated from *Epilobiumangustifolium*, enhances IFNγ production by lymphocytes. *PLOS ONE* 7:e50546, 10 p.

Ramstead, A.G., I.A. Schepetkin, K. Todd, et al. 2015. Aging influences the response of T cells to stimulation by the ellagitannin, oenothein B. *Int. Immunopharmacol.* 26:367–377.

Rieder, S.A., P. Nagarkatti, and M. Nagarkatti. 2012. Multiple anti-inflammatory pathways triggered by resveratrol lead to amelioration of staphylococcal enterotoxin B-induced lung injury. *Br. J. Pharmacol.* 167:1244–1258.

Sajkowska-Kozielewicz, J.J., P. Kozielewicz, N.M. Barnes, I. Wawer, and K. Paradowska. 2016. Antioxidant, cytotoxic, and antiproliferative activities and total polyphenol contents of the extracts of *Geissospermum reticulatum* bark. Oxid. *Med. Cell. Longev.* Article ID 2573580, 8 p.

Salimi, A., M.H. Roudkenar, L. Sadeghi, et al. 2015. Ellagic acid, a polyphenolic compound, selectively induces ROS-mediated apoptosis in cancerous B-lymphocytes of CLL patients by directly targeting mitochondria. *Redox Biol.* 6:461–471.

Santilli, G., J. Anderson, A.J. Thrasher, and A. Sala. 2013b. Catechins and antitumor immunity: Not MDSC's cup of tea. *OncoImmunology* 2:e24443, 3 p.

Santilli, G., I. Piotrowska, S. Cantilena, et al. 2013a. Polyphenol E enhances the antitumor immune response in neuroblastoma by inactivating myeloid suppressor cells. *Clin. Cancer Res.* 19:1116–1125.

Sebastià, N., A. Montoro, D. Hervás, et al. 2014. Curcumin and trans-resveratrol exert cell cycle-dependent radioprotective or radiosensitizing effects as elucidated by the PCC and G2-assay. *Mutat. Res.* 766–767:49–55.

Shanafelt, T.D., T.G. Call, C.S. Zent, et al. 2009. Phase I trial of daily oral polyphenon E in patients with asymptomatic rai stage 0 to II chronic lymphocytic leukemia. *J. Clin. Oncol.* 27:3808–3814.

Shen, M., Q. Sun, J. Wang, W. Pan, and X. Ren. 2016. Positive and negative functions of B lymphocytes in tumors. *Oncotarget* 7:55828–55839.

Shimizu, T., T. Nakazato, M.J. Xian, M. Sagawa, Y. Ikeda, and M. Kizaki. 2006. Resveratrol induces apoptosis of human malignant B cells by activation of caspase-3 and p38 MAP kinase pathways. *Biochem. Pharmacol.* 71:742–750.

Skommer, J., D. Wlodkowic, and J. Pelkonen. 2007. Gene-expression profiling during curcumin-induced apoptosis reveals downregulation of CXCR4. *Exp. Hematol.* 35:84–95.

Spagnuolo, C., M. Russo, S. Bilotto, I. Tedesco, B. Laratta, and G.L. Russo. 2012. Dietary polyphenols in cancer prevention: The example of the flavonoid quercetin in leukemia. *Ann. N. Y. Acad. Sci.* 1259:95–103.

Srinivasa, H., M.S. Bagul, H. Padh, and M. Rajani. 2004. A rapid densitometric method for the quantification of luteolin in medicinal plants using HPTLC. *Chromatographia* 60:131–134.

Sun, S., Y. Cui, K. Ren, et al. 2017. 8-Bromo-7-methoxychrysin reversed M2 polarization of tumor-associated macrophages induced by liver cancer stem-like cells. *Anticancer Agents Med. Chem.* 17:286–293.

Szala, S. 2007. Komórki mikrośrodowiska nowotworowego: Cel terapii przeciwnowotworowej. *Nowotwory J. Oncol.* 57:633–645.

Ting, H., G. Deep, S. Kumar, A.K. Jain, C. Agarwal, and R. Agarwal. 2016. Beneficial effects of the naturally occurring flavonoid silibinin on the prostate cancer microenvironment: Role of monocyte chemotactic protein-1 and immune cell recruitment. *Carcinogenesis* 37:589–599.

Tsao, R. 2010. Chemistry and biochemistry of dietary polyphenols. *Nutrients* 2:1231–1246.

van der Bij, G.J., M. Bögels, S.J. Oosterling, et al. 2008. Tumor infiltrating macrophages reduce development of peritoneal colorectal carcinoma metastases. *Cancer Lett.* 262:77–86.

Wang, Y.L., X. Tan, X.L. Yang, X.Y. Li, K. Bian, and D.D. Zhang. 2015. Total flavonoid from *Glycyrrhizae radix etrhizoma* and its ingredient isoliquiritigenin regulation M2 phenotype polarization of macrophages. *Zhongguo Zhong Yao Za Zhi* 40:4475–4481.

Wright, B. 2013. Forging a modern generation of polyphenol-based therapeutics. *Br. J. Pharmacol.* 169:844–847.

Yi, J., C. Cheng, X. Li, et al. 2017. Protective mechanisms of purified polyphenols from pinecones of *Pinuskoraiensis* on spleen tissues in tumor-bearing S180 mice *in vivo*. *Food Funct.* 8:151–166.

Yu, C.S., K.C. Lai, J.S. Yang, et al. 2010. Quercetin inhibited murine leukemia WEHI-3 cells *in vivo* and promoted immune response. *Phytother. Res.* 24:163–168.

Yu, P., and Y.X. Fu. 2006. Tumor-infiltrating T lymphocytes: Friends or foes? *Lab. Invest.* 86:231–245.

Zhang, H.G., H. Kim, C. Liu, et al. 2007. Curcumin reverses breast tumor exosomes mediated immune suppression of NK cell tumor cytotoxicity. *Biochim. Biophys. Acta* 1773:1116–1123.

Zhao, H., X. Zhang, X. Chen, et al. 2014. Isoliquiritigenin, a f3lavonoid from licorice, blocks M2 macrophage polarization in colitis-associated tumorigenesis through downregulating PGE2 and IL-6. *Toxicol. Appl. Pharmacol.* 279:311–321.

Zou, T., Y. Yang, F. Xia, et al. 2013. Resveratrol inhibits CD4+ T cell activation by enhancing the expression and activity of Sirt1. *PLOS ONE* 8:e75139, 12 p.

CHAPTER 6

Medicinal Plant Product-Based Fabrication Nanoparticles (Au and Ag) and Their Anticancer Effects

Azamal Husen

CONTENTS

ABBREVIATIONS

Au: Gold
Ag: Silver
NPs: Nanoparticles
TEM: Transmission electron microscopy
SEM: Scanning electron microscopy
XRD: X-ray diffraction
FTIR: Fourier transform infrared
AFM: Atomic force microscopy
EDX: Energy-dispersive X-ray
DLS: Dynamic light scattering
SERS: Surface-enhanced Raman scattering
ROS: Reactive oxygen species

6.1 INTRODUCTION

In recent years, nanotechnology has flourished rapidly and emerged as an interdisciplinary science of research with distinguished applications in various sectors (Husen and Siddiqi 2014a, b, c, Siddiqi et al. 2016, 2018a, Siddiqi and Husen 2017a, b, c, d). Among the studied novel metal and/ or metal-oxide nanoparticles (NPs), gold (Au) and silver (Ag) NPs gained much attention because they are used as antioxidants, antimicrobials, catalysts, in sensors including biosensors, and for the destruction of cancer cells, wound treatment, sterilization, food sanitation, X-ray, gene/drug delivery, and so on (Abasi et al. 2016, Chung et al. 2016, Siddiqi and Husen 2016a, b, Husen 2017, Karatoprak et al. 2017, Mattea et al. 2017, Siddiqi et al. 2018a). Commonly, Au- and Ag-NPs are synthesized using various kind of physical and chemical methods. Both methods are considered as fairly expensive, need high energy, and their by-products and wastes are toxic and harmful for the environment and human health; thus, biological methods of NPs synthesis are understood as simple, cost-effective, and environmentally friendly (Husen and Siddiqi 2014b, Husen 2017, Siddiqi et al. 2018b). Several bacteria, fungi, algae, and medical-plant (lower and higher plants) extracts are used as a biological system for the synthesis of NPs (Husen and Siddiqi 2014b, Abasi et al. 2016, Siddiqi and Husen 2016a, b, Karatoprak et al. 2017, Siddiqi et al. 2018d).

Cancer (uncontrolled cell growth) is one of the most dangerous diseases at the global level. It generates numerous pathological and metabolic alterations in the cellular environments. Almost each family is touched by cancer, which is now responsible for nearly one in six deaths globally (WHO 2018). According to WHO (2014a), World Cancer Report ~14 million new cancer cases and 8.2 million deaths were observed in 2012. It has also been stated that lung cancer leads to the greatest mortality (1.5 million deaths), followed by liver (745,000 deaths), stomach (723,000 deaths), colorectal (694,000 deaths), breast (521,000 deaths), and esophageal (400,000 deaths) cancer. In general, cancer cases end up with loss of life (Dite et al. 2010, Smith et al. 2011). During the next two decades, this global problem is likely to rise by 70%, from 14 million to 22 million new cases per year (WHO 2014b). Moten and coworkers (2014) have reported that the populations of Africa, Asia, and Central and South America are linked to 70% of all cancer deaths and 60% of the total new annual cancer cases at the global level.

Nanomaterials have now been commonly used for the effective treatment of human cancers. These particles are unique as they have a high surface area to volume ratio (small size) and are capable of easily entering the cell membranes and/or the biological barriers. These features of NPs increase the effectiveness against the tumor cells at lower concentrations and then decrease the toxicity to surrounding normal cells. Among the used metal NPs, Au- and Ag-NPs have been examined extensively against various cancer cell lines. Studies have also confirmed that the cancer cell growth has been significantly decreased in a concentration dependent manner after the treatment with biogenic synthesized Au- and or Ag-NPs. Thanh and coworkers (2002) have used Au-NPs and reported an exclusive, sensitive, and very specific immunoassay system for antibodies. Thanks to their colloidal nature, Au-NPs transmit drugs to the specific site of the use irrespective of their shape if both of them (Au-NPs and drugs) are biocompatible, stable, and easily form bonds with each other. They are finally deposited in cells and on irradiation with visible light, the heat mediated by Au-NPs destroys the cancerous tissues (Huff et al. 2007, Lowery et al. 2006). Au-NPs have been also found to enhance the intensity of Raman scattering of adjacent molecules and thus are used in surface-enhanced Raman scattering (SERS) for the detection and quantitative analysis of Raman active materials (Zamarion et al. 2008, Dasary et al. 2009, Ding et al. 2013). SERS is used to distinguish cancer cells from normal (noncancerous) cells, when Au nanorods conjugate to anti-epidermal growth factor receptor antibodies (Huang et al. 2007). SERS with Au-NPs has been employed in cancer research to detect tumors (Cai et al. 2008, Huang and El-Sayed 2010). In addition, Au and Ag nanocatalysts are found to be effective in cleaning cancer-causing dyes from water bodies using the electron-relay effects (Joseph and Mathew 2015, Lim et al. 2016). The role of Ag-NPs as an anticancer agent has also been investigated by numerous researchers (Guo et al. 2013, Satapathy et al. 2013, Ortega et al. 2015, Juarez-Moreno et al. 2017, Buttacavoli et al. 2018, Chugh et al. 2018, Dadashpour et al. 2018).

Natural products obtained from various medicinal plants are the most important source of drugs (Nautiyal et al. 2002, Mishra et al. 2011). There are several products from our day-to-day consumption of herbs, fruits, vegetables, tea beverages, etc., whose active ingredients have shown potential health benefits. For instance, Zecca et al. (2004) have reported that the plant-derived polyphenols are strongly involved in preventing neurodegenerative disorders, diabetes, and malignant conditions. In addition, several plant metabolites have revealed cytotoxic efficiency against many forms of cancerous cells. These medicinal plants, namely *Acalypha indica, Andrographis echioides, Catharanthus roseus, Carica papaya, Cassia tora, Clerodendrum phlomidis, Dimocarpus longan, Melia azedarch, M. dubia, Musa paradisiaca, Origanum vulgare, Piper longum, Podophyllum hexandrum, Psidium guajava, Rosa indica, Sida cordifolia*, and *Syzygium aromaticum*, among others, are used for the synthesis and characterization of Au- and Ag-NPs; their anticancer activities have also been reported. Studies have shown that during Au- and Ag-NPs synthesis, the shape, size, and stability of the particles is controlled and/or influenced by several factors such as temperature, pH, incubation time, and plant extract concentrations and that of the metal salt.

The main objective of this chapter is to identify the medicinal-plant extracts (natural and active products-functional groups) involved in the biosynthesis of Au- and Ag-NPs. In addition, the characterization and identification of particle morphology (shape and size) was undertaken with a view to update the nanobiotechnology of NPs and their anticancer activities.

6.2 MEDICINAL PLANTS AND AU- OR AG-NPS SYNTHESIS

Extracts of various parts of different medicinal plant species have been used for synthesis of Au- and/ or Ag-NPs, as summarized in Tables 6.1 and 6.2. In a study, Krishnaraj et al. (2014) have reported the *in vitro* cytotoxic effect of Au-NPs and Ag-NPs (obtained by an *Acalypha indica*-mediated synthesis) against the MDA-MB-231, human breast cancer cells, which could cause significant cytotoxic effects and apoptotic features. Many other research groups have reported anticancer properties of plant-mediated Au-NPs such as those involving *Moringa oleifera* (Anand et al. 2015, Tiloke et al. 2016), *Portulaca grandiflora* (Ashokkumar et al. 2016), *Musa paradisiaca* (Vijayakumar et al. 2017), *Spinacia oleracea* (Ramachandran et al. 2017), and so on. Balasubramani et al. (2015) found the *Antigonon leptopus*-mediated Au-NPs as a free-radical scavenger and an anticancer agent. Mukundan et al. (2015) approached a green route for synthesis of Au-NPs by using the leaf of *Bauhinia tomentosa* and tested *in vitro* anticancer. The distinct UV-Vis absorption peak was noted at 563 nm, which is characteristic of Au-NPs. The reaction mixture color pale yellow immediately changed to a ruby red color, which indicates the reduction of Au^{3+} ions to Au°. Phytoconstituents present in the extract of the leaves acted as the reducing agent during NPs synthesis. FTIR analysis exhibited various peaks at 3,342 (O–H stretch), 1,737 (C=O), 1,639 (N–H bend), 1,436 (C–H bend), 1,365 (C–H rock), and 1,218 cm^{-1} (C–O stretch) spectra which has been reported for the NPs synthesis. Koperuncholan (2015) synthesized Au-NPs by using aqueous leaf extracts of *Hygrophila spinosa* which were examined as an anticancer agent. UV-Vis absorption spectroscopy studies have shown a decrease in the intensity of peak absorbance (at 500 nm). Thereafter, a constant absorbance (at 560 nm) for an hour was observed during the incubation of the reaction mixture. FTIR analysis exhibited a peak at 3,400 cm^{-1} which indicated the presence of OH groups and which was found to be responsible for NPs synthesis. Suganya et al. (2016) approached a green route for the synthesis of Au-NPs by using leaf extracts of *Mimosa pudica* and examined their activity against breast cancer cell lines. A UV-Vis absorption spectroscopy study exhibited an Au-NPs absorption peak at 534 nm within 30 min at 55°C; no change in color intensity was noted after 30 min. FTIR studies confirmed that the peaks at 1,697 cm^{-1} (C=O stretch) and 3,609 cm^{-1} (O–H stretch) corresponded to the reduction and stabilization of Au-NPs. Abel et al. (2016) produced Au-NPs from the *Cassia tora* secondary metabolites conjugate and noted their higher bioavailability and the antioxidant and anticancer effect against the colon cancer cell line (Col320).

Ag-NPs fabricated from various medical plants have also shown anticancer properties due to structurally diverse chemical constituents present in them. They have shown an effective cytotoxic

Table 6.1   Recently Used Medicinal Plant Products as Basis for the Synthesis of Gold Nanoparticles, Their Characterization Techniques, and Their Application in Cancer Cell Treatments

Botanical Name	Family	Plant Part Used	Solvent Used	Synthesis Condition	Characterization Techniques	Shape and Size	Phyto-constituents Responsible for Reduction of Gold Ion	References
Abelmoschus esculentus	Malvaceae	Pulp	Distilled water	Extract mixed with ingredient; reaction at room temperature with continuous stirring for 6 h	XRD, UV-vis, FTIR, RTEM, EDX, DLS	Spherical, triangle, and hexagonal; 4–32 nm	Phytochemicals viz. vitamins and proteins	Rahaman Mollick et al. (2014)
Antigonon leptopus	Polygonaceae	Leaf	Distilled water	Solution based mixing of ingredients; reaction set at particular pH	UV-vis, XRD, FTIR, HRTEM, EDX, SAED, DLS	Spherical; 13–28 nm	Carbonyl, amide, and carboxylic groups	Balasubramani et al. (2015)
Backhousia citriodora	Myrtaceae	Leaf	Distilled water	Solution based mixing of ingredients	UV-vis, XRD, FTIR, TEM	Spherical; 8.40 ± 0.084 nm	Polyphenolic compounds or amines groups	Khandanlou et al. (2018)
Cassia tora	Leguminosae	Leaf	Distilled water	Solution based mixing of ingredients	UV-vis, FTIR, HRTEM, DLS, Zeta Potential	Spherical; mean diam. ~ 5 nm	Amides, alcohols, and aromatic compounds	Abel et al. (2016)
Clerodendrum phlomidis	Lamiaceae	Leaf	Distilled water	Solution based mixing of ingredients	UV-vis, FESEM, TEM, EDX, FTIR	Spherical; 23–42 nm	Hydroxy and amino groups	Sriranjani et al. (2016)
Commelina nudiflora	Commelinaceae	Aqueous extract	Distilled water	Solution based mixing of ingredients	TEM, DLS, Zeta Potential	Spherical, triangular; 24–150 nm	Hydroxy, acids, and amino groups	Kuppusamy et al. (2016)
Cystoseira baccata	Fucaceae	Leaf	Distilled water	Mixing of ingredients with leaf extract; solution kept under stirring for 24 h	UV-vis, XRD, FTIR, HRTEM, EDS, STEM, Zeta Potential	Spherical; ~2.2 ± 8.4 nm	Polysaccharides, phenolic compounds, proteins, vitamins, and terpenoids	González-Ballesteros et al. (2017)

(Continued)

Table 6.1 (Continued) Recently Used Medicinal Plant Products as Basis for the Synthesis of Gold Nanoparticles, Their Characterization Techniques, and Their Application in Cancer Cell Treatments

Botanical Name	Family	Plant Part Used	Solvent Used	Synthesis Condition	Characterization Techniques	Shape and Size	Phyto-constituents Responsible for Reduction of Gold Ion	References
Dendropanax morbifera	Araliaceae	Leaf	Distilled water	Solution based mixing of ingredients	UV-vis, XRD, FETEM, XRD, DLS	Polygonal, few hexagon; 10–20 nm	Phytochemicals and flavonoids	Wang et al. (2016)
Dracocephalum kotschyi	Lamiaceae	Leaf	Distilled water	Solution based mixing of ingredients	UV-vis, TEM-SEAD, SEM-EDAX, XRD, Zeta potential, DLS, FTIR	Triangle, pentagon, hexagon; 7.9–22.63 nm	Amine, amide(II) groups, and alcohols	Dorosti and Jamshidi (2016)
Gymnema sylvestre	Apocynaceae	Leaf	Distilled water	Solution based mixing of ingredients	UV-vis, TEM, EDX, FTIR, DLS	Spherical; 26 nm	Carbonyl groups	Nakkala et al. (2015)
Moringa oleifera	Moringaceae	Flower	Distilled water	Solution based; mild condition	TEM, UV-vis, SEM, EDX, Zeta Potential	Spherical; 3–6 nm	Trace aromatic but abundant aliphatic compounds i.e. proteins and lipids	Anand et al. (2015)
Musa paradisiaca	Musaceae	Peel	Distilled water	Solution based; mixing of ingredients; solution at 353K for 20 min, change in solution color	UV-vis, XRD, FTIR, SEM, EDX, TEM, Zeta Potential	Spherical and triangular; ~50 nm	Phenols and carboxylic acids and amide compounds	Vijayakumar et al. (2017)
Nerium oleander	Apocynaceae	Dried stem bark	Distilled water	Solution based mixing of ingredients	UV-vis, HRTEM XRD, DLS	Spherical; 20–40 nm	Polyphenols including flavonoids, steroids, etc.	Barai et al. (2018)

(Continued)

Table 6.1 (Continued) Recently Used Medicinal Plant Products as Basis for the Synthesis of Gold Nanoparticles, Their Characterization Techniques, and Their Application in Cancer Cell Treatments

Botanical Name	Family	Plant Part Used	Solvent Used	Synthesis Condition	Characterization Techniques	Shape and Size	Phyto-constituents Responsible for Reduction of Gold Ion	References
Nigella sativa	Ranunculaceae	Seeds	Distilled water	Seeds extract in water by soaking and boiling and mixed with ingredient change in color of solution	TEM, UV-vis, FTIR, EDX	Spherical, triangle, and hexagonal; 15–29 nm	Vapors rich in amide, alcohol, phenolic group compounds, and their derivatives	Manju et al. (2016)
Spinacia oleracea	Amaranthaceae	Leaf	Distilled water	Solution based mixing of ingredients	UV-vis, HR-TEM, FE-SEM, XRD	Anisotropic; 5–50 nm	–	Ramachandran et al. (2017)
Vitis vinifera	Vitaceae	Peel	Distilled water	Solution based mixing of ingredients	UV-vis, TEM, XRD, FTIR, Zeta Potential	Spherical; 20–40 nm	Polyphenols	Nirmala et al. (2017)

Table 6.2 Recently Used Medicinal Plant Products as Basis for the Synthesis of Silver Nanoparticles, Their Characterization Techniques, and Their Application in Cancer Cell Treatments

Botanical Name	Family	Part of Plant Used	Solvent Used	Synthesis Conditions	Characterization Techniques	Shape and Size	Phyto-constituents Responsible for Reduction of Silver Ion	References
Andrographis echioides	Acanthaceae	Leaf	Distilled water	Aqueous leaf extract was mixed with AgNO$_3$; color of solution turned brown indicating formation of Ag-NPs; confirmed by UV-Vis absorption	UV-Vis, XRD, FTIR, HRSEM, TEM, EDX, AFM, DLS	Cubic, pentagonal, and hexagonal; 68–91 nm	Extract rich in phenols and their derivatives	Elangovan et al. (2015)
Matricaria chamomilla	Asteraceae	Leaf	Distilled water	Aqueous leaf extract was mixed with AgNO$_3$; colorless solution turned dark brown within 5 min indicating formation of Ag-NPs; confirmed by UV-Vis absorption	UV-Vis, XRD, FTIR, FESEM, TEM, EDX, DLS	Spherical; 45.12 nm	Extract rich in phenolic compound	Dadashpour et al. (2018)
Melia dubia	Meliaceae	Leaf	Distilled water	Aqueous leaf extract was mixed with AgNO$_3$ (28°C) color of solution turned yellow to yellow-red or dark brown indicating formation of Ag-NPs	UV-Vis, XRD, SEM–EDS	Spherical; 7.3 nm	Extract rich in phytochemicals	Kathiravan, Ravi, and Kumar (2014)
Panax ginseng	Araliaceae	Root	–	Aqueous ginseng root extract was mixed with AgNO$_3$ solution, performed ultrasonication (~3 h); colorless solution turned dark yellow; confirmed by UV-Vis absorption	UV-Vis, XRD, FTIR, EDX, FFT, HRTEM	Spherical; 5–15 nm	Aqueous extract rich in ginsenosides (hydroxyl groups), amino acid, and/or protein residues	Sreekanth et al. (2018)
Pimpinella anisum	Apiaceae	Seed	Distilled water	Aqueous seed extract was added into AgNO$_3$ solution at room temperature; after 172 h, yellow color solution turned into dark brown, indicating the formation of Ag-NPs	UV-Vis, FTIR, TEM, EDX	Spherical; 80–85 nm	Extract rich in phytochemicals	Devanesan et al. (2017)

(Continued)

Table 6.2 (Continued) Recently Used Medicinal Plant Products as Basis for the Synthesis of Silver Nanoparticles, Their Characterization Techniques, and Their Application in Cancer Cell Treatments

Botanical Name	Family	Part of Plant Used	Solvent Used	Synthesis Conditions	Characterization Techniques	Shape and Size	Phyto-constituents Responsible for Reduction of Silver Ion	References
Sargassum polycystum	Sargassaceae	Whole plant	Distilled water	Aqueous extract mixed with AgNO$_3$ solution and kept in incubator at 20°C; change in color of the solution indicated the formation of Ag-NPs	UV-Vis, XRD, FTIR, SEM, TEM, EDX	Spherical; ~28 nm	Aqueous extract rich in alkaloids, steroids, flavanoids, protein, terpenoids, amino acids, carbohydrate, quinones, phenols, and tannins	Palanisamy et al. (2017)
Solanum trilobatum	Solanaceae	Unripe fruits	Distilled water	Aqueous unripe fruits extract added into AgNO$_3$ solution; solution color turned yellowish brown indicating the formation of Ag-NPs	UV-Vis, FTIR, SEM, TEM, EDX, XRD	Spherical and polygonal; 12.50–41.90 nm	Hydroxyl/amine groups	Ramar et al. (2014)

effect against several cancer cell lines. Some researchers have proposed that due to the higher concentration of NPs, cell viability is reduced. This was explained on the basis of the active physiochemical interaction of Ag ions with the functional groups of intracellular proteins, along with nitrogenous bases and phosphate groups as well (Mata et al. 2015, Nalavothula et al. 2015, Varghese et al. 2015, Sreekanth et al. 2016). Ag-NPs obtained from petal extract of *Rosa indica* have shown anticancer and anti-inflammatory properties (Manikandan et al. 2015). Those produced from leaf extracts of *Piper longum* and *Melia azedarch* were also shown to have a cytotoxic effect on HEp-2 (Jacob et al. 2012) and HeLa cell lines (Sukirtha et al. 2012). Ag ions can produce reactive oxygen species (ROS) which may cause apoptosis or necrosis in human cancer cells through nuclear condensation and fragmentation. The toxicity of plant-latex-capped Ag-NPs towards human lung carcinoma cells *in vitro* was tested, and authors have suggested that these particles are toxic to A549 cells in a concentration-dependent manner (Valodkar et al. 2011). They also suggested that plant latex can solubilize Ag-NPs in water and act as a potential biocompatible vehicle for transport of Ag-NPs to target tumor cells. Nonetheless, a detailed study of biocompatibility of these plant-latex-mediated NPs is still pending. Ag-NPs have also been observed to cure breast cancer (Franco-Molina et al. 2010, Gurunathan et al. 2013a, b), lung cancer (Foldbjerg et al. 2011), and skin and/or oral carcinoma (Austin et al. 2011). Jeyaraj et al. (2013) have fabricated Ag-NPs from *Podophyllum hexandrum*. Authors have found enhanced anticancer activity in comparison to Cisplatin, the standard anticancer drug. He et al. (2016) used *Dimocarpus longan* peel's aqueous extract to fabricate Ag-NPs, which showed cytotoxicity in a concentration-dependent manner against prostate cancer (PC-3) cells through a decrease of stat 3, bcl-2, and survivin, as well as an increase in caspase-3. He et al. (2016) have proposed that these particles could be used for prostate cancer treatments, although a complete investigation to understand the molecular mechanism and *in vivo* effects of Ag-NPs on prostate cancer is still required. Ag-NPs obtained from several other medicinal plants such as *Allium sativum* (Ahamed et al. 2011), *Annona squamosa* (Vivek et al. 2012), *Citrullus colocynthis* (Satyavani et al. 2011a, b), *Piper longum* (Reddy et al. 2014), *Melia dubia* (Kathiravan, Ravi, and Kumar 2014), *Mentha arvensis* (Banerjee et al. 2017), and *Pimpinella anisum* (Devanesan et al. 2017), among others, have been reported against various human cancer cell lines.

6.3 MECHANISMS OF ACTION

A clear understanding of the interactions of Au- and Ag-NPs with the cancerous cell is vital for the assessment of anticancer activity and for further advanced biomedical applications. Interaction between cancerous cells and Au- or Ag-NPs varies. A possible mechanism is presented in Figure 6.1. These particles have shown cytotoxicity due to ROS and free radicals which cause apoptosis leading to cell death, DNA damage, activation caspase cascade of apoptosis, and dysfunction of mitochondria (Parida et al. 2014; Tiloke et al. 2016; Patil et al. 2017; Bethu et al. 2018; Buttacavoli et al. 2018; Chugh et al. 2018).

6.4 CONCLUSION

Cancer is recognized as a dangerous disease. It is characterized by various signaling mechanisms, for instance, cell proliferation, angiogenesis, and metastasis. Cancerous cells show anomalous metabolic activities in aerobic glycolysis, depletion of DNA in mitochondria, and deviations in the genomic expressions and respiratory chains. In biological sciences, nanomaterials are useful due to their tiny size and effectiveness. Several medicinal plants are rich in secondary metabolites and other products. They have been used for the fabrication (green synthesis) of Au- and/or Ag-NPs. Fabrication of these particles is controlled by factors *viz.* temperature, pH, incubation time, and

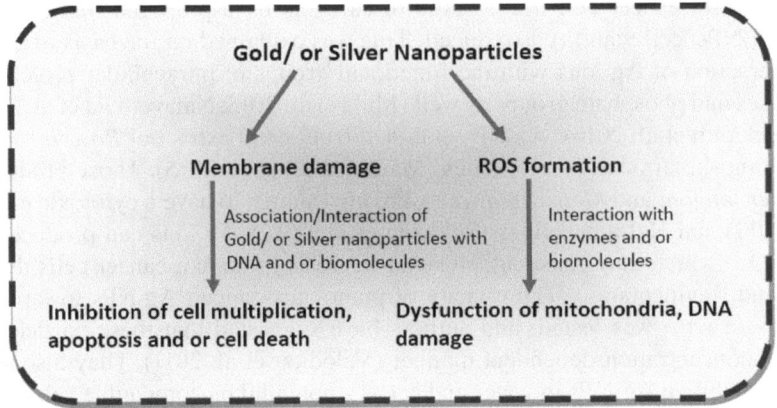

Figure 6.1   Mechanism of action of gold and or silver nanoparticles against cancerous cells.

plant extract and metal salt concentrations. Both Au- and Ag-NPs have shown cytotoxic efficiency against various types of cancerous cells. Studies have confirmed that cancer cell growth has been significantly decreased in a concentration-dependent manner after treatment with biogenic fabricated gold and silver nanoparticles. Numerous mechanisms of action are suggested; nonetheless, it is still essential to explore which specific mechanisms are optimally effective against cancer cells.

REFERENCES

Abasi, E., M. Milani, S.F. Aval, et al. 2016. Silver nanoparticles: Synthesis methods, bio-applications and properties. *Cri. Rev. Microbiol.* 42:173–180.

Abel, E.E., P.R.J. John Poonga, and S.G. Panicker. 2016. Characterization and *in vitro* studies on anticancer, antioxidant activity against colon cancer cell line of gold nanoparticles capped with *Cassia tora* SM leaf extract. *Appl. Nanosci.* 6:121–129.

Ahamed, M., M.A. Majeed-Khan, M.K.J. Siddiqui, M.S. AlSalhi, and S.A. Alrokayan. 2011. Green synthesis: Characterization and evaluation of biocompatibility silver nanoparticles. *Phys. E.* 43:1266–1271.

Anand, K., R.M. Gengan, A. Phulukdaree, and A. Chuturgoon. 2015. Agroforestry waste *Moringa oleifera* petals mediated green synthesis of gold nanoparticles and their anti-cancer and catalytic activity. *J. Ind. Eng. Chem.* 21:1105–1111.

Ashokkumar, T., J. Arockiaraj, and K. Vijayaraghavan. 2016. Biosynthesis of gold nanoparticles using green roof species *Portulaca grandiflora* and their cytotoxic effects against C6 glioma human cancer cells. *Environ. Progsust. Energy.* 35:1732–1740.

Austin, L.A., B.Kang, C.W.Yen, and M.A. El-Sayed. 2011. Plasmonic imaging of human oral cancer cell communities during programmed cell death by nuclear-targeting silver nanoparticles. *J Am Chem Soc.* 133:17594–7.

Balasubramani, G., R. Ramkumar, N. Krishnaveni, et al. 2015. Structural characterization, antioxidant and anti-cancer properties of gold nanoparticles synthesized from leaf extract (decoction) of *Antigononleptopus* Hook. *J. Trace Elem. Med. Biol.* 30:83–89.

Balasubramani, G., R. Ramkumar, R.K. Raja, D. Aiswarya, C. Rajthilak, and P. Perumal. 2017. *Albiziaamara* Roxb. mediated gold nanoparticles and evaluation of their antioxidant, antibacterial and cytotoxic properties. *J. Clust. Sci.* 28:259–275.

Banerjee, P.P., A. Bandyopadhyay, S.N. Harsha, et al. 2017. *Menthaarvensis* (Linn.)-mediated green silver nanoparticles trigger caspase 9-dependent cell death in MCF7 and MDA-MB-231 cells. *Breast Cancer.* 9:265–278.

Barai, A.C., K. Paul, A. Dey, et al. 2018. Green synthesis of *Neriumoleander*-conjugated gold nanoparticles and study of its *in vitro* anticancer activity on MCF-7 cell lines and catalytic activity. *Nano. Converg.* 5:10.

Bethu, M.S., V.R. Netala, L. Domdi, V. Tartte, and V.R. Janapala. 2018. Potential anticancer activity of bio-genic silver nanoparticles using leaf extract of *Rhynchosiasuaveolens*: An insight into the mechanism. *Artif. Cells Nanomed. Biotechnol.* 4:1–11.

Beyene, H.D., A.A. Werkneh, H.K. Bezabh, and T.G. Ambaye. 2017. Synthesis paradigm and applications of silver nanoparticles (AgNPs), a review. *Sus. Mat. Technol.* 13:18–23.

Buttacavoli, M., N.N. Albanese, G.D. Di Cara, et al. 2018. Anticancer activity of biogenerated silver nanopar-ticles: An integrated proteomic investigation. *Oncotarget.* 9:9685–9705.

Cai, W., T. Gao, H. Hong, and J. Sun. 2008. Applications of gold nanoparticles in cancer nanotechnology. *Nanotechnol. Sci. Appl.* 1:17–32.

Chugh, H., D. Sood, I. Chandra, V. Tomar, G. Dhawan, and R. Chandra. 2018. Role of gold and silver nanopar-ticles in cancer nano-medicine. *Artif. Cells Nanomed. Biotechnol.* 13:1–11.

Chung, I.M. III, I. Park, K. Seung-Hyun, M. Thiruvengadam, and G. Rajakumar. 2016. Plant-mediated syn-thesis of silver nanoparticles: Their characteristic properties and therapeutic applications. *Nanoscale Res. Lett.* 11:40.

Dadashpour, M., A. Firouzi-Amandi, M. Pourhassan-Moghaddam, et al. 2018. Biomimetic synthesis of sil-ver nanoparticles using *Matricaria chamomilla* extract and their potential anticancer activity against human lung cancer cells. *Mater. Sci. Eng. C. Mater. Biol. Appl.* 92:902–912.

Dahoumane, S.A., M. Mechouet, K. Wijesekera, et al. 2017. Algae-mediated biosynthesis of inorganic nano-materials as a promising route in nanobiotechnology – A review. *Green Chem.* 19:552–587.

Dasary, S.S.R., A.K. Singh, D. Senapati, H. Yu, and P.C. Ray. 2009. Gold nanoparticle based label-free SERS probe for ultrasensitive and selective detection of trinitrotoluene. *J. Am. Chem. Soc.* 131:13806–13812.

Devanesan, S., M.S. AlSalhi, R. Vishnubalaji, et al. 2017. Rapid biological synthesis of silver nanoparticles using plant seed extracts and their cytotoxicity on colorectal cancer cell lines. *J. Clust. Sci.* 28:595–605.

Ding, X., L. Kong, J. Wang, F. Fang, D. Li, and J. Liu. 2013. Highly sensitive SERS detection of Hg2+ ions in aqueous media using gold nanoparticles/graphene heterojunctions. *ACS Appl. Mater. Interfaces.* 5:7072–7078.

Dite, G.S., A.S. Whittemore, J.A. Knight, et al. 2010. Increased cancer risks for relatives of very early-onset breast cancer cases with and without BRCA1 and BRCA2 mutations. *Br. J. Cancer.* 103:1103–1108.

Dorosti, N., and F. Jamshidi. 2016. Plant-mediated gold nanoparticles by *Dracocephalum kotschyi* as anticho-linesterase agent: Synthesis, characterization, and evaluation of anticancer and antibacterial activity. *J. App. Biomed.* 14:235–245.

Elangovan, K., D. Elumalai, S. Anupriya, R. Shenbhagaraman, P.K. Kaleena, and K. Murugesan. 2015. Phyto mediated biogenic synthesis of silver nanoparticles using leaf extract of *Andrographis echioides* and its bio-efficacy on anticancer and antibacterial activities. *J. Photochem. Photobiol. B. Biol.* 151:118–124.

Foldbjerg, R., D.A. Dang, and H. Autrup. 2011. Cytotoxicity and genotoxicity of silver nanoparticles in the human lung cancer cell line, A549. *Arch. Toxicol.* 85:743–750.

Franco-Molina, M.A., E. Mendoza-Gamboa, C.A. Sierra-Rivera, et al. 2010. Antitumor activity of colloidal silver on MCF-7 human breast cancer cells. *J. Exp. Clin. Cancer Res.* 29:148.

González-Ballesteros, N., S. Prado-López, J.B. Rodríguez-González, M. Lastra, and M.C. Rodríguez-Argüelles. 2017. Green synthesis of gold nanoparticles using brown algae *Cystoseira baccata*: Its activ-ity in colon cancer cells. *Colloids Surf. B. Biointerfaces.* 153:190–198.

Guo, D., L. Zhu, Z. Huang, et al. 2013. Anti-leukemia activity of PVP-coated silver nanoparticles via genera-tion of reactive oxygen species and release of silver ions. *Biomaterials.* 34:7884–7894.

Gurunathan, S., J.W. Han, V. Eppakayala, M. Jeyaraj, and J.H. Kim. 2013a. Cytotoxicity of biologically synthesized silver nanoparticles in MDA-MB-231 human breast cancer cells. *BioMed. Res. Int.* 2013:535796–535805.

Gurunathan, S., J. Raman, S.N. Abd Malek, P.A. John, and S. Vikineswary. 2013b. Green synthesis of silver nanoparticles using *Ganoderma neo-japonicumImazeki*: A potential cytotoxic agent against breast can-cer cells. *Int. J. Nanomed.* 8:4399–4413.

He, Y., Z. Du, S. Ma, et al. 2016. Biosynthesis, antibacterial activity and anticancer effects against prostate cancer (PC-3) cells of silver nanoparticles using *Dimocarpus longan* Lour. peel extract. *Nanoscale Res. Lett.* 11:300.

Huang, X., I.H. El-Sayed, W. Qian, and M.A. El-Sayed. 2007. Cancer cells assemble and align gold nanorods conjugated to antibodies to produce highly enhanced, sharp, and polarized surface Raman spectra: A potential cancer diagnostic marker. *Nano. Lett.* 7:1591–1597.

Huang, X., and M.A. El-Sayed. 2010. Gold nanoparticles: Optical properties and implementations in cancer diagnosis and photothermal therapy. *J. Adv. Res.* 1:13–28.

Huff, T.B., L. Tong, Y. Zhao, M.N. Hansen, J.X. Cheng, and A. Wei. 2007. Hyper thermic effects of gold nanorods on tumor cells. *Nanomedicine.* 2:125–132.

Husen, A., and K.S. Siddiqi. 2014a. Carbon and fullerene nanomaterials in plant system. *J. Nanobiotechnol.* 12:16.

Husen, A., and K.S. Siddiqi. 2014b. Phytosynthesis of nanoparticles: Concept, controversy and application. *Nanoscale Res. Lett.* 9:229.

Husen, A., and K.S. Siddiqi. 2014c. Plants and microbes assisted selenium nanoparticles: Characterization and application. *J. Nanobiotechnol.* 12:28.

Husen, P., and I.A. Solov'yov. 2017. Mutations at the Qo Site of the Cytochrome bc1 Complex Strongly Affect Oxygen Binding. *Phys Chem B.* 121:3308–3317.

Jacob, S.J.P., J.S. Finub, and A. Narayanan. 2012. Synthesis of silver nanoparticles using Piper longum leaf extracts and its cytotoxic activity against Hep-2 cell line. *Coll. Surf. B.: Biointerf.* 91:212–214.

Jeyaraj, M., M. Rajesh, R. Arun, et al. 2013. An investigation on the cytotoxicity and caspase-mediated apoptotic effect of biologically synthesized silver nanoparticles using *Podophyllum hexandrum* on human cervical carcinoma cells. *Colloids Surf. B.: Biointerfaces.* 102:708–717.

Joseph, S., and B. Mathew. 2015. Microwave-assisted green synthesis of silver nanoparticles and the study on catalytic activity in the degradation of dyes. *J. Mol. Liq.* 204:184–191.

Juarez-Moreno, K., E.B. Gonzalez, N. Girón-Vazquez, et al. 2017. Comparison of cytotoxicity and genotoxicity effects of silver nanoparticles on human cervix and breast cancer cell lines. *Hum. Exp. Toxicol.* 36:931–948.

Karatoprak, G.S., G. Aydin, B. Altinsoy, C. Altinkaynak, M. Kos, and I. Ocsoy. 2017. The effect of *Pelargonium endlicherianum* Fenzl. root extracts on formation of nanoparticles and their antimicrobial activities. *Enzy. Micro. Technol.* 97:21–26.

Kathiravan, V., S. Ravi, and S.A. Kumar. 2014. Synthesis of silver nanoparticles from *Melia dubia* leaf extract and their *in vitro* anticancer activity. *Spectrochim. Acta. Part. A. Mol. Biomol. Spectrosc.* 130:116–121.

Khandanlou, R., V. Murthy, D. Saranath, and H. Damani. 2018. Synthesis and characterization of gold-conjugated *Backhousia citriodora* nanoparticles and their anticancer activity against MCF-7 breast and HepG2 liver cancer cell lines. *J. Mater. Sci.* 53:3106–3118.

Koperuncholan, M. 2015. Bioreduction of chloroauric acid (HAuCl4) for the synthesis of gold nanoparticles (GNPs): A special empathies of pharmacological activity. *Int. J. Phytopharm.*5: 72–80.

Krishnaraj, C., P. Muthukumaran, R. Ramachandran, M.D. Balakumaran, and P.T. Kalaichelvan. 2014. Acalypha indica Linn: Biogenic synthesis of silver and gold nanoparticles and their cytotoxic effects against MDA-MB-231, human breast cancer cells. *Biotechnol. Rep.* 4:42–49.

Kuppusamy, P., S.J.A. Ichwan, P.N.H. Al-Zikrim, et al. 2016. In vitro anticancer activity of Au, Ag nanoparticles synthesized using *Commelina nudiflora* L. aqueous extract against HCT-116 colon cancer cells. *Biol. Trace. Elem. Res.* 173:297–305.

Lim, S.H., E.Y. Ahn, and Y. Park. 2016. Green synthesis and catalytic activity of gold nanoparticles synthesized by *Artemisia capillaris* water extract. *Nano. Res. Lett.* 11:474–484.

Lowery, A.R., A.M. Gobin, E.S. Day, N.J. Halas, and J.L. West. 2006. Immunonano shells for targeted photothermal ablation of tumor cells. *Int. J. Nanomed.* 1:149–154.

Manikandan, R., B. Manikandan, T. Raman, et al. 2015. Biosynthesis of silver nanoparticles using ethanolic petals extract of *Rosa indica* and characterization of its antibacterial, anticancer and anti-inflammatory activities. *Spectrochim. Acta. Part. A. Mol. Biomol. Spectrosc.* 138:120–129.

Manju, S., B. Malaikozhundan, S. Vijayakumar, et al. 2016. Antibacterial, antibiofilm and cytotoxic effects of *Nigella sativa* essential oil coated gold nanoparticles. *Microb. Pathog.* 91:129–135.

Mata, R., J. Reddy Nakkala, and S. Rani Sadras. 2015. Catalytic and biological activities of green silver nanoparticles synthesized from *Plumeria alba* (frangipani) flower extract. *Mater. Sci. Eng. C.* 51:216–225.

Mattea, F., J. Vedelago, F. Malano, C. Gomez, M.C. Strumia, and M. Valente. 2017. Silver nanoparticles in X-ray biomedical applications. *Rad. Phys. Chem.* 130:442–450.

Mie, R., M.W. Samsudin, L.B. Din, et al. 2014. Synthesis of silver nanoparticles with antibacterial activity using the lichen *Parmotrema praesorediosum*. *Int. J. Nanomater.* 9:121–127.

Mishra, V.K., R.K. Bacheti, and A. Husen. 2011. Medicinal uses of chlorophyll: A critical overview. In: *Chlorophyll: Structure, Function and Medicinal Uses.* Le, H., and Salcedo, E. (Eds.)–. Nova Science Publishers, Inc., Hauppauge, NY, 177–196.

Moten, A., D. Schafer, P. Farmer, J. Kim, and M. Ferrari. 2014. Redefining global health priorities: Improving cancer care in developing settings. *J. Glob. Health.* 4:010304.

Mukundan, D., R. Mohankumar, and R. Vasanthakumari. 2015. Green synthesis of silver nanoparticles using leaves extract of *Bauhinia tomentosa* Linn and its *in vitro* anticancer potential. *Mater. Today Proc.* 2:4309–4316.

Nakkala, J.R., R. Mata, E. Bhagat, and S.R. Sadras. 2015. Green synthesis of silver and gold nanoparticles from *Gymnema sylvestre* leaf extract: Study of antioxidant and anticancer activities. *J. Nanopart. Res.* 17:151.

Nalavothula, R., J. Alwala, V.B. Nagati, and P.R. Manthurpadigya. 2015. Biosynthesis of silver nanoparticles using *Impatiens balsamina* leaf extracts and its characterization and cytotoxic studies using human cell lines. *Inter. J. Chem. Tech. Res.* 7:2460–2468.

Nautiyal, S., R. Kumar, and A. Husen. 2002. Status of medicinal plants in India: Some latest issue. *Ann. For.* 10:181–190.

Nirmala, J.G., S. Akila, R.T. Narendhirakannan, and S. Chatterjee. 2017. *Vitisvinifera* peel polyphenols stabilized gold nanoparticles induce cytotoxicity and apoptotic cell death in A431 skin cancer cell lines. *Adv. Powder. Technol.* 28:1170–1184.

Ortega, F.G., M.A. Fernández-Baldó, J.G. Fernández, et al. 2015. Study of antitumor activity in breast cell lines using silver nanoparticles produced by yeast. *Int. J. Nanomed.* 10:2021–2031.

Palanisamy, S., P. Rajasekar, G. Vijayaprasath, G. Ravi, R. Manikandan, and N. Marimuthu Prabhu. 2017. A green route to synthesis silver nanoparticles using *Sargassum polycystum* and its antioxidant and cytotoxic effects: An *in vitro* analysis. *Mater. Lett.* 189:196–200.

Parida, U.K., S.K. Biswal, and B.K. Bindhani. 2014. Green synthesis and characterization of gold nanoparticles: Study of its biological mechanism in human SUDHL-4 cell line. *Adv. Biol. Chem.* 4:360–375.

Patil, M.P., D. Ngabire, H.H.P. Thi, M.D. Kim, and G.D. Kim. 2017. Eco-friendly synthesis of gold nanoparticles and evaluation of their cytotoxic activity on cancer cells. *J. Clust. Sci.* 28:119–132.

Rahaman Mollick, M.M.R., B. Bhowmick, D. Mondal, et al. 2014. Anticancer (*in vitro*) and antimicrobial effect of gold nanoparticles synthesized using *Abelmoschus esculentus* (L.) pulp extract via a green route. *RSC Adv.* 4:37838–37849.

Ramachandran, R., C. Krishnaraj, A.S. Sivakumar, et al. 2017. Anticancer activity of biologically synthesized silver and gold nanoparticles on mouse myoblast cancer cells and their toxicity against embryonic zebrafish. *Mat. Sci. Eng. C.* 73:674–683.

Ramar, M., B. Manikandan, P.N. Marimuthu, et al. 2014. Synthesis of silver nanoparticles using *Solanum trilobatum* fruits extract and its antibacterial, cytotoxic activity against human breast cancer cell line MCF 7. *Spectrochim. Acta. A. Mol. Biomol. Spectrosc.* 140:223–228.

Reddy, N.J., D.N. Vali, M. Rani, and S.S. Rani. 2014. Evaluation of antioxidant, antibacterial and cytotoxic effects of green synthesized silver nanoparticles by Piper longum fruit. *Mat. Sci. Eng. C.* 34:115–122.

Santos, EdB., N.V. Madalossi, F.A. Sigoli, and I.O. Mazali. 2015. Silver nanoparticles: Green synthesis, self-assembled nanostructures and their application as SERS substrates. *New J. Chem.* 39:2839–2846.

Satapathy, S.R., P. Mohapatra, R. Preet, et al. 2013. Silver-based nanoparticles induce apoptosis in human colon cancer cells mediated through p53. *Nanomedicine.* 8:1307–1322.

Satyavani, K., S. Gurudeeban, T. Ramanathan, and T. Balasubramanian. 2011a. Biomedical potential of silver nanoparticles synthesized from calli cells of *Citrullus colocynthis* (L.) Schrad. *J. Nanobiotechnol.* 9:43–50.

Satyavani, K., T. Ramanathan, and S. Gurudeekan. 2011b. Green synthesis of silver nanoparticles using stem dried callus extract of bitter apple (*Citrullus colocynthis*). *Dig. J. Nanomater. Biostruct.* 6:1019–1024.

Siddiqi, K.S., and A. Husen. 2016a. Fabrication of metal nanoparticles from fungi and metal salts: Scope and application. *Nanoscale Res. Lett.* 11:98.

Siddiqi, K.S., and A. Husen. 2016b. Fabrication of metal and metal oxide nanoparticles by algae and their toxic effects. *Nanoscale Res. Lett.* 11:363.

Siddiqi, K.S., and A. Husen. 2016c. Engineered gold nanoparticles and plant adaptation potential. *Nanoscale Res. Lett.* 11:400.

Siddiqi, K.S., and A. Husen. 2016d. Green synthesis, characterization and uses of palladium/platinum nanoparticles. *Nanoscale Res. Lett.* 11:482.

Siddiqi, K.S., and A. Husen. 2017a. Plant response to engineered metal oxide nanoparticles. *Nanoscale Res. Lett.* 12:92.

Siddiqi, K.S, and A. Husen. 2017b. Recent advances in plant-mediated engineered gold nanoparticles and their application in biological system. *J Trace Elem Med Biol.* 40:10–23.

Siddiqi, K.S., A. Husen, and R.A.K. Rao. 2018b. A review on biosynthesis of silver nanoparticles and their biocidal properties. *J. Nanobiotechnol.* 16:14.

Siddiqi, K.S., A. Husen, S.S. Sohrab, and M. Osman. 2018c. Recent status of nanomaterials fabrication and their potential applications in neurological disease management. *Nano. Res. Lett.* 13:231.

Siddiqi, K.S., A. Rahman, and A. Husen. 2016. Biogenic fabrication of iron/iron oxide nanoparticles and their application. *Nano. Res. Lett.* 11:498.

Siddiqi, K.S., A. Rahman, and A. Husen. 2018a. Properties of zinc oxide nanoparticles and their activity against microbes. *Nano. Res. Lett.* 13:141.

Siddiqi, K.S., M. Rashid, A. Rahman, Tajuddin, A. Husen, and S. Rehman. 2018d. Biogenic fabrication and characterization of silver nanoparticles using aqueous-ethanolic extract of lichen (*Usnealongissima*) and their antimicrobial activity. *Biomat. Res.* 22:23.

Smith, R.A., V. Cokkinides, D. Brooks, D. Saslow, and O.W. Brawley. 2011. Cancer screening in the United States, 2011: A review of current American Cancer Society guidelines and issues in cancer screening. *CA Cancer. J. Clin.* 60:99–119.

Sreekanth, T.V.M., P.C. Nagajyothi, P. Muthuraman, et al. 2018. Ultra-sonication-assisted silver nanoparticles using *Panax ginseng* root extract and their anti-cancer and antiviral activities. *J. Photochem. Photobiol. B.* 188:6–11.

Sreekanth, T.V.M., M. Pandurangan, M. Jung, Y.R. Lee, and I. Eom. 2016. Eco-friendly decoration of graphene oxide with green synthesized silver nanoparticles: Cytotoxic activity. *Res. Chem. Intermed.* 42:5665–5676.

Srinithya, B., V.V. Kumar, V. Vadivel, B. Pemaiah, S.P. Anthony, and M.S. Muthuraman. 2016. Synthesis of biofunctionalized AgNPs using medicinally important *Sida cordifolia* leaf extract for enhanced antioxidant and anticancer activities. *Mater. Lett.* 170:101–104.

Sriranjani, R., B. Srinithya, V. Vellingiri, et al. 2016. Silver nanoparticle synthesis using *Clerodendrum phlomidis* leaf extract and preliminary investigation of its antioxidant and anticancer activities. *J. Mol. Liq.* 220:926–930.

Suganya, U.S.U., K. Govindaraju, G.G. Kumar, et al. 2016. Anti-proliferative effect of biogenic gold nanoparticles against breast cancer cell lines (MDA-MB-231&MCF-7). *Appl. Surf. Sci.* 371:415–424.

Sukirtha, R., K.M. Priyanka, J.J. Antony, et al. 2012. Cytotoxic effect of green synthesized silver nanoparticles using *Melia azedarach* against *in vitro* HeLa cell lines and lymphoma mice model. *Proc. Biochem.* 47:273–279.

Thanh, N.T.K., and Z. Rosenzweig. 2002. Development of an aggregation-based immunoassay for anti-protein A using gold nanoparticles. *Anal. Chem.* 74:1624–1628.

Tiloke, C., A. Phulukdaree, K. Anand, R.M. Gengan, and A.A. Chuturgoon. 2016. *Moringa oleifera* gold nanoparticles modulate oncogenes, tumor suppressor genes, and caspase-9 splice variants in A549 cells. *J. Cell. Biochem.* 117:2302–2314.

Valodkar, M., R.N. Jadeja, M.C. Thounaojam, R.V. Devkar, and S.Thakore. 2011. In vitro toxicity study of plant latex capped silver nanoparticles in human lung carcinoma cells. *Mat. Sci. Eng. C.* 31:1723–1728.

Varghese, A., P. Anandhi, R. Arunadevi, et al. 2015. Satin leaf (Chrysophyllum oliviforme) extract mediated green synthesis of silver nanoparticles: Antioxidant and anticancer activities. *J. Pharm. Sci. Res.* 7:266–273.

Vijayakumar, S., B. Vaseeharan, B. Malaikozhundan, et al. 2017. Therapeutic effects of gold nanoparticles synthesized using *Musa paradisiaca* peel extract against multiple antibiotic resistant Enterococcus faecalis biofilms and human lung cancer cells (A549). *Microb. Pathog.* 102:173–183.

Vivek, R., R. Thangam, K. Muthuchelian, P. Gunasekaran, K. Kaveri, and S. Kannan. 2012. Green biosynthesis of silver nanoparticles from *Annona squamosa* leaf extract and its *in vitro* cytotoxic effect on MCF-7 cells. *Proc. Biochem.* 47:2405–2410.

Wang, C., R. Mathiyalagan, Y.J. Kim, et al. 2016. Rapid green synthesis of silver and gold nanoparticles using *Dendropanax morbifera* leaf extract and their anticancer activities. *Int. J. Nanomed.* 11:3691–3701.

World Health Organization. 2014a. *Global Battle Against Cancer Won't be Won with Treatment Alone – Effective Prevention Measures Urgently Needed to Prevent Cancer Crisis.* International Agency for Research on Cancer, London, UK.

World Health Organization. 2014b. *International Agency for Research on Cancer – World Cancer Report 2014.* WHO, Geneva, Switzerland.

World Health Organization. 2018. World cancer day 2018. www.who.int/cancer/world-cancer-day/2018/en/ (accessed October 22, 2018).

Zamarion, V.M., R.A. Timm, K. Araki, and H.E. Toma. 2008. Ultrasensitive SERS nanoprobes for hazardous metal ions based on trimercaptotriazine-modified gold nanoparticles. *Inorg. Chem.* 47:2934–2936.

Zecca, L., M.B.H. Youdim, P. Riederer, J.R. Connor, and R.R. Crichton. 2004. Iron, brain ageing and neurodegenerative disorders. *Nat. Rev. Neurosci.* 5:863–873.

CHAPTER **7**

Bladder and Prostate Cancer

Charlie Khoo, Yiannis Philippou, Marios Hadjipavlou, and Abhay Rane

CONTENTS

7.1 INTRODUCTION

There is increasing popularity of and advocacy for the use of plant-based therapies in patients with prostate or bladder cancer. Amongst those undergoing surgery for urological malignancy, one-third use at least one form of complementary and alternative medicine (CAM) before or after surgery (Mani et al. 2015). In particular, plant-based therapies are highly prevalent amongst patients with prostate cancer using CAM (Beebe-Dimmer et al. 2004, Ponholzer et al. 2003, Wilkinson et al. 2008).

There are a number of reasons for this surge in popularity. The incidence of prostate cancer is rising, and the long latency period from intraepithelial malignancy to established prostate cancer gives the patient opportunity to try alternative interventions (Nelson and Montgomery 2003). Use of plant-based therapies may provide patients with a sense of personal control. As 'natural' products, they are perceived as safe. Additionally, when conventional treatments have failed, patients may be more willing to try alternatives, including plant-based treatments (Philippou et al. 2013).

Despite widespread use, the paucity of quality clinical research evaluating the effectiveness of plant-based therapies in managing bladder and prostate cancers makes it difficult for patients and healthcare practitioners to make informed decisions. This chapter summarizes the state of the literature on several promising plant-based therapies that are commonly used in the management of prostate and bladder cancer, discussing mechanisms of action, effectiveness, and highlighting areas for further research.

7.2 METHOD

An English language literature search was performed on Ovid Embase, Ovid Medline, Scopus, and Web of Science to identify studies investigating the potential chemotherapeutic role of various plants and plant extracts against prostate, bladder, and renal cancers. The focus was on agents that have undergone extensive research directed towards understanding their chemotherapeutic mechanisms and on which clinical trials have been undertaken (although a few promising agents with *in vitro* trials have been included). Furthermore, we concentrated primarily on clinical data that relates to treatment rather than prevention of cancer. Search terms included 'prostate cancer', 'bladder cancer', 'plant extract', 'selenium', 'vitamin E', 'pomegranate', 'green tea', 'curcumin', 'resveratrol', 'silibinin', 'gingko', 'modified citrus pectin', *'phellodendron amurense'*, 'red clover', 'Salvia' and 'mistletoe', 'PC-SPES', 'Zyflamend', and 'Profluss'.

7.3 RESULTS

7.3.1 Selenium and Vitamin E

Selenium, named from the Greek σελήνη ('moon'), is a naturally occurring non-metallic trace element essential to human health. It was discovered incidentally in 1817 by the Swedish chemist Jacob Berzelius, who was investigating an impurity slowing production of sulfuric acid at his chemical plant (Trofast 2011). Vitamin E refers to a family of naturally occurring, essential fat-soluble compounds found in a variety of nuts, oils, and leafy green vegetables. In 1922, researchers noted that rats were failing to reproduce when fed solely lard; introducing wheat germ and lettuce to the diet resolved the issue. They determined that these foodstuffs must contain an 'anti-sterility factor', which was first isolated and redubbed 'vitamin E' in 1935 (Evans and Bishop 1992, Oakes 2007).

These compounds are often taken in combination, with several mechanisms proposed to explain their anti-tumorigenic properties. They are both thought to act as antioxidants mopping up free radicals which can attack DNA causing damage and ultimately result in tumorigenesis. Other potential mechanisms derived from *in vitro* studies include inhibition of angiogenesis, cellular proliferation, and induction of apoptosis (Chan et al. 2005).

Until recently, selenium and vitamin E were used extensively by patients both for the prevention and treatment of prostate cancer. Evidence of their potential anti-tumor activity was derived from preclinical studies, epidemiological observations, and clinical trials. In 2009, the Selenium and Vitamin E Cancer Prevention Trial (SELECT), a large-scale, population-based, randomized controlled trial (RCT) reported the results of these agents on the incidence of prostate cancer in North American men (Lippman et al. 2009). SELECT was designed to assess the effect of selenium and vitamin E both alone and in combination as supplements to a normal diet on the prevention of prostate cancer. 35,533 healthy individuals were randomized to receive either selenium or vitamin E alone, in combination, or placebo. The risk of prostate cancer at a median follow-up of seven years was increased by 17% in men randomized to supplementation with vitamin E alone, a deference that started to appear three years after randomization (hazard ratio 1.17). In addition, the results showed no benefit of selenium alone or when combined with vitamin E for prevention of prostate cancer (Klein et al. 2011). Another large multicenter phase III RCT using selenium vs placebo in men with high-grade prostate intraepithelial neoplasia (HGPIN) showed no benefit in the intervention group (Marshall 2010). The results of the above studies now challenge the previous notion of a protective role of selenium and vitamin E supplementation with some studies even suggesting the converse.

Early results of selenium levels on bladder cancer incidence have indicated a protective effect suggesting possible use as a chemopreventive agent. One study reported an increased risk of bladder cancer in people with lower serum selenium concentrations (Kellen et al. 2006). A meta-analysis of bladder cancer incidence which included five observational studies found an inverse association, with an overall risk estimate of 0.67 (95% CI 0.46–0.97), suggesting a protective effect of higher selenium levels against bladder cancer (Dennert et al. 2011). Conversely, secondary analysis of the SELECT trial suggested no effect of selenium and vitamin E alone or in combination on bladder cancer, with a similar incidence occurring in each group (Lotan et al. 2012). Additionally, results from the SELEnium and BLAdder cancer Trial (SELEBLAT) suggest that selenium does not diminish recurrence of bladder cancer in patients who have undergone transurethral resection of bladder tumors (TURBT) for non-invasive transitional cell carcinoma of the bladder (NCT00729287). In this randomized multicenter trial, the intervention group received 200 mg/day selenium-yeast supplementation over three years (with the control group receiving placebo). There was no significant difference in the recurrence-free interval between intervention and control groups (HR = 0.75, 95% CI 0.44–1.28) (Goossens et al. 2015).

7.3.2 Pomegranate

The pomegranate (*Punica granatum* L.) not only has been used medicinally for centuries but is also held sacred in many major religions. Perhaps most famous for the Greek myth of 'Persephone's abduction by Hades', it features on the coats of arms of the British Medical Association, symbolizing 'life' and 'regeneration' (Langley 2000).

The active ingredients of pomegranate juice are polyphenol punicalagins and ellagic acid, which have robust antioxidant properties (Gil et al. 2000). In addition, pomegranate juice extract can inhibit the nuclear factor k-light-chain-enhancer of activated B-cells (NF-κB) pathway in prostate cancer cell line experiments, an inflammatory pathway implicated in the pathogenesis of prostate cancer (Rettig et al. 2008). The first clinical study of pomegranate extract in patients with prostate cancer was conducted by Pantuck et al. This was a phase II clinical trial that included men with rising PSA levels after treatment with surgery or radiotherapy (Pantuck et al. 2006). Eligible patients had a detectable PSA level of > 0.2 and < 5 ng/mL and Gleason score of ≤ 7. Patients were treated with 227 mL of pomegranate juice daily. Clinical endpoints included safety and effect on serum PSA level. The mean PSA doubling time significantly increased with treatment from 15 months at baseline to 54 months after treatment (P < 0.001). In addition, the clinical data was accompanied by *in vitro* experiments, where treatment of prostate cancer cell lines with pomegranate extract showed a decrease of 12% in cell proliferation and a 17% increase in apoptosis (P = 0.005 and P < 0.001, respectively).

Recent studies have provided similar results. The UK NCRD Pomi-T study was a placebo-controlled, double-blind randomized trial of 199 men with localized prostate cancer either being managed with active surveillance or watchful waiting following conventional intervention (Thomas et al. 2014). Participants were randomized to receive either a supplement rich in polyphenol-rich foods (pomegranate, green tea, broccoli, and turmeric) or placebo for six months. Compared to baseline, median PSA levels rose 14.7% in the treatment group, compared to 78.5% in the control group. Paller et al. conducted a randomized, multi-center, double-blind trial of the effects of POMx on participants with histologically confirmed adenocarcinoma of the prostate who had undergone radical therapy with either prostatectomy, external beam radiation therapy, cryotherapy, and/or brachytherapy. Patients with metastasis were excluded. One hundred eight participants were randomized to receive either 1g or 3g of POMx per day. Baseline PSA doubling time (PSADT) was calculated from pre-study baseline PSA results. The trial was finalized 18 months after the last participant enrolled. Both arms experienced a significant increase in PSADT, with no difference observed between the dosing regimens; PSADT in the low-dose group went from 11.9 to 18.8 months (P < 0.001) and 12.2 to 17.5 months in the high-dose group (P < 0.001). Additionally, POMx was well tolerated (Paller et al. 2013).

These results do not correlate with the early results available from randomized trials. Freedland et al. conducted a double-blind, randomized study of 70 men due to undergo radical prostatectomy. Participants received either pomegranate extract (POMx, Pom Wonderful, Los Angeles, CA) or placebo daily for four weeks prior to surgery. Resected samples were analyzed for intra-prostatic Urolithin A (a pomegranate metabolite), 8-oxo-2'-deoxyguanosine (a major marker of oxidative stress), pS6 kinase and NF-κB (biomarkers of prostate cancer inflammation, development, and progression), and Ki-67 (a biomarker of cell proliferation). Urolithin A was detected in 21/33 in the treatment group, and 12/35 in the control (P = 0.031). Although POMx was associated with 16% lower benign tissue 8-OHdG, this was not significant. There were no differences between treatment and control groups in levels of other biomarkers (Freedland et al. 2013).

Although pomegranate juice does seem to increase PSADT, more clinical trials are needed investigating its effect on prostate cancer tissues. It remains an agent of great interest, reflected by the multiple cancer trials registered with ClinicalTrials.gov database investigating the impact of pomegranate supplementation on prostate cancer treatment or prevention.

7.3.3 Green Tea

Green tea, derived from the plant *Camellia sinensis*, has been consumed in the Far East for many centuries. The 'Tea Classic', written around AD 760 by Lu Yu, is recognized as the first of many revered treatises on the subject.

Green tea contains polyphenolic compounds, the most abundant of which is epigallocatechin-3-gallate (EGCG), an antioxidant that is more potent than both vitamin C and E (Cao et al. 2002). Suggested mechanisms of the anti-tumorigenic action of green tea include apoptosis and cell cycle arrest via alterations in mitogen-activated protein kinase, phosphatidylinositol3-kinase (PI3K)/ Akt and protein kinase C pathways, inhibition of inflammatory pathways NF-κB and cyclooxygenase-2 (COX-2), and modulation of the IGF and androgen receptor (AR) axes (Hori et al. 2011). Specifically, EGCG acts as a direct antagonist of androgen action and physically interacts with the ligand binding domain of the AR (Siddiqui et al. 2011).

Preclinical data on the use of green tea for chemoprevention of prostate cancer were generated using the transgenic adenocarcinoma mouse protocol (TRAMP) model, an animal model of mice that are genetically engineered to develop prostate adenocarcinoma. In this study green tea extract (GTE) was administered orally to TRAMP mice for 24 weeks, achieving a 40% reduction in localized prostate tumor development at 20 and 30 weeks. Furthermore, administration of GTE negated the metastatic spread to lymph nodes, liver, lungs, and bone, improving survival by 70% compared with the control mice (Gupta et al. 2001).

Some clinical trials have been undertaken to assess its therapeutic or chemopreventive properties. HGPIN is the main pre-malignant lesion of prostate cancer and will result in a substantial number of cancers in a one-year period (Bishara et al. 2004). A double-blind, placebo-controlled study investigated the effect of consumption of 600 mg GTE/day on progression to prostate cancer in 60 men who were diagnosed with HGPIN. After one year of treatment, only one man was diagnosed with prostate cancer among the 30 men who received GTE daily. In comparison, nine cancers were found among the 30 men treated with placebo. The results suggest a 90% chemoprevention efficacy of GTE in men with a high risk of developing prostate cancer (P < 0.01). Serum PSA levels showed a non-significant decrease in GTE-treated men at 9 and 12 months. Overall, the results of this study suggest that administration of GTE could be an effective therapy of premalignant lesions in high-risk men, allowing intervention before prostate cancer develops (Bettuzzi et al. 2006). A two-year follow-up assessment of the above study was published in 2008 (Brausi et al. 2008). Despite a significant loss of patient follow-up with only nine participants from the placebo arm and 13 from the GTE arm remaining, the cohort underwent a third biopsy. Two further cancer diagnoses appeared in the placebo arm and one in the GTE arm. Overall, even after suspension of the GTE treatment, the GTE arm had an almost 80% reduction in prostate cancer diagnosis compared with the placebo group, though due to the small number of patients these results are not significant.

Some studies have also investigated the role of green tea in patients with confirmed prostate cancer. One of the studies was an open-label, single-arm, two-stage phase II clinical trial that evaluated the effects of four capsules of polyphenon E daily containing a total daily dose of 800 mg EGCG administered during the interval between prostate biopsy and radical prostatectomy (McLarty et al. 2009). Serum was collected before initiation of the drug study and on the day of prostatectomy and analyzed with ELISA for levels of different biomarkers. There was a significant decrease in serum PSA, hepatic growth factor (HGF), vascular endothelial growth factor (VEGF), IGF-1, and the ratio of IGF-1 to IGF binding protein 3 (IGFBP-3). The aforementioned biomarkers are used for ascertaining the metastatic and malignant potential of prostate cancer. However, the results of this study must be interpreted with caution as other trials assessing the effects of green tea on prostate cancer biomarkers and progression have not been as successful. A study from Jatoi et al. used a dosing schedule of 6 g green tea/day in water in six divided doses (Jatoi et al. 2003). Only 1 of 42 patients showed a > 50% reduction in PSA level, and most patients complained of tea toxicity. In conclusion, although green tea appears to have some benefit in the prevention of the progression of HGPIN to prostate cancer, there is no data currently supporting the benefit of increased consumption of green tea amongst men with confirmed prostate cancer.

Evidence supporting the role of green tea in bladder cancer is only restricted to animal models and cell-based studies. Sagara et al. investigated the effect of green tea polyphenol on bladder tumor size and angiogenesis in mice given N-butyl-(-4-hydroxybutyl) nitrosamine (BBN), with and without green tea polyphenol (Sagara et al. 2010). The results showed that green tea polyphenol had no anti-carcinogenic effect, with all mice exposed to BBN developing bladder cancer; however, green tea polyphenol inhibited tumor growth and invasion in mice with established bladder cancer, mostly through the regulation of angiogenesis. Another cell-based study investigated the growth inhibition and cell cycle arrest effects of EGCG on the NBT-II bladder tumor cell line. The results suggest that by down-regulating the cyclin D1, cyclin-dependent kinase 4/6, and retinoblastoma protein machinery, EGCG can inhibit growth and promote cell cycle arrest in NBT-II bladder tumor cell lines (Chen et al. 2004).

Although cellular and animal-based studies are promising, in humans no laboratory studies exist. Nagano et al. conducted a mail study of 38,540 adults living in Hiroshima and Nagasaki and followed them up for over ten years. This large-scale study found no association between consumption of green tea and incidence of bladder cancer (Nagano et al. 2001). A subsequent meta-analysis of the literature conducted by Wu et al. similarly found no association between green tea and bladder cancer (Wu et al. 2013). The potential of green tea as a treatment for bladder cancer in humans requires further evaluation.

7.3.4 Curcumin

Curcumin is derived from turmeric, the spice that makes curry powder yellow. Aside from its use as an ingredient and a dye, has been used in traditional medicine systems for many years. The interest in its antineoplastic properties stems from epidemiological studies correlating prostate cancer incidence and dietary curcumin intake (Hebert et al. 1998).

The major component of curcumin is the lipophilic polyphenol curcumin I (Klempner and Bubley 2012). The anti-cancer activity of curcumin relates to its anti-inflammatory and antiproliferative properties. Curcumin has been shown to suppress the proliferation of both the androgen-dependent prostate cancer cell line LNCaP and the androgen-independent DU145 line (Aggarwal 2008). Its mechanism of antiproliferative action is through its influence on multiple signaling pathways. Curcumin can also inhibit VEGF and angiogenesis *in vivo* (Gururaj et al. 2002). Inflammatory mediators, e.g. COX2 and NF-κB, are also down-regulated by curcumin, expression of which has been associated with a higher prostate tumor grade angiogenesis metastatic potential of prostate cancer (Huang et al. 2001). The ability of curcumin to potentiate the cytotoxicity of chemotherapy agents in prostate cancer cell line experiments has also recently been documented.

Specifically, synergism between curcumin and 5-fluorouracil and paclitaxel has been reported (Hour et al. 2002). Although it is apparent from such experiments that curcumin can exert anticancer properties via multiple targets, little is known about the dose required to achieve the same results *in vivo*. Encouraging animal trials are beginning to emerge (Pan et al. 2017), but more research is needed to realize its potential in humans.

7.3.5 *Resveratrol*

Resveratrol is a polyphenol compound found in the skin of red grapes and other plants. Its role in the 'French Paradox' has been extensively studied, and it is thought to prevent coronary heart disease. Indeed, Louis Pasteur is often attributed as having believed that 'wine is the most healthful and most hygienic of beverages'.

Its potential as a chemopreventive and chemotherapeutic agent has been shown in numerous *in vitro* and *in vivo* studies of human cancers, and like curcumin it can modulate various target and signaling pathways involved in tumorigenesis. The growth inhibitory effects of resveratrol in prostate cancer lie mainly in its effect on the AR, which plays a pivotal role in prostate tumorigenesis. Firstly, resveratrol can regulate AR target gene expression by repressing AR transcriptional activity (Shi et al. 2009), as well as promoting AR degradation (Harad et al. 2007). In addition, resveratrol can repress different classes of androgen-responsive genes, including PSA, human glandular kallikrein-2, AR-specific co-activator, and ARA70 in hormone-responsive cells (Athar et al. 2007), all involved in prostate tumorigenesis and even the development of hormone-refractory prostate cancer (Niu et al. 2008). Furthermore, resveratrol may be able to modulate PI3Ks, which are a family of enzymes involved in cellular functions, e.g. cell growth, proliferation, differentiation, motility, survival, and intracellular trafficking, which in turn are involved in cancer (Chen et al. 2010).

The cellular effects of resveratrol-mediated PI3K inhibition may be partly mediated by inhibition of Forkhead box, class O (FOXO) family tumor suppressor phosphorylation, thereby allowing upregulation of proapoptotic and antiproliferative FOXO targets, e.g. Bcl-2 interacting mediator of cell death (Bim), TNF-related apoptosis inducing ligand (TRAIL), TRAIL death receptors DR5 and DR4, and p27 (Chen et al. 2010).

Interestingly, resveratrol has anticancer activity in prostate cancer, which can be attributed to modulation of microRNAs. MicroRNAs are small non-coding RNAs that negatively regulate gene expression and have recently received much attention as a cancer therapeutic target.

Only a few studies have been published on the effect of resveratrol on bladder cancer cells. These studies, which are mainly cell-based experiments using bladder cancer cell lines, have yielded promising results. In a study by Stocco et al., cells of the bladder cancer cell line ECV304 were incubated with different resveratrol concentrations (Stocco et al. 2012). Resveratrol-induced cell death at high concentrations (> 20 μM) but not at low ones (0.1–20 μM). Pretreatment with 2.5 μM protected the cells from oxidative damage, whereas 50 μM intensified cell death and significantly increased the Bad/Bcl-2 ratio, in this way driving the cells to apoptosis in the presence of a high concentration of the pro-apoptotic protein Bad. Another study by Bai et al. using the T24 bladder cell line confirmed that incubation of bladder cancer cells with resveratrol can induce apoptosis through the Bcl-2 family of pro-apoptotic proteins (Bai et al. 2010).

Only a few *in vivo* animal studies have been completed using either transgenic or xenografted mice (Jasinski et al. 2013). From the available data, it appears that resveratrol may slow prostate tumor formation and progression, but may have the unwanted side-effect of speeding up angiogenesis (Walle et al. 2004).

The use of resveratrol in human trials has proved difficult due to poor oral bioavailability, which is estimated to be 1% with peak plasma concentrations much below the doses used in cell line experiments (Walle et al. 2004). Other grape skin extracts, e.g. muscadine grape (*vitis rotundifolia*) skin extract, which does not contain resveratrol, have also been investigated in prostate cancer. Cell-based experiments have shown that it can induce apoptosis in multiple prostate cancer cell lines by inhibiting Akt signaling (Hudsonet al. 2007). These promising results have led to a phase I/II study of muscadine for men with biochemical recurrence of prostate cancer (NCT01317199).

7.3.6 *Silibinin*

Silibinin is a natural phenol of the flavonolignan family derived from seeds of the milk thistle plant. Milk thistle has been used as a traditional medicine for centuries, with its ability to 'carry the bile' extolled by Pliny the Elder. Throughout the ages, it has been used to treat various ailments of the blood, liver, and kidneys and also to ward off snakes (Grieve 1971).

Evidence exists from animal models of prostate cancer that silibinin can be effective in all stages of tumor progression, not only decreasing the incidence of the tumor itself but decreasing metastatic potential and reducing the size of established prostate tumors in the TRAMP murine model. Experiments by Singh et al. (2008) in the TRAMP model show that *silibinin* treatment decreased proliferation and increased apoptosis in TRAMP tumors. A critical underlying factor involved in the steps towards tumor progression is invasion and angiogenesis. Raina et al. (2008) showed that silibinin can decrease both VEGF and VEGF receptor 2 levels, contributing in this way to the antiangiogenic properties of this agent. In addition, silibinin reduced levels of matrix metalloproteinases-2 and -3 in the TRAMP models, both proteins involved in tumor invasion and metastases.

The chemopreventive and chemotherapeutic effects of silibinin in bladder cancer have also been investigated. Intravesical chemotherapy is often used to reduce the risk of recurrence of superficial bladder tumors after transurethral resection. A recent experiment assessed the effectiveness of intravesical silibinin in preventing the recurrence of superficial bladder tumors in rats. This study yielded positive results, suggesting that silibinin can be an effective and novel intravesical agent for bladder cancer (Zeng et al. 2011).

Although silibinin has shown interesting preventative and anticancer properties in prostate and bladder cancer animal models, no human RCTs investigating the chemotherapeutic effects of silibinin on prostate and bladder cancer have been reported.

7.3.7 *Ginkgo biloba*

The *Ginkgo biloba* tree is native to China. As a living fossil, it symbolizes longevity and vitality in traditional Chinese medicine, amongst others; it is commonly used in the treatment of prostate pathology. It deserves briefly mentioning due to the widespread use; however, currently, there is a dearth of evidence.

Ginkgetin is a biflavonoid found in the leaves of the ginkgo biloba tree. There is a great deal of research ongoing of the effects of dietary flavonoids on human health (the search term 'flavonoid' returns over 100 registered trials on clinicaltrials.gov at time of writing). You et al. investigated the effects of various concentrations of ginkgetin on PC-3 cells, finding that it induced apoptosis in a dose-dependent manner. Western blotting showed decreased concentrations of Bcl-2, Bcl, survivin, and cyclin D1, a cell cycle regulatory protein, suggesting the induction of apoptosis through inhibition of cell survival genes (You et al. 2013). The effect was replicated in human prostate cancer cells DU-145. Signal transducer and activation of transcription (STAT) proteins contribute to regulation of cell growth, survival, and differentiation; STAT3 is constitutively activated in human cancers. A STAT3-dependent assay of natural compounds identified ginkgetin as a potent inhibitor-treated DU-145 arrested at the G0/G1 phase and underwent apoptosis. This effect was replicated in xenograft mice, with tumor volume and weight significantly reduced compared to control after termination at three weeks (Jeon et al. 2015).

Epidemiological evidence suggests this is not replicated in humans. Biggs et al. looked at secondary outcomes from the 'Gingko Evaluation of Memory' study. This prospective study followed 3,069 patients over the age of 75 years randomized to receive either 120 mg of ginkgo extract twice a day or placebo for a median of 6.1 years. The treatment group showed reduced absolute rates of prostate cancer compared to control, but this was not significant (Biggs et al. 2010). In a questionnaire study, over 35,000 men living in Washington self-reported use of supplements. After six years of follow up, there was no association between use of ginkgo and prostate cancer (Brasky et al. 2011). Perhaps as a reflection of this, human *in vivo* studies have not been reported.

7.3.8 Modified Citrus Pectin

Pectin is used as a thickener in jams and marmalades. Citrus pectin is not digestible in humans; it is modified by decreasing the molecular weight, which, it is believed, improves absorbability. It has been promoted as an alternative cancer treatment, but evidence for efficacy is limited.

Pectasol and Pectasol-C (EcoNugenics, Santa Rosa, CA) are commercially available modified citrus pectins (MCPs). Yan and Katz demonstrated cytotoxic efficacy of these preparations on LNCaP and PC3 cells (Yan and Katz 2010); in contrast, another study observed no apoptotic effect of Pectasol on LNCaP cells (Jackson et al. 2007). Jackson et al. did, however, note significant apoptotic effect of fractionated pectin power on LNCaP and LNCaP C4-2 cells (Jackson et al. 2007).

Pienta et al. have shown that MCP inhibits rat prostate cancer metastasis. Rats were injected with rat MAT-LyLu prostate cancer cells. The treatment groups were fed varying concentrations of MCP continuously in their drinking water. At necropsy at day 30, 94% of the control rats had lung metastases compared to 56% of the rats fed a 1% concentration of MCP (P < 0.001) (Pienta et al. 1995). To the author's knowledge, this effect has not been observed with human prostate cancer cell lines *in vivo*.

Thirteen men with newly diagnosed biopsy-confirmed prostate cancer were enrolled to receive 14.4 g of PectaSol capsules per day after appropriate conventional treatment (radical prostatectomy, radiation therapy, cryotherapy, or intermittent androgen-deprivation therapy). Three patients withdrew due to mild side-effects. For six months or more post-procedure, PSA was measured to obtain PSA doubling times (PSA-DT). After participants had returned at least three rising PSA levels, PectaSol was started. PSA levels were then taken to obtain post-PectaSol PSA-DT. Eight of the ten included participants demonstrated increasing PSA-DT after taking PectaSol; seven of these were significant (Guess et al. 2003). Although this study had highly significant results, PSA levels do not necessarily correlate with severity; additionally, sample size was small. To knowledge, no other human studies have been conducted.

7.3.9 *Phellodendron amurense*

More commonly known as the Amur cork tree, the bark of *Phellodendron amurense* (huángbǎi) is one of the 50 traditional herbs in Chinese medicine. It is available commercially as Nexrutine®, an extract developed by Next pharmaceuticals, and is marketed as a remedy for muscle soreness.

The prostaglandins are hormone-like signaling molecules which have a diverse range of effects. Prostaglandin E2 is known to inhibit apoptosis, stimulate angiogenesis, and promote metastasis and immunosuppression. It is produced by the action of cyclooxygenase-2 on arachadonic acid. High levels of Cox-2 have been found in various types of tumor cells, including prostate, where it is proposed chronic inflammation contributes to the development neoplasia. Ghosh et al. studied the effects of Nexrutine on human prostate cancer cell lines PC-3 and LNCaP. It was found to have an antiproliferative effect. Polymerase chain reaction showed that the Cox-2 level was significantly reduced in PC-3 cells treated with Nexrutine®. In LNCaP cells treated with tumor necrosis factor alpha, high levels of PGE2 were observed in response. Nexrutine significantly reduced production of PGE2. They conclude that Nexrutine exerts its antiproliferative effects on prostate cancer cells through inhibition of Cox-2 (Ghosh et al. 2007). These results have been reproduced, with the butanol fraction of Nexrutine shown to be responsible for the observed biological activity; the compound palmatine and its structural analog berberine are thought to be the active agents (Hambright et al. 2015, Muralimanoharan et al. 2009).

These effects have been confirmed in animal studies. Transgenic adenocarcinoma of mouse prostate (TRAMP) mice (which develop prostate cancer with 100% frequency) were fed pellet diets containing varying regimens of Nexrutine. The mice underwent MRI prostate evaluation at weeks 18 and 28, and tissue samples were examined for histology at termination. Mice fed the Nexrutine-containing diet had a significantly reduced tumor size and slower tumor progression compared to control (Kumar et al. 2007). This correlated with results from a subsequent study; additionally, an observed effect of the higher dose Nexrutine regimen in this study was lower rates of metastases (although sample size was small) (Ghosh et al. 2010).

No direct human studies have been undertaken, although its use as an adjunct to conventional therapy has shown it to be well tolerated (Swanson and Jones 2015).

7.3.10 Red Clover

Trifolium pratense (red clover) is commonly used as a fodder crop. It contains many different isoflavones, which are phytoestrogens (so-called 'dietary estrogens'). Epidemiologically, it has been noted that Asian men suffer a lower incidence of prostate cancer compared to their Western cousins; this has been in part attributed to their higher intake of dietary isoflavones. For these reasons, it has been hypothesized that red clover may be of use in the treatment of prostate cancer.

Research to date has mainly focused on biochanin-A and formononetin, two of the major isoflavones found in red clover. Biochanin-A has been shown to induce apoptosis *in vitro* in prostate cancer cell lines (Hempstock et al. 1998, Peterson and Barnes 1993). Tumor necrosis factor related apoptosis inducing ligand (TRAIL) is cleaved from the surface of immune cells. It binds to death receptors (DRs) on the surface of cancer cells, inducing apoptosis. Szliska et al. incubated prostate cancer cell lines with biochanin-A, TRAIL, and the two in combination. Both alone were effective in inducing apoptosis compared to control, however, the combined treatment significantly enhanced apoptosis compared to either agent alone. Biochanin-A was found to up-regulate expression of DRs in prostate cancer cells and also inhibit transcription factor NF-κB (which protects tumor cells from apoptosis) (Szliszka et al. 2013). Formononetin is effective against human prostate cancer cell lines *in vitro* compared to control (Liu et al. 2014, Ye et al. 2012, Zhang et al. 2014). There are a number of proposed mechanisms. Formononetin upregulates expression of RASD1, a gene implicated in regulating cell differentiation (Liu et al. 2014). Western blot analysis has also shown that formononetin inhibits production of Bcl-2 (an anti-apoptotic protein) whilst increasing production of Bax (a pro-apoptotic) (Zhang et al. 2014). Furthermore, it also inactivates the extracellular signal regulated kinase 1/2 (ERK1/2) mitogen-activated protein kinase (MAPK) pathway (known to regular cellular mitosis) (Ye et al. 2012).

Animal studies confirm the studied cellular effects. Prostatic sections of rats fed a diet rich in red clover isoflavones were shown to have significantly increased levels of estrogen receptor β (known to inhibit cell proliferation) and decreased levels of transforming growth factor β1 (which functions to increase proliferation and differentiation) (Slater et al. 2002). Mice implanted with LNCaP xenografts were injected with biochanin-A daily for ten days. At three weeks, tumor incidence in the treatment group was significantly lower; however, by six weeks, incidence was similar. Tumor volume was lower in the treatment group until week nine, when volumes were similar. The authors conjectured that repeated dosing may sustain the observed effects (Rice et al. 2002).

In a small study of men undergoing radical prostatectomy, patients were assigned to consume 160 mg/day of red clover derived isoflavones or to control. Biochemical parameters were compared, as well as prostatic specimens post-surgery. Although there was no significant difference in serum PSA or testosterone pre- or post-surgery, the histological analysis showed significantly higher rates of apoptosis in the treatment arm, specifically in regions of Gleason grades 1–3 cancer. Additionally, the red clover supplement was well tolerated, with no adverse effects (Jarred et al. 2002). These results correspond with the animal studies, suggesting red clover may be of use in delaying progression of and possibly reducing tumor load in low-grade tumors. Its sole use as a treatment agent in humans remains poorly understood; it currently could be considered a useful adjunctive therapy.

7.3.11 *Salvia*

Salvia is the largest genus of plants in the mint family, containing an estimated 1,000 species. Notable members include *Salvia officinialis* (common garden sage) and *Salvia divinorum*, notorious for inducing psychoactive effects. Diterpenes isolated from plants of the Salvia genus have been shown to have (among others) antioxidant, anti-inflammatory, and antitumor effects. Several species of Salvia have been studied for their antiproliferative effects *in vitro*. Fiore et al. examined the effect of crude methanol extracts from six Salvia species on human cancer cell lines; of the studied species, five demonstrated effective antitumor activity, with *S. sclarea* and *S. spinosa* found to be particularly effective against prostate adenocarcinoma (Fiore et al. 2006). (Interestingly, *S. sclarea* has long been suspected to have health-promoting properties; Nicholas Culpeper, in his seventeenth-century treatise 'Complete Herbal', comments that 'the mucilage of the seed (of *S. sclarea*) made with water, and applied to tumours or swellings, dispersed and take them away' (Culpeper 1826).) The essential oils of *S. officinialis* have been shown to be effective *in vitro* against renal cell adeno-carcinoma (Loizzo et al. 2007).

The roots of *S. miltiorrhiza* (more commonly known as 'danshen') have long been prized in traditional Chinese medicine. The major bioactive derivatives from 'danshen' are tanshinones. It has been hypothesized that they inhibit growth and have cytotoxic effects on various cancer cell lines, including prostatic, through androgen receptor antagonism. Zhang et al. examined the effects of tanshinones both on prostatic cancer cell lines and in mice inoculated with LNCaP cells. *In vitro*, tashinones inhibited growth of prostate cancer cells, being more effective in the androgen-dependent stage. *In vivo*, tumor-bearing mice were randomized to receive corn oil, tanshinone IIA, or Casodex (a commonly used anti-androgen) by gavage. It was found that TIIA significantly decreased protein abundance of androgen receptors, whereas Casodex merely decreased PSA concentration. TIIA antagonized androgen receptor signaling, induced apoptosis, and exerted anti-angiogenetic effects (Zhang et al. 2012). Similar results have been found by other research teams, both *in vitro* and *in vivo* (Gong et al. 2011). Other studies confirm this effect, but propose an androgen receptor independent mechanism; Chuang et al. suggest that tanshinones inhibit proteasome activity, which increases endoplasmic reticular stress and subsequently induces apoptosis (Chuang et al. 2011).

Human studies examining the effects of the more promising *Salvia* species are, sadly, lacking. However, this plant genus remains a promising area of research.

7.3.12 Mistletoe

Mistletoe lectins are thought to exert antiproliferative, cytotoxic, and immunomodulatory effects. They induce release of various cytokines as well as activating natural killer cells, lymphocytes, and antibody-dependent cell-mediated cytotoxicity (Urech et al. 2006).

In animal models, evidence for the induction of urinary bladder carcinogenesis by intravesical administration of mistletoe lectin is equivocal. One study treated rats with chemically induced bladder cancer with differing regimens of intravesical recombinant mistletoe lectin. All experimental groups showed significantly lower rates of atypical hyperplasia and neoplastic transformation; this was most significant in the maximum treatment group (Elsasser-Beile et al. 2001). In contrast, Kunze et al. found insignificant differences between rats with chemically induced carcinoma treated with mistletoe lectin compared with control over both 6 and 15 months (Kunze et al. 2000). Of note, these trials differed in their modes of administration; Elasser-Beile et al. treated intravesically, whereas Kunze et al. treated subcutaneously.

Subsequent human studies provide encouraging results, showing use of intravesical mistletoe lectin in varying regimes to be comparable to the standard of intravesical Bacillus Calmette–Guerin in the adjuvant treatment of non-muscle invasive bladder carcinoma (Elsasser-Beile et al. 2005a, Elsasser-Beile et al. 2005b, Rose et al. 2015). Additionally, researchers found it to be well tolerated, even at higher doses. However, results are not all positive—a study of 45 patients with superficial bladder cancer evaluated the efficacy of subcutaneously administered mistletoe lectin as an adjunctive therapy after TURBT. Time to first recurrence and total number of recurrences was similar in both treatment and control groups (Goebell et al. 2002). Intravesical mistletoe lectin shows great promise as a treatment but needs further evaluation.

7.3.13 Combined Therapies

Several CAM agents exist in combination within the same preparation. The most well-known combined CAM therapy that has been used in prostate cancer is PC-SPES. PC-SPES (BotanicLab, Brea, CA, USA) was an herbal mixture introduced in 1996 consisting of chrysanthemum, *isatis*, *liquorice*, *Ganoderma lucidum*, *Panax pseudoginseng*, *Rabdosia rubescens*, saw palmetto, and *scutellaria* (skullcap) (DiPaola et al. 1998). Despite initial convincing laboratory and animal studies indicating potent anticancer properties of PC-SPES that led to the initiation of multiple RCTs, PC-SPES was withdrawn from the market in 2002 due to contamination with prescription drugs, e.g. warfarin. Furthermore, there was a high variation in the composition of the blend between lots. In addition to the contaminants found in PC-SPES, patients reported various adverse effects including gynecomastia (almost universal), loss of libido, erectile dysfunction, leg cramps, nausea, and thromboembolism (Sovak et al. 2002).

A more recent herbal combined therapy familiar to many patients with prostate cancer is Zyflamend (New Chapter Inc., Brattleboro, VT, USA), an herbal blend consisting of extracts from rosemary, turmeric, ginger, holy basil, green tea, huzhang, Chinese goldthread, barberry, oregano, and Chinese skullcap. An *in vitro* study on human prostate cancer cell lines showed inhibition of COX-1 and COX-2 enzymatic activities, cell growth suppression, and induced apoptosis (Bemis et al. 2005). Furthermore, in a mouse model that mimics advanced disease stages, Zyflamend was found to inhibit androgen-dependent and castrate-resistant tumor growth (Huang et al. 2012). Robust evidence on the potential role of Zyflamend as a therapeutic agent comes from a phase I clinical trial assessing the effect of Zyflamend with various dietary supplements in 23 men with HGPIN. The results showed that 48% of the men had a 25–50% reduction in PSA levels after 18 months. Nine of 15 patients biopsied at 18 months were found to have benign disease, and only two had progressed to carcinoma of the prostate. No serious adverse effects were reported with this preparation (Capodice et al. 2009). More recently, Bilen et al. have reported a case series of four patients with worsening prostate cancer after conventional treatment supplemented with Zyflamend, with or without metformin. PSA dramatically improved in all cases. They propose that Zyflamend may target cancer stem cells and the tumor microenvironment to prevent dormant cancer activating, although do not discuss exact mechanisms (Bilen et al. 2015).

Profluss (Konpharma, Rome, Italy), a combined therapy, contains *serenoa repens* (derived from the berries of the saw palmetto tree), selenium, and lycopene (a dietary carotenoid with strong antioxidant activity). In the multicenter 'Flogosis And Profluss in Prostatic and Genital Disease' (FLOG) study, 168 subjects were divided into two groups. The first (108 subjects) had LUTS with prostatic chronic inflammation associated with BPH and high-grade PIN and/or atypical small acinar proliferation (ASAP). This group was further subdivided into a control group (Ic) or a treatment group which received Ser320mg + Se50mcg + Ly5mg per day (I). The second group (60 participants) had LUTS with prostatic chronic inflammation associated with BPH, but no diagnosis of PIN and/or ASAP. As in group 1, group 2 was further subdivided into a control group (IIc) or a treatment group which received Ser320mg + Se50mcg + Ly5mg per day (II). Group 1 (I and Ic) patients underwent repeat biopsies at six months. Group 2 (II and IIc) underwent repeat biopsies at three months followed by TURP. Profluss was shown to reduce histological inflammation in group 1 compared to control. Additionally, there was a statistically significant reduction in the extension and grading of the inflammatory cell in filtrate (Morgia et al. 2013). These results suggest that Profluss may be of use in managing patients with LUTS secondary to PCI associated with BPH and PIN/ASAP.

7.4 CONCLUSION

There is a multitude of plant-based therapies used by patients with prostate and bladder cancer. The majority of these have established cellular mechanisms of action, with some having evidence of efficacy in animal or human trials. However, data is often lacking or inconclusive. With so many plant-based therapies having promising early results, further *in vivo* randomized controlled trials are needed to provide solid evidence of efficacy. Until these studies have been completed, recommendations for plant-based therapies are likely to remain within the realms of level III and level IV evidence.

REFERENCES

Aggarwal, B.B. 2008. Prostate cancer and curcumin: Add spice to your life. *Cancer Biolther.* 7, no. 9:1436–1440.

Athar, M., J.H. Back, X. Tang, et al. 2007. Resveratrol: A review of preclinical studies for human cancer prevention. *Toxicol. Appl. Pharmacol.* 224, no. 3:274–283.

Bai, Y., Q.Q. Mao, J. Qin, et al. 2010. Resveratrol induces apoptosis and cell cycle arrest of human T24 bladder cancer cells *in vitro* and inhibits tumor growth *in vivo. Cancer Sci.* 101, no. 2:488–493.

Beebe-Dimmer, J.L., D.P. Wood Jr., S.B. Gruber, et al. 2004. Use of complementary and alternative medicine in men with family history of prostate cancer: A pilot study. *Urology* 63, no. 2:282–287.

Bemis, D.L., J.L. Capodice, A.G. Anastasiadis, A.E. Katz, and R. Buttyan. 2005. Zyflamend, a unique herbal preparation with nonselective COX inhibitory activity, induces apoptosis of prostate cancer cells that lack COX-2 expression. *Nutr. Cancer* 52, no. 2:202–212.

Bettuzzi, S., M. Brausi, F. Rizzi, G. Castagnetti, G. Peracchia, and A. Corti. 2006. Chemoprevention of human prostate cancer by oral administration of green tea catechins in volunteers with high-grade prostate intraepithelial neoplasia: A preliminary report from a one-year proof-of-principle study. *Cancer Res.* 66, no. 2:1234–1240.

Biggs, M.L., B.C. Sorkin, R.L. Nahin, L.H. Kuller, and A.L. Fitzpatrick. 2010. Ginkgo biloba and risk of cancer: Secondary analysis of the Ginkgo Evaluation of Memory (GEM) Study. *Pharmacoepidemiol. Drug Saf.* 19, no. 7:694–698.

Bilen, M.A., S.H. Lin, D.G. Tang, et al. 2015. Maintenance therapy containing metformin and/or zyflamend for advanced prostate cancer: A case series. *Case Rep. Oncol. Med.* 2015:1–5.

Bishara, T., D.M. Ramnani, and J.I. Epstein. 2004. High-grade prostatic intraepithelial neoplasia on needle biopsy: Risk of cancer on repeat biopsy related to number of involved cores and morphologic pattern. *Am. J. Surg. Pathol.* 28, no. 5:629–633.

Brasky, T.M., A.R. Kristal, S.L. Navarro, et al. 2011. Specialty supplements and prostate cancer risk in the VITamins and Lifestyle (VITAL) cohort. *Nutr. Cancer* 63, no. 4:573–582.

Brausi, M., F. Rizzi, and S. Bettuzzi. 2008. Chemoprevention of human prostate cancer by green tea catechins: Two years later. A follow-up update. *Eur. Urol.* 54, no. 2:472–473.

Cao, Y., R. Cao, and E. Bråkenhielm. 2002. Antiangiogenic mechanisms of diet-derived polyphenols. *J. Nutr. Biochem.* 13, no. 7:380–390.

Capodice, J.L., P. Gorroochurn, A.S. Cammack, et al. 2009. Zyflamend in men with high-grade prostatic intraepithelial neoplasia: Results of a phase I clinical trial. *J. Soc. Integr. Oncol.* 7:43–51.

Carter, L.G., J.A. D'Orazio, and K.J. Pearson. 2014. Resveratrol and cancer: Focus on *in vivo* evidence. *Endocr. Relat. Cancer* 21, no. 3:R209–225.

Chan, J.M., P.H. Gann, and E.L. Giovannucci. 2005. Role of diet in prostate cancer development and progression. *J. Clin. Oncol.* 23, no. 32:8152–8160.

Chen, J.J., Z.Q. Ye, and M.W. Koo. 2004. Growth inhibition and cell cycle arrest effects of epigallocatechin gallate in the NBT-II bladder tumour cell line. *BJU Int.* 93, no. 7:1082–1086.

Chen, Q., S. Ganapathy, K.P. Singh, S. Shankar, and R.K. Srivastava. 2010. Resveratrol induces growth arrest and apoptosis through activation of FOXO transcription factors in prostate cancer cells. *PLOS ONE* 5, no. 12:e15288.

Chuang, M.T., F.M. Ho, C.C. Wu, et al. 2011. 15,16-Dihydrotanshinone I, a compound of Salvia miltiorrhiza Bunge, induces apoptosis through inducing endoplasmic reticular stress in human prostate carcinoma cells. *Evid. Based Complement Alternat. Med.* 2011:1–9.

Culpeper, N. 1826. *Complete Herbal, and English Physician*. J. Gleave and Sons, Manchester.

Dennert, G., M. Zwahlen, M. Brinkman, M. Vinceti, M.P. Zeegers, and M. Horneber. 2011. Selenium for preventing cancer. *Cochrane Database Syst. Rev.* 5:CD005195.

DiPaola, R.S., H. Zhang, G.H. Lambert, et al. 1998. Clinical and biologic activity of an estrogenic herbal combination (PC-SPES) in prostate cancer. *N. Engl. J. Med.* 339, no. 12:785–791.

Elsässer-Beile, U., C. Leiber, U. Wetterauer, et al. 2005a. Adjuvant intravesical treatment with a standardized mistletoe extract to prevent recurrence of superficial urinary bladder cancer. *Anticancer Res.* 25, no. 6C:4733–4736.

Elsässer-Beile, U., C. Leiber, P. Wolf, M. Lucht, U. Mengs, and U. Wetterauer. 2005b. Adjuvant intravesical treatment of superficial bladder cancer with a standardized mistletoe extract. *J. Urol.* 174, no. 1:76–79.

Elsässer-Beile, U., T. Ruhnau, N. Freudenberg, U. Wetterauer, and U. Mengs. 2001. Antitumoral effect of recombinant mistletoe lectin on chemically induced urinary bladder carcinogenesis in a rat model. *Cancer* 91, no. 5:998–1004.

Evans, H.M., and K.S. Bishop. 1922. On the existence of a hitherto unrecognized dietary factor essential for reproduction. *Science* 56, no. 1458:650–651.

Fiore, G., C. Nencini, F. Cavallo, et al. 2006. In vitro antiproliferative effect of six Salvia species on human tumor cell lines. *Phytother. Res.* 20, no. 8:701–703.

Freedland, S.J., M. Carducci, N. Kroeger, et al. 2013. A double-blind, randomized, neoadjuvant study of the tissue effects of POMx pills in men with prostate cancer before radical prostatectomy. *Cancer Prev. Res.* 6, no. 10:1120–1127.

Ghosh, R., G.E. Garcia, K. Crosby, et al. 2007. Regulation of Cox-2 by cyclic AMP response element binding protein in prostate cancer: Potential role for Nexrutine. *Neoplasia* 9, no. 11:893–899.

Ghosh, R., H. Graham, P. Rivas, et al. 2010. Phellodendronamurense bark extract prevents progression of prostate tumors in transgenic adenocarcinoma of mouse prostate: Potential for prostate cancer management. *Anticancer Res.* 30, no. 3:857–865.

Gil, M.I., F.A. Tomás-Barberán, B. Hess-Pierce, D.M. Holcroft, and A.A. Kader. 2000. Antioxidant activity of pomegranate juice and its relationship with phenolic composition and processing. *J. Agric. Food Chem.* 48, no. 10:4581–4589.

Goebell, P.J., T. Otto, J. Suhr, and H. Rübben. 2002. Evaluation of an unconventional treatment modality with mistletoe lectin to prevent recurrence of superficial bladder cancer: A randomized phase II trial. *J. Urol.* 168, no. 1:72–75.

Gong, Y., Y. Li, Y. Lu, et al. 2011. Bioactive tanshinones in Salvia miltiorrhiza inhibit the growth of prostate cancer cells *in vitro* and in mice. *Int. J. Cancer* 129, no. 5:1042–1052.

Goossens, M., M. Zeegers, H. Van Poppel, et al. 2015. Phase III randomised chemoprevention study of Selenium on the recurrence of non-invasive bladder cancer. The SELEnium and BLAdder cancer Trial (SELEBLAT). *Arch. Public Health* 73(Suppl 1): P5.

Grieve, M.T. 1971. Milk. In: *A Modern Herbal.* Dover Publications, Inc., New York.

Guess, B.W., M.C. Scholz, S.B. Strum, R.Y. Lam, H.J. Johnson, and R.I. Jennrich. 2003. Modified citrus pectin (MCP) increases the prostate-specific antigen doubling time in men with prostate cancer: A phase II pilot study. *Prostate Cancer Prostatic Dis.* 6, no. 4:301–304.

Gupta, S., K. Hastak, N. Ahmad, J.S. Lewin, and H. Mukhtar. 2001. Inhibition of prostate carcinogenesis in TRAMP mice by oral infusion of green tea polyphenols. *Proc. Natl. Acad. Sci. USA* 98, no. 18:10350–10355.

Gururaj, A.E., M. Belakavadi, D.A. Venkatesh, D. Marmé, and B.P. Salimath. 2002. Molecular mechanisms of anti-angiogenic effect of curcumin. *Biochem. Biophys. Res. Commun.* 297, no. 4:934–942.

Hambright, H.G., I.S. Batth, J. Xie, R. Ghosh, and A.P. Kumar. 2015. Palmatine inhibits growth and invasion in prostate cancer cell: Potential role for rpS6/NFkappaB/FLIP. *Mol. Carcinog.* 54, no. 10:1227–1234.

Harada, N., Y. Murata, R. Yamaji, T. Miura, H. Inui, and Y. Nakano. 2007. Resveratrol down-regulates the androgen receptor at the post-translational level in prostate cancer cells. *J. Nutr. Sci. Vitaminol. (Tokyo)* 53, no. 6:556–560.

Hebert, J.R., T.G. Hurley, B.C. Olendzki, J. Teas, Y. Ma, and J.S. Hampl. 1998. Nutritional and socioeconomic factors in relation to prostate cancer mortality: A crossnational study. *J. Natl. Cancer Inst.* 90, no. 21:1637–1647.

Hempstock, J., J.P. Kavanagh, and N.J. George. 1998. Growth inhibition of prostate cell lines *in vitro* by phyto-oestrogens. *BJU Int.* 82, no. 4:560–563.

Henning, S.M., P. Wang, and D. Heber. 2011. Chemopreventive effects of tea in prostate cancer: Green tea versus black tea. *Mol. Nutr. Food Res.* 55, no. 6:905–920.

Hori, S., E. Butler, and J. McLoughlin. 2011. Prostate cancer and diet: Food for thought? *BJU Int.* 107, no. 9:1348–1359.

Hour, T.C., J. Chen, C.Y. Huang, J.Y. Guan, S.H. Lu, and Y.S. Pu. 2002. Curcumin enhances cytotoxicity of chemotherapeutic agents in prostate cancer cells by inducing p21WAF1/CIP1 and C/EBP expression-sand suppressing NF-κB activation. *Prostate* 51, no. 3:211–218.

Huang, E.-C., M.F. McEntee, and J. Whelan. 2012. Zyflamend, a combination of herbal extracts, attenuates tumor growth in murine xenograft models of prostate cancer. *Nutr. Cancer* 64, no. 5:749–760.

Huang, S., C.A. Pettaway, H. Uehara, C.D. Bucana, and I.J. Fidler. 2001. Blockade of NF kappaB activity in human prostate cancer cells is associated with suppression of angiogenesis, invasion and metastasis. *Oncogen.* 20:4188–4197.

Hudson, T.S., D.K. Hartle, S.D. Hursting, et al. 2007. Inhibition of prostate cancer growth by muscadine grape skin extract and resveratrol through distinct mechanisms. *Cancer Res.* 67, no. 17:8396–8405.

Jackson, C.L., T.M. Dreaden, L.K. Theobald, et al. 2007. Pectin induces apoptosis in human prostate cancer cells: Correlation of apoptotic function with pectin structure. *Glycobiology* 17, no. 8:805–819.

Jarred, R.A., M. Keikha, C. Dowling, et al. 2002. Induction of apoptosis in low to moderate-grade human prostate carcinoma by red clover-derived dietary isoflavones. *Cancer Epidemiol. Biomark. Prev.* 11, no. 12:1689–1696.

Jasinski, M., L. Jasinska, and M. Ogrodowczyk. 2013. Resveratrol in prostate diseases – A short review. *Cent. Eur. J. Urol.* 66, no. 2:144–149.

Jatoi, A., N. Ellison, P.A. Burch, et al. 2003. A phase II trial of green tea in the treatment of patients with androgen independent metastatic prostate carcinoma. *Cancer* 97, no. 6:1442–1446.

Jeon, Y.J., S.N. Jung, J. Yun, et al. 2015. Ginkgetin inhibits the growth of DU-145 prostate cancer cells through inhibition of signal transducer and activator of transcription 3 activity. *Cancer Sci.* 106, no. 4:413–420.

Kellen, E., M. Zeegers, and F. Buntinx. 2006. Selenium is inversely associated with bladder cancer risk: A report from the Belgian case-control study on bladder cancer. *Int. J. Urol.* 13, no. 9:1180–1184.

Klein, E.A., I.M. Thompson Jr., C.M. Tangen, et al. 2011. Vitamin E and the risk of prostate cancer. *JAMA* 306, no. 14:1549–1556.

Klempner, S.J., and G. Bubley. 2012. Complementary and alternative medicines in prostate cancer: From bench to bedside? *Oncologist* 17, no. 6:830–837.

Kumar, A.P., S. Bhaskaran, M. Ganapathy, et al. 2007. Akt/cAMP-responsive element binding protein/cyclin D1 network: A novel target for prostate cancer inhibition in transgenic adenocarcinoma of mouse prostate model mediated by Nexrutine, a Phellodendronamurense bark extract. *Clin. Cancer Res.* 13, no. 9:2784–2794.

Kunze, E., H. Schulz, M. Adamek, and H.J. Gabius. 2000. Long-term administration of galactoside-specific mistletoe lectin in an animal model: No protection against N-butyl-N-(4-hydroxybutyl)-nitrosamine-i nduced urinary bladder carcinogenesis in rats and no induction of a relevant local cellular immune response. *J. Cancer Res. Clin. Oncol.* 126, no. 3:125–1238.

Langley, P. 2000. Why a pomegranate? *BMJ* 321, no. 7269:1153–1154.

Lippman, S.M., E.A. Klein, P.J. Goodman, et al. 2009. Effect of selenium and vitamin E on risk of prostate cancer and other cancers. *JAMA* 301, no. 1:39–51.

Liu, X.J., Y.Q. Li, Q.Y. Chen, S.J. Xiao, and S.E. Zeng. 2014. Up-regulating of RASD1 and apoptosis of DU-145 human prostate cancer cells induced by formononetin *in vitro*. *Asian Pac. J. Cancer Prev.* 15, no. 6:2835–2839.

Loizzo, M.R., R. Tundis, F. Menichini, A.M. Saab, G.A. Statti, and F. Menichini. 2007. Cytotoxic activity of essential oils from Labiatae and Lauraceae families against *in vitro* human tumor models. *Anticancer Res.* 27, no. 5 (A):3293–3299.

Lotan, Y., P.J. Goodman, R.F. Youssef, et al. 2012. Evaluation of vitamin E and selenium supplementation for the prevention of bladder cancer in SWOG coordinated SELECT. *J. Urol.* 187, no. 6:2005–2010.

Mani, J., E. Jüngel, G. Bartsch, et al. 2015. Use of complementary and alternative medicine before and after organ removal due to urologic cancer. *Patient Pref. Adher.* 9:1407–1412.

Marshall, J.R. 2010. Randomized phase III trial of selenium supplementation to prevent prostate cancer among men with high grade prostatic intraepithelial neoplasia. Paper presented at the American Urological Association (AUA) Annual Meeting, San Francisco, USA.

McLarty, J., R.L.H. Bigelow, M. Smith, D. Elmajian, M. Ankem, and J.A. Cardelli. 2009. Tea polyphenols decrease serum levels of prostate-specific antigen, hepatocyte growth factor, and vascular endothelial growth factor in prostate cancer patients and inhibit production of hepatocyte growth factor and vascular endothelial growth factor *in vitro*. *Cancer Prev. Res.* 2, no. 7:673–682.

Morgia, G., S. Cimino, V. Favilla, et al. 2013. Effects of serenoarepens, selenium and lycopene (Profluss(R)) on chronic inflammation associated with benign prostatic hyperplasia: Results of 'FLOG' (Flogosis and Profluss in prostatic and Genital Disease), a multicentre Italian study. *Int. Braz. J. Urol.* 39, no. 2:214–221.

Muralimanoharan, S.B., A.B. Kunnumakkara, B. Shylesh, et al. 2009. Butanol fraction containing berberine or related compound from nexrutine inhibits NF kappa B signaling and induces apoptosis in prostate cancer cells. *Prostate* 69, no. 5:494–504.

Nagano, J., S. Kono, D.L. Preston, and K.A. Mabuchi. 2001. A prospective study of green tea consumption and cancer incidence, Hiroshima and Nagasaki (Japan). *Cancer Causes Contr.* 12, no. 6:501–508.

Nelson, P.S., and B. Montgomery. 2003. Unconventional therapy for prostate cancer: Good, bad or questionable? *Nat. Rev. Cancer* 3, no. 11:845–858.

Niu, Y., S. Yeh, H. Miyamoto, et al. 2008. Tissue prostate-specific antigen facilitates refractory prostate tumor progression via enhancing ARA70-regulated androgen receptor transactivation. *Cancer Res.* 68, no. 17:7110–7119.

Oakes, E. 2007. Martha Dartt maxwell. In: *Encyclopedia of World Scientists*, E.G. Anderson (Ed.). Facts on File, New York, 494–495.

Paller, C.J., X. Ye, P.J. Wozniak, et al. 2013. A randomized phase II study of pomegranate extract for men with rising PSA following initial therapy for localized prostate cancer. *Prostate Cancer Prostatic Dis.* 16, no. 1:50–55.

Pan, Z.J., N. Deng, and Z. Zou. 2017. The effect of curcumin on bladder tumor in rat model. *Eur. Rev. Med. Pharmacol. Sci.* 21, no. 4:884–889.

Pantuck, A.J., J.T. Leppert, N. Zomorodian, et al. 2006. Phase II study of pomegranate juice for men with rising prostate-specific antigen following surgery or radiation for prostate cancer. *Clin. Cancer Res.* 12, no. 13:4018–4026.

Peterson, G., and S. Barnes. 1993. Genistein and biochanin A inhibit the growth of human prostate cancer cells but not epidermal growth factor receptor tyrosine autophosphorylation. *Prostate* 22, no. 4:335–345.

Philippou, Y., M. Hadjipavlou, S. Khan, and A. Rane. 2013. Complementary and alternative medicine (CAM) in prostate and bladder cancer. *BJU Int.* 112, no. 8:1073–1079.

Pienta, K.J., H. Nailk, A. Akhtar, et al. 1995. Inhibition of spontaneous metastasis in a rat prostate cancer model by oral administration of modified citrus pectin. *J. Natl. Cancer Inst.* 87, no. 5:348–353.

Ponholzer, A., G. Struhal, and S. Madersbacher. 2003. Frequent use of complementary medicine by prostate cancer patients. *Eur. Urol.* 43, no. 6:604–608.

Raina, K., S. Rajamanickam, R.P. Singh, G. Deep, M. Chittezhath, and R. Agarwal. 2008. Stage-specific inhibitory effects and associated mechanisms of silibinin on tumor progression and metastasis in transgenic adenocarcinoma of the mouse prostate model. *Cancer Res.* 68, no. 16:6822–6830.

Rettig, M.B., D. Heber, J. An, et al. 2008. Pomegranate extract inhibits androgen-independent prostate cancer growth through a nuclear factor-KB-dependent mechanism. *Mol. Cancer Ther.* 7, no. 9:2662–2671.

Rice, L., V.G. Samedi, T.A. Medrano, et al. 2002. Mechanisms of the growth inhibitory effects of the isoflavonoidbiochanin A on LNCaP cells and xenografts. *Prostate* 52, no. 3:201–212.

Rose, A., T. El-Leithy, F.V. Dorp, et al. 2015. Mistletoe plant extract in patients with nonmuscle invasive bladder cancer: Results of a phase Ib/IIa Single Group Dose Escalation study. *J. Urol.* 194, no. 4:939–943.

Sagara, Y., Y. Miyata, K. Nomata, T. Hayashi, and H. Kanetake. 2010. Green tea polyphenol suppresses tumor invasion and angiogenesis in N-butyl-(−4-hydroxybutyl) nitrosamine-induced bladder cancer. *Cancer Epidemiol.* 34, no. 3:350–354.

Shi, W.F., M. Leong, E. Cho, et al. 2009. Repressive effects of resveratrol on androgen receptor transcriptional activity. *PLOS ONE* 4, no. 10:e7398.

Siddiqui, I.A., M. Asim, B.B. Hafeez, V.M. Adhami, R.S. Tarapore, and H. Mukhtar. 2011. Green tea polyphenol EGCG blunts androgen receptor function in prostate cancer. *FASEB J.* 25, no. 4:1198–1207.

Singh, R.P., K. Raina, G. Sharma, and R. Agarwal. 2008. Silibinin inhibits established prostate tumor growth, progression, invasion and metastasis, and suppresses tumor angiogenesis and epithelial-mesenchymal transition in transgenic adenocarcinoma of the mouse prostate model mice. *Clin. Cancer. Res.* 14, no. 23:7773–7780.

Slater, M., D. Brown, and A. Husband. 2002. In the prostatic epithelium, dietary isoflavones from red clover significantly increase estrogen receptor β and E-cadherin expression but decrease transforming growth factor β1. *Prostate Cancer Prostatic Dis.* 5, no. 1:16–21.

Sovak, M., A.L. Seligson, and M. Konas. 2002. Herbal composition PC-SPES for management of prostate cancer: Identification of active principles. *Cancer Spectr. Knowledge Environ.* 94, no. 17:1275–1280.

Stocco, B., K. Toledo, M. Salvador, M. Paulo, N. Koyama, and M.R. Torqueti Toloi. 2012. Dose-dependent effect of resveratrol on bladder cancer cells: Chemoprevention and oxidative stress. *Maturitas* 72, no. 1:72–78.

Swanson, G.P., W.E. Jones, C.S. Ha, C.A. Jenkins, A.P. Kumar, and J. Basler. 2015. Tolerance of Phellodendronamurense bark extract (Nexrutine(R)) in patients with human prostate cancer. *Phytother. Res.* 29, no. 1:40–42.

Szliszka, E., Z.P. Czuba, A. Mertas, A. Paradysz, and W. Krol. 2013. The dietary isoflavonebiochanin-A sensitizes prostate cancer cells to TRAIL-induced apoptosis. *Uroloncol.* 31, no. 3:331–342.

Thomas, R., M. Williams, H. Sharma, A. Chaudry, and P. Bellamy. 2014. A double-blind, placebo-controlled randomised trial evaluating the effect of a polyphenol-rich whole food supplement on PSA progression in men with prostate cancer – The UK NCRN Pomi-T study. *Prostate Cancer Prostatic Dis.* 17, no. 2:180–186.

Trofast, J. 2011. Berzelius' discovery of selenium. *Chem. Int.* 33, no. 5:16.

Urech, K., A. Buessing, G. Thalmann, H. Schaefermeyer, and P. Heusser. 2006. Antiproliferative effects of mistletoe (*Viscum album* L.) extract in urinary bladder carcinoma cell lines. *Anticancer Res.* 26, no. 4 (B):3049–3055.

Walle, T., F. Hsieh, M.H. DeLegge, J.E. Oatis Jr., and U.K. Walle. 2004. High absorption but very low bioavailability of oral resveratrol in humans. *Drug Metab. Dispos.* 32, no. 12:1377–1382.

Wilkinson, S., S. Farrelly, J. Low, A. Chakraborty, R. Williams, and S. Wilkinson. 2008. The use of complementary therapy by men with prostate cancer in the UK. *Eur. J. Cancer Care* 17, no. 5:492–499.

Wu, S., F. Li, X. Huang, et al. 2013. The association of tea consumption with bladder cancer risk: A meta-analysis. *Asia Pac. J. Clin. Nutr.* 22, no. 1:128–137.

Yan, J., and A. Katz. 2010. PectaSol-C modified citrus pectin induces apoptosis and inhibition of proliferation in human and mouse androgen-dependent and -independent prostate cancer cells. *Integr. Cancer Ther.* 9, no. 2:197–203.

Ye, Y., R. Hou, J. Chen, et al. 2012. Formononetin-induced apoptosis of human prostate cancer cells through ERK1/2 mitogen-activated protein kinase inactivation. *Horm. Metab. Res.* 44, no. 4:263–267.

You, O.H., S.H. Kim, B. Kim, et al. 2013. Ginkgetin induces apoptosis via activation of caspase and inhibition of survival genes in PC-3 prostate cancer cells. *Bioorg. Med. Chem. Lett.* 23, no. 9:2692–2695.

Zeng, J., Y. Sun, K. Wu, et al. 2011. Chemopreventive and chemotherapeutic effects of intravesical silibinin against bladder cancer by acting on mitochondria. *Mol. Cancer Ther.* 10, no. 1:104–116.

Zhang, X., L. Bi, Y. Ye, and J. Chen. 2014. Formononetin induces apoptosis in PC-3 prostate cancer cells through enhancing the Bax/Bcl-2 ratios and regulating the p38/Akt pathway. *Nutr. Cancer* 66, no.4:656–661.

Zhang, Y., S.H. Won, C. Jiang, et al. 2012. Tanshinones from Chinese medicinal herb Danshen (Salvia miltiorrhiza Bunge) suppress prostate cancer growth and androgen receptor signaling. *Pharm. Res.* 29, no. 6:1595–1608.

CHAPTER 8

Plant Lectins in Cancer Treatment:
The Case of *Viscum album* L.

Vasileios Tsekouras

CONTENTS

Cancer is a fatal disease and current research focuses on natural resources with potential antimalignant properties. Plant lectins are a heterogeneous protein group implicated in several cellular physiological processes. These molecules are characterized by the ability to identify specific carbohydrates on the cellular membranes as well as glycosylated structures resulting from malignancy. Lectins display great potential in cancer therapy due to their antitumor and immunomodulatory functions, and a vast number of assays have been conducted to evaluate their efficacy as antimalignant compounds. In addition, due to their high glycan specificity and selectivity, lectins have been applied in cancer diagnosis for the detection of cellular and histochemical glycosylation. Lectins extracted from the European mistletoe, *Viscum album* L., exhibited significant cytotoxic and immunomodulatory properties against several cancer types. Nowadays, mistletoe preparations are commercially available as adjuvant therapy for the treatment of cancer.

8.1   INTRODUCTION TO LECTIN RESEARCH

Since the end of the nineteenth century, several proteins were detected in nature characterized by their ability to agglutinate erythrocytes. Consequently, they were named hemagglutinins or phytoagglutinins because they were mainly discovered in plant extracts. Lectins can be described as proteins that bind reversibly and non-enzymatically to specific carbohydrates (De Hoff et al. 2009). In a broader definition, lectins are carbohydrate-binding proteins or glycoproteins of non-immune origin which agglutinate or precipitate glycoconjugates or both. The carbohydrate protein binding is selective without altering the bound ligand (Van Damme et al. 2008).

Initially, lectins were characterized as sugar-binding proteins derived from seeds and functioning as storage proteins with protective properties, against microbes and predators (Brewin and Kardailsky 1997). For many decades, only a few of them had been isolated and even less received interest, despite the fact of being in abundance in nature (Sharon and Lis 2004). The lack of advanced and specialized detection methods had been an important inhibitory factor for any advances regarding lectins. In the past, the most suitable method for detecting novel lectins were agglutination assays by blood cells erythrocytes, from several animal species (Van Damme 2014). Even though agglutination inhibition experiments are simple processes, in lectin research it exhibited a significant role in understanding the interaction of lectin binding sites and specific carbohydrate structures. However, the heamagglutination approach is a sufficient method only for the detection of the proteins, and more sophisticated procedures are required for their isolation and purification. Since lectins are characterized by unique binding traits, as they react reversibly and selectively with carbohydrates, this distinctive ability indicates these molecules as suitable for purification by carbohydrate affinity chromatography (Pohleven et al. 2012). The pace of lectin extraction increased significantly by the evolution of isolation techniques regarding affinity chromatography. Scientific advances in topics as molecular biology, genomics, transcriptomics, and bioinformatics lead to novel facts regarding the expression, distribution, biological role, and the evolutionary relationships of lectins as well as enabling the ability to identify and explain protein carbohydrate binds expressed at low levels (Van Damme et al. 2004).

The protein carbohydrate moieties were the fundamental principle for lectin purification procedures and in conjunction with scientific advances led to the tracing and the extraction of a large number of lectins with remarkable biological activities (Kocourek 1986). The evolution in lectin identification and isolation procedures reveals a number of chemical, physiological, and biochemical traits. These qualities have raised a considerable interest for their pharmacological activities especially in the area of biomedical research and cancer therapy (Jiang et al. 2015). Several studies conducted during the last four decades demonstrated that lectins are useful compounds for the detection, isolation, and characterization of glycoconjugates, for the histochemical detection of carbohydrate structures on and inside various cells and tissues, for distinguishing subtle alterations in cellular glycosylation, and for the examination of glycosylation changes occurring on cells, during physiological and pathological processes (Sharon and Lis 2004, Brooks and Hall 2012).

The most studied lectins are those originated in plants, as they compose a group of easily accessible carbohydrate-binding proteins with important biochemical properties and biological functions. Phytolectins are a quite heterogeneous group presenting several forms of variations in their molecular structure, specificity for certain carbohydrate structures and biological activities. Plant lectins are applicable for glycoprotein detection and characterization in cells and tissues as well as the tracing of alterations that can occur in malignancy (Van Damme 2014).

8.2 LECTIN ABUNDANCE AND CLASSIFICATION

Lectins are a diverse group of bioactive proteins existing in almost every kind of organism. A broader, species-based classification separates the group into three major categories: Lectins of Microorganisms, Animal Lectins, and Plant Lectins. Microorganisms like fungi (Wu et al. 2001), bacteria (Alyousef et al. 2018), viruses (Montelaro et al. 1983), and algae (Yamaguchi et al. 1999) compose sugar-specific proteins interfering with cell adhesion and agglutination (Nizet et al. 2017). These biological properties are resulting to the induction of several cell processes such as immunomodulation (Chang et al. 2007), mitogenic activity (Hori et al. 1987), and cell proliferation (Yu et al. 1993).

Animal lectins are characterized by considerable diversity throughout the animal kingdom, existing in both major groups of vertebrates and invertebrates (Drickamer and Taylor 1993). These proteins are an important element of the host defense against microorganisms due to their ability to recognize not only carbohydrates endogenous to the species but also molecules existing in invaders (Dias et al. 2015). Lectins have been isolated from protozoa (Singh et al. 2016) and several invertebrates such as insects (Gomes et al. 1991), arachnids (Vasta and Cohen 1984), mollusks (Adhya and Singha, 2016), crustaceans (Zhang et al. 2018), corals (Kvennefors et al. 2008), and sea sponges (Gardères et al. 2016). In vertebrates, lectins have been isolated and characterized from fishes (Kugapreethan et al. 2018), amphibians (Qu et al. 2015), reptiles (Sartim et al. 2016), birds (Zhang et al. 2017), and mammals, including human species (Yang and T. Liu 2003, Manikandan et al. 2017).

Plants are a rich source of lectins and the main source for detection, isolation, an further analysis of such molecules. Lectins have been found in several plant groups in both classes of monocots and dicots, traced in almost every kind of plants tissue with different cellular localizations and molecular properties (Van Damme et al. 1998a, Dias et al. 2015). Plant lectins have been purified from roots (M. Quinn and Etzler 1987), rhizomes (Bains et al. 2005), tubers (Seo et al. 1990), bulbs (Ooi et al. 2000), barks (Broekaert et al. 1984), stems (Singh and Das 1994), leaves (Jawade et al. 2016), fruits (Raja et al. 2011), flowers (Peumans et al. 1997), and seeds (Sharma et al. 2009).

Plant lectins are a heterogeneous group of proteins varying in several biochemical and physiochemical characteristics such as molecular structure, sugar-binding specificity, and biological properties, and, consequently, there is a difficulty in grouping them in subdivisions. Generally, plant lectins have been classified based on criteria such as structural forms, sugar specificity, and evolutionary relations.

A structural-based division distinguishes the lectins in four major types: the merolectins, hololectins, chimerolectins, and superlectins. The merolectins consist of a single carbohydrate-binding domain while the hololectins have two or more domains, identical or homologous, resulting in binding to structurally similar sugars. The chimerolectins are actually fusion proteins where a carbohydrate-binding domain is combined to unrelated non-binding domains with different biological activities. Finally, superlectins are also fusion proteins consisting of two different glycan-binding domains and possess the ability to recognize structurally unrelated carbohydrates (Van Damme et al. 1998a,b, Lavelle et al. 2000).

Due to the complexity and heterogeneity of their carbohydrate-binding specificity, plant lectins have been initially subdivided in groups, according to terms of sugar-binding domains. According to their specificity, they have been categorized as galactose, N-acetylglucosamine (GlcNAc), N-acetylgalactosamine (GalNAc), glucose, L-fucose, mannose, maltose, and sialic acid specific lectins as well as complex glycan-binding lectins (Goldstein and Poretz 1986, Lis and Sharon 1986, Hashim et al. 2017). This classification is simplified as it is based on the inhibition from monosaccharides in agglutination assays while the binding of a lectin can be extended further from the protein glycan interactions to additional hydrophobic interactions and electrostatic interactions (Roth 2011).

Advances in analytical techniques generate significant progress in lectinology, highlighting the insufficiency of existing classification systems to explain and describe new data. The traditional classifications are being supplanted by homology-based systems that incorporate sequence and structural homology plus evolutionary relations in order to categorize the plant lectins (De Hoff et al. 2009). A vast number of plant lectins have been identified and genome-transcriptome analyses revealed that groups are particularly widespread; thus many proteins contain one or more carbohydrate-binding domains, forming complex structures. Thorough analyses of the lectin sequences distinguish 12 different, evolutionary and structurally related carbohydrate-binding domains identified in plants. Every protein domain is characterized by unique folding with highly specific sugar binding sites. The domains are *Agaricus bisporus* agglutinin homologs, amaranthins, class V chitinase homologs, cyanovirin family, *Euonymus europaeus* agglutinin family, *Galanthus nivalis* agglutinin family, proteins with hevein domains, jacalins, proteins with a legume lectin domain,

Table 8.1   Examples of Plant Lectins with Different Carbohydrate Binding Domains

Lectin Domain	Carbohydrate Specificity	Plant Lectin (Origin)	References
Agaricus bisporus agglutinin family	T-Antigen	MarpoABA (*Marchantia polymorpha*)	Peumans et al. (2007)
Amaranthins	T-Antigen	(*Amaranthus caudatus*) Amaranthin	Rinderle et al. (1989)
Class V chitinase homologs	Blood group B, high-man	(*Nicotiana tabacum*) RobpsCRP	Melchers et al. (1994)
Cyanovirin family	High-Man glycans (Man-8 and Man-9)	(*Ceratopteris richardii*) CrCVNH	Koharudin et al. (2008)
Euonymus europaeus agglutinin family	Blood group B, high-man N-glycans	EEA (*Euonymus europaeus*)	Fouquaert et al. (2008)
Galanthus nivalis agglutinin family	High-Man glycans	GNA (*Galanthus nivalis*)	Fouquaert et al. (2007)
Hevein domain	GlcNAc N-glycans	DSA (*Datura stramonium*)	Nishimoto et al (2015)
Jacalins	Gal	Jacalin (*Artocarpus heterophyllus*)	Kabir (1998)
Legume lectin domain	High-Man N-glycans	ConA (*Canavalia ensiformis*)	Maupin et al. (2012)
LysM domains	GalNAc	LysM CERK1 (*Arabidopsis thaliana*)	Willmann et al. (2011)
Nicotiana tabacum agglutinin domain	GlcNAc-oligomers, high-Man N-glycans	NICTABA (*Nicotiana tabacum*)	Delporte et al. (2015)
Ricin-B domain	Gal/GalNAc	RCA I (*Ricinus communis*)	Wu et al. (2006)

LysM domains, *Nicotiana tabacum* agglutinin family, and ricin-B family (Van Damme et al. 2008). Representative lectins from each family are listed in Table 8.1.

8.3 PLANT LECTINS IN CANCER DIAGNOSIS AND TREATMENT

Lectins display great potential in cancer therapy and diagnosis by being molecules able to specifically recognize carbohydrates on the cellular membrane and their glycosylation resulting from malignancy (Mody et al. 1995). Cancer is a complex process in which genetic alterations affect and modify cell signaling, functioning, replication rhythm, apoptosis, and metastasis. Cancer cells exhibit irregular glycosylation patterns, and this process plays a significant role in cell development, signaling, interaction, proliferation, differentiation, and migration (Estrada-Martínez et al. 2017). During malignant transformation, the glycans express alterations such as over expression of certain structures, loss of expression, the appearance of novel or incomplete structures, and the accumulation of precursors (Varki et al. 2017). Glycan is a term used for any sugar or assembly of sugars, existing in a free form or attached to molecules. Glycans are involved in basic molecular and biological processes occurring in cancer affecting cell signaling, cell–cell and cell–matrix interactions, tumor cell invasion, angiogenesis, immune modulation, and metastasis (Pinho and Reis 2015). The glycosylation of proteins is the key element for a wide variety of biological processes affecting their localization, stability, and folding. Consequently, aberrant glycosylation in malignant cells results in many biological pathways affecting cell signaling, migration, adhesion, and cell death, regulating apoptosis, and autophagy (Korekane and Taniguchi 2015).

Plant lectins are capable of specific recognition and binding to certain oligosaccharide structures in a non-enzymatic action without causing modifications, thus discriminating between subtle variations throughout complex carbohydrate forms (Kim et al. 2009). Due to their high selectivity and specificity against certain glycan structures, lectins have been applied in cancer diagnosis. In addition, a vast number of these compounds are also known for their cytotoxicity, emerging as potential anti-cancer therapeutics. Several lectins exhibited antitumor properties by binding to cancer cell membranes or their receptors, causing cytotoxicity, apoptosis, and inhibition of tumor growth (Coulibaly and Youan 2017). During the last decades, many plant lectins, predominantly galactoside- and galNAc-specific, have been applied in several *in vitro*, *in vivo*, and clinical trials to document their potential applications as drugs for the treatment of cancer (Ernst et al. 2003).

8.3.1 Plant Lectins in Cancer Diagnosis

Cancer biomarkers are important for the detection, diagnosis, and prognosis of the disease in early stages, and there is a need to discover novel biomarkers to develop new diagnostics for use in clinical practice (Carrigan and Krahn 2016). A biomarker is an objectively measured characteristic that indicates biological or pathogenic processes in an organism by analyzing biomolecules such as nucleic acids, proteins, peptides, as well as biochemical modifications (Goossens et al. 2015).

Plant lectins are used in recognition of alterations resulting from physiological and pathological processes especially against cancer. Alterations in glycan moieties of glycoproteins are a common trait of cancer cells, and aberrant O-glycants are thoroughly expressed on cells membranes as Tn, T, Lewis a, Lewis x, and the Forssman pentasaccharide antigens. These forms represent cancer glycomarkers, and highly selective specific lectins can recognize and identify them in histochemical applications (Lowe and Marth 2003, Poiroux et al. 2017).

The lectins *Amaranthus caudatus* agglutinin and *Artocarpus integrifolia* agglutinin can detect specific glycoforms of serum T, Tn, and sialyl-T antigens (Wu et al. 2008). Lens culinaris agglutinin A is used for the analysis of alpha-fetoprotein (AFP), a fetal specific glycoprotein used as a lectin reactive biomarker for the early recognition of hepatocellular carcinoma (Shiraki et al. 1995, Kumada et al. 1999). Normal colorectal mucosa binds strongly to the Dolichos biflorus agglutinin (DBA) but not to peanut agglutinin (PNA), while malignant tissue of colonic carcinomas exhibited strong PNA-binding (Shah et al. 1989). *Pinellia pedatisecta* agglutinin, PPA applied for the recognition of glycosylation patterns in leukemia cells HL60 and Kasumi-1, as well as the solid tumors cell lines human hepatoma BEL-7404, human liver cell Huh7, and human non-small cell lung carcinoma H1299 (Li et al. 2014). *Sambucus nigra* agglutinin and *Maackia amurensis* agglutinin are used in a lectin immunoenzymatic method in order to explore prostate cancer sialylation isoforms (Meany et al. 2009).

Immobilized jacalin has been applied as a selective biorecognition layer, in an impedimetric leukemia sensor, as a strategy to improve the early recognition of circulating tumor cells in blood fluid (Cancino-Bernardi et al. 2015). *Sambucus nigra* agglutinin immobilized on agarose beads exhibited selective binding and isolation of the STn glycan, a pan-carcinoma biomarker, presented in cancer patients serum glycoproteins (Silva et al. 2017). A multilectin affinity chromatography of endometrioid ovarian cancer tissue was conducted to identify potential glycoprotein markers for this cancer. The biotinylated lectins *Phaseolus vulgaris* erythro agglutinin E-PHA, *Aleuria aurantia* AAL, and *Datura stramonium* DSL captured glycoproteins with 47 specific lectin reactive potential tumor markers (Abbott et al. 2010).

The significance of plant lectins in cancer biomarkers detection was improved by lectin microarray technologies. The enzyme-linked lectin assay is a set of different lectins, spotted on microslides for scanning membrane glycoproteins isolated from the tumors. The formation of lectins in microarrays enables multiple and simultaneous discrimination of several selective binding sites for the rapid characterization of glycans (Hu and Wong 2009). The CA125 biomarker is essential for the diagnosis of ovarian cancer but is not cancer-specific, and other gynecological conditions promote its expression. A novel microarray platform consisting of specific aberrant glycoforms succeeded in distinguishing ovarian neoplasms (Chen et al. 2013). A sensitive profiling of various clinical samples was obtained by a lectin microarray assay, performed to highlight the glycosylation differences between metastatic and non-metastatic breast cancer. In this study, *Psophocarpus tetragonolobus* lectin-I (PTL-I) and jacalin, both Tn antigen specific, exhibited significant differences in binding patterns among metastatic and non-metastatic breast tumors (Fry et al. 2011).

8.3.2 Plant Lectins in Cancer Treatment

Lectin anti-cancer activities generate from diverse mechanisms such as immune system modulation, autophagy, apoptosis, and inhibition of tumor growth (Coulibaly and Youan 2017).

Plant lectins are dynamically involved in the activation of the immune system by their interaction with specific glycan moieties existing on the surface of immune cells. Immune surveillance system is expressed by the induction of NK cells, helper T-cells, and natural killer (NKT) cells (Li et al. 2017). The T-lymphocytes are important for organizing the immune response including their capacity to activate the production of cytokines which act as hormonal messengers involved in many immune effects (Berger 2000, Shaw et al. 2018). In cancer treatment, protein carbohydrate moieties can trigger signal transductions, resulting in T lymphocyte proliferation and cytokine production, affecting tumor growth (Souza et al. 2013). ATF1011 is a lectin purified from *Aloe arborescens* Mill that binds to specific receptor sites on normal and MM102 tumor cells, augmenting cytotoxic T-cell response (Yoshimoto et al. 1987). Mistletoes, as *Viscum album coloratum*, contain type-2 RIPs that elaborate immunomodulatory properties, as demonstrated by *in vitro* and *in vivo* studies (Lee et al. 2009). The B-chain of the Korean mistletoe lectin (KML) exhibits immunomodulatory and antimalignant activities, by promoting NK cells activation, production of cytokines, and macrophage activities *in vitro* (Yoon et al. 2003). Jacalin, from *Artocarpus heterophyllus*, is reported to induce production of antiinflammatory cytokines against colon cancer (Pereira da Silva et al. 2013).

Despite the complex immune system responses, tumors can still develop. In this case, cellular autophagy can be significant by promoting the differentiation of immune cells and maintaining the internal homeostasis in immune cells (Li et al. 2017). Autophagy is an evolutionary conserved cellular degradation pathway for the clearance of damaged proteins and organelles. It is an important homeostatic function that ensures organelle quality control and protein maintenance and serves as an alternative source of energy in periods of metabolic stress. Autophagy promotes cell viability; however, in several cases of malignancy, it can induce apoptotic or caspase-independent cell death (Mathew et al. 2007, Liu et al. 2013). The tumor suppressor genes p53 and PTEN are targeted in chemotherapy to damage the autophagic mechanism and cause instability and uncontrolled cell growth (Yau et al. 2015). *Polygonatum cyrtonema* lectin (PCL) is a mannose/sialic acid specific lectin that has been reported to induce autophagy against human melanoma A375 cancer cells through a mitochondria-mediated ROS-p38-p53 pathway (Liu et al. 2009). Similar behavior was reported for the manose-binding *Galanthus nivalis* agglutinin (Wu and Bao 2013). Concanavalin A (Con A), extracted from lectins from Jack bean seeds, binds to mannose moieties on the cell and promotes BNIP3-mediated mitochondria autophagy in Hepatoma Cells (Lei and Chang 2007). Soybean lectin and *Clematis montana* lectin elicited autophagy in HeLa cells via the generation of reactive oxygen species (Peng et al. 2009, Panda et al. 2014). Abrus agglutinin (AGG) is a type II ribosome-inactivating protein that is reported to present autophagy dependent cell death against cervical cancer cell (Panda et al. 2017).

Plant lectins inhibit tumor cell growth by promoting cytotoxic effects and mediating apoptosis. Apoptosis is a highly regulated, controlled process and a vital component of several cell operations such as the immune system functioning, programmed cell death for maintaining cell population, hormone dependent atrophy, and embryonic development (Elmore 2007). Apoptosis can be activated through two main ways: the extrinsic or death receptor pathway and the intrinsic or mitochondrial pathway. Activation of caspase cascade is involved in both pathways, ending up in cell death. The extrinsic pathway is initiated by specific ligands that bind to membrane receptors of tumor necrosis. The intrinsic pathway is activated by intracellular or extracellular stress like DNA damage, growth factors deficiency, and activation of oncogene (Shi et al. 2016).

Ricin, extracted from *Ricinus communis*, is a heterodimer 64 kDa type II ribosome-inactivating protein consisting of two distinct N-glycosylated polypeptide chains joined by a single disulfide bond. Chain B is responsible for cell binding by specifically targeting galactosides residues on cell surfaces. The A chain removes an adenine residue from a loop of 28S ribosomal RNA, preventing protein synthesis (Lord and Spooner 2011). Ricin-induced apoptosis against human cervical cancer cells (HeLa) is mediated by the generation of reactive oxygen species and caspase-3 activation (Rao et al. 2005). In another study, ricin caused apoptotic death to hepatoma cell BEL7404, throughout caspases activation, and PARP cleavage activity (Hu et al. 2001). Concanavalin A is a legume lectin reported to induce mitochondria-mediated apoptosis in murine macrophage PU5-1.8 cells by collapsing mitochondrial membranes, releasing cytochrome c and eventually activating caspases (Liu et al. 2009b). The α-D-galactose-binding native frutalin from breadfruit seeds as well as the recombinant form, frutalin expressed and purified from *Pichia pastoris*, have similar cytotoxic effects on HeLa cells, by inducing cell apoptosis and inhibiting cell proliferation (Oliveira et al. 2011). Two chito-oligosaccharide specific lectins from *Benincasa hispida* (BhL) and *Datura innoxia* (DiL9) possessed antiproliferative activity and induced apoptosis in human pancreatic cancer cells PANC-1 by mitochondrial membrane depolarization leading to the activation of caspases executing cell death (Singh et al. 2016a).

Plant lectins exhibited antitumor effects in several *in vitro* assays. Significant dose-dependent antiproliferative effect was reported by wheat germ (WGA), concanavalin A (Con-A), *Griffonia simplicifolia* (GSA-IA4), and *Phaseolus vulgaris* (PHA-L) agglutinins against melanoma cell lines SK-MEL-28, HT-144, and C32, while the agglutinin, the peanut agglutinin (PNA), exhibited significant stimulatory effect on the C32 cancer cells (Kiss et al. 1997). A panel of lectins, *Phaseolus vulgaris* leukoagglutinin PHA-L, *Griffonia simplifolia* I-A4 agglutinin GSA-IA4, *Arachis hypogaea* agglutinin PNA, *Triticum vulgare* agglutinin WGA, and *Canavalia ensiformis* agglutinin ConA, were tested against three cancer cell lines rhabdomyosarcoma Hs 729, human leiomyosarcoma SK-UT-1, and SK-LMS-1 and demonstrated inhibition of cell proliferation (Remmelink et al. 1999). Wheat germ agglutinin (WGA) is reported to inhibit cell growth on human breast cancer cells lines (Valentiner et al. 2003) and exhibited highly toxic properties towards pancreatic cancer cells *in vitro* (Schwarz et al. 1999). Abrin extracted from *Abrus precatorius* is a type II ribosome-inactivating protein that demonstrated antiproliferative effects and apoptosis induction towards colon cancer cells (Yu et al. 2016). Jacalin showed a non-toxic, dose-dependent inhibition of cell proliferation towards human colon cancer cells HT29 (Yu et al. 2001). Tepary bean lectin, extracted from *Phaseolus acutifolius*, presented dose dependent inhibition of viability, by affecting cell proliferation, colony formation, and DNA synthesis, on human cervix carcinoma cells C33-A and human colon adenocarcinoma cell line Sw480 (Valadez-Vega et al. 2011). A 30kDa mannose-specific lectin, isolated from *Hyacinth bulbs*, suppressed the growth of human colon cancer cells Caco-2 and cervical cancer cells HeLa (Naik et al. 2017). Concomitant administration of Korean mistletoe agglutinin and doxorubicin against human breast cancer cells MCF-and MDA-MB231 demonstrated strong synergistic effect in cell growth inhibition, compared to lectin or doxorubicin alone treatment (Hong et al. 2014).

A number of *in vivo* studies against animal models are conducted regarding plant-derived lectins and their potential anticancer properties. Abrin was administered to nude mice, bearing three different human tumors. The treatment caused the delay of tumor growth in the case of Lewis lung carcinoma and increased life expectancy in mice inoculated with Ehrlich ascites carcinoma and B16 melanoma (Fodstad et al. 1977). Similar results presented in other assays where abrin isoforms exhibited inhibition of tumor growth in mice with sarcoma 180 (Lin et al. 1982), tumor reduction in mice bearing Dalton's lymphoma ascites, and increased animal life span (Ramnath et al. 2002). WGA was found to control tumor growth on lymphoma (Ganguly and Das 1994), and a wheat germ extract inhibits colon cancer in F-344 rats (Zalatnai et al. 2001). Peanut agglutinin (PNA) against mice bearing Dalton's lymphoma caused the reduction of lymphoma proliferation and activation of autophagic and apoptotic procedures (Mukhopadhyay et al. 2014). *Momordica charantia* lectin (MCL), a type II ribosome inactivating protein, exhibited antitumor activity towards athymic nude mice xenograft inoculated by nasopharyngeal carcinoma cell line CNE2 (Fang et al. 2012).

8.4 *VISCUM ALBUM* LECTINS IN CANCER TREATMENT

Mistletoes are semi-parasitic plants from the order Santalales, grouped in many families such as *Loranthaceae, Santalaceae*, and *Viscaceae*. The species are thoroughly distributed all over the world and are named in accordance with their geographical location as European, American, Mexican, Korean, African, Japanese, and Indian mistletoes. These obligate semiparasites grow on a variety of host plants and traditionally, are regarded as medicinal repositories (Patel and Panda 2014).

Mistletoes, especially European *Viscum album* L., are very important herbal drugs for complementary and anthroposophic medicine for cancer treatment, and a number of extracts or mistletoe derived compounds are applied against malignancies (Bussing 2000, Mulsow et al. 2016). Many preparations from mistletoes, derived from various hosts, have been used in several assays, and fermented preparations are commercially available such as Abnobaviscum, Cefaleksin, Eurixor, Helixor, Lektinol, Iscador, Iscucin, and Isorel (Thies et al. 2001, Giacometti 2015). Many anticancer compounds have been identified in *Viscum album* L., such as viscotoxins, flavonoids, triterpene acids, and most important, the mistletoe lectins (Twardziok et al. 2016). Mistletoe contains three galactose and N-acetylgalactosamine specific isolectin groups, MLI, MLII, and MLIII, responsible for anticancer effects. The three isoforms are glycoproteins, classified to the ribosome-inactivating proteins family type II, and they consist of two polypeptide chains, A-chain and B-chain, linked by a disulfide bridge (Voelter et al. 2005). The lectins demonstrate differences in molecular mass and carbohydrate specificity. MLI is a dimer of 115 kDa, formed by two identical subunits, each one consisting of two chains, an A-chain of 29kDa and a galactose-specific B-chain of 34 kDa. The MLII consists of two chains, a 27 kDa A-chain and a N-acetylgalactoseamine specific 32 kDa B-chain, while the MLIII has an A-chain of 25 kDa and a B-chain of 30kDa that binds both sugars (Lavelle et al. 2002, Tonevitsky et al. 2004).

Mistletoe lectins exhibit anticancer properties either by triggering apoptotic death or via the stimulation of immune responses (Elluru et al. 2006). Mistletoe lectins enhance immunomodulatory activity due to B chain-selective binding, which results in cytokine secretion and natural killer cell activation. The potential application of mistletoe lectin isoforms in order to activate the natural immune system was tested in several assays, against malignancies and in healthy individuals as well (Hajto et al. 1997, Hajto et al. 1998, Klein et al. 2002). In clinical trials with patients diagnosed with Hodgkin disease, multiple myeloma, and breast cancer, when treated with the mistletoe extract *Viscum album* Qu (Quercus), several parameters of humoral and cellular immunity, such as leukocytes, lymphocytes, and T-lymphocytes, were increased (Gardin 2009). In another study, patients with mammary carcinoma exhibited a significant amplification of natural killer and helper T-cells after the administration of MLI (Beuth et al. 1995). Glioma patients in standard oncologic treatment presented immunostimulatory effect after the administration of an ML standardized mistletoe preparation that improved their quality of life (Lenartz et al. 1996). In accordance, there are indications that mistletoe lectin administration affects the patient's quality of life due to immune stimulation, demonstrated by another clinical trial on patients with breast cancer receiving chemotherapy (Semiglasov et al. 2004). In a preclinical assay, RAW 117 H 10 lymphosarcoma cells and L-1 sarcoma cells were inoculated into BALB/c-mice to test the activity of aqueous mistletoe extract. Antimetastatic and immunomodulatory activity was reported because the tumor size was significantly reduced and peripheral blood leukocyte and thymocyte were increased (Braun et al. 2002). Lectin from Korean mistletoe *V. album* subsp. *coloratum* against human colon adenocarcinoma Caco-2 cells, indicated up-regulation of the cytokine gene expression resulting in augmented expression of cytokines and tumor necrosis factor-alpha (Monira et al. 2009).

The toxic effect of ML results from the cooperation of both polypeptide chains. The B chain is selectively bound to carbohydrates on cell surfaces, and afterwards, the cytotoxic A-chain inhibits the protein synthesis by targeting and cleaving a specific adenine residue from the 28S ribosomal RNA, causing apoptosis or necrosis (Endo et al. 1988). Mistletoe lectin I when administered in low concentrations towards the human T-cell leukemia line MOLT-4 caused cell death, resulting from apoptotic processes (Möckel et al. 1997). Caspase-derived apoptotic cell death was demonstrated against leukemic T- and B-cell lines treated with ML-I, where the administration of the protein was followed by the activation of caspase-3, caspase-8, and caspase-9 (Bantel et al. 1999). The European mistletoe extract Abnobaviscum F®, from the host tree Fraxinus, induced apoptosis towards *in vitro* cultures of human myeloid leukemia K562, human plasmacytoma RPMI-8226, and murine lymphocytic leukemia L1210 cells. A more detailed analysis of K562 cells highlighted the existence of multiple processes leading to cell death, such as the activation of the intrinsic caspase pathway (caspase-9, JNK-1/2, and p38 MAPK), the phosphorylation of Protein kinase B, the reduction of cellular glutathione, and the augmentation of endoplasmic reticulum stress (Park et al. 2012). The Korean mistletoe lectin induced apoptosis in human hepatic adenocarcinoma SK-Hep-1 and human hepatoma Hep3B cell lines, by regulating caspase 3 processes and telomerase inhibition (Lyu et al. 2002).

Mistletoe preparations and cytotoxic lectins demonstrate antitumor and antiproliferative activities against many types of malignancies, *in vitro* and *in vivo*. The mistletoe lectin I exhibited cytotoxic properties, in three tumor cell Yoshida sarcoma cells, leukemia cells Molt4 (Urech et al. 1995). The treatment of MFM-223 and KPL-1 cell lines with Iscador M, Iscador Q, and Abnobaviscum Fraxini-2 caused proliferation inhibition of cancer cells (Knöpfl-Sidler et al. 2005). The lectin-rich preparation Iscador®M indicated potential use for breast cancer therapy towards the breast cancer cell line MDA-MB-468-HER2, as tumor cell proliferation and migration were affected (Hugo et al. 2007).

An assay involving 16 cell lines representing several human tumor cancer types was conducted, aiming to evaluate antiproliferative effects derived from the administration of three aqueous mistletoe extracts, Iscador M special, Iscador Qu special, and Iscador P. The lectin-poor preparation Iscador P presented no cell antiproliferative activity while the Iscador preparations exhibited significant anticancer properties against most of the cancer cells (Maier and Fiebig 2002). *Viscum album* L. extracts containing lectins inhibited cell proliferation and induced apoptosis, via caspase 8 and 9 pathways, in human acute myeloid leukemia cells *in vitro* and *in vivo* (Delebinski et al. 2015). Another preparation of somaclonal variant mistletoe lectins, derived from callus cultures (Kintzios et al. 2002), demonstrated considerably increased cytotoxic properties *in vitro*, on PC12 pheochromocytoma and RAW 264.7 macrophage cell cultures. Additionally, considerably lower toxicity was observed on the non-cancer, immortalized Vero cells, indicating potential use of variant lectins in cancer treatment (Barberaki et al. 2015). The lectin-rich mistletoe extract ABNOBAviscum Fraxini-2® presented antitumor activity towards the human pancreatic cancer xenograft PAXF 736 (Rostock et al. 2005). Recombinant mistletoe lectins (rML) were injected to SCID mice, bearing transplanted human ovarian cancer cells, indicating that rML possesses anti-tumor activities when administered locally into the peritoneum of the tumor (Schumacher et al. 2000) Patients with non-muscle invasive bladder cancer have been treated with a mistletoe extract to examine the safety, effectiveness, and maximum tolerated dose of the preparation for complementary cancer treatment (Rose et al. 2015).

8.5 CONCLUSION

Malignant cells exhibit altered glycosylation patterns affecting many intracellular pathways and modifying a range of biological properties. These aberrant alterations provide a potential base for the development of new anticancer agents and the discovery of novel cancer biomarkers with selective targeting properties, against particular glycosylation patterns (Gondim et al. 2017). Biotechnological advances improve the knowledge regarding molecular events and changes during malignancies, discovering new pathways for early diagnosis and treatment of cancer. Cell culture, gene mapping, in situ hybridization, and microarray technology are useful methods for more effective and accurate cancer detection and therapy. Nevertheless, cancer still remains a fatal disease, and there is a great need for the discovery of new selective anticancer drugs and diagnostic tools (Ghorbani and Karimi 2015). Current research focuses on natural resources with potential antimalignant properties. Lectins are highly specific glycan-binding proteins, widely distributed in living organisms and involved in several cellular physiological and pathological processes, exhibiting immunomodulatory properties and antitumor functions (Gondim et al. 2017).

Plant lectins are bioactive molecules able to identify and recognize animal cell carbohydrates. Lectins exhibit significant properties for cancer diagnosis and treatment owing to their unique characteristic of distinguishing subtle alterations in cellular glycosylation. Several cellular responses are generated due to specific protein–cell bounding, resulting in a panel of activities ranging from immune system activation to cancer cell death. Lectin chemotherapy presented cytotoxic, apoptotic, autophagic, and antitumor effects, as well as significant immunostimulant and mitogenic properties. In addition to pharmaceutical practicing, plant lectins are also employed in histochemical techniques, towards malignant cells and tissues, aiming at the detection of glycoconjugates identified as cancer markers. The diagnostic potential of plant lectins has been particularly expanded by immobilizing them in microarrays, contributing to rapid and sensitive analysis of glycans. *Viscum album* L., the European mistletoe, is one of the most important plant species with anticancer properties, containing a number of anticancer compounds, in particular three cytotoxic and immunomodulatory lectins. *In vitro* and *in vivo* experiments provide significant indications that mistletoe-derived lectins exhibit anticancer activities, in a selective manner, against several cancer types. Nowadays, lectin-rich mistletoe preparations are at the top of anthroposophic cancer treatment and complementary therapy.

Plant lectins are promising tools for cancer treatment and diagnosis. Nevertheless, extensive research is required for lectins with potential pharmaceutical properties to determine the type of cancer that target the precise mode of action, the method of administration, and the dosage. The progressive transition from the primary *in vitro* evaluation to *in vivo* tests and clinical trials is a necessity to obtain integrated and accurate evaluation of plant lectin anticancer properties.

REFERENCES

Abbott, K.L., J.M. Lim, L. Wells, B.B. Benigno, J.F. McDonald, and M. Pierce. 2010. Identification of candidate biomarkers with cancer-specific glycosylation in the tissue and serum of endometrioid ovarian cancer patients by glycoproteomic analysis. *Proteomics* 10, no. 3:470–481.

Adhya, M., and B. Singha. 2016. Gal/GalNAc specific multiple lectins in marine bivalve Anadara granosa. *Fish and Shellfish Immunology* 50:242–246.

Alyousef, A.A., A. Alqasim, and M.S. Aloahd. 2018. Isolation and characterization of lectin with antibacterial, antibiofilm and antiproliferative activities from Acinetobacter baumannii of environmental origin. *Journal of Applied Microbiology* 124 no. 5:1139–1146.

Bains, J.S., V. Dhuna, J. Singh, S.S. Kamboj, K.K. Nijjar, and J.N. Agrewala. 2005. Novel lectins from rhizomes of two Acorus species with mitogenic activity and inhibitory potential towards murine cancer cell lines. *International Immunopharmacology* 5 no. 9:1470–1478.

Bantel, H., I.H. Engels, W. Voelter, K. Schulze-Osthoff, and S. Wesselborg. 1999. Mistletoe lectin activates caspase-8/FLICE independently of death receptor signaling and enhances anticancer drug-induced apoptosis. *Cancer Research* 59 no. 9:2083–2090.

Barberaki, M., E. Dermitzaki, A.N. Margioris, M. Theodosaki, S. Grafakos, and S. Kintzios. 2015. Protein extracts from somaclonal mistletoe (Viscum album L.) callus with increased tumor cytotoxic activity *in vitro*. *Current Bioactive Compounds* 11:1–1.

Berger, A. 2000. Th1 and Th2 responses: What are they? *BMJ* 321 no. 7258:424.

Beuth, J., B. Stoffel, H. L. Ko, G. Buss, L. Tunggal, and G. Pulverer. 1995. Immunostimulating activity of different dosages of mistletoe lectin-1 in patients with mammary carcinoma. *Arzneimittel-Forschung* 45 no. 4:505–507.

Braun, J.M., H.L. Ko, J.M. Schierholz, and J. Beuth. 2002. Standardized mistletoe extract augments immune response and down-regulates local and metastatic tumor growth in murine models. *Anticancer Research* 22 no. 6C:4187–4190.

Brewin, N.J., and I.V. Kardailsky. 1997. Legume lectins and nodulation by Rhizobium. *Trends in Plant Science* 2 no. 3:92–98.

Broekaert, W.F., M. Nsimba-Lubaki, B. Peeters, and W.J. Peumans. 1984. A lectin from elder (Sambucus nigra L.) bark. *Biochemical Journal* 221 no. 1:163–169.

Brooks, S.A., and D.M.S. Hall. 2012. Lectin histochemistry to detect altered glycosylation in cells and tissues. In: *Metastasis Research Protocols*. Dwek, M., S.A. Brooks, and U. Schumacher (Eds.). Totowa, NJ, Humana Press, 31–50.

Bussing, A. 2000. Biological and pharmacological properties of Viscum album L. from tissue flask to man. In: *Mistletoe: The Genus Viscum*.Bussing, A. (Ed.). Amsterdam, The Netherlands, Harwood Academic Publishers, 123–182.

Cancino-Bernardi, J., V.S. Marangoni, H.A.M. Faria, and V. Zucolotto. 2015. Detection of leukemic cells by using jacalin as the biorecognition layer: A new strategy for the detection of circulating tumor cells. *ChemElectroChem* 2 no. 7:963–969.

Carrigan, P., and T. Krahn. 2016. Impact of biomarkers on personalized medicine. In: *New Approaches to Drug Discovery*. Nielsch, U., U. Fuhrmann, and S. Jaroch (Eds.). Cham, Springer International Publishing, 285–311.

Chang, H.H., P.J. Chien, M.H. Tong, and F. Sheu. 2007. Mushroom immunomodulatory proteins possess potential thermal/freezing resistance, acid/alkali tolerance and dehydration stability. *Food Chemistry*. 105 no. 2:597–605.

Chen, K., A. Gentry-Maharaj, M. Burnell, C. Steentoft, L. Marcos-Silva, U. Mandel, I. Jacobs, A. Dawnay, U. Menon, and O. Blixt. 2013. Microarray glycoprofiling of CA125 improves differential diagnosis of ovarian cancer. *Journal of Proteome Research* 12 no. 3:1408–1418.

Coulibaly, S. F., and B.C. Youan. 2017. Current status of lectin-based cancer diagnosis and therapy. *AIMS Molecular Science*. 4 no. 1:1–27.

De Hoff, P.L., L.M. Brill, and A.M. Hirsch. 2009. Plant lectins: The ties that bind in root symbiosis and plant defense. *Molecular Genetics and Genomics*. 282 no. 1:1–15.

Delebinski, C.I., M. Twardziok, S. Kleinsimon, F. Hoff, K. Mulsow, J. Rolff, S. Jäger, A. Eggert, and G. Seifert. 2015. A natural combination extract of Viscum album L. containing both triterpene acids and lectins is highly effective against AML *in vivo*. *PLOS ONE*. 10 no. 8:e0133892.

Delporte, A., S. Van Holle, N. Lannoo, and E.J.M. Van Damme. 2015. The tobacco lectin, prototype of the family of nictaba-related proteins. *Current protein & peptide science*, no. 16:5–16. doi: 10.2174/13892 03716666150213154107.

Dias, R.O., L.S. Machado, L. Migliolo, and O.L. Franco. 2015. Insights into animal and plant lectins with antimicrobial activities. *Molecules*. 20 no. 1:519–541.

Drickamer, K., and M.E. Taylor. 1993. Biology of animal lectins. *Annual Review of Cell Biology*. 9 no. 1:237–264.

Elluru, S.R., J.P.D. Duong Van Huyen, S. Delignat, F. Prost, J. Bayry, M.D. Kazatchkine, and S.V. Kaveri. 2006. Molecular mechanisms underlying the immunomodulatory effects of mistletoe (Viscum album L.) extracts iscador. *Arzneimittel-Forschung*. 56 no. 6A:461–466.

Elmore, S. 2007. Apoptosis: A review of programmed cell death. *Toxicologic Pathology*. 35 no. 4:495–516.

Endo, Y., K. Tsurugi, and H. Franz. 1988. The site of action of the A-chain of mistletoe lectin I on eukaryotic ribosomes the RNA N-glycosidase activity of the protein. *FEBS Letters*. 231 no. 2:378–380.

Ernst, E., K. Schmidt, and M.K. Steuer-Vogt. 2003. Mistletoe for cancer? A systematic review of randomised clinical trials. *International Journal of Cancer.* 107 no. 2:262–267.

Estrada-Martínez, L.E., U. Moreno-Celis, R. Cervantes-Jiménez, R.A. Ferriz-Martínez, A. Blanco-Labra, and T. García-Gasca. 2017. Plant lectins as medical tools against digestive system cancers. *International Journal of Molecular Sciences.* 18 no. 7:1403.

Fang, E.F., C.Z. Zhang, T.B. Ng, J.H. Wong, W.L. Pan, X.J. Ye, Y.S. Chan, and W.P. Fong. 2012. Momordica charantia lectin, a type II ribosome inactivating protein, exhibits antitumor activity toward human nasopharyngeal carcinoma cells *in vitro* and *in vivo. Cancer Prevention Research.* 5 no. 1:109–121.

Fodstad, O., S. Olsnes, and A. Pihl. 1977. Inhibitory effect of abrin and ricin on the growth of transplantable murine tumors and of abrin on human cancers in nude mice. *Cancer Research.* 37 no. 12:4559–4567.

Fouquaert, E., S.L. Hanton, F. Brandizzi, W.J. Peumans, and E.J.M. Van Damme. 2007. Localization and topogenesis studies of cytoplasmic and vacuolar homologs of the *Galanthus nivalis* Agglutinin. *Plant and Cell Physiology,* no. 48 no. 7:1010–1021. doi: 10.1093/pcp/pcm071.

Fouquaert, E., W.J. Peumans, D.F. Smith, P. Proost, S.N. Savvides, and E.J.M. Van Damme. 2008. The "old" euonymus europaeus agglutinin represents a novel family of ubiquitous plant proteins. *Plant Physiology,* 147 no. 3:1316–1324. doi: 10.1104/pp.108.116764.

Fry, S.A., B. Afrough, H.J. Lomax-Browne, J.F. Timms, L.S. Velentzis, and A.J.C. Leathem. 2011. Lectin microarray profiling of metastatic breast cancers. *Glycobiology.* 21 no. 8:1060–1070.

Ganguly, C., and S. Das. 1994. Plant lectins as inhibitors of tumour growth and modulators of host immune response. *Chemotherapy.* 40 no. 4:272–278.

Gardères, J., I. Domart-Coulon, A. Marie, B. Hamer, R. Batel, W.E.G. Müller, and M.L. Bourguet-Kondracki. 2016. Purification and partial characterization of a lectin protein complex, the clathrilectin, from the calcareous sponge Clathrina clathrus. *Comparative Biochemistry and Physiology: Part B, Biochemistry and Molecular Biology.* 200:17–27.

Gardin, N.E. 2009. Immunological response to mistletoe (Viscum album L.) in cancer patients: A four-case series. *Phytotherapy Research.* 23 no. 3:407–411.

Ghorbani, M., and H. Karimi. 2015. Role of biotechnology in cancer control. *IJSRST.* 5 no. 1:180–185.

Giacometti, J. 2015. Plant lectins in cancer prevention and treatment. *Medicina Fluminensis.* 51:211–229.

Goldstein, I.J., and R.D. Poretz. 1986. 2 – Isolation, physicochemical characterization, and carbohydrate-binding specificity of lectins. In: *The Lectins.* Liener, I.E., N. Sharon, I.J. Goldstein (Eds.). Academic Press, Orlando 33–247.

Gomes, Y.D., A.F. Furtado, and L.B.B. Coelho. 1991. Partial purification and some properties of a hemolymph lectin from panstrongylus megistus (Hemiptera, Reduviidae). *Applied Biochemistry and Biotechnology.* 31 no. 1:97–107.

Gondim, A.C.S., I. Romero-Canelón, E.H.S. Sousa, C.A. Blindauer, J.S. Butler, M.J. Romero, C. Sanchez-Cano, B.L. Sousa, R.P. Chaves, C.S. Nagano, B.S. Cavada, and P.J. Sadler. 2017. The potent anti-cancer activity of Dioclea lasiocarpa lectin. *Journal of Inorganic Biochemistry.* 175:179–189.

Goossens, N., S. Nakagawa, X. Sun, and Y. Hoshida. 2015. Cancer biomarker discovery and validation. *Translational Cancer Research.* 4 no. 3:256–269.

Hajto, T., K. Hostanska, J. Fischer, and R. Saller. 1997. Immunomodulatory effects of Viscum album agglutinin-I on natural immunity. *Anti-Cancer Drugs.* 8, Suppl 1:S43–S46.

Hajto, T., K. Hostanska, K. Weber, H. Zinke, J. Fischer, U. Mengs, H. Lentzen, and R. Saller. 1998. Effect of a recombinant lectin, Viscum album agglutinin on the secretion of interleukin-12 in cultured human peripheral blood mononuclear cells and on NK-cell-mediated cytotoxicity of rat splenocytes *in vitro* and *in vivo. Natural Immunity.* 16 no. 1:34–46.

Hashim, O.H., J.J. Jayapalan, and C.S. Lee. 2017. Lectins: An effective tool for screening of potential cancer biomarkers. *PeerJ.* 5:e3784.

Hong, C.E., A.K. Park, and S.Y. Lyu. 2014. Synergistic anticancer effects of lectin and doxorubicin in breast cancer cells. *Molecular and Cellular Biochemistry.* 394 no. 1–2:225–235.

Hori, K., H. Matsuda, K. Miyazawa, and K. Ito. 1987. A mitogenic agglutinin from the red alga Carpopeltis flabellata. *Phytochemistry.* 26 no. 5:1335–1338.

Hu, R., Q. Zhai, W. Liu, and X. Liu. 2001. An insight into the mechanism of cytotoxicity of ricin to hepatoma cell: Roles of Bcl-2 family proteins, caspases, Ca(2+)-dependent proteases and protein kinase C. *Journal of Cellular Biochemistry.* 81 no. 4:583–593.

Hu, S., and D.T. Wong. 2009. Lectin microarray. *Proteomics: Clinical Applications.* 3 no. 2:148–154.

Figure 9.1 *Podophyllum peltatum.*

Figure 9.2 *Vinca rosea.*

Figure 9.3 *Aconitum fischeri.*

Figure 9.4 *Acronychia* sp.

Figure 9.5 *Brucea.*

Figure 9.6 *Bursera.*

Figure 9.7 *Chamaecyparis.*

Figure 9.9 *Cistus creticus.*

Figure 9.10 *Ficus carica.*

Figure 9.12 *Gossypium herbaceum.*

Figure 9.13 *Hypericum.*

Figure 9.14 *Juniperus virginiana.*

Figure 9.15 *Magnolia virginiana.*

Figure 9.16   *Nerium oleander.*

Figure 9.17   *Nigella.*

Figure 9.18   *Origanum vulgare.*

Figure 9.19   *Paeonia officinalis.* ©2001 Horticopia, Inc.

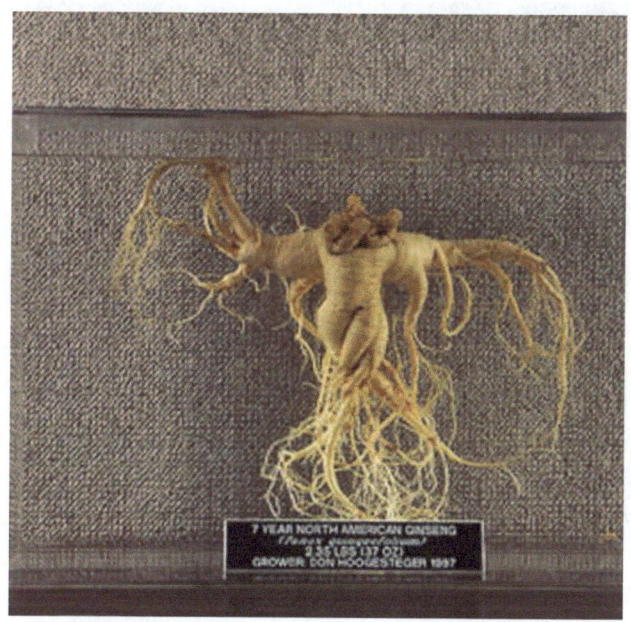

Figure 9.20 *Ginseng/Panax.*

Figure 9.21 *Plumeria.*

Figure 9.22 *Punica granatum.*

Figure 9.23 *Salvia sclarea.*

Figure 9.24 *Sargassum.*

Figure 9.25 *Seseli.*

Figure 9.26 *Tropaeolum.*

Figure 9.27 *Valeriana.*

Figure 9.28 *Wikstroemia indica.*

Hugo, F., S. Schwitalla, B. Niggemann, K.S. Zänker, and T. Dittmar. 2007. Viscum album extracts Iscador®P and Iscador®M counteract the growth factor induced effects in human follicular B-NHL cells and breast cancer cells. *Medicina Fluminensis*. 67:90–96.

Jawade, A.A., S.K. Pingle, R.G. Tumane, A.S. Sharma, A.S. Ramteke, and R.K. Jain. 2016. Isolation and characterization of lectin from the leaves of *Euphorbia tithymaloides* (L.). *Tropical Plant Research*. 3 no. 3:634–641.

Jiang, Q.L., S. Zhang, M. Tian, S.Y. Zhang, T. Xie, D.Y. Chen, Y.J. Chen, J. He, J. Liu, L. Ouyang, and X. Jiang. 2015. Plant lectins, from ancient sugar-binding proteins to emerging anti-cancer drugs in apoptosis and autophagy. *Cell Proliferation*. 48 no. 1:17–28.

Kabir, S. 1998. Jacalin: A jackfruit (*Artocarpus heterophyllus*) seed-derived lectin of versatile applications in immunobiological research. *Journal of immunological methods*, 212:193–211. doi: 10.1016/S0022-1759(98)00021-0.

Kim, Y.S., H.S. Yoo, and J.H. Ko. 2009. Implication of aberrant glycosylation in cancer and use of lectin for cancer biomarker discovery. *Protein and Peptide Letters*. 16 no. 5:499–507.

Kintzios, S., M. Barberaki, P. Tourgielis, G. Aivalakis, and A. Volioti. 2002. Preliminary evaluation of somaclonal variation for the *in vitro* production of new toxic proteins from Viscum album L. *Journal of Herbs, Spices and Medicinal Plants*. 9 no. 2–3:217–221.

Kiss, R., I. Camby, C. Duckworth, R. De Decker, I. Salmon, J.L. Pasteels, A. Danguy, and P. Yeaton. 1997. In vitro influence of Phaseolus vulgaris, Griffonia simplicifolia, concanavalin A, wheat germ, and peanut agglutinins on HCT-15, LoVo, and SW837 human colorectal cancer cell growth. *Gut*. 40 no. 2:253–261.

Klein, R., K. Classen, P.A. Berg, R. Lüdtke, M. Werner, and R. Huber. 2002. *In vivo*-induction of antibodies to mistletoe lectin-1 and viscotoxin by exposure to aqueous mistletoe extracts: A randomised double-blinded placebo controlled phase I study in healthy individuals. *European Journal of Medical Research*. 7 no. 4:155–163.

Knöpfl-Sidler, F., A. Viviani, L. Rist, and A. Hensel. 2005. Human cancer cells exhibit *in vitro* individual receptiveness towards different Mistletoe extracts. *Die Pharmazie*. 60 no. 6:448–454.

Kocourek, J.A.N. 1986. 1 – Historical background. In: *The Lectins*. Liener, I.E., N. Sharon, and I.J. Goldstein (Eds.)., Academic Press, Orlando, 1–32.

Koharudin, L.M.I., A.R. Viscomi, J.G. Jee, S. Ottonello, and A.M. Gronenborn. 2008. The evolutionarily conserved family of cyanovirin-n homologs: Structures and carbohydrate specificity. *Structure*, 16 no. 4:570–584. doi: https://doi.org/10.1016/j.str.2008.01.015.

Korekane, H., and N. Taniguchi. 2015. Glycosylation in cancer: Enzymatic basis for alterations in N-glycan branching. In: *Glycoscience: Biology and Medicine*. Taniguchi, N., Tamao Endo, G.W. Hart, P.H. Seeberger, and C.-H. Wong (Eds.). Tokyo, Japan, Springer, 1349–1356.

Kugapreethan, R., Q. Wan, J. Nilojan, and J. Lee. 2018. Identification and characterization of a calcium-dependent lily-type lectin from black rockfish (Sebastes schlegelii): Molecular antennas are involved in host defense via pathogen recognition. *Developmental and Comparative Immunology*. 81:54–62.

Kumada, T., S. Nakano, I. Takeda, S. Kiriyama, Y. Sone, K. Hayashi, H. Katoh, T. Endoh, T. Sassa, and S. Satomura. 1999. Clinical utility of Lens culinaris agglutinin-reactive alpha-fetoprotein in small hepatocellular carcinoma: Special reference to imaging diagnosis. *Journal of Hepatology*. 30 no. 1:125–130.

Kvennefors, E.C., W. Leggat, O. Hoegh-Guldberg, B.M. Degnan, and A.C. Barnes. 2008. An ancient and variable mannose-binding lectin from the coral Acropora Millepora binds both pathogens and symbionts. *Developmental and Comparative Immunology*. 32 no. 12:1582–1592.

Lavelle, E.C., G. Grant, A. Pusztai, U. Pfüller, O. Leavy, E. McNeela, K.H.G. Mills, and D.T. O'Hagan. 2002. Mistletoe lectins enhance immune responses to intranasally co-administered herpes simplex virus glycoprotein D2. *Immunology*. 107 no. 2:268–274.

Lavelle, E.C., G. Grant, A. Pusztai, U. Pfüller, and D.T. O'Hagan. 2000. Mucosal immunogenicity of plant lectins in mice. *Immunology*. 99 no. 1:30–37.

Lee, C.H., J.K. Kim, H.Y. Kim, S.M. Park, and S.M. Lee. 2009. Immunomodulating effects of Korean mistletoe lectin *in vitro* and *in vivo*. *International Immunopharmacology*. 9 no. 13–14:1555–1561.

Lei, H.Y., and C.P. Chang. 2007. Induction of autophagy by concanavalin A and its application in anti-tumor therapy. *Autophagy*. 3 no. 4:402–404.

Lenartz, D., B. Stoffel, J. Menzel, and J. Beuth. 1996. Immunoprotective activity of the galactoside-specific lectin from mistletoe after tumor destructive therapy in glioma patients. *Anticancer Research*. 16 no. 6B:3799–3802.

Li, C.J., W.T. Liao, M.Y. Wu, and P.Y. Chu. 2017. New insights into the role of autophagy in tumor immune microenvironment. *International Journal of Molecular Sciences.* 18 no. 7:1566.

Li, N., G. Dong, S. Wang, S. Zhu, Y. Shen, and G. Li. 2014. Pinellia pedatisecta agglutinin-based lectin blot analysis distinguishes between glycosylation patterns in various cancer cell lines. *Oncology Letters.* 8 no. 2:837–840.

Lin, J.Y., T.C. Lee, and T.C. Tung. 1982. Inhibitory effects of four Isoabrins on the growth of sarcoma 180 cells. *Cancer Research.* 42 no. 1:276–279.

Lis, H., and N. Sharon. 1986. Lectins as molecules and as tools. *Annual Review of Biochemistry.* 55 no. 1:35–67.

Liu, B., C.Y. Li, H.J. Bian, M.W. Min, L.F. Chen, and J.K. Bao. 2009. Antiproliferative activity and apoptosis-inducing mechanism of concanavalin A on human melanoma A375 cells. *Archives of Biochemistry and Biophysics.* 482 no. 1–2:1–6.

Liu, Z., Y. Luo, T.T. Zhou, and W.Z. Zhang. 2013. Could plant lectins become promising anti-tumour drugs for causing autophagic cell death? *Cell Proliferation.* 46 no. 5:509–515.

Lord, J.M., and R.A. Spooner. 2011. Ricin trafficking in plant and mammalian cells. *Toxins.* 3 no. 7:787–801.

Lowe, J.B.L., and J.D. Marth. 2003. A genetic approach to mammalian glycan function. *Annual Review of Biochemistry.* 72 no. 1:643–691.

Lyu, S.Y., S.H. Choi, and W.B. Park. 2002. Korean mistletoe lectin-induced apoptosis in hepatocarcinoma cells is associated with inhibition of telomerase via mitochondrial controlled pathway independent of p53. *Archives of Pharmacal Research.* 25 no. 1:93–101.

Maier, G., and H.H. Fiebig. 2002. Absence of tumor growth stimulation in a panel of 16 human tumor cell lines by mistletoe extracts *in vitro.* *Anti-Cancer Drugs.* 13 no. 4:373–379.

Manikandan, B., M. Ramar, T. Raman, M. Periasamy, and A. Munusamy. 2017. Purification and characterisation of a pronase-inducible lectin isolated from human serum. *International Journal of Biological Macromolecules.* 99:443–453.

Mathew, R., V. Karantza-Wadsworth, and E. White. 2007. Role of autophagy in cancer. *Nature Reviews. Cancer.* 7 no. 12:961–967.

Maupin, K.A., D. Liden, and B.B. Haab. 2012. The fine specificity of mannose-binding and galactose-binding lectins revealed using outlier motif analysis of glycan array data. *Glycobiology,* 22 no. 1:160–169. doi: 10.1093/glycob/cwr128.

Meany, D.L., Z. Zhang, L.J. Sokoll, H. Zhang, and D.W. Chan. 2009. Glycoproteomics for prostate cancer detection: Changes in serum PSA glycosylation patterns. *Journal of Proteome Research.* 8 no. 2:613–619.

Melchers, L.S., M.A. Groot, J.A. Knaap, A.S. Ponstein, M.B. Sela-Buurlage, J.F. Bol, B.J.C. Cornelissen, P.J.M. Elzen, and H.J.M. Linthorst. 1994. A new class of tobacco chitinases homologous to bacterial exo-chitinases displays antifungal activity. *The Plant Journal,* 5 no. 4:469–480. doi: 10.1046/j.1365-313X.1994.05040469.x.

Möckel, B., T. Schwarz, H. Zinke, J. Eck, M. Langer, and H. Lentzen. 1997. Effects of mistletoe lectin I on human blood cell lines and peripheral blood cells. Cytotoxicity, apoptosis and induction of cytokines. *Arzneimittel-Forschung.* 47 no. 10:1145–1151.

Mody, R., S. Joshi, and W. Chaney. 1995. Use of lectins as diagnostic and therapeutic tools for cancer. *Journal of Pharmacological and Toxicological Methods.* 33 no. 1:1–10.

Monira, P., Y. Koyama, R. Fukutomi, K. Yasui, M. Isemura, and H. Yokogoshi. 2009. Effects of Japanese mistletoe lectin on cytokine gene expression in human colonic carcinoma cells and in the mouse intestine. *Biomedical Research.* 30 no. 5:303–309.

Montelaro, R.C., M. West, and C.J. Issel. 1983. Isolation of equine infectious anemia virus glycoproteins. Lectin affinity chromatography procedures for high avidity glycoproteins. *Journal of Virological Methods.* 6 no. 6:337–346.

Mukhopadhyay, S., P.K. Panda, B. Behera, C.K. Das, M.K. Hassan, D.N. Das, N. Sinha, A. Bissoyi, K. Pramanik, T.K. Maiti, and S.K. Bhutia. 2014. In vitro and *in vivo* antitumor effects of Peanut agglutinin through induction of apoptotic and autophagic cell death. *Food and Chemical Toxicology.* 64:369–377.

Mulsow, K., T. Enzlein, C.I. Delebinski, S. Jaeger, G. Seifert, and M.F. Melzig. 2016. Impact of mistletoe triterpene acids on the uptake of mistletoe lectin by cultured tumor cells. *PLOS ONE.* 11 no. 4:e0153825.

Naik, S., R.S. Rawat, S. Khandai, M. Kumar, S.S. Jena, M.A. Vijayalakshmi, and S. Kumar. 2017. Biochemical characterisation of lectin from Indian hyacinth plant bulbs with potential inhibitory action against human cancer cells. *International Journal of Biological Macromolecules.* 105 no. 1:1349–1356.

Nizet, V., A. Varki, and M. Aebi. 2017. Microbial lectins: Hemagglutinins, adhesins, and toxins. In: *Essentials of Glycobiology (Internet), 3rd edition, 2015–2017* Varki, A., R.D. Cummings, and J.D. Esko (Eds.). Cold Spring Harbor, NY, Cold Spring Harbor Laboratory Press.

Nishimoto, K., K. Tanaka, T. Murakami, H. Nakashita, H. Sakamoto, and S. Oguri. 2015. *Datura stramonium* agglutinin: Cloning, molecular characterization and recombinant production in *Arabidopsis thaliana.* *Glycobiology,* 25 no. 2:157–169. doi: 10.1093/glycob/cwu098.

Oliveira, C., A. Nicolau, J.A. Teixeira, and L. Domingues. 2011. Cytotoxic effects of native and recombinant Frutalin, a plant galactose-binding lectin, on HeLa cervical cancer cells. *Journal of Biomedicine and Biotechnology.* 2011:1–9.

Ooi, L. Sm., T.B. Ng, Y. Geng, and V.E. Ooi. 2000. Lectins from bulbs of the Chinese daffodil Narcissus tazetta (family Amaryllidaceae). *Biochemistry and Cell Biology.* 78 no. 4:463–468.

Panda, P.K., B. Behera, B.R. Meher, D.N. Das, S. Mukhopadhyay, N. Sinha, P.P. Naik, B. Roy, J. Das, S. Paul, T.K. Maiti, R. Agarwal, and S.K. Bhutia. 2017. Abrus Agglutinin, a type II ribosome inactivating protein inhibits Akt/PH domain to induce endoplasmic reticulum stress mediated autophagy-dependent cell death. *Molecular Carcinogenesis.* 56 no. 2:389–401.

Panda, P.K., S. Mukhopadhyay, B. Behera, C.S. Bhol, S. Dey, D.N. Das, N. Sinha, A. Bissoyi, K. Pramanik, T.K. Maiti, and S.K. Bhutia. 2014. Antitumor effect of soybean lectin mediated through reactive oxygen species-dependent pathway. *Life Sciences.* 111 no. 1–2:27–35.

Park, Y.K., Y.R. Do, and B.C. Jang. 2012. Apoptosis of K562 leukemia cells by Abnobaviscum F (R), a European mistletoe extract. *Oncology Reports.* 28 no. 6:2227–2232.

Patel, S., and S. Panda. 2014. Emerging roles of mistletoes in malignancy management. *Biotech.* 4 no. 1:13–20.

Peng, H., H. Lv, Y. Wang, Y.H. Liu, C.Y. Li, L. Meng, F. Chen, and J.K. Bao. 2009. Clematis montana lectin, a novel mannose-binding lectin from traditional Chinese medicine with antiviral and apoptosis-inducing activities. *Peptides.* 30 no. 10:1805–1815.

Pereira da Silva, G., G. Thais, M. Patricia, L. Veronez, L. Rosengela, and G. Sergio. 2013. Jacalin has distinct immunomodulatory effects on early and late stages of experimental colon carcinogenesis. *Frontiers in Immunology.* 4:212–30.

Peumans, W.J., E. Fouquaert, A. Jauneau, P. Rougé, N. Lannoo, H. Hamada, R. Alvarez, B. Devreese, and E.J.M. Van Damme. 2007. The liverwort *Marchantia polymorpha* expresses orthologs of the fungal Agaricus bisporus agglutinin family. *Plant Physiology,* 144 no. 2:637–647. doi: 10.1104/pp.106.087437.

Peumans, W.J., K. Smeets, K. Van Nerum, F. Van Leuven, and E.J.M. Van Damme. 1997. Lectin and alliinase are the predominant proteins in nectar from leek (Allium porrum L.) flowers. *Planta.* 201 no. 3:298–302.

Pinho, S.S., and C.A. Reis. 2015. Glycosylation in cancer: Mechanisms and clinical implications. *Nature Reviews: Cancer.* 15 no. 9:540–555.

Pohleven, J., B. Strukelj, and J. Kos. 2012. Affinity chromatography of lectins. In: *Affinity Chromatography.* Available from: www.intechopen.com/books/affinity-chromatography/carbohydrate-affinity-chromatography.

Poiroux, G., A. Barre, E.J.M. Van Damme, H. Benoist, and P. Rougé. 2017. Plant lectins targeting O-glycans at the cell surface as tools for cancer diagnosis, prognosis and therapy. *International Journal of Molecular Sciences.* 18 no. 6:1232.

Qu, M., C. Tong, L. Kong, X. Yan, O.V. Chernikov, P.A. Lukyanov, Q. Jin, and W. Li. 2015. Purification of a secreted lectin from Andrias davidianus skin and its antibacterial activity. *Comparative Biochemistry and Physiology: Toxicology and Pharmacology.* 167:140–146.

Quinn, J.M., and M.E. Etzler. 1987. Isolation and characterization of a lectin from the roots of Dolichos biflorus. *Archives of Biochemistry and Biophysics.* 258 no. 2:535–544.

Raja, S.B., M.R. Murali, N.K. Kumar, and S.N. Devaraj. 2011. Isolation and partial characterisation of a novel lectin from Aegle marmelos Fruit and Its effect on adherence and invasion of shigellae to HT29 cells. *PLOS ONE.* 6 no. 1:e16231.

Ramnath, V., G. Kuttan, and R. Kuttan. 2002. Antitumour effect of abrin on transplanted tumours in mice. *Indian Journal of Physiology and Pharmacology.* 46 no. 1:69–77.

Rao, P.V. L., R. Jayaraj, A.S. Bhaskar, O. Kumar, R. Bhattacharya, P. Saxena, P.K. Dash, and R. Vijayaraghavan. 2005. Mechanism of ricin-induced apoptosis in human cervical cancer cells. *Biochemical Pharmacology*. 69 no. 5:855–865.

Remmelink, M., F. Darro, C. Decaestecker, N. Bovin, M. Gebhart, H. Kaltner, H.J. Gabius, R. Kiss, I. Salmon, and A. Danguy. 1999. In vitro influence of lectins and neoglycoconjugates on the growth of three human sarcoma cell lines. *Journal of Hepatology*. 125:275–285.

Rinderle, S.J., I.J. Goldstein, K. Matta, and R.M. Ratcliffe. 1989. Isolation and characterization of Amaranthin, a lectin present in the seeds of *Amaranthus caudatus*, that recognizes the T- (or Cryptic T)-antigen. *The Journal of biological chemistry*, 264:16123–16131.

Rose, A., T. El-Leithy, F.V. vom Dorp, A. Zakaria, A. Eisenhardt, S. Tschirdewahn, and H. Rübben. 2015. Mistletoe plant extract in patients with nonmuscle invasive bladder cancer: Results of a phase Ib/IIa Single Group Dose Escalation study. *Journal of Urology*. 194 no. 4:939–943.

Rostock, M., R. Huber, T. Greiner, P. Fritz, R. Scheer, J. Schueler, and H.H. Fiebig. 2005. Anticancer activity of a lectin-rich mistletoe extract injected intratumorally into human pancreatic cancer xenografts. *Anticancer Research*. 25 no. 3B:1969–1975.

Roth, J. 2011. Lectins for histochemical demonstration of glycans. *Histochemistry and Cell Biology*. 136 no. 2:117–130.

Sartim, M., M. Pinto Pinheiro, R.A.P. de Pádua, S. Sampaio, and M. Nonato. 2017. Structural and binding studies of a C-type galactose-binding lectin from Bothrops jararacussu snake venom. *Toxicon*. 126:59–69.

Schumacher, U., S. Feldhaus, and U. Mengs. 2000. Recombinant mistletoe lectin (rML) is successful in treating human ovarian cancer cells transplanted into severe combined immunodeficient (SCID) mice. *Cancer Letters*. 150 no. 2:171–175.

Schwarz, R.E., D.C. Wojciechowicz, A.I. Picon, M.A. Schwarz, and P.B. Paty. 1999. Wheatgerm agglutinin-mediated toxicity in pancreatic cancer cells. *British Journal of Cancer*. 80 no. 11:1754–1762.

Semiglasov, V.F., V.V. Stepula, A. Dudov, W. Lehmacher, and U. Mengs. 2004. The standardised mistletoe extract PS76A2 improves QoL in patients with breast cancer receiving adjuvant CMF chemotherapy: A randomised, placebo-controlled, double-blind, multicentre clinical trial. *Anticancer Research*. 24 no. 2C:1293–1302.

Seo, Y.J., S. Une, I. Tsukamoto, and M. Miyoshi. 1990. The effect of lectin from Taro tuber (Colocasia antiquorum) given by force-feeding on the growth of mice. *Journal of Nutritional Science and Vitaminology*. 36 no. 3:277–285.

Shah, M., S.S. Shrikhande, and V.S. Swaroop. 1989. Lectin binding in colorectal mucosa. *Indian Journal of Gastroenterology*. 8 no. 1:31–33.

Sharma, A., T.B. Ng, J.H. Wong, and P. Lin. 2009. Purification and characterization of a lectin from Phaseolus vulgaris cv. (Anasazi Beans). *Journal of Biomedicine and Biotechnology*. 2009:929568.

Sharon, N., and H. Lis. 2004. History of lectins: From hemagglutinins to biological recognition molecules. *Glycobiology*. 14 no. 11:53R–62R.

Shaw, D.M., F. Merien, A. Braakhuis, and D. Dulson. 2018. T-cells and their cytokine production: The anti-inflammatory and immunosuppressive effects of strenuous exercise. *Cytokine*. 104:136–142.

Shi, Z., R. Sun, T. Yu, R. Liu, L.J. Cheng, J.K. Bao, L. Zou, and Y. Tang. 2016. Identification of novel pathways in plant lectin-induced cancer cell apoptosis. *International Journal of Molecular Sciences*. 17 no. 2:228–228.

Shiraki, K., K. Takase, Y. Tameda, M. Hamada, Y. Kosaka, and T. Nakano. 1995. A clinical study of lectin-reactive alpha-fetoprotein as an early indicator of hepatocellular carcinoma in the follow-up of cirrhotic patients. *Hepatology*. 22 no. 3:802–807.

Silva, L.S., M., C. Gomes, and M.B. Quinaz Garcia. 2017. Flow lectin affinity chromatography – A model with Sambucus nigra agglutinin. *Journal of Glycobiology*. 06 no. 1:1–8.

Singh, R.S., A.K. Walia, and J.R. Kanwar. 2016a. Protozoa lectins and their role in host–pathogen interactions. *Biotechnology Advances*. 34 no. 5:1018–1029.

Singh, R., and H.R. Das. 1994. Purification of lectins from the stems of peanut plants. *Glycoconjugate Journal*. 11 no. 4:282–285.

Singh, R., L. Nawale, D. Sarkar, and C.G. Suresh. 2016b. Two chitotriose-specific lectins show anti-angiogenesis, induces caspase-9-mediated apoptosis and early arrest of pancreatic tumor cell cycle. *PLOS ONE*. 11 no. 1:e0146110.

Souza, M.A., F.C. Carvalho, L.P. Ruas, R. Ricci-Azevedo, and M.C. Roque-Barreira. 2013. The immunomodulatory effect of plant lectins: A review with emphasis on ArtinM properties. *Glycoconjugate Journal.* 30 no. 7:641–657.

Thies, A., U. Pfüller, M. Schachner, H.P. Horny, I. Molls, and U. Schumacher. 2001. Binding of mistletoe lectins to cutaneous malignant melanoma: Implications for prognosis and therapy. *Anticancer Research.* 21 no. 4B:2883–2887.

Tonevitsky, A.G., I.I. Agapov, I.B. Pevzner, N.V. Maluchenko, M.M. Moisenovich, R.A. Palmer, M. Yurkova, K. Pfüller, and U. Pfüller. 2004. A new gene encoding the ribosome-inactivating protein from mistletoe extracts. *Arzneimittel-Forschung.* 54 no. 4:242–249.

Twardziok, M., S. Kleinsimon, J. Rolff, S. Jäger, A. Eggert, G. Seifert, and C.I. Delebinski. 2016. Multiple active compounds from Viscum album L. synergistically converge to promote apoptosis in Ewing sarcoma. *PLOS ONE.* 11 no. 9:e0159749.

Urech, K., G. Schaller, P. Ziska, and M. Giannattasio. 1995. Comparative study on the cytotoxic effect of viscotoxin and mistletoe lectin on tumor cells in culture. *Phytotherapy Research.* 9 no. 1:49–55.

Valadez-Vega, C., G. Alvarez-Manilla, L. Riverón-Negrete, A. García-Carrancá, J.A. Morales-González, C. Zuñiga-Pérez, E. Madrigal-Santillán, J. Esquivel-Soto, C. Esquivel-Chirino, R. Villagómez-Ibarra, M. Bautista, and Á. Morales-González. 2011. Detection of cytotoxic activity of lectin on human colon adenocarcinoma (Sw480) and epithelial cervical carcinoma (C33-A). *Molecules.* 16 no. 3:2107–2118.

Valentiner, U., S. Fabian, U. Schumacher, and A.J. Leathem. 2003. The influence of dietary lectins on the cell proliferation of human breast cancer cell lines *in vitro. Anticancer Research.* 23 no. 2B:1197–1206.

Van Damme, E.J.M. 2014. History of plant lectin research. In: *Lectins: Methods and Protocols.* Hirabayashi, J. (Ed.). New York, NY, Springer, 3–13.

Van Damme, E.J.M. , W.J. Peumans, A. Barre, and P. Rougé. 1998a. Plant lectins: A composite of several distinct families of structurally and evolutionary related proteins with diverse biological roles. *Critical Reviews in Plant Sciences.* 17 no. 6:575–692.

Van Damme, E.J.M., W.J. Peumans, A. Pusztai, and S. Bardocz. 1998b. Plant lectins: A special class of plant proteins. In: *Handbook of Plant Lectins: Properties and Biomedical Applications.* Van Damme, E.J.M., W.J. Peumans, A. Pusztai, S. Bardocz (Eds.). Chichester, UK, John Wiley and Sons3–13.

Van Damme, E.J.M., A. Barre, P. Rougé, and W.J. Peumans. 2004. Cytoplasmic/nuclear plant lectins: A new story. *Trends in Plant Science.* 9 no. 10:484–489.

Van Damme, E.J.M., N. Lannoo, and W.J. Peumans. 2008. Plant lectins. In: *Advances in Botanical Research.* Kader, J.-C., M. Delseny (Eds.). Academic Press, Waltham, MA, USA,107–209.

Varki, A., R. Kannagi, B. Toole, and P. Stanley. 2017. Glycosylation changes in cancer. In: *Essentials of Glycobiology (Internet), 3rd edition*, 2015–2017. Varki, A., R.D. Cummings, J.D. Esko (Eds.). Cold Spring Harbor (NY), Cold Spring Harbor Laboratory Press.

Vasta, G.R., and E. Cohen. 1984. Sialic acid-binding lectins in the 'whip scorpion' (Mastigoproctus giganteus) serum. *Journal of Invertebrate Pathology.* 43 no. 3:333–342.

Voelter, W., R. Wacker, S. Stoeva, R. Tsitsilonis, and C. Betzel. 2005. Mistletoe lectins, structure and function. *Frontiers in Natural Product Chemistry.* 1 no. 1:149–162.

Willmann, R., H.M. Lajunen, G. Erbs, M. Newman, D. Kolb, K. Tsuda, F. Katagiri, J. Fliegmann, J.J. Bono, V.J Cullimore, A.K. Jehle, F. Götz, A. Kulik, A. Molinaro, V. Lipka, A. Gust, and T. Nuernberger. 2011. Arabidopsis lysin-motif proteins LYM1 LYM3 CERK1 mediate bacterial peptidoglycan sensing and immunity to bacterial infection. *Proceedings of the National Academy of Sciences*, 108:19824–19829. doi: 10.1073/pnas.1112862108.

Wu, A.M., J.H. Wu, M.S. Tsai, G.V. Hegde, S.R. Inamdar, B.M. Swamy, and A. Herp. 2001. Carbohydrate specificity of a lectin isolated from the fungus Sclerotium rolfsii. *Life Sciences.* 69 no. 17:2039–2050.

Wu, A.M., J.H. Wu, T. Singh, L.J. Lai, Z. Yang, and A. Herp. 2006. Recognition factors of Ricinus communis agglutinin 1 (RCA1). *Molecular Immunology*, 43 no. 10:1700–1715. doi: https://doi.org/10.1016/j.molimm.2005.09.008.

Wu, A.M., J.H. Wu, Z. Yang, T. Singh, I.J. Goldstein, and N. Sharon. 2008. Differential contributions of recognition factors of two plant lectins – Amaranthus caudatus lectin and Arachis hypogea agglutinin, reacting with Thomsen-Friedenreich disaccharide (Galβ1–3GalNAcα1–Ser/Thr). *Biochimie.* 90 no. 11–12:1769–1780.

Wu, L., and J.K. Bao. 2013. Anti-tumor and anti-viral activities of Galanthus nivalis agglutinin (GNA)-related lectins. *Glycoconjugate Journal.* 30 no. 3:269–279.

Yamaguchi, M., T. Ogawa, K. Muramoto, Y. Kamio, M. Jimbo, and H. Kamiya. 1999. Isolation and character-ization of a mannan-binding lectin from the freshwater cyanobacterium (blue-green algae) Microcystis viridis. *Biochemical and Biophysical Research Communications*. 265 no. 3:703–708.

Yang, R.Y., and F.T. Liu. 2003. Galectins in cell growth and apoptosis. *Cellular and Molecular Life Sciences*. 60 no. 2:267–276.

Yau, T., X. Dan, C.C. Ng, and T.B. Ng. 2015. Lectins with potential for anti-cancer therapy. *Molecules*. 20 no. 3:3791–3810.

Yoon, T.J., Y.C. Yoo, T.B. Kang, S.K. Song, K.B. Lee, E. Her, K.S. Song, and J.B. Kim. 2003. Antitumor activ-ity of the Korean Mistletoe Lectin is attributed to activation of macrophages and NK cells. *Archives of Pharmacal Research*. 26 no. 10:861–867.

Yoshimoto, R., N. Kondoh, M. Isawa, and J. Hamuro. 1987. Plant lectin, ATF1011, on the tumor cell sur-face augments tumor-specific immunity through activation of T cells specific for the lectin. *Cancer Immunology, Immunotherapy*. 25 no. 1:25–30.

Yu, L. G., J.D. Milton, D.G. Fernig, and J.M. Rhodes. 2001. Opposite effects on human colon cancer cell pro-liferation of two dietary Thomsen-Friedenreich antigen-binding lectins. *Journal of Cellular Physiology*. 186 no. 2:282–287.

Yu, L.J., D.G. Fernig, J.A. Smith, J.D. Milton, and J.M. Rhodes. 1993. Reversible inhibition of prolifera-tion of epithelial cell lines by Agaricus bisporus (edible mushroom) lectin. *Cancer Research*. 53 no. 19:4627–4632.

Yu, Y., R. Yang, X. Zhao, D. Qin, Z. Liu, F. Liu, X. Song, L. Li, R. Feng, and N. Gao. 2016. Abrin P2 sup-presses proliferation and induces apoptosis of colon cancer cells via mitochondrial membrane depolar-ization and caspase activation. *Acta Biochimica et Biophysica Sinica*. 48 no. 5:420–429.

Willmann, R., H.M. Lajunen, G. Erbs, M. Newman, D. Kolb, K. Tsuda, F. Katagiri, J. Fliegmann, J.J. Bono, V.J Cullimore, A.K. Jehle, F. Götz, A. Kulik, A. Molinaro, V. Lipka, A. Gust, and T. Nuernberger. 2011. Arabidopsis lysin-motif proteins LYM1 LYM3 CERK1 mediate bacterial peptidoglycan sensing and immunity to bacterial infection. *Proceedings of the National Academy of Sciences*, no. 108:19824–19829. doi: 10.1073/pnas.1112862108.

Wu, A.M., J.H. Wu, T. Singh, L.J. Lai, Z. Yang, and A. Herp. 2006. Recognition factors of Ricinus communis agglutinin 1 (RCA(1)). *Mol Immunol*. 43:1700–15.

Zalatnai, A., K. Lapis, B. Szende, E. Rásó, A. Telekes, Resetár A, and M. Hidvégi. 2001. Wheat germ extract inhibits experimental colon carcinogenesis in F-344 rats. *Carcinogenesis*. 22 no. 10:1649–1652.

Zhang, W., M. Van Eijk, H. Guo, A. Van Dijk, O.B. Bleijerveld, M.H. Verheije, G. Wang, H.P. Haagsman, and E.J.A. Veldhuizen. 2017. Expression and characterization of recombinant chicken mannose binding lectin. *Immunobiology*. 222 no. 3:518–528.

Zhang, X.W., X. Man, X. Huang, Y. Wang, Q.S. Song, K.M. Hui, and H.W. Zhang. 2018. Identification of a C-type lectin possessing both antibacterial and antiviral activities from red swamp crayfish. *Fish and Shellfish Immunology*. 77:22–30.

CHAPTER 9

Plants Species with Anticancer Activity

Evangelia Flampouri, Spyridon Kintzios, and Maria Barberaki

CONTENTS

9.1 INTRODUCTION: GENERAL BOTANICAL ISSUES

In this part of the book, a detailed analysis will be given on each of a number of species (almost 300 species) with documented anticancer properties either *in vitro* or in clinical use. The success stories of mistletoe, periwinkle, and yew that became drugs in the first line of treatment of various cancers are remarkable. Additionally, many plant species have shown activity against cancer, and many others are promising for future use. Before we proceed with the analysis, however, and for the purpose of a better understanding of the description of each species, a brief overview of botanical terms is given in following:

9.1.1 Life Cycle

Plants can be distinguished according to their life cycle (germination, growth, flowering, and seed production) as annuals, biennials, and perennials. *Annuals* complete their life-cycle within a year. *Biennials* grow without flowering in the first year, coming into flowering in the second. Both these groups are herbs, which flower only once, produce seeds, and then die. *Perennials* flower for several or many years in succession.

9.1.2 Plant Anatomy

The stem is made up of *internodes*, separated by *nodes*. The leaves arise at these nodes. The stem is either unbranched or has *side branches* emerging from buds in the leaf axils. The side branches may themselves branch. Shoots continue to grow at the tip and develop new leaves, with buds in the axils, which can grow into branches. The shoot can either be hairless, or it may carry hairs of various kinds, often glandular.

Roots serve to anchor a plant (in the soil or another host plant) and to facilitate the uptake of water and mineral salts. The *main* or *tap root* is normally vertical. From this grow *lateral roots,* which may themselves branch, and in this way, the full root system develops. Many plants have swollen roots which contain stores of food.

A fully developed *leaf* consists of the *blade,* the *leafstalk* (petiole), and *leaf base.* Sometimes there is no leafstalk, in which case the leaf is termed *sessile,* or unstalked; otherwise, it is known as *petiolate,* or stalked. The leaf base is often inconspicuous but sometimes has a *leaf sheath.* The leaf base may have blunt or pointed extensions at either side of the stem (*amplexicaul*) or even completely encircle and fuse with the stem (*perfoliate*). In *decurrent* leaves, the leaf blade extends some distance down the sides of the stem.

Leaves can have different shapes, which often serve as taxonomic characters. They are distinguished in *simple* leaves with undivided blade and *compound* leaves, consisting of several separate leaflets. Some have *parallel* or curved veins, without a central midrib; others have *pinnate* veins, with an obvious midrib and lateral veins. A leaf can have any one of a number of shapes, including *linear, lanceolate, elliptic, ovate, hastate* (spear-shaped), *reniform* (kidney-shaped), *cordate* (heart-shaped), *rhombic, spatulate* or *spathulate* (spoon-shaped), or *sagittate* (arrow-shaped). There are also differences in leaf *margins* including *entire, crenate* (bluntly toothed), *serrate* (serrated), *dentate* (toothed), *sinuate/undulate* (wavy), *pinnately lobed,* or *palmately lobed* leaves. Accordingly, compound leaves can be found as *pinnate* (*imparipinnate* if there is a terminal leaflet and *paripinnate* if not). Leaves grow as lateral appendages of the stem, from nodes. In the case of *alternate* leaves, there is a single leaf at each node, and successive leaves are not directly above each other. *Opposite* leaves are placed as a pair, one at each side of the node. When there are three or more leaves at each node, they are described as *whorls* (in whorls).

The *inflorescence* is the part of the stem which carries the flowers. A *spike* is a flowerhead in which the individual flowers are stalkless. It can be short and dense or long and loose. A *raceme* is similar but consists of stalked flowers. A *panicle* is an inflorescence whose main branches are themselves branched. In an *umbel,* the flower stalks are of equal length and arise from the same point on the stem. A *head* consists of many unstalked or short-stalked flowers growing close together at the end of a stem. The particularly densely clustered head of composites is known as a *capitulum.*

The *flower* is a thickened shoot which carries the reproductive parts of the plant. Its individual parts can be interpreted as modified leaves. The *perianth* consists either of *perianth segments* or of *sepals* and *petals.* More commonly, these are differentiated into an outer ring of usually green sepals (the *calyx*) and an inner ring of usually colored petals (the *corolla*). The male part of the flower (*androecium*) consists of the *stamens*; the female part (*gynoecium*) consists of the *ovary, style,* and *stigma,* together known as the *pistil.* Each stamen consists of a thin *filament* and an *anther,* the latter containing the *pollen.* In the center of the flower is the *pistil* (gynoecium). This consists of at least one *carpel,* often more, either free or fused. The pistil is divided into *ovary, style,* and *stigma.*

The *fruit* develops from the *ovary,* after pollination. It protects the seeds until they are ripe and often also has particular adaptations for seed dispersal. *Dehiscent* fruits open to release the seeds, while *indehiscent* do not.

9.2 SPECIES-SPECIFIC INFORMATION

9.2.1 Success Stories

Plant Species Used in Contemporary Clinical Cancer Treatment

Happy tree (*Camptotheca acuminata*) (Nyssaceae)	**Antitumor**
	Antileukemic
	Antineoplastic
	Tumor inhibitor

Synonyms: Happy tree-xi shu, cancer tree, tree of life.

Location: Central China, sometimes grown in southern California.

Origin: Southern China and Tibet.

Appearance:

Stem: Trunk is usually straight with gray bark and spreading lateral branches.
Leaves: large, glossy, and strongly ribbed; on new growth flushes they are pale pinkish bronze.
Flowers: tree bears stalked, spherical heads of tiny white flowers close to the branch tip.
Fruit: curious yellow-green, sharply angled fruit which finally turn brown before falling.
In bloom: May–July.

Degree of rarity: low (widely cultivated in parts of China).

Tradition: *Camptotheca* trees had been used as fuelwood and as ornamental species in its native China. There were no reports on medicinal uses of *Camptotheca* in China, and thus it was believed the trees had no medicinal value before its antitumor activity was discovered in 1957. *Camptotheca acuminata* is commonly known as happy tree (xi shu) because the trees can be used as folk medicine to cure stubborn phlegm as well as other diseases, thus making patients 'happy'. There are three local names to refer the uses of *C. acuminata* in China. In addition, there are at least 25 other local names for tree morphology and habitat of the species by 35 ethnic groups. The Dong people made paste from fresh leaves or fruits and powder from dry materials from any part of this tree species and mixed some or all with rice wine to cure many stubborn diseases including furunclulosis, skin diseases, and even a liu (probably a kind of cancer).

Biology: medium-sized deciduous trees growing to 20 meters (*Camptotheca acuminata, Camptotheca lowreyana*).

Parts used: bark, stem tissues, leaves, seeds (endosperm and embryo), fruits, and root bark.

Active ingredients: Camptothecin, irinotecan, topotecan, rubitecan, exatecan, CKD-602, gimatecan, karenitecin, diflomotecan.

Particular value: *Camptothecin* was originally isolated by Wall et al. in pioneering studies in 1966 as a novel alkaloidal leukemia and tumor inhibitor. Topoisomerase I was validated as a target for cancer chemotherapy when it was identified as the sole target of camptothecin (CPT). This compound was isolated in 1966 from the Chinese tree *Camptotheca acuminata*, and its therapeutic development was initially limited by its poor solubility and unacceptable toxicity.

Precautions:

The successful clinical introduction was substantially delayed, largely due to problems with insolubility and instability, particularly of the ring E lactone. Late diarrhea (generally occurring more than 24 hours after administration of Camptosar®) can be life threatening since it may be prolonged and may lead to dehydration, electrolyte imbalance, or sepsis. Late diarrhea should be treated promptly with loperamide. Patients with diarrhea should be carefully monitored and given fluid and electrolyte rement if they become dehydrated or antibiotic therapy if they develop ileus, fever, or severe neutropenia. The most common side effects with Hycamtin® (topotecan hydrochloride) are infections. Blood disorders are also common, where reduced levels of certain blood cells may cause anemia, reduced resistance to infections, and increased bruising or bleeding.

Indicative dosage and application:

- The usual dose of topotecan injection is 1.5 mg/m^2 of body surface area per day. (1.5 mg/mVd for five days). This treatment will normally be repeated every three weeks. This treatment may vary, depending on the results of your regular blood tests. (Responsible company GlaxoSmithKline, trade name: Hycamtin®). The product has an FDA approval.
- Topotecan hydrochloride is given into the vein by an intravenous drip over a 30 minute period. Topotecan hydrochloride is given once daily for five days for the treatment of ovarian and small cell lung cancer and once every day for three days for cervical cancer. This is usually repeated every three weeks from the start of each course.

- Topotecan capsules: hard capsules for oral administration, 0.25 mg and 1 mg. The recommended dose of topotecan capsules is 2.3 mg/m^2 once daily for five consecutive days every 21 days. The US Food and Drug Administration has given approval to *topotecan* capsules to be used by patients who had a complete or partial response to first-line chemotherapy and who are at least 45 days from the end of that treatment (2008).
- 50 to 350 mg/m^2 (recommended dose) of irinotecan. Weekly schedule of bolus 5-FU/LV or with an every-two-week schedule of infusional 5-FU/LV. Weekly and once-every-three-week dosage schedules were used for the single agent irinotecan studies. It is approved by FDA (company Pfizer Inc., trade name: Camptosar®).

Documented target cancers:

- Camptothecin derivatives are used in the chemotherapy of glioblastoma, sarcomas, ovarian cancers, recurrent small cell lung cancer, colorectal cancers, and lung cancers.
- Its semisynthetic derivatives, irinotecan and topotecan, are currently extensively used for colorectal and ovarian cancer treatments, as well as being evaluated in numerous other clinical trials.
- Non-camptothecin topoisomerase I poison such as indenoisoquinoline derivatives (indotecan and indimitecan) and Genz-644282 are currently under clinical testing. Topoisomerase II poisons such as etoposide, anthracyclins (daunorubicin and doxorubicin), and mitoxantrone are extensively used for the treatment of different types of cancer.
- Hycamtin® (Topotecan) monotherapy is indicated for the treatment of: patients with metastatic carcinoma of the ovary after failure of first-line or subsequent therapy, patients with relapsed small cell lung cancer (SCLC) for whom re-treatment with the first-line regimen is not considered appropriate. Topotecan in combination with cisplatin is indicated for patients with carcinoma of the cervix recurrent after radiotherapy and for patients with Stage IVB disease. Patients with prior exposure to *cisplatin* require a sustained treatment-free interval to justify treatment with the combination.
- Camptosar® Injection (irinotecan hydrochloride) is indicated as a component of first-line therapy in combination with 5-fluorouracil and leucovorin for patients with metastatic carcinoma of the colon or rectum. Camptosar® is also indicated for patients with metastatic carcinoma of the colon or rectum whose disease has recurred or progressed following initial fluorouracil-based therapy.

Further details:

- Camptothecin is a natural alkaloid isolated from the bark of Chinese tree *Camptotheca acuminata*. Camptothecin (a quinoline alkaloid) is commercially obtained from stem tissues of the 'Happy Tree' (*Camptotheca acuminata*), which is now mainly grown in plantations for this purpose.
- The alkaloid *camptothecin* was first isolated from the Chinese tree *Camptotheca acuminata* by Wani and Wall contemporaneously with their discovery of Taxol®.
- It acts as DNA topoisomerase I poison and stabilizes topoisomerase I-DNA covalent complex by inhibiting the religation of DNA strand. Semisynthetic derivatives of camptothecin such as topotecan (Hycamtin®) and irinotecan (Camptosar®) have been clinically approved by the US Federal Drug Administration.
- Camptothecin is the active agent derived from the bark extract of the *Camptotheca acuminata* tree. Two analogs of camptothecin have been developed that are clinically active and less toxic than the parent compound: irinotecan (CPT-11) and topotecan.
- Topoisomerase inhibitors are divided into TP1 and TP2 inhibitors. TP1 inhibitors stem from the bark of the *Camptotheca acuminata* tree. They interact with topoisomerase enzymes and disrupt DNA duplication, resulting in apoptosis and cell death. Drugs in this class include topotecan and irinotecan. TP2 inhibitors are podophyllotoxin alkaloids found in the American may apple. The active agents are primarily etoposide (VP16, VePesid) and teniposide. Both TP1 and TP2 inhibitors are used in the treatment of solid cancers and to a degree in the treatment of glioma.
- The mechanism of action appears to be stabilization of the covalent adduct between topoisomerase I and the 3′-phosphate group of the DNA backbone.
- Irinotecan and topotecan are structurally related semisynthetic derivatives of the natural product camptothecin. This compound is inactive in the parent form and is converted by intracellular carboxylesterase activity to its active metabolite termed SN-38.

- Camptosar® (irinotecan hydrochloride injection) is an antineoplastic agent of the topoisomerase I inhibitor class. Irinotecan hydrochloride was clinically investigated as CPT-11. Camptosar® is supplied as a sterile, pale yellow, clear, aqueous solution. It is available in two single-dose sizes: 2 mL-fill vials contain 40 mg irinotecan hydrochloride and 5 mL-fill vials contain 100 mg irinotecan hydrochloride.
- CPT-11 is a prodrug and only a weak topo 1 inhibitor, the endogenous bioactivation by carboxylesterase-catalyzed hydrolysis generates the active metabolite SN-38.
- Irinotecan requires de-esterification by carboxylesterases to yield SN-38, a metabolite that is 1,000 times more active at inhibiting topoisomerase I. Topotecan is not a prodrug and binds directly to topoisomerase I without activation.
- Both irinotecan and topotecan are cell cycle-nonspecific agents with activity in all phases of the cell cycle. Irinotecan is indicated for use in patients with colorectal cancers, for whom DNA topoisomerase I is expressed at relatively high levels, and in patients with lung cancers. *Topotecan* is used to treat ovarian cancer, small cell lung cancer, and acute myelogenous leukemia. Adverse reactions for both agents include myelosuppression, diarrhea, vomiting, liver enzyme elevations, asthenia, and fever; for topotecan, side effects also include microscopic hematuria and alopecia. It has been reported that irinotecan toxicity is modified by genetic variations in UDP-Glucuronosyltransferase (UGT1A1), the enzyme that glucuronidates SN-38 and converts it to an inactive product in bile.
- Rubitecan (9-nitrocamptothecin) is a semisynthetic compound, derived from *camptothecin* by direct nitration, and the most advanced of all the new camptothecin analogs. Rubitecan itself induces apoptosis associated with the upregulation of the cytokines cyclin B1 and cdc2, but not of cyclins A, E, and cdk2. A number of Phase II clinical trials have been reported, mostly using oral dosing at 1.5 mg m^{-2} per day for five consecutive days (five days on–two days off) on a continuous basis. Phase III trials are in progress.
- CKD-602 represents an attempt to improve the solubility of camptothecin, by adding an N-aminoethyl group at the 7-position of ring B where there is some bulk tolerance, and was the best of a number of such analogs evaluated. It showed good activity in a range of human tumor xenografts in nude mice (e.g. SKOV-3 ovarian, MX-1, LX-1, and HT29, WIDR, and CX-1 colon lines), with activity broadly comparable to *topotecan* at the same dose. A preliminary report of a Phase II trial of CKD-602, at a dose of 0.6 mg m^{-2} day^{-1} in patients with ovarian cancer, said that 20% showed partial responses, with the major toxicity being neutropenia.
- Gimatecan has an unusual lipophilic butyloxyliminomethyl substituent at the 7-position and was the best of a series of similar compounds evaluated. *Gimatecan* was the most active against neuroblastoma cell lines, suggested due to an ability to cause high levels of DNA breaks. In Phase I clinical studies, oral gimatecan had an acceptable toxicity profile, with myelotoxicity being dose-limiting. It showed a very long terminal half-life and favorable pharmacokinetics, with a significant incidence of tumor responses.
- Karenitecin (BNP1350) is another semisynthetic, lipophilic *camptothecin* derivative, designed to have higher oral bioavailability and greater lactone stability than *camptothecin* by employing an ethyltrimethylsilyl derivative at the 7-position. Karenitecin proved to be a much poorer substrate than topotecan for BCRP, which may be a major reason for its broader-spectrum activity. Karenitecin-resistant A2780 variants over expressed BCRP, had reduced catalytic activity of topo I and showed cross-resistance to many other camptothecins. *In vitro* studies with cloned cytochrome P450 (CYP) isoenzymes suggested that karenitecin is metabolized by CYP3A4, CYP2C8, and CYP2D6 and is an inhibitor of CYP3A4 and CYP2C8. Karenitecin was also active in a wide range of adult and pediatric central nervous system malignancies growing in athymic nude mice, suggesting possible clinical utility in patients with such tumors.
- Diflomotecan is prepared by a totally synthetic and stereospecific route and was selected for development from a series of E ring modified homocamptothecin analogs possessing a 7- rather than a 6-membered hydroxylactone ring. Diflomotecan is more effective than camptothecin in inducing topo I-mediated DNA cleavage, due to a greater ability to stabilize the topo I–DNA cleavage complexes 57. It is also a very potent cytotoxin in cell line assays, with higher overall antiproliferative activity than camptothecin, topotecan, or SN38 against a series of 43 early-passage human colon cancer cell lines in culture. *In vitro* studies suggest it is metabolized in humans primarily by CYP3A4. A Phase I clinical trial of oral diflomotecan in adult patients with solid malignant tumors suggested a maximum tolerated dose of 0.27 mg day^{-1} daily for five days, repeated every three weeks.
- The silatecan7-tert-butyldimethylsilyl-10-hydroxy-camptothecin (DB-67) represents a new generation of camptothecin derivatives that exhibits a potent *in vitro* DNA topoisomerase I (TOP1)-mediated DNA-damaging activity, improved blood stability, and holds significant promise for the

treatment of human cancers. It is a synthetic, highly lipophilic derivative of camptothecin, with potential antineoplastic and radiosensitizing activities. Silatecan DB-67 binds to and stabilizes the topoisomerase I-DNA covalent complex, inhibiting the religation of topoisomerase I-mediated single-stranded DNA breaks and producing lethal double-stranded DNA breaks when encountered by the DNA replication machinery; inhibition of DNA replication and apoptosis follow.

- Cositecan is a CPT analog. Cositecan is in phase III clinical trials by BioNumerik for the treatment of ovarian cancer. Phase II clinical trials are also being conducted for the treatment of malignant melanoma, non-small cell lung cancer, and breast cancer.
- Exatecan is a drug which is a structural analog of camptothecin with antineoplastic activity.
- Lurtotecan is a semisynthetic analog of camptothecin with antineoplastic activity. Lurtotecan selectively stabilizes the topoisomerase I-DNA covalent complex and forms an enzyme–drug–DNA ternary complex. As a consequence of the formation of this complex, both the initial cleavage reaction and religation steps are inhibited and subsequent collision of the replication fork with the cleaved strand of DNA results in inhibition of DNA replication, double strand DNA breakage, and triggering of apoptosis. Independent from DNA replication inhibition, lurtotecan also inhibits RNA synthesis, multi-ubiquitination, and degradation of topoisomerase I, and chromatin reorganization. *Lurtotecan* is under investigation in clinical trial NCT00022594 (liposomal Lurtotecan in treating patients with metastatic or locally recurrent head and neck cancer).
- Belotecan it is a semi-synthetic camptothecin analog indicated for small cell lung cancer and ovarian cancer, approved in South Korea under the trade name Camtobell(R), presented in 2 mg vials for injection. The drug is marketed by ChongKunDang Pharmaceuticals since 2003. Belotecan blocks topoisomerase I with a pIC_{50}, stabilizing the cleavable complex of topoisomerase I-DNA, which inhibits the religation of single-stranded DNA breaks gen.
- Camptothecin (CPT), the third largest anticancer drug, is produced mainly by *Camptotheca acuminata* and *Nothapodytes foetida*. CPT itself is the starting material for clinical CPT-type drugs, but the plant-derived CPT cannot support the heavy demand from the global market. Research efforts have been made to identify novel sources for CPT. In this study, three CPT-producing endophytic fungi, *Aspergillus sp.* LY341, *Aspergillus sp.* LY355, and *Trichoderma atroviride*.

LY357, were isolated and identified from *C. acuminata*.

References

Chen, A.Y., S.J. Shih, L.N. Garriques, M.L. Rothenberg, M. Hsiao, and D.P. Curran. 2005. Silatecan DB-67 is a novel DNA topoisomerase I-targeted radiation sensitizer. *Molecular Cancer Therapeutics* 4, no. 2 (Feb):317–324.

Denny, W.A. 2007. Comprehensive medicinal chemistry II, therapeutic areas II. *Cancer, Infectious Diseases, Inflammation & Immunology and Dermatology* 7:111–128. Editors-in-Chief: John B. Taylor and David J. Triggle:499–510.

Gerson, St.L., P.F. Caimi, M.W. Basem, and R.J. Creger. 2018. Pharmacology and molecular mechanisms of antineoplastic agents for hematologic malignancies. In: *Hematology Basic Principles and Practise*, Seventh Edition. R. Hoffman, E.J. Benz Jr., L.E. Silbersteim, H.E. Heslop, J.I. Weitz, J. Anastasi et al. (Eds). Amsterdam, the Netherlands, Elsevier, 849–912, Chapter 57.

Góngora-Castillo, E., K.L. Childs, G. Fedewa, 2012. Development of transcriptomic resources for interrogating the biosynthesis of monoterpene indole alkaloids in medicinal plant species. *PLOS ONE* 7, no. 12:e52506.

Jain, Ch.K. 2017. Reference module in life sciences: *Topoisomerases*. 18 no. 1:75–92.

Jing-zhi, Gong, Q.j.L., X. Wang, et al. 2018. Floral morphology and morphogenesis in *Camptotheca* (*Nyssaceae*), and its systematic significance. *Annals of Botany* 121, no. 7:1411–1425.

Li, S., and W. Zhan. 2014. Access ethnobotany of *Camptotheca* Decaisne: New discoveries of old medicinal uses. *Pharmaceutical Crops* 5 (Suppl 2: M7):140–145 2210-2906/14 2014 Bentham Open Open.

Lin, C.S., P.C. Chen, C.K. Wang, C.W. Wang, Y.J. Chang, C.J. Tai, and C.J. Tai. 2014. Antitumor effects and biological mechanism of action of the aqueous extract of the *Camptotheca acuminata* fruit in human endometrial carcinoma cells. *Evidence-Based Complementary and Alternative Medicine* 2014:564810.

Lorence, A., F. Medina-Bolivar, and C.L. Nessler. 2004. *Camptothecin* and 10-hydroxycamptothecin from *Camptotheca acuminata* hairy roots. *Plant Cell Reports* 22, no. 6 (January):437–441.

Lorence, A., and C.L. Nessler. 2004. *Camptothecin*, over four decades of surprising findings. *Phytochemistry* 65, no. 20:2735–2749.

Majumder, H.K. 2013. Topoisomerases. In: *Brenner's Encyclopedia of Genetics*. Maloy, S. and Hughes, K. (Eds.). Oxford, U.K. Elsevier, 78–79.

Matsuura, H.N., M.R. Rau, and A.G. Fett-Neto. 2014. Oxidative stress and production of bioactive monoterpene indole alkaloids: Biotechnological implications. *Biotechnology Letters* 36, no. 2 (Feb):191–200.

Nabors, L.B., B. Surboeck, and W. Grisold. 2016. Complications from pharmacotherapy. In: *Handbook of Clinical Neurology*. Vinken, P. and Bruyn. G. (Eds.). Oxford, U.K. Elsevier, 134:235–250, Chapter 14.

Newman, D.J., G.M. Cragg, and D.G.I. Kingston. 2015. Natural products as pharmaceuticals and sources for lead structures. In: *The Practice of Medicinal Chemistry*, 4th Edition. Wermuth, C.G., Aldous, D., and Rognan, D. (Eds.). Academic Press Inc., San Diego, 101–139, Chapter 5.

Newman, D.J., and D.G.I. Kingston. 2015. *The Practice of Medicinal Chemistry, Natural Products as Pharmaceuticals and Sources for Lead Structures*. B. D. ed. Roitberg, sub (4th Edition), Ed. P.D. Cotter.

Newton, H.B. 2006. Clinical pharmacology of brain tumor chemotherapy. In: *Handbook of Brain Tumor Chemotherapy*. Newton, H.B. (Ed.).

Patten, A.M., D.G. Vassão M.P. Wolcott, L.B. Davin, and G.L. Norman. 2010. Volume 3.27. Trees: A Remarkable Biochemical Bounty. In: Comprehensive Natural Products II. Chemistry and Biology. Hung-Wen (Ben) Liu and Lew Mander (Eds). Oxford, sa U.K. Elsevier 1173–1296.

Pu, X., X. Qu, F. Chen, J. Bao, G. Zhang, and Y. Luo. 2013. *Camptothecin*-producing endophytic fungus *Trichoderma atroviride* LY357: Isolation, identification, and fermentation conditions optimization for *camptothecin* production. *Applied Microbiology and Biotechnology* 97, no. 21 (Nov):9365–9375.

Zhang, Z., S. Li, S. Zhang, C. Liang, D. Gorenstein, and R.Sc. Beasley. 2004. New *camptothecin* and *ellagic acid* analogues from the Root Bark of *Camptotheca acuminate*. *Planta Medica* 70, no. 12:1216–1221.

Websites

www.worldbotanical.com

https://en.wikipedia.org/wiki

http://life.nthu.edu.tw

https://academic.oup.com/aob/article-abstract

www.amazon.com/10-seeds-Camptotheca-acuminata-Cancer

www.thegoodscentscompany.com

www.gardensonline.com.au

www.accessdata.fda.gov

http://au.gsk.com/media

https://dtp.cancer.gov

www.sciencedirect.com

www.researchgate.net

www.thieme-connect.com

http://labeling.pfizer.com

www.ema.europa.eu

www.drugbank.ca

www.sps.nhs.uk/medicines

https://pubchem.ncbi.nlm.nih.gov/compound

Catharanthus

See in *Vinca*

Cephalotaxus

See in *Taxus* under Further details.

Mandrake American (*Podophyllum peltatum*) (Berberidaceae)	Antitumor

Synonyms: Wild lemon, Ground lemon, May Apple, Racoonberry

You can find it commonly in the eastern United States and Canada, North America, growing there profusely in wet meadows and in damp, open woods. Podophyllotoxin, a potential natural anticancer agent, is also found in the Himalayan mayapple in much greater amounts, but this plant is endangered in the wild.

Origin: North America.

Appearance:

Stem: solitary, mostly unbranched, 0.3–0.5 m high.

Root: is composed of many thick tubers, fastened together by fleshy fibers, which spread greatly underground.

Leaves: smooth, stalked, peltate in the middle like an umbrella, of the size of the hand, composed of five to seven wedge-shaped divisions.

Flowers: solitary, drooping white, about 2 cm across, with nauseous odor.

Fruit: size and shape of a common rosehip, being 3 to 6 cm long. Yellow in color, sweet in taste.

In bloom: May.

Degree of rarity: low.

Tradition: North American Indians used it as an emetic and vermifuge.

Biology: The rhizome develops underground for several years before a flowering stem emerges (only one shoot per root). The plant can be propagated either by runners or by seed. For cultivation, adequate fertilization is recommended.

Part used: root, resin.

Active ingredients: podophyllotoxin (a neutral crystalline substance), podophylloresin (amorphus resin), diphyllin and aryltetralin (*podophyllum lignan*), etoposide (VP – 16), teniposide (semisynthetic derivative of 4'- demethylepipodophyllotoxin, naturally occurring compounds).

Particular value: It was included in the British Pharmacopoeia in 1864. It is considered as one of the medicines with the most extensive service: it is used for all hepatic complaints, as antibilious, cathartic, hydragogue, and purgative.

Precautions: Leaves and roots are poisonous, podophyllotoxins are classical spindle poisons causing inhibition of mitosis by blocking mitrotubular assembly and should be avoided during pregnancy. *Podophyllum* may cause abnormal heart rhythms, allergic skin symptoms (burning sensation, irritation, itching, pain, stinging, swelling, and ulcers), bad or bitter taste, bile stimulation, birth defects, bowel blockage, changes in cell growth, confusion, decreased electrolytes, decreased muscle tone, dehydration, diarrhea (sometimes bloody or watery), digestive tract pain and inflammation, dizziness, fetal deaths, hallucinations, hair loss, irritation of the stomach and intestines, kidney failure, liver problems, low levels of white blood cells, low potassium levels, metabolic changes, mouth inflammation, movement disorders, muscle problems (fatigue, fiber breakdown, pain, paralysis, soreness, and weakness), nausea and vomiting, reduced nail growth, retinoblastoma (tumor of the retina of the eye), seizures, stomach pain, tenderness of the tongue, urinary problems (inability to empty the bladder), and worsening of mental illness symptoms.

Indicative dosage and application:

- Etoposide is used as etoposide phosphate (Etopophos; Bristol-Myers Squibb Company, Princeton, NJ), and because it is water soluble it can be made up to a concentration of 20 mg/mL, however, it can be given as a five-minute bolus, in high doses in small volumes, and as a continuous infusion. Capsules formulation contain 50 mg and 100 mg etoposide (Vepesid 50 mg and 100 mg capsule, soft).
- Teniposide is a semisynthetic derivative of podophyllotoxin. The chemical name for teniposide is 4'-demethylepipodophyllotoxin9-[4,6-O-(R)-2-thenylidene-β-D-glucopyranoside]. Teniposide differs from etoposide, by the substitution of a thenylidene group on the glucopyranoside ring.
- Teniposide (VUMON®, teniposide injection) (also commonly known as VM-26) is supplied as a sterile nonpyrogenic solution in a nonaqueous medium intended for dilution with a suitable parenteral vehicle prior to intravenous infusion. It is available in 50 mg (5 mL) ampules.

- Penile warts in selected cases can be safely treated with 0.5–2.0% podophyllin self-applied by the patient at a fraction of the cost of commercially available podophyllotoxin.

Documented target cancers:

- Antiproliferative effects on human peripheral blood mononuclear cells and inhibition of *in vitro* immunoglobulin synthesis.
- Etoposide appears to be one of the most active drugs for small cell lung cancer, testicular carcinoma (the Food and Drug Administration approved indication), ANLL, and malignant lymphoma.
- Etoposide Injection is indicated in adults for the management of: resistant non-seminomatous testicular tumors in combination with other chemotherapeutic agents, small cell lung cancer, in combination with other chemotherapeutic agents, acute monoblastic leukemia (AML M5), and acute myelomonoblastic leukemia (AML M4) when standard induction therapy has failed (in combination with other chemotherapeutic agents).
- Etoposide capsules are indicated in combination with other approved chemotherapeutic agents in adults, for the treatment of recurrent or refractory testicular cancer, small-cell lung cancer, Hodgkin's lymphoma, Non-Hodgkin's lymphoma, relapsed or refractory acute myeloid leukemia, and platinum-resistant/refractory epithelial ovarian cancer.
- Etoposide also has demonstrated activity in refractory pediatric neoplasms, hepatocellular, esophageal, gastric, and prostatic carcinoma, ovarian cancer, chronic and acute leukemia, although additional single and combination drug studies are needed to substantiate these data.
- Teniposide, in combination with other approved anticancer agents, is indicated for induction therapy in patients with refractory childhood acute lymphoblastic leukemia.
- Proresid (a mixture of natural extracts from *Podophyllum* sp.) has been used to a triple-drug therapy with high doses of Endoxan and Methotrexat instead of the earlier long-term Endoxan treatment in addition to surgery.
- Teniposide causes dose-dependent single- and double-stranded breaks in DNA and DNAprotein cross-links. The mechanism of action appears to be related to the inhibition of type II topoisomerase activity since teniposide does not intercalate into DNA or bind strongly to DNA.

Further details:

- Podophyllotoxin is a natural product isolated from *Podophyllum peltatum* and *Podophyllum emodi* and has long been known to possess medicinal properties. Etoposide (VP-16), a podophyllotoxin derivative, is currently in clinical use in the treatment of many cancers, particularly small cell lung carcinoma and testicular cancer. This compound arrests cell growth by inhibiting DNA topo-isomerase II, which causes double strand breaks in DNA. VP-16 does not inhibit tubulin polymerization, however, its parent compound, podophyllotoxin, which has no inhibitory activity against DNA topoisomerase II, is a potent inhibitor of microtubule assembly. In addition to these two mechanisms of action, an unknown third mechanism of action has also been proposed for some of the recent modifications of podophyllotoxins. Some of the congeners exhibited potent anti-tumor activity, of which etoposide and teniposide are in clinical use, NK 611 is in phase II clinical trials, and many compounds are in the same line. Recent developments on podophyllotoxins have led structure-activity correlations which have assisted in the design and synthesis of new podophyllotoxin derivatives of potential antitumor activity. Modification of the A-ring gave compounds having significant activity but less than that of etoposide, whereas modification of the B-ring resulted in the loss of activity. One of the modifications in the D-ring produced GP-11 which is almost equipotent with etoposide. E-ring oxygenation did not affect the DNA cleavage which led to the postulation of the third mechanism of action. It has also been observed that free rotation of E-ring is necessary for the antitumor activity. The C4-substituted aglycones have a significant role in these recent developments. Epipodophyllotoxin conjugates with DNA cleaving agents such as distamycin increased the number of sites of cleavage. The substitution of a glycosidic moiety with arylamines produced enhanced activity. Modification in the sugar ring resulted in the development of the agent, NK 611, which is in clinical trial at present.

- Podophyllotoxin is also one of the main ingredients of the Bajiaolian root. Bajiaolian (*Dysosma pleianthum*), one species in the Mayapple family, has been widely used as a general remedy and for the treatment of snake bite, weakness, condyloma accuminata, lymphadenopathy, and tumors in China for thousands of years. The herb was recommended by either traditional Chinese medical doctors or herbal pharmacies for postpartum recovery and treatment of a neck mass, hepatoma, lumbago, and dysmenorrhoea.

- *Podophyllum emodi*: Indian *Podophyllum*, a native of Northern India. The roots are much stouter, more knotty, and twice as strong as the American. It contains twice as much podophyllotoxin. It is official in India and in close countries, and it is used in place of ordinary *podophyllum*.

- A mixture of natural and semisynthetic (modified) glycosides from *Podophyllumemodi* has been used for many years in the treatment of rheumatoid arthritis, but its use is hampered by gastrointestinal side effects. Highly purified podophyllotoxin (CPH86) and a preparation containing two semisynthetic podophyllotoxin glycosides (CPH82) are currently being tested in clinical trials. In this study, these drugs were shown to inhibit *in vitro* [3H]-thymidine uptake of human peripheral blood mononuclear cells stimulated by the mitogens concanavalin A, phytohemagglutinin, and pokeweed mitogen. Complete inhibition was observed with CPH86 in concentrations > or = 20 ng/ml and with CPH82 in concentrations > or = 1 microgram/ml.

- In conclusion, both CPH86 and CPH82 inhibit mitogen-induced lymphocyte proliferation and immunoglobulin synthesis, and the results may be of help in determining optimal dose levels if related to treatment effects.

- Aryltetralin lignan is a constituent of the resins and roots/rhizomes of *Podophyllum hexandrum* and *P. peltatum*. A method confirms that *P. hexandrum* resins and roots/rhizomes contain approximately four times the quantity of lignans as do those of *P. peltatum* and also that there is a significant variation in the lignan content of *P. hexandrum* resins.

- The epipodophyllin etoposide is a semisynthetic drug, prepared from the natural product podophyllotoxin, which is isolated from the plant *Podophyllum peliatum*. It has been widely used in cancer therapy and has sparked intense research into other derivatives, of which teniposide is also a well-established cancer drug and other analogs continue to be developed (Figure 9.1).

Figure 9.1   (See color insert.) *Podophyllum peltatum*.

References

Aguirre-Alvarado, C., A. Segura-Cabrera, I. Velázquez-Quesada, et al. 2016. Virtual screening-driven repositioning of etoposide as CD44 antagonist in breast cancer cells. *Oncotarget* 7, no. 17 (Apr 26):23772–23784.

Atwal, M., E.L. Lishman, C.A. Austin, and I.G. Cowell. 2017. Myeloperoxidase enhances *etoposide* and mitoxantrone-mediated DNA damage: A target for myeloprotection in cancer chemotherapy. *Molecular Pharmacology* 91, no. 1 (Jan):49–57.

Bhattacharya, P.K., A.J. Pappelis, S.C. Lee, J.N. BeMiller, and C.S. Karagiannis. 1996. Nuclear (DNA, RNA, histone and non-histone protein) and nucleolar changes during growth and senescence of may apple leaves. *Mechanisms of Ageing and Development* 92, no. 2–3:83–99.

But, P.P., L. Cheng, and I.M. Kwok. 1997. Instant methods to spot-check poisonous podophyllum root in herb samples of clematis root. *Veterinary and Human Toxicology* 39, no. 6:366.

Chu, B., S. Shi, X. Li, et al. 2016. Preparation and evaluation of *teniposide*-loaded polymeric micelles for breast cancer therapy. *International Journal of Pharmaceutics* 513, no. 1–2 (Nov 20):118–129.

Damayanthi, Y., and J.W. Lown. 1998. *Podophyllotoxins*: Current status and recent developments. *Current Medicinal Chemistry* 5, no. 3:205–252.

Frasca, T., A.S. Brett, and S.D. Yoo. 1997. Mandrake toxicity. A case of mistaken identity. *Archives of Internal Medicine* 157, no. 17:2007–2009.

Goel, H.C., J. Prasad, A. Sharma, and B. Singh. 1998. Antitumour and radioprotective action of *Podophyllum hexandrum*. *Indian Journal of Experimental Biology* 36, no. 6:583–587.

Hatfield, L.A., H.A. Huskamp, and E.B. Lamont. 2016. Survival and toxicity after cisplatin plus etoposide versus carboplatin plus etoposide for extensive-stage small-cell lung cancer in elderly patients. *Journal of Oncology Practice* 12, no. 7 (Jul):666–673.

He, S., Z. Cui, D. Mei, H. Zhang, X. Wang, W. Dai, and Q. Zhang. 2012. A cremophor-free self-microemulsified delivery system for intravenous injection of teniposide: Evaluation *in vitro* and *in vivo*. *AAPS PharmSciTech* 13, no. 3 (Sep):846–852.

He, S., H. Yang, R. Zhang, Y. Li, and L. Duan. 2015. Preparation and *in vitro-in vivo* evaluation of teniposide nanosuspensions. *International Journal of Pharmaceutics* 478, no. 1 (Jan 15):131–137.

Koh, A.J., B.P. Sinder, P. Entezami, L. Nilsson, and L.K. McCauley. 2017. The skeletal impact of the chemotherapeutic agent etoposide. *Osteoporosis International* 28, no. 8 (Aug):2321–2333.

Li, X., Z. Wu, X. An, et al. 2017. Blockade of the LRP16-PKR-NF-κB signaling axis sensitizes colorectal carcinoma cells to DNA-damaging cytotoxic therapy. *ELife* 18, (Aug):6.

McDow, R.A. 1996. Cryosurgery and *podophyllum* in combination for condylomata. *American Family Physician* 53, no. 6:1987–1988.

Mo, L., L. Hou, D. Guo, X. Xiao, P. Mao, and X. Yang. 2012. Preparation and characterization of teniposide PLGA nanoparticles and their uptake in human glioblastoma U87MG cells. *International Journal of Pharmaceutics* 436, no. 1–2 (Oct):815–824.

Nudelman, A., M. Ruse, H.E. Gottlieb, and C. Fairchild. 1997. Studies in sugar chemistry. VII. Glucuronides of *podophyllum* derivatives. *Archiv Der Pharmazie* 330, no. 9–10:285–289.

Papież, M.A., W. Krzyściak, K. Szade, et al. 2016. *Curcumin* enhances the cytogenotoxic effect of *etoposide* in leukemia cells through induction of reactive oxygen species. *Drug Design, Development and Therapy* 10:557–570.

Perkins, J.A., A.F. Inglis Jr, and M.A. Richardson. 1998. Iatrogenic airway stenosis with recurrent respiratory papillomatosis. *Archives of Otolaryngology: Head and Neck Surgery* 124, no. 3:281–287.

Sarin, Y.K., R.S. Kadyan, and R. Simon. 1997. Condyloma accuminata. *Indian Pediatrics* 34, no. 8:741–742.

Schacter, L. 1996. Etoposide phosphate: What, why, where, and how? *Seminars in Oncology* 23, no. 6 (Suppl 13):1–7.

Takeya, T., and S. Tobinaga. 1997. Weitz' aminium salt initiated electron transfer reactions and application to the synthesis of natural products. *Yakugaku Zasshi* 117, no. 6:353–367.

Teicher, B.A., T. Silvers, M. Selby, et al. 2017. Small cell lung carcinoma cell line screen of *etoposide/carboplatin* plus a third agent. *Cancer Medicine* 6, no. 8 (Aug):1952–1964.

White, D.J., C. Billingham, S. Chapman, et al. 1997. Podophyllin 0.5% or 2.0% v podophyllotoxin 0.5% for the self treatment of penile warts: A double blind randomised study. *Genitourinary Medicine* 73, no. 3:184–187.

Yan, J., J. Sun, and Z. Zeng. 2018. *Teniposide* ameliorates bone cancer nociception in rats via the P2X7 receptor. *Inflammopharmacology* 26, no. 2 (Apr):395–402.

Ying-Qian, L., J. Tian, Keduo Qian, et al. 2015. Recent progress on C-4-modified *podophyllotoxin*. Analogs as potent antitumor agents. *Medicinal Research Reviews* 35, no. 1 (Jan):1–62.

Websites

www.sciencedirect.com
www.heathsnaturalfoods.com
www.medicines.org.uk
www.accessdata.fda.gov

Yew (*Taxus baccata*) (Taxaceae and Coniferae)	Antineoplastic **Antimicrotubular**

You can find yews in: Europe, North Africa, and Western Asia.
Appearance:

Stem: a tree 1.2 to 1.5 m high, forming with age a trunk covered with red-brown, peeling bark and topped with a rounded or wide-spreading head of branches.
Leaves: spirally attached to twigs but by twisting of the stalks brought more or less into two opposed ranks, dark, glossy, almost black-green above, grey, pale-green or yellowish beneath, 15 to 45 cm long, 2 to 3 cm wide.
Flowers: unisexual, with the sexes invariably on different trees, produced in spring from the leaf axils of the proceeding summer's twigs. Male, a globose cluster of stamens; female, an ovule surrounded by small bracts, the so-called fruit bright red, sometimes yellow, juicy and encloses the seed.

Biology: can be propagated by seed or cuttings. Seeds may require warm and cold stratification. Mature woodcuttings taken in winter can be rooted under mist.

Degree of rarity: the important clinical efficacy of taxol has led to the drug supply crisis. As a result, NCI (National Cancer Institute) has developed plans to avert similar supply crisis in the future by initiating exploratory research projects for large-scale production.

Tradition: no tree is more associated with the history and legends of Great Britain. Before Christianity, it was a sacred tree favored by the Druids, who built their temples near these trees—a custom followed by the early Christians. The association of the tree with worship still prevails. The wood was formerly much valued in archery for the making of long bows. The wood is said to resist the action of water and is very hard.

Part used: stem segments, needles 1 to 2 cm long, and roots.
Active ingredients:

- Taxane diterpenes, among them paclitaxel (earlier known as taxol), cephalomannine.
- Key precursors: baccatin III, 10-desacetylbaccatin III, 9-dihydrobaccatin III, 13-Acetyl-9-dihydrobaccatin III, baccatin VI.
- Related compounds, such as taxotere.

Particular value: taxol research is being carried out on ovarian cancer, breast cancer, colon and gastric cancers, arthritis, Alzheimer's, as an aid in coronary and heart procedures, and as an antiviral agent. The uses of yew in any form for any medical or health reason should only be done after consulting a health care professional.

The status of taxus application in cancer therapy:

Taxol (containing paclitaxel) is an anticancer drug, it was originally isolated from the Pacific Yew tree in the early 1960s, was approved in 1992 by the Food & Drug Administration for use against ovarian cancer, and has also shown activity against breast, lung, and other cancers. This drug was also registered in Poland in 1996.

In 1958 the U.S. National Cancer Institute (NCI) initiates a program to screen 35,000 plant species for anticancer activity. In 1963, Drs Monroe Wall and M.C. Wani of Research Triangle Institute, North Carolina, subsequently find that an extract or the bark of Pacific yew tree has anti-tumor activity. Since that time its use as an anticancer drug has become well established.

Human trials started in 1983. Despite a few deaths caused by unforeseen allergic reactions due to the form in which the drug was administered, great promise was shown for women with previously incurable ovarian cancer. This led the NCI to issue a contract with Bristol Myers-Squibb (BMS), a pharmaceutical company based in the United States, for the clinical development of taxol. FDA approved Taxol (Paclitaxel) Injection in 1998 (6/30/1998) (Bristol-Myers Squibb Company).

In 2004 Taxotere (Docetaxel) Injection Concentrate 20MG/0.5ML was approved by FDA (Approval Date: 8/18/2004) [Responsible Company: Aventis Pharmaceuticals, Inc. Indications: breast cancer (60–100 mg/m^2), non small cell lung cancer (75–100 mg/m^2), and prostate cancer (75 mg/m^2)].

Intense research on finding alternatives to *taxol* extracted from the bark of the Pacific yew is ongoing. Taxol has been chemically synthesized and semi-synthetic versions have been developed using needles and twigs from other yew species grown in agricultural settings. This is reducing the pressure on natural strands of Pacific yew, but bark is still being used for taxol production.

Precautions:

- Poisonous. Many cases of poisoning amongst cattle have resulted from eating parts of it. The fruit and seeds seem to be the most poisonous parts of the tree.
- In the treatment of cancer: reduction in white and red blood cells counts and infection. Other common side effects include hair loss, nausea and vomiting, joint and muscle pain, nerve pain, numbness in the extremities, and diarrhea. Severe hypersensitivity can also occur, demonstrated by symptoms of shortness of breath, low blood pressure, and rash. The likelihood of these reactions is lowered by the use of several kinds of medications that are given before the taxol infusion.

Indicative dosage and application: the doses of taxol given to most patients are:

- 110 mg/m^2 in 22%
- 135 mg/m^2 in 48%
- 170 mg/m^2 in 22%

These doses are significantly lower, because of limited hematopoietic tolerance, than those previously demonstrated to be safe in minimally pre-treated or untreated patients (200–250 mg/m^2).

- Injectable solution: 6 mg/ml (paclitaxel), 20 ml/4ml (docetaxel), 80 ml/4ml (docetaxel), 20 ml/4ml (docetaxel), 160 ml/8ml (docetaxel).

More specifically, for ovarian cancer, various regimens exist: 135–175 mg/m^2 IV over three hours in three weeks, for breast cancer 175 mg/m^2 IV over three hours in three weeks, for non-small cell lung cancer 135 mg/m^2 IV over 24 hours in three weeks. There are several off-label indications that are being investigated, including: pancreatic cancer, head/neck cancer, small-cell lung cancer, upper GI adenocarcinoma, hormone-refractory prostate cancer, NHL, urothelium transitional cell carcinoma, and stage IIB–IV melanoma (to be confirmed).

Documented target cancers:

- Activity against the P-388, P-1534, and L-1210 murine leukemia models.
- Strong activity against the B16 melanoma system.
- Cytotoxic activity against KB cell culture system, Walker 256 carcinosarcoma, sarcoma 180, and Lewis lung tumors.
- Significant activity against several human tumor xenograft systems, including the MX-1 mammary tumor.
- Introduced to all ovarian cancer patients (meeting defined disease criteria).
- Responses in patients with metastatic breast cancer and in patients with other forms of advanced malignancy including lung cancer, cancer of the head and neck region, and lymphomas.
- It is used together with another drug to treat non-small cell lung cancer.
- In a large international clinical trial (phase III) to support the use of immunotherapy in patients with breast cancer and establish a new standard of care in PD-L1+ patients, results showed that patients with an aggressive form of advanced breast cancer can benefit from immunotherapy when used in combination with chemotherapy (taxol) as first-line treatment.

Further details:

- The antitumorous properties of paclitaxel are based on the ability to bind and to stabilize microtubules and block cell replication in the late G2-M phase of the cell cycle. In 1979 it was demonstrated that taxol affects the tubulin-microtubule equilibrium: it decreases both the critical concentration of tubulin (to almost 0–1 mg/ml) and the induction time for polymerization, either in the presence or absence of GTP, MAPs, and magnesium. Taken in conjunction with observations showing that *taxol* promotes the end-to-end joining of microtubules, these results point to a rather complex mechanism of action for taxol that is not yet completely understood.
- Early studies with HeLa cells and BALB/c mouse fibroblasts treated with low concentrations of taxol (0.25 µmol/L), which produce minimal inhibition of DNA, RNA, and protein synthesis, demonstrated that taxol blocks cell cycle traverse in the mitotic phases. Recently, taxol has been demonstrated to prevent transition from the G_0 phase to the S phase in fibroblasts during stimulation of DNA synthesis by growth factors and to delay traverse of sensitive leukemia cells in nonmitotic phases of the cell cycle. These findings indicate that the integrity of microtubules may be critical in the transmission of proliferative signals from cell-surface receptors to the nucleus. Proposed explanations that at least in part account for taxol's inhibitory effects in non-mitotic phases include disruption of tubulin in the cell membrane and/or direct inhibition of the disassembly of the interphase cytoskeleton, which may upset many vital cell functions such as locomotion, intracellular transport, and transition of proliferative transmembrane signals.
- The plum yews (*Cephalotaxus harringtonia* family: Cephalotaxaceae (plum yew family)) are similar to, and closely related to, the yews, family Taxaceae. Common Names: Japanese plum yew, Harrington plum yew, cow-tail pine, plum yew. The plum yews are evergreen, coniferous shrubs or small trees with flat, needle-like leaves arranged in two ranks on the green twigs, and fleshy, plum-like seeds borne only on female plants. Japanese plum yew is a shrub or small tree, but most cultivars are quite a bit smaller. Japanese plum yew is native to Japan, Korea, and eastern China, where it grows in the forest understory. Japanese plum yew has the potential to be a very useful landscape plant in the southern US. It is more tolerant of heat than the true yews (*Taxus*). It produces cephalomannine, a promising agent for cancer therapy.
- *Taxus brevifolia* is the first source of taxol. It is common in the Olympic Peninsula in Washington and in Vancouver Island in British Columbia. The taxol supply needs for preclinical and early clinical studies were easily met by bark collections in Oregon between 1976 and 1985, from the bark of the tree. In 1988 it was demonstrated that the precursor, 10-desacetylbaccatin III, isolated from the needles of the tree, can be converted to taxol and related active agents by a relatively simple semi-synthetic procedure, and alternatively, more efficient processes for this conversion have recently been reported.
- The taxol content of fresh needles of 35 different *Taxus* cultivars from different locations within the US has been analyzed. At least six contain amounts comparable to or higher than those found in the

dried bark of *T. brevifolia*. These observations have resulted in the initiation of a study of the nursery cultivar, *Taxus* X *media Hicksii*, as a potential renewable large-scale source of taxol.

- Taxol has been shown to inhibit steroidogenesis in human Y-1 adrenocortical tumors and in MLTC-1 Leydig tumors by decreasing the intracellular transport of cholesterol to cholesterol side-chain cleavage enzymes. This effect appears to be related to perturbations in microtubule dynamics.

- Taxol has also been shown to inhibit specific functions in many nonmalignant tissues, which may be mediated through microtubule disruption. For example, in human neutrophils, taxol inhibits relevant morphological and biochemical processes, including chemotaxis, migration, cell spreading, polarization, generation of hydrogen peroxide, and killing of phagocytosed microorganisms. Taxol also antagonizes the effects of microtubule-disrupting drugs on lymphocyte function and adenosine 3', 5'- cyclic monophosphate metabolism and inhibits the proliferation of stimulated human lymphocytes, but blast transformation is not affected during lymphocyte activation. Taxol has also been found to mimic the effects of endotoxic bacterial lipopolysaccharide on macrophages, resulting in a rapid decrement of receptors for tumor factor-α and tumor necrosis factor-α release. This finding suggests that an intracellular target affected by taxol may be involved in the actions of lipopolysacccharide on macrophages and other cells. Interestingly, taxol inhibits chorioretinal fibroblast proliferation and contractility in an *in vitro* model of proliferative vitreoretinopathy, a fact that may be relevant to the treatment of traction retinal detachment and proliferative vitreoretinopathy. Taxol also inhibits the secretory functions of many specialized cells. Examples include insulin secretion in isolated rat islets of Langerhans, protein secretion in rat hepatocytes, and the nicotinic receptor-stimulated release of catecholamines from chromaffin cells of the adrenal medulla.

- NCI and Program Resources, in collaboration with various organizations, are undertaking analytical surveys of needles of a number of *Taxus* species. They include *T. baccata* from the Black Sea–Caucasus region of Georgia and Ukraine and *T. cuspidata* from Siberian regions of Russia; *Taxus canadensis* from the Gaspe Peninsula of Quebec; *Taxus globosa* from Mexico; *Taxus sumatriensis* from the Philippines; and various *Taxus* species from the US. In a number of samples, the taxol content of the needles is comparable to that of the dried bark of *T. brevifolia*.

- Taxotere is a highly promising synthetic analog of taxol. It promotes the assembly and stability of microtubules with potency approximately twice that of taxol. Recently, taxol and taxotere have been shown to complete for the same binding site. While most of the effects of taxotere mirror those of taxol, it appears that the microtubules formed by taxotere induction are structurally different from those formed by taxol induction. Taxotere is currently produced by attaching a synthetic side chain to 10-deacetyl baccatin III, which is readily available from the European yew *T. baccata*, in yields approaching 1 kg of three from 3 kg of needles.

- Paclitaxel is extracted from the bark of the Pacific yew and docetaxel from the needles of the English yew. Taxenes promote and stabilize microtubular assembly, which seems to inhibit fast axonal transport leading to axonal dysfunction. Paclitaxel is used for the treatment of breast, ovarian, and lung cancers. About one-half of patients treated with taxol in doses over 200 mg/m^2 develop a sensory neuropathy involving all sensory modalities.

- Docetaxel (Taxotere®) is a semisynthetic taxoid derived from a precursor, extracted from the needles from the European yew, *Taxus baccata*. Like paclitaxel, docetaxel is a potent inhibitor of cell replication. It promotes the *in vitro* assembly of stable microtubules and induces microtubule bundle formation in cells. Docetaxel is a more potent inhibitor of cell replication than paclitaxel. Response rates of up to 35% have been reported in non-small cell lung cancer and platinum-refractory advanced ovarian cancer. Hematological toxicity is encountered frequently. Nonhematological toxicities include alopecia, skin rash, fluid retention, diarrhea, hypersensitivity reactions, and peripheral neuropathy. The drug is usually administered intravenously in a one-hour infusion time, usually 100 mg/m^2, once in every 21 days.

- Cell culture has already been used to produce ^{14}C labeled taxol from ^{14}C sodium acetate. The USDA (United States Department of Agriculture) has received a patent for the production of taxol from cultured callus cells of *T. brevifolia*. They have licensed this process to Phyton Catalytic, who estimates that they will begin commercial production soon. The advantage of this system is that the major secretion product of the cells is taxol, which reduces the purification to an ether extraction of the medium. ESCA genetics has also announced technology for producing high levels of taxol in plant cell cultures, and they project large-scale production in the near future. Additionally, callus cultures of *T. cuspitata* and *T. canadensis* have been substained in a taxol-producing system for over two months. A fungus indigenous to *T. brevifolia* that produces small amounts of taxol has recently been isolated and cultured.

- As a target for chemical synthesis, taxol presents a plethora of potential problems. Perhaps most obvious is the challenge presented by the central B ring, an eight- membered carbocycle. Such rings are notoriously difficult to form because of both entropic and enthalpic factors. The normally high transannular strain of an eight-membered ring is further increased in this case by the presence of the geminal dimethyl groups, which project into the interior of the B ring. Then the transfused C ring with its angular methyl group and another ring (A ring), which is a 1,3 C3 bridge, must be introduced. The A ring includes a somewhat problematic bridgehead alkene formally forbidden in a six-membered ring by Bredt's rule. If assembling the carbon skeleton alone is not a daunting enough task, one should consider the high degree of oxygenation that must be introduced in a manner which allows the differential protection of five alkoxy groups in a minimum of three orthogonal classes. Additionally, some of the functionality is quite sensitive to environmental conditions. The oxetane ring, for example, will open under acidic or nucleophilic conditions, and the 7-hydroxyl group, if left unprotected, will epimerize under basic conditions. Despite the many attempts to synthesize taxol, the molecule still remains inaccessible by total synthesis.

- Taxol is supplied as a sterile solution of 6 mg/mL in 5-mL ampoules (30 mg per ampoule). Because of taxol's aqueous insolubility, it is formulated in 50% cremophor EL and 50% dehydrated alcohol. The contents of the ampoule must be diluted further in either 0.9% sodium chloride or 5% dextrose. During early phase I and II studies, taxol was diluted to final concentrations of 0.003–0.60 mg/mL. These concentrations were demonstrated to be stable for 3–24 hours in early stability studies. This short stability period required the administration of large volumes of fluids and/or drug preparation at frequent intervals for patients receiving higher doses. In recent studies, concentrations of 0.3–1.2 mg/mL in either 5% dextrose or normal saline solution have demonstrated both chemical and physical stability for at least 12 hours.

- *Pestalotiopsis microspora* (an endophytic fungus) was isolated from the inner bark of a small limb of Himalayan yew, *Taxus wallachiana*, which has been shown to produce taxol in mycelial culture. Fungal taxol was evaluated in the standard 26 cancer cell line test and for its ability, when compared to authentic taxol, to inhibit cell division. The fungal compound was found to be identical to authentic taxol (methods used: NMR, UV absorption, and electrospray mass spectroscopy). It showed a pattern of activity comparable to that produced by standard authentic taxanes in the 26 cancer cell line test. In addition, its ability to induce mitotic arrest at a concentration of 37 ng/ml was consistent with a tubulin-stabilizing mode of action. The discovery that fungi make taxol increasingly adds to the possibility that horizontal gene transfer may have occurred between *Taxus* spp. and its corresponding endophytic organisms. This demonstration supports the idea that certain endophytic microbes of *Taxus* species may make and tolerate taxol in order to better compete and survive in association with these trees. Since *Taxus* spp. grow in areas that are generally damp and shaded certain plant-pathogenic fungi (water molds) also prefer this niche.

- *Taxus marei* Hu ex Liu is a native Taiwan species sparsely distributed in mountainous terrain. Many are giant trees with a diameter at breast height greater than 100 cm and an estimated age of more than 1,000 years. Taxol accumulates in the needles of these trees and selected superior trees with respect to high taxol and 10-deacetyl baccatin III concentrations. It was found that rooted cutting (steckling) ramets of these trees also exhibited high taxol concentrations in mature needles, confirming that taxol yield is a heritable trait. Young needles from vegetatively propagated elite yew trees can serve as a rewable and economic tissue source for increasing taxol production. Micropropagation of mature *Taxus marei* was achieved using bud explants derived from approximately 1,000-year-old field grown trees. It might be a very useful tool for the mass propagation of superior yew trees and the production of high-quality (orthotropic) plantlets for nursery operation.

- Taxol and its relatives are emerging as yet another class of naturally occurring substances, like the enediyne antitumor antibiotics and the macrocyclic immunophilin ligands, that combine novel molecular architecture, important biological activity, and fascinating mode of action.

- Patients with an aggressive form of advanced breast cancer can benefit from immunotherapy when used in combination with chemotherapy (taxol) as first-line treatment, according to the results of a large international phase III clinical trial.

References

Attila, T., V. Adsay, and D.O. Faigel. 2019. The efficacy and safety of endoscopic ultrasound-guided ablation of pancreatic cysts with alcohol and paclitaxel: A systematic review. *European Journal of Gastroenterology and Hepatology* 31, no. 1 (Nov 2):1–9.

Chang, S.H., C.K. Ho, Z.Z. Chen, and J.Y. Tsay. 2001. Micropropagation of *Taxus mairei* from mature trees. *Plant Cell Reports* 20, no. 6:496–502.

Cragg, G.M. 1998. *Paclitaxel (taxol): A Success Story with Valuable Lessons for Natural Product Drug Discovery and Development*. Medicinal research reviews, 18(5), 315–331.

Cragg, G.M., S.A. Schepartz, M. Suffness, and M.R. Grever. 1993. The taxol supply crisis. New NCI policies for handling the large-scale production of novel natural product anticancer and anti-HIV agents. *Journal of Natural Products* 56, no. 10:1657–1668.

Diker, S., and Ö. Diker. 2018. Optic atrophy after cabazitaxel treatment in a patient with castration-resistant prostate cancer: A case report. *Scottish Medical Journal* (Nov 5):36933018810653.

Furmanowa, M., K. Glowniak, K. Syklowska-Baranek, G. Zgórka, and A. Józefczyk. 1997. Effect of picloram and methyl jasmonate on growth and *taxane* accumulation in callus culture of *Taxus x media var. Hatfieldii. Plant Cell, Tissue and Organ Culture* 49, no. 1:75–79.

Grieve, M. 1994. *A Modern Herbal*. Edited and Introduced by Mrs. C.F. Leyel. Tiger Books International, London.

Helfferich, C. 1993. taxol *Revisited Article, Alaska Science Forum* 1126.

Hirasuna, T.J., L.J. Pestchanker, V. Srinivasan, and M.L. Shuler. 1996. taxol production in suspension cultures of Taxus baccata. *Plant Cell, Tissue and Organ Culture* 44, no. 2:95–102.

Ketchum, R.E.B., and D.M. Gibson. 1996. Pactitaxel production in suspension cell cultures of Taxus. *Plant Cell, Tissue and Organ Culture* 46, no. 1:9–16.

Ketchum, R.E.B., D.M. Gibson, and L.G. Gallo. 1995. Media optimization for maximum biomass production in cell cultures of pacific yew. *Plant Cell, Tissue and Organ Culture* 42, no. 2:185–193.

Luo, J.P., Q. Mu, and Y.-H. Gu. 1999. Protoplast culture and paclitaxel production by Taxus yunnanensis. *Plant Cell, Tissue and Organ Culture* 59, no. 1:25–29.

Nicolaou, K.C., W.M .Dai, and R.K. Guy. 1994. Chemistry and biology of taxol . *Angewandte Chemie International Edition in English* 33, no. 1:15–44.

Postma, T.J., and J.J. Heimans. 2012. Handbook of clinical neurology 2012. In: *Neurological Complications of Chemotherapy to the Peripheral Nervous System* 105. https://doi.org/10.1016/B978-0-444-53502-3.00 032-X. Elsevier B.V.:917–936: Chapter 60.

Reeves, P.M., M.A. Abbaslou, F.R.W. Kools, and M.C. Poznansky. 2017. CXCR4 blockade with AMD3100 enhances taxol chemotherapy to limit ovarian cancer cell growth. *Anti-Cancer Drugs* 28, no. 9 (Oct):935–942.

Rowinsky, E.K., L.A. Cazenave, and R.C. Donehower. 1990. taxol : A novel investigational antimicrotubule agent. *Journal of the National Cancer Institute* 82, no. 15 (August 1):1247–1259.

Samuelsson, G. 1992. *Drugs of Natural Origin – A Textbook of Pharmacognosy. Third Revised, Enlarged and Translated Edition*. Swedish Pharmaceutical Press. Stockholm, Sweden.

Schmid, P., S. Adams, H.S. Rugo, et al. 2018. *Atezolizumab* and *nab-paclitaxel* in advanced triple-negative breast cancer. *New England Journal of Medicine*. 379.22: 2108–2121.

Strobel, G., X. Yang, J. Sears, R. Kramer, R.S. Sidhu, and W.M. Hess. 1996. taxol from *Pestalotiopsis microspora*, an endophytic fungus of *Taxus wallachiana. Microbiology* 142, no. 2:435–440.

Wang, Y., C. Zhang, S. Zhang, et al. 2017. Kanglaite sensitizes colorectal cancer cells to taxol via NF-κB inhibition and connexin 43 upregulation. *Scientific Reports* 7, no. 1 (Apr 28):1280.

Weaver, B.A. 2014. How taxol /paclitaxel kills cancer cells. *Molecular Biology of the Cell* 25, no. 18 (Sep 15):2677–2681.

Whiteley, W., and R. Grant. Neurology and clinical neuroscience 2007. In: *Neurological Complications of Cancer Treatments*, ed. A.H.V. Schapira, associate ed. Z. K. Wszolek https. Elsevier B.V.:1353–1360: Chapter 100.

Websites

www.gi.alaska.edu/ScienceForum/ASF11/1126.html
http://.epnws1.ncifcrf.gov.2345/dis3d/dtp.htpl
www.pfc.cfs.nrcan.gc.ca/ecosystem/yew/ taxol .html
www. taxol .com/timeli.html
https://floridata.com
www.ncbi.nlmnih..../query.fcgi?cmd
www.ncbi.nlm.nih.gov/pubmed
www.nejm.org/doi
https://reference.medscape.com
www.sciencedirect.com
www.accessdata.fda.gov

Periwinkle (*Vinca rosea*, Linn) (Apocynaceae)	Immunomodulator Cytotoxic Antitumor Antineoplastic

Madagascar periwinkle is a modern-day success story in the search for naturally occurring anticancer drugs.

Synonyms: *Catharanthus roseus* (G. Don), *Lochnera rosea* (Reichb.), Madagascar periwinkle, and rose periwinkle.

You can find it in: East Indies, Madagascar, and America.

Origin: Madagascar, tropical Africa, and generally in the Tropics.

Appearance:

Stem: small under-shrub up to 40–80 cm high in its native habitat. The broken stem of Madagascar periwinkle exudes a milky latex sap.
Leaves: retains its glossy leaves throughout the winter. Are always found in pairs on the stem.
Flowers: springing from their axils, five-petaled flowers are typically rose pink, but among the many cultivars are those with pink, red, purple, and white flowers. The flowers are tubular, with a slender corolla tube about an inch long that expands to about 25 mm and a half across. They are borne singly throughout most of the summer.
In bloom: summer.

Biology: it propagates itself by long, trailing, and rooting stems and by their means not only extends itself in every direction but succeeds in obtaining an almost exclusive possession of the soil. Because of the dense mass of stems, the periwinkle deprives the weaker plants of light and air.

Degree of rarity: low. It has escaped cultivation and is naturalized in most of the tropical world where it often becomes a rampant weed. Over the past hundreds of years, the periwinkle has been widely cultivated and can now be found growing wild in most warm regions of the world, including the Southern United States. It is established in several areas in the southern US. Madagascar periwinkle is grown commercially for its medicinal uses in Australia, Africa, India, and southern Europe.

Tradition: it was one of the plants believed to have power to exorcize evil spirits. Apuleius, in his *Herbarium* (printed 1480), writes: 'this wort is of good advantage for many purposes, first against devil sinks and demoniacal possessions and against snakes and wild beasts and against poisons and for various wishes and for envy and for terror ...' 'The periwinkle is a great binder', said an old herbalist, and both Dioscorides and Galen commended it against fluxes. It was considered a good remedy for cramps. An ointment prepared from the bruised leaves with lard has been largely used in domestic medicine and is reputed to be both soothing and healing in all inflammatory ailments of the skin and an excellent remedy for bleeding piles. In India, juice from the leaves was used to treat wasp stings. In Hawaii, the plant was boiled to make a poultice to stop bleeding, and throughout the Caribbean, an extract from the flowers was used to make a solution to treat eye irritation and infections. In France, it is considered an emblem of friendship.

Part used: leaves, stems, flower buds.

Active ingredients:

- Ajmalicine, vindoline, catharanthine.
- Vindoline is enzymatically coupled with catharanthine to produce the powerful cytotoxic dimeric alkaloids: vinblastine (VB), vincristine (VC), leurosidine, anhydrovinblastine.
- Vinorelbine (semisynthetic vinca alkaloid).
- Vinflunine (gem-difluoromethylenated derivative of vinca alkaloid).

Particular value: cure for diabetes, anticancer drug. The plant has been used for centuries to treat diabetes, high blood pressure, asthma, constipation, and menstrual problems. In the 1950s researchers learned of a tea that Jamaicans had been drinking to cure diabetes. A native who had been drinking the tea sent a small envelope full of leaves to researchers explaining that the leaves came from a plant known as the Madagascar periwinkle. The native explained that the tea was used in the absence of insulin treatment and apparently already had a worldwide reputation and was being sold as a remedy under the name Vinculin.

The status of vinca application in cancer therapy: the plant was used in traditional medicine. When tested in scientific studies it was demonstrated that it could be used in diabetes and anticancer research with great advantages. In the 1950s, Dr. Johnston, who had been practicing in the Jamaica area, was quite convinced that his diabetic patients had received some benefit from drinking extracts of the periwinkle leaves. Therefore, it was decided among the researchers to send these leaves to Dr. Collip at the University of Western Ontario. The doctor had already been working with another group on insulin derived from a hormone, so it seemed logical to send the leaves of the periwinkle to him. Dr. Collip decided to make a water extract to determine if, when given orally, they would lower blood sugar levels. These extracts were given to animals but were not found to have any effect on the blood sugar or the disease. One of Dr. Collip's colleagues, Dr. McAlpine, decided to give the water extract to a few of his diabetic patients, who had volunteered to try it. There was no effect except in one mildly diabetic woman. In the absence of oral activity and as a final resort, Dr. Collip decided to give the most concentrated dose to a few rats by intraperitoneal injection. The rats survived for about five days but then died rather unexpectedly from diffuse multiple abscesses. This intrigued the doctor because the extracts that had been given had been sterilized. Dr. Collip became very excited because another colleague of his had published that overdoses of cortisone in rats also led to their death from multiple abscesses. Dr. Collip wondered if perhaps the periwinkle plant might be a source of cortisone. Unfortunately, it was found that the two had very different mechanisms involved. In cortisone, lymphocytes are destroyed resulting in its well-known immunosuppressive effect; in the case of the periwinkle extracts, it was found that after a single injection there was a rapid but transient depression of the WBC count, which was traced to the destruction of the bone marrow. In view of the dramatic effect on the bone marrow, it looked like there might be one or more compounds present in the periwinkle that might be useful in the treatment of cancers of the hematopoietic system such as lymphomas and leukemias. Therefore, it was decided to try and identify and isolate the component in the extracts responsible for the effects on the WBC counts and bone marrow.

In 1954 Dr. Charles T. Beer came to work in Dr. Collip's laboratory on a one-year fellowship. He looked at the problem of isolating active compounds from the periwinkle plant. When he started working on the project, the supply of periwinkle leaves was a problem. Dr. Johnston in Jamaica was still convinced that the research was headed in the wrong direction. He felt like researchers should look for a cure for diabetes instead of a cure for cancer. So he decided to continue to supply Dr. Beer with dried periwinkle leaves. Unfortunately, it took so many leaves to make the extract that he decided to grow the periwinkle himself in Ontario. After working on the project for a year, Dr. Beer finally isolated a small amount of unknown alkaloid. In rats, this alkaloid was highly active, and there was a dramatic decrease in the WBC counts and a marked depletion of the bone marrow. He decided to name the alkaloid vincaleukoblastine (the name was shortened later to vinblastine). He found upon further observation of the plant that periwinkle contained tons of useful alkaloids (70 in all at last count). Some of the alkaloids isolated contained properties that lowered blood sugar levels, others lowered blood pressure, some acted as hemostatics. Upon further investigation of vinblastine, Dr. Beer also noted some activity but in smaller amounts. The related alkaloid was vincristine (VCR), but was present in an amount insufficient for isolation in the laboratory. VCR was later isolated in crystalline form by chemists at the Eli Lilly Co.

Later, there were isolated about 100 alkaloids, but they were not all suitable for clinical use. The most of the part of this investigation was done by the American pharmaceutical company Eli Lilly, and the responsible professor was Dr. Gordon H. Svoboda. The *C. roseus* bisindoles, vinblastine and vincristine, were the first plant products to be approved by the FDA for cancer treatment in the early 1970s and are still currently used. The needs of production of final product for medical use are high, without satisfactory cover, because of the low concentration of vinblastine and vincristine in *C. roseus*, in spite of its being cultivated in several tropical countries.

Vinorelbine is a semisynthetic vinca alkaloid. It is a third-generation vinca alkaloid. The introduction of third-generation drugs (vinorelbine, gemcitabine, taxanes) in platinum combination improved survival of patients with advanced NSCLC, with very similar results from the various drugs. EMEA in 1989 approved the Vinorelbine and FDA in 1995.

In 1996, a pharmaceutical development process was initiated for a new molecule vinflunine, and phase I clinical trials started two years later, followed by phase II trials in 2000 in patients with a variety of solid tumors. A phase III trial in advanced non-small cell lung cancer was initiated in 2003, and the results were reported in 2007. Two other phase III studies in advanced bladder and metastatic breast cancers are ongoing. Institut de Recherche Pierre Fabre support the development of vinflunine, and in 2009 EMEA approved the use of vinflunine (trade name: Javlor 25 mg/mL concentrate for solution for infusion) for the treatment of patients with advanced or metastatic transitional cell carcinoma of urothetial tract after failure of a prior platinum-containing regimen.

Precautions: Madagascar periwinkle is poisonous if ingested or smoked. It has caused poisoning in grazing animals. Even under a doctor's supervision for cancer treatment, products from Madagascar periwinkle produce undesirable side effects.

The principal dose-limiting toxicity of Vincristine is peripheral neurotoxicity. In the beginning, only symmetrical sensory impairment and parasthesias may be encountered. However, neuritic pain and motor dysfunction may occur with continued treatment. Loss of deep tendon reflexes, foot and wrist drop, ataxia, and paralysis may also be observed with continued use. These effects are almost always symmetrical and may persist for weeks to months after discontinuing the drug. These effects usually begin in adults who have received a cumulative dose of 5 to 6 mg, and the toxicity may occasionally be profound after a cumulative dose of 15 to 20 mg. Children generally tolerate this toxicity better than adults do, and the elderly are particularly susceptible. Other toxicities involving VCR are gastrointestinal with symptoms such as constipation, abdominal cramps, diarrhea, etc. Cardiovascular symptoms include hypertension and hypotension and a few reports of massive myocardial infarction. The principle toxicity of vinblastine is myelosuppression or in particular neutropenia. Neurotoxicity occurs much less commonly with VBL than VCR and is generally observed in patients who have received protracted therapy. Hypertension is the most common cardiovascular toxicity of VBL. Sudden and massive myocardial infarctions and cerebrovascular events have also been associated with the use of single-agent VBL and multi-agent regimens. Pulmonary toxicities include acute pulmonary edema and acute bronchospasm. Pregnant women and people with neuromuscular disorders should steer clear of these drugs. With pregnant women, VCR and VBL have been found to cause severe birth defects. Vinflunine is a cytotoxic anticancer medicinal product and, as with other potentially toxic compounds, caution should be exercised in handling it. Procedure for proper handling and disposal of anticancer medicinal products should be considered.

Indicative dosage and application:

- VCR is routinely given to children as a bolus intravenous injection at doses of 2.0 mg/m² weekly. For children weighing 10 kg or less, the starting dose should be 0.05 mg/kg administered as a weekly intravenous injection.
- For adults, the conventional weekly dose is 1.4 mg/m² to 1.5 mg/m² up to a maximum weekly dose of 2 mg.
- A restriction of the absolute single dose of VCR to 2.0 mg/m² has been adopted by many clinicians over the last several decades, mainly because of reports that show an increasing neurotoxicity at higher doses.
- Vinblastine has been given by several schedules. The most common schedule involves weekly bolus doses of 6 mg/m² incorporated into combination chemotherapy regimens such as ABVD (adriamycin, bleomycin, VBL, dacarbazine) and the MOPP-AVB hybrid regimen (nitrogen mustard, VCR, prednisone, procarbazine, adriamycin, bleomycin, VBL).
- Vinorelbine is approved as 10 mg/ml concentrate for solution for infusion. Vinorelbine is usually given at 25–30 mg/m² once weekly. The maximum tolerated dose per administration: 35.4 mg/m² body surface area and the maximum total dose per administration: 60 mg. There are also soft capsules of 20 mg and 30 mg.
- The novel microtubule inhibitor, vinflunine, has a unique mechanism of action that differs from other members of the vinca alkaloid class in terms of tubulin-binding affinity, microtubule dynamics, spiral formation, and intracellular accumulation. Vinflunine has shown significant activity *in vivo*, which involves its antimitotic, antiangiogenic, and antivascular properties. The recommended dose is 320 mg/m² vinflunine as a 20-minute intravenous infusion every three weeks.

Documented target cancers: extracts from Madagascar periwinkle have been shown to be effective in the treatment of various kinds of: leukemia, skin cancer, lymph cancer, breast cancer, and Hodgkin's disease.

Vincristine is used against childhood leukemia, malignant lymphomas, including Hodgkin's disease and non-Hodgkin's lymphomas, leukemias, including acute lymphocytic leukemia, chronic lymphocytic leukemia, acute myelogenous leukemia, and blastic crisis of chronic myelogenous leukemia, multiple myeloma, solid tumors, including breast carcinoma, small cell bronchogenic carcinoma, head and neck carcinoma, and soft tissue sarcomas, paediatric solid tumors, including Ewing's sarcoma, embryonal rhabdomyosarcoma, neuroblastoma, Wilms' tumor, retinoblastoma and medulloblastoma, idiopathic thrombocytopenic purpura. Patients with true ITP refractory to splenectomy and short-term treatment with adrenocortical steroids may respond to vincristine, but the medicinal product is not recommended as primary treatment of this disorder. Recommended weekly doses of vincristine given for three to four weeks have produced permanent remissions in some patients. If patients fail to respond after three to six doses, it is unlikely that there will be any beneficial results with additional doses. In 2012, vincristine sulfate liposome injection (Marqibo) was approved by FDA for the treatment of adult patients with Philadelphia chromosome-negative (Ph–) acute lymphoblastic leukemia.

VBL is mainly used for the treatment of Hodgkin's disease (generalized Hodgkin's disease: stages III and IV, Ann Arbor modification of Rye staging system), testicular cancer, breast cancer, Kaposi's sarcoma, and other lymphomas (lymphocytic lymphoma: nodular and diffuse, poorly and well differentiated, histiocytic lymphoma), mycosis fungoides (advanced stages), advanced carcinoma of the testis, letterer-Siwe disease (histiocytosis X), choriocarcinoma resistant to other chemotherapeutic agents, carcinoma of the breast, unresponsive to appropriate endocrine surgery and hormonal therapy.

Vinorelbine is indicated as a single agent or in combination for the first line treatment of stage 3 or 4 non-small cell lung cancer and for the treatment of advanced breast cancer stage 3 and 4 relapsing after or refractory to an anthracycline containing regimen.

Vinflunine was identified in preclinical studies as having marked antitumor activity *in vivo* against a large panel of experimental tumor models, with tumor regressions being recorded in human renal and small cell lung cancer tumor xenografts. Vinflunine is indicated in monotherapy for the treatment of adult patients with advanced or metastatic transitional cell carcinoma of the urothelial tract after failure of a prior platinum-containing regimen.

Further details:

- *Vinca major* (Apocynaceae family), with common names: large periwinkle, big periwinkle, is a big periwinkle fast-growing herbaceous perennial groundcover with evergreen foliage and pretty blue flowers. It is native to France and Italy and eastward through the Balkans to northern Asia Minor and the western Caucasus. *V. major* and *V. minor* are the most commonly cultivated. Herbalists have long used it for curing diabetes because it can be an efficient substitute for insulin. It is used in herbal practice for its astringent and tonic properties in menorrhagia and in hemorrhages generally. For obstructions of mucus in the intestines and lungs, diarrhea, congestions, hemorrhages, etc., periwinkle tea is a good remedy. In cases of scurvy and for relaxed sore throat and inflamed tonsils, it may also be used as a gargle. It may be applied externally on bleeding piles.
- Apparently, all the vincas are poisonous if ingested. Numerous alkaloids, some useful to man, have been isolated from big and common periwinkle.
- Common periwinkle (*Vinca minor*) is similar to but has smaller leaves (less than 5 cm long) and smaller flowers (2.5 cm or less across) than *Vinca major* and is more cold hardy and more tolerant of shade. It is used for producing catharanthus alkaloids. Also, a homoeopathic tincture is prepared from the fresh leaves of it and is given medicinally for the milk-crust of infants as well as for internal hemorrhages. Its flowers are gently purgative but lose their effect on drying. If gathered in the spring and made into a syrup, they will impart thereto all their virtues, and this is excellent as a gentle laxative for children and also for overcoming chronic constipation in grown persons.
- Vinblastine and vincristine are dimeric catharanthus alkaloids isolated from *vincas* plants. Both vinblastine and vincristine are large, dimeric compounds with similar but complex structures. They are composed of an indole nucleus and a dihydroindole nucleus. They are structurally identical with the exception of the substitutent attached to the nitrogen of the vindoline nucleus where VCR possesses a formyl group and VBL has a methyl group. However, VCR and VBL differ dramatically in their antitumor spectrum and clinical toxicities. Both alkaloids are therapeutically proven to be effective in the treatment of various neoplastic diseases. Consequently, the determinations of these compounds in plant samples, as well as biological fluids, are of interest to many scientists. Many gas and high-performance liquid chromatographic (HPLC) and mass spectrometric methods have been developed for the determination of VC and VB in either plant samples or biological systems. The potential use of information-rich detectors such as mass spectrometry with capillary zone electrophoresis (CZE) has made this a more attractive separation method.
- Vinblastine and vincristine, which belong to the group of *Vinca* alkaloids, induce cytotoxicity by direct contact with tubulin, which is the basic protein subunit of microtubules. Other biochemical effects that have been associated with VBL and VCR include: competition for transport of amino acids into cells; inhibition of purine biosynthesis; inhibition of RNA, DNA, and protein synthesis; inhibition of glycolysis; inhibition of release of histamine by mast cells and enhanced release of epinephrine; and disruption in the integrity of the cell membrane and membrane functions. Microtubules are present in eukaryotic cells and are vital to the performance of many critical functions including maintenance of cell shape, mitosis, meiosis, secretion, and intracellular transport. VBL and VCR exert their antimicrotubule effects by binding to a site on tubulin that is distinctly different from the binding sites of others. They have a binding constant of 5.6×10^{-5} M and initiate a sequence of events that lead to disruption of microtubules. The binding of VBL and VCR to

tubulin, in turn, prevents the polymerization of these subunits into microtubules. The net effects of these processes include the blockage of the polymerization of tubulin into microtubules, which may eventually lead to the inhibition of vital cellular processes and cell death. Although most evidence suggests that mitotic arrest is the principal cytotoxic effect of the alkaloids, there is also evidence that the lethal effects of these agents may be attributed in part to effects on other phases of the cell cycle. The alkaloids also appear to be cytotoxic to nonproliferating cells *in vitro* and *in vivo* in both G1 and S cell cycle phases. In other words, VBL and VCR work by inhibiting mitosis in metaphase.

- Vinorelbine binds to tubulin and prevents formation of the mitotic spindle, resulting in the arrest of tumor cell growth in metaphase. This agent may also interfere with amino acid, cyclic AMP, and glutathione metabolism; calmodulin-dependent Ca^{2+}-transport ATPase activity; cellular respiration; and nucleic acid and lipid biosynthesis.

- Vinflunine is a new *Vinca* alkaloid uniquely fluorinated, by the use of superacid chemistry, in a little exploited region of the catharanthine moiety. *In vitro* investigations have confirmed the mitotic-arresting and tubulin-interacting properties of vinflunine shared by other *Vinca* alkaloids. Vinflunine induced smaller spirals with a shorter relaxation time, effects which might be associated with reduced neurotoxicity. Vinflunine appears to participate in P-glycoprotein-mediated drug resistance mechanisms.

- Vinflunine binds to tubulin at or near to the *Vinca* binding sites inhibiting its polymerization into microtubules, which results in treadmilling suppression, disruption of microtubule dynamic, mitotic arrest, and apoptosis. *In vivo*, vinflunine displays significant antitumor activity against a broad spectrum of human xenografts in mice both in terms of survival prolongation and tumor growth inhibition.

- A new approach in cancer treatment is PEGylated liposomes. PEGylated liposomes are important drug carriers that can passively target tumor by enhanced permeability and retention (EPR) effect in neoplasm lesions. Until now, most studies remain at mouse models.

- Studies with germinating seedlings have suggested that alkaloid biosynthesis and accumulation are associated with seedling development. Studies with mature plants also reveal this type of developmental control. Furthermore, alkaloid biosynthesis in cell suspension cultures appears to be coordinated with cytodifferentiation. Vindoline biosynthesis in *Catharanthus roseus* also appears to be under this type of developmental control. Vindoline as well as the dimeric alkaloids are restricted to leaves and stems, whereas catharanthine is distributed equally throughout the aerial and underground tissues. The developmental regulation of vindoline biosynthesis has been well documented in *C. roseus* seedlings, in which it is light inducible. This is in contrast to catharanthine, which also accumulates in etiolated seedlings. Furthermore, cell cultures that accumulate catharanthine but not vindoline recover this ability upon redifferentiation of shoots. These observations suggest that the biosynthesis of catharanthine and vindoline is differentially regulated and that vindoline biosynthesis is under more rigid tissue-development- and environment-specific control than is that of catharanthine. The early stages of alkaloid biosynthesis in *C. roseus* involve the formation of tryptamine from tryptophan and its condensation with secologanin to produce the central intermediate strictosidine, the common precursor for the monoterpenoid indole alkaloids. The enzymes catalyzing these two reactions are tryptophan decarboxylase (TDC) and strictosidine synthase (STR1), respectively. Strictosidine is the precursor for both the *Iboga* (catharanthine) and *Aspidosperma* (tabersonine and vindoline) types of alkaloids. The condensation of vindoline and catharanthine leads to the biosynthesis of the bisindole alkaloid vinblastine.

- A successful attempt of production of *Indole* alkaloids by selected hairy root lines of *Catharanthus roseus* has been done. Approximately 150 hairy root clones from four varieties were screened for their biosynthetic potential. Two key factors affecting productivity, growth rate, and specific alkaloid yield. The detection of vindoline in these clones may potentially present a new source for the *in vitro* production of vinblastine. Production of vindoline and catharanthine by plant tissue culture and subsequent catalytic coupling *in vitro* is a possible alternative to using tissue culture alone to produce vinblastine and vincristine. Recently, enzyme-catalyzed techniques have been developed for the conversion of vindoline and catharanthine to bisindole alkaloids. Catharanthine is readily produced in cell suspension and hairy root cultures in amounts equal to or above that found in intact plant.

Figure 9.2 (See color insert.) *Vinca rosea*.

- The ethnobotanical significance of *periwinkle* is exemplified by its international usage as a traditional remedy for abundant ailments and not only for cancer. Antitumor terpenoid indole alkaloids (TIAs) are present only in micro quantities in the plant and are highly poisonous per se rendering a challenge for researchers to increase yield and reduce toxicity (Figure 9.2).

References

Awada, A., L. Dirix, L. Manso Sanchez, et al. 2013. Safety and efficacy of neratinib (HKI-272) plus vinorelbine in the treatment of patients with ErbB2-positive metastatic breast cancer pretreated with anti-HER2 therapy. *Annals of Oncology* 24, no. 1 (Jan):109–116.

Bhadra, R., S. Vani, and J.V. Shanks. 1993. Production of indole alkaloids by selected hairy root lines of *Catharanthus roseus*. *Biotechnology and Bioengineering* 41, no. 5:581–592.

Braguer, D., J.M. Barret, H. McDaid, and A. Kruczynski. 2008. Antitumor activity of vinflunine: Effector pathways and potential for synergies. *Seminars in Oncology* 35, no. 3 Suppl 3 (Jun):S13–S21.

Canellos, G.P., J.R. Anderson, K.J. Propert et al. 1992. Chemotherapy of Advanced Hodgkin's disease with MOPP, BVD, or MOPP alternating with ABVD. *New England Journal of Medicine* 327, no. 21:1478–1484.

Capasso, A. 2012. *Vinorelbine* in cancer therapy. *Current Drug Targets*. Review 13, no. 8 (Jul):1065–1071.

Chen-His, H., K.S. Chao, H.-F. Liao, and Y.-J. Chen. 2013. Norcantharidin, derivative of cantharidin, for cancer stem cells. *Evidence-Based Complementary and Alternative Medicine* 2013:838651, 11 pages. Article ID.

Chu, I., J.A. Bodnar, E.L. White, and R.N. Bowman. 1996. Quantification of vincristine and vinblastine in Catharanthus roseus plants by capillary zone electrophoresis. *Journal of Chromatography A* 755, no. 2:281–288.

Danieli, B., G. Lesma, M. Martinelli, D. Passarella, A. Silvani, B. Pyuskyulev, and M.N. Tam. 1998. *Vinblastine*-type antitumor *alkaloids*: A method for creating new C17 modified analogues. *Journal of Organic Chemistry* 63, no. 23:8586–8588.

De Sanctis, V., M. Alfò, A. Di Rocco, et al. 2017. Second cancer incidence in primary mediastinal B-cell lymphoma treated with methotrexate with leucovorin rescue, doxorubicin, cyclophosphamide, vincristine, prednisone, and bleomycin regimen with or without rituximab and mediastinal radiotherapy: Results from a monoinstitutional cohort analysis of long-term survivors. *Hematological Oncology* 35, no. 4 (Dec):554–560.

Deyell, R.J., B. Wu, S.R. Rassekh, et al. 2018. Phase I study of vinblastine and temsirolimus in pediatric patients with recurrent or refractory solid tumors: Canadian Cancer Trials Group Study IND.. *Pediatric Blood and Cancer* 218 (Nov):4:e27540.

Diaby, V., R. Tawk, V. Sanogo, H. Xiao, and A.J. Montero. 2015. A review of systematic reviews of the cost-effectiveness of hormone therapy, chemotherapy, and targeted therapy for breast cancer. *Breast Cancer Research and Treatment* 151, no. 1 (May):27–40.

Fahy, J., P. Hellier, F. Breillout, and C. Bailly. 2008. *Vinflunine*: Discovery and synthesis of a novel microtubule inhibitor. *Seminars in Oncology* 35, no. 3 Suppl 3 (June):S3–S5.

Garnier, F., Ph. Label, D. Hallard, J.C. Chénieux, M. Rideau, and S. Hamdi. 1996. Transgenic *periwinkle* tissues overproducing *cytokinins* do not accumulate enhanced levels of *indole alkaloids*. *Plant Cell, Tissue and Organ Culture* 45, no. 3:223–230.

Gourmelon, C., H. Bourien, P. Augereau, A. Patsouris, J.S. Frenel, and M. Campone. 2016. Vinflunine for the treatment of breast cancer. *Expert Opinion on Pharmacotherapy* 17, no. 13 (Sep):1817–1823.

Grieve, M. 1994. *A Modern Herbal. Edited and Introduced by Mrs. C.F. Leyel.* Tiger books international, London.

Grossi, F., J. Bennouna, L. Havel, M. Hochmair, and T. Almodovar. 2016. Oral vinorelbine plus cisplatin versus pemetrexed plus cisplatin as first-line treatment of advanced non-squamous non-small-cell lung cancer: Cost minimization analysis in 12 European countries. *Current Medical Research and Opinion* 32, no. 9 (Sep):1577–1584.

Gurr, S.J. 1996. The hidden power of plants. *The Garden* 121:262–264.

Imai, H., T. Shukuya, R. Yoshino, et al. 2013. Efficacy and safety of platinum combination chemotherapy re-challenge for relapsed patients with non-small-cell lung cancer after postoperative adjuvant chemotherapy of cisplatin plus vinorelbine. *Chemotherapy* 59, no. 4:307–313.

Isambert, N., J.P. Delord, J.M. Tourani, et al. 2014. How to manage intravenous *vinflunine* in cancer patients with renal impairment: Results of a pharmacokinetic and tolerability phase I study. *British Journal of Clinical Pharmacology* 77, no. 3 (Mar):498–508.

Jagetia, G.C., H. Krishnamurthy, and P. Jyothi. 1996. Evaluation of cytotoxic effects of different doses of *vinblastine* on mouse spermatogenesis by flow cytometry. *Toxicology* 112, no. 3:227–236.

Jordan, M.A., D. Thrower, and L. Wilson . 1992. *Journal of Cell Science* 102, no. 3:401–416.

Jordan, M.A., D. Thrower, and L. Wilson. 1991. Mechanism of inhibition of cell proliferation by *Vinca alkaloids*. *Cancer Research* 51, no. 8:2212–2222.

Joyce, C. 1992. What past plants hunts produced. *BioScience* 42:402.

Kallio, M., T. Sjöblom, and J. Lähdetie. 1995. Effects of *vinblastine* and *colchicine* on male rat meiosis *in vivo*: Disturbances in spindle dynamics causing micronuclei and metaphase arrest. *Environmental and Molecular Mutagenesis* 25, no. 2:106–117.

Kruczynski, A., and B.T. Hill. 2001. Vinflunine, the latest vinca alkaloid in clinical development. A review of its preclinical anticancer properties. *Critical Reviews in Oncology/Hematology* 40, no. 2 (Nov):159–173.

Lin, Y.Y., H.W. Kao, J.J. Li, et al. 2013. Tumor burden talks in cancer treatment with pegylated liposomal drugs. *PLOS ONE* 8, no. 5 (May10):e63078.

Lobert, S., and C. Puozzo. 2008. Pharmacokinetics, metabolites, and preclinical safety of vinflunine. *Seminars in Oncology* 35, no. 3 Suppl 3 (Jun):S28–S33.

Madoc-Jones, H., and F. Mauro. 1968. Interphase action of *vinblastine* and *vincristine*: Differences in their lethal actions through the mitotic cycle of cultured mammalian cells. *Journal of Cellular Physiology* 72, no. 3:185–195.

Montagna, E., A. Palazzo, P. Maisonneuve, et al. 2017. Safety and efficacy study of metronomic *vinorelbine*, cyclophosphamide plus capecitabine in metastatic breast cancer: A phase II trial. *Cancer Letters* 400, no. 400 (Aug):276–281.

Mukai, H., N. Masuda, H. Ishiguro, et al. 2015. Phase I trial of afatinib plus vinorelbine in Japanese patients with advanced solid tumors, including breast cancer. *Cancer Chemotherapy and Pharmacology* 76, no. 4 (Oct):739–750.

Naghmeh, N., A. Valdiani, D. Cahill, Y.-H. Tan, M. Maziah, and R. Abiri. 2015. Ornamental exterior versus therapeutic interior of Madagascar periwinkle (Catharanthus roseus): The two faces of a versatile herb. *Scientificworldjournal* 2015[10.1155/2015/982412].

Ngan, V.K., K. Bellman, D. Panda, B.T. Hill, M.A. Jordan, and L. Wilson. 2000. Novel actions of the antitumor drugs *vinflunine* and *vinorelbine* on microtubules. *Cancer Research* 60, no. 18 (Sep 15):5045–5051.

Noble, R.L. 1990. The discovery of the *vinca alkaloids* – Chemotherapeutic agents. *Biochemistry and Cell Biology = Biochimie et Biologie Cellulaire* 68, no. 12:1344–1351.

Peter Kutney, J., L. Siu Leung Choi, J. Nakano, H. Tsukamoto, M. McHugh, and C. Andre Boulet. 1988. A highly efficient and commercially important synthesis of the antitumor *Catharanthus* alkaloids vinblastine and *leurosidine* from *catharanthine* and *vindoline*. *Heterocycles* 27, no. 8:1845–1853.

Pollner, F. 1990. Chemo edging up on four so-far intractable tumors: U.S. and European teams report the first clinical successes—Some dramatic—From novel attacks. *Medical World News* 31:13–16.

Powell, J. 1991. Senior seminar presentation. *Biology*: Fall 4900.

Rowinsky, E.K., and R.C. Donehower. 1991. The clinical pharmacology and use of antimicrotuble agents in cancer therapeutics. *Pharmacology and Therapeutics* 52, no. 1:35–84.

Samuelsson, G. 1992. *Drugs of Natural Origin – A Textbook of Pharmacognosy. Third Revised, Enlarged and Translated Edition.* Swedish Pharmaceutical Press. Stockholm, Sweden.

Shirahama, T., T. Kohno, T. Kaijima, Y. Nagaoka, D. Morimoto, K. Hirata, and S. Uesato. 2006. Stereoselective conversion of anhydrovinblastine into vinblastine utilizing an anti-vinblastine monoclonal antibody as a chiral mould. *Chemical and Pharmaceutical Bulletin. Tokyo* 54, no. 5 (May):665–668.

St-Pierre, B., F.A. Vazquez-Flota, and V. De Luca V. 1999. Multicellular compartmentation of *Catharanthus roseus* alkaloid biosynthesis predicts intercellular translocation of a pathway intermediate. *Plant Cell* 11, no. 5:887–900.

Xue, Zhang, Cong-Cong Lin, Wai-Kei-Nickie Chan, Kang-Lun Liu, Zhi-Jun Yang, and Hong-Qi Zhang. 2017. Augmented anticancer effects of Cantharidin with liposomal encapsulation: *In vitro* and *in vivo* evaluation. *Molecules* 22:1052.

Zhang, Y., S.H. Yang, and X.L. Guo. 2017. New insights into *vinca alkaloids* resistance mechanism and circumvention in lung cancer. *Biomedicine and Pharmacotherapy = Biomedecine and Pharmacotherapie* 96, Dec:659–666.

Websites

http://biotech.icmb.utexas.edu/botany
https://floridata.com
www.ncbi.nlm.nih.gov/pubmed
www.ema.europa.eu
www.drugs.com
www.hindawi.com
https://pubchem.ncbi.nlm.nih.gov
www.accessdata.fda.gov

Mistletoe (*Viscum album*) (Loranthaceae)	Immunomodulator Cytotoxic

You can find it throughout Europe, Asia, N. Africa.

Appearance:

Stem: yellowish-greenish, branched, forming bushes 0.6–2 m in diameter.

Root: non-existent. The plant is a semi-parasitic evergreen shrub growing on branches of various tree hosts, mostly apple, poplar, ash, hawthorn, and lime, more rarely on oak and pear.

Leaves: opposite, tongue-shaped, yellowish-greenish.

Flowers: small, inconspicuous, clustered in groups of three.

Fruit: globular, pea-sized white berry, ripening in December.

In bloom: March–May.

Biology: mistletoe is propagated exclusively by seed, which is carried distantly with the aid of birds (mostly the thrush). According to host specificity three different races can be distinguished. The plant is dioecious with very reduced male and female flowers. The life cycle of *V. album* is described starting from seed germination to the development of the leaves. The parasitism affords special adaptation to mineral nutrition.

Degree of rarity: low, though not in abundant numbers.

Tradition: following their visions, the Druids used to cut mistletoe from trees with a golden knife at the beginning of the year. They held that the plant protected its possessor from all evil. According to a Scandinavian legend, Balder, the god of Peace, was slain with an arrow made of mistletoe. Later, however, mistletoe was rendered an emblem of love rather than hate. Its poisonous nature has been further exploited for the construction of knifes as a defensive weapon.

Parts used: leaves and young twigs.

Active ingredients: viscotoxin, mistletoe alkaloids, and three lectins (lactose-specific lectin, galactose-specific lectin, N-acetylgalactosamine-specific lectin).

Particular value: mistletoe preparations are well-tolerated with no significant toxicities observed so far. Studies show that mistletoe therapy has a positive effect on the patient's immune system and quality of life and can improve tolerance of standard therapies—such as chemotherapy—without decreasing their efficacy.

The status of mistletoe application in cancer therapy: mistletoe was introduced in the treatment of cancer in 1917. Rudolf Steiner (1861–1925), founder of the Society for Cancer Research, in Arlesheim (Switzerland), was the first to mention the immunoenhancing properties of mistletoe, suggesting its use as an adjutant therapy in cancer treatment.

Therapy of cancer with a Viscum extract has been carried out in Europe for over ten decades in thousands of patients. Extracts from the plant are used mainly as injections.

Currently, there is a number of mistletoe preparations used in many countries against different kinds of cancer:

Iscador and Helixor are licensed medications made from plants growing on different host trees, like oak, apple, pine, and fir, and administered in different kinds of cancer therapy. Some Iscador preparations also include metal, e.g. silver, mercury, and copper. Iscador is usually given by injection. However, it can also be taken orally. The injection treatment typically lasts 14 days with one injection each day. It has been approved for use in Austria, Switzerland, and Germany; it apparently is also being used in France, Holland, Eastern Europe, Britain, and Scandinavia. Proponents of the treatment claim that in 1978 almost 2,000,000 ampules were sold in countries where Iscador is prescribed and that about 30,000 patients are treated with it each year. Iscador is manufactured by the Verein fuer Krebsforschung (Cancer Research Association), a non-profit organization in Arlesheim, Switzerland.

Since 1980, a total of 26 clinical trials have been conducted on Helixor mistletoe products. Also, white-berry mistletoe's effects on various malignant diseases have been researched and proven in over 150 clinical trials, 30 of which used Helixor products.

Mistletoe therapy has been studied in patients with: various stages of breast cancer, lung cancer, various stages of colon cancer, chronic myeloid leukemia, ovarian cancer, pleural carcinosis (effusion in the pleural cavity compressing the lung), malignant lymphoma, malignant melanoma, pancreatic cancer. Most trials sought to determine whether mistletoe cancer treatment can prolong survival and/or improve quality of life.

Mistletoe products from Helixor are aqueous extracts of fresh mistletoe plants. The extract from fir mistletoe (fir in Latin = *abies*; Helixor® A) is primarily suited for patients just starting therapy, as adjunctive treatment for chemotherapy or radiation, and for patients who are prone to allergies or highly sensitive. The extract from apple tree mistletoe (apple in Latin = *malus*, Helixor® M) strongly stimulates the immune system and warmth organism. It has proved especially beneficial for tumors of the lower abdomen (bladder, uterus, and ovaries) and abdomen (stomach, colon, liver, pancreas), as well as for breast cancer after menopause. The extract from pine mistletoe (pine in Latin = *pinus*, Helixor® P) is particularly rich in mistletoe lectin, which has a strong immunomodulating and tumor-inhibiting effect. It is primarily used for skin cancer (malignant melanoma), bone cancer (sarcoma), kidney cancer, and testicular cancer, as well as breast cancer prior to menopause.

Iscucin-Viscum preparations contain mistletoe from eight different host-trees and are produced according to a particular 'rhythmic' procedure and additionally 'potentialized'. Sterilization is achieved by the addition of oligodynamic silver. The indications given are: precancerous conditions, postoperative tumor prevention, operable tumors, and inoperable tumors. Each of the eight preparations (according to host-tree) has its own list of indications. Iscucin is supposed to be injected close to the tumor between 5 and 7 p.m.; the dosage and the frequency depend on body temperature. However, no preclinical studies have been published on Iscucin. In the clinical field, only individual case histories are available, four of which have minimal documentation, and results that can be explained without iscucin. Iscucin is produced and distributed by Wala-Heilmittel GmbH, Eckwalden.

Isorel is an aqueous extract from whole shoots of mistletoe, the subspecies fir (Isorel A), apple (Isorel M), and pine (Isorel P) in each case. The preparation is injected hypodermically. It is usually applied for the medicative treatment of malignant tumors, postoperative and recidivation and prophylaxis of metastases, malignant illness of the hemopoietic system and defined precancerous stages. Isorel A is used principally for the treatment of male patients, while Isorel M is the respective preparation for female patients. Isorel is produced and distributed by Novipharm, Austria.

The use of mistletoe therapy in cancer diseases is permitted by the National Health Service (NHS) (Germany), and mistletoe is currently the most prescribed anticancer medication used in northern and central Europe.

There are more preparations of mistletoe, such as Eurixor and Plenosol. Eurixor, Isorel, and Vysorel are no longer available on the market for sale. Iscador, Isorel, and Plenosol are also sold as Iscar, Vysorel, and Lektinol.

New experimental data demonstrate that the special extract preparation Lektinol® (Madaus AG, Cologne, Germany), standardized for bioactive mistletoe lectin (ML), has antitumoral potencies *in vitro* and in animal tumor models.

AbnobaVISCUM is a standardized manufacture product (performed by using a patented pressing procedure). Using this manufacturing process to produce the pressed juice ensures that the entire range of plant ingredients is transferred to the extract. The plant extract contains 75% of the plant material used. With the help of a buffer solution, the various concentrations of AbnobaVISCUM are then produced.

AbnobaVISCUM uses plant material from both a summer and a winter. These two raw materials are processed separately to pressed juice and are subsequently mixed in another patented procedure.

The particular form of extracting AbnobaVISCUM affects the formation of liposomes in AbnobaVISCUM. Liposomes are cell-like structures in or to which one can include or attach active substances. Liposomes can thus enable targeted transport of the active substances to the 'site of action', i.e. the tumor. The medical term for this process is 'drug targeting. Formulations include: Solution for injection (for the 20 mg, 2 mg, 0.2 mg, 0.02 mg strengths) and liquid dilution for injection (for potency levels D 6, D 10, D 20, D 30) Distinguished according to the type of mistletoe host tree, formulations are classified as: Abietis (fir), Aceris (maple), Amygdali (almond tree), Betulae (birch), Crataegi (hawthorn), Fraxini (ash), Mali (apple tree), Pini (pine), Quercus (oak). Active substance is provided in the following concentration range: 20 mg, 2 mg, and 0.2 mg, with each ampoule containing respectively 1 ml, 0.1 ml, 0.01 ml, and 0.001 ml extract of fresh mistletoe herb from the respective host tree.

Therapeutic indications: AbnobaVISCUM is used according to the anthroposophical understanding of man and nature to stimulate the forming and integrative forces for the elimination and re-assimilation of growth processes which have become independent in adults, e.g. in malignant tumor diseases, also with accompanying disorders of the hematopoietic organs, in benign tumor diseases as prophylaxis against relapse following tumor surgery, and in the preliminary stages of certain cancers (defined precancerous conditions).

AbnobaVISCUM has been authorized in Korea for pleurodesis therapy.

Viscum album is listed in the *Homeopathic Pharmacopoeia* of the United States, which is the officially recognized compendium for homeopathic drugs in this country, but not sold as a drug in the United States.

Precautions: it is generally recommended that treatment be stopped during menstrual period and pregnancy. According to a report by the Swiss Cancer League, fermented Iscador products contain large numbers of both dead and live bacteria and some yeast.

- Home-made mistletoe preparations can be very poisonous. Reported minor side-effects (for Isorel) include a small increase in temperature of 1 to 1.5°C which disappear after one to two days. For Helixor, if the dosage is increased too rapidly, temperature rises of 1–1.5°C and headache may occur. Several clinical studies of the fermented form of Iscador have noted that patients experience moderate fever (a rise of 2.3 to 2.4°C) on the day of the injections. Local reactions around the injection site, temporary headaches, and chills are also associated with the fever. It is recommended to wait for the normalization of the temperature before a new injection is administered. In the case of hyperthyroidism, it is recommended to start with low doses and increase them gradually. High-dose *Viscum album* treatment may have interrupted frequently recurring tumors in individual patients with recurrent bladder cancer. In this study, patients were treated by Viscum Salicis A ampule (Box of 10 × 1-ml ampoules: Ingredients: 1 gm contains: *Viscum album* – Salicis (Willow tree mistletoe) 13X, 93.7%, Water, Salt, Uriel Company).

Indicative dosage and application:

- In all 11 melanoma cell lines tested: lectins isolated from *Viscum album* showed an antiproliferative effect at concentrations of 1–10 ng/ml, viscotoxin's antiproliferative effect rises at concentrations of 0.5–1 µg/ml, and alkaloids' antiproliferative effect begins at 10 µg/ml.
- Lectins ML I, ML II, and ML III were able, at concentrations from 0.02 to 20/pg ml, to enhance the secretion of the cytokines tumor necrosis factor (TNF) alpha, interleukin (IL)-1 alpha, IL-1 beta, and IL-6 by human monocytes.
- At the treatment with AbnobaVISCUM recommend that individual doses can be obtained with the 0.02 mg formulation. Otherwise, the dose should be increased in increments to 0.2 mg, 2 mg, or 20 mg, given in each case as –two to three injections a week.

Documented target cancers:

- Viscumin, a galactoside-binding lectin, is a powerful inflammatory mediator able to stimulate the immune system.
- A purified lectin (MLI) from *Viscum album* has immunomodulating effects in activating monocytes/macrophages for inflammatory responses.
- *Viscum album* L. extracts have been shown to provide a DNA stabilizing effect.
- Since Iscador stimulates the production of the natural killer cells, it can be applied in order to stabilize the number of T4 cells and thus the clinical condition of HIV positive persons. Laboratory tests suggested that the progress of the HIV infection was inhibited.
- Iscador has an increased action against breast cancer cells and colon cancer cells.
- In most patients (but healthy individuals, as well) the quality of life increased remarkably.
- Water-soluble polysaccharides of *Viscum album* exert a radioprotective effect, which could be a valuable complement to radiotherapy of cancer.
- Iscador therapy proved to be clinically and immunologically effective and well tolerated in immunocompromised children with recurrent upper respiratory infections, due to the Chernobyl accident.
- When whole mistletoe preparations are employed, the effect is host tree-specific.

- In 'Non-Small Cell Lung Cancer' cells (NSCLC cells) treatment with *Viscum album* Extracts (VAE), data indicate that VAE targets Axl to suppress cell proliferation and to circumvent cisplatin- and erlotinib-resistance.
- In a randomized phase II study where mistletoe was used as complementary treatment in patients with advanced non-small-cell lung cancer, severe non-hematological side-effects and hospitalizations were less frequent in patients treated with Iscador, warranting further investigation of Iscador as a modifier of chemotherapy-related toxicity.

Further details:

- Dendritic cell (DC) and *Viscum album*: the results of this study provide further evidence for a mode-of-action of the anti-cancer effects of mistletoe preparations via stimulation of DC maturation. Furthermore, the results obtained with this *in vitro* assay point to a possible role for mistletoe lectins in counteracting the tumor-induced immunosuppression of DC. Further experiments should clarify whether or not this effect is due to a specific interaction of ML with DC and focus on identifying the molecular mechanisms which are involved.
- The Chinese herb *Viscum alniformosanae* is the source of a conditioned medium (CM), designated as 572-CM, which is capable of stimulating mononuclear cells. This CM has the capacity to induce the promyelocytic cell line HL-60 to differentiate into morphologically and functionally mature monocytoid cells. Investigations have shown that 572-CM does not contain IFN-r, TNF, IL-1, and IL-2.
- A galactose-specific lectin from *Viscum album* (VAA) was found to induce the aggregation of human platelets in a dose- and sugar-dependent manner. Small non-aggregating concentrations of VAA primed the response of platelets to known aggregants (ADP, arachidonic acid, thrombin, ristocetin, and A23187). VAA-induced platelet aggregation was completely reversible by the addition of the sugar inhibitor lactose, and the platelets from disrupted aggregates maintained the response to other aggregants. The lectin-induced aggregation of washed platelets was more resistant to metabolic inhibitors than thrombin- or arachidonic acid-dependent cell interaction.
- The Korean mistletoe extract possesses antitumor activity *in vivo* and *in vitro*. Antiproliferative activities have been attributed to *Viscum album* C, *Viscum album* Qu, and *Viscum album* M (trade name Iscador) on melanoma cell lines. *Viscum album* C contains viscotoxin, alkaloids, and lectins. *Viscum album* Qu was extracted by Medac (Germany). *Viscum album* M is a preparation by the Institute Hiscia (Switzerland). The antiproliferative effect of the extracts on 11 melanoma cell lines obtained through the EORTC-MCG were tested in monolayer proliferation tests. In most of the melanoma cell lines tested, there was a significant antiproliferative effect of *Viscum album* C at a concentration of 100 µg/ml, whereas *Viscum album* M showed an antiproliferative effect at 1000 µg/ml. The lectins isolated from *Viscum album* C, when compared with each other showed almost in all 11 melanoma cell lines tested a similar antiproliferative effect. It was seen at concentrations of 1–10 ng/ml. The antiproliferative effect of viscotoxin rises at concentrations of 0.5–1 µg/ml, whereas the antiproliferative effect of alkaloids begins at 10 µg/ml.
- Mistletoe lectin I from *Viscum album* applied *in vitro* for 1 h in appropriate doses, caused irreversible inhibition of leukemic L1210 cell proliferation. The toxin appeared to be cytotoxic to normal bone marrow progenitor cells, as well as observed to the P388 and L1210 leukemia cells.
- Partially and highly purified lectins from *Viscum album* cause a dose-dependent decrease of viability of human leukemia cell cultures, MOLT-4, after 72 h treatment. The LC50 of the partially purified lectin was 27.8 ng/ml, of the highly purified lectin 1.3 ng/ml. Compared to the highly purified lectin a 140-fold higher protein concentration of an aqueous mistletoe drug was required to obtain similar cytotoxic effects on MOLT-4 cells. The cytotoxicity of the highly purified lectin was preferentially inhibited by D-galactose and lactose; cytotoxicity of the mistletoe drug and the partially purified lectin were preferentially inhibited by lactose and N-acetyl-D-galactosamine (GalNAc).
- Two lectin fractions with almost the same cytotoxic activity on MOLT-4 cells but with different carbohydrate affinities were isolated by affinity chromatography from the mistletoe drug: mistletoe lectin I with an affinity to D-galactose and GalNAc and mistletoe lectin II with an affinity to

GalNAc. The lectin fractions and the mistletoe drug inhibited protein synthesis of MOLT-4 cells stronger than DNA synthesis.

- The 5-bromo-2'-deoxyuridine-induced sister chromatid exchange (SCE) frequency of amniotic fluid cells (AFC) remained stable after the addition of a therapeutical concentration of *Viscum album* (Iscador P) but decreased significantly after administration of high drug doses. As the proliferation index remained stable, even at extremely high drug concentrations, this effect could not be ascribed to a reduction of proliferation. No indications of cytogenetic damage or effects of mutagenicity were seen after the addition of the preparation. In addition, increasing concentrations of *Viscum album* L. extracts were shown to significantly reduce sister chromatid exchange (SCE) frequency of phytohae-magglutinin (PHA)-stimulated peripheral blood mononuclear cells (PBMC) of healthy individuals.

- The three mistletoe lectins. ML I, ML II, and ML III were able, at concentrations from 0.02 to 20 pg/ml (100–10,000-fold lower than those showing toxic effects) to enhance the secretion of the cytokines tumor necrosis factor (TNF) alpha, interleukin (IL)-1 alpha, IL-1 beta, and IL-6 by human monocytes several-fold over control values were observed. The immunoactivating concentrations by the three lectins were found different for each donor. At toxic concentrations, the amounts of IL- 1 alpha, IL-1 beta, and to a less extent of TNF alpha in monocytes supernatants were particularly high.

- Application of an aqueous extract from *Viscum album coloratum*, a Korean mistletoe significantly inhibited lung metastasis of tumor metastasis produced by highly metastatic murine tumor cells, B16-BL6 melanoma, colon 26-M3.1 carcinoma, and L5178Y-ML25 lymphoma cells in mice. The antimetastatic effect resulted from the suppression of tumor growth and the inhibition of tumor-induced angiogenesis by inducing TNF-alpha.

- A peptide isolated from the *Viscum album* extract (Iscador) stimulated macrophages *in vitro* and *in vivo*, and activated macrophages were found to have cytotoxic activity towards L-929 fibroblasts.

- Iscador inhibited 20-methylcholanthrene-induced carcinogenesis in mice. Intraperitoneal administration of Iscador (1 mg/dose) twice weekly for 15 weeks could completely inhibit 20-methylcholanthrene-induced sarcoma in mice and protect these animals from tumor-induced death. Iscador was found to be effective even at lowered doses. After administration of 0.166, 0.0166, and 0.00166 mg/dose, 67, 50, and 17% of animals respectively did not develop sarcoma.

- Patients with advanced breast cancer who were treated parenterally with Iscador showed an improvement in repair, possibly due to a stimulation of repair enzymes by lymphokines or cytokines secreted by activated leukocytes or an alteration in the susceptibility to exogenous agents resulting in less damage.

- Iscador Pini, an extract derived from *Viscum album* L. grown on pines and containing a non-lectin associated antigen, strongly induced proliferation of peripheral blood mononuclear cells.

- The mistletoe lectin ML-A inactivates rat liver ribosomes by cleaving a N-glycosidic bond at A-4324 of 28 S rRNA in the ribosomes, as it is characteristic of the common ribosome-inactivating proteins (RIPs).

- Macrophages from mice treated with *Viscum album* extract were shown to be active in inhibiting the proliferation of tumor cells in culture. These activated macrophages have now been shown to protect mice from dying of progressive tumors when injected intraperitoneally into the animals. Prophylactic as well as multiple treatments with macrophages activated with *Viscum album* extract seemed more effective than a single treatment. Thus, in addition to a direct cytotoxic effect of *Viscum album* extract, the activation of macrophages may contribute to the overall antitumor activity of the drug.

- During a phase I/II study to determine the effect of *Viscum album* (Iscador) in HIV infection, 40 HIV-positive patients (with CD4-lymphocyte count > 200) were injected with 0.01 mg up to 10 mg subcutaneously twice a week over a period of 18 weeks. The extract was well tolerated and suggested to have anti-HIV activities.

- Iscador was found to be cytotoxic to animal tumor cells such as Dalton's lymphoma ascites cells (DLA cells) and Ehrlich ascites cells *in vitro* and inhibited the growth of lung fibroblasts (LB cells), Chinese hamster ovary cells (CHO cells), and human nasopharyngeal carcinoma cells (KB cells) at very low concentrations. Moreover, administration of Iscador was found to reduce ascites tumors and solid tumors produced by DLA cells and Ehrlich ascites cells. The effect of the drug could be seen when the drug was given either simultaneously, after tumor development, or when given prophylactically, indicating a mechanism of action very different from other chemotherapeutic drugs. Iscador was not found to be cytotoxic to lymphocytes.

- The ML-I lectin from *Viscum album* has been shown to increase the number and cytotoxic activity of natural killer cells and to induce antitumor activity in animal models. The same lectin inhibits cell growth and induces apoptosis (programmed cell death) in several cell types.

- Polysaccharides are possibly involved in the pharmacological effects of *Viscum album* extracts, which are used in cancer therapy. The main polysaccharide of the green parts of Viscum is a highly esterified galacturonan whereas in Viscum 'berries' a complex arabinogalactan is predominant and interacting with the galactose-specific lectin (ML I).

- In mice, an increased number of plaque-forming cells to sheep red blood cells (SRBC) followed the injection of Isorel (Novipharm, Austria) together with SRBC. Further, survival time of a foreign skin graft was shortened if Isorel was applied at the correct time. Finally, suppressed immune reactivity in tumorous mice recovered following Isorel injection. Isorel was further shown to be cytotoxic to tumor cells *in vitro*. Its application to tumor-bearing mice could prolong their life but without any therapeutic effect. However, a combination of local irradiation and Isorel was very effective: following 43 Gy of local irradiation to a transplanted methylcholanthrene-induced fibrosarcoma (volume about 240 mm^3) growing in syngeneic CBA/HZgr mice, the tumor disappeared in about 25% of the animals; the addition of Isorel increased the incidence of cured animals to over 65%. The combined action of Isorel, influencing tumor viability on the one hand and the host's immune reactivity on the other, seems to be favorable for its antitumor action *in vivo*.

- Water-soluble polysaccharides of *Viscum album* were shown to exert a radioprotective effect which was a function of both the radiation dose and the drug dose and time of its injection. The maximum radioprotective efficacy of polysaccharides was observed after their injection 15 min before irradiation.

- Iscador was found to reduce the leukocytopenia produced by radiation and cyclophosphamide treatment in animals. Weight loss due to radiation was considerable, whereas weight loss due to cyclophosphamide was not altered. Hemoglobin levels also were not affected, indicating that treatment with the extract reduces lymphocytopenia and hence could be used along with chemotherapy and radiation therapy.

- Hexanoic extracts of *Viscum cruciatum* Sieber parasitic on *Crataegus monogyna* Jacq. (I), *Crataegus monogyna* Jacq. parasitized with *Viscum cruciatum* Sieber (II), and *Crataegus monogyna* Jacq. non - parasitized (III), and of a triterpenes enriched fractions isolated from I, II, and III (CFI, CFII, CFIII) respectively demonstrated significant cytotoxic activity against cultured larynx cancer cells (HEp-2 cells).

- Intratumoral (IT) injection of European mistletoe (*Viscum album* L.) preparations might induce local tumor response through combined cytotoxic and immunomodulatory actions of the preparations. It is possible that immune-related ADRs (adverse drug reactions), such as pyrexia and local inflammatory reactions, might be critical for tumor response. In light of these results, IT mistletoe therapy seems to be safe, and prospective trials are recommended.

- An interventional (clinical trial) with 550 patients (estimated enrollment) is in process: randomized, open-label, active-controlled, prospective, multinational phase III confirmative study with 2 treatment groups and an adaptive design (Bauer and Köhne 1994). The study is designed to compare the efficacy of treatment with abnobaVISCUM® 900 with Mitomycin C (MMC). Patients with completely resected superficial bladder carcinoma (Stage Ta) with an intermediate risk classification according to the European Association of Urology (EAU, update 2013) and with one immediately post operative MMC 40 mg intravesical instillation will be eligible for inclusion in the study. Estimated study completion date is 2021.

- The Gorter Model is a novel and creative approach in the treatment of cancer patients and other chronic diseases. The goal of the treatment is to improve the immune system ('immune restoration'). This program was developed by Robert Gorter, MD, PhD. *Viscum album* is used to improve effectively the T-cell function and the functioning of the warmth organization of the patient. Approximately 85% of patients receive mistletoe. Mistletoe is provided as a simple subcutaneous injection twice a week, preferably in the early morning. Prolonged survival of patients with breast cancer, ovarian, uterine, and cervical cancers.

Useful addresses:

Camphill Wellbeing Trust
www.mistletoetherapy.org.uk

Cancer Research UK
www.cancerhelp.org.uk
Helpline: 0808 800 404 (free, only UK)

Macmillan Cancer Support
www.macmillan.org.uk
Helpline: 0808 808 0000 (free, only UK)

NHSinform. Cancer Information
Online at your fingerTIPS (Tailored information for the People of Scotland)
www.nhsinform.co.uk/cancer/TIPS

Medical Center Cologne
MD, PhD Robert Gorter
info@gorter-model.org

ABNOBA GmbH
Hohenzollernstraße 16 75177 Pforzheim I Germany
www.abnoba.de

Helixor Heilmittel GmbH
Fischermühle 1
72348 Rosenfeld
Germany
www.helixor.com/company

NOVIPHARM Klagenfurter Str. 164
9210 Pörtschach am Wörther Se, Austria
Tel : Phone number +43 4272 2751
Business website: www.novipharm.com

Verein fuer Krebsforschung
Association for Cancer Research
Kirschweg 9, CH-4144 Arlesheim
Tel: +41 61 706 72 72
Fax: +41 61 706 72 00
Web: www.vfk.ch

Uriel Pharmacy
N8464 Sterman Road
East Troy, WI 53120
USA
Phone: 1.866.642.2858 (toll-free)
Fax: 1.262.642.8780
www.urielpharmacy.com

References

Barney, C.W., F.G. Hawksworth, and B.W. Geils. 1998. *Eur. J*. Hosts of Viscum album. *Forest Pathology* 28, no. 3:187–208.

Bar-Sela, G., A. Gershony, and N. Haim. 2006. Mistletoe (Viscum album) preparations: An optional drug for cancer patients. *Harefuah* 145, no. 1 :42–46, 77, 77.

Bar-Sela, G., M., L. Wollner, A. Hammer, and E. Agbarya. Dudnik, and N. Haim. 2013. *Mistletoe as Complementary Treatment in Patients with Advanced Non-Small-Cell Lung Cancer treated with Carboplatin-Based Combinations: A Randomised phase II Study. Eur J Cancer* 49 , no. 5:1058–1064.

Büssing, A. 2000. *Mistletoe the Genus Viscum*. Harwood Academic Publishers. Amsterdam.

Büssing, A., G. Schaller, and U. Pfüller. 1998. Generation of reactive oxygen intermediates (ROI) by the thionins from *Viscum album L.*. *Anticancer Research* 18, no. 6A:4291–4296.

Büssing, A., and M. Schietzel. 1999. Apoptosis-inducing properties of *Viscum album* L. extracts from different host trees, correlate with their content of toxic Mistletoe lectins. *Anticancer Research* 19, no. 1A:23–28.

CA (anonymous). 1983. Unproven methods of cancer management: *Iscador*. *CA: A Cancer Journal for Clinicians* 33:186–188.

Congress, U.S., and Office of Technology Assessment 1990. *Unconventional Cancer Treatments*. D.C. Sept, Washington. U.S. Government Printing Office:81–86.

Fink, J.M. 1988. Third opinion. *An International Directory to Alternative Therapy Centers for the Treatment and Prevention of Cancer and Other Degenerative Diseases. Garden City Park, 137. New York: A Very Publishing Group Inc* (2nd edition).

Franz, H. 1986. Mistletoe lectins and their A and B chains. *Oncology* 43, no. 1:23–26.

Gaafar, R., A.R. Abdel Rahman, F. Aboulkasem, and A. El Bastawisy. 2014. Mistletoe preparation (*Viscum Fraxini-2*) as palliative treatment for malignant pleural effusion: A feasibility study with comparison to bleomycin. *Ecancermedicalscience* 8:424.

Gómez, M.A., M.T. Sáenz, M.D. García, M.C. Ahumada, and R. De La Puerta. 1997. Cytostatic activity against Hep-2 cells of methanol extracts from Viscum cruciatum Sieber parasitic on Crataegus monogyna Jacq. and two isolated principles. *Phytotherapy Research* 11, no. 3:240–242.

Grieve, M. 1994. *A Modern Herbal. Edited and Introduced by Mrs. C.F. Leyel*. Tiger books international, London.

Hauser, S., and A. Kast. 1990. *Iscucin* - Preparations for pre- and postoperative treatment of malignant tumours. *BCCA Cancer Information Centre SEARCH File* 701.

Hauser, S.P. 1993. Unproven methods in cancer treatment. *Current Opinion in Oncology* 5, no. 4:646–654.

Health professional version. 2018. In: *Mistletoe Extracts (PDQ®) Cover of PDQ Cancer Information Summaries PDQ Cancer Information Summaries www.ncbi.nlm.nih.gov/books/NBK66054/*.

Heiny, B.M., and J. Beuth. 1994. Mistletoe extract standardized for the galactoside-specific *lectin* (ML-1) induces b-endorphin release and immunopotentiation in breast cancer patients. *Anticancer Research* 14, no. 3B:1339–1342.

Hekal, I.A., T. Samer, and E.I. Ibrahim. *Viscum fraxini* 2, as an adjuvant therapy after resection of superficial bladder cancer: Prospective clinical randomized study 2008. Proceedings of the 43rd Annual Congress of the Egyptian Urological Association in Conjunction with the European Association of Urology, Novarum. *Abstract* 120:10–14. Hurghada, Egypt:8.

Horneber, M.A., G. Bueschel, R. Huber, K. Linde, and M. Rostock. 2008. Mistletoe therapy in oncology. *Cochrane Database of Systematic Reviews* 16, no. 2 (Apr):CD003297.

Kienle, G.S., R. Grugel, and H. Kiene. 2011. Safety of higher dosages of *Viscum album L* in animals and humans—Systematic review of immune changes and safety parameters. *BMC Complementary and Alternative Medicine* :11:72.

Kienle, G.S., and H. Kiene. 2010. Review article: Influence of *Viscum album L* (European mistletoe) extracts on quality of life in cancer patients: A systematic review of controlled clinical studies. *Integrative Cancer Therapies* 9, no. 2 (Jun):142–157.

Kim, S., K.C. Kim, and C. Lee. 2017. Mistletoe (*Viscum album*) extract targets Axl to suppress cell proliferation and overcome cisplatin- and erlotinib-resistance in non-small cell lung cancer cells. *Phytomedicine* 1, no. 36 (Dec):183–193.

Olsnes, S., F. Stirpe, K. Sandvig, and A. Pihl. 1982. Isolation and characterization of *viscumin*, a toxic lectin from *Viscum album L*. *(Mistletoe)*. *Journal of Biological Chemistry* 257, no. 22:13263–13270.

Ontario Breast Cancer Information Exchange Project. 1994. *Guide to Unconventional Cancer Therapies* (1st edition). Ontario Breast Cancer Information Exchange Project, Toronto:76–79.

Renatus, Z. 2009. Mistletoe preparation *iscador*: Are there methodological concerns with respect to controlled clinical trials. *Evidence-Based Complementary and Alternative Medicine* 6, no. 1 (Mar):19–30.

Ruebben, H. 2014. Intravesical mistletoe extract in superficial bladder cancer: A phase III efficacy study (clinical trial study record). Internet. National Institutes of Health, Bethesda, MD: US updated 2015 Mar; cited 2015 Apr 24. https://clinicaltrials.gov/ct2/show/NCT02106572.

Samuelsson, G. 1992. *Drugs of Natural Origin – A Textbook of Pharmacognosy. Third Revised, Enlarged and Translated Edition*. Swedish Pharmaceutical Press. Stockholm, Sweden.

Steele, M.L., J. Axtner, A. Happe, M. Kröz, H. Matthes, and F. Schad. 2015. Use and safety of intratumoral application of European mistletoe (*Viscum album L*) preparations in Oncology. *Integrative Cancer Therapies* 14, no. 2 (Mar):140–148.

Steinborn, C., A.M. Klemd, A.S. Sanchez-Campillo, et al. 2017. *Viscum album* neutralizes tumor-induced immunosuppression in a human *in vitro* cell model. *PLOS ONE* 12, no. 7:e0181553.

Sverrisson, E.F., P.N. Espiritu, and P.E. Spiess. 2013. New therapeutic targets in the management of urothelial carcinoma of the bladder. *Research and Reports in Urology* 5, no. 5 (Mar):53–65.

Sweeney, E.C., A.G. Tonevitsky, R.A. Palmer et al. 1998. Mistletoe *lectin I* forms a double trefoil structure. *FEBS Letters* 431, no. 3 (Jul 24):367–370.

Tröger, W., D. Galun, M. Reif, A. Schumann, N. Stanković, and M. Milićević. 2013. *Viscum album* [L] extract therapy in patients with locally advanced or metastatic pancreatic cancer: A randomised clinical trial on overall survival. *European Journal of Cancer* 49, no. 18 (Dec):3788–3797.

Tröger, W., D. Galun, M. Reif, A. Schumann, N. Stanković, and M. Milićević. 2014. Quality of life of patients with advanced pancreatic cancer during treatment with mistletoe: A randomized controlled trial. *Deutsches Ärzteblatt International* 21, (Jul) 111, 29–30:493–502.

von Schoen-Angerer, T.. 2015. High-dose Viscum album extract treatment in the prevention of recurrent bladder cancer: A retrospective case series. Perm Jornal. Fall 19, no. 4:76–83.

Wilson, B.R. *Cancer quackery primer 1985*. In: Dallas, Oregon: The Author.

Websites

www.clinicaltrials.gov
www.vedamsbooks.com/no13147.htm
www.colciencias.gov.co/simbiosis/ingles/projects.html
www.sare.org/htdocs/hypermail/html-home/3-html/0469.html
http://ericir.syr.edu/plweb-cgi/fastweb?getdoc+ericdb+972182+11+wAAA+Botany
https://floridata.com
www.ncbi.nlm.nih.gov/pubmed

Sho-saiko-to (SST), Juzen-taiho-to (JTT)

Sho-saiko-to (SST) and *Juzen-taiho-to* (JTT) are not plants but Japanese modified Chinese herbal medicines, or *Kampo*.

Sho-saiko-to (SST) has been used for decades in Japan as a key alternative health solution to support the liver. It is also known as Xiao-Chai-Hu-Tang (XCHT) and minor bupleurum decoction. *Sho-Saiko-To* (SST) is a highly researched herbal solution, with over 200 scientific research papers published (these statements have not been evaluated by the Food and Drug Administration). The synergetic botanical formula includes *Bupleurum* root, *Pinellia* tuber, *Scutellaria* root, *Ginseng*, *Jujube*, *Licorice*, and *Ginger*. Sho-saiko-to Herbal Extract is made in Japan and encapsulated and packaged in the United States (Honso® SST).

Sho-saiko-to is used to treat fever, malaria, gastrointestinal disorders, and chronic liver diseases. *Sho-saiko-to* appears to prevent liver injury and stimulate the immune system in laboratory and animal studies. *Sho-saiko-to* and its components may reduce or stop the spread of liver cancer cells and have little effect on normal human white blood cells. (One clinical trial showed that *sho-saiko-to* was associated with a decreased risk of developing liver cancer in patients with cirrhosis. Clinical trials are now being performed to determine whether *sho-saiko-to* can increase survival in patients with liver cancer).

Sho-saiko-to and its components demonstrate marked antiproliferative effects on hepatoma cell lines and ovarian cancer cell lines. Morphological analysis of cells grown in the presence of *Sho-saiko-to* show evidence of apoptosis. *Sho-saiko-to* has been shown to prevent liver injury and promote liver regeneration in animal models and to enhance various aspects of immune function, including effects on killer cells, interleukins, interferon, and macrophage. *Sho-saiko-to* may cause interstitial pneumonitis, a potentially fatal condition. Concurrent use of interferon may increase this risk. *Sho-saiko-to*-related pneumonitis has been reported in 74 patients (approximately 1 in 20,000). Liver injuries and hepatitis have been associated with the use of *sho-saiko-to*.

In an attempt to know the mechanism of action of XCHT, Fujian University of Traditional Chinese Medicine made a study and proved that XCHT reduced tumor size and weight, as well as significantly decreased cell viability both *in vivo* and *in vitro*. XCHT suppressed the expression of the proliferation marker Ki-67 in HCC tissues and inhibited Huh7 colony formation. XCHT induced apoptosis in HCC tumor tissues and in Huh7 cells. Finally, XCHT altered the expression of Bax, Bcl-2, CDK4 and cyclin-D1, which halted cell proliferation and promoted apoptosis (potent anticancer activity). Aslo, Xiaochaihu decoction and Ginseng-Chinese Date-Licorice Root group can markedly inhibit the growth of H22 mouse solid liver cancer and improve the immune function of tumor-bearing mice. The inhibiting effect is probably closely related with improving the tumor-bearing host immune function. There is a hint that compatibility Ginseng-Chinese Date-Licorice Root of strengthening body resistance is the core of inhibiting effect.

Tokushima University School of Medicine, in Japan, provided evidence that *sho-saiko-to* functions as a potent anti-fibrosuppressant via the inhibition of oxidative stress in hepatocytes and hepatic stellate cells and that its active components are baicalin and baicalein. In addition, *sho-saiko-to* has anti-carcinogenic properties in that it inhibits chemical hepatocarcinogenesis in animals, acts as a biological response modifier, and suppresses the proliferation of hepatoma cells by inducing apoptosis and arrests the cell cycle. Among the active components of *sho-saiko-to*, baicalin, baicalein, and saikosaponin-a have the ability to inhibit cell proliferation. It should be noted that baicalin and baicalein are flavonoids with chemical structures very similar to silybinin, which shows anti-fibrogenic activities. This may provide valuable information on the search for novel anti-fibrogenic agents.

Recommended use/supplement facts: serving size three packs (one pack each time, three times daily)—4,200 mg.

Juzen-taiho-to was formulated by Taiping Hui-Min Ju (Public Welfare Pharmacy Bureau) in Chinese Song Dynasty in AD 1200. It is prepared by extracting a mixture of ten medical herbs (*Rehmannia glutinosa, Paeonia lactiflora, Liqusticum wallichii, Angelica sinesis, Glycyrrhiza uralensis, Poria cocos, Atractylodes macrocephala, Panax ginseng, Astragalus membranaceus* and *Cinnamomum cassia*) that tone the blood and vital energy and strengthen health and immunity. This potent and popular prescription has traditionally been used against anemia, anorexia, extreme exhaustion, fatigue, kidney and spleen insufficiency, and general weakness, particularly after illness. Juzen-taiho-to is the most effective biological response modifier among 116 Chinese herbal formulates. Animal models and clinical studies have revealed that it demonstrates extremely low toxicity (LD50 > 15 g/kg op murine), self-regulatory and synergistic actions of its components in immunomodulatory and immunopotentiating effects (by stimulating hemopoietic factors and interleukins production in association with NK cells, etc.), potentiates therapeutic activity in chemotherapy (mitomycin, cisplatin, cyclophosphamide, and fluorouracil) and radiotherapy, inhibits the recurrence of malignancies, prolongs survival, as well as ameliorating and/or preventing adverse toxicities (GI disturbances such as anorexia, nausea, vomiting, hematotoxicity, immunosuppression, leukopenia, thrombocytopenia, anemia, and nephropathy, etc.) of many anticancer drugs.

Liver metastasis: the effect of the medicine was assayed after the inoculation of a liver-metastatic variant (L5) of murine colon 26 carcinoma cells into the portal vein. Oral administration of Juzen-taiho-to for seven days before tumor inoculation resulted in dose-dependent inhibition of liver tumor colonies and significant enhancement of survival rate as compared with the untreated control, without side effects. Juzen-taiho-to significantly inhibited the experimental liver metastasis of colon 26-L5 cells in mice pretreated with anti-asialo GM1 serum and untreated normal mice, whereas it did not inhibit metastasis in 2-chloroadenosine-pretreated mice or T-cell-deficient nude mice. Oral administration of Juzen-taiho-to activated peritoneal exudate macrophages (PEM) to become cytostatic against the tumor cells. These results show that oral administration of Juzen-taiho-to inhibited liver metastasis of colon 26-L5 cells, possibly through a mechanism mediated by the activation of macrophages and/or T-cells in the host immune system. Thus, Juzen-taiho-to may be efficacious for the prevention of cancer metastasis. It was suggested that Juzen-taiho-to may prolong the survival time of SV40 T antigen TG mice by improving their nutritional condition, inhibiting the MAPK pathway, and strengthening the immune system without causing hepatic toxicity.

Both *sho-saiko-to* (SST) and *Juzen-taiho-to* (JTT) suppressed the activities of thymidylate synthetase and thymidine kinase involved in de novo and salvage pathways for pyrimidine nucleotide synthesis, respectively, in mammary tumors of SHN mice with the reduction of serum prolactin level. These results indicate that SST and JTT may have the antitumor effects on mammary tumors.

Juzen-taiho-to also improves the general condition of cancer patients receiving chemotherapy and radiation therapy. Oral administration of TJ-48 accelerates recovery from hemopoietic injury induced by radiation and the anti-cancer drug mitomycin C. The effects are found to be due to its stimulation of spleen colony-forming units. It has been suggested that the administration of TJ-48 should be of benefit to patients receiving chemotherapy, radiation therapy, or bone marrow transplantation.

In combination with an anticancer drug UFT (5-fluorouracil derivative), it prevented the body weight loss and the induction of the colonic cancer in rats treated with a chemical carcinogen 1,2-dimethylhydrazine (DMH) and suppressed markedly the activity of thymidylate synthetase (TS) involved in the de novo pathway of pyrimidine synthesis in colonic cancer induced by DMH.

The combination of TJ-48 and mitomycin C (MMC) produced significantly longer survival in p-388 tumor-bearing mice than MMC alone, and TJ-48 decreased the diverse effects of MMC such as leukopenia, thrombopenia, and weight loss.

Imunostimulation: in mice, TJ-48 augmented antibody production and activated macrophage by oral administration of TJ-48 but reduced the MMC-induced immunosuppression in mice. TJ-48 showed a mitogenic activity in splenocytes but not in thymocytes, and an anti-complementary activity was also observed. Anti-complementary activity and mitogenic activity were both observed in high-molecular polysaccharide fraction but not in low-molecular weight fraction. Of several polysaccharide fractions in TJ-48, only pectic polysaccharide fraction (F-5-2) showed potent mitogenic activity. F-5-2 was also shown to have the highest anti-complementary activity. However, the poly-galacturonan region is essential for the expression of the mitogenic activity, but the contribution of poly-galacturonan region to the anti-complementary activity is less. F-5-2 activates complement via the alternative complement pathway and induces the proliferation of B cells but does not differentiate those cells from antibody-producing cells.

Contribution to the prevention of the lethal and marked side effects of recombinant human.

TNF (rhTNF) and lipopolysaccharide (LPS) without impairing their antitumor activity: these drugs are thought to decrease the oxygen radicals and stabilize the cell membranes, with a deep relation to the arachidonic cascade. The release of prostaglandins and leukotriene B4 was suppressed by pretreatment with *sho-saiko-to*. Thromboxane B2 was transiently increased, followed by suppression. After pretreatment with Hochu-ekki-to or Juzen-taiho-to, suppression of leukotriene B4 could not be observed. The release of prostaglandin D2 was suppressed in mice pretreated with *sho-saiko-to*, Juzentaiho-to, or Ogon (*Scutellariae Radix*) but it increased following pretreatment with Hochu-ekki-to. Chemicals that could prevent the lethality of rhTNF and LPS also revealed suppression of prostaglandins, leukotriene B4, and thromboxane B2. In general, drugs that prevented the lethality of rhTNF and LPS without impairing the antitumor activity could inhibit the release of leukotriene B4 and/or prostaglandin D2. rhTNF could activate the arachidonic cascade in combination with LPS. The lethality of rhTNF and LPS could be prevented by pretreatment with Japanese modified traditional Chinese medicines and the crude drug, Ogon.

In BDF1-mice which were implanted with P-388 leukemic cells, JTX prolonged significantly the average survival days of the MMC-treated group. In tumor-free BDF1-mice, JTX improved the leukopenia and the body weight loss which were caused by MMC. Additionally, JTX delayed the appearance of deaths by lethal dosis of MMC. These results indicate that JTX enhances the anti-tumor activity of MMC and lessens the adverse effects of it. JTX may be useful for patients undertaking MMC treatment.

TJ-48 has the capacity to accelerate recovery from hematopoietic injury induced by radiation and the anti-cancer drug mitomycin C (MMC). The effects are found to be due to its stimulation of spleen colony-forming unit (CFU-S) counts on day 14.

Compound isolation: the n-hexane extract from TJ-48 shows a significant immunostimulatory activity. The extract is further fractionated by silica gel chromatography and HPLC in order to identify its active components. 1H-NMR and GC-EI-MS indicate that the active fraction is composed of free fatty acids (oleic acid and linolenic acid). When 27 kinds of free fatty acids (commercially available) were tested using the HSC proliferating assay, oleic acid, elaidic acid, and linolenic acid were found to have potent activity. The administration of oleic acid to MMC-treated mice enhanced CFU-S counts on days 8 and 14 to twice the control group. These findings strongly suggest that fatty acids contained in TJ-48 actively promote the proliferation of HSCs. Although many mechanisms seem to be involved in the stimulation of HSC proliferation, we speculate that at least one of the signals is mediated by stromal cells, rather than any direct interaction with the HSCs.

The inhibitory effect of Juzen-taiho-to on progressive growth of a mouse fibrosarcoma is partly associated with prevention of gelatin sponge-elicited progressive growth, probably mediated by endogenous factors including antioxidant substances, in addition to the augmentation of host-mediated antitumor activity.

Juzen-taiho-to could be an effective drug for protecting against the side effects (nephrotoxicity, immunosuppression, hepatic toxicity, and gastrointestinal toxicity) induced by carboplatin in the clinic as well as by cisplatin.

Sodium L-malate, $C_4H_4Na_2O_5$, was found to exhibit protective effects against both nephrotoxicity (ED50: 0.4 mg/kg, p.o.) and bone marrow toxicity (ED50: 1.8 mg/kg, p.o.), without reducing the antitumor activity of CDDP. These findings indicate that *Angelicae Radix* and its constituent sodium L-malate could provide significant protection against CDDP-induced nephrotoxicity and bone marrow toxicity without reducing the antitumor activity.

Water-soluble ingredients of the herbal medicine *sho-saiko-to* dose-dependently inhibited the proliferation of a human hepatocellular carcinoma cell line (KIM-1) and a cholangiocarcinoma cell line (KMC-1). Fifty percent effective doses on day three of exposure to *sho-saiko-to* were 353.5 +/– 32.4 micrograms/ml for KIM-1 and 236.3 +/– 26.5 micrograms/ml for KMC-1. However, almost no suppressive effects were detected in normal human peripheral blood lymphocytes or normal rat hepatocytes. *Sho-saiko-to* suppressed the proliferation of the carcinoma cell lines significantly more strongly than did each of its major ingredients, i.e. saikosaponin a, c, and d, ginsenoside Rb1 and Rg1, glycyrrhizin, baicalin, baicalein, and wogonin, or another herbal medicine, juzen-taiho-to (P < 0.05 or 0.005). Because such ingredients are barely soluble in water, there could be synergistic or additive effects of the ingredients in *sho-saiko-to*. Morphological, DNA, and cell cycle analyses revealed two possible modes of action of *sho-saiko-to* to suppress the proliferation of carcinoma cells: (a) it induces apoptosis in the early period of exposure, and (b) it induces arrest at the G0/G1 phase in the late period of exposure.

The effect of Shi-Quan-Da-Bu-Tang (TJ-48) on hepatocarcinogenesis induced by N-nitrosomorpholine (NNM) was investigated in male Sprague-Dawley rats. Rats were given drinking water containing NNM for eight weeks and also from the start of the experiment, regular chow pellets containing 2.0 or 4.0% TJ-48 until the end of the experiment. Preneoplastic and neoplastic lesions staining for the ntal type of glutathione-S-transferase (GST-P) or gamma-glutamyl transpeptidase (GGT) were examined histochemically. In week 15, quantitative histological analysis showed that prolonged administration of either 2.0 or 4.0% TJ-48 in the diet significantly reduced the size, volume, and/or number of GST-P-positive and GGT-positive hepatic lesions. This treatment also caused a significant increase in the proportion of interleukin-2 receptor-positive lymphocytes among the lymphocytes infiltrating the tumors as well as a significant decrease in the labelling index of preneoplastic lesions. These findings indicate that TJ-48 inhibits the growth of hepatic enzyme-altered lesions and suggests that its effect may be in part due to activation of the immune system.

Juzen-taiho-to, a Kampo formula, originally consists of a mixture of *Shimotsu-to* and *Shikunshi-to* formulas together with two other crude ingredients. *Shimotsu-to* exerts an inhibitory effect on estrogen-induced expression of c-fos, interleukin (IL)-1alpha, and tumor necrosis factor (TNF)-alpha in uteri of ovarectomized mice. In another study, short- and long-term experiments were designed to determine the effects of *Juzen-taiho-to* and *Shimotsu-to* on the estrogen-related endometrial carcinogenesis in mouse uteri, associated with the expression of cyclooxygenase (COX)-1 and -2. *Shimotsu-to* is responsible for the preventive effects of *Juzen-taiho-to* on estrogen-related endometrial carcinogenesis in mice, through the inhibition of estrogen-related COX-2 as well as c-fos, IL-1alpha, and TNF-alpha expressions.

References

Aburada, M., S. Takeda, E. Ito, M. Nakamura, and E. Hosoya. 1983. Protective effects of *juzentaihoto*, dried decoctum of 10 Chinese herbs mixture, upon the adverse effects of *mitomycin C* in mice. *Journal of Pharmacobio-Dynamics* 6, no. 12:1000–1004.

Bensky, D., and A. Gamble. 1993. *Chinese Herbal Medicine*: Materia medica. Eastland Press, Seattle.

Chen, Y.W., M.Y. Tsai, H.B. Pan, H.H. Tseng, Y.T. Hung, and C.P. Chou. 2014. Gadoxetic acid-enhanced MRI and sonoelastography: Non-invasive assessments of chemoprevention of liver fibrosis in thioacetamide-induced rats with *Sho-Saiko-To. PLOS ONE* 9, no. 12 (Dec 9):e114756.

Deng, G., R.C. Kurtz, and A. 2011. Deng, G., R.C. Kurtz, A. Vickers, N. Lau, K.S. Yeung, J. Shia, and B. Cassileth. 2011. A single arm phase II study of a Far-Eastern traditional herbal formulation (*sho-sai-ko-to* or *xiao-chai-hu-tang*) in chronic hepatitis C patients. *Journal of Ethnopharmacology* 136, no. 1 (Jun 14):83–87.

Hisha, H., H. Yamada, M.H. Sakurai et al. 1997. Isolation and identification of hematopoietic stem cell-stimulating substances from Kampo (Japanese herbal) medicine, *Juzen-taiho-to. Blood* 90, no. 3:1022–1030.

Honso Professional Catalog. 2002. Honso USA, Inc. Tempe, Arizona.

Horie, Y., K. Kato, S. Kameoka, and K. Hamano. 1994. Bu ji (hozai) for treatment of postoperative gastric cancer patients. *American Journal of Chinese Medicine* 22, no. 3–4:309–319.

Horii, A., M. Kyo, M. Asakawa, R. Yasumoto, and M. Maekawa. 1991. Multidisciplinary treatment for bladder carcinoma–Biological response modifiers and kampo medicines. *Urologia Internationalis* 47, no. 1:108–112.

Hsu, L.M., Y.S. Huang, S.H. Tsay, F.Y. Chang, and S.D. Lee. 2006. Acute hepatitis induced by Chinese hepatoprotective herb, *xiao-chai-hu-tang. Journal of the Chinese Medical Association* 69, no. 2 (Feb):86–88.

Ikehara, S., H. Kawamura, Y. Komatsu, et al. 1992. Effects of medicinal plants on hemopoietic cells. *Advances in Experimental Medicine and Biology* 319:319–330.

Kang, H., T.W. Choi, K.S. Ahn, et al. 2009. Upregulation of interferon-gamma and interleukin-4, Th cell-derived cytokines by So-Shi-Ho-Tang (*Sho-Saiko-To*) occurs at the level of antigen presenting cells, but not CD4 T cells. *Journal of Ethnopharmacology* 123, no. 1 (May 4):6–14.

Kim, A., M. Im, and J.Y. Ma. 2016. Sosiho-tang ameliorates cachexia-related symptoms in mice bearing colon 26 adenocarcinoma by reducing systemic inflammation and muscle loss. *Oncology Reports* 35, no. 3 (Mar):1841–1850.

Li, J., M. Xie, and Y. Gan. 2008. Effect of Xiaochaihu decoction and different herbal formulation of component on inhibiting H22 liver cancer in mice and enhancing immune function. *Zhongguo Zhong Yao Za Zhi = Zhongguo Zhongyao Zazhi = China Journal of Chinese Materia Medica* 33, no. 9 (May):1039–1044.

Liu, X., and N. Li. 2012. Regularity analysis on clinical treatment in primary liver cancer by traditional Chinese medicine. *Zhongguo Zhong Yao Za Zhi = Zhongguo Zhongyao Zazhi = China Journal of Chinese Materia Medica* 37, no. 9 (May):1327–1331.

Miyanishi, K., T. Hoki, S. Tanaka, and J. Kato. 2015. *World Journal of Hepatology* 7, no. 3 (Mar 27):593–599.

Ohnishi, Y., H. Fujii, Y. Hayakawa, et al. 1998. Oral administration of a Kampo (Japanese herbal) medicine *Juzen-taiho-to* inhibits liver metastasis of colon 26-L5 carcinoma cells. *Japanese Journal of Cancer Research* 89, no. 2:206–213.

Ohnishi, Y., H. Fujii, F. Kimura, et al. 1996. Inhibitory effect of a traditional Chinese medicine, *Juzen-taiho-to*, on progressive growth of weakly malignant clone cells derived from murine fibrosarcoma. *Japanese Journal of Cancer Research* 87, no. 10:1039–1044.

Oka, H., S. Yamamoto, T. Kuroki, et al. 1995. Prospective study of chemoprevention of hepatocellular carcinoma with *Sho-saiko-to* (TJ-9). *Cancer* 76, no. 5:743–749.

Onishi, Y., T. Yamaura, K. Tauchi, et al. 1998. Expression of the anti-metastatic effect induced by *Juzen-taiho-to* is based on the content of Shimotsu-to constituents. *Biological and Pharmaceutical Bulletin* 21, no. 7:761–765.

Rino, Y., N. Yukawa, and N. Yamamoto. 2015. Does herbal medicine reduce the risk of hepatocellular carcinoma? *World Journal of Gastroenterology* 21, no. 37 (Oct 7):10598–10603.

Sakamoto, S., R. Furuichi, M. Matsuda, et al. 1994. Effects of Chinese herbal medicines on DNA-synthesizing enzyme activities in mammary tumors of mice. *American Journal of Chinese Medicine* 22, no. 1:43–50.

Sakamoto, S., H. Kudo, K. Kuwa, et al. 1991. Anticancer effects of a Chinese herbal medicine, *juzen-taiho-to*, in combination with or without 5-fluorouracil derivative on DNA-synthesizing enzymes in 1,2-dimethylhydrazine induced colonic cancer in rats. *American Journal of Chinese Medicine* 19, no. 3–4:233–241.

Sato, A., M. Toyoshima, A. Kondo, K. Ohta, H. Sato, and A. Ohsumi. 1997. Pneumonitis induced by the herbal medicine *sho-saiko-to* in Japan. *Nihon Kyobu Shikkan Gakkai Zasshi* 35, no. 4:391–395.

Satomi, N., A. Sakurai, F. Iimura, R. Haranaka, and K. Haranaka. 1989. Japanese modified traditional Chinese medicines as preventive drugs of the side effects induced by tumor necrosis factor and lipopolysaccharide. *Molecular Biotherapy* 1, no. 3:155–162.

Shimizu, I. 2000. *Sho-saiko-to*: Japanese herbal medicine for protection against hepatic fibrosis and carcinoma. *Journal of Gastroenterology and Hepatology (Mar 15)* 15 Suppl:D84–D90.

Song, K.H., Y.H. Kim, and B.Y. Kim. 2014. *Sho-saiko-to*, a traditional herbal medicine, regulates gene expression and biological function by way of microRNAs in primary mouse hepatocytes. *BMC Complementary and Alternative Medicine* 14, no. 1 (Jan 11):14.

Sugiyama, K., H. Ueda, and Y. Ichio. 1995. Protective effect of *juzen-taiho-to* against carboplatin-induced toxic side effects in mice. *Biological and Pharmaceutical Bulletin* 18, no. 4 :544–548.

Sugiyama, K., H. Ueda, Y. Ichio, and M. Yokota. 1995. Improvement of cisplatin toxicity and lethality by *juzen-taiho-to* in mice. *Biological and Pharmaceutical Bulletin* 18, no. 1:53–58.

Sugiyama, K., H. Ueda, Y. Suhara, Y. Kajima, Y. Ichio, and M. Yokota. 1994. Protective effect of sodium L-malate, an active constituent isolated from *Angelicae radix*, on cis-diamminedichloroplatinum(II)-Induced toxic side effect. *Chemical and Pharmaceutical Bulletin* 42, no. 12:2565–2568.

Tagami, K., K. Niwa, Z. Lian, J. Gao, H. Mori, and T. Tamaya. 2004. Preventive effect of *Juzen-taiho-to* on endometrial carcinogenesis in mice is based on *Shimotsu-to* constituent. *Biological and Pharmaceutical Bulletin* 27, no. 2 (Feb):156–161.

Takaoka, A., M. Iacovidou, T.H. Hasson, D. Montenegro, X. Li, M. Tsuji, and A. Kawamura. 2014. Biomarker-guided screening of *Juzen-taiho-to*, an oriental herbal formulation for immunostimulation. *Planta Medica* 80, no. 4 (Mar):283–289.

Tatsuta, M., H. Iishi, M. Baba, A. Nakaizumi, and H. Uehara. 1994. Inhibition by shi-quan-da-bu-tang (TJ-48) of experimental hepatocarcinogenesis induced by N-nitrosomorpholine in Sprague-Dawleyrats. *European Journal of Cancer* 30, no. 1:74–78.

Yamada, H. 1989. Chemical characterization and biological activity of the immunologically active substances in Juzen-taiho-to. *Gan to Kagaku Ryoho. Cancer and Chemotherapy* 16, no. 4 Pt 2–2:1500–1505.

Yano, H., A. Mizoguchi, K. Fukuda, M. Haramaki, S. Ogasawara, S. Momosaki, and M. Kojiro. 1994. The herbal medicine sho-saiko-to inhibits proliferation of cancer cell lines by inducing apoptosis and arrest at the G0/G1 phase. *Cancer Research* 54, no. 2:448–454.

Zee-Cheng, R.K. 1992. Shi-quan-da-bu-tang (ten significant tonic decoction), SQT. A potent Chinese bio-logical response modifier in cancer immunotherapy, potentiation and detoxification of anticancer drugs. *Methods and Findings in Experimental and Clinical Pharmacology* 14, no. 9:725–736.

Zhao, J., L. Liu, Y. Zhang, Y. Wan, and Z. Hong. 2017. The Herbal mixture Xiao-Chai-Hu Tang (XCHT) induces apoptosis of human hepatocellular carcinoma HUH7 cella *in vitro* and *in vivo*. *African Journal of Traditional, Complementary, and Alternative Medicines* 14, no. 3 (Mar 1):231–241.

Zheng, H.C., A. Noguchi, K. Kikuchi, T. Ando, T. Nakamura, and Y. Takano. 2014. Gene expression profil-ing of lens tumors, liver and spleen in α-crystallin/SV40 T antigen transgenic mice treated with *Juzen-taiho-to*. *Molecular Medicine Reports* 9, no. 2 (Feb):547–552.

Zhu, K., I. Fukasawa, M. Furuno, et al. 2005. Inhibitory effects of herbal drugs on the growth of human ovarian cancer cell lines through the induction of apoptosis. *Gynecologic Oncology* 97, no. 2:405–409.

Websites

http://sho-saiko-to.com
www.ncbi.nlm.nih.gov
https://europepmc.org
www.accessdata.fda.gov

9.2.2 Species with Anticancer Active Ingredients

Acacia catechu (Willd.) (Catechu) (Leguminosae) Antitumor

Synonyms: *Catechu nigrum (Leguminosae),* catechu black, cutch
  Location: Burma, India
  Appearance:

Stem: handsome trees
Leaves: compoundly pinnate
Flowers: are arranged in rounded or elongated clusters

Degree of rarity: low
  Tradition: is sold under the name of Catechu. It occurs in commerce in black, shining pieces or cakes.
  Parts used: leaves, young shoots
  Active ingredients: Proteins: Concanavalin A, abrin B chain and trypsin inhibitor (ACTI) (*Acacia confusa*)
  Particular value: it is used as an astringent to overcome relaxation of mucous membranes in general. An infusion can be employed to stop nose-bleeding and is also employed as an injection for uterine haemorrhage leucorrhoea and gonorrhoea. Externally, it is applied in the form of powder, to boils, ulcers, and cutaneous eruptions.
  Documented target cancers:

- Sarcoma 180 cells and Hela cell culture (mice)
- Human Breast Adenocarcinoma
- Human epithelial carcinoma cell line A431

Further details:

- Synthetic chimeric protein (ANB-ACTI) of abrin B chain and trypsin inhibitor
- Synthetic chimeric protein (Con A-ACTI) of Concanavalin A and trypsin inhibitor

Mode of action: abrin B chain of chimeric protein may act as a vector to carry ACTI into the tumor cells. ACTI in the chimeric protein potentiates its antitumor activity as well as its resistance to tryptic digestion.

References

Ghate, N.B., B. Hazra, R. Sarkar, and N. Mandal. 2014. Heartwood extract of *Acacia catechu* induces apoptosis in human breast carcinoma by altering bax/bcl-2 ratio. *Pharmacognosy Magazine* 10, no. 37:27–33.

Lin, J.Y., Y.S. Hsieh, and S.C. Chu. 1989. Chimeric protein: Abrin B chain-trypsin inhibitor conjugate as a new antitumor agent. *Biochemistry International* 19, no. 2:313–323.

Lin, J.Y., and L.L. Lin. 1985. Antitumor lectin-trypsin inhibitor conjugate. *Journal of the National Cancer Institute* 74, no. 5:1031–1036.

Lo, Y.L., C.Y. Hsu, and J.D. Huang. 1998. Comparison of effects of surfactants with other MDR reversing agents on intracellular uptake of epirubicin in Caco-2 cell line. *Anticancer Research* 18, no. 4C:3005–3009.

Monga, J., C.S. Chauhan, and M. Sharma. 2011. Human epithelial carcinoma cytotoxicity and inhibition of DMBA/TPA induced squamous cell carcinoma in Balb/c mice by *Acacia catechu* heartwood. *Journal of Pharmacy and Pharmacology* 63, no. 11:1470–1482.

Monga, J., C.S. Chauhan, and M. Sharma. 2013. Human breast adenocarcinoma cytotoxicity and modulation of 7,12-dimethylbenz [a] anthracene-induced mammary carcinoma in Balb/c mice by *Acacia catechu* (Lf) wild heartwood. *Integrative Cancer Therapies* 12, no. 4:347–362.

Aconitum napellus L. (Aconite) (Ranunculaceae)   **Poisonous**

Location: lower mountain slopes of north portion of Eastern Hemisphere. From Himalayas through Europe to Great Britain
   Appearance:

Stem: 3 ft. high
Root: fleshy, spindle-shaped, pale-colored when young, dark brown skin when mature
Leaves: dark, green glossy, deeply divided in palmate manner
Flowers: in erect clusters of a dark blue or white color
In bloom: late Spring–early Summer

Degree of rarity: not rare
   Tradition: one of the most useful drugs. It is used for many years as an anodyne, diuretic, and diaphoretic. It was used, also, for poisoning arrows. It is mentioned by Dioscorides that arrows tipped with the juice would kill wolves.
   Parts used: the whole plant, but especially the root (Aconiti tuber)
   Active ingredients: alkaloids: aconitine, aconine, benzaconine (picraconitine)
   Particular value: it produces highly toxic alkaloids, so all procedures must be done carefully.
   Precautions: keep away from children, even in gardens. In the dose of 3–6 mg it can cause death.
   Documented target cancers: novel derivatives prepared from *Aconitum* alkaloids are reported to show significantly suppressive effects in the cancer human cell lines A172, A549, HeLa, and Raji.
   Further details: the whole plant contains diterpene alkaloids (N-deethylaconotine) (*Aconitum napellus* and *Aconitum napellus* ssp. *neomontanum*) (Figure 9.3).

Figure 9.3 (See color insert.) *Aconitum fischeri.*

References

Ameri, A., and T. Simmet. 1999. Interaction of the structurally related aconitum alkaloids, aconitine and 6-benzyolheteratisine, in the rat hippocampus. *European Journal of Pharmacology* 386, no. 2–3:187–194.

Been, A. 1992. Aconitum: Genus of powerful and sensational plants. *Pharmacy in History* 1, no. 1:35–37.

Cole, C.T., and M.A. Kuchenreuther. 2001. Molecular markers reveal little genetic differentiation among *Aconitum noveboracense* and *A. columbianum* (*Ranunculaceae*) populations. *American Journal of Botany* 88, no. 2:337–347.

Colombo, M.L., M. Bravin, and F. Tome. 1988. A study of the diterpene alkaloids of *Aconitum napellus* ssp. neomontanum during its onthogenetic cycle. *Pharmacological Research Communications* 20 Suppl 5:123–128.

Fico, G., A. Braca, N. De Tommasi, F. Tomè, and I. Morelli. 2001. Flavonoids from *Aconitum napellus* subsp. neomontanum. *Phytochemistry* 57, no. 4:543–546.

Grieve, M. 1994. *A Modern Herbal*. Leyel, C.F. (Ed.). Tiger books international, London.

Hazawa, M., K. Wada, K. Takahashi, T. Mori, N. Kawahara, and I. Kashiwakura. 2009. Suppressive effects of novel derivatives prepared from Aconitum alkaloids on tumor growth. *Investigational New Drugs* 27, no. 2:111–119.

Imazio, M., R. Belli, F.E. Pomari, et al. 2000. Malignant ventricular arrhythmias due to *Aconitum napellus* seeds. *Circulation* 102, no. 23:2907–2908.

Kim, D.K., H.Y. Kwon, K.R. Lee, D.K. Rhee, and O.P. Zee. 1998. Isolation of a multidrug resistance inhibitor from *Aconitum* pseudo-laeve var. erectum. *Archives of Pharmacal Research* 21, no. 3:344–347.

Li, Z.B., and F.P. Wang. 1998. Two new diterpenoid alkaloids, beiwusines A and B, from *Aconitum kusnezoffii*. *Journal of Asian Natural Products Research* 1, no. 2:87–92.

Marchenko, M.M., H.P. Kopyl'chuk, and O.V. Hrygor'ieva. 2000. Activity of cytoplasmic proteinases from rat liver in Heren's carcinoma during tumor growth and treatment with medicinal herbs. *Ukrains'Kyĭ Biokhimichnyĭ Zhurnal* 72, no. 3:91–94.

Peng, C.S., F.P. Wang, and X.X. Jian. 2000. New norditerpenoid alkaloids from *Aconitum hemsley anum var. pengzhouense*. *Journal of Asian Natural Products Research* 2, no. 4:245–249.

Ulubelen, A., A.H. Meriçli, F. Meriçli, N. Kilinçer, A.G. Ferizli, M. Emekci, and S.W. Pelletier. 2001. Insect repellent activity of diterpenoid alkaloids. *Phytotherapy Research* 15, no. 2:170–171.

Wang, F.P., C.S. Peng, X.X. Jian, and D.L. Chen. 2001. Five new norditerpenoid alkaloids from *Aconitum sinomontanum*. *Journal of Asian Natural Products Research* 3, no. 1:15–22.

Yamanaka, H., A. Doi, H. Ishibashi, and N. Akaike. 2002. Aconitine facilitates spontaneous transmitter release at rat ventromedial hypothalamic neurons. *British Journal of Pharmacology* 135, no. 3:816–822.

Acronychia baueri (Rutaceae) (Sapindales)   **Cytotoxic**

Location: tropical forests of Australia, but also in China, South Asia, India, Indonesia, Myanmar, and the Pacific Ocean islands

Appearance: shrub or a small to moderately stinging tree that reaches up to 18 meters high and has a diameter of 33 cm

Stem: long and pale stem with a length of up to 4 cm
Leaves: elliptical in shape and length 5 to 13 cm, shiny dark green color at the upper part while lighter at the bottom
Flowers: five petals and eight stamens
In bloom: March–September

Tradition:

- Used by indigenous populations in Asia and Australia for the treatment of microbial and fungal infections and as anti-spasmodic, antipyretic, and anti-hemorrhagic agents.
- The fruits are used in salads and spices, while the oils in the flowers and the leaves are used in cosmetology.
- Its wood is excellent for the manufacture of chisels and chisel sticks.

Parts used: bark, stem, fruit

Active ingredients: alkaloids of acridone, acronycin, melicopine, 4-geranyloxyferulic acid

Figure 9.4   (See color insert.) *Acronychia* sp.

Documented target cancers:

- Acronychine has been reported to show significant activity against adenocarcinoma 755, leukemia C-1498, and myeloma X-5563.
- B82 leukemia, AKR melanomas, L5178Y, S-91, Shionogi carcinoma 115, and Ridgeway sarcoma (Figure 9.4).

References

Epifano, F., S. Fiorito, and S. Genovese. 2013. Phytochemistry and Pharmacognosy of the genus Acronychia. *Phytochemistry* 95:12–18.

Epifano, F., S. Fiorito, V.A. Taddeo, and S. Genovese. 2015. 4′-Geranyloxyferulic acid: An overview of its potentialities as an anti-cancer and anti-inflammatory agent. *Phytochemistry Reviews* 14, no. 4:607–612.

Svoboda, G.H., G.A. Poore, P.J. Simpson, and G.B. Boder. 1966. Alkaloids of *Acronychia baueri* Schott I: Isolation of the alkaloids and a study of the antitumor and other biological properties of acronycine. *Journal of Pharmaceutical Sciences* 55, no. 8:758–768.

Acronychia oblongifolia (Rutaceae) (Sapindales)	Cytotoxic

Location: In all types of rainforest
   Appearance:

Stem: 12 m high
Leaves: 4–12 cm long and emit a pleasant smell when crushed. Oil dots are visible and numerous, and the leaf blade is very glossy.
Flowers: produced on the bare stems and behind the foliage

Parts used: bark, stem
   Active ingredients:

- Flavonols: 5,3′-dihydroxy-3,6,7,8,4′-pentamethoxyflavone, 5-hydroxy-3,6,7,8,3′,4′-hexamethoxy flavone, digicitrin, 3-O-demethyldigicitrin, 3,5,3′-trihydroxy-6,7,8,4′-tetramethoxyflavone, and 3,5-dihydroxy-6,7,8,3′,4′-pentamethoxyflavone
- Alkaloids: 1,2,3-trimethoxy-10-methyl-acridone, 1,3,4-trimethoxy-10-methyl-acridone, des-N-methyl acronycine, normelicopine, and noracronycine

Documented target cancers:

- Human nasopharyngeal carcinoma
- Tubulin inhibitor

Further details:

- *Acronychia porteri* contains various flavonols (see above) which showed activity against (KB) human nasopharyngeal carcinoma cells (IC_{50} 0.04 micrograms/ml) and inhibited tubulin assembly into microtubules (IC_{50} 12 μM).
- *Acronychia pedunculata*: the bark contains acrovestone and bauerenol, two crystallin substances.
- *Acronychia baueri* (*Rutaceae*): the bark contains the alkaloids, 1,2,3-trimethoxy-10-methyl-acridone, 1,3,4-trimethoxy-10-methyl-acridone, des-N-methylacronycine, normelicopine, and noracronycine.
- *Acronychia laurifolia* BL: contains acronylin, a phenolic compound.
- *Acronychia haplophylla*: This plant contains the alkaloids acrophylline and acrophyllidine.

References

Biswas, G.K., and A. Chatterjee. 1970. Isolation and structure of acronylin: A new phenolic compound from *Acronychia laurifolia* BL. *Chemistry and Industry* 20, no. 20:654–655.

Chowrashi, B.K., B. Mukherjea, and S. Sikder. 1976. Some central effects of *Acronychia laurifolia* Linn. (Letter). *Indian Journal of Physiology and Pharmacology* 20, no. 4:250–251.

Epifano, F., S. Fiorito, and S. Genovese. 2013. Phytochemistry and pharmacognosy of the genus Acronychia. *Phytochemistry* 95:12–18.

Funayama, S., and G.A. Cordell. 1984. Chemistry of acronycine IV. Minor constituents of acronine and the phytochemistry of the genus *Acronychia*. *Journal of Natural Products* 47, no. 2:285–291.

Lahey, F.N., and M. McCamish. 1968. Acrophylline and acrophyllidine. Two new alkaloids from *Acronychia haplophylla*. *Tetrahedron Letters* 12:1525–1527.

Lichius, J.J., O. Thoison, A. Montagnac, et al. 1994. Antimitotic and cytotoxic flavonols from *Zieridium pseudobtusifolium* and *Acronychia porteri*. *Journal of Natural Products* 57, no. 7:1012–1016.

Svoboda, G.H., G.A. Poore, P.J. Simpson, and G.B. Boder. 1966. Alkaloids of *Acronychia baueri* Schott I. Isolation of the alkaloids and a study of the antitumor and other biological properties of acronycine. *Journal of Pharmaceutical Sciences* 55, no. 8:758–768.

Wu, T.S., M.L.Wang, T.T. Jong, A.T. McPhail, D.R. McPhail, and K.H. Lee. 1989. X-ray crystal structure of acrovestone, a cytotoxic principle from *Acronychia pedunculata*. *Journal of Natural Products* 52, no. 6:1284–1289.

Zhou, F.X., and Z.D. Min. 1989. Studies on the chemical constituents of *Acronychia pedunculata* (L.) Mig. *Chung Kuo Chung Yao Tsa Chih* 14, no. 2(30–31): 62.

***Ardisia gigantifolia* (Primulaceae) (Ericales)**   **Anti-cancer**

Location: tropical areas
  Appearance: shrub or half-leaved tree

  Leaves: elliptical, blade-like, alternating
  Flowers: white or pink
  Fruit: round and reddish

Tradition: the dried rhizome of *Ardisia gigantifolia* is mainly used as Chinese folk medicine in the south of China for the treatment of rheumatism, pain of muscles, and bones.
  Parts used: dried rhizome
  Active ingredients: triterpenoid saponins
  Documented target cancers:
  • A triterpenoid saponin-rich extract of *Ardisia gigantifolia* is reported active on human breast adenocarcinoma cells *in vitro* and *in vivo*

References

Chen, Z., Y. Ling, Q. Zhang, and H. Wan. 2017. Characterization of the major chemical constituents in *Ardisia gigantifolia* by high performance liquid chromatography coupled to electrospray ionization and quadrupole time-of-flight mass spectrometry. *Anal. Methods* 9, no. 39:5816–5825.

Mu, L.H., L. Bai, X.Z. Dong, F.Q. Yan, D.H. Guo, X.L. Zheng, and P. Liu. 2014. Antitumor activity of triterpenoid saponin-rich *Adisia gigantifolia* extract on human breast adenocarcinoma cells *in vitro* and *in vivo*. *Biological and Pharmaceutical Bulletin* 37, no. 6:1035–1041.

Mu, L.H., X.W. Huang, D.H. Guo, X.Z. Dong, and P. Liu. 2013. A new triterpenoid saponin from *Ardisia gigantifolia*. *Journal of Asian Natural Products Research* 15, no. 10:1123–1129.

Aerva lanata (Amaranthaceae) (Caryophyllales)	**Anti-leukemic**

Location: India, Africa, Asia
  Appearance: woody, prostrate, or succulent, perennial herb

  Stem: annual with a branching, straggling and sprawling, sometimes as much as 6 feet (1.8 m) in length.
  Leaves: simple, alternating, rounded, tapered at the base, hairy on the upper surface
  Flowers: tiny clusters of two or three flowers, greenish to white

Tradition: used in religious ceremonies, especially during the Pongal festival as a talisman against evil spirits. Traditional medicine against snake bites.
  Parts used: whole plant
  Active ingredients: ervine, methylervine, ervoside, aervine, methylaervine, aervoside, ervolanine and aervolanine, kaempferol, quercetin, isorhamnetin, persinol, persinosides A and B, methyl grevillate, lupeol, lupeol acetate benzoic acid, β-sitosteryl acetate, and tannic acid
  Documented target cancers:

• DLA (Dalton's Lymphoma Ascites)
• Leukemia cells
• Gut, lung, and cervix cancer

Related species:
Aerva javanica, traditionally used for the treatment of wounds, ulcer, cough, diarrhea, and hyperglycaemia. Anticancer potential of leaf and callus methanolic extracts from Aerva javanica has been reported *in vitro* on human breast cancer cell line MCF7.

References

Adepu, A., S. Narala, A. Ganji, and S.A. Chilvalvar. 2013. Review on natural plant: *Aerva lanata. International Journal of Pharmaceutical Sciences* 3:398–402.

Alsherif, E.A., A.M. Ayesh, A.S. Allogmani, and S.M. Rawi. 2012. Exploration of wild plants wealth with economic importance tolerant to difficult conditions in Khulais Governorate, Saudi Arabia. *Scientific Research and Essays* 7, no. 45:3903–3913.

Bhanot, A., R. Sharma, S. Singh, M.N. Noolvi, and S. Singh. 2013. *In vitro* anti cancer activity of ethanol extract fractions of *Aerva lanata* L. *Pakistan Journal of Biological Sciences* 16, no. 22:1612–1617.

Ghazanfar, S.A. 1994. *Handbook of Arabian Medicinal Plants*. CRC Press/Taylor and Francis, Bosa Roca, USA.

Goyal, M., A. Pareek, B.P. Nagori, and D. Sasmal. 2011. Aerva lanata: A review on phytochemistry and pharmacological aspects. *Pharmacognosy Reviews* 5, no. 10:195–198.

Kamalanathan, D., and D. Natarajan. 2014. Antiproliferative and antioxidant potential of leaf and leaf derived callus extracts of *Aerva lanata* (L.) Juss. Ex Schult. Against human breast cancer (MCF-7) cell lines. *Natural Products Journal* 4, no. 4:271–279.

Mamykova, R.U., T.S. Ibragimov, and R.K. Pernebekova. 2014. Study of onthogenesis of *Aerva lanata* (L.) Juss. *Life Science Journal* 11, no. 2:128–131.

Siveen, K.S., and G. Kuttan. 2011. Immunomodulatory and antitumor activity of *Aerva lanata* ethanolic extract. *Immunopharmacology and Immunotoxicology* 33, no. 3:423–432.

Zhao, Y., D. Kumar, D.N. Prasad, R.K. Singh, and Y. Ma. 2015. Morphoanatomic, physicochemical, and phytochemical standardization with HPTLC fingerprinting of aerial parts of *Aerva lanata* (Linn) Juss ex Schult. *Journal of Traditional Chinese Medical Sciences* 2, no. 1:39–44.

Agave angustifolia (Asparagaceae) (Asparagales)   **Anti-cancer**

Location: Mexico, central America
  Appearance: large succulent plant

Stem: flowering stem of up to 5 m (usually 2.5 m)
Leaves: stiff, straight, hollow, and sword-shaped. It is blue-green with wide white marginal bands and
  tiny brown horns as well as such a conical thorn at the end of the leaf.
Flowers: greenish-yellow to white in extreme inflorescences (fear) and then transformed into small
  plants, the bulbils, which then fall into the ground and roots.

Tradition: ornamental plant: especially in rock gardens (using many rocks) and in hedgerows.
Edible: used for mescal drink (tequila), burnt leaves are used to give the barbeque flavor. Flowering
shoots are cooked and juices extracted, fermented, and distilled to an alcoholic beverage. Also, the
juice can be concentrated in a sweet syrup known as 'Agave Syrup' or 'Agave Nectar'.
  Parts used: inflorescences and flowers, young stem of flowers, shoots, leaf bases, fruits
  Active ingredients: saponins: natural glycosides of steroids, triterpenes, branched fructans
  Precautions: pregnant women should not use *Agave americana* internally. Large quantities of
the herb can irritate the digestive system and even cause liver damage. The plant can trigger an
allergic reaction in some people and cause irritation and rashes. Caution is advised when the herb is
collected and handled due to the sharp blades at the tip of the leaves.
  Indicative dosage and application: as of today, there is no proven safe or effective dose for agave
as a herbal medicine but as with most things, moderation is always good.
  Documented target cancers:

- Colon cancer

Related species:

- *Agave americana*: methanol extract of the leaves exhibited an IC_{50} of 826.1 µg/mL against Human
  MCF-7 breast carcinoma cells and an IC_{50} of 8.455 µg/mL against VERO (non cancerous, African
  green monkey kidney cell line).
- *Agave sisalana*: homoisoflavonoids from *Agave sisalana* have been reported to have immunopharmacological activity as they significantly inhibited the production of IL-2 and IFN-γ in activated
  peripheral blood mononuclear cells in a concentration-dependent manner.
- *Agave intermixta* L.: aqueous extract has been reported to exhibit a cytostatic activity on HeLa
  contaminant Carcinoma (HEp-2 cells).

Related compounds:
Inulin-type prebiotics and in particular inulin-type fructans (ITF) are emerging as a food ingredient
which can positively influence the intestinal microbiota composition and metabolism and have been
shown to beneficially alter biomarkers of colon cancer in both *in vitro* and *in vivo* studies.

References

Allsopp, P., S. Possemiers, D. Campbell, I.S. Oyarzábal, C. Gill, and I. Rowland. 2013. An exploratory study
  into the putative prebiotic activity of fructans isolated from *Agave angustifolia* and the associated anticancer activity. *Anaerobe* 22:38–44.
Anajwala, C.C., R.M. Patel, S.L. Dakhara, and J.K. Jariwala. 2010. *In vitro* cytotoxicity study of *Agave americana*, *Strychnos nuxvomica* and *Areca catechu* extracts using MCF-7 cell line. *Journal of Advanced
  Pharmaceutical Technology and Research* 1, no. 2:245–252.

Chen, P.Y., Y.C. Kuo, C.H. Chen, Y.H. Kuo, and C.K. Lee. 2009. Isolation and immunomodulatory effect of homoisoflavones and flavones from *Agave sisalana* Perrine ex Engelm. *Molecules* 14, no. 5:1789–1795.

Hernández-Valle, E., M. Herrera-Ruiz, G.R. Salgado, et al. 2014. Anti-inflammatory effect of 3-O-[(6'-O-palmitoyl)-β-D-glucopyranosyl sitosterol] from *Agave angustifolia* on ear edema in mice. *Molecules* 19, no. 10:15624–15637.

Kajla, M., K. Bhattacharya, K. Gupta, U. Banerjee, P. Kakani, L. Gupta, and S. Kumar. 2015. Identification of the temperature induced larvicidal efficacy of *Agave angustifolia* against Aedes, Culex, and Anopheles larvae. *Frontiers in Public Health* 3:286.

Lacaille-Dubois, M.A. 2005. Bioactive saponins with cancer related and immunomodulatory activity: Recent developments. *Studies in Natural Products Chemistry* 32:209–246.

Macfarlane, G.T., and S. Macfarlane. 2011. Fermentation in the human large intestine: Its physiologic consequences and the potential contribution of prebiotics. *Journal of Clinical Gastroenterology* 45:S120–S127.

Sáenz, M.T., M.D. Garcia, A. Quilez, and M.C. Ahumada. 2000. Cytotoxic activity of *Agave intermixta* L. (Agavaceae) and *Cissus sicyoides* L. (Vitaceae). *Phytotherapy Research* 14, no. 7:552–554.

Ageratum houstonianum (Asteraceae) (Asterales)   Anti-cancer (*in silico*)

Location: (native) Mexico, Central America
    Appearance: annual herb

    Leaves: slightly aromatic and dentate, light green, soft, hairy, oval to triangular in shape and arranged mostly opposite
    Flowers: flowers are arranged on the bumps at the end of the shoots and have a color of pale purple-blue, purple, or red

Tradition:

- Ornamental plant: in gardens, flowerbeds, pots, rocky gardens, like chill-pots, etc.
- Pharmaceutical use: its juices are used externally for healing cuts and wounds and as an antiseptic. It also reduces pain.
- It has insect repellent properties in immature stages and adult mosquitoes.
- It has antimicrobial properties (against certain bacteria). Research into its ability to be used against diabetes, dermatophytes, and mites.
- Investigations of their effect on the preference of where mosquitoes lay their eggs, as well as their antioxidant capabilities.

Parts used: whole plant, but mostly the leaves
    Active ingredients: terpene mixtures (mostly precocene I, precocene II)
    Precautions: it contains enough alkaloids, which can cause hepatic injury. In general, however, it is safe to use.
    Documented target cancers:

- Osteosarcoma (in-silico results)
- Colorectal cancer, endometrial carcinoma, breast, esophageal, and gastric cancer, pancreatic, ovarian, carcinoma of the scalp and throat

References

Kumar, N. 2014. Biological potential of a weed *Ageratum houstonianum* Mill: A review. *Indo American Journal of Pharmaceutical Research* 4, no. 6:2683–2689.

Lu, X.N., X.C. Liu, Q.Z. Liu, and Z.L. Liu. 2014. Isolation of insecticidal constituents from the essential oil of *Ageratum houstonianum* Mill. against Liposcelis bostrychophila Badonnel. *Journal of Chemistry* 2014:1–6.

Ming, L.C. 1999. Ageratum conyzoides: A tropical source of medicinal and agricultural products. *Perspectives on New Crops and New Uses* 469–473.

Rizvi, S.M.D., S. Shakil, M. Zeeshan, et al. 2014. An enzoinformatics study targeting polo-like kinases-1 enzyme: Comparative assessment of anticancer potential of compounds isolated from leaves of *Ageratum houstonianum*. *Pharmacognosy Magazine* 10 Suppl 1:S14–S21.

Verma, A., S.M.D. Rizvi, S. Shaikh, et al. 2014. Compounds isolated from *Ageratum houstonianum* inhibit the activity of matrix metalloproteinases (MMP-2 and MMP-9): An oncoinformatics study. *Pharmacognosy Magazine* 10, no. 37:18–26.

Agrimonia pilosa (Agrimony) (Rosaceae)	**Immunomodulator**
	Cytotoxic

Location: everywhere, though not very far northward. On hedge-banks, meadows, open woods, and roadsides.

Origin: Chinese

Appearance:

Stem: erect and cylindrical, hairy, 50–150 cm high, mostly unbranched
Root: long, woody, and black
Leaves: 7.7–20 cm long, pinnate to other leaflets
Flowers: small, yellow, on terminal spikes, emitting an apricot-like odor. Fruits bear hairy spines. Fruit deeply grooved.
In bloom: June–September.

Degree of rarity: low

Tradition: one of the most famous 'magic' herbs, it has been used against wounds of various causes and for the prevention and cure of liver disorders. The Chinese *A. pilosa* is known as *xian he cao*.

Parts used: roots
Active ingredients: agrimoniin (tannin), unidentified components of methanolic extract
Particular value: its use presents a relatively low risk of side effects.
Precautions: avoid use in case of constipation.
Indicative dosage and application: agrimoniin: intraperitoneal injection with 10 mg/kg
Documented target cancers:

- The ethanol extract is reported to induce programmed cell death (apoptosis) in the human liver cancer HepG2 cells.
- Agrimoniin is capable of inducing interleukin-1.
- The methanol extract from roots of the plant helps to prolong the life span of mammary carcinoma-bearing mice while inhibiting tumor growth.
- Is cytotoxic to tumor cells, normal cells are far less affected.

Further details:

- An antimutagenic activity against benzo [a]pyrene (B[a]P) was marked in the presence of *Agrimonia pilosa* extracts (boiled for 2 h in a water bath) whereas that against 1,6-dinitropyrene (1,6-diNP) and 3,9-dinitrofluoranthene (3,9-diNF) varied from 20 to 86%. The observed differences in inhibition might be due to the inactivation of metabolic enzymes.

- A significant amount of interleukin-1 (IL-1) beta in the culture supernatant of the human peripheral blood mononuclear cells was stimulated with agrimoniin. Agrimoniin induced IL-1 beta secretion dose- and time-dependently. The adherent peritoneal exudate cells from mice intraperitoneally injected with agrimoniin (10 mg/kg) also secreted IL-1 four days later. These results suggested that agrimoniin is a novel cytokine inducer.
- To evaluate the antitumor activity of *Agrimonia pilosa*, the effects of the methanol extract from roots of the plant (AP-M) on several transplantable rodent tumors were investigated. AP-M inhibited the growth of S-180 solid type tumors. On the other hand, the prolongation of life span induced by AP-M on S-180 ascites type tumor-bearing mice was markedly minimized or abolished by the pretreatment with cyclophosphamide. AP-M showed considerably strong cytotoxicity on MM-2 cells *in vitro*, but the effect was diminished to one-tenth by the addition of serum to the culture. Against the host animals, the peripheral white blood cells in mice were significantly increased from two to five days after the i.p. injection of AP-M. On day four after the injection of AP-M, the peritoneal exudate cells, which possessed the cytotoxic activity on MM-2 cells *in vitro*, were also increased about five-fold relative to those in the non-treated control. The spleen of the mice was enlarged, and the spleen cells possessed the capacity to uptake 3H-thymidine. However, AP-M did not show direct migration activity like other mitogens against spleen cells from non-treated mice. These results indicate that the roots of *Agrimonia pilosa* contain some antitumor constituents, and possible mechanisms of the antitumor activity may include host-mediated actions and direct cytotoxicity.

References

Horikawa, K., T. Mohri, Y. Tanaka, and H. Tokiwa. 1994. Moderate inhibition of mutagenicity and carcinogenicity of benzo[a]pyrene, 1,6-dinitropyrene and 3,9-dinitrofluoranthene by Chinese medicinal herbs. *Mutagenesis* 9, no. 6:523–526.

Kimura, Y., M. Takido, and S. Yamanouchi. 1968. Studies on the standardization of crude drugs. XI. Constituents of *Agrimonia pilosa var. japonica. Yakugaku Zasshi* 88, no. 10:1355–1357.

Koshiura, R., K. Miyamoto, Y. Ikeya, H. Taguchi. 1985. Antitumor activity of methanol extract from roots of Agrimonia pilosa Ledeb. *Japanese Journal of Pharmacology* 38, no. 1:9–16.

Min, B.S., Y.H. Kim, M. Tomiyama, N. Nakamura, H. Miyashiro, T. Otake, and M. Hattori. 2001. Inhibitory effects of Korean plants on HIV-1 activities. *Phytotherapy Research* 15, no. 6:481–486.

Miyamoto, K., N. Kishi, and R. Koshiura. 1987. Antitumor effect of agrimoniin, a tannin of *Agrimonia pilosa* Ledeb., on transplantable rodent tumors. *Japanese Journal of Pharmacology* 43, no. 2:187–195.

Miyamoto, K., N. Kishi, T. Murayama, T. Furukawa, and R. Koshiura. 1988. Induction of cytotoxicity of peritoneal exudate cells by agrimoniin, a novel immunomodulatory tannin of *Agrimonia pilosa* Ledeb. *Cancer Immunology, Immunotherapy* 27, no. 1:59–62.

Murayama, T., N. Kishi, R. Koshiura, K. Takagi, T. Furukawa, and K. Miyamoto. 1992. Agrimoniin, an antitumor tannin of *Agrimonia pilosa* Ledeb., induces interleukin-1. *Anticancer Research* 12, no. 5:1471–1474.

Nho, K.J., J.M. Chun, and H.K. Kim. 2011. Agrimonia pilosa ethanol extract induces apoptotic cell death in HepG2 cells. *Journal of Ethnopharmacology* 138, no. 2:358–363.

Agrocybe aegerita (Strophariaceae) (Agaricales)	Anti-cancer
	Anti-angiogenic

Location: Chile, China (tea-oil trees), US, Japan, Korea, Australia
  Appearance: large smooth mushroom

Stem: creamy color

Tradition: as a flavoring in sauces, salads, pasta, chic, soup. It is said to have bacon flavor.
  Parts used: fruiting body

Active ingredients: polysaccharides (AG-HN1, AG-HN2), lectins (AAL), secondary metabolites (cylindan, agrocybenine), phenols (flavonoids), alkaloids, glucans, and sterols (ergosterol)

Particular value: in Chinese medicine, it is used as a diuretic and for stomach ulcers. Anti-inflammatory, antifungal, antioxidant, antibiotic properties, lowers blood sugar.

Precautions: reported hepatic toxicity after overconsumption by animals

Documented target cancers:

- Tumor angiogenesis model using both Caco-2 (epithelial tumor colon cell line) and HUVEC (endothelial cell line)
- Human SK-N-BE(2)-C neuroblastoma cells, murine C6, and human U-251 glioma cells
- HeLa (cervical cancer cell line), SW480 (colon cancer cell line), SGC-7901 (gastric cancer cell line), MGC80-3 (gastric adenocarcinoma cell line), BGC-823 (gastric cancer cell line), HL-60 (promyelocytic leukemia cells), and murine S-180 sarcoma cell line
- HepG2 (hepatocellular carcinoma) and SH-SY5Y (neuroblastoma) cell lines
- Leukemic U937 cell line

Related species:

Agrocybe cylindracea has been shown to exhibit HIV-1 reverse transcriptase inhibitory activity, antiproliferative activity against HepG2 and MCF7 cells

Related compounds:

Laccases that have been isolated from a number of mushrooms, like lectins and antifungal proteins which are defense or antipathogenic proteins, have been shown to exhibit antiproliferative activity toward tumor cells.

References

Chen, Y., S. Jiang, Y. Jin, et al. 2012. Purification and characterization of an antitumor protein with deoxyribonuclease activity from edible mushroom *Agrocybe aegerita*. *Molecular Nutrition and Food Research* 56, no. 11:1729–1738.

Guest, T.C., and S. Rashid. 2016. Anticancer laccases: A review. *Journal of Clinical and Experimental Oncology* 5, no. 1:2.

Hu, D.D., R.Y. Zhang, G.Q. Zhang, H.X. Wang, and T.B. Ng. 2011. A laccase with antiproliferative activity against tumor cells from an edible mushroom, white common *Agrocybe cylindracea*. *Phytomedicine* 18, no. 5:374–379.

Landi, N., S. Pacifico, S. Ragucci, et al. 2017. Purification, characterization and cytotoxicity assessment of Ageritin: The first ribotoxin from the basidiomycete mushroom *Agrocybe aegerita*. *Biochimica et Biophysica Acta (BBA) – General Subjects* 1861, no. 5:1113–1121.

Liang, Y., H.H. Liu, Y.J. Chen, and H. Sun. 2014. Antitumor activity of the protein and small molecule component fractions from *Agrocybe aegerita* through enhancement of cytokine production. *Journal of Medicinal Food* 17, no. 4:439–446.

Liang, Y., and H. Sun. 2015. The tumor protection effect of high-frequency administration of whole tumor cell vaccine and enhanced efficacy by the protein component from *Agrocybe aegerita*. *International Journal of Clinical and Experimental Medicine* 8, no. 5:6914–6925.

Lin, S., L.T. Ching, K. Lam, and P.C.K. Cheung. 2017. Anti-angiogenic effect of water extract from the fruiting body of *Agrocybe aegerita*. *LWT* 75:155–163.

Muthu, N., and I. Shanmugasundaram. 2015. Comparative study of phytochemicals in aqueous and silver nanoparticles extracts of *Agrocybe aegerita*, (V. Brig.) singer black poplar mushroom. *International Journal of Pharmacy and Biological Sciences* 6:190–197.

Ou, H.T., C.J. Shieh, J.Y.J. Chen, and H.M. Chang. 2005. The antiproliferative and differentiating effects of human leukemic U937 cells are mediated by cytokines from activated mononuclear cells by dietary mushrooms. *Journal of Agricultural and Food Chemistry* 53, no. 2:300–305.

Pei, Y.H., X.Li, T.R. Zhu, and L.J. Wu. 1990. Studies on the structure of a new flavanonol glucoside of the root-sprouts of *Agrimonia pilosa* Ledeb. *Yao Xue Xue Bao* 5, no. 4:267–270.
Zhao, C., H. Sun, X. Tong, and Y. Qi. 2003. An antitumour lectin from the edible mushroom *Agrocybe aegerita*. *Biochemical Journal* 374, no. 2:321–327.

Allanblackia floribunda (Clusiaceae) (Malpighiales)   **Cytotoxic**

Location: wet and tropical zone of Western, Central, and Eastern Africa (from Sierra Leone to Tanzania)

Appearance: evergreen tree, reaching up to 30 m

Stem: stump is loose and dark brown. It carries slim red scarves, exuding a sticky yellowish juice. Branches are thin, fallen, and sometimes form a propeller.
Leaves: green, oppositely arranged, elliptically elongated, and biconvex or rounded at the edge
Flowers: pink or reddish with round and convex petals

Tradition: commonly planted to provide shade for the seedlings in cacao establishments. Fruit fat, known as 'allanblackia oil' (stearic acid 45–58% and oleic acid 40–51%) is used for margarines or similar substitutions of dairy products. The nectar and pollen that honeybees collect from the tree is an important honey source. The sticky, yellow resin from the bark is used for building walls, pit-props, window frames, etc. in homes in Nigeria.

Parts used: leaves, fruits, seeds, bark, and bark juice

Active ingredients: triglycerides (palmitic, stearic, oleic), saponins, alkaloids, flavanatins, steroids, terpenoids, cardiac glycosides, difluanoids, anthraquinones, xanthones, gallic acid, polysaccharides, benzophenones

Documented target cancers:

- Human nasopharyngeal epidermoid carcinoma (KB) cells
- BT-549 cancer cell line (human breast carcinoma)
- (Low effect) BT-20, PC-3 (prostate carcinoma)
- CW-480 (colon carcinoma)

References

Adebayo, J.O., and A.U. Krettli. 2011. Potential antimalarials from Nigerian plants: A review. *Journal of Ethnopharmacology* 133, no. 2:289–302.
Boudjeko, T., R. Megnekou, A.L. Woguia, F.M. Kegne, J.E.K. Ngomoyogoli, C.D.N. Tchapoum, and O. Koum. 2015. Antioxidant and immunomodulatory properties of polysaccharides from *Allanblackia floribunda* Oliv stem bark and Chromolaena odorata (L.) King and HE Robins leaves. *BMC Research Notes* 8, no. 1:759.
Boudjeko, T., J.E.K. Ngomoyogoli, A.L., Woguia, and N.N. Yanou. 2013. Partial characterization, antioxidative properties and hypolipidemic effects of oilseed cake of Allanblackia floribunda and Jatropha curcas. *BMC Complementary and Alternative Medicine* 13, no. 1:352.
Fadeyi, S.A., O.O. Fadeyi, A.A. Adejumo, C. Okoro, and E.L. Myles. 2013. *In vitro* anticancer screening of 24 locally used Nigerian medicinal plants. *BMC Complementary and Alternative Medicine* 13, no. 1:79.
Fobane, L., E.N. Ndam, and M.J. Mbolo. 2014. Population structure and natural regeneration of *Allanblackia floribunda* oliv. (Clusiaceae) in a forest concession of East Cameroon. *Journal of Biodiversity and Environmental Sciences* 4:403–408.
Folarin, O.M., A.S. Oreniyi, and O.G. Oladipo. 2017. Physicochemical and kinetics parameters of *Allanblackia floribunda* Seed oil and some of its metal carboxylates. *Science International* 5, no. 2:56–62.

Kuete, V., A.G. Azebaze, A. Mbaveng, E.L. Nguemfo, E.T. Tshikalange, P. Chalard, and A.E. Nkengfack. 2011. Antioxidant, antitumor and antimicrobial activities of the crude extract and compounds of the root bark of *Allanblackia floribunda*. *Pharmaceutical Biology* 49, no. 1:57–65.

Lenta, B.N., C. Vonthron-Sénécheau, B. Weniger, et al. 2007. Leishmanicidal and cholinesterase inhibiting activities of phenolic compounds from *Allanblackia monticola* and *Symphonia globulifera*. *Molecules* 12, no. 8:1548–1557.

Nkengfack, A.E., G.A. Azebaze, J.C. Vardamides, Z.T. Fomum, and F.R. van Heerden. 2002. A prenylated xanthone from *Allanblackia floribunda*. *Phytochemistry* 60, no. 4:381–384.

Allium cepa L. (Amaryllidaceae) (Asparagales)   **Cytotoxic**

Location: anthropogenic soil, meadows, and fields
  Appearance: evergreen bulb, reaching up to 0.6 cm

  Leaves: its leaves are cylindrical, yellowish-green, fleshy, hollow, and alternating at the base of the plant. In autumn the leaves wither and the outer flakes of the bulb dry and become brittle.
  Flowers: the inflorescence is a round shade and is brought to the end of a large, hollow sprout. The flowers are white and 5–10 per inflorescence. The seeds are shiny black and triangular.

Tradition:

- Flavors: the bulb is eaten raw or boiled in soups or as a flavor additive, salad leaves, and even grains.
- Cosmetology: dyeing (peeling from the skin), for removal of parchment, for hair growth in the hair.
- Insect repellent, the juice is also used to prevent metal rust or as glass and copper polish.
- Anthelmintic, anti-inflammatory, antiseptic, antispasmodic, for anorexia, diuretic, for excretion of gases, for abrasions or fungal problems in the skin. Known to accelerate healing.

Parts used: bulb, seed, bark, leaves, flowers
  Active ingredients: alkylcysteine sulphoxides, gamma-glutamylpeptides, sulfur-containing gamma-glutamic acids, alliinase, flavonoids (quercetin) and respective glycosides, phenolic derivatives (photocatechic acid, ferulic acid, phloroglucin), vitamin B1, vitamin C, thiopropanal-5-oxide (which causes tears)
  Precautions: bleeding disorder: onion might slow blood clotting. Diabetes: onion might lower blood sugar.
  Documented target cancers:

- Ovarian, skin, and gastrointestinal cancer
- Colon and kidneys cancer
- Leukemic cancer cells
- B16F10 melanoma cell line
- Breast cancer, SAS oral cells
- APH (atypical prostatic hyperplasia)

Related species: *Allium porrum L., Allium schoenoprasum L., Allium sativum L.*

References

Elberry, A.A., S. Mufti, J. Al-Maghrabi, et al. 2014. Immunomodulatory effect of red onion (*Allium cepa* Linn) scale extract on experimentally induced atypical prostatic hyperplasia in Wistar rats. *Mediators of Inflammation* 2014:640746.

Han, M.H., W.S. Lee, J.H. Jung, et al. 2013. Polyphenols isolated from *Allium cepa* L. induces apoptosis by suppressing IAP-1 through inhibiting PI3K/Akt signaling pathways in human leukemic cells. *Food and Chemical Toxicology* 62:382–389.

Lai, W.W., S.C. Hsu, F.S. Chueh, et al. 2013. Quercetin inhibits migration and invasion of SAS human oral cancer cells through inhibition of NF-κB and matrix metalloproteinase-2/–9 signaling pathways. *Anticancer Research* 33, no. 5:1941–1950.

Levy, D., Z. Ben-Herut, N. Albasel, F. Kaisi, and I. Manasra. 1981. Growing onion seeds in an arid region: Drought tolerance and the effect of bulb weight, spacing and fertilization. *Scientia Horticulturae* 14, no. 1:1–7.

Mohammad Nabavi, S., S. Habtemariam, M. Daglia, and S. Fazel Nabavi. 2015. Apigenin and breast cancers: From chemistry to medicine. *Anti-Cancer Agents in Medicinal Chemistry (Formerly Current Medicinal Chemistry-Anti-Cancer Agents)* 15, no. 6:728–735.

Nelsen, C.E., and G.R. Safir. 1982. Increased drought tolerance of mycorrhizal onion plants caused by improved phosphorus nutrition. *Planta* 154, no. 5:407–413.

Shilpa, S. 2010. Tumor inhibition and cytotoxicity assay by aqueous extract of onion (*Allium cepa*) & garlic (*Allium sativum*): An in-vitro analysis. *International Journal of Phytomedicine* 2, no. 1:80–84.

Syed, N., D., V.M. Adhami, M.I. Khan, and H. Mukhtar. 2013. Inhibition of Akt/MTOR signaling by the dietary flavonoid fisetin. *Anti-Cancer Agents in Medicinal Chemistry (Formerly Current Medicinal Chemistry-Anti-Cancer Agents)* 13, no. 7:995–1001.

Allium cepa var. Aggregatum G. Don. (Amaryllidaceae) (Asparagales)   Anti-cancer, anti-inflammatory

---

Location: anthropogenic soil, meadows, and fields
   Appearance: bulb, growing up to 0.3 cm.

   Leaves: forms a cluster of plant-siblings with their head consisting of many cloves. It has a cluster of green narrow leaves.
   Flowers: the inflorescence is a round shade and is brought to the end of a large, hollow sprout. The flowers are white and 5 - 10 per inflorescence. The seeds are shiny black and triangular.

Tradition:

- Flavors: (in particular, bulbs are consumed raw, boiled, pickled, cooked, or fried).
- Pharmaceuticals: source of antioxidants, antibacterial, antimycotic, antiviral, helps to lower blood sugar and cholesterol, anti-inflammatory, anthelmintic, antispasmodic, diuretic, expectorant, heals wounds, astringent, antipyretic, antipyretic, tonic.

Parts used: leaves, flowers, bulbs
   Active ingredients: quercetin, isorhamnetin, and their glycosides, furostanol saponins, allicin, kaempferol, diallyl disulfide, diallyl trisulfide
   Documented target cancers:

- Colon cancer cells
- Jurkat, Wehi 164, and K562 cancer cell lines
- HeLa human carcinoma cell lines
- MCF-7 cell line
- HL-60 leukemic cell line

References

Fattorusso, E., M. Iorizzi, V. Lanzotti, and O. Taglialatela-Scafati. 2002. Chemical composition of shallot (Allium ascalonicum Hort.) *Allium ascalonicum* Hort. *Journal of Agricultural and Food Chemistry* 50, no. 20:5686–5690.

Guo, J., A. Liu, H. Cao, Y. Luo, J.M. Pezzuto, and R.B. van Breemen. 2008. Biotransformation of the chemo-preventive agent isoliquiritigenin by UDP-glucuronosyltransferases. *Drug Metabolism and Disposition: The Biological Fate of Chemicals* 36, no. 10:2104–2112.

Hsu, Y.L., C.C. Chia, P.J. Chen, S.E. Huang, S.C. Huang, and P.L. Kuo. 2009. Shallot and licorice constituent isoliquiritigenin arrests cell cycle progression and induces apoptosis through the induction of ATM/p53 and initiation of the mitochondrial system in human cervical carcinoma HeLa cells. *Molecular Nutrition and Food Research* 53, no. 7:826–835.

Mohammadi-Motlagh, H.R., A. Mostafaie, and K. Mansouri. 2011. Anticancer and anti-inflammatory activities of shallot (*Allium ascalonicum*) extract. *Archives of Medical Science* 7, no. 1:38–44.

Alocasia cucullata (Lour.) Schott (Araceae) (Alismatales)   Cytotoxic

---

Location: east Asia, southeast China, India, Sri Lanka, south Vietnam, Laos, south Thailand

Appearance: evergreen plant with thick, straight shoots that branch off from the base and reach up to 1 m in height

Leaves: it creates a sturdy cluster of large leaves, located on long-standing green stems. Its glossy dark green leaves are wide, heart-shaped, and have a curved edge, and its limbs are raised.

Flowers: the inflorescence is a patch with a greenish to pale yellow sword on top and is 'cocooned' on a modified leaf. It is brought to the end of a long and thick stalk.

Tradition: it is used as an ornamental plant. It is thought to bring luck, so it is planted in Buddhist temples. Edible: swamp plants are consumed in parts of India (very well cooked). Pharmaceutical: snake bites, abscesses and infections, body pains, rheumatism, gout.

Parts used: spadix, stem, leaves

Active ingredients: calcium oxalate crystals, alkaloids, steroids, phenolic components, N-acetyl-D: -lactosamine (LacNAc) specific lectin, amino acids, polysaccharides

Precautions: the presence of calcium oxalate crystals in spade, stem, and leaves requires good cooking before consumption to solubilize them as they cause discomfort in the oral cavity and a burning and stinging sensation. It is not poisonous but some deaths have been recorded after consumption (similar to cyanogenic glycoside poisoning).

Documented target cancers:

- Cervical cancer, breast cancer (SiHa cell line)
- Gastric cancer cells

Related species: *Alocasia macrorrhizos, Alocasia odora*

References

Goonasekera, C.D.A., V.W.J.K. Vasanthathilake, N. Ratnatunga, and C.A.S. Seneviratne. 1993. Is Nai Habarala (*Alocasia cucullata*) a poisonous plant? *Toxicon* 31, no. 6:813–816.

Kaur, A., S.S. Kamboj, J. Singh, A.K. Saxena, and V. Dhuna. 2005. Isolation of a novel N-acetyl-D-lactosamine specific lectin from *Alocasia cucullata* (Schott.). *Biotechnology Letters* 27, no. 22:1815–1820.

Peng, Q., H. Cai, X. Sun, X. Li, Z. Mo, and J. Shi. 2013. *Alocasia cucullata* exhibits strong antitumor effect *in vivo* by activating antitumor immunity. *PLOS ONE* 8, no. 9:e75328.

Wei, P., C. Zhiyu, T. Xu, and Z. Xiangwei. 2015. Antitumor effect and apoptosis induction of *Alocasia cucullata* (Lour.) G. Don in human gastric cancer cells *in vitro* and *in vivo*. *BMC Complementary and Alternative Medicine* 15, no. 1: https://doi.org/10.1186/s12906-015-0554-2.

Alocasia macrorrhiza **(Araceae) (Alismatales)**   **Cytotoxic, anti-tumor**

Location: tropical forests
    Appearance: rhizomatous evergreen perennial

Leaves: heart-shaped with intense green color

Tradition: roots and leaves are used in folk medicine of some countries as a topical patch for burns. The juice from fresh cut stems is used as an antidote to pain, redness, stinging, and itching caused by contact with plants.
    Parts used: all parts
    Active ingredients: alocasin B, hyrtiosin B, α-monopalmitin, 3-*epi*-betulinic acid, 3-*epi*-ursolic acid (8), β-sitosterol (9), and β-sitosterol 3-*O*-β-D-glucoside
    Precautions: contact with the sap can irritate the sensitive skin.
    Documented target cancers:
- The aqueous extract of the plant is reported to exhibit cytotoxic and apoptotic effect on human hepatocellular carcinoma cells *in vitro* and additionally to inhibit the development of hepatoma *in vivo*.

References

Fang, S., C. Lin, Q. Zhang, L. Wang, P. Lin, J. Zhang, and X. Wang. 2012. Anticancer potential of aqueous extract of *Alocasia macrorrhiza* against hepatic cancer *in vitro* and *in vivo*. *Journal of Ethnopharmacology* 141, no. 3:947–956.

Zhu, L.H., X.S. Huang, W.C. Ye, and G.X. Zhou. 2012. Study on chemical constituents of *Alocasia macrorrhiza* (L.) Schott. *Chinese Pharmaceutical Journal* 47:1029–1031.

Angelica archangelica **L. (Angelica) (Umbelifereae)**   **Cytotoxic**

Location: native in cold and moist areas in Scotland and in countries further north (Lapland, Iceland)
    Origin: Syria
    Appearance:

Stem: stout, fluted, 1.3–2 m high and hollow
Root: long, spindle-shaped, thick, and fleshy with large heavy specimens
Leaves: bright green, composed of numerous small leaflets, divided into three principal groups each of which is subdivided into three lesser groups. Edges are finely toothed or serrated.
Flowers: small and numerous, yellowish or greenish, grouped into large, globular umbels
In bloom: July

Degree of rarity: not rare (as it is largely cultivated in some areas)
    Tradition: it was well known for its protection against contagion, for purifying the blood, and for curing every conceivable malady, such as poisons, agues, and all infectious maladies.
    Parts used: root, leaves, seeds
    Active ingredients:

- *Pyranocoumarins*: decursin, archangelici, and 8(S),9(R)-9-angeloyloxy-8,9-dihydrooroselol
- *Chalcones*: 4-hydroxyderricin, xanthoangelol, and ashitaba-chalcone
- *Polysaccharide*: uronic acid

Precautions: should not be given to patients who have tendency towards diabetes, because it increases sugar in the urine

Documented target cancers:

- Cytotoxic on PANC-1 human pancreas cancer cells and Crl mouse breast cancer cells
- Skin cancer (mouse), Ehrlich tumors (mouse), and the stimulation of the uptake of tritiated thymidine into murine and human spleen cells

Further details:

- *Pyranocoumarins*: *decursin*: cytotoxic against various human cancer cell lines, possibly due to protein kinase C activation. Relatively low cytotoxicity against normal fibroblasts
- *Polysaccharide*: cytotoxic, immunostimulating
- *Angelica gigas*: roots contain the cytotoxic pyranocoumarin decursin (also found in *A. decursiva*) Fr. et Sav.
- *Angelica sinensis*: the rhizome contains a low molecular weight (3 Kd) polysaccharide composed partly of uronic acid. It shows strong anti-tumor activity on Ehrlich Ascites tumor-bearing mice. It also exhibits immunostimulating activities, both *in vitro* and *in vivo*.
- *Angelica keiskei*: roots contain two angular furanocoumarins, archangelicin and 8(S),9(R)-9-angeloyloxy-8,9-dihydrooroselol as well as three chalcones, 4-hydroxyderricin, xanthoangelol, and ashitaba-chalcone which can suppress 12-O-tetradecanoylphorbol-13-acetate (TPA)-stimulated 32Pi-incorporation into phospholipids of cultured cells. In addition, 4-hydroxyderricin and xanthoangelol have anti-tumor-promoting activity in mouse skin carcinogenesis induced by 7,12-dimethylbenz[a]anthracene (DMBA) plus TPA, possibly due to the modulation of calmodulin involved systems.
- *Angelica acutiloba* is one of the main components of the oriental Kampo-prescription, Shi-un-kou, (which other two constituents are *Lithospermum erythrorhizon* and *Macrotomia euchroma*). The drug exhibits inhibitory activity on Epstein–Barr virus activation and skin tumor formation in mice. Roots contain an immunostimulating polysaccharide (AIP) consisting of uronic acid, hexose, and peptide.
- *Angelica radix* is another oriental herb whose administration in mice is associated with an increased production of the tumor necrosis factor (TNF), possibly through stimulation of the reticuloendothelial system (RES).

References

Ahn, K.S., W.S. Sim, and I.H. Kim. 1996. Decursin: A cytotoxic agent and protein kinase C activator from the root of *Angelica gigas*. *Planta Medica* 62, no. 1:7–9.

Choy, Y.M., K.N. Leung, C.S. Cho, C.K. Wong, and P.K. Pang. 1994. Immunopharmacological studies of low molecular weight polysaccharide from *Angelica sinensis*. *American Journal of Chinese Medicine* 22, no. 2:137–145.

Haranaka, K., N. Satomi, A. Sakurai, R. Haranaka, N. Okada, and M. Kobayashi. 1985. Antitumor activities and tumor necrosis factor producibility of traditional Chinese medicines and crude drugs. *Cancer Immunology, Immunotherapy* 20, no. 1:1–5.

Konoshima, T., M. Kozuka, H. Tokuda, and M. Tanabe. 1989. Anti-tumor promoting activities and inhibitory effects on Epstein-Barr virus activation of Shi-un-kou and its constituents. *Yakugaku Zasshi* 109, no. 11:843–846.

Kumazawa, Y., K. Mizunoe, and Y. Otsuka. 1982. Immunostimulating polysaccharide separated from hot water extract of *Angelica acutiloba* Kitagawa (Y. Tohki). *Immunology* 47, no. 1:75–83.

Okuyama, T., M. Takata, J. Takayasu, et al. 1991. Anti-tumor-promotion by principles obtained from *Angelica keiskei*. *Planta Medica* 57, no. 3:242–246.

Sigurdsson, S., H.M. Ögmundsdóttir, and S. Gudbjarnason. 2005. The cytotoxic effect of two chemotypes of essential oils from the fruits of Angelica archangelica L. *Anticancer Research* 25, no. 3B:1877–1880.
Sigurdsson, S., H.M. Ögmundsdóttir, J. Hallgrimsson, and S. Gudbjarnason. 2005. Antitumour activity of Angelica archangelica leaf extract. *In vivo* 19, no. 1:191–194.

Annona cherimola (Annona) (Annonaceae)	Cytotoxic

Location: Central America (Ecuador, Colombia, and Bolivia)
  Appearance:

Stem: 5–10 m high, erect, low brunched
Leaves: briefly deciduous, alternate, two-ranked, with minutely hairy petioles 0.8–1.5 cm long, ovate to elliptic or ovate-lanceolate
Flowers: fragrant, solitary or in groups of two or three, on short hairy stalks along the branches, three outer greenish petals and three smaller, inner pinkish petals
In bloom: Spring, summer, autumn, winter

Parts used: fruit
  Active ingredients: Annonaceous acetogenins (lactones), alkaloids
  Documented target cancers:

* Human nasopharyngeal carcinoma cells
* High cytotoxicity against tumor cell lines: MCF-7 (human breast adenocarcinoma), ME-180 (epidermal carcinoma of the cervix), K562 (chronic myeloid leukemia)
* Prostate adenocarinoma and pancreatic carcinoma cell line (human), sarcoma

Further details:

* *Annona muricata*: leaves contain two Annonaceous acetogenins, muricoreacin and murihexocin C., showing significant cytotoxicities among human tumor cell lines with selectivities to the prostate adenocarinoma (PC-3) and pancreatic carcinoma (PACA-2) cell lines.
* *Annona senegalensis* is used against sarcomas.
* *Annona purpurea* contains alkaloids.
* *Annona reticulata*: seeds contain the cytotoxic gamma-lactone acetogenin, cis-/trans-isomurisolenin, along with annoreticuin, annoreticuin-9-one, bullatacin, squamocin, cis-/trans-bullatacinone, and cis-/trans-murisolinone.
* The bark of *Annona squamosa* yielded three new mono-tetrahydrofuran (THF) ring acetogenins, each bearing two flanking hydroxyls and a carbonyl group at the C-9 position. These compounds were isolated using the brine shrimp lethality assay as a guide for the bioactivity-directed fractionation. (2,4-cis and trans)-Mosinone A is a mixture of ketolactone compounds bearing a threo/trans/threo ring relationship and a double bond two methylene units away from the flanking hydroxyl. The other two new acetogenins differ in their stereochemistries around the THF ring; mosin B has a threo/trans/erythro configuration across the ring, and mosin C possesses a threo/cis/threo relative stereochemistry. Also found was annoreticuin-9-one, a known acetogenin that bears a threo/trans/threo ring configuration and a C-9 carbonyl and is new to this species. The structures were elucidated based on spectroscopic and chemical methods. Compounds 1–4 all showed selective cytotoxic activity against the human pancreatic tumor cell line, PACA-2, with potency 10–100 times that of Adriamycin.
* Activity-guided fractionation of the stem bark of *Annona senegalensis* gave four bioactive ent-kaurenoids. Compound 2 showed selective and significant cytotoxicity for MCF-7 (breast cancer) cells (ED50 1.0 microgram/ml), and 3 and 4 exhibited cytotoxic selectivity for PC-3 (prostate cancer) cells but with weaker potencies (ED50 17–18 micrograms/ml). The structure of the new compound, 3, was deduced from spectral evidence.

- The bark extracts of *Annona squamosa* yielded a new bioactive acetogenin, *squamotacin* (1), and the known compound, molvizarin, which is new to this species. Compound 1 is identical to the potent acetogenin, bullatacin, except that the adjacent bistetrahydrofuran (THF) rings and their flanking hydroxyls are shifted two carbons toward the gamma-lactone ring. Compound 1 showed cytotoxic selectively for the human prostate tumor cell line (PC-3), with a potency of over 100 million times that of Adriamycin.

- Bioactivity-directed fractionation of the seeds of *Annona muricata* L. (*Annonaceae*) resulted in the isolation of five new compounds: cis-annonacin, cis-annonacin-10-one, cis-goniothalamicin, arianacin, and javoricin. Three of these are among the first cis mono-tetrahydrofuran ring acetogenins to be reported. NMR analyses of published model synthetic compounds, prepared cyclized formal acetals, and prepared Mosher ester derivatives permitted the determinations of absolute stereochemistries. Bioassays of the pure compounds, in the brine shrimp test, for the inhibition of crown gall tumors, and in a panel of human solid tumor cell lines for cytotoxicity, evaluated relative potencies. Compound 1 was selectively cytotoxic to colon adenocarcinoma cells (HT-29) in which it was 10,000 times the potency of adriamycin.

- In a continuing activity-directed search for new antitumor compounds, using brine shrimp lethality (BST), mixtures of three additional pairs of bis-THF ketolactone acetogenins were isolated from the ethanol extract of the bark of *Annona bullata* Rich. (*Annonaceae*). Compared with (2,4-cis and trans)-bullatacinone, these new compounds each have one more aliphatic OH group at a different position on the hydrocarbon chain, and, thus, were named (2,4-cis and trans)-10-hydroxybullatacinone (1 and 2), (2,4-cis and trans)-12-hydroxybullatacinone (3 and 4), and (2,4-cis and trans)-29-hydroxybullatacinone. These mixtures all showed potent activities in the brine shrimp lethality test (BST) and exhibited cytotoxicities comparable to those of adriamycin against human solid tumor cells in culture with selectivities exhibited especially toward the breast cancer cell line (MCF-7).

- From *Annona bullata*, three more pairs of new ketolactone Annonaceous acetogenins were isolated by bioactivity-directed isolation. They are hydroxylated adjacent bistetrahydrofuran (THF) acetogenins and are named (2,4-cis and trans)-32-hydroxybullatacinone (1 and 2), (2,4-cis and trans)-31-hydroxybullatacinone (3 and 4), and (2,4-cis and trans)-30-hydroxybullatacinone. The structures were elucidated by analysis of the 1H- and 13C-nmr spectra of 1-6 and their acetates and the ms of their tri-trimethylsilyl (TMSi) derivatives as compared with bullatacinone. This is the first time that Annonaceous acetogenins with OH groups at successive positions near the end of the aliphatic chain have been reported. All of the new compounds showed potent activities in the brine shrimp lethality test and against human solid tumor cells in culture, with selectivities exhibited especially toward the colon cancer cell line (HT-29).

- Structural work and chemical studies are reported for several cytotoxic agents from the plants *Annona densicoma*, *Annona reticulata*, *Claopodium crispifolium*, *Polytrichum ohioense*, and *Psorospermum febrifugum*. Studies are also reported based on development of a mammalian cell culture benzo[a]pyrene metabolism assay for the detection of potential anticarcinogenic agents from natural products.

References

Cassady, J.M., W.M. Baird, and C.J. Chang. 1990. Natural products as a source of potential cancer chemotherapeutic and chemopreventive agents. *Journal of Natural Products* 53, no. 1:23–41.

Chang, F.R., J.L. Chen, H.F. Chiu, M.J. Wu, and Y.C. Wu. 1998. Acetogenins from seeds of *Annona reticulata*. *Phytochemistry* 47, no. 6:1057–1061.

Durodola, J.I. 1975. Viability and transplantability of developed tumour cells treated *in vitro* with antitumour agent C/M2 isolated from a herbal cancer remedy – Of *Annona senegalensis*. *Planta Medica* 28, no. 4:359–362.

Durodola, J.I. 1975. Antitumour effects against sarcoma 180 ascites of fractions of *Annona senegalensis*. *Planta Medica* 28, no. 1:32–36.

Fatope, M.O., O.T. Audu, Y. Takeda, L. Zeng, G. Shi, H. Shimada, and J.L. McLaughlin. 1996. Bioactive ent-kaurene diterpenoids from *Annona senegalensis*. *Journal of Natural Products* 59, no. 3:301–303.

Gu, Z.M., X.P. Fang, Y.H. Hui, and J.L. McLaughlin. 1994. 10-, 12-, and 29-hydroxybullatacinones: New cytotoxic Annonaceous acetogenins from *Annona bullata* Rich (Annonaceae). *Natural Toxins* 2, no. 2:49–55.

Gu, Z.M., X.P. Fang, L.R. Miesbauer, D.L. Smith, and J.L. McLaughlin. 1993. 30-, 31-, and 32-hydroxybullatacinones: Bioactive terminally hydroxylated annonaceous acetogenins from *Annona bullata*. *Journal of Natural Products* 56, no. 6:870–876.

Hopp, D.C., L. Zeng, Z. Gu, and J.L. McLaughlin. 1996. Squamotacin: An annonaceous acetogenin with cytotoxic selectivity for the human prostate tumor cell line (PC-3). *Journal of Natural Products* 59, no. 2:97–99.

Hopp, D.C., L. Zeng, Z.M. Gu, J.F. Kozlowski, and J.L. McLaughlin. 1997. Novel mono-tetrahydrofuran ring acetogenins, from the bark of *Annona squamosa*, showing cytotoxic selectivities for the human pancreatic carcinoma cell line, PACA-2. *Journal of Natural Products* 60, no. 6:581–586.

Kim, G.S., L. Zeng, F. Alali, L.L. Rogers, F.E. Wu, S. Sastrodihardjo, and J.L. McLaughlin. 1998. Muricoreacin and murihexocin C, mono-tetrahydrofuran acetogenins, from the leaves of *Annona muricata*. *Phytochemistry* 49, no. 2:565–571.

Quispe-Mauricio, A., R. Callacondo, J. Rojas-Camayo, C. Zavala, R. Posso, and W. Vaisberg. 2009. Cytotoxic effect of Annona cherimola seeds on cell cultures of cervical and breast cancer and chronic myeloid leukemia. *Acta Médica Peruana* 26, no. 3:156–161.

Rieser, M.J., Z.M. Gu, X.P. Fang, L. Zeng, K.V. Wood, and J.L. McLaughlin. 1996. Five novel mono-tetrahydrofuran ring acetogenins from the seeds of *Annona muricata*. *Journal of Natural Products* 59, no. 2:100–108.

Sonnet, P.E., and M. Jacobson. 1971. Tumor inhibitors. II. Cytotoxic alkaloids from *Annona purpurea*. *Journal of Pharmaceutical Sciences* 60, no. 8:1254–1256.

Wélé, A., Y. Zhang, J.P. Brouard, J.L. Pousset, and B. Bodo. 2005. Two cyclopeptides from the seeds of *Annona cherimola*. *Phytochemistry* 66, no. 19:2376–2380.

***Annona sylvatica* (Annonaceae) (Magnoliales)** **Cytotoxic**

Location: Brazil
  Appearance: deciduous tree

  Leaves: simple and glossy
  Flowers: creamy to yellow

Tradition: used as antitussive, spasmolytic, antipyretic, and for the treatment of dysentery, abdominal pain, and sore throat
  Parts used: fruit, leaves, bark-wood
  Active ingredients: hinesol, z-karyophyllene, β-maaliene, γ-gurjunene, silphiperfol-5-en-3-ol, ledol, cubecol-1-epi, and muurola-3,5-diene
  Documented target cancers:
  • The essential oil from the leaves of *Annona sylvatica* has been reported to show growth inhibitory activity against U251 (glioma, CNS), UACC-62 (melanoma), MCF-7 (breast), NCI-ADR/RES (ovarian expressing phenotype multiple drug resistance), 786-0 (renal), NCI-H460 (lung, nonsmall cells), PC-3 (prostate), OVCAR-03 (ovarian), and HT29 (colon) cell lines

References

Biba, V.S., A. Amily, S. Sangeetha, and P. Remani. 2014. Anticancer, antioxidant and antimicrobial activity of Annonaceae family. *World Journal of Pharmacy and Pharmaceutical Sciences* 3, no. 3:1595–1604.

Formagio, A.S., Mdo C. Vieira, L.A. dos Santos, et al. 2013. Composition and evaluation of the anti-inflammatory and anticancer activities of the essential oil from *Annona sylvatica* A. St.-Hil. *Journal of Medicinal Food* 16, no. 1:20–25.

| *Anthocephalus cadamba* (Rubiaceae) (Gentianales) | Anti-tumor |

Location: south and south-east Asia
   Appearance: large tree, up to 45 m tall, with a diameter of 100–160 cm

   Leaves: glossy green, opposite, simple more or less sessile to petiolate, ovate to elliptical
   Flowers: bisexual, 5-merous, calyx tube funnel-shaped, corolla gamopetalous saucer-shaped with a
   narrow tube, the narrow lobes imbricate in bud

Tradition: ornamental plants with religious significance in India, Java, and Malaysia. It is considered sacred to Lord Krishna. Used in ayurvedic medicine.
   Parts used: stem, fruit, flowers, leaves
   Active ingredients: ethercadamine and isocadamamine, chlorogenic acid, lupeol, betulinic acid-type triterpenes, cadambagenic acid, quinovic acid, and β-sitosterol
   Documented target cancers:
   • Defatted methanol extract has shown direct cytotoxicity on the Ehrlich ascites carcinoma cell line in
     a dose dependant manner. It has also exhibited significant decrease in the tumor volume, viable cell
     count, tumor weight, and elevated the life span of Ehrlich ascites carcinoma tumor-bearing mice.

References

Dolai, N., I. Karmakar, R.B. Suresh Kumar, B. Kar, A. Bala, and P.K. Haldar. 2012. Evaluation of antitumor
   activity and *in vivo* antioxidant status of *Anthocephalus cadamba* on Ehrlich ascites carcinoma treated
   mice. *Journal of Ethnopharmacology* 142, no. 3:865–870.
Dwevedi, A., K. Sharma, and Y.K. Sharma. 2015. Cadamba: A miraculous tree having enormous pharmacological implications. *Pharmacognosy Reviews* 9, no. 18:107–113.

| *Anthriscus sylvestris* (Apiaceae/Umbelliferae) (Apiales) | Cytotoxic |

Location: Europe, Western Asia, and Northwest Africa
   Appearance: herbaceous biennial that grows up to 1 m in height

   Stem: grows to a height of 60–170 cm, it is hollow and covered in hairs.
   Leaves: the leaves are 15–30 cm long and have a triangular form. The leaflets are ovate and subdivided.
   Flowers: white with five notched petals

Tradition: natural insect repellent when applied directly to the skin. The root is impregnated for several days in rice rinses and then cooked with other foods as a tonic for general weakness. The leaves are eaten raw, cooked, or used as a flavor. Used to treat headaches and also as a tonic, as an antitussive, as antipyretic, as an analgesic, and as a diuretic. Hollow stems of the plant are traditional materials from which children make peashooters.
   Parts used: leaves, flowers, seeds, root, shoot
   Active ingredients: deoxypodophyllotoxin (lignin)
   Particular value: the leaves and the stem are used for green pigment extraction.
   Precautions: can be mistaken for several similar-looking poisonous plants, among them *Conium maculatum* and *Aethusa cynapium*. Can be confused with *Heracleum mantegazzianum*, the sap of which can cause severe burns after coming in contact with the skin.
   Documented target cancers:

   • Deoxypodophyllotoxin, isolated from *Anthriscus sylvestris*, has been reported to arrest the cell
     cycle of HeLa human cervix carcinoma cells.

- Anthricin, isolated from *Anthriscus sylvestris*, has been reported to induce apoptosis in two breast cancer cell lines, MCF7 (estrogen receptor positive) and MDA-MB-231 (estrogen receptor, progesterone receptor, and Her2/Neu receptor negative).

Related compounds: the lignan family of natural products includes compounds with important antineoplastic and antiviral properties such as podophyllotoxin and two of their semisynthetic derivatives, etoposide and teniposide. The mechanism by which podophyllotoxin blocks cell division is related to its inhibition of microtubule assembly in the mitotic apparatus. However, etoposide and teniposide were shown not to be inhibitors of microtubule assembly which suggested that their antitumor properties were due to another mechanism of action, via their interaction with DNA and inhibition of DNA topoisomerase II.

References

Gordaliza, M., M.A. Castro, J.M.M. del Corral, and A.S. Feliciano. 2000. Antitumor properties of podophyllotoxin and related compounds. *Current Pharmaceutical Design* 6, no. 18:1811–1839.

Jung, C.H., H. Kim, J. Ahn, et al. 2013. Anthricin isolated from *Anthriscus sylvestris* (L.) Hoffm. Inhibits the growth of breast cancer cells by inhibiting Akt/mTOR signaling, and its apoptotic effects are enhanced by autophagy inhibition. *Evidence-Based Complementary and Alternative Medicine* 2013:385219.

Yong, Y., S.Y. Shin, Y.H. Lee, and Y. Lim. 2009. Antitumor activity of deoxypodophyllotoxin isolated from *Anthriscus sylvestris*: Induction of G2/M cell cycle arrest and caspase-dependent apoptosis. *Bioorganic and Medicinal Chemistry Letters* 19, no. 15:4367–4371.

Antogonon leptopus (Polygonaceae) (Caryophyllales)   **Cytotoxic**

Location: Mexico, Central America
  Appearance: fast-growing climbing vine

Stem: 6–10 m high, sometimes up to 15 m. Older stems may be brown and woody to the base of the plant, while the younger ones are reddish-brown or green.
Leaves: hairless or somewhat hairy, especially along their nerves, on their underside. Light to dark green in color, strongly wrinkled in appearance, lined with a vein network.
Flowers: clustered, with a hairy stalk, usually end up in vines

Tradition: use of *A. leptopus* tea as an ethnomedicine to ameliorate inflammatory pain. In Chinese traditional medicine, it is used for the treatment of nephritis, hepatitis, and colitis.
  Parts used: seeds, fruit, shoot, leaves
  Active ingredients: phenolic compound 2,3,4-trihydroxy benzaldehyde, hydroxylated analogs of benzaldehyde, steroid, tannins, terpenoids
  Particular value: the aerial portion, including flowers, is used in the preparation of tea used as a cold remedy. Plant extracts are used to treat thrombosis and have analgesic, anti-inflammatory, and anti-diabetic activity.
  Documented target cancers:

Human adenocarcinoma breast cancer (MCF-7) cell line

Further details: gold nanoparticles synthesized from *A. leptopus* extracts revealed important biological properties. Cytotoxicity study of synthesized nanoparticles exhibited a growth inhibitory property (GI_{50}) at the concentration of 257.8 µg/mL for human adenocarcinoma breast cancer (MCF-7) cells after 48 h.

Related species: gold nanoparticles of *Acalypha indica* have also shown anticancer activities against breast cancer cell lines.

References

Balasubramani, G., R. Ramkumar, N. Krishnaveni, A. Pazhanimuthu, T. Natarajan, R. Sowmiya, and P. Perumal. 2015. Structural characterization, antioxidant and anticancer properties of gold nanoparticles synthesized from leaf extract (decoction) of *Antigonon leptopus* Hook. & Arn. *Journal of Trace Elements in Medicine and Biology* 30:83–89.

Mulabagal, V., R.L. Alexander-Lindo, D.L. DeWitt, and M.G. Nair. 2011. Health-beneficial phenolic aldehyde in *Antigonon leptopus* Tea. *Evidence-Based Complementary and Alternative Medicine (eCAM)* 8, no. 1:http://dx.doi.org/10.1093/ecam/nep041

Swaroopa Rani, V., S. Sujatha, G.K. Mohan, and B.R. Kumar. 2010. Antidiabetic effect of *Antigonon leptopus* Hook & Arn. leaf on streptozotocin-induced diabetic rats. *Pharmacologyonline* 2:922–931.

Vanisree, M., R.L. Alexander-Lindo, D.L. DeWitt, and M.G. Nair. 2008. Functional food components of *Antigonon leptopus* tea. *Food Chemistry* 106, no. 2:487–492.

***Aralia elata* (Araliaceae) (Apiales)**   **Cytotoxic**

Location: Russia, China, Korea, Japan

Appearance: thorny, deciduous tree with a low, oval dome. It normally gets to 5 m (18 ft) tall and about 3 m (10 ft) wide.

Stem: spiny, pithy, usually unbranched stems which are crowned at the top with a spreading umbrella-like canopy of huge, showy, bi-pinnate leaves.

Leaves: very large bipinnate leaves at the tips of stout, spiny stems. They are ellipsoid and barbed and have hair. There is one sheet per node in each stem.

Flowers: white, with radial symmetry, hermaphrodite, pollinated by insects

In bloom: every summer

Tradition: it is used to treat rheumatoid arthralgia, cough, diabetes, jaundice, stomach ulcers, and stomach cancer.

Parts used: stems, leaves, roots

Active ingredients: Saponins, Araloside A, aralin

Documented target cancers:

Aralin, a novel cytotoxic protein, isolated from the shoots of *A. elata* has exhibited potent cytotoxic activity against various types of human cancer cell lines; cervical carcinoma cells (HeLa) proved the most sensitive, with an IC_{50} value of 0.08 ng/ml.

Related species: *Aralia chinensis* is used as a warming painkilling herb in the treatment of rheumatoid arthritis. The root is also considered to be useful in the treatment of diabetes. *Aralia nudicaulis* is used internally in the treatment of rheumatism and stomach aches. Closely related to the American species *Aralia spinosa*, with which it is easily confused.

References

Kim, S.J., W.S. Yoo, H. Kim, et al. 2015. *Aralia elata* prevents neuronal death by downregulating tonicity response element binding protein in diabetic retinopathy. *Ophthalmic Research* 54, no. 2:85–95.

Tomatsu, M., M. Ohnishi-Kameyama, and N. Shibamoto. 2003. Aralin, a new cytotoxic protein from *Aralia elata*, inducing apoptosis in human cancer cells. *Cancer Letters* 199, no. 1:19–25.

***Araucaria angustifolia* (Araucariaceae) (Pinales)** **Cytotoxic**

Location: North American warm deserts
  Appearance: tall evergreen tree, 20–50 m

Leaves: sharply acute, needle-like, lanceolate, usually in pairs, dark green or glaucous

Tradition: in Brazilian folk medicine various parts of the tree are used to treat various illnesses, particularly diseases of the respiratory tract.
  Parts used: bark, needles, and resin
  Active ingredients: dodecanoic and hexadecanoic acids, phenolic compounds
  Documented target cancers:
* *Angustifolia* extract is reported to induce a significant cytotoxicity in laryngeal carcinoma HEp-2 cells compared to the non-tumor human epithelial (HEK-293) cells, indicating a selective activity of the plant for the cancer cells.

Reference

Branco, Cdos S., É.D. de Lima, T.S. Rodrigues, et al. 2015. Mitochondria and redox homoeostasis as chemotherapeutic targets of *Araucaria angustifolia* (Bert.) O. Kuntze in human larynx HEp-2 cancer cells. *Chemico-Biological Interactions* 231:108–118.

Arctium lappa (Asteraceae) (Asterales) **Cytotoxic**

Location: most of Europe, including Britain, east to northern Asia
  Appearance: herbaceous, biennial plant, growing to 2 m by 1 m

Stem: erect stem, upper part with longitudinal edges, multi-branched
Leaves: the lower leaves are very large, furrowed above, heart-shaped, and of a gray color on their under surfaces. The upper leaves are much smaller, more egg-shaped and less densely arranged.
Flowers: spherical reddish purple or pink, hermaphrodite, and are pollinated by bees, lepidoptera, self.
In bloom: from June to September

Tradition: dried burdock roots are used in folk medicine as a diuretic, diaphoretic, and a blood purifying agent. In Britain, its root has been traditionally used as a flavoring in the herbal beverage 'Dandelion and burdock', which is still commercially produced.
  Parts used: leaves, flowers, shoots, fruits
  Active ingredients: taraxaterol, pantothenic acid, palmitate
  Documented target cancers:

* The methanolic extract of the seeds of *Arctium lappa* was found to show inhibitory activity against the LNCaP prostate cancer cell line.
* Arctigenin (a lignan from *Arctium lappa* L) has been reported to show cytotoxicity to lung cancer (A549), liver cancer (HepG2), and stomach cancer (KATO III) cell lines, while no cytotoxicity to several normal cell lines.

Related species: *Arctium minus* is one of the foremost detoxifying herbs in herbal medicine. It is antibacterial, antifungal, and carminative. It has soothing, mucilaginous properties and is said to be one of the most certain cures for many types of skin diseases, burns, bruises, etc. It is used in the treatment of herpes, eczema, acne, impetigo, ringworm, boils, bites, etc. It is used in the treatment of colds with sore throat and cough, measles, pharyngitis, acute tonsillitis, and abscesses.

References

Chan, Y.S., L.N. Cheng, J.H. Wu, et al. 2011. A review of the pharmacological effects of *Arctium lappa* (burdock). *Inflammopharmacology* 19, no. 5:245–254.

Ming, D.S., E. Guns, A. Eberding, and G.H. Towers. 2004. Isolation and characterization of compounds with anti-prostate cancer activity from *Arctium lappa* L. using bioactivity-guided fractionation. *Pharmaceutical Biology* 42, no. 1:44–48.

Susanti, S., H. Iwasaki, Y. Itokazu, M. Nago, N. Taira, S. Saitoh, and H. Oku. 2012. Tumor specific cytotoxicity of arctigenin isolated from herbal plant *Arctium lappa* L. *Journal of Natural Medicines* 66, no. 4:614–621.

Zhao, F., L. Wang, and K. Liu. 2009. *In vitro* anti-inflammatory effects of arctigenin, a lignan from *Arctium lappa* L., through inhibition on iNOS pathway. *Journal of Ethnopharmacology* 122, no. 3:457–462.

Ardisia brevicaulis (Primulaceae) (Ericales)   **Pro-apoptotic Cytotoxic**

Location: America, Asia, Australia, and the Pacific Islands
   Appearance: small shrub

Stem: 10–15 cm
Leaves: narrowly ovate to elliptic
Flowers: pink, 4–5 mm. The defining characteristic of the genus is the small tube formed at the center of the flower by the stamens, which are joined at their bases.

Tradition: its root is used as a folk medicine for irregular menstruation, postpartum anemia, rheumatism, sciatica, traumatism, contusions, and stomachache.
   Parts used: root, leaves, and fruits
   Active ingredients: 4-hydroxy-2-methoxy-6-[(8Z)-decapent-8-en-1-yl] phenolic acid, 5-r (8Z)-decapent-8-en-1-yl] resorcinol, Ardisiphenol D
   Documented target cancers:

- Inhibitory activities against the proliferation of PANC-1, A549, SGC7901, MCF-7, and PC-3 cancer cells, selective cytotoxic activity against the NCI-H460 cancer cell line
- Alkylphenols from the roots of *Ardisia brevicaulis* induce G1 arrest and apoptosis through endoplasmic reticulum stress pathway in human non-small-cell lung cancer cells.
- Antitumor effect of resorcinol derivatives from the roots of *Ardisia brevicaulis* by inducing apoptosis.

Related species:

- *Ardisia maculosa*: cytotoxic activity against human cancer cells
- *Ardisia crispa*: an antimetastatic and cytostatic substance, was isolated from *Ardisia crispa* and identified as 2-methoxy-6-tridecyl-1,4-benzoquinone
- *Ardisia virens*: cytotoxic alkyl benzoquinones and alkyl phenols

References

Chang, C.P., H.S. Chang, C.F. Peng, S.J. Lee, and I.S. Chen. 2011. Antitubercular resorcinol analogs and benzenoid C-glucoside from the roots of *Ardisia cornudentata*. *Planta Medica* 77, no. 1:60–65.

Chang, H.S., Y.J. Lin, S.J. Lee, C.W. Yang, W.Y. Lin, I.L. Tsai, and I.S. Chen. 2009. Cytotoxic alkyl benzoquinones and alkyl phenols from *Ardisia virens. Phytochemistry* 70, no. 17–18:2064–2071.

Chen, L.P., F. Zhao, Y. Wang, L.L. Zhao, Q.P. Li, and H.W. Liu. 2011. Antitumor effect of resorcinol derivatives from the roots of *Ardisia brevicaulis* by inducing apoptosis. *Journal of Asian Natural Products Research* 13, no. 8:734–743.

Pu, L.X., X.H. Yuan, and T.J. Tang. 2009. Analysis of volatile oil from *Ardisia brevicaulis. Zhong Yao Cai =* *Zhongyaocai = Journal of Chinese Medicinal Materials* 32, no. 11:1694–1697.

Zhao, F., Y. Hu, C. Chong, et al. 2014. Ardisiphenol D, a resorcinol derivative identified from *Ardisia brevicaulis*, exerts antitumor effect through inducing apoptosis in human non-small-cell lung cancer A549 cells. *Pharmaceutical Biology* 52, no. 7:797–803.

Zheng, Y., and F.E. Wu. 2007. Resorcinol derivatives from *Ardisia maculosa. Journal of Asian Natural Products Research* 9, no. 6–8:545–549.

Zhu, G.Y., B.C.K. Wong, A. Lu, et al. 2012. Alkylphenols from the roots of *Ardisia brevicaulis* induce G1 arrest and apoptosis through endoplasmic reticulum stress pathway in human non-small-cell lung cancer cells. *Chemical and Pharmaceutical Bulletin* 60, no. 8:1029–1036.

Aristolochia elegans (Aristolochia) (Aristolochiaceae)

Location: South America, with Brazil being its home territory
Appearance:

Stem: slender woody stems twine gracefully in tight coils around fence wire and other supports to lift the vine to heights of 10 or 12 feet.
Root: short horizontal rhizome with numerous long, slender roots below
Leaves: rich glossy green, about 3 inches long by 2 inches wide which grow closely, creating a dense mass of foliage.
Flowers: light green and covered with purple brown spots on the flared lips of the blossom in a pattern reminiscent of calico fabric
In bloom: summer

Parts used: dried rhizome and roots
Active ingredients: *sesquiterpene lactone versicolactone A.*
Indicative dosage and application: it is still under tests.
Documented target cancers: mutagenic activity in the Ames test
Further details:

- *Aristolochia versicolar*: roots contain the sesquiterpene lactone *versicolactone A.*
- Other species*: Aristolochia tagala, Aristolochia rigida.* Two aristolochia acids and a flavonol glycoside have been isolated from *A. rigida*. Only *Aristolochic acid IV* has shown a weak direct mutagenic activity in the Ames test.

References

Konigsbauer, H. 1968. On the usability of *Aristolochia tagala* Cham. in Königsbauer. *Zeitschrift Fur Haut- und Geschlechtskrankheiten* 43, no. 4:159–163.

Pistelli, L., E. Nieri, A.R. Bilia, A. Marsili, and R. Scarpato. 1993. Chemical constituents of *Aristolochia rigida* and mutagenic activity of aristolochic acid IV. *Journal of Natural Products* 56, no. 9:1605–1608.

Zhang, J., L.X. He, H.Z. Xue, R. Feng, and Q.L. Pu. 1991. The structure of versicolactone A from *Aristolochia versicolar S.M. Hwang. Yao Xue Xue Bao = Acta Pharmaceutica Sinica* 26, no. 11:846–851.

Aristolochia mollissima (Aristolochiaceae) (Piperales)   **Cytotoxic**

Location: tropical climate
  Appearance: climbing shrubs, perennial, herbaceous

  Stem: thin, with a series of longitudinal embossments, dense and gray
  Leaves: blade cordate to ovate-cordate, 3.5–10 × 2.5–8 cm, papery, abaxially densely gray to white
    tomentose, adaxially densely strigose, veins palmate, two to three pairs from base, base cordate or
    auriculate, sinus 1–2 cm deep, apex obtuse or acut
  Flowers: single, d in stems 2–4cm long
  In bloom: June to August

Tradition: the root and fruit have been used in traditional Chinese medicine as analgesic, anticancer,
antimalarian, and anti-inflammatory agents and also for the treatment of stomachache, abdominal
pain, and rheumatism.
  Parts used: leaves, flowers, shoots, fruits
  Active ingredients: allantoin, aristolactone, mollislactone, β-sitosterol, aristolochic acid A,
9-ethoxyaristololactam, and 9-ethoxyaristo-lactone
  Precautions: *Aristolochia mollissima* Hance is reported in the literature to have a relatively
strong renal toxicity.
  Documented target cancers: osteosarcoma, the cytotoxicity of the rhizome oil on four cancer
cell lines (ACHN, Bel-7402, Hep G2, and HeLa) is reported to be significantly strong.
  Related species: *Aristolochia ringens* Vahl. (Aristolochiaceae): the ethanolic and dichloromethane: methanol root extracts possess significant anticancer activities *in vitro* and *in vivo*.

References

Akindele, A.J., Z. Wani, G. Mahajan, et al. 2015. Anticancer activity of *Aristolochia ringens* Vahl.
    (Aristolochiaceae). *Journal of Traditional and Complementary Medicine* 5, no. 1:35–41.
Wu, T.S., Y.Y. Chan, and Y.L. Leu. 2000. The constituents of the leaves of *Aristolochia mollissima* Hance.
    Journal of the Chinese Chemical Society 47, no. 4B:957–960.

Aristolochia rotunda (Aristolochiaceae) (Piperales)   **Cytotoxic**

Location: native to Southern Europe
  Appearance: herbaceous perennial plant, characterized as underground rhizomes

  Leaves: heart- or needle-shaped
  Flowers: tubular, yellowish or red

Tradition: used in ancient Egypt as a medicine for snake bite
  Parts used: root and aerial parts
  Active ingredients: aristolochic acid, aristolocin, magnoborin, aristolactams
  Precautions:
  Aristolochic acid, besides being nephrotoxic, is also carcinogenic and mutagenic, and some of
the people who escaped fatal nephrotoxicity have subsequently developed urothelial malignancies.
The use of herbal remedies containing *Aristolochia* species is widespread in Taiwan, and Taiwan
also has the highest incidence of urothelial cancer in the world. All herbal remedies containing

members of this genus should be considered toxic and carcinogenic. *Aristolochia* is banned in Germany, Austria, France, Great Britain, Belgium, and Japan.

Documented target cancers:

- Aristolactam III has been reported to exhibit cytotoxic activity against three kinds of human cancer cell lines (A-549, SK-MEL-2, and SK-OV-3).

References

Kumar, V., Poonam, A.K. Prasad, and V.S. Parmar. 2003. Naturally occurring aristolactams, aristolochic acids and dioxoaporphines and their biological activities. *Natural Product Reports* 20, no. 6:565–583.
Kuo, P.C., Y.C. Li, and T.S. Wu. 2012. Chemical constituents and pharmacology of the Aristolochia (馬兜鈴 mǎdōu ling) species. *Journal of Traditional and Complementary Medicine* 2, no. 4:249–266.

Artemisia argyi (Asteraceae) (Asterales)	Anti-tumor cytotoxic

Location: China, Japan, and the Far East

Appearance: standing, grayish, herbaceous plant about one meter tall, with short branches and rhizomes

Leaves: ovate. The thinner leaves are about six inches long, feathered with wide bony lobes. The upper leaves are smaller and tripartite.

Flowers: the flowers are hermaphrodite and are pollinated by insects. Single flowers are light yellow in color and are tubular.

In bloom: July–October

Tradition: used as antiseptic, expectorant, and antipyretic. Increases blood supply to the pelvic area and stimulates menstruation, helping to treat infertility, dysmenorrhea, asthma, and cough.

Parts used: leaves and fruit

Active ingredients: 2-methylbutanol, 2-hexanal, tricyclene and artemic alcohol, essential oils (quinole and thujone), bitter substances, tannin, resin, insulin, gluconate, vitamins A, B1, B2, C, water-soluble polysaccharide (FAAP-02)

Particular value: the fine fibers of the plant have been used as a binding agent for the seal stamping paste.

Documented target cancers:

- A water-soluble polysaccharide (FAAP-02), isolated from *Artemisia argyi*, has been shown to significantly inhibit the growth of the transplanted tumors and prolonged the survival time of Sarcoma 180 (S180) tumor-bearing mice.
- HT-29 human colon cancer cells
- Cervical cancers associated with the human papilloma virus (SiHa and CaSki cells)
- Precautions: although no reports of toxicity have been seen for this species, skin contact with some members of this genus can cause dermatitis or other allergic reactions in some people.

Further details:

- The antitumor activity of FAAP-02 might be associated with its immunostimulatory effects.
- Essential oil extracted from *A. argyi* leaves decreased melanin production in B16F10 cells and showed potent antioxidant activity.

Related species: *Artemisia princeps* var. *orientalis*: an antioxidant and anticancer agent in hepatocarcinoma cell lines. *Artemisia vulgaris*: all parts of the plant are anthelmintic, antiseptic, antispasmodic, carminative, cholagogue, diaphoretic, digestive, emmenagogue, expectorant, nervine, purgative, stimulant, slightly tonic.

References

Bao, X., H. Yuan, C. Wang, J. Liu, and M. Lan. 2013. Antitumor and immunomodulatory activities of a polysaccharide from *Artemisia argyi*. *Carbohydrate Polymers* 98, no. 1:1236–1243.

Choi, E.J., and G.H. Kim. 2013. Antioxidant and anticancer activity of Artemisia princeps var. orientalis extract in HepG2 and Hep3B hepatocellular carcinoma cells. *Chinese Journal of Cancer Research = Chung-Kuo Yen Cheng Yen Chiu* 25, no. 5:536–543.

Huang, H.C., H.F. Wang, K.H. Yih, L.Z. Chang, and T.M. Chang. 2012. Dual bioactivities of essential oil extracted from the leaves of *Artemisia argyi* as an antimelanogenic versus antioxidant agent and chemical composition analysis by GC/MS. *International Journal of Molecular Sciences* 13, no. 11:14679–14697.

Lee, H.G., K.A. Yu, W.K. Oh, et al. 2005. Inhibitory effect of jaceosidin isolated from *Artemisia argyi* on the function of E6 and E7 oncoproteins of HPV 16. *Journal of Ethnopharmacology* 98, no. 3:339–343.

Arum maculatum (Araceae) (Alismatales)   **Cytotoxic**

Location: tropical and temperate regions
    Appearance: perennial, herbaceous plant, emerging from bulbs

    Leaves: arrow-shaped, deep green, shiny, with red-brown dots
    Flowers: red inflorescences

Tradition: the leaves of *Arum maculatum* are a known folkloric medicine in Jordan. Preparations are used as contraceptive, diuretic, laxative, aphrodisiac, anti-rheumatic, and abortive
    Parts used: dried root (all parts are poisonous)
    Active ingredients: polysaccharides, saponites
    Precautions: there have been reports of *Arum maculatum* poisoning after wild tuber consumption with suicidal intention.
    Documented target cancers:
- Treatment with *A. maculatum* aqueous extract has been reported to remarkably lower the mitotic index in bone marrow cells of Swiss male mice at all exposure times and in almost all concentrations tested compared to controls.

References

Abu-Darwish, M.S., and T. Efferth. 2018. Medicinal plants from Near east for cancer therapy. *Frontiers in Pharmacology* 9:56.

Alencar, V.B., N.M., Alencar, A.M. Assreuy, et al. 2005. Pro-inflammatory effect of *Arum maculatum* lectin and role of resident cells. *International Journal of Biochemistry and Cell Biology* 37, no. 9:1805–1814.

Nabeel, M., S. Abderrahman, and A. Papini. 2008. Cytogenetic effect of *Arum maculatum* extract on the bone marrow cells of mice. *Caryologia* 61, no. 4:383–387.

Prakash Raju, K.N.J., K. Goel, D. Anandhi, V.R. Pandit, R. Surendar, and M. Sasikumar. 2018. Wild tuber poisoning: *Arum maculatum*–A rare case report. *International Journal of Critical Illness and Injury Science* 8, no. 2:111–114.

Brucea antidysenterica (Brucea) (Simaroubaceae)	Cytotoxic

Location: China, Japan
  Parts used: stem
  Active ingredients:

- Cytotoxic: Bruceoside C, bruceanic acid A and its methyl ester 2 (new), bruceanic acid B, C, and D.
- Quassinoid glucosides: bruceosides D, E, and F, bruceantinoside C and yadanziosides G and N, bruceanic acids.
- Alkaloids: 1,11-dimethoxycanthin-6-one, 11-hydroxycanthin-6-one, and canthin-6-one.

Indicative dosage and application: tested in human carcinoma cells at:

- 250 micrograms/ml showed 42% growth inhibition.
- 500 micrograms/ml showed 56% growth inhibition.

The 50% of the results are visible after the first seven hours.
  Documented target cancers: leukemia and non-small cell lung, colon, CNS, melanoma, and ovarian cancer.

- Bruceanic acid D is cytotoxic against P-388 lymphocytic leukemia cells.
- Bruceanic acid A against KB and TE 671 tumor cells, brain metastasis, in lung cancer with radiotherapy.
- Bruceoside C is used against KB, A-549, RPMI, and TE-671 tumor cells.
- The three above mentioned alkaloids are cytotoxic and are used as antileukemic alkaloids.
- *In vivo* studies using RPMI 8226 human-SCID xenografts report that bruceantin induces regression in early as well as advanced tumors.

Further details:

- The fruit of *Brucea javanica* contains quassinoid glucosides, which show selective cytotoxicity in the leukemia and non-small cell lung, colon, CNS, melanoma, and ovarian cancer cell lines with log GI50 values in the range of −4.14 to −5.72. A fruit-derived emulsion inhibited human squamous cell carcinoma cells. At a dose of 250 micrograms/ml at 96 hrs after drug exposure, it showed 42% growth inhibition and at 500 micrograms/ml inhibited 56% of the cell growth. The effect of more than 50% of the growth inhibition was evident at more than 7 h after drug exposure. In the analysis of the mechanism of the drug using a flow cytometry, the arrest in G1 phase of cell cycle was found during incubation of cancer cells with drug.
- The 10% *Brucea javanica* emulsion has synergetic with radiotherapy in treating brain metastasis in lung cancer. Median survival (15 months) of the patients treated was prolonged for 50%.
- In addition, the venous emulsion of BJOE had strong action against the elevation of intracranial pressure produced by SNP (P < 0.01) while oral emulsion had mild action against it, which was similar to the clinical observation exhibiting improvement of clinical manifestations after application of BJOE on intracranial hypertension caused by brain metastasis from lung cancer.
- The stem of *Brucea antidysenterica* contains bruceanic acid A and its methyl ester 2, as well as the bruceanic acids B, C, and D. It also contains three cytotoxic, quassinoid glycosides, bruceantinoside C, and the yadanziosides G and N.
- These species also contain three cytotoxic antileukemic alkaloids, 1,11-dimethoxycanthin-6-one, 11-hydroxycanthin-6-one, and canthin-6-one (Figure 9.5).

Figure 9.5   (See color insert.) *Brucea.*

References

Cuendet, M., and J.M. Pezzuto. 2004. Antitumor activity of bruceantin: An old drug with new promise. *Journal of Natural Products* 67, no. 2:269–272.

Fukamiya, N., M. Okano, T. Aratani, K. Negoro, A.T. McPhail, M. Ju-ichi, and K.H. Lee. 1986. Antitumor agents, 79. Cytotoxic antileukemic alkaloids from *Brucea antidysenterica. Journal of Natural Products* 49, no. 3:428–434.

Fukamiya, N., M. Okano, M. Miyamoto, K. Tagahara, and K.H. Lee. 1992. Antitumor agents, 127. Bruceoside C, a new cytotoxic quassinoid glucoside, and related compounds from *Brucea javanica. Journal of Natural Products* 55, no. 4:468–475.

Fukamiya, N., M. Okano, K. Tagahara, T. Aratani, Y. Muramoto, and K.H. Lee. 1987. Antitumor agents, 90. Bruceantinoside C, a new cytotoxic quassinoid glycoside from *Brucea antidysenterica. Journal of Natural Products* 50, no. 6:1075–1079.

Kupchan, S.M., R.W. Britton, M.F. Ziegler, and C.W. Sigel. 1973. Bruceantin, a new potent antileukemic simaroubolide from *Brucea antidysenterica. Journal of Organic Chemistry* 38, no. 1:178–179.

Lu, J.B., S.Y. Shu, and J.Q. Cai. 1994. Experimental study on effect of *Brucea javanica* oil emulsion on rabbit intracranial pressure. *Chung Kuo Chung HSI i Chieh Ho Tsa Chih* 14, no. 10:610–611.

Ohnishi, S., N. Fukamiya, M. Okano, K. Tagahara, and K.H. Lee. 1995. Bruceosides D, E, and F, three new cytotoxic quassinoid glucosides from *Brucea javanica. Journal of Natural Products* 58, no. 7:1032–1038.

Okano, M., K.H., Lee, and I.H. Hall. 1981. Antitumor agents. 39. Bruceantinoside-A and -B, novel antileukemic quassinoid glucosides from Brucea antidysenterica. *Journal of Natural Products* 44, no. 4:470–474.

Phillipson, J.D., and F.A. Darwish. 1979. TLX-5 lymphoma cells in rapid screening for cytotoxicity in *Brucea* extracts. *Planta Medica* 35, no. 4:308–315.

Phillipson, J.D., and F.A. Darwish. 1981. Bruceolides from Filjian *Brucea javanica. Planta Medica* 41, no. 3:209–220.

Sakaki, T., S. Yoshimura, T. Tsuyuki, T. Takahashi, and T. Honda. 1986. Yadanzioside P, a new antileukemic quassinoid glycoside from *Brucea javanica* (L.) Merr with the 3-O-(beta-D-glucopyranosyl) bruceantin structure. *Chemical and Pharmaceutical Bulletin* 34, no. 10:4447–4450.

Toyota, T., N. Fukamiya, M. Okano, K. Tagahara, J.J. Chang, and K.H. Lee. 1990. Antitumor agents, 118. The isolation and characterization of bruceanic acid A, its methyl ester, and the new bruceanic acids B, C, and D, from *Brucea antidysenterica. Journal of Natural Products* 53, no. 6:1526–1532.

Wang, Z.Q. 1992. Combined therapy of brain metastasis in lung cancer. *Chung Kuo Chung HSI i Chieh Ho Tsa Chih* 12, no. 10:609–610, 581.

Xuan, Y.B., S. Yasuda, K. Shimada, S. Nagai, and H. Ishihama. 1994. Growth inhibition of the emulsion from to *Brucea javanica* cultured human carcinoma cells. *Gan to Kagaku Ryoho. Cancer and Chemotherapy* 21, no. 14:2421–2425.

Bursera simaruba **(Bursera) (Burseraceae)**   **Cytotoxic**

Location: Central and northern South America
    Appearance:

Stem: 6–17 m high, with reddish bark that reveals a smooth and sinuous gray underbark, thick trunk, large irregular branches

Figure 9.6   (See color insert.) *Bursera.*

Leaves: 10–28 cm long with three to seven oval or elliptic leaflets, each 2.5–5 cm long
Flowers: small, inconspicuous, with three to five greenish petals, blooming in elongate racemes
In bloom: winter

Parts used: stem, leaves

Active ingredients: lignans: deoxypodophyllotoxin, beta-peltatin methyl ether, picro-beta-peltatin methyl ether, and dehydro-beta-peltatin methyl ether.

Documented target cancers: lymphocytic leukemia, human epidermoid carcinoma of the nasopharynx. Lignans: deoxypodophyllotoxin (KB, PS test systems), 5′- desmethoxydeoxypodophyllotoxin (morelensin) (KB test system). Sapelins A and B: PS system. Acetone and methanol extracts of Bursera simaruba are reported to be moderately toxic to breast cancer cells ($100\ \mu g/ml < IC_{50} < 200\ \mu g/ml$).

Further details:

* The stem of *Bursera permollis* contains four cytotoxic lignans: deoxypodophyllotoxin, beta-peltatin methyl ether, picro-beta-peltatin methyl ether, and dehydro-beta-peltatin methyl ether. Deoxypodophyllotoxin and another lignan, 5′-desmethoxydeoxypodophyllotoxin, were also isolated from the dried exudate of *Bursera morelensis*. *B. microphylla* also contains deoxypodophyllotoxin.
* The leaves of *Bursera klugii* contain non-polar substances, such as sapelins A and B, which showed activity against two test systems, the P-388 lymphocytic leukemia (3PS) and the human epidermoid carcinoma of the nasopharynx (9KB).
* The isolation and identification of deoxypodophyllotoxin and a new lignan, 5′-desmethoxydeoxypodophyllotoxin, from (Burseraceae) are reported.
* The existence of lignans with antitumor activity in *Bursera schlechtendalii* has been reported (Figure 9.6).

References

Bianchi, E., M.E. Caldwell, and J.R. Cole. 1968. Antitumor agents from *Bursera microphylla* (Burseraceae) I. Isolation and characterization of deoxypodophyllotoxin. *Journal of Pharmaceutical Sciences* 57, no. 4:696–697.

Cates, R.G., B. Prestwich, A. Innes, et al. 2013. Evaluation of the activity of Guatemalan medicinal plants against cancer cell lines and microbes. *Journal of Medicinal Plants Research* 4, no. 35:2616–2627.

Jolad, S.D., R.M. Wiedhopf, and J.R. Cole. 1977a. Cytotoxic agents from *Bursera klugii* (Burseraceae) I: Isolation of sapelins A and B. *Journal of Pharmaceutical Sciences* 66, no. 6:889–890.

Jolad, S.D., R.M. Wiedhopf, and J.R. Cole. 1977b. Cytotoxic agents from Bursera klugii (Burseraceae) I: isolation of sapelins A and B. *Journal of Pharmaceutical Sciences* 66, no. 6:892–893.

McDoniel, P.B., and J.R. Cole. 1972. Antitumor activity of *Bursera schlechtendalii* (Burseraceae): Isolation and structure determination of two new lignans. *Journal of Pharmaceutical Sciences* 61, no. 12:1992–1994.

Wickramaratne, D.B., W. Mar, H. Chai, et al. 1995. Cytotoxic constituents of *Bursera permollis*. *Planta Medica* 61, no. 1:80–81.

Capsicum annuum (Solanaceae) (Solanales)   **Anti-tumor**

Location: native to southern North America and northern South America
  Appearance: shrubby perennial herb with height of 50–75 cm

  Leaves: small, pale green
  Flowers: white, grown individually in groups of two or three

Tradition: source of sweet peppers and hot chili and popular spices such as cayenne, chili, and paprika
  Parts used: fruit
  Active ingredients: capsaicin, capsanthin 3′-ester, capsanthin 3,3′-diester, dihydrocapsaicin, kerketin, luteolin
  Documented target cancers:
  • Capsanthin, capsanthin 3′-ester, and capsanthin 3,3′-diester, major carotenoids in paprika, have been reported to exhibit potent anti-tumor-promoting activity in an *in vivo* mouse skin two-stage carcinogenesis assay using 7, 1 dimethylbenz[a]anthracene as an initiator and TPA as a promoter.

References

Maoka, T., K. Mochida, M. Kozuka, et al. 2001. Cancer chemopreventive activity of carotenoids in the fruits of red paprika *Capsicum annuum* L. *Cancer Letters* 172, no. 2:103–109.

Srinivasan, K. 2016. Biological activities of red pepper (*Capsicum annuum*) and its pungent principle capsaicin: A review. *Critical Reviews in Food Science and Nutrition* 56, no. 9:1488–1500.

Carica papaya (Caricaceae) (Brassicales)   **Cytotoxic**

Location: originated in Central America and is now grown in tropical areas world-wide
  Appearance: giant herbaceous plant that resembles a tree, 5–6m high

  Leaves: limited to the top of the trunk, arranged in coils, large, with a diameter of 50–70 cm
  Flowers: male or female, white
  Fruit: up to 5 kg

Tradition: used in folk medicine to treat hair loss, open wounds, and hypertension. Also used as an aphrodisiac, anti-inflammatory, analgesic, and for reducing gastrointestinal gas.
  Active ingredients: gluco tropaeolin benzyl isothiocyanate, papain, chemopapain

Documented target cancers:

- Papaya juice is reported to cause cell death in the liver cancer cell line Hep G2 with the half maximal inhibitory concentration (IC_{50}) of 20 mg/mL and 22.8 g/mL.
- n-hexane extract from papaya seed and pulp is reported to inhibit the viability of acute promyelotic leukemia HL-60 cells.
- MCF-7, MDA-MB-231, and T47D breast cancer cells
- T-cell lines (H9, Jurkat, Molt-4, CCRF-CEM, and HPB-ALL)
- Burkitt's lymphoma cell lines (Ramos and Raji)
- Chronic myelogenous leukemia cell line (K562)
- Cervical carcinoma cell line (Hela)
- Hepatocellular carcinoma cell lines (HepG2 and Huh-7)
- Lung adenocarcinoma cell line (PC14)
- Pancreatic epithelioid carcinoma cell line (Panc-1)
- Mesothelioma cell lines (H2452, H226, and MESO-4)
- Plasma cell leukemia cell line (ARH77)
- Anaplastic large cell lymphoma cell line (Karpas-299)
- Mesothelioma cell line (JMN)
- Pancreatic adenocarcinoma cell line (Capan-1)

References

Krishna, K.L., M. Paridhavi, and J.A. Patel. 2008. Review on nutritional, medicinal and pharmacological properties of Papaya (*Carica papaya* Linn.). *Indian Journal of Natural Products and Resources* 7, no. 4:364–373.

Nguyen, T.T., P.N. Shaw, M.O. Parat, and A.K. Hewavitharana. 2013. Anticancer activity of *Carica papaya*: A review. *Molecular Nutrition and Food Research* 57, no. 1:153–164.

Vij, T., and Y. Prashar. 2015. A review on medicinal properties of *Carica papaya* Linn. *Asian Pacific Journal of Tropical Disease* 5, no. 1:1–6.

Casearia sylvestris Sw. (Casearia) (Flacourtiaceae)   **Antitumor**

Parts used: leaves

Active ingredients: Clerodane diterpenes: casearins A-F

Documented target cancers:

- Reported elective cytotoxicity against HeLa, A-549, and HT-29 tumor cells.
- Extracts from *Casearia sylvestris* are cytotoxic to MCF-7 cells at high concentrations, while on lower concentrations they have antiproliferative effects.

Further details:
- The structures of the active ingredients mentioned before have been completely elucidated by two-dimensional nuclear magnetic resonance, circular dichroism spectroscopy, X-ray analysis, and chemical evidence.

References

Bou, D.D., J.H.G. Lago, C.R. Figueiredo, A.L. Matsuo, R.C. Guadagnin, M.G. Soares, and P. Sartorelli. 2013. Chemical composition and cytotoxicity evaluation of essential oil from leaves of *Casearia sylvestris*, its main compound α-zingiberene and derivatives. *Molecules* 18, no. 8:9477–9487.

Da Silva, S.L., Jda S. Chaar, and T. Yano. 2009. Chemotherapeutic potential of two gallic acid derivative compounds from leaves of *Casearia sylvestris* Sw (Flacourtiaceae). *European Journal of Pharmacology* 608, no. 1–3:76–83.

Felipe, K.B., M.R. Kviecinski, F.O. da Silva, et al. 2014. Inhibition of tumor proliferation associated with cell cycle arrest caused by extract and fraction from *Casearia sylvestris* (Salicaceae). *Journal of Ethnopharmacology* 155, no. 3:1492–1499.

Itokawa, H., N. Totsuka, H. Morita, K. Takeya, Y. Iitaka, E.P., Schenkel, and M. Motidome. 1990. New antitumor principles, casearins A-F, for *Casearia sylvestris* Sw. (Flacourtiaceae). *Chemical and Pharmaceutical Bulletin* 38, no. 12:3384–3388.

Oberlies, N.H., J.P. Burgess, H.A. Navarro, et al. 2002. Novel bioactive clerodane diterpenoids from the leaves and twigs of *Casearia sylvestris*. *Journal of Natural Products* 65, no. 2:95–99.

Silva, S.Ld, JdS. Chaar, PdM.S. Figueiredo, and T. Yano. 2008. Cytotoxic evaluation of essential oil from *Casearia sylvestris* Sw on human cancer cells and erythrocytes. *Acta Amazonica* 38, no. 1:107–112.

Cassia acutifolia (Cassia, Senna) (Leguminosae)   **Cytotoxic**

Location: Egypt, Nubia, Arabia, and Sennar
  Appearance:

  Stem: erect, smooth, pale green with long spreading branches, 0.70 m high
  Leaves: bearing leaflets in four or five pairs, 1 inch long, lanceolate or obovate, brittle, grayish-green, of a faint, peculiar odor, and mucilaginous, sweetish taste
  Flowers: small, yellow

Parts used: dried leaflets, pods
  Active ingredients: Bitetrahydroanthracene derivative: torosaol-III. Pyranosides. Polysaccharides. Piperidine
  Documented target cancers: KB cells, solid Sarcoma-180 (mice)
  Further details:

- It has been found to contain ingredients that are cytotoxic and DNA damaging.

Related species:

- *Cassia torosa* Cav.: the flowers contain torosaol-III, physcion, 5,7'-physcionanthrone-physcion, 5,7'-biphyscion, torosanin-9,10-quinone, 5,7-dihydroxy-chromone, naringenin, and chrysoeriol. Dimeric tetrahydroanthracenes exhibited cytotoxic activity against KB cells in the tissue culture.
- *Cassia angustifolia* L.: the leaves contain water-soluble polysaccharides, including L-rhamnose, L-arabinose, D-galactose, D-galacturonic acid, and derivatives thereof, exhibiting activity against the solid Sarcoma-180 in CD1 mice.
- *Cassia leptophylla* contains the DNA-damaging compound piperidine.

References

Kitanaka, S., and M. Takido. 1994. Bitetrahydroanthracenes from flowers of *Cassia torosa* Cav. *Chemical and Pharmaceutical Bulletin* 42, no. 12:2588–2590.

Kwon, B.M., S.H., Lee, S.U. Choi, et al. 1998. Synthesis and *in vitro* cytotoxicity of cinnamaldehydes to human solid tumor cells. *Archives of Pharmacal Research* 21, no. 2:147–152.

Lee, C.W., D.H. Hong, S.B. Han, S.H. Park, H.K. Kim, B.M. Kwon, and H.M. Kim. 1999. Inhibition of human tumor growth by 2'-hydroxy- and 2'-benzoyloxycinnamaldehydes. *Planta Medica* 65, no. 3:263–266.

Messana, I., F. Ferrari, M.S. Cavalcanti, and G. Morace. 1991. An anthraquinone and three naphthopyrone derivatives from *Cassia pudibunda*. *Phytochemistry* 30, no. 2:708–710.

Müller, B.M., J. Kraus, and G. Franz. 1989. Chemical structure and biological activity of water-soluble polysaccharides from *Cassia angustifolia* leaves. *Planta Medica* 55, no. 6:536–539.

Chamaecyparis lawsonianna (Cypress hinoki) (Cupressaceae)   **Anti-leukemic**

Location: eastern Asia and North America
    Origin: southern Japan and the island of Taiwan
    Appearance:

> *Stem*: reddish evergreen conifer with attractive soft and stringy brown bark, cypress can get over 3 m
> tall with a trunk diameter of 12 cm
> *Leaves*: drooping flat frondlike branchlets bearing small scalelike leaves. Has two kinds of leaves: adult
> leaves are like closely adpressed overlapping scales; leaves on juvenile branchlets and young plants
> don't overlap and are shaped more like tiny awls or broad needles. The scalelike leaves are borne in
> pairs of two unequal sizes and shapes.

Degree of rarity: low. The typical form of the Hinoki false cypress is rarely cultivated, and most
gardeners are more familiar with one or more of the many dwarf cultivars selected for size, form,
and foliage color.
    Tradition: is used as specimens and for hedging, screening, and windbreaks
    Active ingredients: Alkaloids; hinokitiol, tropolone
    Documented target cancers: high potency in the P388 leukemia assay
    Further details:
- Tropolone derivatives prepared from hinokitiol, which naturally occurs in the plants of
  Chamaecyparis species, show high potency in the P388 leukemia assay. It preferentially inhibits the
  soluble guanylate cyclase from leukemic lymphocytes. This inhibition correlates with its preferential cytotoxic effects for these same cells, since cyclic GMP is thought to be involved in lymphocytic
  cell proliferation and leukemogenesis and, in general, the nucleotide is elevated in leukemic vs
  normal lymphocytes and changes have been reported to occur during remission and relapse of this
  disease (Figure 9.7).

Figure 9.7   (See color insert.) *Chamaecyparis.*

References

Takemoto, D.J., C. Dunford, D. Vaughn, K.J. Kramer, A. Smith, and R.G. Powell. 1982. Guanylate cyclase activity in human leukemic and normal lymphocytes. Enzyme inhibition and cytotoxicity of plant extracts. *Enzyme* 27, no. 3:179–188.

Yamato, M., K. Hashigaki, N. Kokubu, T. Tashiro, and T. Tsuruo. 1986. Synthesis and antitumor activity of tropolone derivatives. 3. *Journal of Medicinal Chemistry* 29, no. 7:1202–1205.

Chelidonium majus L. (Chelodonium, Celandine) (Papaveraceae)   **Immunomodulatory**

Location: found by old walls, on waste ground, and in hedges, nearly always in the neighborhood of human habitations

Appearance:

Stem: slender, round, and slightly hairy, 0.5–1 m high, much branched

Root: thick, fleshy

Leaves: yellowish-green, much paler, almost grayish below, graceful in form and slightly hairy, 15–30 cm long, 5–7.5 cm wide, deeply divided as far as the central rib, so as to form usually two pairs of opposite leaflets with rounded teeth edges

Flowers: arranged at the ends of the stems in loose umbels

In bloom: summer

Tradition: it was used as a drug plant since the Middle Ages, and Dioscorides and Pliny mention it. It was used to take away specks from the eye and to stop incipient suffusions. It is useful, also, as alterative, diuretic, purgative, in jaundice, eczema, and scrofulous diseases.

Parts used: the whole herb

Active ingredients: Alkaloids: chelidonine and its semi-synthetic compound: Tris(2-([5bS-(5ba,6b,12ba)]-5b,6,7,12b,13,14-Hexahydro-13-methyl][1,3]-benzodioxolo[5,6-c]-1,3-dioxolo[4, 5- i] phenanthridinium-6-ol]-Ethaneaminyl) Phosphinesulfide 6HCl (Ukrain).

Particular value: although Ukrain of high concentrations is cytostatic for malignant cells and may suppress the growth of cancer, it is not cytostatic in normal concentrations.

Indicative dosage and application:

- Every second day in a dose of 10 mg per injection. Each patient receives 300 mg of the drug (30 injections).
- In lung cancer it is used in an intravenous injection every three days. One course consists of ten applications of 10 mg each.

Documented target cancers:

- It has been reported that the herb extract of *Chelidonium majus* showed preventive effects on glandular stomach tumor development in rats treated with N-methyl-N'-nitro-N nitrosoguanidine (MNNG) and hypertonic sodium chloride.
- The incidence of forestomach neoplastic lesions (papillomas and squamous cell carcinomas) also showed a tendency to decrease with the herbal extract treatment.
- The extract has been reported to induce apoptosis on human epidermoid carcinoma A431 cells.

Further details:

- Ukrain is a semi-synthetic thiophosphoric acid compound of alkaloid chelidonine isolated from *Chelidonium majus* L. Its full chemical name is Tris(2-([5bS-(5ba,6b,12ba)]-5b,6,7,12b,13,14- Hexah ydro-13-methyl][1,3]-benzodioxolo[5,6-c]-1,3-dioxolo[4,5-i]phenanthridinium-6ol]Ethaneaminyl),

Phosphinesulfide6HCl. Ukrain causes a regression of tumors and metastases in many oncological patients. More than 400 documented patients with various carcinomas in different stages of development have been treated with Ukrain. J.W. Nowicky produced Ukrain for the first time in 1978. (Austrian Patent No. 354644, Vienna, 25 January 1980). Ukrain can be immunologically effective in lung cancer patients and can improve human cellular response.

- Ukrain was applied as an intravenous injection every three days on nine men (aged 42–68 years, mean 57 years) with histologically proven lung cancer, previously untreated. One course consisted of ten applications of 10 mg each. The treatment was generally well tolerated. The results showed an increase in the proportion of total T-cells and a significant decrease in the percentage of T-suppressor cells. There were no signs of activation of NK, T-helper, and B-cells. The restoration of cellular immunity was accompanied by an improvement in the clinical course of the disease. This effect was particularly pronounced in patients who responded to further chemotherapy. Objective tumor regression was seen in 44.4% of treated patients. Four out of nine patients (44.4%) died of progressive disease during the course of this study.

- Thirty-six stage III cancer patients were treated with Ukrain. The drug was injected intravenously every second day in a dose of 10 mg per injection. Each patient received 300 mg of the drug (30 injections). The cytostatic effect of Ukrain was monitored clinically and by ultrasonography (USG) and computer tomography (CT), as well as by determination of CEA and CA-125 in the sera of patients with rectal and ovarian cancers, respectively. The influence of Ukrain on immune parameters was evaluated by monoclonal antibodies (MAb) to CD2, CD4, CD8, and CD22. The influence of Ukrain on immune parameters in cancer patients was matched with its effect on these parameters in 20 healthy volunteer controls. The results obtained indicate that Ukrain, in a concentration not cytostatic in normal cells, is cytostatic for malignant ones and may suppress the growth of cancer. The compound also has immunoregulatory properties, regulating the T lymphocyte subsets.

- The effect of Ukrain on the growth of Balb/c syngenic mammary adenocarcinoma was assessed. Intravenous, but not subcutaneous or intraperitoneal, administration of this drug was found to be effective in delaying tumor growth in an actual therapeutic protocol initiated five days after tumor implantation. No untoward side effects were observed using these *in vivo* treatment modalities. Ukrain's *in vivo* effects against the development of mammary tumors may be due, at least in part, to its ability to restore macrophage cytolytic function.

- For the treatment of AIDS patients with Kaposi's sarcoma, Ukrain was injected i.v. in the dose of 5 mg every other day for a total of ten injections. During treatment the Kaposi's sarcoma lesions diminished in size, showed decoloration, and no lesion appeared in the 30-day interval after the beginning of treatment. Both patients tolerated Ukrain well and showed an improved immunohaematological status: an increase in total leukocytes, T-lymphocytes, and T-suppressor numbers. In one case T-helper lymphocytes were also increased.

- Ukrain is an effective biological response modifier augmenting, by up to 48-fold, the lytic activity of splenic lymphocytes obtained from alloimmunized mice. The lytic activities of interleukin-2 (IL-2) treated spleen cells and peritoneal exudate lymphocytes were also significantly increased by the addition of Ukrain to the cell mediated lysis (CML) assay medium. The highest Ukrain-induced enhancement of splenic lymphocytolytic activity *in vitro* was found to occur at day 18 after alloimmunization and was dose-dependent and specific for the immunizing P815 tumor cells. Since *Ukrain* was present only during the CML assays, its mode of action is thought to be via direct activation of the effector cells' lytic mechanism(s). The effect of Ukrain on the growth of Balb/c syngenic mammary adenocarcinoma was also evaluated. Intravenous, but not subcutaneous or intraperitoneal, administration of this drug was found to be effective in delaying tumor growth in an actual therapeutic protocol initiated five days after tumor implantation. No deleterious side effects were observed using these *in vivo* treatment modalities. The role of macrophages in the observed retardation of tumor development was investigated, using peritoneal exudate macrophages (PEM) in cytotoxicity assays. Previous studies showed that PEM of mammary tumor-bearing mice lose their capacity to kill a variety of tumor target cells including the *in vitro* cultured homologous tumor cells (DA-3). Pretreatment of PEM from normal mice with 2.5 µM Ukrain for 24 h, followed by stimulation with either IFN-gamma or with lipopolysaccharide (LPS) plus IFN-gamma enhanced their cytotoxic activity. Treatment of PEM from tumor-bearing mice with 2.5 µM Ukrain and LPS results in a reversal of their defective cytotoxic response against DA-3 target cells. Furthermore, Ukrain alone, in the absence of a secondary signal, induced the activation of tumoricidal function of PEM from tumor-bearing, but not from normal, mice. These data indicate that Ukrain's *in vivo* effects against the development of mammary tumors may be due, at least in part, to its ability to restore macrophage cytolytic function.

References

Brüller, W. 1992. Studies concerning the effect of ukrain *in vivo* and *in vitro*. *Drugs under Experimental and Clinical Research* 18:13–16.

Ciebiada, I., E. Korczak, J.W. Nowicky, and A. Denys. 1996. Estimation of direct influence of ukrain preparation on influenza viruses and the bacteria *E. coli* and *S. aureus*. *Drugs under Experimental and Clinical Research* 22, no. 3–5:219–223.

Ciebiada, I., E. Korczak, J.W. Nowicky, and A. Denys. 1996. Does the ukrain preparation protect mice against lethal doses of bacteria? *Drugs under Experimental and Clinical Research* 22, no. 3–5:207–211.

Ebermann, R., G. Alth, M. Kreitner, and A. Kubin. 1996. Natural products derived from plants as potential drugs for the photodynamic destruction of tumor cells. *Journal of Photochemistry and Photobiology: B, Biology* 36, no. 2:95–97.

Ernst, E., and K. Schmidt. 2005. Ukrain–a new cancer cure? A systematic review of randomised clinical trials. *BMC Cancer* 5, no. 1.

Kim, D.J., B. Ahn, B.S. Han, and H. Tsuda. 1997. Potential preventive effects of *Chelidonium majis* L. (Papaveraceae) herb extract on glandular stomach tumor development in rats treated with N-methyl-N'-nitro-N nitrosoguanidine (MNNG) and hypertonic sodium chloride. *Cancer Letters* 112, no. 2:203–208.

Liepins, A., and J.W. Nowicky. 1992. Activation of spleen cell lytic activity by the alkaloid thiophosphoric acid derivative: Ukrain. *International Journal of Immunopharmacology* 14, no. 8:1437–1442.

Liepins, A., and J.W. Nowicky. 1996. Modulation of immune effector cell cytolytic activity and tumour growth inhibition *in vivo* by ukrain (NSC 631570). *Drugs under Experimental and Clinical Research* 22, no. 3–5:103–113.

Lohninger, A., and F. Hamler. 1992. Chelidonium *majus* L. (ukrain) in the treatment of cancer patients. *Drugs under Experimental and Clinical Research* 18:73–77.

Malaveille, C., M. Friesen, A.M. Camus, et al. 1982. Mutagens produced by the pyrolysis of opium and its alkaloids as possible risk factors in cancer of the bladder and oesophagus. *Carcinogenesis* 3, no. 5:577–585.

Nowicky, J.W., G. Manolakis, D. Meijer, V. Vatanasapt, and W.J. Brzosko. 1992. Ukrain both as an anti cancer and immunoregulatory agent. *Drugs under Experimental and Clinical Research* 18:51–54.

Nowicky, J.W., A. Staniszewski, W. Zbroja-Sontag, B. Slesak, W. Nowicky, and W. Hiesmayr. 1991. Evaluation of thiophosphoric acid alkaloid derivatives from *Chelidonium majus* L. ('ukrain') as an immunostimulant in patients with various carcinomas. *Drugs under Experimental and Clinical Research* 17, no. 2:139–143.

Park, S.W., S.R. Kim, Y. Kim, J.H. Lee, H.J. Woo, Y.K. Yoon, and Y.I. Kim. 2015. Chelidonium majus L. extract induces apoptosis through caspase activity via MAPK-independent NF-κB signaling in human epidermoid carcinoma A431 cells. *Oncology Reports* 33, no. 1:419–424.

Ranadive, K.J., S.V. Gothoskar, and B.U. Tezabwala. 1972. Carcinogenicity of contaminants in indigenous edible oils. *International Journal of Cancer* 10, no. 3:652–666.

Ranadive, K.J., S.V. Gothoskar, and B.U. Tezabwala. 1973. Testing carcinogenicity of contaminants in edible oils. II. Argemone oil in mustard oil. *Indian Journal of Medical Research* 61, no. 3:428–434.

Shi, G.Z. 1992. Blockage of *Glyrrhiza uralensis* and *Chelidonium majus* in MNNG induced cancer and mutagenesis. *Chung-Hua Yu Fang i Hsueh Tsa Chih* 26, no. 3:165–167.

Slesak, B., J.W. Nowicky, and A. Harłozinska. 1992. *In vitro* effects of *Chelidonium majus* L. alkaloid thiophosphoric acid conjugates (ukrain) on the phenotype of normal human lymphocytes. *Drugs under Experimental and Clinical Research* 18:17–21.

Sotomayor, E.M., K. Rao, D.M. Lopez, and A. Liepins. 1992. Enhancement of macrophage tumouricidal activity by the alkaloid derivative ukrain. *In vitro* and *in vivo* studies. *Drugs under Experimental and Clinical Research* 18:5–11.

Staniszewski, A., B. Slesak, J. Kołodziej, A. Harłozińska-Szmyrka, and J.W. Nowicky. 1992. Lymphocyte sub-sets in patients with lung cancer treated with thiophosphoric acid alkaloid derivatives from *Chelidonium majus* L. (ukrain). *Drugs under Experimental and Clinical Research* 18:63–67.

Steinacker, J., T. Kroiss, O.B. Korsh, and A. Melnyk. 1996. Ukrain therapy in a frontal anaplastic grade III astrocytoma (case report). *Drugs under Experimental and Clinical Research* 22, no. 3–5:275–277.

Voltchek, I.V., A. Liepins, J.W. Nowicky, and W.J. Brzosko. 1996. Potential therapeutic efficacy of ukrain (NSC 631570) in AIDS patients with Kaposi's sarcoma. *Drugs under Experimental and Clinical Research* 22, no. 3–5:283–286.

Xian, M.S., K. Hayashi, J.P. Lu, and M. Awai. 1989. Efficacy of traditional Chinese herbs on squamous cell carcinoma of the esophagus: Histopathologic analysis of 240 cases. *Acta Medica Okayama* 43, no. 6:345–351.

Chrysanthemum

See in *Glycyrriza* under further details.

Cinnamomum camphora (Cinnamomum) (Lauraceae)	Cytotoxic
	Immunomodulator

Location: East Asia
Appearance:

Stem: 20–40 m, many branched, evergreen
Leaves: evergreen with oval oblong blades
Flowers: white, small, and clustered
In bloom: spring

Degree of rarity: low, as it can be cultivated successfully in sub-tropical countries

Tradition: Chinese use the Camphor oil exudes in the process of extracting Camphor for many centuries. It was mentioned by Marko Polo in the thirteenth century and Camoens in 1571, who called it the 'balsam of disease'. Very useful in complaints of stomach and bowels, in spasmodic cholera, and flatulent colic.

Parts used: gum

Active ingredients: Cinnamaldehydes: 2'-Hydroxycinnamaldehyde (HCA) and 2'-benzoxy-cinnamaldehyde (BCA)

Precautions: in large doses it is very poisonous. Should be used cautiously in certain heart disease.

Documented target cancers: human cancer cells lines, SW-620 human tumor xenograft
Further details:

- The species is cytotoxic (the key functional group of the cinnamaldehyde-related compounds for their antitumor activity is the propenal group).
- Immunomodulation is effected due to the inhibition of farnesyl protein transferase. RAS activation, which is accompanied with its farnesylation, has been known to be important in immune cell activation as well as in carcinogenesis. Extracts inhibit the lymphoproliferation and induce a T-cell differentiation through the blockade of early steps in signaling pathway leading to cell growth.
- *Cinnamomum cassia* Blume (Lauraceae): the bark contains 2'-hydroxycinnamaldehyde which reacts with benzoyl chloride in order to give 2'-benzoyloxycinnamaldehyde. Both compounds strongly inhibited *in vitro* growth of 29 kinds of human cancer cells and *in vivo* growth of SW-620 human tumor xenograft without the loss of body weight in nude mice.

- Two kinds of cinnamaldehyde derivative, 2'-hydroxycinnamaldehyde (HCA) and 2'-benzoxy-cin-namaldehyde (BCA), were studied for their immunomodulatory effects. These compounds were screened as anticancer drug candidates from stem bark of *Cinnamomum cassia* for their inhibitory effect on activity. Treatment of these cinnamaldehydes to mouse splenocyte cultures induced suppression of lymphoproliferation following both Con A and LPS stimulation in a dose-dependent manner. A dose of I μM of HCA and BCA inhibited the Con A-stimulated proliferation by 69% and 60% and the LPS-induced proliferation by 29% and 21%, respectively. However, the proliferation induced by PMA plus ionomycin was affected by neither HCA nor BCA treatment. Decreased levels of antibody production by HCA or BCA treatment were observed in both SRBC-immunized mice and LPS-stimulated splenocyte cultures. The exposure of thymocytes to HCA or BCA for 48 h accelerated T-cell differentiation from CD4 and CD8 double positive cells to CD4 or CD8 single positive cells. The inhibitory effect of cinnamaldehyde on lymphoproliferation was specific to the early phase of cell activation, showing the strongest inhibition of Con A- or LPS-stimulated proliferation when added concomitantly with the mitogens. In addition, the treatment of HCA and BCA to splenocyte cultures attenuated the Con A-triggered progression of cell cycle at G1 phase with

Cinnamomum camphora (L.) J. Presl

Figure 9.8 *Cinnamomum camphora.*

no inhibition of S to G2/M phase transition. Although cinnamaldehyde treatment had no effect on the IL-2 production by splenocyte cultures stimulated with Con A, it inhibited markedly and dose-dependently the expression of IL-2Ralpha and interferon-gamma. Taken together, the results in this study suggest both HCA and BCA (Figure 9.8).

References

Balachandran, B., and V.M. Sivaramkrishnan. 1995. Induction of tumours by Indian dietary constituents. *Indian Journal of Cancer* 32, no. 3:104–109.

Chen, C.H., S.W. Yang, and Y.C. Shen. 1995. New steroid acids from *Antrodia cinnamomea*, a fungal parasite of *Cinnamomum micranthum*. *Journal of Natural Products* 58, no. 11:1655–1661.

Choi, J., K.T. Lee, H. Ka, W.T. Jung, H.J. Jung, and H.J. Park. 2001. Constituents of the essential oil of the *Cinnamomum cassia* stem bark and the biological properties. *Archives of Pharmacal Research* 24, no. 5:418–423.

Haranaka, K., N. Satomi, A. Sakurai, R. Haranaka, N. Okada, and M. Kobayashi. 1985. Antitumor activities and tumor necrosis factor producibility of traditional Chinese medicines and crude drugs. *Cancer Immunology, Immunotherapy* 20, no. 1:1–5.

Haranaka, R., R. Hasegawa, S. Nakagawa, A. Sakurai, N. Satomi, and K. Haranaka. 1988. Antitumor activity of combination therapy with traditional Chinese medicine and OK432 or MMC. *Journal of Biological Response Modifiers* 7, no. 1:77–90.

Ikawati, Z., S. Wahyuono, and K. Maeyama. 2001. Screening of several Indonesian medicinal plants for their inhibitory effect on histamine release from RBL-2H3 cells. *Journal of Ethnopharmacology* 75, no. 2–3:249–256.

Kwon, B.M., S.H. Lee, S.U. Choi, et al. 1998. Synthesis and *in vitro* cytotoxicity of cinnamaldehydes to human solid tumor cells. *Archives of Pharmacal Research* 21, no. 2:147–152.

Lee, C.W., D.H. Hong, S.B. Han, S.H. Park, H.K. Kim, B.M. Kwon, and H.M. Kim. 1999. Inhibition of human tumor growth by 2-hydroxy- and 2-benzoyloxycinnamaldehydes. *Planta Medica* 65, no. 3:263–266.

Ling, J., and W.Y. Liu. 1996. Cytotoxicity of two new ribosome-inactivating proteins, cinnamomin and camphorin, to carcinoma cells. *Cell Biochemistry and Function* 14, no. 3:157–161.

Mihail, R.C. 1992. Oral leukoplakia caused by cinnamon food allergy. *Journal of Otolaryngology* 21, no. 5:366–367.

Moayedi, Y., S.A. Greenberg, B.A. Jenkins, et al. 2018. Camphor white oil induces tumor regression through cytotoxic T cell-dependent mechanisms. *Molecular Carcinogenesis*.https://doi.org/10.1101/386789

Sakamoto, S., H. Yoshino, Y. Shirahata, K. Shimodairo, and R. Okamoto. 1992. Pharmacotherapeutic effects of kuei-chih-fu-ling-wan (keishi-bukuryo-gan) on human uterine myomas. *American Journal of Chinese Medicine* 20, no. 3–4:313–317.

Sedghizadeh, P.P., and C.M. Allen. 2002. White plaque of the lateral tongue. *Journal of Contemporary Dental Practice* 3, no. 3:46–50.

Westra, W.H., J.S. McMurray, J. Califano, P.W. Flint, and R.L. Corio. 1998. Squamous cell carcinoma of the tongue associated with cinnamon gum use: A case report. *Head and Neck* 20, no. 5:430–433.

Zee-Cheng, R.K. 1992. Shi-quan-da-bu-tang (ten significant tonic decoction), SQT. A potent Chinese biological response modifier in cancer immunotherapy, potentiation and detoxification of anticancer drugs. *Methods and Findings in Experimental and Clinical Pharmacology* 14, no. 9:725–736.

***Cistus creticus* (Cistaceae) (Malvales)**   **Cytotoxic**

Location: Mediterranean region
  Appearance: bush, 30–50 cm

  Leaves: opposite, deep green, speared with wavy fringes
  Flowers: purple, on the peaks of the tender twigs, with five wide petals, long and always tatty

Figure 9.9   (See color insert.) *Cistus creticus.*

Tradition: used as antimicrobial, antiseptic, controls bleeding, remedy for snuffles and diarrhea
    Parts used: leaves
    Active ingredients: labdane type diterpenes
    Documented target cancers:
    Shoot extract of *Cistus creticus* is reported to have cytotoxic activity on HeLa (cervix), MDA-MB-453 (breast), and FemX (melanoma) cancer cells (Figure 9.9).

Reference

Skorić, M., S. Todorović, N. Gligorijević, R. Janković, S. Živković, M. Ristić, and S. Radulović. 2012. Cytotoxic activity of ethanol extracts of *in vitro* grown *Cistus creticus* subsp. creticus L. on human cancer cell lines. *Industrial Crops and Products* 38:153–159.

Citrus maxima **(Rutaceae) (Sapindales)**   **Anti-tumor**

Location: South and Southeast Asia
    Appearance: evergreen fruit tree

    Leaves: oval, elliptical or extended, 5–20 cm
    Flowers: fragrant, white or yellow, 1.5–3.5 cm diameter, hairy

Parts used: blossom, seeds, fruit, leaves
  Tradition:

- The fruit is widely used in confectionery. In the native countries, leaf, flower, and seed extracts are used as tranquilizers and painkillers in cases of epilepsy, dyskinesia, and extensive cough.
- In Vietnam, its flowers are used to make perfumes. Its wood is suitable for the construction of tools.
- The pulp is used as appetizer, antitoxic, cardiac stimulant, and stomach tonic.

Active ingredients: neohesperidin and naringin, α-pinene, sabenine, geranyl formate, Z-citral, E-citral, geranyl acetate, β-farnesene
  Indicative dosage and application:

- 200 and 400 mg/kg significantly decreased tumor parameters in Ehrlich's ascites carcinoma in Swiss albino mice.

Documented target cancers:
- Methanol extract of *Citrus maxima* leaves significantly decreased tumor parameters such as tumor volume, viable tumor cell count in Ehrlich's ascites carcinoma in Swiss albino mice.

Reference

Kundusen, S., M. Gupta, U.K. Mazumder, P.K. Haldar, P. Saha, and A. Bala. 2011. Antitumor activity of *Citrus maxima* (Burm.) Merr. Leaves in Ehrlich's ascites carcinoma cell-treated mice. *ISRN Pharmacology* 2011:138737.

Citrus unshiu (Rutaceae) (Sapindales)	Anti-tumor	Antiproliferative

Location: Japan, China
  Appearance: mandarin tree
  Parts used: fruit, peel
  Active ingredients: flavonoids, hesperedin,3′,4′,5,6,7,8-hexamethoxyflavone (nobiletin), and naringin
  Tradition: it has been used as traditional medicine in Korea, China, and Japan. It is divided as immature peel or mature peel. *C. unshiu* immature peel (Cheong pi or Chung-pi in Korea, Qing pi in China, Jyōhi in Japan) has been used for relief of pleuralgia caused by liver disease and remedy of indigestion. On the other hand, *C. unshiu* mature peel (Jin pi in Korea, Chen pi in china, Chinpi in Japan) has been used for improving bronchial, asthmatic conditions, cardiac, and blood circulation in these countries.
  Documented target cancers:

- *Citrus unshiu* pulp is reported to inhibit chemically induced colon, tongue, and lung tumorigenesis in animal models.
- The methanol extract has been shown to have inhibitory activity on cancer cells *in vitro*. The most important observation of its antiproliferative activity has been made in human gastric adenocarcinoma.
- The water extract of *C. unshiu* peel has been shown to suppress breast cancer (MCF-7) cell proliferation by activating the intrinsic and extrinsic apoptosis pathways through ROS-dependent AMPK pathway activation.

References

Kim, M.Y., H.H. Bo, E.O. Choi, et al. 2018. Induction of apoptosis by *Citrus unshiu* Peel in human breast cancer MCF-7 cells: Involvement of ROS-dependent activation of AMPK. *Biological and Pharmaceutical Bulletin* 41, no. 5:713–721.

Song, Y.W., S. Shrestha, R. Gyawali, D.S. Lee, and S.K. Cho. 2015. *Citrus unshiu* leaf extract containing phytol as a major compound induces autophagic cell death in human gastric adenocarcinoma AGS cells. *Journal of the Korean Society for Applied Biological Chemistry* 58, no. 2:257–265.

Tanaka, T., M. Tanaka, T. Kuno, and T. Kuno. 2012. Cancer chemoprevention by citrus pulp and juices containing high amounts of β-cryptoxanthin and hesperidin. *Journal of Biomedicine and Biotechnology* 2012:516981.

***Cladonia furcata* (Cladoniaceae) (Lecanorales)**   **Cytotoxic**

Location: North America
  Appearance: lichen
  Parts used: thalli
  Active ingredients: vulpinic acid, lecanoric acid, gyrophoric acid, salazinic acid, lobaric acid, evernic acid, usnic acid, protolichesterinic acid
  Documented target cancers:

- Human melanoma FemX and human colon carcinoma LS174 cell lines
- Human leukemia K562 and HL-60 cells

References

Lin, X., Y.J. Cai, Z.X. Li, Q. Chen, Z.L. Liu, and R. Wang. 2003. Structure determination, apoptosis induction, and telomerase inhibition of CFP-2, a novel lichenin from *Cladonia furcata*. *Biochimica et Biophysica Acta* 1622, no. 2:99–108.

Lin, X., Y.J. Cai, Z.X. Li, Z.L. Liu, S.F. Yin, and J.C. Zhao. 2001. *Cladonia furcata* polysaccharide induced apoptosis in human leukemia K562 cells. *Acta Pharmacologica Sinica* 22, no. 8:716–720.

Ranković, B.R., M.M. Kosanić, and T.P. Stanojković. 2011. Antioxidant, antimicrobial and anticancer activity of the lichens *Cladonia furcata*, *Lecanora atra* and *Lecanora muralis*. *BMC Complementary and Alternative Medicine* 11, no. 1:97.

***Clerodendrum infortunatum* (Lamiaceae) (Lamiales)**   **Anti-tumor**

Location: Asia, Bangladesh, India, Myanmar, Pakistan, Thailand, Malaysia, the Andaman Islands, and Sri Lanka
  Appearance: perennial shrub

Stem: 0.5–5 m
Leaves: simple, hairy, elliptic, serrated, 3.5–20 cm width, and 6–25 length
Flowers: white, pink, purple, hairy
In bloom: April to August

Tradition: in traditional Ayurvedic and Siddha medicine, the leaves and roots of *Clerodendrum* are used as a treatment for hair loss, asthma, cough, diarrhea, rheumatism, fever, and various skin diseases. It is also known for its tampering and antimicrobial action. A mixture of leaves and roots is applied externally to skin diseases, mainly fungal infections. Fresh leaves help in diarrhea, headaches, and liver disorders.

Parts used: leaves, roots

Active ingredients: clerodinin A, oleanolic acid

Documented target cancers:

- *In vivo* investigations in mice suffering from Elrich's carcinoma have shown that the presence of Clerodinin A and Oleanolic acid compounds in *Clerodendrum infortunatum* methanol extract resulted in a significant tumor reduction. The life expectancy of the treated mice was also increased by about 27 days.

Reference

Sannigrahi, S., U.K. Mazumder, D. Pal, and S.L. Mishra. 2012. Terpenoids of methanol extract of *Clerodendrum infortunatum* exhibit anticancer activity against Ehrlich's ascites carcinoma (EAC) in mice. *Pharmaceutical Biology* 50, no. 3:304–309.

Cocculus hirsutus (Menispermaceae) (Ranunculales)	Anti-lymphomatic

Location: India

Appearance: climbing scandent shrub with hairy sepals

Leaves: simple, alternate, hairy, with three lobes, obtuse
Flowers: hairy
In bloom: February to March

Tradition: the root is bitter and used as a laxative, soothing, tonic, diuretic, antipyretic, against malaria and joint pain, as ointment in skin diseases, and kidney problems. The juice of the leaves coagulates in the water and forms a substance that is used for eye problems, eczema, infections, and dyspepsia. In combination with sugar it is used for acute gonorrhea. Portions of the root are mixed with pepper for chronic rheumatism and cachexia. Combination of plant roots of the same genus, dissolved in water, relieves abdominal pains. The roots are also used as aphrodisiac, tonic, and drugs for stomach upsets especially in children.

Parts used: roots, leaves

Active ingredients: essential oil, β-sitosterol, ginnol, glycosides, sterols, and alkaloids, D-trilobine, DL-coclaurine, isotrilobine, syringaresinol, protoquericitol

Documented target cancers:

- Dalton lymphoma cells, both *in vitro* and *in vivo* (mice infected with lymphoma). Significant reduction in viability and growth of cancer cells was observed, as well as increased life expectancy of infected mice.

Related species:

- In folk medicine of India, combination of plant roots of the *Cocculus* genus, dissolved in water, relieves abdominal pains.

References

Marya, H., and B. Bothara. 2011. Ethnopharmacological properties of *Cocculus hirsutus* (L.) Diels – A review. *International Journal of Pharmaceutical Sciences Review and Research* 7, no. 1:108–112.
Thavamani, B.S., M. Mathew, and D.S. Palaniswamy. 2014. Anticancer activity of *Cocculus hirsutus* against Dalton's lymphoma ascites (DLA) cells in mice. *Pharmaceutical Biology* 52, no. 7:867–872.

***Codonopsis pilosula* (Campanulaceae) (Asterales)**   **Cytotoxic**

Location: native to Asia
   Appearance: flowering plant

  Stem: twin stems up to 2 meters high
  Leaves: alternating, oval, up to 7.3 cm long, hairy
  Flowers: on the edge of the branches, hermaphrodite, insect-pollinated, bell-shaped, length up to 2 cm,
    and yellow-green with pink pearls inside
  In bloom: June to August

Parts used: roots
  Active ingredients: codonopyrrolidium B, hesperidine, alpha-sinasterol, 5-hydroxymethylfurfu-
ral, perlolyrine, taraxerol, taraxeryl acetate, squalene sitosterol, and beta-daucosterol
  Tradition: it is widely used in traditional Chinese medicine.
  Documented target cancers:

- *In vitro* studies in human ovarian cancer cells (human epithelial ovarian cancer HO-8910 cell line)
  have shown that treatment with an acidic polysaccharide derived from the root of *C. pilosula* has led
  to a reduction in invasive capacity as well as the ability of cancer cells to migrate.
- A pectic polysaccharide from *C. pilosula* has been reported to exhibit obvious cytotoxicity to human
  lung adenocarcinoma A549 cells in a dose- and time-dependent manner.

References

Xin, T., F. Zhang, Q. Jiang, et al. 2012. The inhibitory effect of a polysaccharide from *Codonopsis pilosula*
   on tumor growth and metastasis *in vitro*. *International Journal of Biological Macromolecules* 51, no.
   5:788–793.
Yang, C., Y. Gou, J. Chen, J. An, W. Chen, and F. Hu. 2013. Structural characterization and antitumor activity
   of a pectic polysaccharide from *Codonopsis pilosula*. *Carbohydrate Polymers* 98, no. 1:886–895.

***Colchicum autumnale* (Meadow saffron) (Liliaceae)**   **Cytotoxic**

Synonyms: Autumn Crocus, Naked Ladies
  Location: Southern and Central Europe, in meadows and deciduous woods
  Appearance:

  Root: scaly corm, up to 7 cm
  Leaves: basal, linear-lanceolate, up to 40 cm long
  Flowers: long-tubed purple or white, directly emerging from the underground corm. They share a
    resemblance to the flowers of *Crocus sativus*, but they possess six anthers.
  Fruit: oval capsule
  In bloom: August–October

Tradition: considered to be the Hermodactyls of the Arabians, it has been used against rheumatism
and gout.
  Parts used: root, seeds
  Active ingredients: colchicine (alkaloid) and related compounds, such as thiocolchicine and
thioketones

Particular value: it is used as anti-rheumatic, cathartic, emetic.

Precautions: extremely poisonous! Colchicine acts upon all secretive organs, such as the bowels and kidneys.

Documented target cancers:

- Colchicine and several of its analogs show good antitumor effect in mice infected with P388 lymphocytic leukemia
- High antitubulin effects of derivatives of 3-demethylthiocolchicine, methylthio ethers of natural colchicinoids, and thioketones derived from thiocolchicine
- Treatment of esophageal cancer with colchamine

Further details:

- *Colchicum autumnale*: it is considered to have, also, cytostatic effects.
- Esterification of the phenolic group in 3-demethylthiocolchicine and exchange of the N-acetyl group with other N-acyl groups or a N-carbalkoxy group afforded many compounds which showed superior activity over the parent drug as inhibitors of tubulin polymerization and of the growth of L1210 murine leukemia cells in culture. A comparison of naturally occurring Colchicum alkaloids with thio isosters, obtained by replacing the OMe group at C(10) with a SCH3 group, showed the thio ethers to be invariably more potent in these assays. The comparison included 3-demethylthiodemecolcine prepared from 3-demethylthiocolchicine by partial synthesis. Thiation of thiocolchicine with Lawesson's reagent afforded novel thiotropolones which exhibited high antitubulin activity. Their structures are fully secured by spectral data. Colchicine and several of its analogs show good antitumor effect in mice infected with P388 lymphocytic leukemia, and all of them show high affinity for tubulin and inhibit tubulin polymerization at low concentration. Consequently, antitubulin assays with this class of compounds can serve as valuable prescreens for the initial evaluation of potential antitumor drugs.
- Colchicine can cause induction of chromosome (loss and gain): the fruit fly *Drosophila melanogaster* is one of the standard systems used for mutagen screening. The colchicine-containing drugs Colchicum-Dispert and Colchysat Burger were fed at extremely low concentrations (1:300,000 and 1:50,000 respectively) to *Drosophila* females. Among their offspring a remarkably high frequency of aneuploid individuals (XO and XXY flies) were found. These aneuploids correspond karyotypically to the human Ullrich-Turner (XO) and Klinefelter's (XXY) syndromes and result from chromosome loss (XO) and chromosome gain (XXY). The maximum aneuploidy frequency observed after colchicine feeding was 24 times the control value. Depending on their size the aneuploidy frequencies are as great as those obtained by X-irradiation with some hundred or some thousand R.
- *Colchicum speciosum*: it concerns as a tumor inhibitor, with antileukemic activity.

References

Brncić, N., I. Visković, R. Perić, A. Dirlić, D. Vitezić, and D. Cuculić. 2001. Accidental plant poisoning with *Colchicum autumnale*: Report of two cases. *Croatian Medical Journal* 42, no. 6:673–675.

Danel, V.C., J.F. Wiart, G.A. Hardy, F.H. Vincent, and N.M. Houdret. 2001. Self-poisoning with *Colchicum autumnale L.* flowers. *Journal of Toxicology: Clinical Toxicology* 39, no. 4:409–411.

Haupt, H. 1996. Toxic and less toxic plants 29. *Kinderkrankenschwester* 15, no. 9:337–338.

Klintschar, M., C. Beham-Schmidt, H. Radner, G. Henning, and P. Roll. 1999. Colchicine poisoning by accidental ingestion of meadow saffron (*Colchicum autumnale*): Pathological and medicolegal aspects. *Forensic Science International* 106, no. 3:191–200.

Kupchan, S.M., R.W. Britton, C.K. Chiang, N. NoyanAlpan, and M.F. Ziegler. 1973. Tumor inhibitors. 88. The antileukemia principles of Colchicum speciosum. *Lloydia* 36, no. 3:338–340.

Lindholm, P., J. Gullbo, P. Claeson, et al. 2002. Selective cytotoxicity evaluation in anticancer drug screening of fractionated plant extracts. *Journal of Biomolecular Screening* 7, no. 4:333–340.

M. Akram, M., O. Alam, K. Usmanghani, and N. Akhter. 2012. *Colchicum autumnale*: A review. *Journal of Medicinal Plants Research* 6, no. 8: 1489 – 1491.

Muzaffar, A., A. Brossi, C.M. Lin, and E. Hamel. 1990. Antitubulin effects of derivatives of 3-demethylthio-colchicine, methylthio ethers of natural colchicinoids, and thioketones derived from thiocolchicine. Comparison with colchicinoids. *Journal of Medicinal Chemistry* 33, no. 2:567–571.

Rueffer, M., and M.H. Zenk. 1998. Microsome-mediated transformation of O-methylandrocymbine to dem-ecolcine and colchicine. *FEBS Letters* 30, no. 438 (1–2):111–113.

Schrader, A., O. Schulz, H. Volker, and H. Puls. 2001. Recent plant poisoning in ruminants of northern and eastern Germany. *Communication from the Practice for the Practice Berlin Munchener Tierärztliche Wochenschrift* 114, no. 5–6:218–221.

Traut, H., and U. Sommer. 1976. The induction of chromosome loss and gain by colchicines. *MMW Munchener Med. Wochenscr* 3, no. 118 (36):1113–1116.

Van and Os, F.H. 1970. Plants with cytostatic effect Farmaco. *Science* 25, no. 6:455–483.

Vitkin, B.S. 1969. Treatment of esophageal cancer with colchamine. *Voprosy Onkologii* 15, no. 11:90–92.

Weinberger, A., and J. Pinkhas. 1980. The history of colchicine. *Korot* 7, no. 11–12:760–763.

Yamada, M., Y. Kobayashi, H. Furuoka, and T. Matsui. 2000. Comparison of enterotoxicity between autumn crocus (*Colchicum autumnale L.*) and colchicine in the guinea pig and mouse: Enterotoxicity in the guinea pig differs from that in the mouse. *Journal of Veterinary Medical Science* 62, no. 8:809–813.

Yamada, M., T. Matsui, Y. Kobayashi, H. Furuoka, M. Haritani, M. Kobayashi, and M. Nakagawa. 1999. Supplementary report on experimental autumn crocus (*Colchicum autumnale L.*) poisoning in cattle: Morphological evidence of apoptosis. *Journal of Veterinary Medical Science* 61, no. 7:823–825.

Yamada, M., M. Nakagawa, M. Haritani, M. Kobayashi, H. Furuoka, and T. Matsui. 1998. Histopathological study of experimental acute poisoning of cattle by autumn crocus (*Colchicum autumnale L.*). *Journal of Veterinary Medical Science* 60, no. 8:949–952.

Coleus forskohlii (Labiatae/Lamiaceae) (Lamiales)   **Anti-tumor**

Location: India

Appearance: herbaceous plant with annual stem and perennial root stock

Stem: grows at about 45–60 cm high and is often hairy
Leaves: bright green, 7–12 cm long and 3 to 5 cm wide
Flowers: the inflorescence is snout, 15–30 cm in length, and the flowers are 2 to 2.5 cm in size, usually perfect, and the cup is scaly on the inside, with an oval lip. The crown is blue or lilac.

Tradition: the leaves and root of the plant were traditional medicine in India for digestive problems, as well as heart and lung problems. Extracts of the root regulate blood pressure, colic, chronic cough, dysuria, epilepsy, fever, and indigestion.

Parts used: root

Active ingredients: the diterpene forskolin derived from the root of the plant, deactylforskolin, 9-deoxyforskolin, 1, 9-deoxyforskolin, 1, 9-dideoxy-7-deacetylforskolin

Documented target cancers:

Forskolin is reported to reduce tumor colonization of the highly metastatic melanoma cell line (B16- F10) in the lungs.

References

Agarwal, K.C., and R.E. Parks Jr. 1983. Forskolin: A potential antimetastatic agent. *International Journal of Cancer* 32, no. 6:801–804.

Kavitha, C., K. Rajamani, and E. Vadivel. 2010. *Coleus forskohlii* A comprehensive review on morphology, phytochemistry and pharmacological aspects. *Journal of Medicinal Plants Research* 4, no. 4:278–285.

Sapio, L., M. Gallo, M. Illiano, E. Chiosi, D. Naviglio, A. Spina, and S. Naviglio. 2017. The natural cAMP elevating compound forskolin in cancer therapy: Is it time? *Journal of Cellular Physiology* 232, no. 5:922–927.

Colocasia esculenta (Araceae) (Alismatales)	Cytotoxic

Location: (native) Malaysia, India, Australia, Turkey, southeastern US
  Appearance: fast-growing herbaceous plant

  Leaves: up to 40 × 24.8 cm, dark green, triangular-ovate, sub-rounded and mucronate at apex, tip of the basal lobes rounded or sub-rounded
  Flowers: monogenic, small, without a perianth, the male flowers in the upper part of the spadix have stamens completely merged, while the female flowers are at the base of the spadix.

Parts used: bulbs, leaves
  Documented target cancers:
  • Colonic adenocarcinoma cells

References

Brown, A.C., J.E. Reitzenstein, J. Liu, and M.R. Jadus. 2005. The anti-cancer effects of poi (*Colocasia esculenta*) on colonic adenocarcinoma cells *in vitro*. *Phytotherapy Research* 19, no. 9:767–771.
He, X., S.C. Miyasaka, M.M. Fitch, and Y.J. Zhu. 2015. Taro (Colocasia esculenta (L.) Schott). In *Agrobacterium Protocols*. Wang, E. (Ed.). Springer, New York, NY, 97–108.

Colocasia gigantea (Araceae) (Alismatales)	Cytotoxic

Location: Vietnam, Japan, southeast tropical Asia, Polynesia, China
  Appearance: fast-growing herbaceous plant

  Leaves: huge, velvety or glossy texture, dark green, heart-shaped, lined with darker ribs and succulent stems
  Flowers: large aromatic with yellow-white swords, often hidden from the foliage when they bloom

Parts used: entire plant, petioles
  Active ingredients: lectins, tannins, phenolic compounds
  Tradition: in the native habitat, edible leaves and stems are sometimes eaten as vegetables or used as feed for pigs. They are also planted in gardens as cultivars. Sometimes it is used as a soup ingredient, chanpuru, and sushi. The term zuiki refers to the leaf and stem of the species *C. gigantea* and *C. esculenta*. Higo-zuiki (dried stem) is produced exclusively in the county of Kumamoto, and it contains saponins that are thought to affect sexual pleasure. It is also considered cyanogenic.
  Documented target cancers:
  • Cervical cancer cells: HeLa cell line

Reference

Meng, A.P., and M.D. Amornpun Sereemaspun. 2015. Anticancer activity of selected *Colocasia gigantia* fractions. *Journal of the Medical Association of Thailand* 98, no. 1:S98–S106.

Conium maculatum (Apiaceae) (Apiales)   **Cytotoxic**

Location: Europe, north and southern Africa, western Asia
  Appearance: extremely poisonous flowering weed

  Leaves: opposite, complex, soft with no hair
  Flowers: small, white, clustered in umbels

Tradition: this is believed to be the plant used for the execution of Socrates in 399 BC.
  Parts used: fruit
  Active ingredients: pyridine alkaloids like coniine, N-methylconiine, conhydrine, pseudoconhydrine, and gamma-coniceine
  Documented target cancers:
  • The ethanolic extract of *Conium maculatum* is reported to reduce cell viability and colony formation in cervix cancer HeLa cells *in vitro*.

Reference

Mondal, J., A.K. Panigrahi, and A.R. Khuda-Bukhsh. 2014. Anticancer potential of *Conium maculatum* extract against cancer cells *in vitro*: Drug-DNA interaction and its ability to induce apoptosis through ROS generation. *Pharmacognosy Magazine* 10 Suppl 3:S524–S533.

Conyza triloba (Asteraceae) (Asterales)   **Anti-proliferative**

Location: Morocco, central to northeastern Africa, Madagascar, Asia, India, Himalayas
  Appearance: flowering plant

  Flowers: inflorescence fear, consisting of 20–80 (–200) heads and triangular outline. The flowers are yellow, and the ducks are in white or bright straw color.

Active ingredients: compounds such as terpenoids and flavonoids (quercetin 3-glycoside, rutin, and pinostrobin) and quinones. Also sterol (alpha-spinasterol), beta-caryophyllene 4,5-alpha-oxide, and two triterpenic erythrodiol and 3-beta-tridecanooxy-28-hydroxyolean-12-ene.
  Documented target cancers:

  • Hepa1c1c7 and H4IIE1, A549, HT29, and PC3 cell lines
  • Colorectal carcinogenesis

References

El-Sayed, W.M., W.A. Hussin, A.A. Mahmoud, and M.A.Al. Fredan. 2013. The *Conyza triloba* extracts with high chlorophyll content and free radical scavenging activity had anticancer activity in cell lines. *BioMed Research International* 2013:945638.

Mahmoud, A.A., M.A.Al. Fredan, and W.M. El-Sayed. 2016. Isolation, characterization and anticancer activity of seven compounds from the aerial parts of *Conyza triloba*. *International Journal of Pharmacognosy and Phytochemical Research* 8, no. 12:2071–2079.

Mata, R., A. Rojas, L. Acevedo, et al. 1997. Smooth muscle relaxing flavonoids and terpenoids from *Conyza filaginoides*. *Planta Medica* 63, no. 1:31–35.

Coptis chinensis (Ranunculaceae) (Ranunculales)	Anti-proliferative Cytotoxic

Location: forests of Mishmi Hills of Arunachal Pradesh in India
  Appearance: perennial herb

  Leaves: palmate or maple-like in appearance with the blade oval-triangular, 10 cm wide
  Flowers: one to two in number, equal to or more than the length of the leaves. Intersectal inflorescence
    with three to eight flowers.

Tradition: *C. chinensis* rhizomes are well-known as antimicrobial herbs with a long history of use
in traditional Chinese medicine for the treatment of gastrointestinal problems, gallbladder inflammation, abdominal cramps, and for the control of excessive bleeding.
  In alternative medicine, *C. chinensis* helps treat the following: diabetes, diarrhea, ear infections,
heart disease, high blood pressure, high cholesterol, psoriasis, respiratory infections.
  Parts used: root
  Active ingredients: berberine, coptisine, epiberberine, berberrubine, palmatine, columbamine,
jatrorrhizine, worenine, magnoflorine, φερουλικό οξύ, obakunone, and obakulactone
  Documented target cancers:

- Liver cancer cell lines (SK-Hep1, HepG2, Hep3B, PLC/PRF/5)
- Leukemia cancer cell lines (K562, U937, P3H1)

References

Kong, W., J. Wei, P. Abidi, et al. 2004. Berberine is a novel cholesterol-lowering drug working through a unique
    mechanism distinct from statins. *Nature Medicine* 10, no. 12:1344–1351.
Xiang, K.L., S.D. Wu, S.X. Yu, et al. 2016. The first comprehensive phylogeny of Coptis (Ranunculaceae) and
    its implications for character evolution and classification. *PLOS ONE* 11, no. 4:e0153127.

Corydalis ochotensis (Papaveraceae) (Ranunculales)	Anti-cancer

Location: northern hemisphere, high mountains of tropical eastern Africa, China, Himalayas
  Appearance: biennial growing to 1 m

  Leaves: basal leaves long petiolate; blade glaucous abaxially, broadly ovate, or deltoid, triternate, pri-
    mary petiolules long. Secondary petiolules short, leaflets deeply or slightly divided, obovate, rhom-
    bic-obovate, or ovate, obtuse, mucronate. Lower cauline leaves long petiolate, upper ones shortly
    petiolate, like basal leaves.
  Flowers: racemes 3–5 cm, at fruiting to 9 cm, four–eight-flowered, bracts ovate to broadly ovate, 0.5–
    1.4 cm, entire, partly concealing flowers. Pedicel 2–4 mm, recurved in fruit. Sepals reniform, to
    0.5 mm, dentate. Corolla yellow, keels of outer petals often with a purplish gray tint, outer petals
    acuminate, with short obtuse crest extended clearly beyond petal apex.

Tradition: it is used as a popular medicine in China as an antipyretic, analgesic, and for its diuretic
properties. It also inhibits allergic reactions.
  Active ingredients: fumariline, sanguinarine, tetrahydroberberine, berberine, ochotensine, chei-
lanthifoline, and lienkonine

Precautions: pregnancy and breast-feeding: not safe during pregnancy. It might induce menstrual flow and uterus contractions. This could cause a miscarriage.

Documented target cancers:

Inhibitory activity on cancer cell lines Hep2 (human epidermoid carcinoma), HepG2 (human liver cancer cell line), SW480 (colon carcinoma), and A549 (lung carcinoma)

References

Gao, X.F., Y.L. Peng, M. Lidén, and Y.W. Wang. 2008. Three new species of Corydalis (Fumariaceae) from Northwestern Sichuan, China. *Novon* 18, no. 3:330–335.

Lidén, M., T. Fukuhara, and T. Axberg. 1995. Phylogeny of Corydalis, ITS and morphology. In: *Systematics and Evolution of the Ranunculiflorae*. Jensen, U., and J.W. Kadereitpp (Eds.). Springer, Vienna, 183–188.

Yu, J.J., D.L. Cong, Y. Jiang, Y. Zhou, Y. Wang, and C.F. Zhao. 2014. Study on alkaloids of *Corydalis ochotensis* and their antitumor bioactivity. *Zhong Yao Cai = Zhongyaocai = Journal of Chinese Medicinal Materials* 37, no. 10:1795–1798.

Zhang, M.-L., Z.-Y. Su, and M. Lidén. 2008. Corydalis. In: *Flora of China*. Wu, Z.Y., and P.H. Raven (Eds.). Science Press, Beijing, China & Missouri Botanical Garden Press, Saint Louis, USA, 545–652.

Couroupita guianensis (Lecythidaceae) (Ericales)   **Cytotoxic**

Location: south India, Malaysia, (native) Hondura

Appearance: large evergreen tree with a height of 20 meters

Leaves: alternate, oblong-anthoid, up to 20 cm long, slightly toothed and hairy under the nerves

Flowers: the inflorescence is botryoid, and stems from the trunk and other large branches. The flowers are reddish with a yellow tint on the outside.

Tradition: shamans of South America used the plant for malaria. Fruit pulp is used to disinfect wounds while young leaves are used for toothache. According to Buddhist tradition, Maya held onto the branch of a blossoming sal tree while she was giving birth to the Lord Buddha. Because of this, the sal tree is revered by many Buddhist people around the world.

Parts used: juice, leaves, fruit

Active ingredients: alkaloids, phenolics and flavonoids, isatin and indirubin (vital to their own antimicrobial activity), flavonoids: 2′, 4′dihydroxy-6′-methoxy-3′, 5′-dimethylchalcone, 7-hydroxy, 5-methoxy-6,8-dimethylflavonoids, and phenolic acid 4-hydroxybenzoic acid.

Precautions: fruit may cause toxicity and a serious allergic reaction.

Documented target cancers:

• Cytotoxic against human promylocytic leukemia (HL60) cells

References

Martínez, A., E. Conde, A. Moure, H. Domínguez, and R.J. Estévez. 2012. Protective effect against oxygen reactive species and skin fibroblast stimulation of *Couroupita guianensis* leaf extracts. *Natural Product Research* 26, no. 4:314–322.

Premanathan, M., S. Radhakrishnan, K. Kulangiappar, G. Singaravelu, V. Thirumalaiarasu, T. Sivakumar, and K. Kathiresan. 2012. Antioxidant and anticancer activities of isatin (1H-indole-2,3-dione), isolated from the flowers of *Couroupita guianensis* Aubl. *Indian Journal of Medical Research* 136, no. 5:822–826.

Crassocephalum crepidioides (Asteraceae) (Asterales)	Anti-tumor

Location: tropical and subtropical regions, especially tropical Africa
    Appearance: orthocladic succulent herb, reaching 180 centimeters, smooth or hairy

    Leaves: laminated and are elliptical, they end up oval in the contour
    Flowers: cylindrical, appear on the top with red petals

Tradition: especially popular in southwestern Nigeria to treat indigestion. In the Democratic Republic of Congo the leaf is given to treat stomach upsets and as a treatment for fresh wounds. A leaf lotion or decoction is used to treat headache in Nigeria and Tanzania, and a mixture of *C. crepidioides* and *Cymbopogon giganteus* is used orally and externally to treat epilepsy.
    Parts used: whole plant, leaves
    Active ingredients: the essential oil obtained from the leaf hydrodeposition consists mainly of monoterpenes (myrcene, limonene, and alpha-copaene).
    Documented target cancers:
- *C. crepidioides* extract has been shown to delay tumor growth in S-180-bearing mice (*in vivo*), while supernatant of cultured *C. crepidioides*-stimulated RAW264.7 macrophages was cytotoxic to murine sarcoma S-180 cells (*in vitro*).

References

Denton, O.A., R.R. Schippers, and L.P.A. Oyen. 2004. Plant resources of tropical Africa. *Vegetables*. Backhuys, Netherlands.

Tomimori, K., S. Nakama, R. Kimura, K. Tamaki, C. Ishikawa, and N. Mori. 2012. Antitumor activity and macrophage nitric oxide producing action of medicinal herb, *Crassocephalum crepidioides*. *BMC Complementary and Alternative Medicine* 12, no. 1:78.

Crataegus sanguinea (Rosaceae) (Rosales)	Anti-proliferative

Location: north America (US), native to southern Siberia, Mongolia, and the extreme north of China
    Appearance: deciduous shrub that reaches a height of 6 m

    Leaves: oval, tooth in black green. The layout of the leaves is alternate.

Tradition: used in folk medicine as heart tonic
    Parts used: fruit
    Active ingredients: flavonoids, polyphenols, terpenoids
    Documented target cancers:
- *In vitro* experiments have shown no inhibition of growth or increase in cell death on normal mouse fibroblasts but a stronger inhibition of cell growth and an increase in cell death of Hep-2 and MGC-803 tumor cells.

Reference

Sun, J., G. Gao, Y. Gao, L. Xiong, X. Li, J. Guo, and Y. Zhang. 2013. Experimental research on the *in vitro* antitumor effects of Crataegus sanguinea. *Cell Biochemistry and Biophysics* 67, no. 1:207–213.

***Cratoxylum formosum* (Hypericaceae) (Malpighiales)**   **Cytotoxic**

Location: Singapore
   Appearance: deciduous small tree

   Leaves: simple, opposite, elongated
   Flowers: light pink

Tradition: a mixture of bark and leaf of the plant in combination with coconut oil is used in folk medicine to cure skin diseases. The bark is also used for the treatment of diarrhea in domesticated animals.
   Parts used: stems, leaves
   Active ingredients: formosumone A, toxyloxanthone B, vismione D, gallic acid, and quercetin
   Documented target cancers:

* *Cratoxylum formosum* extracts are reported cytotoxic against the cervical cancer cell lines C-33A, HeLa, and SiHa.
* The ethyl acetate extract of the plant is reported to significantly inhibit the growth of the oral cancer cell lines ORL-48 and ORL-136 but not the normal renal cell line Vero.
* *Cratoxylum formosum* is reported to selectively induce cell death in the human hepatocellular carcinoma HepG2 compared with the non-cancerous African green monkey kidney epithelial cell line (Vero).

References

Nonpunya, A., N. Weerapreeyakul, and S. Barusrux. 2014. *Cratoxylum formosum* (Jack) Dyer ssp. pruniflorum (Kurz) Gogel.(Hóng yá mù) extract induces apoptosis in human hepatocellular carcinoma HepG2 cells through caspase-dependent pathways. *Chinese Medicine* 9, no. 1: https://doi.org/10.1186/1749-8546-9-12

Promraksa, B., J. Daduang, P. Chaiyarit, et al. 2015. Cytotoxicity of *Cratoxylum formosum* subsp. Pruniflorum Gogel extracts in oral cancer cell lines. *Asian Pacific Journal of Cancer Prevention* 16, no. 16:7155–7159.

Promraksa, B., J. Daduang, T. Khampitak, et al. 2015. Anticancer potential of *Cratoxylum formosum* Subsp. Pruniflorum (Kurz.) Gogel extracts against cervical cancer cell lines. *Asian Pacific Journal of Cancer Prevention* 16, no. 14:6117–6121.

***Crinum asiaticum* (Crinum) (var. *toxicarium* (Hubert))**   **(Liliaceae) Inhibitor**

Location: wild in low, humid spots in various parts of India and on the coast of Ceylon
   Appearance: evergreen bulb-forming perennial

   Leaves: simple, spiral, thin, and relatively elliptical. The blade length of the leaf is over 36 inches and has a green color.
   Flowers: white with pleasant odor
   In bloom: summer

Degree of rarity: low, as it is cultivated in Indian gardens
   Tradition: ornamental plant used in gardens. In Polynesia, it is used as jewelry (earrings). The trunk is used in deep-sea fishing and the shiny part of the leaves as fish bait.
   Parts used: leaves, root, and bulb

Active ingredients: N-(3,4-dioxo-benzyl)-4-O-phenylethylamine, pyrrole phenanthridine, asi-aticumines A and B, criasiaticidine A, lycorine

Particular value: the bulb was admitted to the Pharmacopoeia of India as a valuable emetic.

Documented target cancers:

- Lycorine inhibits not only induction of MM46 cell death by calprotectin but also inhibits the suppressive effect of calprotectin on target DNA synthesis at a half effective concentration of 0.1–0.5 microgram/ml.
- Lycorine has been reported to possess inhibitory activity against protein translation.
- Colon cancer cell line SW620, lung in the cell line (CHAGO), HepG2, gastric cancer in the cell line (KATOIII), and breast cancer in the cell line (BT474).
- Meth-A (mouse sarcoma), Lewis lung carcinoma (mouse lung carcinoma) tumor cell lines.

Further details:
- It has been demostrated that calprotectin, an abundant calcium-binding protein complex in polymorphonuclear leukocytes (PMNs), has the capacity to induce growth inhibition and apoptotic cell death against a variety of tumor cell lines and normal cells such as fibroblasts. Therefore, calprotectin which is released to extracellular spaces might cause tissue destruction in severe inflammatory conditions. Using MM46 mouse mammary carcinoma cells as targets, hot water extracts of *Crinum asiaticum* (lycorine is the active inhibitory molecule) showed strong inhibition of calprotectin-induced cytotoxicity *in vitro*. The dose–response relationship between the inhibitory effects of lycorine on calprotectin action and target protein synthesis showes that lycorine inhibition for calprotectin cytotoxicity is not solely due to its inhibitory effect on protein synthesis.

References

Ji, Y.B., P. Tian, Q.C. Dai, S.T. Wang, and N. Chen. 2013. The present research situation of *Crinum asiaticum* alkaloids active ingredient. *Applied Mechanics and Materials* 411–414:3181–3186.

Min, B.S., J.J. Gao, N. Nakamura, Y.H. Kim, and M. Hattori. 2001. Cytotoxic alkaloids and a flavan from the bulbs of *Crinum asiaticum* var. japonicum. *Chemical and Pharmaceutical Bulletin* 49, no. 9:1217–1219.

Yui, S., M. Mikami, M. Kitahara, and M. Yamazaki. 1998. The inhibitory effect of lycorine on tumor cell apoptosis induced by polymorphonuclear leukocyte-derived calprotectin. *Immunopharmacology* 40, no. 2:151–162.

Crocus sativus (Saffron) (Iridaceae)　　　　　　　　　　　　　　　**Cytotoxic Chemopreventive**

Synonyms: Crocus, Saffron Crocus, Krokos (Greek), Zaffer (Arabian)

Location: from Europe to Asia, in meadows or (mostly) in cultivation

Origin: wild forms are found in Italy, Greece, the Balkans, and Eastern Asia (mainly Iran).

Appearance:

Root: corm

Leaves: short and linear, with a white-pale central nerve, up to 30 cm long

Flowers: long-tubed pale violet, directly emerging from the underground corm, with three yellow anthers and red-orange styles, up to 10 cm long

In bloom: September–November

Degree of rarity: considerable as wild form, though widely cultivated.

Biology: the plant is perennial, with five forms existing in the wild state. Fruit setting requires cross-fertilization. Corms must not be left to grow in the same ground for too long (longer than three years).

Tradition: already known in ancient times, saffron is referred to as Karkom in the Song of Solomon (iv. 14). The luxury yellow dye traditionally derived from the plant has been mentioned in various Greek myths, along with its scent and flavor.

Parts used: flower stigmas

Active ingredients: crocin, crocetin, picrocrocin, and safranal (carotenoids)

Particular value: it is used as carminative, emmenagogue, diaphoretic for children, and for chronic haemorage of the uterus in adults.

Indicative dosage and application:

- Oral administration of 200 mg/kg body weight of the extract increased the life span of S-180, EAC, DLA tumor-bearing mice to 111.0, 83.5, and 112.5%, respectively. The same extract was found to be cytotoxic to P38B, S-180, EAC, and DLA tumor cells *in vitro* (potential use of saffron as an anticancer agent).
- Intraperitoneal administration of *Nigella sativa* (100 mg/kg body wt) and oral administration of *Crocus sativus* (100 mg/kg body wt) 30 days after subcutaneous administration of MCA (745 nmol × two days) restricted tumor incidence to 33.3 and 10%, respectively, compared with 100% in MCA-treated controls.

Documented target cancers:

- Crocin, safranal, and picrocrocin inhibit the growth of human cancer cells *in vitro*.
- Saffron extract (dimethyl-crocetin) posseses anticarcinogenic, anti-mutagenic, and immunomodulating effects: dose-dependent cytotoxic effect to carcinoma, sarcoma, and leukemia cells *in vitro*, delayed ascites tumor growth, and increased the life span of the treated mice compared to untreated controls by 45–120%. In addition, it delayed the onset of papilloma growth and decreased incidence of squamous cell carcinoma and soft tissue sarcoma in treated mice.
- Crocetin has a dose-dependent inhibitory effect on DNA and RNA synthesis in isolated nuclei and suppressed the activity of purified RNA polymerase II. Also, crocetin causes a dose-dependent inhibition of nucleic acid and protein synthesis. (Cell lines: HeLa (cervical epitheloid carcinoma), A549 (lung adenocarcinoma), and VA13 (SV-40 transformed fetal lung fibroblast) cells.)
- Antitumor activity against intraperitoneally transplanted sarcoma-180 (S-180), Ehrlich ascites carcinoma (EAC), and Dalton's lymphoma ascites (DLA) tumors in mice.

Further details:

- Doses inducing 50% cell growth inhibition (LD50) on HeLa cells were 2.3 mg/ml for an ethanolic extract of saffron dry stigmas, 3 mM for crocin, 0.8 mM for safranal, and 3 mM for picrocrocin. Crocetin did not show any cytotoxic effect.
- Cells treated with crocin exhibited wide cytoplasmic vacuole-like areas, reduced cytoplasm, cell shrinkage, and pyknotic nuclei, suggesting apoptosis induction. Considering its water-solubility and high inhibitory growth effect, crocin is the more promising saffron compound to be assayed as a cancer therapeutic agent.
- Saffron (dimethyl-crocetin) disrupts DNA-protein interactions, e.g. topoisomerases II, important for cellular DNA synthesis (significant inhibition in the synthesis of nucleic acids but not protein synthesis).
- The effects of carotenoids of *Crocus sativus* L. (saffron) on cell proliferation and differentiation of HL-60 cells have been studied and compared with those of all-trans retinoic acid. Results demonstrated that the doses inducing 50% inhibition of cell growth were 0.12 µM for all-trans retinoic acid (ATRA) and for carotenoids of saffron 0.8 µM for dimethylcrocetin (DMCRT), 2 µM for crocetin CRT, and 2 µM for crocins (CRCs). At 5 µM, all these compounds induced differentiation of HL-60 cells, at 85% for ATRA, 70% for DMCRT, 50% for CRT, and 48% for CRCs. In these experiments, leukemic cells were cultured for five days in the absence or in the presence of up to 5 µM ATRA or seminatural and natural carotenoids. Since retinoids have a potential application as

chemopreventive agents in humans, their toxicity as an important limiting factor for their use in treatment should be extensively explored. The seminatural (DMCRT and CRT) and natural carotenoids (CRCs) of *Crocus sativus* L. are not provitamin A precursors and could therefore be less toxic than retinoids, even at high doses.

- Topical application of *Nigella sativa* and *Crocus sativus* extracts (common food spices) inhibited two-stage initiation/promotion [dimethylbenz[a]anthracene (DMBA)/croton oil] skin carcinogenesis in mice. A dose of 100 mg/kg body wt of these extracts delayed the onset of papilloma formation and reduced the mean number of papillomas per mouse, respectively. The possibility that these extracts could inhibit the action of 20-methylcholanthrene (MCA)-induced soft tissue sarcomas was evaluated by studying the effect of these extracts on MCA-induced soft tissue sarcomas in albino mice.

References

Abdullaev, F.I. 1994. Inhibitory effect of crocetin on intracellular nucleic acid and protein synthesis in malignant cells. *Toxicology Letters* 70, no. 2:243–251.

Aung, H.H., C.Z. Wang, M. Ni, et al. 2007. Crocin from *Crocus sativus* possesses significant anti-proliferation effects on human colorectal cancer cells. *Experimental Oncology* 29, no. 3:175–180.

Escribano, J., G.L. Alonso, M. Coca-Prados, and J.A. Fernandez. 1996. Crocin, safranal and picrocrocin from saffron (*Crocus sativus* L.) inhibit the growth of human cancer cells *in vitro*. *Cancer Letters* 27, no. 100 (1–2):23–30.

Nair, S.C., S.K. Kurumboor, and J.H. Hasegawa. 1995. Saffron chemoprevention in biology and medicine a review. *Cancer Biotherapy* 10, no. 4:257–264.

Nair, S.C., B. Pannikar, and K.R. Panikkar. 1991. Antitumour activity of saffron (*Crocus sativus*). *Cancer Letters* 57, no. 2:109–114.

Salomi, M.J., S.C. Nair, and K.R. Panikkar. 1991. Inhibitory effects of *Nigella sativa* and saffron (*Crocus sativus*) on chemical carcinogenesis in mice. *Nutrition and Cancer* 16, no. 1:67–72.

Samarghandian, S., and A. Borji. 2014. Anticarcinogenic effect of saffron (Crocus sativus L.) and its ingredients. *Pharmacognosy Research* 6, no. 2:99–107.

Tarantilis, P.A., H. Morjani, M. Polissiou, and M. Manfait. 1994. Inhibition of growth and induction of differentiation of promyelocytic leukemia (HL-60) by carotenoids from *Crocus sativus* L. *Anticancer Research* 14, no. 5 (A):1913–1918.

Croton lechleri (Euphorbiaceae) (Malpighiales)   Cytotoxic

Location: northwestern South America
    Appearance: large to medium sized tree

    Leaves: large, heart-shaped leaves
    Flowers: green-white flowers

Parts used: bark, resin, tree juice
    Tradition: it is used as an ointment to stop bleeding, dandruff, rashes, dermatitis, germs. It is considered antiseptic. It aids wound healing, has anti-inflammatory action, and relieves itching (from insect bites, poison ivy, or allergic skin reactions). It heals intestinal problems.
    Active ingredients: taspine, dimethylcedrusine
    Documented target cancers:

- Essential oil of the trunk from the plant was found to be antimatabolic against heterocyclic amines (HCAs), which appear to induce liver and colon cancer.
- Taspine alkaloid is reported to show inhibition against human melanoma SK23 and colon carcinoma HT29 cell lines.

- The plant is reported to exhibit moderate toxic effects *in vivo* and cytotoxic effects on HeLa cervical cancer cells.

Further details:
- The SP-303 (large proanthocyanidin oligomer isolated from the latex of the plant species *Croton lechleri*) has been approved by the FDA for fine diarrhea in people with HIV.

References

Alonso-Castro, A.J., E. Ortiz-Sánchez, F. Domínguez, G. López-Toledo, M. Chávez, Ade J. Ortiz-Tello, and A. García-Carrancá. 2012. Antitumor effect of *Croton lechleri* Mull. Arg.(Euphorbiaceae). *Journal of Ethnopharmacology* 140, no. 2:438–442.

Montopoli, M., R. Bertin, Z. Chen, J. Bolcato, L. Caparrotta, and G. Froldi. 2012. *Croton lechleri* sap and isolated alkaloid taspine exhibit inhibition against human melanoma SK23 and colon cancer HT29 cell lines. *Journal of Ethnopharmacology* 144, no. 3:747–753.

Ubillas, R., S.D. Jolad, R.C. Bruening, et al. 1994. SP-303, an antiviral oligomeric proanthocyanidin from the latex of *Croton lechleri* (Sangre de Drago). *Phytomedicine* 1, no. 2:77–106.

Cryptolepis sanguinolenta (Apocynaceae) (Gentianales) **Cytotoxic**

Location: West Africa
  Appearance: climbing shrub

Leaves: smooth, elongated, elliptical, greenish, and rounded at the top
Flowers: yellow color

Tradition: it is an established antimalarial in West African ethnomedicine.

Additionally, the root extract is used to treat hepatitis, hypertension, incontinence, upper respiratory tract infections, colic, stomach problems, amniotic dysentery and diarrhea, wounds, measles, hernia, snake bites, rheumatism, and insomnia.

The rooting of the roots serves as an aphrodisiac.

Parts used: root, stems, and leaves

Active ingredients: cryptolepine, pryptolepine, hydroxyptolepine, neocryptolepine, biscyptlepine, and quindoline

Documented target cancers:

- Studies in which *in vitro* toxicity of *C. sanguinolenta* and cryptolepine toxicity was examined in HCT116 human colon adenocarcinoma, SKOV3 human ovary adenocarcinoma, MCF7 human breast adenocarcinoma, and MDA MB 361 human breast adenocarcinoma cell lines report a dose-dependent and time reduction in cell viability.
- Cryptolepine, a major alkaloid isolated from the roots of *Cryptolepis sanguinolenta* is reported to inhibit the growth of skin cancer cell lines SCC-13 and A43, through inhibition of topoisomerase and induction of DNA damage.

References

Ansah, C., and N.J. Gooderham. 2002. The popular herbal antimalarial, extract of *Cryptolepis sanguinolenta*, is potently cytotoxic. *Toxicological Sciences* 70, no. 2:245–251.

Ansah, C., and K.B. Mensah. 2013. A review of the anticancer potential of the antimalarial herbal Cryptolepis sanguinolenta and its major alkaloid cryptolepine. *Ghana Medical Journal* 47, no. 3:137–147.

Pal, H.C., and S.K. Katiyar. 2017. Cryptolepine a plant alkaloid, inhibits the growth of nonmelanoma skin cancer cells through inhibition of topoisomerase and induction of DNA damage. *Cancer Research* 2017:77.

Curcuma caesia (Zingiberaceae) (Zingiberales)	Anti-tumor	Anti-mutagenic

Location: North-East and Central India
   Appearance:

   Leaves: elongated lobes, smooth, with a purple line in the central nerve of the lamina
   Flowers: light yellow, red on the outside

Tradition: used in Ayurveda as anti-asthmatic and anti-inflammatory. Dried roots and leaves are used to treat leprosy, asthma, trauma (wounds), fertility, sore throat, chest pain, and allergies. Decoction from the fresh rhizome is used as an anti-diarrhea, for relief of stomach pains, epilepsy, as an anemythmic, aphrodisiac, ointment in snake and scorpion bites, for blood test, and rapid healing.
   Parts used: rhizomes and leaves
   Active ingredients: camphor oil, ar-tumerone, (Z) -b-ocimene, ar-curcumene, 8-cineoles, β-elemene, borneol, bornyl acetate, and γ-curcumene
   Documented target cancers:

- Ethanolic extracts of rhizomes are reported to exert antitumor activity *in vivo* against diethylnitrosamine induced hepatocellular carcinoma in mice.
- Rhizome extracts of *Curcuma caesia* are reported to show antimutagenicity in dose dependent manner.
- Hexane extracts of dried *C. caesia* rhizome have been shown to exhibit a dose-dependent inhibition only in cancer cells, with notable activity against the human liver cancer cell line HepG2.

References

Devi, H.P., P.B. Mazumder, and L.P. Devi. 2015. Antioxidant and antimutagenic activity of *Curcuma caesia* Roxb. rhizome extracts. *Toxicology Reports* 2:423–428.

Hadem, K.L.H., R.N. Sharan, and L. Kma. 2015. Phytochemicals of Aristolochia tagala and *Curcuma caesia* exert anticancer effect by tumor necrosis factor-α-mediated decrease in nuclear factor kappaB binding activity. *Journal of Basic and Clinical Pharmacy* 7, no. 1:1–11.

Mukunthan, K.S., R.S. Satyan, and T.N. Patel. 2017. Pharmacological evaluation of phytochemicals from South Indian Black Turmeric (*Curcuma caesia* Roxb.) to target cancer apoptosis. *Journal of Ethnopharmacology* 209:82–90.

Curcuma longa (Zingiberaceae) (Zingiberales)	Anti-tumor

Location: native to India and Indonesia
   Appearance: herbaceous plant with a large oval rhizome and cylindrical tubers

   Fruit: capsule-divided into three sections
   Leaves: large, hairless, mounted in clusters
   Flowers: yellow and white

Tradition: in Ayurvedic herbal medicine it is considered an alternative antiparasitic, microbicidal, stimulating, tonic, anti-almintant, and dissolves stomach gases. In Bihar (eastern India), the flower is used against fertility and the leaves for the treatment of colds, fever and pneumonia.

Parts used: rhizomes, flowers, and leaves

Active ingredients: curcumin, bisdemethoxycurcumin, demethoxycurcumin, turmerone, aur-turmerone, zingiberene, curlone

Documented target cancers:

- The antitumor capacity of curcumin loaded nanoparticles in the treatment of lung cancer with little toxicity to normal tissues has been reported.
- Micromolar concentrations of curcumin combined with genistein are reported to inhibit the growth of MCF-7 breast cells.

References

Chang, H.B., and B.H. Chen. 2015. Inhibition of lung cancer cells A549 and H460 by curcuminoid extracts and nanoemulsions prepared from *Curcuma longa* Linnaeus. *International Journal of Nanomedicine* 10:5059–5080.

Chen, Y.C., and B.H. Chen. 2018. Preparation of curcuminoid microemulsions from *Curcuma longa* L. to enhance inhibition effects on growth of colon cancer cells HT-29. *RSC Advances* 8, no. 5:2323–2337.

Donipati, P., and S.H. Sreeramulu. 2015. *In vitro* anticancer activity of *Curcuma longa* against human breast cancer cell line MCF-7. *Journal of Pharmacy and Pharmaceutical Sciences* 4, no. 11:1188–1193.

Wilken, R., M.S. Veena, M.B. Wang, and E.S. Srivatsan. 2011. Curcumin: A review of anti-cancer properties and therapeutic activity in head and neck squamous cell carcinoma. *Molecular Cancer* 10, no. 1:12.

Curcuma zedoaria (Zingiberaceae) (Zingiberales)   **Cytotoxic**

Location: native to India and Indonesia

Appearance: perrenial herb

Leaves: oblique, with a characteristic purple color in the middle vein of the leaf blade

Flowers: formed in clusters above the ground

Tradition: antidote to the Indian Cobra. The rhizome is thought to aid in digestion, cleanses blood, provides colic relief, and treats cold and infections. The essential oil extracted from the plant is an active ingredient in antibacterial preparations. In India, the rhizome is chewed or taken in the form of a beverage to treat the defectiveness resulting from childbirth. Bitter root tincture is used to prevent recurrence of malaria and for ulcer treatment.

Parts used: rhizomes and leaves

Active ingredients: isocurcumenol, curcumin, curcumol, curdione, dehydrocurdione, furanodi-enone, curzerenone, socurcumenol, and zederone

Documented target cancers:

- Research on the essential oil from *C. zedoaria* Roscoe has shown effective cytotoxicity in non-small cell lung cancer causing cellular apoptosis.
- Other studies have shown that a tangy plant extract is a promising agent against esophageal cancer.
- Curdione, isolated from the plant, has been found to inhibit the production of prostaglandin E2 and cyclooxygenase 2 (cox2), both of which are involved in inflammation and carcinogenic processes. This substance was found to inhibit the proliferation of breast MCF-7 cancer cells by induction of apoptosis.
- Zeboarin, kurdiona, and kurkumol are reported to have antineoplastic function by inhibiting ribosomal activity of the cancer cells.
- Elemene was found to have anticancer activity against promyelocytic leukemia cells HL-60.

- Curcumin and crurcumenol are reported to inhibit the growth of U-14 cell sarcoma cells in mice.
- Extracts from the plant showed inhibition of the MDA-MB-231 human breast cancer cell line.
- Isocurcumenol is shown to significantly inhibit the proliferation of various types of cancer cells such as lung cancer, leukemia, and nasopharyngeal carcinoma.

References

Carvalho, F.R., R.C. Vassão, M.A. Nicoletti, and D.A. Maria. 2010. Effect of *Curcuma zedoaria* crude extract against tumor progression and immunomodulation. *Journal of Venomous Animals and Toxins Including Tropical Diseases* 16, no. 2:324–341.

Gao, X.F., Q.L. Li, H.L. Li, et al. 2014. Extracts from *Curcuma zedoaria* inhibit proliferation of human breast cancer cell MDA-MB-231 *in vitro*. *Evidence-Based Complementary and Alternative Medicine* 2014:730678.

Khaing, S.L., S.Z. Omar, C.Y. Looi, A. Arya, N. Mohebali, and M.A. Mohd. 2017. Identification of active extracts of *Curcuma zedoaria* and their real-time cytotoxic activities on ovarian cancer cells and HUVEC cells. *Biomedical Research* 28, no. 21:9182–9187.

Lakshmi, S., G. Padmaja, and P. Remani. 2011. Antitumour Effects of Isocurcumenol Isolated from Curcuma zedoaria Rhizomes on Human and Murine Cancer Cells. *International Journal of Medicinal Chemistry* 2011:253962.

Shin, Y., and Y. Lee. 2013. Cytotoxic activity from *Curcuma zedoaria* through mitochondrial activation on ovarian cancer cells. *Toxicological Research* 29, no. 4:257–261.

Cyathula prostrata (Amaranthaceae) (Caryophyllales)   **Anti-tumor Anti-leukemic**

Location: India

Appearance: shrub up to one meter

Leaves: opposed, diamond-ovate to elliptical, acute at the tip
Flowers: light pink to purple

Tradition: in Nigeria it is traditionally used for the treatment of various inflammations and pain associated with diseases. It has also been used in Ayurvedic medicine for the treatment of asthma, liver disease, anemia, diabetes, and malaria.

Parts used: whole plant, roots

Active ingredients: alkaloids, flavonoids, glycosides, steroids, and tannins

Documented target cancers:

- The methanolic extract of *Cyathula prostrata* is reported to increase the survival period of animals with Dalton's lymphoma ascites cells induced tumors.
- The methanolic extract of the plant reversed the tumor-induced alterations in DNA fragmentation in Ehrlich ascites carcinoma (EAC)-bearing mice.
- An ethanolic *C. prostrata* extract is reported to show dose–response effects on cervical cancer HeLa and the malignant U937 cell lines, while cytotoxicity of the plant extract was not evident when treating normal peripheral blood mononuclear cells.

References

Mayakrishnan, V., P. Kannappan, K. Shanmugasundaram, and N. Abdullah. 2014. Anticancer activity of *Cyathula prostrata* (Linn) Blume against Dalton's lymphomae in mice model. *Pakistan Journal of Pharmaceutical Sciences* 27, no. 6:1911–1917.

Priya, K., S. Krishnakumari, and M. Vijayakumar. 2013. *Cyathula prostrata*: A potent source of anticancer agent against daltons ascites in Swiss albino mice. *Asian Pacific Journal of Tropical Medicine* 6, no. 10:776–779.

Schnablegger, G.E., L. Venables, T.C. Koekemoer, A.A. Sowemimo, and M. Van de Venter. 2013. *Cyathula prostrata* ethanol extract activates the extrinsic pathway of apoptosis in HeLa and U937 cell lines. *South African Journal of Botany* 88:380–387.

Cyclocarya paliurus (Rosaceae) (Rosales)   **Cytotoxic**

Location: native to eastern and central China
    Appearance: deciduous tree, reaches 30 meters height

    Leaves: feathered leaves 20–25 cm long with 5 to 11 leaflets. The leaflets are 5–14 cm long and 2–6 cm wide.

Tradition: used in traditional Chinese medicine as an antihyperlipidemic
    Parts used: fruit
    Active ingredients: proteins, polysaccharides, saponins, flavonoids, steroids, and phenolics
    Documented target cancers:

- A *Cyclocarya paliurus* polysaccharide is reported to show strong inhibition of human cancer gastic cell growth.
- The polysaccharide is reported to significantly inhibit the proliferation of cancerous HeLa cells, while not influencing the growth of normal HUVEC cells.

References

Liu, X., S.Q. Wang, M.Y. Xie, J.H. Xie, D.F. Huang, and Y.F. Tang. 2007. Effects of polysaccharide from *Cyclocarya paliurus* (Batal.) on growth of HeLa cells and human umbilical vein endothelial cells. *Food Science* 10:135.

Xie, J.H., X. Liu, M.Y. Shen, et al. 2013. Purification, physicochemical characterisation and anticancer activity of a polysaccharide from *Cyclocarya paliurus* leaves. *Food Chemistry* 136, no. 3–4:1453–1460.

Cymbopogon citratus (Poacea) (Poales)   **Anti-cancer**

Location: South and Southeast Asia
    Appearance: aromatic, evergreen, clump-forming, perennial grass

    Leaves: straplike with graceful, tilting edges. They are bright blue-green and release a citrus aroma when crushed.

Tradition: used as a spice in eastern cuisine. Provides protection from microbial infections of the urinary and respiratory system, the colon, stomach, and open wounds. It reduces fever. Helps the body get rid of undesirable toxic substances, cleanses the liver, digestive system, kidneys, pancreas, and bladder, helping to stay healthy.
    Used for immediate relief from toothache, joint pain, muscle pain, and generally fights any kind of inflammation associated with pain.
    Parts used: leaves, stem, roots

Active ingredients: luteolin 7-O-neohesperidoside (a promising compound to fight inflammatory skin conditions), b-d-Xylofuranose (anti-cancer and anti-inflammatory action)

Documented target cancers:

Polysaccharide fractions for the plant exhibihted potential cytotoxic and apoptotic effects on Siha (uterus carcinoma) and LNCap (prostate cancer) cell lines, as they induced apoptosis through the events of up-regulation of caspase-3, down-regulation of bcl-2 family genes followed by cytochrome *c* release.

References

Pereira, R.P., R. Fachinetto, A. de Souza Prestes, et al. 2009. Antioxidant effects of different extracts from *Melissa officinalis*, *Matricaria recutita* and *Cymbopogon citratus*. *Neurochemical Research* 34, no. 5:973–983.

Thangam, R., M., Sathuvan, A. Poongodi, et al. 2014. Activation of intrinsic apoptotic signaling pathway in cancer cells by Cymbopogon citratus polysaccharide fractions. *Carbohydrate Polymers* 107:138–150.

Cynodon dactylon (Poacea) (Poales)   **Anti-cancer**

Location: middle East (origin), Australia, Africa, southern Europe

Appearance: stoloniferous grass

Leaves: the blades are small, gray-green, usually 2–15 cm.

Tradition: it is considered one of the most useful medicinal plants and is included in many combinations for the treatment of the prostate. Natural antibiotic and one of the most famous diuretics. It cleanses the body of toxins and reduces blood cholesterol. Helps with liver colic, bile stones, and cellulite. It acts as antiseptic and anti-inflammatory in urinary infections, such as cystitis, uranitis, prostatitis (in excellent combination with Achilles), rheumatism, and arthritic skin diseases and hepatitis. It is beneficial for nephrolithiasis and slamming (kidney sand).

Parts used: herb and root stalk

Active ingredients: apigenin, luteolin, horridin, and carotenoids: β-carotene, neoxanthin, glycosides, saponins, and volatile oils

Precautions: the hybrid variety Tifton 85 produces cyanide under certain conditions, and has been involved in several animal deaths.

Documented target cancers:

- The methanolic extract of roots of *Cynodon dactylon* protected mice from diethyl nitrosamine induced hepatic carcinoma. Additionally, methanolic extracts have been shown to reverse the adverse effects of Ehrlich's ascites carcinoma (EAC) in Swiss albino mice. *C. dactylon* was also found to induce apoptotic cell death in COLO 320 DM (colon adenocarcinoma) cells.
- Indicative dosage and application: the dose of 500 mg/kg bw was identified as the most effective dose (regarding its hypoglycemic and antidiabetic effects). Forms: juice, paste, crushed

References

Albert-Baskar, A., and S. Ignacimuthu. 2010. Chemopreventive effect of *Cynodon dactylon* (L.) Pers. extract against DMH-induced colon carcinogenesis in experimental animals. *Experimental and Toxicologic Pathology* 62, no. 4:423–431.

Kowsalya, R., J. Kaliaperumal, M. Vaishnavi, and E. Namasivayam. 2015. Anticancer activity of *Cynodon dactylon* L. root extract against diethyl nitrosamine induced hepatic carcinoma. *South Asian Journal of Cancer* 4, no. 2:83–87.

Marappan, S., and A. Subramaniyan. 2012. Antitumor activity of methanolic extract of *Cynodon dactylon* leaves against Ehrlich ascites induced carcinoma in mice. *Journal Advances Scient Resource* 3:105–108.

Singh, S.K., P.K. Rai, D. Jaiswal, and G. Watal. 2008. Evidence-based critical evaluation of glycemic potential of *Cynodon dactylon*. *Evidence-Based Complementary and Alternative Medicine* 5, no. 4:415–420.

Cynomorium coccineum (Cynomoriacea)   **Anti-cancer**

Location:

- *Cynomorium* var. *coccineum coccineum* is found in the Mediterranean region, the Canary Islands and Mauritania.
- *Cynomorium* var. *coccineum songaricum* is located in Central Asia and Mongolia, where it grows at high altitudes.
- Several authorities believe that *C. songaricum*, which is called 'suo yang' in China, is a distinct species.

Appearance: parasitic perennial flowering plant

Stem: fleshy, unbranched (most of which is underground)
Leaves: scale-like, membranous
Flowers: low-growing inflorescence (in spring). Dark-red or purplish, the inflorescence consists of a dense, erect, club-shaped mass of minute scarlet flowers, which may be male, female, or hermaphrodite.

Tradition: the Knights of St. John, in some of their writings, claim that the juice of the plant stops bleeding, heals the wounds, and restores the natural forces. Generally, it is believed to strengthen bones and tendons and to revive blood while stimulating vigor.

Parts used: dry parts for medical use, fresh as a snack, root

Active ingredients: contains anthocyanic glycosides, triterpenoid saponins, and lignans. *Cynomorium coccineum* var. *coccineum* from Sardinia was found to contain gallic acid and cyanidin-3-O-glucoside as the main constituents.

Indicative dosage and application: supplements, extract, pills, and powder

Documented target cancers:

- Colon adenocarcinoma
- Melanoma cells

References

Abdel-Magied, E.M., H.A. Abdel-Rahman, and F.M. Harraz. 2001. The effect of aqueous extracts of Cynomorium coccineum and Withania somnifera on testicular development in immature Wistar rats. *Journal of Ethnopharmacology* 75, no. 1:1–4.

Cui, Z., Z. Guo, J. Miao, Z. Wang, Q. Li, X. Chai, and M. Li. 2013. The genus Cynomorium in China: An ethnopharmacological and phytochemical review. *Journal of Ethnopharmacology* 147, no. 1:1–15.

Meng, H.C., S. Wang, Y. Li, Y.Y. Kuang, and C.M. Ma. 2013. Chemical constituents and pharmacologic actions of Cynomorium plants. *Chinese Journal of Natural Medicines* 11, no. 4:321–329.

Rosa, A., M. Nieddu, A. Piras, A. Atzeri, D. Putzu, and A. Rescigno. 2015. Maltese mushroom (*Cynomorium coccineum* L.) as source of oil with potential anticancer activity. *Nutrients* 7, no. 2:849–864.

Rosa, A., A. Rescigno, A. Piras, et al. 2012. Chemical composition and effect on intestinal Caco-2 cell viability and lipid profile of fixed oil from *Cynomorium coccineum* L. *Food and Chemical Toxicology* 50, no. 10:3799–3807.

Cynomorium songaricum (Cynomoriacea)	Anti-cancer

Note: *Cynomorium songarium* or *Cynomorium coccineum* var. *songaricum* are considered by some as different species.

Location:

- *Cynomorium* var. *coccineum songaricum* is located in Central Asia and Mongolia, where it grows at high altitudes.
- *Cynomorium* var. *coccineum coccineum* is found in the Mediterranean region, the Canary Islands, and Mauritania.
- *C. songaricum* is called 'suo yang' in China.

Appearance: parasitic perennial flowering plant

Stem: fleshy, cylindrical, mostly buried in the ground, and only with the small top on the ground and a slightly expanded base that is in a diameter of 3 to 6 cm.

Leaves: interchangeable, oval, triangular or triangular-ovate, 0.5 to 1 cm long, less than 1 cm wide with acute peak.

Flowers: dark purple, polygamous

Parts used: root

Active ingredients: cynoterpene, acetylursolic acid, ursolic acid, palmitic acid, oleic acid, and linoleic acid, sterol contains β-sitosterol, campesterol, β-sitosterol palmitate, and daucosterol

Tradition: the combination of *Cynomorium*, Ba Ji Tian (Morinda Root), and other herbs has remarkable therapeutic function on sexual dysfunction. According to the Chinese Materia Medica, it is sweet in flavor and warm in nature. Essential functions are invigorating the kidney, strengthening Yang, and relaxing bowel. Main uses and indications include impotence due to kidney deficiency, nocturnal emission, premature ejaculation, weak lower limbs, and deficiency-type constipation.

Documented target cancers: lung cancer cell line A549 and also see *Cynomorium coccineum*.

Reference

Yang, F., P.W. Zhao, P. Sun, L.J. Ma, and P.W. Zhao. 2016. Effect of *Cynomorium songaricum* polysaccharide on telomere of lung cancer A549 cells. Zhongguo Zhong yao za zhi- Zhongguo Zhongyao zazhi. *Zhongguo Zhong Yao za Zhi = Zhongguo Zhongyao Zazhi = China Journal of Chinese Materia Medica* 41, no. 5:917–921.

Cyrtomium fortumei (Cyrtomium) (Dryopteridaceae)	Anti-melanoma

Location: Asia (Japan, Korea, and China)

Appearance: evergreen fern with erect 'shuttlecocks' of pinnate fronds, height 30–60 cm

Leaves: about 12–26 pairs of narrowly sickle-shaped, deep dull green leaflets, pinnate fronds to 1.2 m tall

Tradition: ornamental: decoration of garden and interior

Documented target cancers:

- The methanolic extract of the *Cyrtomium fortunei* J. Smith root has been shown to inhibit tyrosinase activity and melanin production in melanoma cells.

References

Choi, S.Y. 2013. Inhibitory effects of *Cyrtomium fortunei* J. Smith root extract on melanogenesis. *Pharmacognosy Magazine* 9, no. 35:227–230.

Yang, S.J., M.C. Liu, N. Liang, H.M. Xiang, and S. Yang. 2013. Chemical constituents of *Cyrtomium fortumei* (J.) Smith. *Natural Product Research* 27, no. 21:2066–2068.

Dasymaschalon trichophorum (Dasymaschalon) (Annonaceae)   **Anti-cancer**

Location: China
   Appearance: tree to 3.5 m tall
   Leaves: densely, hairy, elliptic to obovate

   Flowers: axillary or terminal on young growth, sepals 7–16 × 4–9 mm

Parts used: stems, leaves
   Active ingredients: five aristolactam alkaloids, which are 10-amino-3,6-dihydroxy-2,4-dim ethoxyphenanthrene-1-carboxylic acid lactam, enterocarpam-II, oldhamactam, goniopedaline, and stigmalactam.
   Documented target cancers:
   The essential oil from the fruits exhibited strong activity against lung adenocarcinoma SPCA-1 cell line.

References

Li, X.B., G.Y. Chen, X.P. Song, C.R. Han, and C.J. Zheng. 2013. Composition and antitumor activities of essential oil from the fruits of *Dasymaschalon trichophorum*. *Zhong Yao Cai = Zhongyaocai = Journal of Chinese Medicinal Materials* 36, no. 11:1786–1788.

Liu, Y.L., D.K., Ho, J.M. Cassady, et al. 1992. Dasytrichone, a novel flavone from *Dasymaschalon trichophorum* with cancer chemopreventive potential. *Natural Product Letters* 1, no. 3:161–165.

Zhou, X., J. Wu, J. Bai, X. Hu, E. Li, N. Shi, and Y. Pei. 2013. A new aristolactam alkaloid from the stems of *Dasymaschalon trichophorum*. *Chinese Journal of Natural Medicines* 11, no. 1:81–83.

Daucus carota (Daucus) (Apiaceae)   **Anti-cancer Anti-leukemic**

Location: native to temperate regions of Europe and southwest Asia, and naturalized to North America and Australia
   Appearance: biennial plant that grows between 30 and 60 cm tall

   Leaves: coming from the top of the root with a long stem
   Flowers: pink, yellowish, or white, they grow the second year

Tradition: used as diuretic, emmenagogue, and stimulant and are used for dropsy, chronic dysentery, kidney ailments, and worms. A tea made from the plant was believed to help maintain low blood sugar.

Parts used: root, flowers, leaves

Active ingredients: falcarinol, carotenoids, vitamins (A, C, K, B1, B3, B6)

Precautions: can cause a slight intoxication to grazing mammals, like cattle and horses. The wildflower can be toxic to humans. Skin contact with wet *Daucus carota* can cause irritation and vesication occur. The leaves may cause phytophotodermatitis.

Documented target cancers:

- Different fractions from carrot juice extract showed cytotoxicity against human lymphoid leukemia cell lines.
- Pentane-based fractions of *Daucus carota* oil showed cytotoxicity against human breast adenocarcinoma cell lines MDA-MB-231 and MCF-7.
- Ethylacetate partition layer of root of *Daucus carota* L. showed strong cytotoxic effects on HepG2, HeLa, C6, and NB41A3 cell lines.

References

Diab-Assaf, M., S., El-Sharif, and M. Mroueh. 2007. Evaluation of anti-cancer effect of *Daucus carota* on the human promyelocytic leukemia HL-60 cells. AACR International Conference on Molecular Diagnostics in Cancer Therapeutic Development, Atlanta, USA.

Han, E.J. 2000. Cytotoxicity of Daucus carota L. on various cancer cells. *Journal of the Korean Society of Food Science and Nutrition* 29:153.

Shebaby, W.N., M. Mroueh, K. Bodman-Smith, A. Mansour, R.I. Taleb, C.F. Daher, and M. El-Sibai. 2014. *Daucus carota* pentane-based fractions arrest the cell cycle and increase apoptosis in MDA-MB-231 breast cancer cells. *BMC Complementary and Alternative Medicine* 14, no. 1:387.

Zaini, R., K.R. Brandt, M., Clench, and C.L. Le Maitre. 2012. Effects of bioactive compounds from carrots (Daucus carota L.), polyacetylenes, beta-carotene and lutein on human lymphoid leukaemia cells. *Anti-Cancer Agents in Medicinal Chemistry (Formerly Current Medicinal Chemistry-Anti-Cancer Agents)* 12, no. 6:640–652.

Dendrobium formosum (Dendrobium) (Orchidaceae)   **Anti-cancer**

Location: Himalayas, northern Indochina, and the Andaman Islands

Appearance: orchid species

Leaves: short, ovate leaves grow alternately over the whole length of the stems.

Flowers: large, sepal, and petals pure white, lip white with an orbicular yellow marking on its disc extended to the base in the form of a ridge. Sepals very narrow, oblong-lanceolate, acute, spreading, the lateral ones very slightly keeled. Petals much larger than the petals, sub-orbicular.

Parts used: stems

Active ingredients: rich in flavonoids, glycosides, carbohydrates

Documented target cancers:

- The ethanolic extract of *Dendrobium formosum* has demonstrated potent anticancer activity in both *in vitro* and *in vivo* studies against Dalton's lymphoma cells.

References

Ng, T.B., J. Liu, J.H. Wong, X. Ye, S.C. Wing Sze, Y. Tong, and K.Y. Zhang. 2012. Review of research on Dendrobium, a prized folk medicine. *Applied Microbiology and Biotechnology* 93, no. 5:1795–1803.

Prasad, R., and B. Koch. 2014. Antitumor activity of ethanolic extract of *Dendrobium formosum* in T-cell lymphoma: An *in vitro* and *in vivo* study. *BioMed Research International* 2014:753451.

Dendropanax arboreus (Dendropanax) (Araliaceae)   **Cytotoxic**

Appearance:

Stem: spines absent
Root: stilt roots absent
Leaves: spiral, not scale-like, simple, trinerved at base, coriaceous, symmetric at the base, palmately lobed, smooth margined
Flowers: bisexual, stalked, round

Active ingredients:

- Falcarinol, dehydrofalcarinol, diyenne, falcarindiol, dehydrofalcarindiol, and
- Two novel polyacetylenes: dendroarboreols A and B

Further details:
- The major compound responsible for the *in vitro* cytotoxicity was falcarinol. Several other known compounds were isolated and found to be cytotoxic, including dehydrofalcarinol, a diyenne, falcarindiol, and dehydrofalcarindiol. In addition, two novel polyacetylenes, dendroarboreols A and B, were isolated and characterized by standard and inverse-detected NMR methods.

Reference

Bernart, M.W., J.H. Cardellina, M.S. Balaschak, M.R. Alexander, R.H. Shoemaker, and M.R. Boyd. 1996. Cytotoxic falcarinol oxylipins from *Dendropanax arboreus*. *Journal of Natural Products* 59, no. 8:748–753.

Dendropanax morbifera (Araliaceae) (Apiales)   **Cytotoxic**

Location: subtropical regions and southern regions of Korea
Appearance: evergreen tree, reaches 15 meters high

Leaves: oval
Flowers: in bloom from August to September

Tradition: in traditional Chinese medicine various parts of the plant are used for a wide range of diseases, mainly improvement of the function of the liver, prevention of old-age diseases (hypertension, diabetes) as the root contains saponins, immune enhancement, blood purification, antimicrobial action, antioxidant, regeneration of hard tissue (bones and teeth), enhancement of brain activity, sedative action.

Parts used: leaf, stem, root, and yellow resin
Active ingredients: chlorogenic acid, rutin, saponins
Documented target cancers:
- Treatment with the extracts of debarked stems, green leaves, and yellow leaves is reported to cause an increase of apoptotic or senescent cells in human hepatoma cells Huh-7.

Reference

Hyun, T.K., M.O. Kim, H. Lee, Y. Kim, E. Kim, and J.S. Kim. 2013. Evaluation of anti-oxidant and anti-cancer properties of *Dendropanax morbifera* Léveille. *Food Chemistry* 141, no. 3:1947–1955.

Dendrophthoe falcata (Loranthaceae) (Santalales)   **Anti-tumor**

Location: tropical Africa, Asia, Australia, India

Appearance: parasitic plant with a gray smooth bark and small branches (2 mm to 2.5 cm in thickness)

Leaves: simple, plump, oval, thick, leathery, and their arrangement is usually opposite, alternating or in coils.

Flowers: flowers are spikly, pentameric, orange-red, axillary, in small spread with thick bunches. The cup of cauliflower is about 4 mm long, and the crown of the flower is 2.5 to 5 cm long, divided into the back. The stamens are six, long thread, and the anthers are monotonous.

Tradition: the whole plant is used for medical purposes. It is mainly used for treatment of pulmonary tuberculosis, asthma, menstrual disorders, and impotence, swelling, sores, ulcers, kidney stones, and bladder stones. It is also a diuretic, aphrodisiac, and narcotic. Also important is its action on wound healing, anti-microbial, antioxidant, and hepato-protective activity. In the traditional medicine system, *D. falcata* is recommended for the treatment of epilepsy.

Parts used: leaves, flowers, aerial parts

Active ingredients: flavonoids such as quercetin and tannins, β-amyrin, oleanolic acid, β-sitosterol, and stigmasterol. The root contains catechins and the cortex leucocynidin. Tannins have mainly been found in the fruit.

Precautions:

In high doses the plant can be toxic, while its decoction has anti-fertility action.

Documented target cancers:

• Cytotoxic effect against human breast adenocarcinoma MCF-7 cells

Ethanolic and aqueous extracts showed significant decrease in tumor volume, viable cell count, tumor weight, and elevated the life span of Ehlrich ascites carcinoma tumor-bearing mice.

References

Dashora, N., V. Sodde, K. Prabhu, and R. Lobo. 2011. Antioxidant activities of *Dendrophthoe falcata* (Lf) Etting. *Pharmaceutical Crops* 2, no. 1:24–27.

Dashora, N., V. Sodde, K.S. Prabhu, and R. Lobo. 2011. *In vitro* cytotoxic activity of *Dendrophthoe falcata* on human breast adenocarcinoma Cells-MCF-7. *International Journal of Cancer Research* 7, no. 1:47–54.

Karthikeyan, A., R. Rameshkumar, N. Sivakumar, I.S. Al Amri, S. Karutha Pandian, and M. Ramesh. 2012. Antibiofilm activity of *Dendrophthoe falcata* against different bacterial pathogens. *Planta Medica* 78, no. 18:1918–1926.

Sathishkumar, G., C. Gobinath, A. Wilson, and S. Sivaramakrishnan. 2014. Dendrophthoe falcata (Lf) Ettingsh (Neem mistletoe): A potent bioresource to fabricate silver nanoparticles for anticancer effect against human breast cancer cells (MCF-7). *Spectrochimica Acta Part A* 128:285–290.

Singh, R.B., and P.K. Gupta. 2013. Morphotaxonomy, medicinal use and new host range of *Dendrophthoe falcata* var. coccinia in Champaran, its cause and consequences. *Indian Journal of Life Sciences* 2, no. 2:39.

Sinoriya, P., R. Irchhaiya, B. Sharma, G. Sahu, and S. Kumar. 2011. Anticonvulsant and muscle relaxant activity of the ethanolic extract of stems of *Dendrophthoe falcata*. (Linnean F.) in mice. *Indian Journal of Pharmacology* 43, no. 6:710.

Sinoriya, P.O.O.J.A., R.E.N.D.R.A. Sharma, and A.R.T.I. Sinoriya. 2011. A review on Dendrophthoe falcata (Linn. F.). *Asian Journal of Pharmaceutical and Clinical Research* 4:1–5.

Dimocarpus longan (Sapindaceae) (Sapindales)   **Anti-cancer**

Location: southeastern Asia, Puerto Rico, Florida, Thailand, Hawaii, Vietnam, Australia, Taiwan
   Appearance: tree about 15 meters high, with dense foliage, symmetrical crown, and rough bark

   Leaves: composite up to 30 cm in length, with six to nine pairs
   Flowers: equestrian, female and hermaphrodite, small, whitish

Tradition:

- Crushed seeds produce foam that is used as a shampoo.
- Wood is used in the manufacture of furniture and other items.
- The tree is also planted as ornamental.

Active ingredients: water soluble polysaccharide LP1 (longan 1 polysaccharide (LP1)) and phenolic compounds: gallic acid, corilagin, and ellagic acid
   Documented target cancers:

- Polyphenol-rich longan seed extract induced S phase arrest and apoptotic death in three human colorectal carcinoma cell lines.
- A water-soluble polysaccharide extracted from *Dimocarpus longan* pulp had significantly high antitumor activity against SKOV3 and HO8910 ovarian carcinoma cell lines.
- High pressure-assisted extract of longan significantly inhibited cancer cell growth of SGC 7901 (gastric carcinoma) and A549 (lung adenocarcinoma) cell lines.

References

Chung, Y.C., C.C. Lin, C.C. Chou, and C.P. Hsu. 2010. The effect of Longan seed polyphenols on colorectal carcinoma cells. *European Journal of Clinical Investigation* 40, no. 8:713–721.

Meng, F.Y., Y.L. Ning, J. Qi, et al. 2014. Structure and antitumor and immunomodulatory activities of a water-soluble polysaccharide from *Dimocarpus longan* pulp. *International Journal of Molecular Sciences* 15, no. 3:5140–5162.

Prasad, K.N., J., Hao, J. Shi, et al. 2009. Antioxidant and anticancer activities of high pressure-assisted extract of longan (*Dimocarpus longan* Lour.) fruit pericarp. *Innovative Food Science and Emerging Technologies* 10, no. 4:413–419.

Dioscorea bulbifera (Dioscoreaceae) (Dioscoreales)   **Anti-tumor**

Location: Florida, southern Asia, Caribbean, Australia, northern America, US, Hawaii, Puerto Rico
   Appearance: dioecious, herbaceous perennial vine

   Leaves: protruding into long stems, alternating, generally heart-shaped, and up to 20 cm long
   Flowers: rarely flowering, small with pale green color, aromatic, arising from the axillary areas of the leaves

Tradition:

- It is used both in Indian and Chinese traditional medicine to treat gastric cancer and rectal and goiter carcinomas. The extract is said to be antihyperlipidemic, anti-tumor, antioxidant, analgesic, anti-inflammatory, antidiabetic, and anti-hyperglycemic. Dried, it is known to possess toxins for treatment of poisonous infections and is also considered suitable against dysentery and diarrhea. Additionally, it has been used against dog bites, snake bites, and food poisoning in China.
- The extract of *D. bulbifera* tubers is used in the treatment of goitre. *D. bulbifera* serves as a well-known ophthalmic remedy. In particular, purulent ophthalmia is treated with sap from stem as eye drops in Congo.

Parts used: tuber, bulbils

Active ingredients: ethyl-O-β-D-fructo-pyranoside and butyl-O-β-D-fructopyranoside, a-glucones, flavonol glycosides, myricetin, hyperoside, methyl-O-α-d-fructofuranoside, and butyl-O-α-D-fructofuranoside.

Precautions: the roots of *D. bulbifera* from China have been reported as toxic and therefore have limited use in traditional medicine. Dioscine and diosbulbin B derived from its roots are responsible for liver toxicity, nausea, abdominal pain, coma, and even death.

Documented target cancers:

- The ethyl acetate soluble fraction of 75% ethanol extract of the rhizomes of *Dioscorea bulbifera* L. showed an inhibitory effect against the tumor promotion of JB6 cells.
- *Dioscorea bulbifera* extract exhibited cytotoxic activity against HeLa breast cancer cells.
- The ethanol and ethylacetate extracts were found to have the ability to decrease tumor weight in S180 and H22 tumor cell-bearing mice.
- Extracts induced significant inhibition in human liver cancer, colon cancer, and other tumor cells.
- Tubers petroleum ether extract had strong inhibitory effects on the formation of ascites of the HepA tumor cell lines on mice *in vivo*.
- *D. bulbifera* decoction could inhibit cell growth in the human squamous cell carcinoma cell line SiHa, in the human cervical cancer cells HeLa, and in the human hepatoma cells HepG2, in a dosage- and time-dependent manner.
- SMMC7721 human hepatocellular carcinoma cells

References

Gao, H., M. Kuroyanagi, L. Wu, N. Kawahara, T. Yasuno, and Y. Nakamura. 2002. Antitumor-promoting constituents from *Dioscorea bulbifera* L. in JB6 mouse epidermal cells. *Biological and Pharmaceutical Bulletin* 25, no. 9:1241–1243.

Ghosh, S., V.S. Parihar, P. More, D.D. Dhavale, and B.A. Chopade. 2015. Phytochemistry and therapeutic potential of medicinal plant: *Dioscorea bulbifera*. *Medicinal Chemistry* 5, no. 4:154–159.

Guan, X.R., L. Zhu, Z.G. Xiao, Y.L. Zhang, H.B. Chen, and T. Yi. 2017. Bioactivity, toxicity and detoxification assessment of *Dioscorea bulbifera* L.: A comprehensive review. *Phytochemistry Reviews* 16, no. 3:573–601.

Itharat, A.and B. Ooraikul. 2007. Research on Thai medicinal plants for cancer treatment. In: Advances in Medicinal Plant Research, ed. S.N. Acharya, and J.E. Thomas. Research Signpost, India..

Kosuge, T., M. Yokota, K. Sugiyama, T. Yamamoto, M.Y. Ni, and S.C. Yan. 1985. Studies on antitumor activities and antitumor principles of Chinese herbs. I. Antitumor activities of Chinese herbs. *Yakugaku Zasshi* 105, no. 8:791–795.

Li, J.H., X.H. Zhang, F.W. Zheng, Y.X. Zhou, and J. Gao. 1999. Study on the antitumor constituents and biological activity of *D. bulbifera*. *China Pharmacy* 8:40–41.

Murray, R.D.H., Z.D. Jorge, N.H. Khan, M. Shahjahan, and M. Quaisuddin. 1984. Diosbulbin D and 8-epidiosbulbin E acetate, norclerodane diterpenoids from *Dioscorea bulbifera* tubers. *Phytochemistry* 23, no. 3:623–625.

***Dracaena draco* (Asparagaceae) (Asparagales)**   **Anti-leukemic**

Location: Portugal, Madeira, Spain (Canary islands), Morocco
  Appearance: evergreen tree

  Leaves: green and red in their base, arranged in dense rosettes at the ends of the branches
  Flowers: very fragrant, they form large greenish clusters with white petals.

Parts used: resin
  Active ingredients: cytotoxic steroids-saponins
  Tradition: the resin of *D. draco* is used from antiquity for artistic purposes, as well as for traditional medicine as a haemostatic, for ulcer, antimicrobial, antiviral, wound healing, anticancer, anti-inflammatory, antioxidant, etc.
  Precautions: it may cause allergic reactions such as: itching, edema, skin rashes.
  Documented target cancers:
* A natural steroidal saponin, icogenin, isolated from *Dracaena draco*, was found cytotoxic against the myeloid leukemia cell line HL-60.

References

Di Stefano, V.D., R. Pitonzo, and D. Schillaci. 2014. Phytochemical and anti-staphylococcal biofilm assessment of *Dracaena draco* L. spp. draco resin. *Pharmacognosy Magazine* 10 Suppl 2:S434–S440.

González, A.G., F. León, L. Sánchez-Pinto, J.I. Padrón, and J. Bermejo. 2000. Phenolic compounds of Dragon's Blood from Dracaena d raco. *Journal of Natural Products* 63, no. 9:1297–1299.

Gupta, D., B., Bleakley, and R.K. Gupta. 2008. Dragon's blood: Botany, chemistry and therapeutic uses. *Journal of Ethnopharmacology* 115, no. 3:361–380.

Hernández, J.C., F. León, J. Quintana, F. Estévez, and J. Bermejo. 2004. Icogenin, a new cytotoxic steroidal saponin isolated from *Dracaena draco*. *Bioorganic and Medicinal Chemistry* 12, no. 16:4423–4429.

Krawczyszyn, J., and T. Krawczyszyn. 2016. Photomorphogenesis in *Dracaena draco*. *Trees* 30, no. 3:647–664.

***Dracunculus vulgaris* (Araceae) (Alismatales)**   **Cytotoxic**

Location: Balkan region
  Appearance: tuberous herbaceous perennial

  Leaves: palmate, wide and large, crimson-brown on the inside and green on the outside
  Flowers: the inflorescence is surrounded by a dark purple leaf bract called spathe, which can grow up to 125 cm in height (usually 40–50 cm).

Tradition: in folk medicine, tubers of this plant are used externally in the treatment of rheumatism and hemorrhoids.
  Parts used: root collected in spring or fall
  Active ingredients: saponin and conicine alkaloids, estragole, phelandrine, methyl cavicol, iodine, rutin, and coumarin
  Documented target cancers:
* Ethyl acetate, methanol, and water (infusion and decoction) extracts from *D. vulgaris* tubers are reported to exhibit high cytotoxic, DNA damaging, and apoptotic activities on MCF-7 breast cancer cell line.

Reference

Aslantürk, Ö.S., and T.A. Çelik. 2013. Potential antioxidant activity and anticancer effect of extracts from Dracunculus vulgaris Schott. Tubers on MCF-7 breast cancer cells. *International Journal of Research in Pharmaceutical and Biomedical Sciences* 4, no. 2:394–402.

Dregea volubilis (Apocynaceae) (Gentianales)   **Anti-leukemic**

Location: India, Sri Lanka, Myanmar, Thailand, Indonesia, China
  Appearance: big bush

Leaves: oval, partially rounded, opposed, occasionally spiral, rather leathery, with a length of 7.5–15 cm and a width of 5–10 cm
Flowers: green or yellowish green, aromatic, spiral, hermaphrodite, and width about 1 cm

Tradition:

- Shoots and flowers are used in cooking, mainly in Thai cuisine as a vegetable or as a seasoning.
- Beverages from the leaves have multiple uses. It is said to be extensively used to treat inflammation, sore throat, eczema, asthma, as an antidote to poisons as well as urinary tract disorders.
- Roots and stems are used as emesis and expectorant, and also extracts from root fragments are used, with insertion into the nose, to induce sneezing.

Parts used: leaves, roots, juice
  Active ingredients: steroids, triterpenoids, flavonoids, biflavonoids, tannins, and glycosides
  Documented target cancers:
- A pentacyclic triterpenoid compound isolated from the fruits of *D. volubilis* showed anti-tumor activity on the K562 leukemic cell line.

References

Aydiv, Ç., Ö.Z.A.Y. Cennet, O. Dusen, R. Mammadov, and F. Orphan. 2017. Total phenolics, antioxidant, antibacterial and cytotoxic activity studies of ethanolic extracts *Arisarum vulgare* O. Targ. Tozz. and *Dracunculus vulgaris* Schott. *International Journal of Secondary Metabolite* 4, no. 2:114–122.

Biswas, M., P.K. Haldar, and A.K. Ghosh. 2010. Antioxidant and free-radical-scavenging effects of fruits of *Dregea volubilis*. *Journal of Natural Science, Biology, and Medicine* 1, no. 1:29–34.

Hossain, E., A. Rawani, G. Chandra, S.C. Mandal, and J.K. Gupta. 2011. Larvicidal activity of *Dregea volubilis* and *Bombax malabaricum* leaf extracts against the filarial vector Culex quinquefasciatus. *Asian Pacific Journal of Tropical Medicine* 4, no. 6:436–441.

Moulisha, B., M.N. Bikash, P. Partha, G.A. Kumar, B. Sukdeb, and H.P. Kanti. 2009. *In vitro* anti-leishmanial and anti-tumour activities of a pentacyclic triterpenoid compound isolated from the fruits of *Dregea volubilis* Benth Asclepiadaceae. *Tropical Journal of Pharmaceutical Research* 8, no. 2:127–131.

Dryopteris erythrosora (Dryopteridaceae) (Polypodiales)   **Cytotoxic**

Location: southeastern Asia, Japan, China, Korea, Philippines
  Appearance: medium and large-sized evergreen fern

Leaves: glowing orange color (emerging leaves), with variations of rich red color, light green, and finally dark green for the mature leaves, dentate, 30–130cm long and 20–40cm wide.
Flowers: sterile

Tradition: *Dryopteris erythrosora* is mainly used as ornamental in gardens and in shaded areas.

Active ingredients: gliricidin 7-O-hexoside, apigenin 7-O-glucoside, quercetin 7-O-rutinoside, quercetin 7-O-galactoside, kaempferol 7-O-gentiobioside, kaempferol-3-O-rutinoside, myricetin 3-O-rhamnoside, quercitrin, (glucoside of flavanol), 3-deoxyanthocyanines

Documented target cancers:

- Flavonoids extract from *D. erythrosora* showed obvious cytotoxic effects on human non-small cell lung cancer A549 cells.

References

Cao, J., X. Xia, X. Chen, J. Xiao, and Q. Wang. 2013. Characterization of flavonoids from *Dryopteris erythrosora* and evaluation of their antioxidant, anticancer and acetylcholinesterase inhibition activities. *Food and Chemical Toxicology* 51:242–250.

Xie, Y., Y. Zheng, X. Dai, Q. Wang, J. Cao, and J. Xiao. 2015. Seasonal dynamics of total flavonoid contents and antioxidant activity of *Dryopteris erythrosora*. *Food Chemistry* 186:113–118.

Echinacea purpurea (Asteraceae) (Asterales)   **Anti-proliferative**

Location: native to North America, distributed throughout the eastern and central US and southern Canada

Appearance: herbaceous perennial

Stem: erect, stout, branched, hirsute or glabrous, 60–180 cm high
Leaves: basal leaves ovate to ovate-lanceolate, acute, coarsely or sharply serrate
Flowers: daisy-like head units, which are a conglomeration of many tiny florets

Tradition: used for infections, such as syphilis and septic wounds, but also as an 'anti-toxin' for snakebites and blood poisoning. Traditionally, it is described as an 'anti-infective' and was indicated in bacterial and viral infections, mild septicaemia, furunculosis, and other skin conditions, including boils, carbuncles, and abscesses.

Parts used: the fresh or dried flowering tops and the fresh pressed juice from the flowering tops

Active ingredients: alkamides, caffeic acid derivatives, polysaccharides, and alkenes (such as polyenes)

Precautions:

- Individuals with asthma or atopy may be predisposed to allergic reactions from oral/topical use of echinacea, consistent with IgE-mediated hypersensitivity.
- Individuals sensitive to members of the Asteraceae/Compositae plant family may be more likely to experience allergic responses (ragweed, chrysanthemums, marigolds, daisies, etc.).

Documented target cancers:

- *Echinacea* root extracts have cytotoxic effects on pancreatic cancer cell line in a concentration- and time-dependent manner.
- Daily consumption of *Echinacea* has been reported to be prophylactic and significantly alleviate leukemia in a mouse model.
- *E. purpurea* flower extract caused significant inhibition of proliferation in human colon cell lines.

References

Barnes, J., L.A. Anderson, S. Gibbons, and J.D. Phillipson. 2005. Echinacea species (*Echinacea angustifolia* (DC.) Hell., *Echinacea pallida* (Nutt.) Nutt., *Echinacea purpurea* (L.) Moench): A review of their chemistry, pharmacology and clinical properties. *Journal of Pharmacy and Pharmacology* 57, no. 8:929–954.

Basch, E., C. Ulbricht, S. Basch, et al. 2005. An evidence-based systematic review of Echinacea (*E. angustifolia* DC, *E. pallida*, *E. purpurea*) by the Natural Standard Research Collaboration. *Journal of Herbal Pharmacotherapy* 5, no. 2:s57–s88.

Chicca, A., B. Adinolfi, E. Martinotti, et al. 2007. Cytotoxic effects of Echinacea root hexanic extracts on human cancer cell lines. *Journal of Ethnopharmacology* 110, no. 1:148–153.

Miller, S.C. 2005. Echinacea: A miracle herb against aging and cancer? Evidence *in vivo* in mice. *Evidence-Based Complementary and Alternative Medicine* 2, no. 3:309–314.

Mistríková, I., and Š. Vaverková. 2007. Morphology and anatomy of *Echinacea purpurea*, *E. angustifolia*, *E. pallida* and *Parthenium integrifolium*. *Biologia* 62, no. 1:2–5.

Tsai, Y.L., C.C. Chiu, J.Y.F. Yi-Fu Chen, K.C. Chan, and S.D. Lin. 2012. Cytotoxic effects of Echinacea purpurea flower extracts and cichoric acid on human colon cancer cells through induction of apoptosis. *Journal of Ethnopharmacology* 143, no. 3:914–919.

Echinophora platyloba (Apiaceae) (Apiales)	**Anti-proliferative**

Location: native to Iran

  Appearance: spiny plant

  Stem: branched from bottom, with grooved branches, yellowish or bluish-green, thick, sturdy, and stiff, highly branched, with tentacles distorted

  Leaves: radical broad leaves

  Flowers: 'glowing' plant with yellow flowers and cylindrical fruit, each in a different warp

Tradition:

- In Iranian traditional medicine it is used as an antifungal agent for preventing fungi contamination of the dairy products and foods.
- In addition, it is also used as an aromatic food in the production of cheese and yoghurt.

Parts used: aerial parts (leaves and stem), essential oil

  Active ingredients: stigmasterol, sitosterol, stigmasterol-β-D-glycoside, quercetin, betulinic acid, and ursolic acid

  Documented target cancers:

- *Echinophora platyloba* methanolic extract was found to time- and dose-dependently inhibit the proliferation of mouse the fibrosarcoma cell line WEHI-164.
- *E. platyloba* extract severely repressed division of human leukemia cells in *in vitro* studies on Acute Promyelocytic Leukemia cell line (NB4).

References

Avijgan, M., and M. Mahboubi. 2015. *Echinophora platyloba* DC. as a new natural antifungal agent. Echinophora platyloba DC. As a new natural antifungal agent. *Asian Pacific Journal of Tropical Disease* 5, no. 3:169–174.

Delaram, M., S. Kheiri, and M.R. Hodjati. 2011. Comparing the effects of *Echinophora platyloba*, Fennel and bo on pre-menstrual syndrome. *Journal of Reproduction and Infertility* 12, no. 3:221.

Entezari, M., F.H. Dabaghian, and M. Hashemi. 2014. The comparison of antimutagenicity and anticancer activities of Echinophora platyloba DC on acute promyelocytic leukemia cancer cells. *Journal of Cancer Research and Therapeutics* 10, no. 4:1004–1007.

Hosseini, Z., Z. Lorigooini, M. Rafieian-Kopaei, H.A. Shirmardi, and K. Solati. 2017. A review of botany and pharmacological effect and chemical composition of Echinophora species growing in Iran. *Pharmacognosy Research* 9, no. 4:305–312.

Majid, A., M. Mohaddesse, D. Mahdi, S. Mahdi, S. Sanaz, and N. Kassaiyan. 2010. Overview on *Echinophora platyloba*, a synergistic anti-fungal agent candidate. *Journal of Yeast and Fungal Research* 1, no. 5:88–94.

Shahneh, F.Z., S. Valiyari, A. Azadmehr, et al. 2013. Inhibition of growth and induction of apoptosis in fibrosarcoma cell lines by *Echinophora platyloba* DC: *In vitro* analysis. *Advances in Pharmacological Sciences* 2013:512931.

Eclipta alba (Asteraceae) (Asterales)   **Anti-tumor**

Location: widespread across much of the world

Appearance: annual herbareous plant, reaching a height of 60 cm

Stem: straight or cylindrical cylindrical shaft

Leaves: elongated to oblique. The inflorescence may be terminal or axillary about 1 cm wide.

Flowers: white or creamy and hermaphrodite

Tradition: used by traditional Chinese medicine. All parts of the plant have diuretic, obstructive, analgesic, antipyretic, and tonic properties. The oil is used as a treatment against hair loss, athlete's foot, eczema, and dermatitis as well as healing wounds. The leaves are still used as an antidote for snake and scorpion bites in Korea.

Parts used: leaves and young shoots

Active ingredients: edelolactone, eclalbasaponins, ursolic acid, oleanolic acid, luteolin, and apigenin

Documented target cancers:

- *Eclipta alba* extract is reported as anticancer and potent MDR reversal agent in diethylnitrosamine) and 2-acetylaminofluorene liver cancer induced animal models.
- HeLa cells
- The alcoholic extract of *Eclipta alba* is found cytotoxic against various cancer cell lines but significantly induced apoptosis in human breast cancer cell lines by disrupting mitochondrial membrane potential and DNA damage.
- The alcoholic extract is also reported to inhibit migration in both MCF 7 and MDA-MB-231 breast cancer cells in a dose-dependent manner.

References

Chaudhary, H., P.K. Jena, and S. Seshadri. 2014. *In vivo* evaluation of *Eclipta alba* extract as anticancer and multidrug resistance reversal agent. *Nutrition and Cancer* 66, no. 5:904–913.

Jahan, R., A. Al-Nahain, S. Majumder, and M. Rahmatullah. 2014. Ethnopharmacological significance of *Eclipta alba* (L.) hassk. (Asteraceae). *International Scholarly Research Notices* 2014:385969.

Yadav, N.K., R.K. Arya, K. Dev, et al. 2017. Alcoholic extract of *Eclipta alba* shows *in vitro* antioxidant and anticancer activity without exhibiting toxicological effects. *Oxidative Medicine and Cellular Longevity* 2017:9094641.

Eriophyllum

See in *Eupatorium* under active ingredients.

Ervatamia divaricata (Ervatamia) (Apocynaceae)   Cytotoxic

Location: Southeast Asia
Appearance:

Stem: round, many branches, 0.5–3 m
Leaves: single, green, the surfaces of which are smooth and with raised veins. The length is 6–15 cm,
   and its width is 2–4 cm.
Flowers: snow white, 1–5 cm diameter, fragrant. The flower stalk protrudes from the leaves and bears
   one or two flowers.

Parts used: root, steam, and leaf
   Active ingredients: Vinca alkaloids (conophylline)
   Documented target cancers:

* The dimeric indole alkaloids of the plant have been reported to show cytotoxic activities against
  P-388 murine leukemia and A-549 human lung carcinoma cells.
* Conophylline inhibits the growth of K-ras-NRK cells, but this inhibition is reversible.
* The alkaloid also inhibits the growth of K-ras-NRK and K-ras-NIH3T3 tumors transplanted into
  nude mice.
* On the other hand, it shows no effect on survival of the mice loaded with L1210 leukemia.

Further details:

* *Ervatamia heyneana*: the whole plant contains unidentified factors with anticancer properties.
* *Ervatamia microphylla* contains conophylline, a vinca alkaloid, isolated from the plant.

References

Chitnis, M.P., D.D. Khandalekar, M.K. Adwankar, and M.B. Sahasrabudhe. 1971. Anticancer activity of the
   extracts of root, stem & leaf of *Ervatamia heyneana*. *Indian Journal of Experimental Biology* 9, no.
   2:268–270.
Johnson, R.K., M.P. Chitnis, W.M. Embrey, and E.B. Gregory. 1978. *In vivo* characteristics of resistance and
   cross-resistance of an adriamycin-resistant subline of P388 leukemia. *Cancer Treatment Reports* 62, no.
   10:1535–1547.
Umezawa, K., T. Taniguchi, M. Toi, et al. 1996. Growth inhibition of K-ras-expressing tumours by a new vinca
   alkaloid, conophylline, in nude mice. *Drugs under Experimental and Clinical Research* 22, no. 2:35–40.
Zhang, H., X.N. Wang, L.P. Lin, J. Ding, and J.M. Yue. 2007. Indole alkaloids from three species of the
   Ervatamia genus: *E. officinalis*, *E. divaricata*, and *E. divaricata Gouyahua*. *Journal of Natural Products*
   70, no. 1:54–59.

Eugenia jambolana (Myrtaceae) (Myrtales)   Cytotoxic

Location: native to the Indian subcontinent, Southeast Asia, and China
   Appearance: tropical berry tree with height of up to 30 meters, it can live up to 100 years.

Leaves: young leaves are pink and later become leathery, dark green, elongated, oval to elliptical, and
      6 to 12 cm long.

Tradition: the seed is used in traditional diabetes medicine. Leaves and bark are used to control
both pressure and gingivitis. Fruit pulp as well as bark and seed extract greatly help to reduce blood
glucose and urine.

Active ingredients: oleanolic acid, vetulinic acid, 1.8-cineole, quercetin, amphoferol, maldivin,
delphinidin-3,5-diglucoside, and petunidin-3,5-diglucoside

Documented target cancers:
- The fruit extract has been reported to exhibit cytotoxic properties against MCF-7 and MDA-MB-231
  breast cancer cell lines.

References

Donepudi, A.C., L.M. Aleksunes, M.V. Driscoll, N.P. Seeram, and A.L. Slitt. 2012. The traditional ayurvedic
      medicine, E ugenia jambolana (J amun fruit), decreases liver inflammation, injury and fibrosis during
      cholestasis. *Liver International* 32, no. 4:560–573.
Li, L., L.S. Adams, S. Chen, C. Killian, A. Ahmed, and N.P. Seeram. 2009. Eugenia jambolana Lam. berry
      extract inhibits growth and induces apoptosis of human breast cancer but not non-tumorigenic breast
      cells. *Journal of Agricultural and Food Chemistry* 57, no. 3:826–831.

Eupatorium cannabinum (Agrimony (Hemp)) (Compositae)   **Antitumor**

Synonyms: Holy Rope, St. John's Herb
      Location: common on the banks of rivers, sides of ditches, at the base of cliffs on the seashore,
and in other damp areas in most parts of Britain and Europe
      Appearance:

Stem: round, growing from 60 to 150 cm, with short branches, reddish in color, covered with downy
      hair, woody below
Root: woody
Leaves: the root-leaves are on long stalks. The stem-leaves have only very short footstalks. All the
      leaves bear distinct, short hairs.
Flowers: flower heads being arranged in crowded masses of a dull lilac color at the top of the stem or
      branches. Each little composite head consists of about five or six florets.
In bloom: late in summer and autumn

Degree of rarity: low
      Tradition: it has the reputation of being a good wound herb, whether bruised or made into an
ointment with lard. They used it as a strong purgative and emetic and for curing dropsy.
      Parts used: herb
      Active ingredients:

- Sesquiterpene lactones: eupatoriopicrin (EUP) (*E. cannabinum*), eupaserrin and deacetyleupaserrin
  (*E. semiserratum*), eupacunin (*E. cuneifolium*), lactones from *E. rotundifolium*
- Gamma-lactones: germacranolides (*E. semiserratum* and *Eriophyllum confertiflorum*)
- Eupatolide (*E. formosanum* HAY)
- Flavones: *eupatorin* and 5-hydroxy-3′,4′,6,7-tetramethoxyflavone (*E. altissimum*)

Particular value: herbalists recognize its cathartic, diuretic, and anti-scorbutic properties and consider it a good remedy for purifying the blood.

Precautions: cytotoxicity
Indicative dosage and application:

- Growth inhibition of the Lewis lung carcinoma and the F10 26 fibrosarcoma was found after i.v. injection of 20 or 40 mg/kg EUP (in mice C57B1), at a tumor volume of 500 microliters.

Documented target cancers:

- Antileukemic: eupaserrin and deacetyleupaserrin, germacranolides, flavones
- Antitumor: eupatoriopicrin, eupatolide, flavones
- Cytotoxic: flavones
- The ethanolic extract has been reported to exert cytotoxic activity on HT29 colon cancer cells leading to mitotic disruption and non-apoptotic cell death.

Further details:

- The sesquiterpene lactone eupatoriopicrin (EUP) from *Eupatorium cannabinum* L. has been shown to be cytotoxic in a glutathione (GSH)-dependent way, through the induction of DNA damage in tumor cells. The amount of EUP required to demonstrate DNA damage after a 24-hr post-incubation period lay within the concentration range that was effective in the clonogenic assay (1–10 micrograms/ml). Glutathione (GSH) depletion of the cells to about 99%, by use of buthionine sulphoximine (BSO), enhanced the extent of DNA damage.
- Germacranolides: the alpha,beta-unsaturated ester side chain adjacent to the gamma-lactone and either a primary or secondary allylic alcohol or both demonstrates an *in vivo* antileukemic activity.
- Flavones showed confirmed activity in the P-388 lymphocytic leukemia assay in mice, and the chloroform solubles showed both cytotoxic activity in the 9KB carcinoma of the nasopharynx cell culture assay and antitumor activity in the P-388 lymphocytic leukemia assay.
- *E. rotundifolium* is a native of New England and Virginia.

References

Dobberstein, R.H., M. Tin-wa, H.H. Fong, F.A. Crane, and N.R. Farnsworth. 1977. Flavonoid constituents from Eupatorium altissimum L. (Compositae). *Journal of Pharmacological Sciences* 66, no. 4:600–602.

Elsässer-Beile, U., W. Willenbacher, H.H. Bartsch, H. Gallati, J. Schulte Mönting, and S. von Kleist. 1996. Cytokine production in leukocyte cultures during therapy with Echinacea extract. *Journal of Clinical Laboratory Analysis* 10, no. 6:441–445.

Kupchan, S.M., J.W. Ashmore, and A.T. Sneden. 1978. Structure–activity relationships among *in vivo* active germacranolides. *Journal of Pharmaceutical Sciences* 67, no. 6:865–867.

Kupchan, S.M., T. Fujita, M. Maruyama, and R.W. Britton. 1973. The isolation and structural elucidation of eupaserrin and deacetyleupaserrin, new anti-leukemic sesquiterpene lactones from *Eupatorium semiserratum*. *Journal of Organic Chemistry* 38, no. 7:1260–1264.

Kupchan, S.M., J.E. Kelsey, M. Maruyama, J.M. Cassady, J.C. Hemingway, and J.R. Knox. 1969. Tumor inhibitors. XLI. Structural elucidation of tumor-inhibitory sesquiterpene lactones from *Eupatorium rotundifolium*. *Journal of Organic Chemistry* 34, no. 12:3876–3883.

Kupchan, S.M., M. Maruyama, R.J. Hemingway, et al. 1971. Eupacunin, a novel anti-leukemic sesquiterpene lactone from *Eupatorium cuneifolium*. *Journal of the American Chemical Society* 93, no. 19:4914–4916.

Lee, K.H., H.C. Huang, E.S. Huang, and H. Furukawa. 1972. Antitumor agents. II. Eupatolide, a new cytotoxic principle from *Eupatorium formosanum* HAY. *Journal of Pharmaceutical Sciences* 61, no. 4:629–631.

Ribeiro-Varandas, E., F. Ressurreição, W. Viegas, and M. Delgado. 2014. Cytotoxicity of Eupatorium cannabinum L. ethanolic extract against colon cancer cells and interactions with bisphenol A and doxorubicin. *BMC Complementary and Alternative Medicine* 14, no. 1:264.

Woerdenbag, H.J., W. Lemstra, T.M. Malingré, and A.W. Konings. 1989. Enhanced cytostatic activity of the sesquiterpene lactone eupatoriopicrin by glutathione depletion. *British Journal of Cancer* 59, no. 1:68–75.

Woerdenbag, H.J., J.C. van der Linde, H.H. Kampinga, T.M. Malingré, and A.W. Konings. 1989. Induction of DNA damage in Ehrlich ascites tumour cells by exposure to eupatoriopicrin. *Biochemical Pharmacology* 38, no. 14:2279–2283.

Euphorbia fischeriana (Euphorbiaceae) (Euphorbiales)   **Anti-tumor**

Location: northeastern China in the meadows and on the lower, dry slopes of the mountains. It is also found in pine forests at 100–600 m.

Appearance: perennial herbaceous flowering plant

Stem: single, 5–7 mm thick

Leaves: leaves alternate, stipules absent, basal scale-leaves ovate-oblong, 1–2 cm × 4–6 mm, petiole absent, leaf blades gradually larger upward, to oblong, 4–6.5 × 1–2 cm, base subtruncate, apex rounded or acute, lateral veins inconspicuous. Primary involucral leaves usually five, similar to normal leaves, primary rays five, 4–6 cm, secondary involucral leaves usually three, ovate, ca. 4 × 2 cm, cyathophylls two, triangular-ovate, ca. 2 × 2 cm, base subtruncate, apex acute

Flowers: cyathium sessile, involucre campanulate, ca. 4 × 4–5 mm, white pubescent, lobes rounded, white pubescent, glands four, pale brown, rounded. Male flowers many, exserted from involucre. Female flower: ovary pedicel 3–5 mm, exserted from cup, ovary densely white pubescent, styles connate below middle, persistent, style arms unlobed, slightly emarginate at middle

Tradition:

- The roots have often been used in traditional Chinese medicine to treat a wide range of ailments, such as edema, ascites, and cancer.

Documented target cancers:

- Extracts of *E. fischeriana* have been proven to be effective against several types of cancer, including malignant melanoma, Lewis lung carcinoma, and ascitic hepatoma, in mice.

Parts used: root

Active ingredients: emonoterpenes, flavonoids, anthraquinones, tannins, cereburosides, glycerols, phenolics, diterpenates, triterpenids, and steroids. Identified compounds: jolkinolide B, 12-deoxyphorbol-13-acetate (commonly known as prostratin, a protein kinase C activator), 17-hydroxyisoquinolide A and B

Precautions:

The fresh herb can cause skin irritation or allergic reactions.

References

Barrero, R.A., B. Chapman, Y. Yang, et al. 2011. *De novo* assembly of *Euphorbia fischeriana* root transcriptome identifies prostratin pathway related genes. *BMC Genomics* 12, no. 1:600.

Dong, M.H., Q. Zhang, Y.Y. Wang, B.S. Zhou, Y.F. Sun, and Q. Fu. 2016. *Euphorbia fischeriana* Steud inhibits malignant melanoma via modulation of the phosphoinositide-3-kinase/Akt signaling pathway. *Experimental and Therapeutic Medicine* 11, no. 4:1475–1480.

Geng, Z.F., Z.L. Liu, C.F. Wang, et al. 2011. Feeding deterrents against two grain storage insects *from Euphorbia fischeriana*. *Molecules* 16, no. 1:466–476.

Sun, Y.X., and J. Liu. 2011. Chemical constituents and biological activities of *Euphorbia fischeriana* Steud. *Chemistry and Biodiversity* 8, no. 7:1205–1214.

Wang, L., H. Duan, Y. Wang, et al. 2010. Inhibitory effects of Lang-du extract on the *in vitro* and *in vivo* growth of melanoma cells and its molecular mechanisms of action. *Cytotechnology* 62, no. 4:357–366.

Euphorbia helioscopia (Euphorbiaceae) (Euphorbiales) | **Anti-proliferative**

Location: dry, well-drained soils, native to most of Europe, northern Africa, and eastward through Asia
Appearance: herbaceous annual plant, can reach the height of 10–50 cm

Stem: single, upright, hairless stem, which branches at the top
Leaves: oval, broadly near the tip, 1.5–3 cm long, with a thin, toothed margin
Flowers: small, yellow-green, with two to five basic bracelets, which are similar to leaves, but more yellow

Tradition:

- Young shoots are used cooked, however, due to the toxicity of the plant, it should be used with caution. In addition, the new leaves are used as a tea substitute. Finally, this genus has been distinguished as a potential source of latex (for rubber production).
- It is anti-periodic. Leaves and stems are antipyretic and anthelmintic. In addition, the plant has a laxative effect. The milky juice is applied externally to cutaneous rashes, while the seeds are mixed with roasted pepper and used to treat cholera.

Parts used: leaves, stems, root, milky juice, seeds, and seed oil; four esters can be isolated from the fresh aerial parts of the plant.
Active ingredients:

- *E. helioscopia* contains the jatrophone-type diterpenoids euphoheliosnoid A, B, C, and D and other toxic diterpenes such as euphoscopins, epieuphoscopins euphornins, cuphohelioscopins, and euphohelionone.

Precautions:

- All species belonging to the genus *Euphorbia* are poisonous and must be treated with caution.
- The sap can cause severe inflammation and blistering if it comes in direct contact with the skin.
- If the herb is ingested it can lead to a sharp local irritation of the mucous membranes in the mouth and throat, dizziness, acute gastrointestinal disorders with vomiting, diarrhea, cramping, and severe pain.

Documented target cancers:

- The ethyl acetate extract (EAE) was found to markedly inhibit the proliferation of SMMC-7721 cells in a time- and dose-dependent manner.
- Application of aquatic extract the root of *Euphorbia helioscopia* against 7721, HeLa, and MKN-45 cancer cell lines indicated obvious antitumor activity.
- Aquatic extract of the root of *Euphorbia helioscopia* showed antitumor effect in S180- and H22-bearing mice and prolonged their life-span.
- It has shown significant antiproliferation potential against different types of cancer cells (human hepatocellular carcinoma cell lines SMMC-7721, BEL-7402, HepG2, gastric carcinoma cell line SGC-7901, and colorectal cancer cell line SW-480), mediating G1-phase arrest of cell cycle.

References

Jiwajinda, S., V. Santisopasri, A. Murakami, et al. 2002. *In vitro* anti-tumor promoting and anti-parasitic activities of the quassinoids from *Eurycoma longifolia*, a medicinal plant in Southeast Asia. *Journal of Ethnopharmacology* 82, no. 1:55–58.

Kanchanapoom, T., R. Kasai, P. Chumsri, Y. Hiraga, and K. Yamasaki. 2001. Canthin-6-one and β-carboline alkaloids from *Eurycoma harmadiana*. *Phytochemistry* 56, no. 4:383–386.

Kardono, L.B., C.K. Angerhofer, S. Tsauri, K. Padmawinata, J.M. Pezzuto, and A.D. Kinghorn. 1991. Cytotoxic and antimalarial constituents of the roots of *Eurycoma longifolia*. *Journal of Natural Products* 54, no. 5:1360–1367.

Kuo, P.C., L.S. Shi, A.G. Damu, et al. 2003. Cytotoxic and antimalarial β-carboline alkaloids from the roots of *Eurycoma longifolia*. *Journal of Natural Products* 66, no. 10:1324–1327.

Tee, T.T., and H.L.P. Azimahtol. 2005. Induction of apoptosis by *Eurycoma longifolia* Jack extracts. *Anticancer Research* 25, no. 3(B):2205–2213.

***Eurycoma longifolia* (Simaroubaceae)**   **Cytotoxic**

Location: Indonesia
   Parts used: roots
   Active ingredients:

- Four canthin-6-one alkaloids: 9-methoxycanthin-6-one, 9-methoxycanthin-6-one-N-oxide, 9-hydroxycanthin-6-one, and 9-hydroxycanthin-6-one-N-oxide, and
- One quassinoid: eurycomanone

Documented target cancers:

- Canthin-6-ones 1–4 were found to be active with all cell lines tested: breast, colon, fibrosarcoma, lung, melanoma, KB, and murine lymphocytic leukemia (P-388).
- Eurycomanone was significantly active against the human cell lines tested [breast, colon, fibrosarcoma, lung, melanoma, KB, and KB-V1 (a multi-drug resistant cell line derived from KB)] but was inactive against murine lymphocytic leukemia (P-388).
- Cytotoxic against human lung cancer (A-549) and human breast cancer (MCF-7) cell lines.

Further details:
- Two additional isolates from the roots of *Eurycoma longifolia*, the beta-carboline alkaloids beta-carboline-1-propionic acid and 7-methoxy-beta-carboline-1-propionic acid, were not significantly active with these cultured cells. However, they were found to demonstrate significant antimalarial activity as judged by studies conducted with cultured *Plasmodium falciparum* strains.

References

Jiwajinda, S., V. Santisopasri, A. Murakami, et al. 2002. *In vitro* anti-tumor promoting and anti-parasitic activities of the quassinoids from *Eurycoma longifolia*, a medicinal plant in Southeast Asia. *Journal of Ethnopharmacology* 82, no. 1:55–58.

Kanchanapoom, T., R. Kasai, P. Chumsri, Y. Hiraga, and K. Yamasaki. 2001. Canthin-6-one and β-carboline alkaloids from *Eurycoma harmadiana*. *Phytochemistry* 56, no. 4:383–386.

Kardono, L.B., C.K. Angerhofer, S. Tsauri, K. Padmawinata, J.M. Pezzuto, and A.D. Kinghorn. 1991. Cytotoxic and antimalarial constituents of the roots of *Eurycoma longifolia*. *Journal of Natural Products* 54, no. 5:1360–1367.

Kuo, P.C., L.S. Shi, A.G. Damu, et al. 2003. Cytotoxic and antimalarial β-carboline alkaloids from the roots of *Eurycoma longifolia*. *Journal of Natural Products* 66, no. 10:1324–1327.

Tee, T.T., and H.L.P. Azimahtol. 2005. Induction of apoptosis by *Eurycoma longifolia* Jack extracts. *Anticancer Research* 25, no. 3(B):2205–2213.

Fagara macrophylla (Fagara) (Rutaceae) **Cytotoxic Anti-leukemic**

Location: Africa
  Parts used: roots
  Active ingredients:

- Alkaloids: nitidine chloride, 6-oxynitidine, 6-methoxy-5,6-dihydronitidine (*Fagara macrophylla*)
- Fagaronine (Fine) (*Fagara xanthoxyloides*)

Indicative dosage and application:

- Alkaloids: nitidine chloride and 6-methoxy-5,6-dihydronitine are used at doses of 30–50 mg/kg.
- Fagaronine is used at a concentration of 3×10^{-6} mol/l at day four.

Documented target cancers:

- The alkaloids nitidine chloride and 6-methoxy-5,6-dihydronitine are about equipotent in P-388 mouse leukemia, giving high T/C values of 240–260%.
- Fagaronine (Fine) inhibits cell proliferation of human erythroleukemia K562 cells by 50% at a concentration of $3 \ 10^{-6}$ mol/l at day four (more information in further details).

Further details:

- The known alkaloids nitidine chloride (1), 6-oxynitidine (2), and 6-methoxy-5,6-dihydronitidine (3) have been isolated from *Fagara macrophylla*. Compound 3 was the major product and was shown to be an artifact. The alkaloids 1 and 3 have been interconverted by treatment of 1 under basic conditions or 3 under acidic conditions. On sublimation 1 and 3 formed 8,9-dimethoxy-2,3-methylenedioxy benzo[c]phenanthridine which could then be converted to 5,6-dihydronitidine. The alkaloids 1 and 3 are about equipotent in P-388 mouse leukemia, giving high T/C values of 240–260% at doses of 30–50 mg/kg. The other compounds were inactive. The structural requirement for antitumor activity in the phenanthridine series is the ability to form a C-6 iminium ion.
- Fagaronine (Fine) is an antileukemic drug extracted from the root of *Fagara xanthoxyloides* Lam. (Rutaceae). Fine inhibits cell proliferation of human erythroleukemia K562 cells by 50% at a concentration of 3×10^{-6} mol/l at day four. It stimulates incorporation of labeled macromolecular thymidine on day one, but decreases incorporation on days two, three, and four. Fine induces a cell accumulation in G2 and late-S phases.

References

Comoë, L., Y. Carpentier, B. Desoize, and J.C. Jardillier. 1988. Effect of fagaronine on cell cycle progression of human erythroleukemia K562 cells. *Leukemia Research* 12, no. 8:667–672.

Comoë, L., J. Kouamouo, P. Jeannesson, B. Desoize, R. Dufour, E.A. Yapo, and J.C. Jardillier. 1987. Cytotoxic effects of root extracts of *Fagara zanthoxyloides* Lam. (Rutaceae) on the human erythroleukemia K 562 cell line. *Annales Pharmaceutiques Francaises* 45, no. 1:79–86.

Kuete, V., L.P. Sandjo, A.T. Mbaveng, J.A. Seukep, B.T. Ngadjui, and T. Efferth. 2015. Cytotoxicity of selected Cameroonian medicinal plants and Nauclea pobeguinii towards multi-factorial drug-resistant cancer cells. *BMC Complementary and Alternative Medicine* 15, no. 1:309.

Messmer, W.M., M. Tin-Wa, H.H. Fong, C. Bevelle, N.R. Farnsworth, D.J. Abraham, and J. Trojánek. 1972. Fagaronine, a new tumor inhibitor isolated from *Fagara zanthoxyloides* Lam. (Rutaceae). *Journal of Pharmaceutical Sciences* 61, no. 11:1858–1859.

Wall, M.E., M.C. Wani, and H. Taylor. 1987. Plant antitumor agents, 27. Isolation, structure, and structure activity relationships of alkaloids from *Fagara macrophylla*. *Journal of Natural Products* 50, no. 6:1095–1099.

Ferula dissecta (Apiaceae) (Apiales)   **Anti-cancer**

Location: Washington, British Columbia, and Southern California. In addition, it appears in many Canadian plains.

Appearance: robust perennial

Leaves: both basal and cauline leaves large and somewhat roughened, ternate-pinnately dissected into small, narrow ultimate segments up to 1 cm long

Flowers: brownish-purple or yellow (the two colors are rarely found together), some of them always sterile

Parts used: roots, leaves

Tradition:

- It provides a significant amount of food for wildlife and for pets. It is considered to be a valuable breeding stock because of its high stature and high level of biomass production.
- The dried plant roots have been used to treat a wide range of diseases, including edema, ascites, stomach infusions, and liver and lung cancer.

Active ingredients: terpenoid coumarins and sesquiterpene derivatives

Documented target cancers:

- Cervical cancer HeLa cell line
- PC3 cells (prostate cancer)

References

Alam, M., A. Khan, A. Wadood, et al. 2016. Bioassay-guided isolation of sesquiterpene coumarins from *Ferula narthex* Bioss: A new anticancer agent. *Frontiers in Pharmacology* 7:26.

Fan, C., X. Li, J. Zhu, J. Song, and H. Yao. 2015. Endangered Uyghur medicinal plant Ferula identification through the second internal transcribed spacer. *Evidence-Based Complementary and Alternative Medicine* 2015:479879.

Huang, J., H.Y. Han, G.Y. Li, et al. 2013. Two new terpenoid benzoates with antitumor activity from the roots of *Ferula dissecta. Journal of Asian Natural Products Research* 15, no. 10:1100–1106.

Tilley, D., L.S. John, D. Ogle, N. Shaw, and J. Cane 2010. *Plant Guide: Fernleaf Biscuitroot (Lomatium dissectum [Nutt.] Mathias & Constance).* US Department of Agriculture, Natural Resources Conservation Service, Aberdeen Plant Materials Center, Aberdeen, ID.

Wang, Z.Q., C. Huang, J. Huang, H. Han, G. Li, J. Wang, and T. Sun. 2014. The stereochemistry of two monoterpenoid diastereomers from *Ferula dissecta. RSC Advances* 4, no. 28:14373–14377.

Zhou, Y., F. Xin, G. Zhang, H. Qu, D. Yang, and X. Han. 2017. Recent advances on bioactive constituents in ferula. *Drug Development Research* 78, no. 7:321–331.

Ficus carica L. (Ficus) (Urticaceae)   **Anti-leukemic**

Location: indigenous to Persia, Asia Minor, and Syria, wild in most of the Mediterranean countries

Appearance:

Stem: 6–7 m high

Root: free from stagnant water, sheltered from cold

Leaves: broad, rough, deciduous, deeply lobed

Flowers: concealed within the body of the fruit
In bloom: July–August

Parts used: seeds, fruit
  Active ingredients: Lectins *(Ficus cunia)*
  Documented target cancers:

- It is used for different types of leukemia (chronic myeloid leukemia, acute myeloblastic leukemia, acute lymphoblastic leukemia, and chronic lymphocytic leukemia).
- *Ficus carica* latex was found to inhibit growth of a benz-[a]-pyrene-induced sarcoma and resulted in the disappearance of small tumors in albino rats. This work inspired the isolation and structure elucidation of a mixture of 6-O-acyl--d-glucosyl--sitosterol isoforms from the latex of *Ficus carica* that demonstrated anti-proliferative activity in several tumor cell lines.
- A mixture of 6-O-acyl-β-d-glucosyl-β-sitosterols, isolated from *Ficus carica* latex and soybeans, was found to inhibit the growth of several cancer cell lines: Raji Burkitt B cell lymphoma cells, DG-75 Burkitt B-cell lymphoma cells, Jurkat T-cell leukemia cells, HD-MAR T-cell leukemia cells, DU-145 prostate cancer cells, and MCF-7 mammary cancer cells.
- Extracts for the fog latex were found to inhibit the proliferation of a human gastric adenocarcinoma cell line.

Further details:

- The seeds of *Ficus cunia* contain a lectin with a molecular weight of 3,300 to 3,500, which can agglutinate white blood cells (leukocytes and mononuclear cells) from patients with different types of leukemia (as mentioned above).
- A lectin, isolated from the seeds of *Ficus cunia* and purified by affinity chromatography on fetuin-Sepharose, was homogeneous in PAGE, GPC, HPLC, and immunodiffusion and had mol wt of 3,200–3,500. In SDS-PAGE and HPLC in the absence and presence of 2-mercaptoethanol, the lectin gave a single band or peak corresponding to M(r) 3,300–3,500, thus indicating it to be a monomer. The lectin agglutinated human erythrocytes regardless of blood group, bound to Ehrlich ascites cells and to human rat spermatozoa, and was thermally stable; Ca^{2+} enhanced its activity. The lectin is a metalloprotein that was inactivated by dialysis with EDTA followed by acetic acid, but reactivated by the addition of Ca^{2+}. The lectin contained 2.0% of carbohydrates, large proportions of acidic amino acids, but little methionine. In hapten-inhibition assays, chitin oligosaccharides linked beta-GlcNAc] and N-acetyl-lactosamine were inhibitors of which N,N'-tetra-acetylchitotetraose was the most potent. Among the macromolecules tested that contain either multiple N-acetyl-lactosamine and/or linked beta-GlcNAc, asialofetuin glycopeptide was the most potent inhibitor. Thus, an N-acetyl group and substitution at C-1 of D-GlcN are necessary for binding.
- Semipurified saline extracts of seeds from *Crotolaria juncea, Cassia marginata, Ficus racemosa, Cicer arietinum* (L-532), *Gossipium indicum* (G-27), *Melia composita, Acacia lenticularis, Meletia ovalifolia, Acacia catechu,* and *Peltophorum ferrenginium* were tested for leukoagglutinating activity against whole leukocytes and mononuclear cells from patients with chronic myeloid leukemia, acute myeloblastic leukemia, acute lymphoblastic leukemia, chronic lymphocytic leukemia, various lymphoproliferative/haematologic disorders, and normal healthy subjects. In addition, bone marrow cells from three patients undergoing diagnostic bone marrow aspiration and activated lymphocytes from mixed lymphocyte cultures (MLC) were also tested. All the seed extracts agglutinated white blood cells from patients with different types of leukemia. But none of them reacted with peripheral blood cells of normal individuals, patients with various lymphoproliferative/haematologic disorders, or cells from MLC. Leukoagglutination of leukemic cells with each of the seed extracts was inhibited by simple sugars. Only in one instance, cells from bone marrow of an individual who had undergone diagnostic bone marrow aspiration for a non-malignant condition were agglutinated. It is felt that purification of these seed extracts may yield leukemia-specific lectins (Figure 9.10).

Figure 9.10   (See color insert.) *Ficus carica.*

References

Agrawal, S., and S.S. Agarwal. 1990. Preliminary observations on leukaemia specific agglutinins from seeds. *Indian Journal of Medical Research* 92:38–42.

Guyot, M., M., Durgeat, and E. Morel. 1986. Ficulinic acid A and B, two novel cytotoxic straight-chain acids from the sponge *Ficulina ficus. Journal of Natural Products* 49, no. 2:307–309.

Hashemi, S.A., S. Abediankenari, M. Ghasemi, M. Azadbakht, Y. Yousefzadeh, and A.A. Dehpour. 2011. The effect of fig tree latex (*Ficus carica*) on stomach cancer line. *Iranian Red Crescent Medical Journal* 13, no. 4:272–275.

Kapoor, L.D. 2017. *Handbook of Ayurvedic Medicinal Plants. Herbal Reference Library.* CRC Press, New York, NY.

Lansky, E.P., H.M. Paavilainen, A.D. Pawlus, and R.A. Newman. 2008. Ficus spp. (fig): Ethnobotany and potential as anticancer and anti-inflammatory agents. *Journal of Ethnopharmacology* 119, no. 2:195–213.

Ray, S., H. Ahmed, S., Basu, and B.P. Chatterjee. 1993. Purification, characterisation, and carbohydrate specificity of the lectin of *Ficus cunia. Carbohydrate Research* 242, no. 242:247–263.

Rubnov, S., Y. Kashman, R. Rabinowitz, M. Schlesinger, and R. Mechoulam. 2001. Suppressors of cancer cell proliferation from fig (*Ficus carica*) resin: Isolation and structure elucidation. *Journal of Natural Products* 64, no. 7:993–996.

Ficus religiosa **(Moraceae) (Rosales)**   **Cytotoxic**

Location: (sub) tropical (warm and humid) climate, India (native)

Appearance: large, deciduous tree during the dry season or semi-evergreen tree, which reaches a height of up to 30 meters

Stem: the diameter of its trunks reaches up to 3 meters.

Leaves: the leaves are heart-shaped and have a distinctive, extended tip, which is 10–17 cm long and 8–12 cm wide, with 6–10 cm stem.

Tradition:

- Respected as a symbol of prosperity, happiness, good fortune, and longevity.
- Ornamental: it is grown as an ornamental plant in gardens and parks for its shade, in areas that are free of frost. It can also be cultivated as a Bonsai pot plant.
- Pharmaceutical: it is used in traditional medicine for about 50 types of disorders, including asthma, diabetes, diarrhea, epilepsy, gastric problems, inflammatory disorders, and finally infectious and sexual disorders.

Parts used: leaves, bark, fruit, branches
    Active ingredients: phytosterolin, β-sitosterol, its glycoside, albuminoids
    Documented target cancers:

- Human cervical cancer cell lines (SiHa, HeLa)
- Breast cancer cell lines

References

Chandrasekar, S.B., M. Bhanumathy, A.T. Pawar, and T. Somasundaram. 2010. Phytopharmacology of *Ficus religiosa*. *Pharmacognosy Reviews* 4, no. 8:195–199.

Choudhari, A.S., S.A. Suryavanshi, and R. Kaul-Ghanekar. 2013. The aqueous extract of *Ficus religiosa* induces cell cycle arrest in human cervical cancer cell lines SiHa (HPV-16 positive) and apoptosis in HeLa (HPV-18 positive). *PLOS ONE* 8, no. 7:e70127.

Haneef, J., M. Parvathy, S.K. Thankayyan R, H. Sithul, and S. Sreeharshan. 2012. Bax translocation mediated mitochondrial apoptosis and caspase dependent photosensitizing effect of *Ficus religiosa* on cancer cells. *PLOS ONE* 7, no. 7:e40055.

Fissistigma cavaleriei (Annonaceae) (Magnoliales)   **Anti-tumor**

Location: native to Asia-Tropical, New Caledonia, and Asia-Temperate
    Appearance: perennial shrub

Stem: climbers to 8 m tall, most parts reddish pubescent
Leaves: petiole 6–8 mm, leaf blade oblong-lanceolate to oblong-elliptic. Inflorescences opposed or alternate, cymose, one- to five-flowered

Tradition: used as a folklore medicine for treatment of inflammation, arthritis, and tuberculosis
    Parts used: fresh leaves (essential oil)
    Active ingredients: essential oil's major compounds: β-Phellandrene (24.3%), Germacrene-D (8.4%), β-Caryophyllene (6.1%), and Caryophyllene Oxide (4.5%)
    Documented target cancers:
- Sarcoma 180-bearing mice

Reference

Yang, Z., W. Lu, X. Ma, and D. Song. 2012. Bioassay-guided isolation of an alkaloid with antiangiogenic and antitumor activities from the extract of *Fissistigma cavaleriei* root. *Phytomedicine* 19, no. 3–4:301–305.

***Foeniculum vulgare* (Apiaceae, Umbelliferae) (Apiales)**   **Cytotoxic**

Location: north Europe, US, south Canada, Asia, Australia
    Appearance: upright, perennial–umbelliferous herb

Stem: green with hollow stalks that reach a height of 2.5 m
Leaves: leaves reach a length of 40 cm and are divided thinly into distal filamentous pieces, approximately 0.5 mm wide.
Flowers: 5–15 cm wide. Each shade has 20–50 tiny yellow flowers with small stems.

Tradition:

- It is one of the main ingredients of absinthe drink. Also, the essential oil from the seeds is added to perfumes, soaps, and pharmaceutical and cosmetic products.
- In traditional medicine, fennel was used as an aphrodisiac as well as to encourage menstruation and lactation, although according to some sources great attention was needed during the last use because of its possible infant toxicity. In addition, it has been approved by the German Commission for short-term treatment of indigestion, swelling, and reflux of the upper respiratory tract. Furthermore, the oil appears to be antioxidant, antimicrobial, anticonvulsant, and to stimulate gastrointestinal motility.

Parts used: leaves, fruit, stem, seeds
    Active ingredients: phenylpropenes estragol, trans-anethole, phenolic glycosides, phenolic acids. Flavonoids like eriodictyol-7-rutinoside, quercetin-3-rutinoside and rosmarinic acid
    Precautions: Estragole (Methylchavicol), one of the main components of the essential oil of *F. vulgare,* has been reported to be associated with the development of malignant tumors in rodents. This was the basis for the recommendations of the Scientific Committee on Food (SCF) of the European Union to restrict the use of this substance.
    Some people can have allergic skin reactions to fennel. People who are allergic to plants such as celery, carrot, and mugwort are more likely to also be allergic to fennel. Fennel can also make the skin extra sensitive to sunlight and make it easier to get a sunburn.
    Documented target cancers:

- B16F10 melanoma cell line
- Breast cancer cell line (MCF7)
- Liver cancer cell line (Hepg-2)

References

Badgujar, S.B., V.V. Patel, and A.H. Bandivdekar. 2014. *Foeniculum vulgare* Mill: A review of its botany, phytochemistry, pharmacology, contemporary application, and toxicology. *BioMed Research International* 2014:842674.

Pradhan, M., S. Sribhuwaneswari, D. Karthikeyan, et al. 2008. In-vitro cytoprotection activity of *Foeniculum vulgare* and *Helicteres isora* in cultured human blood lymphocytes and antitumour activity against B16F10 melanoma cell line. *Research Journal of Pharmacy and Technology* 1, no. 4:450–452.

Rather, M.A., B.A. Dar, S.N. Sofi, B.A. Bhat, and M.A. Qurishi. 2016. Foeniculum vulgare: A comprehensive review of its traditional use, phytochemistry, pharmacology, and safety. *Arabian Journal of Chemistry* 9:S1574–S1583.

Sharopov, F., A. Valiev, P. Satyal, I. Gulmurodov, S. Yusufi, W.N. Setzer, and M. Wink. 2017. Cytotoxicity of the essential oil of Fennel (Foeniculum vulgare) from Tajikistan. *Foods* 6, no. 9.

Galium aparine (Rubiaceae) (Gentianales)	Cytotoxic Anti-leukemic

Location: native to a wide region of Europe, North Africa, and Asia from Britain and the Canary Islands to Japan, US, Canada, Mexico, Central America, South America, Australia, New Zealand, some oceanic islands, and scattered locations in Africa

Appearance: vigorous, scrambling annual

Leaves: narrow, small, in rosettes
Flowers: hermaphrodite, white-yellow, small, in stumps

Tradition: used in skin conditions, particularly in psoriasis. It is also used as a lymphotonic with smoothing and diuretic action.

Parts used: aerial parts and fresh juice
Active ingredients: β-sitosterol, daucosterol, and dibutyl phthalate
Documented target cancers:

- *G. aparine* extract has been reported as cytotoxic in MCF-7 and MDA-MB-231 human breast cancer cell lines in a concentration- and time-dependent manner.
- β-sitosterol, daucosterol, and dibutyl phthalate, contained in *G. aparine* petroleum ether phase, have been reported to inhibit the proliferation of leukemia cell K562 in a dose- and time-response manner.

References

Atmaca, H., E. Bozkurt, M. Cittan, and H.D. Dilek Tepe. 2016. Effects of *Galium aparine* extract on the cell viability, cell cycle and cell death in breast cancer cell lines. *Journal of Ethnopharmacology* 186:305–310.

Shi, G., J. Liu, W. Zhao, Y. Liu, and X. Tian. 2016. Separation and purification and *in vitro* anti-proliferative activity of leukemia cell K562 of *Galium aparine* L. petroleum ether phase. *Saudi Pharmaceutical Journal* 24, no. 3:241–244.

Galium verum (Rubiaceae) (Gentianales)	Cytotoxic

Location: native to Eurasia and Africa

Appearance: perennial herbaceous plant with golden yellow flowers that are 2–3 mm in diameter and grouped in many-flowered panicles

Stem: height 20–60 cm (8–25 in.), stem lower part four-edged, upper part round, fine-haired
Leaves: regular (actinomorphic), usually eight whorled leaves, stalkless. Blade needle-like, bristle-tipped, top shiny, underside white-haired, with entire margin, clearly revolute
Flowers: corolla wheel-shaped, dark yellow, 3 mm (0.12 in.) broad, fused, four-lobed. Calyx lacking. Stamens four. Pistil of two fused carpels, styles two. Inflorescence broad, narrowly conical, densely flowered cyme. Flower with mild honey fragrance

Tradition:

- Roasted seed acts as a substitute for coffee. Also, this plant can be used as rennet for milk coagulation. The dried plant has an odor of freshly harvested straw and was used to fill layers.
- It has a long history as a medicinal plant although it is little used in modern medicine. It is mainly used as a diuretic, as well as for treatment of skin diseases. Leaves, stems, and flowering shoots act as anticonvulsive, astringent, diuretic, for foot care and healing. Also, the plant is used as a treatment

for urinary tract stone. Powder prepared from the fresh plant is used to soothe irritated skin and reduce inflammation, while the plant is also used as a healing agent for cuts, skin infections, and when there is slow wound healing.

Parts used: leaves, seeds, flower, stem
    Active ingredients: 1-methoxy-2-hydroxyanthraquinones
    Documented target cancers:

- Laryngeal carcinoma cell lines (Hep-2 and HLaC79, Hep2-Tax, and HLaC79-Tax)
- Head and neck cancer

References

Banthorpe, D.V., and J.J. White. 1995. Novel anthraquinones from undifferentiated cell cultures of *Galium verum*. *Phytochemistry* 38, no. 1:107–111.

Lakić, N., N. Mimica-Dukić, J. Isak, and B. Božin. 2010. Antioxidant properties of *Galium verum* L. (Rubiaceae) extracts. *Open Life Sciences* 5, no. 3:331–337.

Schmidt, M., C. Polednik, J. Roller, and R. Hagen. 2014. *Galium verum* aqueous extract strongly inhibits the motility of head and neck cancer cell lines and protects mucosal keratinocytes against toxic DNA damage. *Oncology Reports* 32, no. 3:1296–1302.

Schmidt, M., C.J. Scholz, G.L. Gavril, C. Otto, C. Polednik, J. Roller, and R. Hagen. 2014. Effect of *Galium verum* aqueous extract on growth, motility and gene expression in drug-sensitive and-resistant laryngeal carcinoma cell lines. *International Journal of Oncology* 44, no. 3:745–760.

Garcinia hombroniana (Garcinia) (Guttifereae)	Cytotoxic	Antitumor

Location: riverine and coastal alluvial regions. Malaysia, Brunei
    Appearance:

Stem: 10 m high, numerous branches
Leaves: tertiary branches hold much of the leaves
Flowers: in clusters of not more than five small flowers

Tradition: very powerful drastic hydragogue, cathartic, very useful in dropsical conditions
    Parts used: gum resin
    Active ingredients: Garonolic acids
    Precautions: full dose is rarely given alone, as it causes vomiting, nausea, and griping. In high dose it can cause death.
    Documented target cancers:

- *Garcinia hunburyi* (Gamboge), when steam processed (0.15 MPa, $126°C$ for 30 min), is cytotoxic on K562 tumor cells.
- Compounds isolated from the bark are reported to display good cytotoxic effects against DBTRG glioma cells.

Further details:

- The technology for processing steamed *Garcinia hunburyi* with high pressure was synthetically selected by using orthogonal experimental design, based on the indexes of anti-inflammatory, bacteriocidal, anti-tumor effects, and gambagic acid content. The result shows that the best way is to steam for 0.5 h at 126°C.

Figure 9.11 *Garcinia* fruit.

- The cytotoxicity of different processed products of Gamboge on K562 tumor cells was observed. The result showed that the antitumor action of *Garcinia hunburyi* processed by steaming (0.15 MPa, 126°C for 30 min) was the strongest.
- However, there is a possibility that the Nigerian cola plant (*Garcinia*) may be a cause of human cancer in countries where kola nuts are widely consumed as stimulants (e.g. via chewing), because of their content of primary and secondary amines and their relative methylating potential due to nitrosamide formation (Figure 9.11).

References

Atawodi, S.E., P. Mende, B. Pfundstein, R. Preussmann, and B. Spiegelhalder. 1995. Nitrosatable amines and nitrosamide formation in natural stimulants: *Cola acuminata*, *C. nitida* and *Garcinia cola*. *Food and Chemical Toxicology* 33, no. 8:625–630.

Jamila, N., M. Khairuddean, N.S. Yaacob, N.N. Kamal, H. Osman, S.N. Khan, and N. Khan. 2014. Cytotoxic benzophenone and triterpene from *Garcinia hombroniana*. *Bioorganic Chemistry* 54:60–67.

Lu, Y., G. Wang, and D. Ye. 1996. Comparison of cytotoxicity of different processed products of gamboge on K562 tumor cells. *Chung Kuo Chung Yao Tsa Chih* 21, no. 2(90–91):127.

Ye, D., and L. Kong. 1996. Selection of technology for processing steamed *Garcinia hunburyi* with high pressure. *Chung Kuo Chung Yao Tsa Chih* 21, no. 8:472–473.

***Gentiana lutea* (Gentianaceae) (Gentianales)**   **Cytotoxic**

Location: central and southern Europe
   Appearance: perennial, herbaceous plant

   Leaves: wide, elliptical, contrasting, light green
   Flowers: large, bright yellow

Tradition: used in Serbian and Peruvian folk medicines to treat digestive disorders

Parts used: blooming peak, root

Active ingredients: gentiopicroside, gentisin, bellidifolin-8-O-glucoside, demethylbellidifo-lin-8-O-glucoside, isovitexin, swertiamarin, and amarogentin

Precautions:

The highly toxic *Veratrum album* can be mistaken for *Gentiana lutea* and has caused accidental poisoning when used in homemade preparations.

Documented target cancers:

- *Gentiana lutea* leaf extract is reported to exhibit moderate cytotoxic effects toward cervical cancer HeLa cells, while gentiopicrin, mangiferin, and isogentisin exert strong activity.

References

Balijagić, J., T. Janković, G. Zdunić, et al. 2012. Chemical profile, radical scavenging and cytotoxic activity of yellow gentian leaves (*Genitaneae luteaefolium*) grown in northern regions of Montenegro. *Natural Product Communications* 7, no. 11:1487–1490.

Kesavan, R., U.R. Potunuru, B. Nastasijević, A. T, G. Joksić, and M. Dixit. 2013. Inhibition of vascular smooth muscle cell proliferation by *Gentiana lutea* root extracts. *PLOS ONE* 8, no. 4:e61393.

Mirzaee, F., A. Hosseini, H.B. Jouybari, A. Davoodi, and M. Azadbakht. 2017. Medicinal, biological and phytochemical properties of Gentiana species. *Journal of Traditional and Complementary Medicine* 7, no. 4:400–408.

Glaucium flavum (Papaveraceae) (Ranunculales)   **Cytotoxic**

Location: native to Algeria, Macronesia, Western Asia, Caucasus, as well as Europe

Appearance: perennial summer flowering plant that reaches a height of 60 cm

Leaves: thick, leather-like, and blue-gray

Flowers: yellow, similar in shape to red poppy

Tradition: used in Algerian folk medicine to treat warts (benign tumors)

Active ingredients: soquinoline alkaloid protopine, bacconolin

Documented target cancers:

- *G. flavum* root crude alkaloid extract is reported to inhibitproliferation of breast cancer cells (MDA-MB-435, MDA-MB-231, and Hs578T), causing G2/M phase apoptosis, without affecting normal cells.

- Bocconoline strongly inhibits the viability of breast cancer cells with an IC_{50} of 7.8 μM, with a low cytotoxic effect against normal human cells.

Further details:

Glaucine is the main alkaloid component in *Glaucium flavum*. Glaucine has bronchodilator and anti-inflammatory effects, acting as a PDE4 inhibitor and calcium channel blocker, and is used medically as an antitussive in some countries. Glaucine may produce side effects such as sedation, fatigue, and a hallucinogenic effect characterized by colorful visual images and as a recreational drug.

References

Bournine, L., S. Bensalem, P. Peixoto, et al. 2013. Revealing the anti-tumoral effect of Algerian *Glaucium flavum* roots against human cancer cells. *Phytomedicine* 20, no. 13:1211–1218.

Bournine, L., S. Bensalem, J.N. Wauters, et al. 2013. Identification and quantification of the main active anti-cancer alkaloids from the root of *Glaucium flavum*. *International Journal of Molecular Sciences* 14, no. 12:23533–23544.

Glycyrrhiza glabra L. (Glycyrrhiza, Liquorice) (Leguminosae)	Antitumor

Location: South-east Europe, South-west Asia
   Appearance:

Stem: graceful, with light, spreading, pinnate foliage, presenting an almost feathery appearance from a distance
Root: double: the one part consisting of a vertical or tap root, often with several branches penetrating to a depth of 1–1.5 m, the other of horizontal rhizomes or stolons, thrown out of the root below the surface of the ground
Leaves: leaflets
Flowers: from the axils of the leaves spring racemes or spikes of papilionaceous small pale blue, violet, yellowish-white, or purplish, followed by small pods.
In bloom: summer

Degree of rarity: not rare. It is cultured in Spain, Italy, UK, and US.
   Tradition: very common in use in South Italy for stomach disorders, cough, and also as a sweetener.
   Parts used: root and stolons
   Active ingredients: Glycyrrhizic acid, glycyrrhetinic acid, flavonoids, triterpenoids
   Documented target cancers:

- Prevention, skin cancer, leukemia
- Some triterpenoids from *Glycyrrhiza* spp. were effective against adriamycin (ADM)-resistant P388 leukemia cells (P388/ADM), which were resistant to multiple anticancer drugs.
- Ehrlich ascites tumor cells
- Androgen-independent PC-3 prostate cancer cells

Further details:

- *Glycyrrhiza uralensis* is one of the main ingredients of Hua-sheng-ping (*Chrysanthemum morifolium, Glycyrrhiza uralensis, Panax notoginseng*), which has many medicinal uses.
- *Glycyrrhizic acid*, the active ingredients in licorice, and its metabolite carbenoxolone are members of short-chain dehydrogenase reductase (SDR) enzymes. The SDR family includes over 50 proteins from human, mammalian, insect, and bacterial sources.
- *Glycyrrhiza uralensis*: extracts have strong antimutagenic properties, indicated for syndromes such as Spleen-Stomach Asthenic Cold, and have been proven to be an effective prescription for pre-cancerous lesions. An important component is glycyrrhetinic acid, which can protect against rapid DNA damage and decrease the unscheduled DNA synthesis induced by benzo(alpha)pyrene.
- *Glycyrrhiza inflata*: extracts contain six flavonoids with significant antioxidant effects, showing anti-promoting effects on two-stage carcinogenesis in mouse skin induced by DMBA plus croton oil. The TPA enhanced 32Pi-incorporation into phospholipid fraction in HeLa cells was inhibited, and the micronuclei in mouse bone marrow cells induced by cytoxan were also depressed.

References

Agarwal, R., Z.Y. Wang, and H. Mukhtar. 1991. Inhibition of mouse skin tumor-initiating activity of DMBA by chronic oral feeding of glycyrrhizin in drinking water. *Nutrition and Cancer* 15, no. 3–4:187–193.

Biglieri, E.G. 1995. My engagement with steroids: A review. *Steroids* 60, no. 1:52–58.

Chen, X., and R. Han. 1995. Effect of glycyrrhetinic acid on DNA damage and unscheduled DNA synthesis induced by benzo(alpha)pyrene. *Chinese Medical Sciences Journal = Chung-Kuo i Hsueh k'o Hsueh Tsa Chih* 10, no. 1:16–19.

Chen, X.G., and R. Han. 1994. Effect of glycyrrhetinic acid on DNA damage and unscheduled DNA synthesis induced by benzo (a) pyrene. *Yao Hsueh Hsueh Pao* 29, no. 10:725–729.

Duax, W.L., and D. Ghosh. 1997. Structure and function of steroid dehydrogenases involved in hypertension, fertility, and cancer. *Steroids* 62, no. 1:95–100.

Fu, N., Z. Liu, and R. Zhang. 1995. Anti-promoting and anti-mutagenic actions of G9315. *Chung Kuo i Hsueh Ko Hsueh Yuan Hsueh Pao* 17, no. 5:349–352.

Fu, Y., T.C. Hsieh, J. Guo, J. Kunicki, M.Y. Lee, Z. Darzynkiewicz, and J.M. Wu. 2004. Licochalcone-A, a novel flavonoid isolated from licorice root (Glycyrrhiza glabra), causes G2 and late-G1 arrests in androgen-independent PC-3 prostate cancer cells. *Biochemical and Biophysical Research Communications* 322, no. 1:263–270.

Hasegawa, H., J.H. Sung, S. Matsumiya, M. Uchiyama, Y. Inouye, R. Kasai, and K. Yamasaki. 1995. Reversal of daunomycin and vinblastine resistance in multidrug-resistant P388 leukemia *in vitro* through enhanced cytotoxicity by triterpenoids. *Planta Medica* 61, no. 5:409–413.

Horn, B. 1986. Rarities in family practice. Consequences for education and continuing education. *Schweizerische Rundschau für Medizin Praxis* 75, no. 44:1323–1327.

Liu, X.R., W.Q. Han, and D.R. Sun. 1992. Treatment of intestinal metaplasia and atypical hyperplasia of gastric mucosa with xiao wei yan powder. *Chung Kuo Chung HSI i Chieh Ho Tsa Chih* 12, no. 10(602–603):580.

Montanari, G., F. Zanaletti, G.C. Quadrelli, et al. 1988. Arterial hypertension with hypokalemia. *Minerva Medica* 79, no. 3:209–214.

Shackleton, C.H. 1993. Mass spectrometry in the diagnosis of steroid-related disorders and in hypertension research. *Journal of Steroid Biochemistry and Molecular Biology* 45, no. 1–3:127–140.

Sheela, M.L., M.K. Ramakrishna, and B.P. Salimath. 2006. Angiogenic and proliferative effects of the cytokine VEGF in Ehrlich ascites tumor cells is inhibited by Glycyrrhiza glabra. *International Immunopharmacology* 6, no. 3:494–498.

Shi, G.Z. 1992. Blockage of *Glyrrhiza uralensis* and *Chelidonium majus* in MNNG induced cancer and mutagenesis. *Chung-Hua Yu Fang i Hsueh Tsa Chih* 26, no. 3:165–167.

Takahashi, K., K. Yoshino, T. Shirai, A. Nishigaki, Y. Araki, and M. Kitao. 1988. Effect of a traditional herbal medicine (shakuyaku-kanzo-to) on testosterone secretion in patients with polycystic ovary syndrome detected by ultrasound. *Nippon Sanka Fujinka Gakkai Zasshi* 40, no. 6:789–792.

Wang, Z.Y., R. Agarwal, W.A. Khan, and H. Mukhtar. 1992. Protection against benzo[a]pyrene- and N-nitrosodiethylamine-induced lung and forestomach tumorigenesis in A/J mice by water extracts of green tea and licorice. *Carcinogenesis* 13, no. 8:1491–1494.

Webb, T.E., P.C. Stromberg, H. Abou-Issa, R.W. Curley Jr, and M. Moeschberger. 1992. Effect of dietary soybean and licorice on the male F344 rat: An integrated study of some parameters relevant to cancer chemoprevention. *Nutrition and Cancer* 18, no. 3:215–230.

White, P.C., T. Mune, and A.K. Agarwal. 1997. 11 Beta-hydroxysteroid dehydrogenase and the syndrome of apparent mineralocorticoid excess. *Endocrine Reviews* 18, no. 1:135–156.

Yamashiki, M., A. Nishimura, H. Suzuki, S. Sakaguchi, and Y. Kosaka. 1997. Effects of the Japanese herbal medicine 'Sho-saiko-to' (TJ-9) on *in vitro* interleukin-10 production by peripheral blood mononuclear cells of patients with chronic hepatitis C. *Hepatology* 25, no. 6:1390–1397.

Yu, X.Y. 1993. A prospective clinical study on reversion of 200 precancerous patients with hua-sheng-ping. *Chung Kuo Chung HSI i Chieh Ho Tsa Chih* 13, no. 3:147–149.

Zee-Cheng, R.K. 1992. Shi-quan-da-bu-tang (ten significant tonic decoction), SQT. A potent Chinese biological response modifier in cancer immunotherapy, potentiation and detoxification of anticancer drugs. *Methods and Findings in Experimental and Clinical Pharmacology* 14, no. 9:725–736.

Glyptopetalum sclerocarpum (Celastraceae)   **Cytotoxic**

Location: India, China

Appearance: evergreen tree or shrub, 2–12 meters high, often climbing

Leaves: leathery or thick leathery, elongated to elliptical, rare oval, wedge-shaped to rounded and edge
Flowers: yellow-white with petals slightly thick, oval

Tradition: the stems are used to cure malaria and skin infection.

Parts used: bark, pericarp

Active ingredients: glyptopetolide, 20-hydroxytryptone, sclerocarpic acid, isoarborinol, and cangoronine

Documented target cancers:

- Has been tested against: P-388 lymphocytic leukemia, KB carcinoma of the nasopharynx, and a number of human cancer cell types, i.e. HT-1080 fibrosarcoma, LU-1 lung cancer, COL-2 colon cancer, MEL-2 melanoma, and BC-1 breast cancer.
- Intense but non-specific cytotoxic activity was observed when 22-hydroxytingenone was evaluated with a battery of cell lines comprised of P-388 lymphocytic leukemia, KB carcinoma of the nasopharynx, and a number of human cancer cell types, i.e. HT-1080 fibrosarcoma, LU-1 lung cancer, COL-2 colon cancer, MEL-2 melanoma, and BC-1 breast cancer.

Further details:

- 22-hydroxytingenone was isolated from *Glyptopetalum sclerocarpum* M. Laws, and its unambiguous ^{13}C-NMR assignments were accomplished through the use of APT, HETCOR, and selective INEPT spectroscopy.
- Intense, but nonspecific cytotoxic activity was observed when this substance was evaluated with a battery of cell lines comprised of the P-388 lymphocytic leukemia, KB carcinoma of the nasopharynx, and a number of human cancer cell types, i.e. HT-1080 fibrosarcoma, LU-1 lung cancer, COL-2 colon cancer, MEL-2 melanoma, and BC-1 breast cancer.

References

Bavovada, R., G. Blaskó, H.L. Shieh, J.M. Pezzuto, and G.A. Cordell. 1990. Spectral assignment and cytotoxicity of 22-hydroxytingenone from *Glyptopetalum sclerocarpum*. *Planta Medica* 56, no. 4:380–382.
Sotanaphun, U., V. Lipipun, and R. Bavovada. 2004. Constituents of the pericarp of *Glyptopetalum sclerocarpum*. *Fitoterapia* 75, no. 6:606–608.
Sotanaphun, U., R. Suttisri, V. Lipipun, and R. Bavovada. 1998. Quinone-methide triterpenoids from *Glyptopetalum sclerocarpum*. *Phytochemistry* 49, no. 6:1749–1755.
Sotanaphun, U., R. Suttisri, V. Lipipun, and R. Bavovada. 2000. A new 3,4-seco-ursane triterpenoid from Glyptopetalum sclerocarpum. *Chemical and Pharmaceutical Bulletin* 48, no. 9:1347–1349.

Goniothalamus sp. (Annonaceae) **Cytotoxic (Agrimony) (Rosaceae)**

Location: Malaysia, China

Active ingredients:

- Acetogenins: gardnerilins A and B
- Styrylpyrone(SPD), goniodiol-7-monoacetate
- Acetogenin lactones: goniothalamicin, annonacin

Indicative dosage and application: doses used in rat mammary tumors with good effects were: 2, 10, and 50 mg/kg.

Documented target cancers:

- Antiestrogen (mice), breast cancer, cytotoxic, 9ASK (astrocytoma), and weakly active against 3PS murine leukemia

- Human large cell lung carcinoma (COR-L23) cells
- MCF-7 breast cancer, cervical cancer cells (HeLa), HT-29 colon cancer, and CEM-ss leukemia cell lines

Further details:

- *Goniothalamus gardneri*: the roots contain the C35 acetogenins gardnerilins A and B.
- *Goniothalamus amuyon* and other *Goniothalamus* species contain the styrylpyrone, goniodiol-7-monoacetate [6R-(7R,8R-dihydro-7-acetoxy-8-hydroxystyryl)-5, 6-dihydro-2-pyrone].
- The estrogen antagonism: agonism ratio for SPD is much higher than Tamoxifen, which is indicative of the breast cancer antitumor activity as seen in compounds such as MER-25. Pretreatment assessment on 1 mg/kg BW SPD and Tam showed that SPD is not a very good estrogen antagonist compared to Tam, as it was unable to revert the estrogenicity effect of estradiol benzoate (EB) on immature rat uterine weight. Antitumor activity assessment for SPD exhibited significant tumor growth retardation in 7,12-dimethyl benzanthracene (DMBA) induced rat mammary tumors at all doses employed (2, 10, and 50 mg/kg) compared to the controls. This compound was found to be more potent than Tam (2 and 10 mg/kg) and displayed greater potency at a dose of 10 mg/kg. It caused complete remission of 33.3% of tumors but failed to prevent onset of new tumors. However, SPD administration at 2 mg/kg caused 16.7% complete remission and partial remission. It also prevented the onset of new tumors throughout the experiment.
- Goniodiol-7-monoacetate showed potent (ED50 values less than 0.1 microgram/ml) cytotoxicities against KB, P-388, RPMI, and TE671 tumor cells.
- The stem bark of *Goniothalamus giganteus* Hook. Thomas contains the gamma-lactone goniothalamicin, a tetrahydroxy-mono-tetrahydrofuran fatty acid, along with annonacin.
- Goniothalamicin is cytotoxic and insecticidal and inhibits the formation of crown gall tumors on potato discs. Annonacin, the only other reported mono-tetrahydrofuran acetogenin, was also isolated, which is active against 9ASK (astrocytoma) and weakly active against 3PS murine leukemia.

References

Abdel-Wahab, S.I., A.B. Abdul, S. Mohan, A.S. Al-Zubairi, M.M. Elhassan, and A.S. Al-Zubairi. 2009. Oncolysis of breast, liver and leukemia cancer cells using ethyl acetate and methanol extracts of Goniothalamus umbrosus. *Research Journal of Biological Sciences* 4, no. 2:209–215.

Alabsi, A.M., R. Ali, A.M. Ali, et al. 2013. Induction of caspase-9, biochemical assessment and morphological changes caused by apoptosis in cancer cells treated with goniothalamin extracted from *Goniothalamus macrophyllus*. *Asian Pacific Journal of Cancer Prevention* 14, no. 11:6273–6280.

Alkofahi, A., J.K. Rupprecht, D.L. Smith, C.J. Chang, and J.L. McLaughlin. 1988. Goniothalamicin and annonacin: Bioactive acetogenins from *Goniothalamus giganteus* (Annonaceae). *Experientia* 44, no. 1:83–85.

Chen, Y., Z. Jiang, R.R. Chen, and D.Q. Yu. 1998. Two linear acetogenins from *Goniothalamus gardneri*. *Phytochemistry* 49, no. 5:1317–1321.

Hawariah, A., and J. Stanslas. 1998. Antagonistic effects of styrylpyrone derivative (SPD) on 7,12-dimethyl-benzanthracene-induced rat mammary tumors. *In vivo* 12, no. 4:403–410.

Tantithanaporn, S., C. Wattanapiromsakul, A. Itharat, and N. Keawpradub. 2011. Cytotoxic activity of acetogenins and styryl lactones isolated from Goniothalamus undulatus Ridl. root extracts against a lung cancer cell line (COR-L23). *Phytomedicine* 18, no. 6:486–490.

Wu, Y.C., C.Y. Duh, F.R. Chang, et al. 1991. The crystal structure and cytotoxicity of goniodiol-7-monoacetate from *Goniothalamus amuyon*. *Journal of Natural Products* 54, no. 4:1077–1081.

Gossypium herbaceum L. (Gossypium, Cotton root) (Malvaceae)	Cytotoxic, Antitumor

Location: Asia Minor, cultivated in US and Egypt, Mediterranean, India

Appearance:

Stem: 0.5–2 m high, branching stems
Root: the root bark consists of thin flexible bands covered with a brownish yellow periderm, odor not strong, tastes slightly acid.
Leaves: palmate, hairy, green, lobes lanceolate and acute
Flowers: yellow with a purple spot in the center
In bloom: August–September

Degree of rarity: low

Tradition: one of the well-known Chinese medicines used in Chinese medicine as an anticancer crude drug

Parts used: bark of root

Active ingredients: Catechin, Gossypol

Indicative dosage and application: it is used as crude extract, mixed with other herbs, usually oral intake.

Documented target cancers:

- Murine B16 melanoma and L1210 lymphoma cells
- Human head and neck squamous cell carcinoma *in vivo*
- Embryonal and alveolar rhabdomyosarcoma cell lines

Figure 9.12 *Gossypium herbaceum.*

Further details:

- *Gossypium indicum*: has a moderate antimutagenic activity against benzo[a]pyrene. Its aqueous-alcoholic extracts from unripe cotton balls are well known for their antitumor activity. The hydrophilic fractions contain certain amounts of *catechin* and its derivatives, which are responsible for the antitumor activities of the herb (Figure 9.12).

References

Choi, J.J., K.N. Yoon, S.K. Lee, et al. 1998. Antitumor activity of the aqueous-alcoholic extracts from unripe cotton ball of *Gossypium indicum*. *Archives of Pharmacal Research* 21, no. 3:266–272.

Fuh, B.R., C.L. Chen, S. Wang, and J. Lin. 2006. (–)Gossypol inhibits cell proliferation and induces apoptosis in rhabdomyosarcoma *in vitro*. *Proceedings of the American Association for Cancer Research* 47:1111.

Lee, H., and J.Y. Lin. 1988. Antimutagenic activity of extracts from anticancer drugs in Chinese medicine. *Mutation Research* 204, no. 2:229–234.

Stipanovic, R.D., L.S. Puckhaber, A.A. Bell, A.E. Percival, and J. Jacobs. 2005. Occurrence of (+)- and (–)-gossypol in wild species of cotton and in *Gossypium hirsutum* var. Marie-galante (Watt) Hutchinson. *Journal of Agricultural and Food Chemistry* 53, no. 16:6266–6271.

Wolter, K.G., S.J. Wang, B.S. Henson, et al. 2006. (–)-Gossypol inhibits growth and promotes apoptosis of human head and neck squamous cell carcinoma *in vivo*. *Neoplasia* 8, no. 3:163–172.

Gracilaria edulis (Gracilariaceae) (Florideophyceae)   **Cytotoxic**

Location: Asia, South America, Africa, and Oceania, southern England and northwestern France
  Appearance: seaweed

  Stem: red algae, grows upright up to 20 cm in length. The bifurcations are intense and diverge; they are 1–1.5 mm in diameter with a slight stenosis in the base.

Tradition: this herb is used in Japanese, Hawaiian, and Filipino cuisine. It is antimicrobial and antibacterial, and it addresses decisively the phenomenon of eutrophication in lakes.
  Active ingredients: source of significant phenols
  Documented target cancers:

- Ehrlich ascites carcinoma-bearing mice
- Human lung adenocarcinoma cell line A549
- Human prostate cancer cell line: zinc oxide (ZnO) nanoparticles using the extracts of *Gracilaria edulis*

References

Priyadharshini, R.I., G. Prasannaraj, N. Geetha, and P. Venkatachalam. 2014. Microwave-mediated extracellular synthesis of metallic silver and zinc oxide nanoparticles using macro-algae (*Gracilaria edulis*) extracts and its anticancer activity against human PC3 cell lines. *Applied Biochemistry and Biotechnology* 174, no. 8:2777–2790.

Sakthivel, R., S. Muniasamy, G. Archunan, and K.P. Devi. 2016. *Gracilaria edulis* exhibit antiproliferative activity against human lung adenocarcinoma cell line A549 without causing adverse toxic effect *in vitro* and *in vivo*. *Food and Function* 7, no. 2:1155–1165.

Sundaram, M., S. Patra, and G. Maniarasu. 2012. Antitumor activity of ethanol extract of *Gracilaria edulis* (Gmelin) Silva on Ehrlich ascites carcinoma-bearing mice. *Zhong Xi Yi Jie He Xue Bao = Journal of Chinese Integrative Medicine* 10, no. 4:430–435.

***Guatteria friesiana* (Annonaceae) (Magnoliopsida)** **Cytotoxic**

Location: Brazilian and Colombian Amazon basin
    Appearance: small tree known as 'envireira' or 'envira'

    Leaves: large and interchangeable
    Flowers: flower consists of six petals and has a pale yellowish color.

Tradition: *Guatteria friesiana* is renowned for its antioxidant and antimicrobial action and its anti-malaria and insect-repellent properties.
    Parts used: essential oil from the leaves
    Active ingredients: alcanoids, terpenoids, main components in the herb's extract of essential oil, having antitumor activity are isomers of eudesmol (α-, β-, γ-eudesmol)
    Documented target cancers:

- B16-F10 (mouse melanoma)
- HepG2 (human hepatocellular carcinoma)
- HL-60 (human promyelocytic leukemia)
- K562 (human chronic myelocytic leukemia)

References

Costa, E.V., P.E.O.D. Cruz, M.L.B. Pinheiro, et al. 2013. Aporphine and tetrahydroprotoberberine alkaloids from the leaves of *Guatteria friesiana* (Annonaceae) and their cytotoxic activities. *Journal of the Brazilian Chemical Society* 24, no. 5:788–796.

Costa, E.V., M.L.B. Pinheiro, B.H.L. Maia, et al. 2016. 7,7-Dimethylaporphine and other alkaloids from the bark of *Guatteria friesiana*. *Journal of Natural Products* 79, no. 6:1524–1531.

***Gymnema sylvestre* (Apocynaceae) (Gentianales)** **Anti-proliferative**

Location: tropical forests of Southern and Central India and Sri Lanka
    Appearance: perennial woody vine

    Stem: climbing plant with scattered shoots
    Leaves: elongate oval shaped
    Flowers: small yellow umbelliferous inflorescence which is produced throughout the year. The arils are smooth, biconvex with a length of more than 3 inches. The sepals are long and oval.

Tradition: used in folk, Ayurvedic, and homeopathic systems of medicine used as diuretic and antitussive. It heals jaundice, constipation, bruising, heart disease, and malaria. It is used for weight loss and to reduce cholesterol and blood sugar. It is also used against asthma, eye problems, inflammation, and snake bites.
    Active ingredients: triterpenoids, saponins: ginsenosides, soyasaponins, saikosaponins, gymnemagenol

Documented target cancers:

- HeLa (human cervical carcinoma) cell line
- Gland cancer (main effect is the inhibition of proliferation of cells)
- Human lung adenocarcinoma cell line (A549) and human breast carcinoma cell line (MCF7)

Further details:
- It is hepato-protective and antimicrobial, while it prevents caries and reduces the percentage of triglycerides.

References

Arunachalam, K.D., L.B. Arun, S.K. Annamalai, and A.M. Arunachalam. 2015. Potential anticancer properties of bioactive compounds of Gymnema sylvestre and its biofunctionalized silver nanoparticles. *International Journal of Nanomedicine* 10:31–41.

Kanetkar, P., R. Singhal, and M. Kamat. 2007. Recent advances in Indian herbal drug research guest editor: Thomas Paul Asir Devasagayam Gymnema sylvestre: A memoir. *Journal of Clinical Biochemistry and Nutrition* 41, no. 2:77–81.

Khanna, V.G., and K. Kannabiran. 2009. Anticancer-cytotoxic activity of saponins isolated from the leaves of Gymnema sylvestre and Eclipta prostrata on HeLa cells. *International Journal of Green Pharmacy* 3, no. 3:227–229.

Kumar, P.M., M.V. Venkataranganna, K. Manjunath, G.L. Viswanatha, and G. Ashok. 2016. Methanolic leaf extract of Gymnema sylvestre augments glucose uptake and ameliorates insulin resistance by upregulating glucose transporter-4, peroxisome proliferator-activated receptor-gamma, adiponectin, and leptin levels *in vitro*. *Journal of Intercultural Ethnopharmacology* 5, no. 2:146–152.

Srikanth, A.V., S. Maricar, M.N. Lakshmi, P.R. Kumar, and B. Madhava Reddy. 2010. Anticancer activity of *Gymnema sylvestre*. R. Br. *International Journal of Pharmaceutical Sciences and Nanotechnology* 3, no. 1:2–4.

Gynostema pentaphyllum (Cucurbitaceae) (Cucurbitales)	Cytotoxic

Location: China, India, Korea, Japan, southeast Asia
Appearance: herbaceous climbing vine

Leaves: serrated and usually form groups of five leaflets

Tradition: widely used in Chinese medicine for the treatment of several diseases, including hepatitis, diabetes, and cardiovascular disease. In particular, it improves blood circulation, stimulates the function of the liver, strengthens the immune and nervous systems, and reduces blood sugar and cholesterol levels. It also has soothing effects, relieving spasms and lowering blood pressure. It is used in the treatment of asthma, peptic ulcer, and bronchitis. It improves insulin sensitivity.

Parts used: whole plant

Active ingredients: compounds belonging to the terpenoid group, and more particularly in the triterpene group, including saponins. These saponins, called gypenosides, have a strong anti-knock action.

Documented target cancers:

- HL-60 (human promyelocytic leukemia cells)
- Colon 205 (human colon cancer cells)

- Du145 (human prostate carcinoma cells)
- A549 human lung cancer cells
- U87 glioblastoma cells
- HepG2 human hepatocellular carcinoma
- B16 melanoma cells

References

Cui, J.F., P. Eneroth, and J.G. Bruhn. 1999. Gynostemma pentaphyllum: Identification of major sapogenins and differentiation from Panax species. *European Journal of Pharmaceutical Sciences* 8, no. 3:187–191.

Li, Y., W. Lin, J. Huang, Y. Xie, and W. Ma. 2016. Anti-cancer effects of *Gynostemma pentaphyllum* (Thunb.) Makino (Jiaogulan). *Chinese Medicine* 11, no. 1:43.

Liu. 1992. Therapeutic effect of jiaogulan on leukopenia due to irradiation and chemotherapy. *Zhong Guo Yi Yao Xue Bao* 7, no. 2:99.

Nagai, M., et al. Nov. 1976. Abstracts of Papers. The 23rd Meeting of the Japanese Society of Pharmacognosy. Jpn. 37.

Tanner, M.A., X. Bu, J.A. Steimle, and P.R. Myers. 1999. The direct release of nitric oxide by gypenosides derived from the herb Gynostemma pentaphyllum. *Nitric Oxide* 3, no. 5:359–365.

Winston, D., and S. Maimes 2007. *Adaptogens: Herbs for Strength, Stamina, and Stress Relief*, Rochester, Vermont

Zhou, Y.N. 1993. Effects of a gypenosides-containing tonic on the pulmonary function in exercise workload. *Journal of Guiyang Medical College* 18, no. 4:261.

Hannoa chlorantha (Hannoa) (Simaroubaceae) | **Anti-leukemic**

Location: Africa

Tradition: *Hannoa chlorantha* and *Hannoa klaineana* (Simaroubaceae) are used in traditional medicine of Central African countries against fevers and malaria.

Parts used: stem bark, root bark

Active ingredients:

Quassinoids (15-desacetylundulatone), 14-hydroxychaparrinone, chaparrinone

15-O-beta-D-glucopyranosyl-21-hydroxy-glaucarubolone was found to be more toxic while 6 alpha-tigloyloxy-glaucarubol and 21-hydroxyglaucarubolone were found inactive.

Documented target cancers: P-388 cells mouse lymphocytic leukemia, colon 38 adenocarcinoma

Further details:

- *Hannoa chlorantha* and *Hannoa klaineana*: apart from their documented antimalaria activity, stem bark extracts from *H. klaineana* and *H. chlorantha* are also cytotoxic against P-388 cells mouse lymphocytic leukemia cells. This activity is due to the presence of 14-hydroxychaparrinone (and, in a lesser degree, chaparrinone) from *H. klaineana*. In addition, the quassinoid 15-desacetylundulatone, isolated from the root bark of *Hannoa klaineana*, was found active against P388 and colon 38 adenocarcinoma, while 15-O-beta-D-glucopyranosyl-21-hydroxy-glaucarubolone was found to be more toxic, while 6 alpha-tigloyloxy-glaucarubol and 21-hydroxyglaucarubolone were found inactive.

References

François, G., C. Diakanamwa, G. Timperman, et al. 1998. Antimalarial and cytotoxic potential of four quassinoids from *Hannoa chlorantha* and *Hannoa klaineana*, and their structure-activity relationships. *International Journal for Parasitology* 28, no. 4:635–640.

Lumonadio, L., G. Atassi, M. Vanhaelen, and R. Vanhaelen-Fastre. 1991. Antitumor activity of quassinoids from *Hannoa klaineana*. *Journal of Ethnopharmacology* 31, no. 1:59–65.

Hedyotis diffusa (Rubiaceae) (Gentianales)   **Anti-leukemic**   **Cytotoxic**

Location: China, Japan, Nepal
  Appearance: slender, spreading or ascending annual herb (synonym *Oldenlandia diffusa*)

  Stem: weed, perennial, grows in tropical climate and exceeds 30 cm in length, the stem is malleable
  and soft.
  Leaves: elongate, oblong, no stem with two small leaflets on the base
  Flowers: either in pairs or alone, small and white

Tradition: well-known traditional Chinese herbal medicine, has long been widely applied in the
treatment of inflammation-related diseases, such as appendicitis, urethritis, and bronchitis. It is
renowned for its high efficacy against snake bites. It is analgesic, diuretic, antipyretic, sedative and
is used to treat pains related to ulcers, dysentery and pneumonia. It also strengthens the immune
system, relieves strain urine, and improves blood circulation.
  Parts used: whole plant
  Active ingredients: anthraquinones such as 2-methyl-3-methoxyanthraquinone and 2-methyl-
3-hydroxyanthraquinone and flavonoids such as quercetin
  Documented target cancers:

- Methanol extract cell death of n human lymphoid leukemia cells HL-60 in a dose-dependent manner
- Water extract was able to inhibit the growth of: B16-F10 murine melanoma, A5-49 human lung
  epithelial carcinoma, C-33A human cervix carcinoma, 8.5 MCF-7 human breast adenocarcinoma,
  MDA-MB-453 human breast adenocarcinoma, Ln-Cap human prostate carcinoma, Tsu-Pr1 human
  prostate carcinoma, 5 DU-145 human prostate carcinoma.

References

Cai, Q., J. Lin, L. Wei, et al. 2012. *Hedyotis diffusa* Willd inhibits colorectal cancer growth *in vivo* via inhibition
      of STAT3 signaling pathway. *International Journal of Molecular Sciences* 13, no. 5:6117–6128.
Chen, R., J. He, X. Tong, L. Tang, and M. Liu. 2016. The *Hedyotis diffusa* willd. (Rubiaceae): A review on
      phytochemistry, pharmacology, quality control and pharmacokinetics. *Molecules* 21, no. 6:710.
Gupta, S., D. Zhang, J. Yi, and J. Shao. 2004. Anticancer activities of *Oldenlandia diffusa*. *Journal of Herbal
      Pharmacotherapy* 4, no. 1:21–33.
Kasai, R., T. Shingu, R.Y. Wu, I.H. Hall, and K.H. Lee. 1982. Antitumor agents 57. The isolation and structural
      elucidation of microhelenin-E, a new anti-leukemic nor-pseudoguaianolide, and microhelenin-F from
      Helenium microcephalum. *Journal of Natural Products* 45, no. 3:317–320.
Niu, Y., and Q.X. Meng. 2013. Chemical and preclinical studies on Hedyotis diffusa with anticancer potential.
      Journal of Asian Natural Products Research 15, no. 5:550–565.
Yadav, S.K., and S.C. Lee. 2006. Evidence for Oldenlandia diffusa-evoked cancer cell apoptosis through
      superoxide burst and caspase activation. *Zhong Xi Yi Jie He Xue Bao = Journal of Chinese Integrative
      Medicine* 4, no. 5:485–489.

Helenium microcephalum (Sneezeweed) (Compositae)   **Cytotoxic**

Other common names: smallhead sneezeweed, red and gold sneezeweed
  Location: in mountain meadows and moist areas
  Origin: North America
  Appearance:

Stem: stout, 20–90 cm high
Leaves: alternate, lance-shaped, up to 2–2.5 cm long
Flowers: yellow-orange flowerheads
In bloom: June–September or generally during the warm season of the year

Degree of rarity: medium, since it can be easily cultivated

Biology: *Helenium* is a perennial plant, growing well on moist but well drained soil and requiring full sun. It can be easily propagated by seed and by dividing clumps every three to four years.

Tradition: species of the genus *Helenium* are long valued daisy-like ornamentals used in cutting and butterfly gardens for late summer color ('Helen's Flower').

Parts used: whole plant

Active ingredients:

Helenalin (a sesquiterpene lactone), microhelenin-E (1) and -F (2) (nor-pseudoguaianolides)

Precautions:

The plant and related species (such as *H. hoopesii*) are very poisonous. Helenalin has a documented acute toxicity. Reported effects on liver, kidney, and lung include depression, appetite loss, weak irregular pulse, weakness, stiffness, nasal discharge, bloat, 'spewing sickness', vomiting, foaming at the mouth, coughing, green nasal discharge, diarrhea, and photosensitization. Death may occur rapidly, 4 to 24 hours after ingestion or over a longer period in chronic cases.

Indicative dosage and application:

* The oral median lethal dose of helenalin for five mammalian species is between 85 and 105 mg/kg.
* In a study, they used a single ip dose of helenalin in male mice 43 mg/kg, and they continued for the next three days with ip injection of 25 mg helenalin/kg.

Documented target cancers:

* Helenalin, a sesquiterpene lactone found in species of the plant genus *Helenium*, inhibits the proliferation of cancer cells.
* Microlenin acetate, a dimeric sesquiterpene lactone, has a significant antileukemic activity.

Further details:

* Helenalin causes a marked potentiation of the increases in intracellular free Ca^{2+} concentration ($[Ca^{2+}]i$) produced by mitogens such as vasopressin, bradykinin, and platelet-derived growth factor in Swiss mouse 3T3 fibroblasts. Removing external Ca^{2+} partly attenuated the increased $[Ca^{2+}]i$ responses caused by helenalin. The increased $[Ca^{2+}]I$ responses occurred at concentrations of helenalin that inhibited cell proliferation. At higher concentrations, helenalin inhibited the $[Ca^{2+}]I$ responses. No change in resting $[Ca^{2+}]I$ was caused by helenalin even at high concentrations. Other helenalin analogues also increased the $[Ca^{2+}]I$ response. Helenalin did not inhibit protein kinase C (PKC), and PKC appeared to play a minor role in the effects of helenalin on $[Ca^{2+}]I$ responses in intact cells. Studies with saponin-permeabilized HT-29 human colon carcinosarcoma cells indicated that helenalin caused an increased accumulation of Ca^{2+} into nonmitochondrial stores and that the potentiating effect of helenalin on mitogen-stimulated $[Ca^{2+}]I$ responses was due in part to an increase in the inositol-(1,4,5)-trisphosphate-mediated release of Ca^{2+} from these stores.
* Two new nor-pseudoguaianolides, microhelenin-E (1) and -F (2), were isolated from Texas *Helenium microcephalum* and their structures elucidated on the basis of physicochemical data and spectral evidence. Microhelenin-E demonstrated significant *in vitro* and *in vivo* cytotoxic and antileukemic activities against KB tissue cell culture (ED50 = 1.38 microgram/ml) P-388 lymphocytic leukemia growth in BDF1 male mice (T/C-166% at 8 mg/kg/day), respectively.

- The antitumor sesquiterpene lactones microhelenins-A, B, and C, microlenin acetate, and plenolin were isolated from *Helenium microcephalum*. The structures and stereochemistry of these lactones were determined by physical methods as well as by chemical transformations and correlations. Microlenin acetate is probably the first novel dimeric sesquiterpene lactone demonstrated to have significant antileukemic activity.
- The known compound isohelenalin and a new antileukemic sesquiterpene lactone, isohelenol, were isolated from *Helenium microcephalum*.
- Studies with smallhead sneezeweed indicated that helenalin is the only significant toxic constituent present. The oral median lethal dose of helenalin for five mammalian species was between 85 and 105 mg/kg.
- The acute toxicity of helenalin was examined in male BDF1 mice. The 14-day LD50 for a single ip dose of helenalin in male mice was 43 mg/kg. A single ip injection of 25 mg helenalin/kg increased serum alanine aminotransferase (ALT), lactate dehydrogenase (LDH), urea nitrogen (BUN), and sorbitol dehydrogenase within 6 h of treatment. Multiple helenalin exposures, ip injection of 25 mg helenalin/kg for three days, increased differential polymorphonuclear leukocyte counts and decreased lymphocyte counts. Serum ALT, BUN, and cholesterol levels were also increased by multiple helenalin exposures at 25 mg helenalin/kg/day. Helenalin significantly reduced liver, thymus, and spleen relative weights, and histologic evaluation revealed substantial effects of multiple helenalin exposures on lymphocytes of the thymus, spleen, and mesenteric lymph nodes. No helenalin-induced histologic changes were observed in the liver or kidney. Multiple helenalin exposures (25 mg/kg/day) significantly inhibited hepatic microsomal enzyme activities (aminopyrine demethylase and aniline hydroxylase) and decreased microsomal cytochromes P-450 and b5 contents. Three concurrent days of diethyl maleate (DEM) pretreatment (3.7 mmol DEM/kg, 0.5 h before helenalin treatment) significantly increased the toxicity of helenalin exposure. These results indicate that the hepatic microsomal drug metabolizing system and lymphoid organs are particularly vulnerable to the effects of helenalin. In addition, helenalin toxicity is increased by DEM pretreatments, which have been shown to decrease glutathione concentrations.
- Helenalin (25 mg/kg) administered to immature male ICR mice caused a rapid decrease in hepatic glutathione levels and was lethally toxic to greater than 60% of the animals within 6 d. L-2 Oxothiazolidine 4-carboxylate (OTC), a compound that elevates cellular glutathione levels, administered to mice 6 or 12 h before helenalin, protected against hepatic glutathione depletion and the lethal toxicity of these toxins. OTC administered at the same time as the sesquiterpene lactones was not protective, suggesting that the critical events against which glutathione is protective occur within the first 6 h. In primary rat hepatocyte cultures, helenalin (4–16 µM) caused a rapid lethal injury as determined by the release of lactate dehydrogenase. Cotreatment of cultures with N-acetylcysteine at high concentrations (4 mM) afforded significant protection against lethal injury by both toxins. In contrast, BCNU, which inhibits glutathione reductase, or diethylmaleate, which depletes hepatocellular glutathione, potentiated the hepatotoxicity of helenalin in monolayer rat hepatocytes. These studies suggest that the *in vivo* and *in vitro* toxicity of helenalin is strongly dependent on hepatic glutathione levels, which helenalin rapidly depletes at very low concentrations.

References

Lee, K.H., Y. Imakura, and D. Sims 1976. Antitumor Agents XVII; structure and stereochemistry of *microhelenin-A*, a new antitumor sesquiterpene lactone from *Helenium microcephalum*. *Journal of Pharmacological Sciences* 65, no. 9:1410–1412.

Merrill, J.C., H.L. Kim, S. Safe, C.A. Murray, and M.A. Hayes. 1988. Role of glutathione in the toxicity of the sesquiterpene lactones hymenoxon and helenalin. *Journal of Toxicology and Environmental Health* 23, no. 2:159–169.

Sims, D., K.H. Lee, and R.Y. Wu. 1979. Antitumor agents 37. The isolation and structural elucidation of isohelenol, a new anti-leukemic sesquiterpene lactone, and isohelenalin from *Helenium microcephalum*. *Journal of Natural Products* 42, no. 3:282–286.

Helichrysum zivojinii (Asteraceae)	Anti-leukemic Cytotoxic

Location: Republic of Macedonia
  Appearance: flowering herb

Stem: endemic plant reaching the height of 15–30 cm
Leaves: gray-white fluff and schematically resemble the depiction of a spatula, 6–7 cm long
Flowers: yellow

Tradition: *Helichrysum zivojinii* is used for the treatment of headaches and colds. It generally strengthens the immune system. It is also used as a perfume additive to cosmetic products.

Active ingredients: tomoroside A (1) and tomoroside B (2) and eight known flavonoid glycosides that act against cancer cells
  Particular value: *Helichrysum* oil is used primarily for cosmetic and medicinal purposes.
  Documented target cancers:

- Appears to treat myelogenous leukemia *in vitro.*
- Exerted selective dose-dependent cytotoxic actions against: human cervix adenocarcinoma HeLa, human melanoma Fem-x, human breast adenocarcinoma MDA-MB-361, and human chronic myelogenous leukemia K562.

Reference

Matić, I.Z., I. Aljančić, Ž. Žižak, V. Vajs, M. Jadranin, S. Milosavljević, and Z.D. Juranić. 2013. *In vitro* antitumor actions of extracts from endemic plant *Helichrysum zivojinii*. *BMC Complementary and Alternative Medicine* 13, no. 1:https://doi.org/10.1186/1472-6882-13-36.

Hibiscus sabdariffa (Malvaceae) (Mangoliopsida)	Anti-tumor

Location: West Africa, China, Thailand
  Appearance: annual, erect, bushy, herbaceous subshrub

Leaves: three to five in the shape of a lobster that alternate left and right on the stem
Flowers: diameter of 8–10 cm, white or pale yellow with a characteristic red patch and a fleshy cup at
  the base of each petal

Tradition: it has diuretic effects, helps in the treatment of fever, and is anticorrosive. Its anthocyanin-containing flowers can reduce blood viscosity, reduce blood pressure, and stimulate the intestinal system. The seeds are used to treat body weakness and atony. The plant is also considered as antiseptic, aphrodisiac, digestive, and laxative. It is traditionally used in the treatment of abscess, cough, cancer, indigestion, hypertension, scurvy, and headache. The flower bud is eaten raw in salads or cooked and used as a flavor in sauces, puddings, and soups. The calyx is rich in citric acid and pectin, so it is useful in making jams, jellies, and others. It is used to add color and flavor to tea; it can also be boiled and used as a substitute for coffee. Tender young leaves and strain are used in salads, which give it an acidic flavor. The root is edible but has no particular taste. A strong fiber obtained from the stem is used in the preparation of linen, twine, and cord. Finally, it helps to deal with diabetes and liver problems.

  Parts used: stem, root, calyx, flower bud

  Active ingredients: neochlorogenic acid, chlorogenic acid, cryptochlorogenic acid, protocatechuic acid, caffeoyl shikimic acid, and flavonoids such as quercetin, kaempferol, and their derivatives

Documented target cancers:

- *Hibiscus* protocatechuic acid inhibited 12-O-tetradecanolyphorbol-13-acetate (TPA)-induced skin tumor formation in CD1-mice and inhibited the survival of human promyelocytic leukemia HL-60 cells.
- Human prostate cancer cells *in vitro* and *in vivo*. Leaf extracts inhibited the growth of prostate tumor xenograft in athymic nude mice.
- *Hibiscus* protocatechuic acid inhibits the survival of human promyelocytic leukemia HL-60 cells in a concentration- and time-dependent manner.
- Human gastric carcinoma (AGS) cells

References

Adegunloye, B.J., J.O. Omoniyi, O.A. Owolabi, O.P. Ajagbonna, O.A. Sofola, and H.A. Coker. 1996. Mechanisms of the blood pressure lowering effect of the calyx extract of Hibiscus sabdariffa in rats. *African Journal of Medicine and Medical Sciences* 25, no. 3:235–238.

Aguwa, C.N., O.O. Ndu, C.C. Nwanma, P.O. Udeogaranya, and N.O. Akwara. 2004. Verification of the folkloric diuretic claim of Hibiscus sabdariffa L. Petal extract. *Nigerian Journal of Pharmaceutical Research* 3, no. 1:1–8.

Akindahunsi, A.A., and M.T. Olaleye. 2003. Toxicological investigation of aqueous-methanolic extract of the calyces of Hibiscus sabdariffa L. *Journal of Ethnopharmacology* 89, no. 1:161–164.

Da-Costa-Rocha, I., B. Bonnlaender, H. Sievers, I. Pischel, and M. Heinrich. 2014. *Hibiscus sabdariffa* L.–A phytochemical and pharmacological review. *Food Chemistry* 165:424–443.

Lin, H.H., H.P. Huang, C.C. Huang, J.H. Chen, and C.J. Wang. 2005. Hibiscus polyphenol-rich extract induces apoptosis in human gastric carcinoma cells via p53 phosphorylation and p38 MAPK/FASL cascade pathway. *Molecular Carcinogenesis* 43, no. 2:86–99.

Tseng, T.H., T.W. Kao, C.Y. Chu, F.P. Chou, W.L. Lin, and C.J. Wang. 2000. Induction of apoptosis by hibiscus protocatechuic acid in human leukemia cells via reduction of retinoblastoma (RB) phosphorylation and Bcl-2 expression. *Biochemical Pharmacology* 60, no. 3:307–315.

Hicriopteris glauca (Gleicheniaceae) (Gleicheniales)   **Anti-tumor**

Location: South-east China, Oceania, and in Hawaii
    Appearance: fern

Leaves: spores are created at the back of the leaves for sexual reproduction.

Tradition: the roots and shoots of the plant are used in traditional Chinese medicine to treat bruises and wounds.
    Parts used: roots and shoots
    Active ingredients: lunularic acid 4′-glucoside and 2,9-dihydroxy-4,7-megastigmadiene-3-O-β-glucoside
    Documented target cancers:
- *Hicriopteris glauca* n-BuOH extract is reported to show antitumor activity against human melanoma A375 cells with IC_{50} values of 0.80 µg/ml.

Reference

Fang, X., X. Lin, S. Liang, W.D. Zhang, Y. Feng, and K.F. Ruan. 2012. Two new compounds from *Hicriopteris glauca* and their potential antitumor activities. *Journal of Asian Natural Products Research* 14, no. 12:1175–1179.

Hypericum perforatum L. (Hypericum (St John's Wort)) (Hypericaceae)	Cytotoxic

Location: Britain and throughout Europe and Asia
    Appearance:

Stem: 0.3–1 m high, erect, branching in the upper part
Leaves: pale green, sessile, and oblong, with pellucid dots or oil glands
Flowers: bright cheery yellow in terminal corymb
In bloom: June–August

Tradition: its name has been connected with many ancient superstitions. It was used as aromatic, astringent, resolvent, expectorant, and nervine.

    Parts used: herb tops, flowers

    Active ingredients: aromatic polycyclic diones (pseudohypericin and hypericin)

    Documented target cancers: photodynamic cancer therapy, human cancer cell lines (breast, colon, lung, melanoma, bladder), antiretroviral

    Further details:

- Pseudohypericin and hypericin, the major photosensitizing constituents of *Hypericum perforatum*, have been proposed as a photosensitizer for photodynamic cancer therapy. The presence of foetal calf serum (FCS) or albumin extensively inhibits the photocytotoxic effect of pseudohypericin against A431 tumor cells and is associated with a large decrease in cellular uptake of the compound. These results suggest that pseudohypericin, in contrast to hypericin, interacts strongly with constituents of FCS, lowering its interaction with cells. Since pseudohypericin is two to three times more abundant in *Hypericum* than hypericin and the bioavailabilities of pseudohypericin and hypericin after oral administration are similar, these results suggest that hypericin, and not pseudohypericin, is likely to be the constituent responsible for hypericism. Moreover, the dramatic decrease of photosensitizing activity of pseudohypericin in the presence of serum may restrict its applicability in clinical situations.
- Hypericin and pseudohypericin have potent antiretroviral activity and are highly effective in preventing viral-induced manifestations that follow infections with a variety of retroviruses *in vivo* and *in vitro*. Pseudohypericin and hypericin probably interfere with viral infection and/or spread by direct inactivation of the virus or by preventing virus shedding, budding, or assembly at the cell membrane. These compounds have no apparent activity against the transcription, translation, or

Figure 9.13 (See color insert.) *Hypericum.*

transport of viral proteins to the cell membrane and also no direct effect on the polymerase. This property distinguishes their mode of action from that of the major antiretro-virus group of nucleoside analogs. Hypericin and pseudohypericin have low *in vitro* cytotoxic activity at concentrations sufficient to produce dramatic antiviral effects in murine tissue culture model systems that use radiation leukemia and Friend viruses. Administration of these compounds to mice at the low doses sufficient to prevent retroviral-induced disease appears devoid of undesirable side effects. This lack of toxicity at therapeutic doses extends to humans, as these compounds have been tested in patients as antidepressants with apparent salutary effects. These observations suggest that pseudohypericin and hypericin could become therapeutic tools against retroviral-induced diseases such as acquired immunodeficiency syndrome (AIDS).

- Hexane extracts of *Hypericum drummondii* showed significant cytotoxic activity on cultured P-388, KB, or human cancer cell lines (breast, colon, lung, melanoma) (Figure 9.13).

References

Cott, J. 1995. NCDEU update. Natural product formulations available in Europe for psychotropic indications. *Psychopharmacology Bulletin* 31, no. 4:745–751.

Jayasuriya, H., J.D. McChesney, S.M., Swanson, and J.M. Pezzuto. 1989. Antimicrobial and cytotoxic activity of rottlerin-type compounds from *Hypericum drummondii*. *Journal of Natural Products* 52, no. 2:325–331.

Kacerovská, D., K. Pizinger, F. Majer, and F. Šmíd. 2008. Photodynamic therapy of nonmelanoma skin cancer with topical Hypericum perforatum extract—A pilot study. *Photochemistry and Photobiology* 84, no. 3:779–785.

Meruelo, D., G. Lavie, and D. Lavie. 1988. Therapeutic agents with dramatic antiretroviral activity and little toxicity at effective doses: Aromatic polycyclic diones hypericin and pseudohypericin. *Proceedings of the National Academy of Sciences of the United States of America* 85, no. 14:5230–5234.

Müller, W.E., and R. Rossol. 1994. Effects of Hypericum extract on the expression of serotonin receptors. *Journal of Geriatric Psychiatry and Neurology* 7, no. 1:S63–S64.

Stavropoulos, N.E., A. Kim, U.U. Nseyo, et al. 2006. Hypericum perforatum L. extract–novel photosensitizer against human bladder cancer cells. *Journal of Photochemistry and Photobiology: B, Biology* 84, no. 1:64–69.

Vandenbogaerde, A.L., A. Kamuhabwa, E. Delaey, B.E. Himpens, W.J. Merlevede, and P.A. de Witte. 1998. Photocytotoxic effect of pseudohypericin versus hypericin. *Journal of Photochemistry and Photobiology: B, Biology* 45, no. 2–3:87–94.

Jatropha gossypifolia **(Euphorbiaceae) (Malpighiales)**   **Anti-proliferative**

Location: tropical regions of America, the Florida region, Mexico, hotspots of Australia (Queensland), the Hawaiian Islands, Africa, and Asia

Appearance: small deciduous shrub

Leaves: trilobate, purple and sticky when young, while green and hairy when mature
Flowers: dark red in yellow in the center, clustered and followed by cherry-sized pods containing poisonous brown seeds

Tradition: parts of the plant are widely used in traditional folk medicine in the western regions of Africa.

Parts used: roots, stems, leaves, seeds, and fruits

Active ingredients: jatrophine, curcain, cyclogossine A, octapeptide, cyclogossine B

Documented target cancers:

- A new compound isolated from the root of *J. gossypifolia* has been reported to show potent proliferation inhibitory activity against A-549 human cancer cell line.

References

Falodun, A., Q. Sheng-Xiang, G. Parkinson, and S. Gibbons. 2012. Isolation and characterization of a new anticancer diterpenoid from *Jatropha gossypifolia*. *Pharmaceutical Chemistry Journal* 45, no. 10:636–639.

Félix-Silva, J., R.B. Giordani, A.A.D. Silva Jr, S.M. Zucolotto, and M.D.F. Fernandes-Pedrosa 2014. *Jatropha gossypiifolia* L. (Euphorbiaceae): A review of traditional uses, phytochemistry, pharmacology, and toxicology of this medicinal plant. Evidence-Based Complementary and Alternative Medicine http://dx.doi.org/10.1155/2014/369204.

Juniperus virginiana L. (Juniperus (red cedar)) (Conifereae)   **Tumor inhibitor**

Location: North America, Europe, North Africa, and North Asia. It is known as the American Juniper of Bermuda and also as 'Pencil Cedar'.

Appearance:

Stem: 1.5 m high, erect trunk, spreading branches covered with a shreddy bark
Leaves: straight and rigid, awl-shaped, 0.8–1.5 cm long, with sharp, prinkly points
Flowers: in short cones
In bloom: April–May

Tradition: it is used in the preparation of insecticides, in making liniments and other medicinal preparations, and perfumed soaps. The leaves have diuretic properties.

Parts used: ripe, carefully dried fruits, leaves

Active ingredient: Podophyllotoxin

Further details:

- Podophyllotoxin, the active ingredient of *Juniperus virginiana*, is a tumor inhibitor. However, in mice the use of cedar shavings as bedding increased significantly the incidence of spontaneous tumors of the liver and mammary gland and also reduced the average time at which tumors appeared.
- Both antitumor-promoting and antitumor activities have been attributed to the crude extract from the leaves of *Juniperus chinensis* (Figure 9.14).

Figure 9.14 **(See color insert.)** *Juniperus virginiana.*

References

Ali, A.M., M.M. Mackeen, I. Intan-Safinar, M. Hamid, N.H. Lajis, S.H. el-Sharkawy, and M. Murakoshi. 1996. Antitumour-promoting and antitumour activities of the crude extract from the leaves of *Juniperus chinensis*. *Journal of Ethnopharmacology* 53, no. 3:165–169.

Gawde, A.J., C.L. Cantrell, and V.D. Zheljazkov. 2009. Dual extraction of essential oil and podophyllotoxin from Juniperus virginiana. *Industrial Crops and Products* 30, no. 2:276–280.

Kupchan, S.M., J.C., Hemingway, and J.R. Knox. 1965. Tumor inhibitors. VII. Podophyllotoxin, the active principle of *Juniperus virginiana*. *Journal of Pharmaceutical Sciences* 54, no. 4:659–660.

Sabine, J.R. 1975. Exposure to an environment containing the aromatic red cedar, *Juniperus virginiana*: Procarcinogenic, enzyme-inducing and insecticidal effects. *Toxicology* 5, no. 2:221–235.

Kigelia pinnata (Kigelia) (Bigoniaceae)   **Tumor inhibitor**

Parts used: stembark, fruits

Active ingredients: Lapachol

Documented target cancers: effects against four melanoma cell lines, a renal cell carcinoma line (Caki-2), and a rhabdomyosarcoma RD human cancer cell line

Further details:

- Significant inhibitory activity was shown by the dichloromethane extract of the stembark and lapachol (continuous exposure). Moreover, activity was dose-dependent, the extract being less active after 1 h exposure. Chemosensitivity of the melanoma cell lines to the stembark was greater than that seen for the renal adenocarcinoma line. In marked contrast, sensitivity to lapachol was similar amongst the five cell lines. Lapachol was not detected in the stembark extract.

References

Atolani, O., G.A. Olatunji, O.A. Fabiyi, A.J. Adeniji, and O.O. Ogbole. 2013. Phytochemicals from *Kigelia pinnata* leaves show antioxidant and anticancer potential on human cancer cell line. *Journal of Medicinal Food* 16, no. 10:878–885.

Houghton, P.J., A. Photiou, S. Uddin, P. Shah, M. Browning, S.J. Jackson, and S. Retsas. 1994. Activity of extracts of *Kigelia pinnata* against melanoma and renal carcinoma cell lines. *Planta Medica* 60, no. 5:430–433.

Jackson, S.J., P.J. Houghton, S. Retsas, and A. Photiou. 2000. *In vitro* cytotoxicity of norviburtinal and isopinnatal from *Kigelia pinnata* against cancer cell lines. *Planta Medica* 66, no. 8:758–761.

Koelreuteria henryi (Sapindaceae)   **Tumor inhibitor**

Synonyms: varnish tree

Location: eastern Asia

Origin: China and Korea

Appearance:

Stem: fast-growing, deciduous tree reaching about –17 m in height. At maturity, it has a rounded crown, with a spread equal to or greater than the height.

Leaves: compound leaves that give it an overall lacy appearance. The leaves turn yellow before falling.

Flowers: large clusters of showy yellow flowers

Degree of rarity: low, as it can be cultivated

Active ingredients: protein-tyrosine kinase inhibitors: anthraquinone, stilbene, and flavonoid

Particular value: in cooler zones, use as a free-standing tree where it can be seen in all its glory! It is also good as a small shade tree where space is limited. Golden rain tree should be used more often as a street and park tree.

Documented target cancers:

- Anthraquinone inhibitor, emodin, displayed highly selective activities against src-Her-2/neu and ras-oncogenes.
- Austrobailignan-1, a natural lignan derivative isolated from *Koelreuteria henryi* Dummer is reported to inhibit cell growth of human non-small cell lung cancer A549 and H1299 cell lines.

Further details:
- Protein kinases encoded or modulated by oncogenes were used to prescreen the potential antitumor activity of medicinal plants. Protein-tyrosine kinase-directed fractionation and separation of the crude extracts of Polygonum cuspidatum and *Koelreuteria henryi* have led to the isolation of three different classes of protein-tyrosine kinase inhibitors, anthraquinone, stilbene, and flavonoid.

References

Chang, C.J., C.L. Ashendel, R.L. Geahlen, J.L. McLaughlin, and D.J. Waters. 1996. Oncogene signal transduction inhibitors from medicinal plants. *In vivo* 10, no. 2:185–190.

Chiang, Y.Y., S.L. Wang, C.L. Yang, et al. 2013. Extracts of *Koelreuteria henryi* Dummer induce apoptosis and autophagy by inhibiting dihydrodiol dehydrogenase, thus enhancing anticancer effects. *International Journal of Molecular Medicine* 32, no. 3:577–584.

Lee, T.H., Y.H. Chiang, C.H. Chen, P.Y. Chen, and C.K. Lee. 2009. A new flavonol galloylrhamnoside and a new lignan glucoside from the leaves of *Koelreuteria henryi* Dummer. *Journal of Natural Medicines* 63, no. 2:209–214.

Wu, C.C., K.F. Huang, T.Y. Yang, Y.L. Li, C.L. Wen, S.L. Hsu, and T.H. Chen. 2015. The topoisomerase 1 inhibitor Austrobailignan-1 isolated from *Koelreuteria henryi* induces a G2/M-phase arrest and cell death independently of p53 in non-small cell lung cancer cells. *PLOS ONE* 10, no. 7:e0132052.

Landsburgia quercifolia (Gleicheniaceae) (Gleicheniales)   **Anti-leukemic**

Location: endemic to New Zealand's aquatic ecosystems, mainly located around the subtropical islands of Three Kings

Appearance: brown algae that can reach the height of 1.5 meters

Parts used: thallus

Active ingredients: deoxylapachol,2-(3-methyl-2-butenyl), 2,3-epoxide of deoxylapachol

Documented target cancers:
- Leukemia cells P-388 (IC_{50} 0.6 microgm/ml)

Reference

Perry, N.B., J.W. Blunt, and M.H. Munro. 1991. A cytotoxic and antifungal 1,4-naphthoquinone and related compounds from a New Zealand brown alga, *Landsburgia quercifolia*. *Journal of Natural Products* 54, no. 4:978–985.

Larrea tridentata (Zygophyllaceae) (Zygophyllales)   **Cytotoxic**

Location: North American warm deserts

Appearance: evergreen bush

Leaves: dark green, aromatic
Flowers: five yellow petals, 25 mm in diameter

Tradition: in Mexico the tea is used traditionally as a treatment of kidney and gallbladder stones.
   Parts used: leaves and branches
   Active ingredients: nordihydroguaiaretic acid, guaiaretic acid
   Documented target cancers:
Several lignans isolated from the flowering tops of *Larrea tridentata* have shown growth inhibitory activity against human breast cancer, human colon cancer, and human melanoma cell lines.

References

Chan, J.K.W., S. Bittner, A. Bittner, et al. 2018. Nordihydroguaiaretic acid, a lignan from *Larrea tridentata* (creosote bush), protects against American lifestyle-induced obesity syndrome diet-induced metabolic dysfunction in mice. *Journal of Pharmacology and Experimental Therapeutics* 365, no. 2:281–290.

Lambert, J.D., S. Sang, A. Dougherty, C.G. Caldwell, R.O. Meyers, R.T. Dorr, and B.N. Timmermann. 2005. Cytotoxic lignans from Larrea *tridentata*. *Phytochemistry* 66, no. 7:811–815.

Laurus nobilis (Lauraceae) (Laurales)   **Cytotoxic**

Location: Mediterranean climate
   Appearance: evergreen shrub with orthodontic growth

   Leaves: alternating, lanceolate, sharp, dark green

Parts used: leaves, fruits
   Active ingredients: reticulin, 1,8-quinole, α- and β-pinene, costunolide, zaluzanin D, cineol, eugenol, and methyleugenol
   Tradition: the plant extract is considered suitable for rheumatism as well as for bruising and bronchitis. It is also analgesic and haemostatic.
   Documented target cancers:

- *In vitro* studies demonstrated inhibition of cancer cell growth by processed *L. nobilis* leaf products in HT-29, HCT-116, Caco-2, and SW-480 human cancer cell lines, accompanied by variable levels of elevated apoptosis.
- *Laurus nobilis* hexan extract has shown a cytotoxic effect on AMN3 (mouse mammary adenocarcinoma), REF (rat embryo fibroblast), and HeLa (cervical human carcinoma) cell lines.

References

Bennett, L., M. Abeywardena, S. Burnard, et al. 2013. Molecular size fractions of bay leaf (*Laurus nobilis*) exhibit differentiated regulation of colorectal cancer cell growth *in vitro*. *Nutrition and Cancer* 65, no. 5:746–764.

Gany, Z.S.A. 2010. Cytotoxoic effect of Laurus *nobilis* extract on different cancer cell lines. *Journal of the College of Basic Education* 15, no. 64:195–200.

Sayyah, M., J. Valizadeh, and M. Kamalinejad. 2002. Anticonvulsant activity of the leaf esential oil of *Laurus nobilis* against pentylenetetrazole-and maximal electroshock-induced seizures. *Phytomedicine* 9, no. 3:212–216.

Macaranga denticulata (Euphorbiaceae) (Malpighiales)	Anti-proliferative

Location: Asia: China, Thailand, Indonesia, Nepal, Bangladesh, India, Malaysia, Africa, Australia, the Pacific Islands, and the Indian Ocean
  Appearance: evergreen, dioecious tree

  Leaves: simple, alternating, greenish, smooth or with thin hairs and yellow spots
  Flowers: axillary or grow just below the leaves

Parts used: leaves, flowers
  Active ingredients: 3-epitaraxerol, eraxerone, β-sitosterol, aleuritolic acid 3-p-hydroxybenzoate, cleomiscosin B, 3,3-Di-O-methylellagic acid, mulberrin, β-D-glucopyranoside of methyl salicylate, α-onocerin
  Tradition: the leaf juice is used to clean wounds while flower juice and leaves are used to treat constipation and colic.
  Documented target cancers:
  Prenylated flavonoids from *Macaranga denticulata* are reported to inhibit the proliferation of the lung cancer A-549 cell line.

References

Fan, Q.F., J.Y. Zheng, Q.S. Song, Z. Na, and H.B. Hu. 2015. Chemical constituents of the roots of *Macaranga denticulata*. *Chemistry of Natural Compounds* 51, no. 3:586–587.
Sutthivaiyakit, S., S. Unganont, P. Sutthivaiyakit, and A. Suksamrarn. 2002. Diterpenylated and prenylated flavonoids from *Macaranga denticulata*. *Tetrahedron* 58, no. 18:3619–3622.
Yang, D.S., Z.L. Li, W.B. Peng, et al. 2015. Three new prenylated flavonoids from *Macaranga denticulata* and their anticancer effects. *Fitoterapia* 103:165–170.

Machilus mushaensis (Lauraceae) (Laurales)	Cytotoxic

Location: Philippines, Taiwan, India, China, Korea, Hawaii, and Malaysia
  Appearance: large tree, evergreen

  Stem: bud scales are densely covered by brown hairs. Bud sizes of *M. mushaensis* are from 0.8 to 1.2
    cm in diameter and from 1 to 2 cm in length.
  Leaves: leaf blade deep green to blue-green when alive
  Fruits: 9–10 mm in diameter

Parts used: leaves, flowers
  Active ingredients: n-decanal (61.0%), a-cadinol (20.8%), z-trimenal (3.7%), viridiflorene (2.9%), β-caryophyllene (1.7%), and globulol (1.3%)
  Documented target cancers:

- The oil has exhibited cytotoxic activity against human oral, liver, lung, colon, melanoma, and leukemic cancer cells.
- α-Cadinol is reported to be cytotoxic against three human cancer cell lines, including A-549, MCF-7, and HT-29. The presence of α-cadinol significantly contributed to the anticancer activities of *M. mushaensis* leaf oil.
- Melanoma
- Leukemic cancer cells

References

Chen, C.Y., M.J. Cheng, Y.J. Chiang, et al. 2009. Chemical constituents from the leaves of *Machilus zuihoensis* Hayata var. mushaensis (Lu) YC Liu. *Natural Product Research* 23, no. 9:871–875.

Lu, S.Y., and S.Y. Chen. 1996. Geographical distribution and taxonomic study of Machilus zuihoensis and *Machilus mushaensis*. *Taiwan Journal of Forest Science* 11:239–244.

Pan, L.Y., T.C. Chiang, Y.C. Weng, et al. 2015. Taxonomy and biology of a new ambrosia gall midge Daphnephila urnicola sp. nov. (Diptera: Cecidomyiidae) inducing urn-shaped leaf galls on two species of Machilus (Lauraceae) in Taiwan. *Zootaxa* 3955, no. 3:371–388.

Su, Y.C., and C.L. Ho. 2013. Composition, in-vitro anticancer, and antimicrobial activities of the leaf essential oil of *Machilus mushaensis* from Taiwan. *Natural Product Communications* 8, no. 2:273–275.

Yang, S.Y., M.Y. Chen, and J.T. Yang. 2002. Application of cecidomyiid galls to the systematics of the genus Machilus (Lauraceae) in Taiwan. *Botanical Bulletin of Academia Sinica* 43:31–35.

Macrothelypteris torresiana (Thelypteridaceae) (Polypodiales)   Cytotoxic

---

Location: northern Western Australia
   Appearance: pteridophyte

Stem: robust with a short creeping rhizome
Leaves: fronds are three-pinnate, deeply divided, up to 1.2 m long, with a stalk 0.5 m. Sori are circular but not terminal on the veins.

Parts used: entire plant
   Active ingredients: cytotoxic flavonoids and flavonoid glycosides from this species (a.k.a. *Thelypteris torresiana*), flavotorresin, a flavonoid diglycoside, multiflorin C and five known compounds, a drimane sesquiterpene, thelypterene , multiflorin A, two phenols, and one steroid
   Documented target cancers:
   * Protoapigenone, a novel flavonoid, has exhibited significant cytotoxic activity against five human cancer cell lines: liver cancer cells (Hep G2 and Hep 3B), breast cancer cells (MCF-7, MDA-MB-231), and lung cancer cells (A549).

References

Dudani, S.N., M.K. Mahesh, M.S. Chandran, and T.V. Ramachandra. 2014. Pteridophyte diversity in wet ever-green forests of Sakleshpur in Central Western Ghats. *Indian Journal PltSci* 3, no. 1:28–39.

Lin, A.S., F.R. Chang, H.F. Yen, H. Björkeborn, P. Norlén, and Y.C. Wu. 2007. Novel flavonoids of *Thelypteris torresiana*. *Chemical and Pharmaceutical Bulletin* 55, no. 4:635–637.

Lin, A.S., K. Nakagawa-Goto, F.R. Chang, et al. 2007. First total synthesis of protoapigenone and its analogues as potent cytotoxic agents. *Journal of Medicinal Chemistry* 50, no. 16:3921–3927.

Liu, H., Y. Xiao, C. Xiong, A. Wei, and J. Ruan. 2011. Apoptosis induced by a new flavonoid in human hepatoma HepG2 cells involves reactive oxygen species-mediated mitochondrial dysfunction and MAPK activation. *European Journal of Pharmacology* 654, no. 3:209–216.

Magnolia officinalis (Magnoliaceae) (Magnoliales)   Anti-leukemic

---

Location: China (Anhui, Chongqing, Fujian, Gansu—Present—Origin Uncertain, Guangdong—Present—Origin Uncertain, Guangxi, Guizhou, Hubei, Hunan, Jiangxi, Shaanxi, Sichuan, Zhejiang)
   Appearance: deciduous tree growing up to 20 m in height

Leaves: alternate, simple, and sometimes lobed
Flowers: white, fragrant, insect pollinated

Parts used: dried bark, root-bark, stem, and branch-bark

Active ingredients: polyphenolic neolignans: Honokiol, Magnolol, Magnoloside A–Methylhonokiol 4-0

Tradition:

Magnolia (Hou Po) has been used in traditional Chinese medicine (TCM) for at least 2,000 years. The parts used medicinally are the buds, flowers, and bark. The bark is used to treat a variety of gastrointestinal disorders, as an analgesic, antiedematous, diuretic, expectorant, hypotensive, for loss of appetite, diarrhea, in the treatment of bronchitis, emphysema, and acute paroxysmal cough. In traditional Japanese medicine (Kampo), magnolia bark is an ingredient of herbal mixture Hange-koboku-do, which consists of five plant extracts and formulas Saiboku-to, which consists of ten plant extracts. These extracts are used to reduce anxiety, nervous tension, and in the treatment of depression and to improve the quality of sleep.

Documented target cancers:

- Human prostate cancer cells (PC-3, LNCaP, and C4-2) and xenographs
- SVR angiosarcoma cells
- Chronic lymphocytic leukemia (CLL) in human patients

References

Arora, S., S. Singh, G.A. Piazza, C.M. Contreras, J. Panyam, and A.P. Singh. 2012. Honokiol: A novel natural agent for cancer prevention and therapy. *Current Molecular Medicine* 12, no. 10:1244–1252.

Battle, T.E., J. Arbiser, and D.A. Frank. 2005. The natural product honokiol induces caspase-dependent apoptosis in B-cell chronic lymphocytic leukemia (B-CLL) cells. *Blood* 106, no. 2:690–697.

Fried, L.E., and J.L. Arbiser. 2009. Honokiol, a multifunctional antiangiogenic and antitumor agent. *Antioxidants and Redox Signaling* 11, no. 5:1139–1148.

Hahm, E.R., J.A. Arlotti, S.W. Marynowski, and S.V. Singh. 2008. Honokiol, a constituent of oriental medicinal herb *Magnolia officinalis*, inhibits growth of PC-3 *xenografts in vivo* in association with apoptosis induction. *Clinical Cancer Research* 14, no. 4:1248–1257.

Magnolia virginiana L. (Magnolia) (Magnoliaceae) | **Tumor inhibitor**

Location: North America

Appearance:

Stem: 8 or more ft high, 3–5 ft diameter, smooth gray trunk
Leaves: simple, oval, 6 inches long by 3 broad, silvery, and slightly hairy underneath
Flowers: large, white
In bloom: spring

Tradition: it is used in rheumatism and malaria and is contra-indicated in inflammatory symptoms.

Parts used: bark of stem and root

Active ingredients:

Neolignans: magnolol, honokiol, and monoterpenylmagnolol

Parthenolide

Indicative dosage and application: is still being tested

Documented target cancers:

Figure 9.15   (See color insert.) *Magnolia virginiana.*

- Epstein–Barr virus, skin tumor (mice)
- Human lung carcinoma and breast carcinoma cell lines

Further details:

- *Magnolia officinalis*: the bark contains the neolignans magnolol, honokiol, and monoterpenylmagnolol. The MeOH extract of this plant and *magnolol* exhibited remarkable inhibitory effects on mouse skin tumor promotion in an *in vivo* two stage carcinogenesis test.
- Another tumor inhibitory agent, parthenolide, has been isolated from *Magnolia grandiflora* I.P (Figure 9.15).

References

Celle, G., V. Savarino, A. Picciotto, M.R. Magnolia, P. Scalabrini, and M. Dodero. 1988. Is hepatic ultrasonography a valid alternative tool to liver biopsy? Report on 507 cases studied with both techniques. *Digestive Diseases and Sciences* 33, no. 4:467–471.

Farag, M.A., and D.A. Al-Mahdy. 2013. Comparative study of the chemical composition and biological activities of *Magnolia grandiflora* and Magnolia virginiana flower essential oils. *Natural Product Research* 27, no. 12:1091–1097.

***Mallotus philippinensis* (Mallotus (Kamala) (Euphorbiaceae)**   **Tumor inhibitor**

Location: India, Malay Archipelago, Orissa, Bengal, Bombay, Southern Arabia, China, Australia
   Appearance:

Stem: 7–10 m high, 1–1.5 cm in diameter
Leaves: alternate, articulate petioles 1–2 inches long, ovate with tow obscure glands at base
Flowers: dioecious, covered with ferrugineous tomentosum
In bloom: November–January

Tradition: the root of the tree is used in dyeing and for cutaneous eruptions. It was used by the Arabs internally for leprosy and in solution to remove freckles and pustules.

Parts used: pericarps
Active ingredients:

- Maytansinoid tumor inhibitors: rottlerin, mallotojaponin, phloroglucinol derivatives: mallotolerin, mallotochromanol, mallotophenone, mallotochromene
- Ent-kaurane and rosane diterpenoids

Documented target cancers:

- CaM kinase III inhibitor. Cytotoxic (glioblastomas-human, mice)
- Skin tumor (mice), human larynx (HEp-2), and lung (PC-13) carcinoma cells as well as mouse B16 melanoma, leukemia P388, and L5178Y cells
- The ethanolic extract of the fruit is reported as cytotoxic against breast, colon, cervix, lung, and ovarian cancer cell lines.
- Triterpenoids from the bark are reported to inhibit mouse skin tumor promotion.

Further details:

- *Mallotus phillippinensis*: pericarps contain rottlerin, a 5,7-dihydroxy-2,2-dimethyl-6-(2,4,6-trih ydroxy-3-methyl-5-acetylbenzyl)-8-cinnamoyl-1,2-hromene which has been shown to be an effective CaM kinase III inhibitor. Rottlerin decreased growth and induced cytotoxicity in rat (C6) and two human gliomas (T98G and U138MG) at concentrations that inhibited the activity of CaM kinase III *in vitro* and *in vivo*. Far less demonstrable effects were observed on other Ca^{2++}/CaM-sensitive kinases. Incubation of glial cells with rottlerin produced a block at the G1-S interface and the appearance of a population of cells with a complement of DNA. In addition, rottlerin induced changes in cellular morphology such as cell shrinkage, accumulation of cytoplasmic vacuoles, and packaging of cellular components within membranes.
- The pericarps of *Mallotus japonicus* (Euphorbiaceae) contain mallotojaponin, which inhibited the action of tumor promoter *in vitro* and *in vivo*; it inhibited tumor promoter-enhanced phospholipid metabolism in cultured cells and also suppressed the promoting effect of 12-O-etradecanoylphorbol-13-acetate on skin tumor formation in mice initiated with 7,12-dimethylbenz-[a]anthracene.
- In addition, pericarps contain a variety of phloroglucinol derivatives which were proved to be significantly cytotoxic in culture against human larynx (HEp-2) and lung (PC-13) carcinoma cells as well as mouse B16 melanoma, leukemia P388, the KB system, and L5178Y cells. These phloroglucinol derivatives are: mallotolerin (3-(3-methyl-2-hydroxybut-3-enyl)-5-(3-acetyl-2,4-dihydroxy-5-met hyl-6-methoxybenxyl)-phlorbutyrophenone), mallotochromanol (8-acetyl-5, 7-dihydroxy-6-(3-acetyl-2, 4-dihydroxy-5-methyl-6-methoxybenzyl) 2, 2-dimethyl-3-hydroxychroman), allotophenone (5-methylene-bis-2, 6-dihydroxy-3-methyl-4-methoxyacetophenone), mallotochromene (8-acetyl-5, 7-dihydroxy-6-(3-acetyl-2, 4-dihydroxy-5-methyl-6-methoxybenzyl)2, 2-dimethylchromene), 3-(3,3-dimethylallyl)-5-(3-acetyl-2, 4-dihydroxy-5-methyl-6-methoxybenzyl)-phloracetophenone, and 2,6-dihydroxy-3-methyl-4-methoxyacetophenone.
- *Mallotus anomalus* Meer et Chun contains ent-kaurane and rosane diterpenoids.

References

Arisawa, M., A. Fujita, N. Morita, and S. Koshimura. 1990. Cytotoxic and antitumor constituents in pericarps of *Mallotus japonicus*. *Planta Medica* 56, no. 4:377–379.

Arisawa, M., A. Fujita, M. Saga, T. Hayashi, N. Morita, N. Kawano, and S. Koshimura. 1986. Studies on cytotoxic constituents in pericarps of *Mallotus japonicus*, Part II. *Journal of Natural Products* 49, no. 2:298–302.

Arisawa, M., A. Fujita, R. Suzuki, T. Hayashi, N. Morita, N. Kawano, and S. Koshimura. 1985. Studies on cytotoxic constituents in pericarps of *Mallotus japonicus*, Part I. *Journal of Natural Products* 48, no. 3:455–459.

Parmer, T.G., M.D. Ward, and W.N. Hait. 1997. Effects of rottlerin, an inhibitor of calmodulin-dependent protein kinase III, on cellular proliferation, viability, and cell cycle distribution in malignant glioma cells. *Cell Growth and Differentiation* 8, no. 3:327–334.

Satomi, Y., M. Arisawa, H. Nishino, and A. Iwashima. 1994. Antitumor-promoting activity of mallotojaponin, a major constituent of pericarps of *Mallotus japonicus*. *Oncology* 51, no. 3:215–219.

Sharma, V. 2011. A polyphenolic compound rottlerin demonstrates significant *in vitro* cytotoxicity against human cancer cell lines: Isolation and characterization from the fruits of Mallotus philippinensis. *Journal of Plant Biochemistry and Biotechnology* 20, no. 2:190–195.

Marsdenia tenacissima (Apocynaceae) (Gentianales)   Anti-cancer

---

Location: Himalayas, from Kumaun to Sikkim, Bengal, Ceylon, Indo-China, Western China.
  Appearance: large climbing shrub

  Bark: grey or pale brown, corky, deeply furrowed
  Leaves: broadly ovate
  Flowers: greenish-yellow, in paniculate cymes

Tradition: it is used as herbal medicine for the treatment of asthma, cancer, trachitis, tonsillitis, pharyngitis, cystitis, and pneumonia. *Marsdeniae tenacissimae* Caulis is a Chinese herbal medicine used for the treatment of coughs, rheumatism, carbuncles, and tumors. A water soluble extract of the plant is used as raw material for Xiao-Ai-Ping injection.

  Used for fever, hyperliposis, polyuria, dryness of mouth, worm infestation, heart disease, pruritus, bleeding piles (blood purifier and purgative), and excessive thirst. Also used for skin diseases, toxemia, prutirus, and vomiting.

  Parts used: bark of young plants, stem, stem (milky juice), leaves, dried roots
  Active ingredients: phenolic acid and C21 steroidal glycosides
  Particular value:
  Chinese herbal drug approved for the treatment of cancer if co-administered with anticancer agents such as platinum drugs. It is reported to be able to enhance the clinical effects of chemotherapy in the treatment of lung cancer, liver cancer, and gastric cancer, etc. Chinese medicine suggests that *M. tenacissima* is beneficial for treating patients with cancers such as esophageal cancer, gastric cancer, and lung cancer and has no significant side-effects.

  Documented target cancers:

- Human gastric carcinoma
- Anti-angiogenesis agent
- Human U937, HL60 leukemic cells
- Jurkat (acute T cell leukemia), Raji (Burkitt's lymphoma), and RPMI8226 (human myeloma) cells
- Total aglycones of the plant showed the ability to sensitize KB-3-1, HeLa, HepG2, and K562 cells to paclitaxel treatment.
- Commercially available plant extract restored gefitinib efficacy in resistant non-small cell lung cancer lines.

Indicative dosage and application: (Xiaoaiping injection) intravenous injection: instillation after it is diluted by 5 or 10% glucose injection, 20–100 ml once, once a day
  Side effects: occasional low grade fever, rash, sweaty, wandering muscle joint pain, local injection to stimulate the pain and discomfort

References

Chen, B., C. Li, J. Chen, and J. Ou-Yang. 2009. Effect of extract from *Marsdenia tenacissima* on Jurkat, Raji and RPMI8226 cells *in vitro*. *Chinese Journal of Biochemical Pharmaceutics* 30, no. 3:174–177.7.

Han, S.Y., H.R. Ding, W. Zhao, F. Teng, and P.P. Li. 2014. Enhancement of gefitinib-induced growth inhibition by *Marsdenia tenacissima* extract in non-small cell lung cancer cells expressing wild or mutant EGFR. *BMC Complementary and Alternative Medicine* 14, no. 1:165.

Han, S.Y., M.B. Zhao, G.B. Zhuang, and P.P. Li. 2012. *Marsdenia tenacissima* extract restored gefitinib sensitivity in resistant non-small cell lung cancer cells. *Lung Cancer* 75, no. 1:30–37.

Huang, Z., H. Lin, Y. Wang, Z. Cao, W. Lin, and Q. Chen. 2013. Studies on the anti-angiogenic effect of *Marsdenia tenacissima* extract *in vitro* and *in vivo*. *Oncology Letters* 5, no. 3:917–922.

Khare, C.P. 2015. *Ayurvedic Pharmacopoeial Plant Drugs: Expanded Therapeutics*. CRC Press, New York, NY.

Lee, H., and J.Y. Lin. 1988. Antimutagenic activity of extracts from anticancer drugs in Chinese medicine. *Mutation Research* 204, no. 2:229–234.

Li, D., J. Ouyang, C.P. Li, and J.H. Chen 2008. *Marsdensia tenacissima* induces apoptosis of human U937, HL60 leukemic cells. *Chinese Journal of Biochemical Pharmaceutics* 29, no. 1:33.

Li, M.Q., J.H. Shen, B. Xu, and J. Chen. 2001. The Mechanism of Laboratory Research for Xiaoaiping treating SGC-7901 gastric carcinoma cellular strains. *Journal of Interventional Radiology* 10:228–231.

Sun, J., J.H. Shen, M.H. Zhang, C.H. Li, and Z.Z. Fan. 2000. Experimental study on therapeutic function of 'carcinoma-eliminating injection' on cell of human gastric carcinoma. *Acta Acad Shanghai Tradit Chin Med* 14:41–43.

Zhu, R.J., X.L. Shen, L.L. Dai, X.Y. Ai, R.H. Tian, R. Tang, and Y.J. Hu. 2014. Total aglycones from *Marsdenia tenacissima* increases antitumor efficacy of paclitaxel in nude mice. *Molecules* 19, no. 9:13965–13975.

Maytenus boaria (Maytenus) (Celastraceae)	Cytotoxic

Location: mountains of South America
  Appearance:

Stem: 34 m
Leaves: alternate, simple, narrow, elliptic to lanceolate, tiny teeth, pointed tip

Active ingredients: Maytenin
  Ansa macrolide (maytansine)
  Documented target cancers:
  – Basic cellular carcinoma, Kaposi's sarcomatosis, leukemia
  – *In vitro* activity against human colon adenocarcinoma and large cell lung carcinoma cell lines
  Further details:

- Maytenin demonstrates a low irritant action and late antineoplastic properties.
- Some more species of the same genus appear to have a cytotoxic effect against cancer tumors such as: *Maytenus guangsiensis* Cheng et Sha (antileukemic), *Maytenus ovatus* (antileukemic) (maytansine), *Maytenus senegalensis*.
- *Maytenus wallichiana* Raju et Babu and *Maytenus emarginata* Ding Hou (lymphocytic leukemia)
- Biotechnology: plant tissue cultures of *Maytenus wallichiana* Raju et Babu and *Maytenus emarginata* Ding Hou were initiated. Growth conditions of the callus and the optimum medium composition have been established. Increments of callus wet mass and dynamics of callus growth were determined. Morphological and microscopic observations were also performed. The most efficient growth of the callus, resulting in increments of its wet mass up to 6,460%, was obtained on the modified Murashige and Skoog medium. Extracts of the callus were found to be inactive against microorganisms but proved cytotoxic for lymphocytic leukemia.

References

Dymowski, W., and M. Furmanowa. 1989. The search for cytostatic substances in the tissues of plants of the genus *Maytenus molina in vitro* culture. I. Callas culture and biological studies of its extracts. *Acta Poloniae Pharmaceutica* 46, no. 1:81–89.

Kupchan, S.M., Y. Komoda, W.A. Court, et al. 1972. Maytansine, a novel antileukemic ansa macrolide from *Maytenus ovatus. Journal of the American Chemical Society* 94, no. 4:1354–1356.

Melo, A.M., M.L. Jardim, C.F. De Santana, et al. 1974. First observations on the topical use of Primin, Plumbagin and Maytenin in patients with skin cancer. *Revue Institucionais Antibiotic (Recife)* 14, no. 1–2:9–16.

Monks, N.R., S.A.L. Bordignon, A. Ferraz, et al. 2002. Anti-tumour screening of Brazilian plants. *Pharmaceutical Biology* 40, no. 8:603–616.

Qian, X., C. Gai, and S. Yao. 1979. Studies on the antileukemic principle of *Maytenus guangsiensis. Cheng et Sha. Yao Hsueh Hsueh Pao. Yao Xue Xue Bao = Acta Pharmaceutica Sinica* 14, no. 3:182.

Sneden, A.T., and G.L. Beemsterboer. 1980. Normaytansine, a new antileukemic ansa macrolide from *Maytenus buchananii. Journal of Natural Products* 43, no. 5:637–640.

Tin-Wa, M., N.R. Farnsworth, H.H. Fong, et al. 1971. Biological and phytochemical evaluation of plants. IX. Antitumor activity of *Maytenus senegalensis* (Celastraceae) and a preliminary phytochemical investigation. *Lloydia* 34, no. 1:79–87.

Melaleuca armillaris (Myrtaceae) (Myrtales)   **Cytotoxic**

Location: native to south-eastern Australia (eastern New South Wales, eastern Victoria, and Tasmania). Beyond its native range: southern Victoria and the coastal districts of south-western Australia

Appearance: shrub/small tree, perennial

Bark: hard or corky
Leaves: leaves alternate, linear, 12–25 mm long, c. 1 mm wide, apex recurved, acute, glabrous; petiole 1–2 mm long
Flowers: inflorescences many-flowered dense spikes 3–7 cm long, borne low on the branchlets, rachis ± glabrous to woolly. Flowers solitary within each bract, white or rarely pink. Petals ovate, 2–3 mm long. Stamens 16–18 per bundle, claw 5–6 mm long

Tradition:

- Used as ornamental plant, as fuel, shelter, and for its essential oils
- Widely applied in low concentrations in cosmetics and skin washes

Parts used: whole plant for various uses
Active ingredients: essential oil: 1,8-cineole (85.8%), camphene (5.05%), and α-pinene (1.95%). Methyl eugenol
Documented target cancers:
- MCF7 human breast cancer cells (IC(50), 12 ± 1 mg/L)

References

Ali, B., N.A. Al-Wabel, S. Shams, A. Ahamad, S.A. Khan, and F. Anwar. 2015. Essential oils used in aromatherapy: A systemic review. *Asian Pacific Journal of Tropical Biomedicine* 5, no. 8:601–611.

Amri, I., E. Mancini, L. De Martino, et al. 2012. Chemical composition and biological activities of the essential oils from three Melaleuca species grown in Tunisia. *International Journal of Molecular Sciences* 13, no. 12:16580–16591.

Barbosa, L.C.A., C.J. Silva, R.R. Teixeira, R.M.S.A. Meira, and A.L. Pinheiro. 2013. Chemistry and biological activities of essential oils from Melaleuca L. species. *Agriculturae Conspectus Scientificus* 78, no. 1:11–23.

Chabir, N., M. Romdhane, A. Valentin, et al. 2011. Chemical study and antimalarial, antioxidant, and anticancer activities of *Melaleuca armillaris* (Sol Ex Gateau) Sm essential oil. *Journal of Medicinal Food* 14, no. 11:1383–1388.

Melia azedarach (Melia) (Meliaceae)   **Cytotoxic**

Location: Northern India, China, Himalayas
   Appearance:

Stem: 10–17 m high, reddish brown bark
Leaves: bipinnate, 1–2 in long. The individual leaflets, each about 2 cm long, are pointed at the tips and have toothed edges.
Flowers: large branches of lilac, fragrant, star shaped flowers, that arch or droop in 8 cm panicles.
In bloom: spring–early summer

Parts used: the bark of the root and trunk, seed
   Active ingredients:

* Limonoids: toosendanal, 28-deacetyl sendanin, 12-O-methylvolkensin, meliatoxin B1, trichillin H, and toosendanin, 12-deacetyltrichilin I 1-acetyltrichilin H, 3-deacetyltrichilin H, 1-acetyl-3-deacetyltrichilin H, 1-acetyl-2-deacetyltrichilin H, meliatoxin B1, trichilin H, trichilin D, and 1,12-diacetyltrichilin B
* Meliavolkinin, melianin C, 3-diacetylvilasinin, and melianin B

Documented target cancers:

* KB cells (meliatoxin B1 and toosendanin)
* P388 cells (limonoids of *Melia azedarach*)
* Human prostate (PC-3) and pancreatic (PACA-2) cell lines (3, 23,24-diketomelianin B)
* Leukemia (HL-60) (IC_{50}: 0.016 μM) and stomach (AZ521) (IC_{50}: 0.035 μM) cancer cell lines
* Cancer cell lines A549, H460, and U251

Further details:

* The root bark of *Melia azedarach* contains the trichilin-type limonoids 12-deacetyltrichilin I 1-acetyltrichilin H, 3-deacetyltrichilin H, 1-acetyl-3-deacetyltrichilin H, 1-acetyl-2-deacetyltrichilin H, meliatoxin B1, trichilin H, trichilin D, and 1,12-diacetyltrichilin B.
* The limonoid compound (28-deacetyl sendanin) isolated from the fruit of *Melia toosendan* Sieb. et Zucc. was evaluated on anticancer activity. It has been proved that 28-deacetyl sendanin has more sensitive and selective inhibitory effects on *in vitro* growth of human cancer cell lines in a comparison with adriamycin.
* The fruits of *Melia toosendan* Sieb. et Zucc. contain the limonoids toosendanal, 12-O-methylvolkensin, meliatoxin B1, trichillin H, toosendanin, and 28-deacetyl sendanin.
* The root bark of *Melia volkensii* contains meliavolkinin, melianin C, 1,3-diacetylvilasinin, and melianin B, which all showed marginal cytotoxicities against certain human tumor cell lines. Jones oxidation of melianin B4 gave 3, 23,24-diketomelianin B, which showed selective cytotoxicities for the human prostate (PC-3) and pancreatic (PACA-2) cell lines with potencies comparable to those of adriamycin.

References

Itokawa, H., Z.S. Qiao, C. Hirobe, and K. Takeya. 1995. Cytotoxic limonoids and tetranortriterpenoids from *Melia azedarach*. *Chemical and Pharmaceutical Bulletin* 43, no. 7:1171–1175.

Kikuchi, T., X. Pan, K. Ishii, et al. 2013. Cytotoxic and apoptosis-inducing activities of 12-O-acetylazedarachin B from the fruits of *Melia azedarach* in human cancer cell lines. *Biological and Pharmaceutical Bulletin* 36, no. 1:135–139.

Rogers, L.L., L. Zeng, J.F. Kozlowski, H. Shimada, F.Q. Alali, H.A. Johnson, and J.L. McLaughlin. 1998. New bioactive triterpenoids from *Melia volkensii*. *Journal of Natural Products* 61, no. 1:64–70.

Sumarawati, T., - Israhnanto, and D. Fatmawati. 2017. Anticancer mechanism of Melia azedarach, doxorubicin and Cyclosphamide combination against breast cancer in mice. *Bangladesh Journal of Medical Science* 16, no. 3:428–432.

Tada, K., M. Takido, and S. Kitanaka. 1999. Limonoids from fruit of *Melia toosendan* and their cytotoxic activity. *Phytochemistry* 51, no. 6:787–791.

Takeya, K., Z.S. Quio, C. Hirobe, and H. Itokawa. 1996. Cytotoxic trichilin-type limonoids from *Melia azedarach*. *Bioorganic and Medicinal Chemistry* 4, no. 8:1355–1359.

Wu, S.B., Y.P. Ji, J.J. Zhu, Y. Zhao, G. Xia, Y.H. Hu, and J.F. Hu. 2009. Steroids from the leaves of Chinese Melia azedarach and their cytotoxic effects on human cancer cell lines. *Steroids* 74, no. 9:761–765.

Melia dubia (Meliaceae) (Sapindales)   **Cytotoxic**

Location: from India to tropical Africa, southern China, and Taiwan and through Malaya to tropical Australia

Appearance: deciduous tree growing 6 to 15 meters high

Stem: panicles in the upper axils shorter than the leaves and many flowered

Leaves: crowded, long-stalked, 30 to 90 centimeters long, usually bipinnate. Leaflets are in two to five pairs, ovate to ovate-lanceolate, 4 to 8 centimeters long.

Flowers: flowers are numerous, violet and white, fragrant, about 8 millimeters long, borne on the upper axils of the leaves. Petals are hairy.

Tradition:

- In India, the fruit, with its bitter and nauseating taste, is used for colic. In Concan, the juice of green fruit mixed with sulfur and curds, heated in a copper pot, is used as application for scabies and sores infested by maggots.
- Ornamental: fruit stones used as beads in making necklaces and rosaries.

Documented target cancers:

- Breast cancer (KB cell line)
- Breast cancer MCF-7 cell line: plant extract synthesized nanoparticles (IC$_{50}$ value was 31.2 µg/ml)

Parts used: fruit, leaves, wood

Active ingredients: camphene is a major constituent of the leaf essential oils.

References

Chabir, N., M. Romdhane, A. Valentin, et al. 2011. Chemical study and antimalarial, antioxidant, and anticancer activities of *Melaleuca armillaris* (Sol Ex Gateau) Sm essential oil. *Journal of Medicinal Food* 14, no. 11:1383–1388.

Gopal, V., G.P. Yoganandam, and P. Manju. 2015. A concise review on Melia dubia Cav. (Meliaceae). *EURO J Environmentalica Ecologia* 2:57–60.

Kathiravan, V., S. Ravi, and S. Ashokkumar. 2014. Synthesis of silver nanoparticles from *Melia dubia* leaf extract and their *in vitro* anticancer activity. *Spectrochimica Acta: Part A, Molecular and Biomolecular Spectroscopy* 130:116–121.

Leela, G.D.J., S.I. Monisha, A.A. Immaculate, and J.R. Vimala. 2016. Studies on phytochemical, nutritional analysis and screening of *in vitro* biological activities of melia dubia leaf extract. *International Journal of Scientific and Engineering Research* 7, no. 8:56–68.

Murugesan, S., N. Senthilkumar, C. Rajeshkannan, and K.B. Vijayalakshmi. 2013. Phytochemical characterization of *Melia dubia* for their biological properties. *Chemica Sinica* 4, no. 1:36–40.

Melissa officinalis (Abiatae/Lamiaceae) (Lamiales)   **Anti-tumor**

Location: native to southern Europe and northern Africa; today it can be found growing wildly throughout North America, Europe, Asia, and in the Mediterranean.

Appearance: perennial bushy plant, upright, reaching a height of about 1 m

Leaves: the leaf surface is coarse and deeply veined, and the leaf edge is scalloped or toothed.

Flowers: flowers white or pale pink consisting of small clusters of 4 to 12 blossom in the summer. It is commonly referred to as 'lemon balm' because of its lemon-like flavor and fragrance.

Tradition: it has been traditionally used for different medical purposes as tonic, antispasmodic, carminative, diaphoretic, surgical dressing for wounds, sedative hypnotic strengthening the memory, and relief of stress-induced headache. It is currently used for the relief of stress-induced headache, as a mild sedative-hypnotic, and as an antiviral to improve healing of herpes simplex cold sores.

Parts used: oil, leaves, flowers

Active ingredients: the main constituents of the essential oil are: citral (geranial and neral), citronellal, geraniol, beta-pinene, alpha-pinene, beta-caryophyllene, linalool, borneol l. In addition, fresh herb of lemon balm contains total phenolic, L-ascorbic acid, and carotenoids.

Documented target cancers:

- Very effective against human colon cancer cell line (HCT-116)
- The plant extract inhibited the proliferation of HT-29 and T84 colon carcinoma cells with an inhibitory concentration (IC_{50}) of 346 and 120 µg/ml, respectively.
- The plant extract induced a significant reduction in the viability of MCF-7, MDA-MB-468, and MDA-MB-231 breast cancer cells in a dose-dependent manner.
- *In vivo* studies showed that in induced mammary carcinogenesis in rats the mean tumor volume inhibition ratio in treated group was 40% compared with the untreated rats.
- Indicative dosage and application: the lowest active supplemental dose appears to be 300 mg, and supplementation above this dose appears to confer dose-dependent effects although results are not very reliable (i.e. one study says that 1,200 mg gives thrice as much benefit as 300mg while another suggests 1.4× benefit).

References

Basar, S.N., and R. Zaman. 2013. An overview of badranjboya (Melissa officinalis). *International Research Journal of Biological Sciences* 2:107–109.

Encalada, M.A., K.M. Hoyos, S. Rehecho, et al. 2011. Anti-proliferative effect of Melissa officinalis on human colon cancer cell line. *Plant Foods for Human Nutrition* 66, no. 4:328–334.

Saraydin, S.U., E. Tuncer, B. Tepe, et al. 2012. Antitumoral effects of Melissa officinalis on breast cancer *in vitro* and *in vivo*. *Asian Pacific Journal of Cancer Prevention* 13, no. 6:2765–2770.

Melothria heterophylla (Cucurbitaceae) (Cucurbitales)   **Cytotoxic**

Location: India, Southeast Asia, South China, Taiwan
   Appearance: scandent dioecious perennial with several tuberous roots and with slender branched furrowed stems bearing simple tendrils

   Leaves: amplexicaul, 5–6 cm, oval but of variable shapes
   Flowers: tiny, may be less than 1 cm. Axillary often in pira or cluster of three to four flowers

Tradition:

   • The plant is used as 'Kapha' and 'Vata' in Ayurvedic medicine. The plant is used as 'jangli-kakri' in Himachal Pradesh as a folk medicine.
   • The juice of the roots mixed with cumin and sugar helps in cases of seminal weakness, inflammation, eruptions, and other complications caused with bhallatakaphala svarasa.
   • The seeds of the plant are used for their purgative action. The leaves are used in allergic inflammations. In Konkan, the juice of the roots mixed with cumin and sugar is given in cold milk as a remedy for spermatorrhoea. The plant is also used in the treatment of diabetes, gonorrhea, and dysuria.

Parts used: leaves, fruit, roots (the whole plant)
   Active ingredients: 1, 2, 4, 6-Tetra-O-galloyl--glucopyranose, and gallic acid
   Documented target cancers:
   • Ehrlich ascites carcinoma cells

References

Mondal, A., T.K. Maity, and D. Pal. 2012. Oxygen-dependent-regulation of Ehrlich's ascites carcinoma cell respiration by gallic acid and Rutin isolated from *Melothria heterophylla* (Lour.) Cogn. *Asian Journal of Chemistry* 24, no. 6:2648–2650.
Mondal, A., T. Singha, T.K. Maity, and D. Pal. 2013. Evaluation of antitumor and antioxidant activity of *Melothria heterophylla* (Lour.) Cogn. *Indian Journal of Pharmaceutical Sciences* 75, no. 5:515–522.

Menispermum dauricum (Menispermacea) (Ranunculales)   **Cytotoxic**

Location: North East, North, and East China, Japan, Korea, and South Siberia
   Appearance: dioecious climbing woody vine

   Flowers: small
   Fruit: in clusters, similar to grapes

Tradition: used in folk medicine to treat skin disorders, rheumatism, and cervical cancer. The dry root has detoxifying and dehydrating action and is mainly used for the treatment of sore throat, enteritis, diarrhea, and rheumatism.
   Parts used: leaves
   Active ingredients: daurioxoisoporphines A-D, dechloroacutumidine, 1-epidechloroacutumine, acutumidine, acutumine, dechloroacutumine, and (+)-strigol
   Documented target cancers:
   • Two oxoisoaporphine alkaloids, isolated from the rhizomes of *Menispermum dauricum* are reported as significantly active against the MCF-7 human breast cancer cell line.

References

Chen, J., Y. Xie, T. Zhou, and G. Qin. 2012. Chemical constituents of *Menispermum dauricum. Chinese Journal of Natural Medicines* 10, no. 4:292–294.

Yu, B.W., J.Y., Chen, Y.P. Wang, K.F. Cheng, X.Y. Li, and G.W. Qin. 2002. Alkaloids from *Menispermum dauricum. Phytochemistry* 61, no. 4:439–442.

Yu, B.W., L.H. Meng, J.Y. Chen, T.X. Zhou, K.F. Cheng, J. Ding, and G.W. Qin. 2001. Cytotoxic Oxoisoaporphine alkaloids from *Menispermum dauricum. Journal of Natural Products* 64, no. 7:968–970.

Zhang, X., W. Ye, S. Zhao, and C.T. Che. 2004. Isoquinoline and isoindole alkaloids from *Menispermum dauricum. Phytochemistry* 65, no. 7:929–932.

Millettia leucantha (Fabaceae) (Fabales)   **Cytotoxic**

Location: Thailand, Myanmar, Laos
   Appearance: deciduous tree

   Flowers: white

Parts used: leaves, stem bark
   Active ingredients: chalcones
   Tradition:

- Cooking: sathon sauce is a flavoring sauce used in Isaan cuisine. The leaves of two species of Millettia are used for making sathon sauce: *Millettia utilis* and *Millettia leucantha* var. buetoides. This sauce used for cooking is the only OTOP product made from the sathon tree.

Documented target cancers:

- BCA-1 breast cancer cell line
- Human lung cancer cell line NCI-H460

References

Phrutivorapongkul, A., V. Lipipun, N. Ruangrungsi, et al. 2003. Studies on the chemical constituents of stem bark of *Millettia leucantha*: Isolation of new chalcones with cytotoxic, anti-herpes simplex virus and anti-inflammatory activities. *Chemical and Pharmaceutical Bulletin* 51, no. 2:187–190.

Rayanil, K.O., P. Bunchornmaspan, and P. Tuntiwachwuttikul. 2011. A new phenolic compound with anticancer activity from the wood of *Millettia leucantha. Archives of Pharmacal Research* 34, no. 6:881–886.

Mimusops elengi (Sapotaceae) (Ericales)   **Cytotoxic**

Location: tropical forests in South Asia, Southeast Asia, and northern Australia
   Appearance: large glabrous evergreen trees 12–15 m high, with a compact leafy head and short erect trunk, bark smooth, scaly, and gray

   Leaves: leaves 6.3–10 by 3.2–5 cm, elliptic shortly acuminate, glabrous, base acute or rounded, petioles 1.3–2.5 cm long
   Flowers: flower white, fragrant, nearly 2.5 cm across solitary, buds ovoid, acute, pedicels 6.20 mm long. Calyx 1 cm long, stamens eight, opposite to the inner circle of lobes.

Tradition:

- Distilled extract from the flowers is used amongst the natives of Southern India as a perfume.
- The bark is acrid and sweet; cooling, cardiotonic, alexipharmic, stomachic, anthelmintic, astringent; cures biliousness and diseases of the gum and teeth. The leaves are well known as analgesic and antipyretic.

Parts used: bark, flowers, seeds, root, fruits, the pulp of the ripe fruits, leaves
   Active ingredients:

- Bark: taraxerone, taraxerol, betulinic acid, and spinasterol
- Fruit and seed: quercitol, ursolic acid, dihydro quercetin, quercetin, β - d glycosides of β sitosterol, alphaspinasterol after Saponification
- Leaves, heartwood, and roots: hentriacontane, carotene, and lupeol

Documented target cancers:

- Human laryngeal carcinoma (Hep-2)
- Human hepatcarcinoma (HepG2)
- Cholangiocarcinoma (CL-6)

References

Kadam, P.V., K.N. Yadav, R.S. Deoda, R.S. Shivatare, and M.J. Patil. 2012. *Mimusops elengi*: A review on ethnobotany, phytochemical and pharmacological profile. *Journal of Pharmacognosy and Phytochemistry* 1, no. 3:64–74.

Mahavorasirikul, W., V. Viyanant, W. Chaijaroenkul, A. Itharat, and K. Na-Na-Bangchang. 2010. Cytotoxic activity of Thai medicinal plants against human cholangiocarcinoma, laryngeal and hepatocarcinoma cells *in vitro*. *BMC Complementary and Alternative Medicine* 10, no. 1:55.

Mitracarpus frigidus (Rubiaceae) (Gentianales)   **Cytotoxic**

Location: Brazil
   Appearance: small, dense, mound-forming shrub growing up to 20 centimeters in height

Flowers: the inflorescence is a rounded head of tiny white flowers.

Tradition: is widely employed in traditional medicine in West Africa for headaches, toothache, amenorrhea, dyspepsia, hepatic diseases, and leprosy. Among those folkloric uses, the juice of the plant is applied topically for the treatment of skin diseases.
   Parts used: aerial parts
   Active ingredients:
   Linalool (29.29%) and eugenol acetate (15.85%) are the major constituents of the essential oil, followed by 5-hydroxy-isobornyl isobutyrate (8.41%), 5-methyl-1-undecene (7.69%), and methyl salicylate (6.55%).
   Documented target cancers:

- HL60—human leukemia cell line
- Jurkat—T-cell leukemia
- Breast cancer cell line MCF-7
- HCT—colon carcinoma cell line

References

822. Fabri, R.L., E.S. Coimbra, A.C. Almeida, E.P. Siqueira, T.M. Alves, C.L. Zani, and E. Scio. 2012. Essential oil of *Mitracarpus frigidus* as a potent source of bioactive compounds. *Anais da Academia Brasileira de Ciências* 84, no. 4:1073–1080.

Fabri, R.L., R.A. Garcia, J.R. Florêncio, et al. 2014. Pentacyclic triterpenoids from *Mitracarpus frigidus* (Willd. ex Roem. & Schult.) K. Shum: *In vitro* cytotoxic and leishmanicidal and *in vivo* anti-inflammatory and antioxidative activities. *Medicinal Chemistry Research* 23, no. 12:5294–5304.

Fabri, R.L., M.S. Nogueira, F.G. Braga, E.S. Coimbra, and E. Scio. 2009. *Mitracarpus frigidus* aerial parts exhibited potent antimicrobial, antileishmanial, and antioxidant effects. *Bioresource Technology* 100, no. 1:428–433.

Morinda citrifolia (Rubiaceae) (Gentianales)	Anti-tumor

Location: native from Southeastern Asia to Australia
  Appearance: small evergreen trees or shrubs 3–6 m tall, stems

Stem: four-angled, glabrous
Leaves: glossy, membranous, elliptic to elliptic-ovate, 20–45 cm long, 7–25 cm wide, glabrous, petioles stout, 1.5–2 cm long, stipules connate or distinct, 10–12 mm long, apex entire or two- to three-lobed.
Flowers: ca. 75–90 in ovoid to globose heads, peduncles 10–30 mm long; calyx a truncate rim; corolla white, five-lobed, the tube greenish white, 7–9 mm long, the lobes oblong-deltate, ca. 7 mm long; stamens five, scarcely exserted; style ca. 15 mm long

Tradition: in Vietnam roots serve to treat stiffness and tetanus and have been proven to combat arterial tension. Elsewhere they are used as febrifuge, tonic, and antiseptic. The fruits are used as a diuretic, a laxative, an emollient, and as an emmenagogue, for asthma and other respiratory problems, as a treatment for arthritic and comparable inflammations, in cases of leucorrhoea and sapraemia, and for maladies of inner organs. Roots, leaves, and fruits may have anthelmintic properties. In traditional medicine the parts used are administered raw or as juices and infusions or in ointments and poultices.
  Parts used: roots, stems, bark, leaves, flowers, and fruits
  Active ingredients:

- Major components: scopoletin, octoanoic acid, potassium, vitamin C, terpenoids, alkaloids, anthraquinones (such as nordamnacanthal, morindone, rubiadin, and rubiadin-1-methyl ether, anthraquinone glycoside), -sitosterol, carotene, vitamin A, flavone glycosides, linoleic acid, Alizarin, amino acids, acubin, L-asperuloside, caproic acid, caprylic acid, ursolic acid, rutin, and a putative proxeronine
- Several new flavonol glycosides: an iridoid glycoside, a trisacharide fatty acid ester, and an asperulosidic acid from the fruit. Two novel glycosides and a new unusual iridoid named citrifolinoside
- It contains a natural precursor for Xeronine, named Proxeronine. Proxeronine is converted, inside the body, to the alkaloid Xeronine by the enzyme proxeroninase.

Precautions: is reported to cause hepatotoxicity
  Documented target cancers:
- Anthracene (DMBA) induced mammary breast cancer

References

Wang, M.Y., L. Peng, G. Anderson, and D. Nowicki. 2013. Breast cancer prevention with *Morinda citrifolia* (noni) at the initiation stage. *Functional Foods in Health and Disease* 3, no. 6:203–222.

Yu, E.L., M. Sivagnanam, L. Ellis, and J.S. Huang. 2011. Acute hepatotoxicity after ingestion of *Morinda citrifolia* (N. Berry) juice in a 14-year-old boy. *Journal of Pediatric Gastroenterology and Nutrition* 52, no. 2:222–224.

***Moringa oleifera* (Moringaceae) (Brassicales)**   **Cytotoxic**

Location: native to the southern foothills of the Himalayas in northwestern India and widely cultivated in tropical and subtropical areas

Appearance: small to medium-sized deciduous tree, can reach a height of 10–12 m (32–40 ft), and the trunk can reach a diameter of 45 cm.

Stem: young shoots have purplish or greenish-white, hairy bark

Leaves: build up a feathery foliage of tripinnate leaves

Flowers: fragrant and bisexual, surrounded by five unequal, thinly veined, yellowish-white petals. The flowers are about 1.0–1.5 cm (1/2 in) long and 2.0 cm (3/4 in) broad. They grow on slender, hairy stalks in spreading or drooping later flower clusters which have a length of 10–25 cm.

Tradition: widely used in indigenous systems of medicine, particularly in South Asia

Moringa trees have been used to combat malnutrition, especially among infants and nursing mothers. Since moringa thrives in arid and semiarid environments, it may provide a versatile, nutritious food source throughout the year.

The leaves, roots, seed, bark, fruit, flowers, and immature pods can act as cardiac and circulatory stimulants and possess antipyretic, antiepileptic, anti-inflammatory, antiulcer, antispasmodic, diuretic, antihypertensive, cholesterol-lowering, antioxidant, antidiabetic, hepatoprotective, antibacterial, and antifungal activities.

Parts used: leaves, pods, and/or its kernels for oil extraction and water purification, bark, sap, roots, seeds, and flowers

Active ingredients: taxanes (docetaxel, paclitaxel), vinca alkaloids (vindesine, vinblastine, vincristine), anthracyclines (idarubicin, daunorubicin, epirubicin), and others

Indicative dosage and application:

- Most current research has been conducted on rat models, employing water extracts of the leaves. It appears that 150–200 mg/kg oral intake is deemed as optimal (greater potency than higher and lower doses), and in this case a preliminary human dose can be estimated at 1,600–2,200 mg for a 150 lb person, 2,100–2,900 mg for a 200 lb person, and 2,700–3,600 mg for a 250 lb person, respectively.

Documented target cancers:

- Aqueous leaf extract increased the cytotoxic effect of chemotherapy in pancreatic cancer cells.
- Breast and colorectal cancer cell lines: MDA-MB-231 and HCT-8

References

Al-Asmari, A.K., S.M. Albalawi, M.T. Athar, A.Q. Khan, H. Al-Shahrani, and M. Islam. 2015. *Moringa oleifera* as an anti-cancer agent against breast and colorectal cancer cell lines. *PLOS ONE* 10, no. 8:e0135814.

Berkovich, L., G. Earon, I. Ron, A. Rimmon, A. Vexler, and S. Lev-Ari. 2013. *Moringa oleifera* aqueous leaf extract down-regulates nuclear factor-kappaB and increases cytotoxic effect of chemotherapy in pancreatic cancer cells. *BMC Complementary and Alternative Medicine* 13, no. 1:212.

| *Mormodica charantia* (Mormodica (Bitter melon) (Cucurbitaceae) | **Anti-leukemic** |

Location: East India
  Appearance:

Stem: thin, crawly
Leaves: dark, green, and deeply lobed
Flowers: dioecous, yellow

Parts used: the fruit deprived of the seeds
  Active ingredients: protein (molecular weight of 11,000 daltons)
  Documented target cancers: the fruit and seeds of the bitter melon (*Momordica charantia*) have been reported to have anti-leukemic and antiviral activities:

- Antitumor (mice)
- Antiviralantileukemic (human, selective)
- Immunostimulating (mice)
- Breast cancer, skin tumor, prostatic cancer, squamous carcinoma of tongue and larynx
- Human bladder carcinomas and Hodgkin's disease
- Human T-cell leukemia Jurkat, human HL-60 cell leukemia cell line
- Skin papillomas

Further details:

- This anti-leukemic and antiviral action was associated with an activation of murine lymphocytes. This activity is associated with a single protein component with an apparent molecular weight of 11,000 daltons. The factor is not sensitive to boiling or to pretreatments with trypsin, ribonuclease (RNAse), or deoxyribonuclease (DNAse). As determined by radioactive precursor uptake studies, the purified factor preferentially inhibits RNA synthesis in intact tissue culture cells. Some inhibition of protein synthesis and DNA synthesis also occurs. The factor is preferentially cytostatic for IM9 human leukemic lymphocytes when compared to normal human peripheral blood lymphocytes. In addition, it preferentially inhibits the soluble guanylate cyclase from leukemic lymphocytes. This inhibition correlates with its preferential cytotoxic effects for these same cells, since cyclic GMP is thought to be involved in lymphocytic cell proliferation and leukemogenesis and, in general, the nucleotide is elevated in leukemic vs normal lymphocytes, and changes have been reported to occur during remission and relapse of this disease.
- At least part of the anti-leukemic activity of the bitter melon extract is due to the activation of NK cells in the host organism (mouse), i.e. *in vivo* enhancement of immune functions may contribute to the antitumor effects of the bitter melon extract. In humans, the extract has both cytostatic and cytotoxic activities and can kill leukemic lymphocytes in a dose-dependent manner while not affecting the viability of normal human lymphocyte cells at these same doses. These activities are not due to the presence of the lectins from bitter melon seeds, as these purified proteins had no activity against human lymphocytic cells.

References

Cunnick, J.E., K. Sakamoto, S.K. Chapes, G.W. Fortner, and D.J. Takemoto. 1990. Induction of tumor cytotoxic immune cells using a protein from the bitter melon (*Momordica charantia*). *Cellular Immunology* 126, no. 2:278–289.

Grossmann, M.E., N.K. Mizuno, M.L. Dammen, T. Schuster, A. Ray, and M.P. Cleary. 1909. Eleostearic acid inhibits breast cancer proliferation by means of an oxidation-dependent mechanism. *Cancer Prevention Research* 2(10): 879–886

Grover, J.K., and S.P. Yadav. 2004. Pharmacological actions and potential uses of Momordica charantia: A review. *Journal of Ethnopharmacology* 93, no. 1:123–132.

Jia, S., M. Shen, F. Zhang, and J. Xie. 2017. Recent advances in Momordica charantia: Functional components and biological activities. *International Journal of Molecular Sciences* 18, no. 12:2555.

Jilka, C., B. Strifler, G.W. Fortner, E.F. Hays, and D.J. Takemoto. 1983. *In vivo* antitumor activity of the bitter melon (*Momordica charantia*). *Cancer Research* 43, no. 11:5151–5155.

Kobori, M., M. Ohnishi-Kameyama, Y. Akimoto, C. Yukizaki, and M. Yoshida. 2008. α-Eleostearic acid and its dihydroxy derivative are major apoptosis-inducing components of bitter gourd. *Journal of Agricultural and Food Chemistry* 56, no. 22:10515–10520.

Lin, J.Y., M.J. Hou, and Y.C. Chen. 1978. Isolation of toxic and non-toxic lectins from the bitter pear melon *Momordica charantia* Linn. *Toxicon* 16, no. 6:653–660.

Porro, G., A. Bolognesi, P. Caretto, et al. 1993. *In vitro* and *in vivo* properties of an anti-CD5—Momordin immunotoxin on normal and neoplastic T lymphocytes. *Cancer Immunology, Immunotherapy* 36, no. 5:346–350.

Singh, A., S.P. Singh, and R. Bamezai. 1998. Momordica charantia (Bitter gourd) peel, pulp, seed and whole fruit extract inhibits mouse skin papillomagenesis. *Toxicology Letters* 94, no. 1:37–46.

Takemoto, D.J., R. Kresie, and D. Vaughn. 1980. Partial purification and characterization of a guanylate cyclase inhibitor with cytotoxic properties from the bitter melon (*Momordica charantia*). *Biochemical and Biophysical Research Communications* 94, no. 1:332–339.

Takemoto, D.J., C. Dunford, and M.M. McMurray. 1982. The cytotoxic and cytostatic effects of the bitter melon (*Momordica charantia*) on human lymphocytes. *Toxicon* 20, no. 3:593–599.

Takemoto, D.J., C. Dunford, D. Vaughn, K.J. Kramer, A. Smith, and R.G. Powell. 1982. Guanylate cyclase activity in human leukemic and normal lymphocytes. Enzyme inhibition and cytotoxicity of plant extracts. *Enzyme* 27, no. 3:179–188.

Takemoto, D.J., C. Jilka, and R. Kresie. 1982. Purification and characterization of a cytostatic factor from the bitter melon *Momordica charantia*. *Preparative Biochemistry* 12, no. 4:355–375.

Myrica cerifera (Myricaceae) (Fagales)   **Pro-apoptotic Anti-proliferative**

Location: native to North and Central America and the Caribbean
   Appearance: small tree or large shrub

Leaves: glandular, long, with leathery texture and serrated edges
Flowers: Male flowers have three or four stamens and are surrounded by short bracts. Female flowers develop into fruit, which are globular and surrounded by a natural wax-like layer.

Tradition:

   It is a popular herbal remedy in North America where it is employed to increase the circulation, stimulate perspiration, and keep bacterial infections in check. The plant should not be used during pregnancy. It is harvested in the autumn, thoroughly dried then powdered and kept in a dark in an airtight container. It is used internally in the treatment of diarrhea, irritable bowel syndrome, jaundice, fevers, colds, influenza, catarrh, excessive menstruation, vaginal discharge, etc. Externally, it is applied to indolent ulcers, sore throats, spongy gums, sores, itching skin conditions, dandruff, etc. The wax is astringent and slightly narcotic. It is regarded as a sure cure for dysentery and is also used to treat internal ulcers.

   The tea made from the leaves is used in the treatment of fevers and externally as a wash for itchy skin.

   Parts used: root bark, leaves, fruits, fruits' wax extract
   Active ingredients: diarylheptanoids (myricanone), terpenoids, flavonoids, tannins, and phenols
   Documented target cancers:

- Myricanone, the principal bioactive compound of the plant, has been reported to have apoptosis-promoting ability. Studies have shown that myricanone inhibited the growth of HepG2 cells (human liver cancer cell line) in a dose-dependent manner, whereas it had little or no cytotoxicity on normal liver cells (WRL-68).

Reference

Paul, A., J. Das, S. Das, A. Samadder, and A.R. Khuda-Bukhsh. 2013. Anticancer potential of myricanone, a major bioactive component of Myrica cerifera: Novel signaling cascade for accomplishing apoptosis. *Journal of Acupuncture and Meridian Studies* 6, no. 4:188–198.

***Nauclea orientalis* (Rubiaceae)**   **Antiproliferative**

Parts used: leaves

Active ingredients: nine angustine-type alkaloids were isolated from ammoniacal extracts of *Nauclea orientalis* (10-hydroxyangustine, two diastereoisomeric 3,14-dihydroangustolines).

Documented target cancers:

- The compounds have been found to exhibit *in vitro* anti-proliferative activity against the human bladder carcinoma T-24 cell line and against EGF (epidermal growth factor)-dependent mouse epidermal keratinocytes.
- An indole alkaloid, naucleaoral A, is reported to exhibit significant cytotoxicity against cervical cancer HeLa cells.

Further details:
- The structures of the isolates were determined with spectroscopic methods, mainly 1D- and 2D-NMR spectroscopy. By using overpressure layer chromatography, it was shown that minor quantities of these alkaloids occur in dried *Nauclea orientalis* leaves. The use of ammonia in the extraction process results in a significant increase in the formation of angustine-type alkaloids from strictosamide-type precursors.

References

Erdelmeier, C.A., U. Regenass, T. Rali, and O. Sticher. 1992. Indole alkaloids with *in vitro* antiproliferative activity from the ammoniacal extract of *Nauclea orientalis*. *Planta Medica* 58, no. 1:43–48.
Sichaem, J., S. Surapinit, P. Siripong, S. Khumkratok, J. Jong-aramruang, and S. Tip-pyang. 2010. Two new cytotoxic isomeric indole alkaloids from the roots of Nauclea orientalis. *Fitoterapia* 81, no. 7:830–833.

***Neolitsea variabillima* (Lauraceae) (Laurales)**   **Cytotoxic**

Location: sub-tropical, tropical climate zone

Appearance: small tree

Leaves: leaves alternate or clustered toward apex of branchlet, subverticillate; petiole ca. 1 cm, pubescent; leaf blade ovate-lanceolate, obovate-lanceolate, or long obovate, 8–15 × 3–5 cm

Flowers: umbels two- or three-fascicled, axillary or lateral, four- or five-flowered. Pedicel ca. 2 mm, densely pubescent. Perianth segments four, lanceolate. Female flowers: staminodes six, glabrous, of third whorls each with two stipitate glands at base; ovary ovoid; style pubescent

Parts used: bark, leaf oil

Active ingredients: L-hernovine, L-nandigerine, and L-N-methylhernovine

Documented target cancers:

- The leaf essential oil has been evaluated against: human oral squamous cancer OEC-M1 cells, human hepatocellular carcinoma J5 cells, human lung adenocarcinoma A549 cells, human colon cancer HT-29 cells, human melanoma UACC-62 cells, and human leukemic cell K562 cells melanoma. The IC_{50} values were 38.9, 42.6, 36.9, 16.8, 8.8, and 8.6 µg/mL, respectively.

References

Bayala, B., I.H. Bassole, R. Scifo, C. Gnoula, L. Morel, J.M. Lobaccaro, and J. Simpore. 2014. Anticancer activity of essential oils and their chemical components-a review. *American Journal of Cancer Research* 4, no. 6:591–607.

Cao, Y., X.L. Gao, G.Z. Su, X.L. Yu, P.F. Tu, and X.Y. Chai. 2015. The genus Neolitsea of Lauraceae: A phytochemical and biological progress. *Chemistry and Biodiversity* 12, no. 10:1443–1465.

Lu, S.T., and T.L. Su. 1973. Studies on the alkaloids of Formosan Lauraceous Plants. xVI. Alkaloids of *Neolitsea variabillima* (Hay.) Kanehira et Sasaki. *Journal of the Chinese Chemical Society* 20, no. 2:75–81.

Su, Y.C., K.P. Hsu, E.I., Wang, and C.L. Ho. 2013. Composition and *in vitro* anticancer activities of the leaf essential oil of *Neolitsea variabillima* from Taiwan. *Natural Product Communications* 8, no. 4:531–532.

Nerium oleander (Apocynaceae) (Gentianales)   **Anti-cancer**

Location: Mediterranean region, southern China, Indian subcontinent

Appearance: evergreen shrub

Leaves: in groups of three with narrow lobes
Flowers: tubular with five lobes, usually red or pink

Tradition: in traditional medicine, its leaves have been used to treat heart disease, as a diuretic and antibacterial, but also as an antidote to snakebite.

Parts used: leaves, roots

Active ingredients: oleandrin, oleandrigenin, gentiobioside, digitoxigenin, adyregenin, oleanderocioicacid

Figure 9.16   (See color insert.) *Nerium oleander.*

Documented target cancers:

- A hydroalcoholic extract from the leaves of *Nerium oleander* is reported to induce selective killing of lung cancer cells.
- Anvirzel, an extract of *Nerium oleander*, is reported to induce cell killing in human cancer cells, but not in murine cancer cells.
- Two pentacyclic triterpenoids from the leaves of *N. oleander* are reported to exhibit moderate cytotoxicity against the KB (epidermal carcinoma of the mouth) cell line (Figure 9.16).

References

Calderón-Montaño, J.M., E. Burgos-Morón, M.L. Orta, S. Mateos, and M. López-Lázaro. 2013. A hydroalcoholic extract from the leaves of *Nerium oleander* inhibits glycolysis and induces selective killing of lung cancer cells. *Planta Medica* 79, no. 12:1017–1023.

Pathak, S., A.S. Multani, S. Narayan, V. Kumar, and R.A. Newman. 2000. AnvirzelTM, an extract of *Nerium oleander*, induces cell death in human but not murine cancer cells. *Anti-Cancer Drugs* 11, no. 6:455–463.

Siddiqui, B.S., S. Begum, S. Siddiqui, and W. Lichter. 1995. Two cytotoxic pentacyclic triterpenoids from *Nerium oleander*. *Phytochemistry* 39, no. 1:171–174.

Siddiqui, B.S., N. Khatoon, S. Begum, A.D. Farooq, K. Qamar, H.A. Bhatti, and S.K. Ali. 2012. Flavonoid and cardenolide glycosides and a pentacyclic triterpene from the leaves of *Nerium oleander* and evaluation of cytotoxicity. *Phytochemistry* 77:238–244.

Neurolaena lobata (Neurolaena) (Asteraceae)   **Cytotoxic**

Location: Guatemala

Active ingredients: sesquiterpene lactones: of the germacranolide and furanoheliangolide type

Documented target cancers:

- The dichloromethane extract is reported to inhibit proliferation of anaplastic large cell lymphoma lines of human and mouse origin.
- Sesquiterpenes isolated from the plant are reported to have noteworthy antiproliferative activities against human tumor cell lines A2780, A431, HeLa, and MCF7.

Further details:

- Aqueous and lipophilic extracts of *Neurolaena lobata* were tested against human carcinoma cell lines with cytotoxic effects. In addition to that, they were tested, also, against *Plasmodium falciparum in vitro*. Sesquiterpene lactones, isolated from *N. lobata*, were shown to be active against *P. falciparum in vitro* (antiplasmodial activity).

References

François, G., C.M. Passreiter, H.J. Woerdenbag, and M. Van Looveren. 1996. Antiplasmodial activities and cytotoxic effects of aqueous extracts and sesquiterpene lactones from *Neurolaena lobata*. *Planta Medica* 62, no. 2:126–129.

Passreiter, C.M., S.B. Stoeber, A. Ortega, E. Maldonado, and R.A. Toscano. 1999. Gemacranolide type sesquiterpene lactones from *Neurolaena macrocephala*. *Phytochemistry* 50, no. 7:1153–1157.

Unger, C., R. Popescu, B. Giessrigl, et al. 2013. The dichloromethane extract of the ethnomedicinal plant *Neurolaena lobata* inhibits NPM/ALK expression which is causal for anaplastic large cell lymphomagenesis. *International Journal of Oncology* 42, no. 1:338–348.

Nigella sativa L. (Nigella (Fennel flower)) (Ranunculaceae)   **Cytotoxic**

Location: Asia
  Appearance:

Stem: stiff, erect, branching
Leaves: bears deeply cut grayish-green
Flowers: grayish blue
In bloom: early summer

Tradition: in India, the seeds are believed to increase the secretion of milk and are considered as stimulant, diaphoretic. They also used it in tonic medicines. Romans used it in cooking (Roman Coriander). French used it as a substitute for pepper.

  Parts used: seed, herb
  Active ingredients: thymoquinone and dithymoquinone, alpha-hederin, fatty acids
  Documented target cancers:

- Multi-drug resistant (MDR) human tumor cell lines, Ehrlich ascites carcinoma (EAC), Dalton's lymphonia ascites (DLA), and Sarcoma-180 (S-180) cells. Skin cancer (mice)
- Alpha-hederin and thymoquinone exhibit dose- and time-dependent effects on A549 (lung carcinoma), HEp-2 (larynx epidermoid carcinoma), HT-29 (colon adenocarcinoma), and MIA PaCa-2 (pancreas carcinoma) cell lines.

Further details:

- *Nigella sativa*: Seeds contain thymoquinone (TQ) and dithymoquinone (DIM), which were cytotoxic *in vitro* against multi-drug resistant (MDR) human tumor cell lines (IC_{50}s 78 to 393 μM). Both the parental cell lines and their corresponding MDR variants, over ten-fold more resistant to the standard antineoplastic agents doxorubicin (DOX) and etoposide (ETP), as compared to their respective parental controls, were equally sensitive to TQ and DIM. The inclusion of the competitive MDR modulator quinine in the assay reversed MDR Dx-5 cell resistance to DOX and ETP by 6- to 16-fold, but had no effect on the cytotoxicity of TQ or DIM. Quinine also increased MDR Dx-5 cell accumulation of the P-glycoprotein substrate 3H- taxol in a dose-dependent manner. However, neither TQ nor DIM significantly altered cellular accumulation of 3H- taxol. The inclusion of 0.5% v/v of the radical scavenger DMSO in the assay reduced the cytotoxicity of DOX by as much as 39%, but did not affect that of TQ or DIM. These studies suggest that TQ and DIM, which are cytotoxic for several types of human tumor cells, may not be MDR substrates and that radical generation may not be critical to their cytotoxic activity.
- *Nigella sativa* seeds also contain certain fatty acids which are cytotoxic *in vitro* against Ehrlich ascites carcinoma (EAC), Dalton's lymphonia ascites (DLA), and Sarcoma-180 (S-180) cells. *In vitro* cytotoxic studies showed 50% cytotoxicity to Ehrlich ascites carcinoma, Dalton's lymphoma ascites, and Sarcoma-180 cells at a concentration of 1.5 micrograms, 3 micrograms, and 1.5 micrograms, respectively, with little activity against lymphocytes. The cell growth of KB cells in culture was inhibited by the active principle while K-562 cells resumed near control values on day two and day three. Tritiated thymidine incorporation studies indicated the possible action of an active principle at DNA level. *In vivo* EAC tumor development was completely inhibited by the active principle at the dose of 2 mg/mouse per day × 10.
- Topical application of *Nigella sativa* inhibited two-stage initiation/promotion [dimethylbenz[a] anthracene (DMBA)/croton oil] skin carcinogenesis in mice. A dose of 100 mg/kg body wt of these extracts delayed the onset of papilloma formation and reduced the mean number of papillomas per mouse, respectively. The possibility that these extracts could inhibit the action of 20-methylcholanthrene (MCA)-induced soft tissue sarcomas was evaluated by studying the effect of these extracts on MCA-induced soft tissue sarcomas in albino mice. Intraperitoneal administration of *Nigella sativa*

Figure 9.17   (See color insert.) *Nigella.*

(100 mg/kg body wt) and oral administration of *Crocus sativus* (100 mg/kg body wt) 30 days after subcutaneous administration of MCA (745 nmol × 2 days) restricted tumor incidence to 33.3 and 10%, respectively, compared with 100% in MCA-treated controls (Figure 9.17).

References

Randhawa, M.A., and M.S. Alghamdi. 2011. Anticancer activity of Nigella sativa (black seed)—A review. *American Journal of Chinese Medicine* 39, no. 6:1075–1091.

Rooney, S., and M.F. Ryan. 2005. Effects of alpha-hederin and thymoquinone, constituents of Nigella sativa, on human cancer cell lines. *Anticancer Research* 25, no. 3B:2199–2204.

Salomi, M.J., S.C. Nair, and K.R. Panikkar. 1991. Inhibitory effects of *Nigella sativa* and saffron (*Crocus sativus*) on chemical carcinogenesis in mice. *Nutrition and Cancer* 16, no. 1:67–72.

Salomi, N.J., S.C. Nair, K.K. Jayawardhanan, C.D. Varghese, and K.R. Panikkar. 1992. Antitumour principles from Nigella sativa seeds. *Cancer Letters* 63, no. 1:41–46.

Worthen, D.R., O.A. Ghosheh, and P.A. Crooks. 1998. The *in vitro* anti-tumor activity of some crude and purified components of blackseed, *Nigella sativa* L. *Anticancer Research* 18, no. 3(A):1527–1532.

***Ophiopogon japonicus* (Asparagaceae) (Asparagales)**   **Cytotoxic**

Location: East Asia—China, Korea, Japan, Vietnam
   Appearance: evergreen, perennial plant

Leaves: elongated, 20–40 cm long
Flowers: bell-shaped, white or soft lilac, forming a stump
Fruit: blue berry

Tradition: according to Chinese herbal medicine, the herb enters the heart, the lungs, and the stomach and nourishes the spleen, stomach, heart, lungs and cleanses the 'heart of heat' and suppresses irritability. It is a particularly useful antiseptic for the healing of the wounds of the mouth. Its soothing properties provide relief for insomnia, agony, tachycardia, and anxiety.

   Parts used: root, rhizomes, fibrous root nodules

   Active ingredients: ruscogenin, ophiopogonin D, spicatoside A, ophiopogonol, nolinospiroside F, ruscogenin or glycosides based off of ruscogenin, polysaccharides

   Documented target cancers:

- Monocot mannose-binding lectins from *Phiopogon japonicus* were found to bear remarkable inhibitory effects on the growth of breast cancer MCF-7 cells.

- Several steroidal saponins isolated from the tuberous roots of *Ophiopogon japonicus* have been reported cytotoxic against five human tumor cell lines (HepG2, HLE, BEL7402, BEL7403, and HeLa).

References

Li, N., L. Zhang, K.W. Zeng, Y. Zhou, J.Y. Zhang, Y.Y. Che, and P.F. Tu. 2013. Cytotoxic steroidal saponins from *Ophiopogon japonicus*. *Steroids* 78, no. 1:1–7.

Liu, B., H. Peng, Q. Yao, J. Li, E. Van Damme, J. Balzarini, and J.K. Bao. 2009. Bioinformatics analyses of the mannose-binding lectins from *Polygonatum cyrtonema*, *Ophiopogon japonicus* and *Liparis noversa* with antiproliferative and apoptosis-inducing activities. *Phytomedicine* 16, no. 6–7:601–608.

Origanum vulgare, O. majorana (Oregano (marjoram)) (Lamiaceae)   **Anticancer**

Location: Mediterranean region of Europe and Asia

  Appearance:

Stem: bushy, semi-woody sub-shrub with upright or spreading stems and branches

Leaves: aromatic, oval-shaped, about 4 cm long, and usually pubescent

Flowers: throughout the summer oregano bears tiny (0.3 cm long) purple tube-shaped flowers that peek out of whorls of purplish-green leafy bracts about an inch long.

In bloom: summer

Tradition: it was used from the very ancient years for its medicinal properties, as a remedy for narcotic poisons, convulsions, and dropsy. The whole plant has a strong fragrant, balsamic odor and an aromatic taste.

  Parts used: herb, oil, leaves

  Active ingredients: flavonoids, galangin and quercetin, water-alcoholic extracts, and of isolated compounds (arbutin, methylarbutin and their aglycons—hydroquinone and hydroquinone monomethyl ether)

  Antitumor-promoting activity or *in vitro* cytotoxic effects towards different tumor cell lines were attributed also to *Origanum majorana* extracts or their constituents. When studying cytotoxic activity of *O. majorana* water-alcoholic extracts and of isolated compounds (arbutin, methylarbutin and their aglycons—hydroquinone and hydroquinone monomethyl ether) towards cultured rat hepatoma cells (HTC line), a high dose-dependent HTC cytotoxicity of hydroquinone

  Indicative dosage and application: at 300 μM hydroquinone caused 40% cellular mortality after 24 h of incubation.

  Documented target cancers:

- Antitumor-promoting activity or *in vitro* cytotoxic effects towards different tumor cell lines (rat hepatoma cells (HTC line))
- Immunostimulant
- Antimutagenic
- Oregano extract is reported to lead to growth arrest and cell death in colon adenocarcinoma Caco2 cells
- Human lymphoblastic leukemia cell line Jurkat and MDA-MB 231 breast cancer cells

Further details:

- Some studies have shown that oregano extracts or herbal mixtures with *Origanum* spp. possess *in vitro* antiviral activity or have immunostimulating effects both *in vitro* and *in vivo*. However, little knowledge has been attained so far on mechanisms of immunomodulating activity or underlying active compounds. It has been shown that ethanol extracts of *Origanum vulgare* inhibited intracellular propagation of $ECHO_9$ Hill virus and also showed interferon inducing activity *in vitro*. Flavonoid luteoline, a constituent of Origani herba, has been considered as responsible for the induction of an interferon-like substance. A mixture of herbal preparation containing rosemary, sage, thyme, and oregano (*Origanum vulgare*) showed radical scavenging activity and inhibition of the human immunodeficiency virus (HIV) infection at very low concentrations. It was suggested that the main active compounds of herbal preparations were carnosol, carnosic acid, carvacrol, and thymol. Significant inhibitory effects of *Origanum vulgare* extracts against HIV-1 induced cytopathogenicity in MT-4 cells were also observed. According to Krukowski and co-workers, an increase in immunoglobulin (IgG) levels was observed in reared calves, fed with a conventional concentrate supplemented by a mineral–herbal mixture containing *Origanum majorana*.

- A strong and dose-dependent capacity of inactivating dietary mutagen Trp-P-1 in the *Salmonella typhimurium* TA 98 assay was observed in *Origanum vulgare* water extracts, that exhibited significant antimutagenic effects *in vitro*. *Origanum majorana* aqueous extracts were also able to suppress the mutagenicity of liver-specific carcinogen Trp-P-2. When studying the mechanism of suppressing the mutagenicity of Trp-P-2 in *Origanum vulgare*, it was found that two flavonoids, galangin and *quercetin*, acted as Trp-P-2 specific desmutagens, which neutralized this mutagen during or before mutating the bacteria (*Salmonella typhimurium* TA 98). The amounts of galangin and quercetin required for 50% inhibition (IC_{50}) against 20 ng of Trp-P-2 were 0.12 μg and 0.81 μg, respectively. It was also found that quercetin acted as a mutagen at high concentrations (> 10 μg/plate) but was a desmutagen when applied at low (> 0.1 < 10 μg/plate) concentrations. It was also found that isolated phenolic compounds from different spice plants, including *Origanum vulgare*, strongly inhibited pyrazine cation free radical formation in the Maillard reaction and the formation of mutagenic and carcinogenic amino-imidazoazarene in creatinine containing model systems.

- In a literature survey, referring to the anticancer activity of *Origanum* genus, different approaches, testing systems, and cell lines have been used by different authors when assessing the carcinogenic potential of plants or their isolated compounds. However, there are no available data on practical/ clinical use of oregano in cancer prevention. In 1966 an international project was performed with the aim of screening the native plants of former Yugoslavia for their potential agricultural use in the US and Yugoslavia. In the frame of this project 1,466 samples of 754 plant species were analysed for chemical and antitumor activity. According to the results of the Cancer Chemotherapy National Service Center Screening Laboratories (Washington, DC) a high carvacrol (60–85%) containing *Origanum heracleoticum* (= *O. vulgare* spp. *hirtum* (Link)Ietswaart) was reported to show high antitumor activity. The essential oil of *Origanum vulgare* fed to mice induced the activity of glutathione S-transferase (GST) in various tissues. The GST enzyme system is involved in detoxification of chemical carcinogens and plays an important role in prevention of carcinogenesis, which would explain the anticancer potential of *O. vulgare* essential oil. This oil exhibited high levels of cytotoxicity (at dilutions of up to 1:10,000) against four permanent eukaryotic cell lines including two derived from human cancers (epidermoid larynx carcinoma: Hep-2 and epitheloid cervix carcinoma: HeLa). Other studies that refer to *in vitro* cytotoxic and/or anti-proliferative effects of *Origanum vulgare* extracts or isolated compounds (carvacrol, thymol) include those of Bocharova and He, who observed moderate suppressing activities of *O. vulgare* extracts (CE_{50} = 220 mg/ml) on human ovarian carcinoma cells (CaOv) or of isolated carvacrol and thymol (IC_{50} = 120 μmol/l) on Murine B 16(F10) melanoma cells—a tumor cell line with high metastatic potential.

- Antitumor-promoting activity or *in vitro* cytotoxic effects towards different tumor cell lines were attributed also to *Origanum majorana* extracts or their constituents. When studying cytotoxic activity of *O. majorana* water-alcoholic extracts and of isolated compounds (arbutin, methylarbutin, and their aglycons—hydroquinone and hydroquinone monomethyl ether) towards cultured rat hepatoma cells (HTC line), a high dose-dependent HTC cytotoxicity of hydroquinone was observed, whilst *arbutin* was not active. At 300 μM hydroquinone caused 40% cellular mortality after 24 h of incubation, but no cells remained viable after 72 h. It has been established that this well-known antiseptic of the urinary tract was a more potent cytotoxic compound towards rat hepatoma cells than many classic antitumor agents like azauridin or colchicin, but less than valtrate, a monoterpenic ester of *Valeriana* spp. (Figure 9.18).

Figure 9.18 (See color insert.) *Origanum vulgare.*

References

Abdel-Massih, R.M., R. Fares, S. Bazzi, N. El-Chami, and E. Baydoun. 2010. The apoptotic and anti-proliferative activity of *Origanum majorana* extracts on human leukemic cell line. *Leukemia Research* 34, no.
8:1052–1056.

Adam, K., A. Sivropoulou, S. Kokkini, T. Lanaras, and M. Arsenakis. 1998. Antifungal activities of *Origanum
vulgare* subsp. *hirtum*, *Mentha spicata*, *Lavandula angustifolia*, and *Salvia fruticosa* essential oils against
human pathogenic fungi. *Journal of Agricultural and Food Chemistry* 46, no. 5:1739–1745.

Al Dhaheri, Y., A. Eid, S. AbuQamar, et al. 2013. Mitotic arrest and apoptosis in breast cancer cells induced
by *Origanum majorana* extract: Upregulation of TNF-α and downregulation of survivin and mutant p53.
PLOS ONE 8, no. 2:e56649.

Al-Kalaldeh, J.Z., R. Abu-Dahab, and F.U. Afifi. 2010. Volatile oil composition and antiproliferative activity of
Laurus nobilis, *Origanum syriacum*, *Origanum vulgare*, and *Salvia triloba* against human breast adenocarcinoma cells. *Nutrition Research* 30, no. 4:271–278.

Aruoma, O.I., J.P.E. Spencer, R. Rossi, et al. 1996. An evaluation of the antioxidant and antiviral action of
extracts of rosemary and Provencal herbs. *Food and Chemical Toxicology* 34, no. 5:449–456.

Assaf, M.H., A.A. Ali, M.A. Makboul, J.P. Beck, and R. Anton. 1987. Preliminary study of phenolic glycosides
from *Origanum majorana*, quantitative estimation of arbutin, cytotoxic activity of hydroquinone. *Planta
Medica* 53, no. 4:343–345.

Bocharova, O.A., R.V. Karpova, N.N. Kasatkina, L.G. Polunina, T.S. Komarova, and M.A. Lygenkova. 1999.
The antiproliferative activity for tumor cells is important to compose the phytomixture for prophylactic
oncology. *Farmacevtski Vestnik* 50:378–379.

Kanazawa, K., H. Kawasaki, K. Samejima, H. Ashida, and G. Danno. 1995. Specific desmutagens (antimutagens) in Oregano against dietary carcinogen, Trp-P-2, are galangin and quercetin. *Journal of Agricultural
and Food Chemistry* 43, no. 2:404–409.

Mayer, E., V. Sadar, and J. Spanring. 1971. *New* Crops Screening of Native Plants of Yugoslavia of Potential Use
in the Agricultures *of the USA and SFRJ*. University of Ljubljana, Biotechical Faculty, Final Technical
Report. Printed by Partizanska Knjiga Ljubljana.

Milic, B.L., and N.B. Milic. 1998. Protective effects of spice plants on mutagenesis. *Phytotherapy Research*
12 Suppl 1:S3–S6.

Savini, I., R. Arnone, M.V. Catani, and L. Avigliano. 2009. *Origanum vulgare* induces apoptosis in human
colon cancer $CaCO_2$ cells. *Nutrition and Cancer* 61, no. 3:381–389.

Sivropoulou, A., E. Papanikolaou, C. Nikolaou, S. Kokkini, T. Lanaras, and M. Arsenakis. 1996. Antimicrobial
and cytotoxic activities of *Origanum* essential oils. *Journal of Agricultural and Food Chemistry* 44, no.
5:1202–1205.

Skwarek, T., Z. Tynecka, K. Glowniak, and E. Lutostanska. 1994. Plant inducers of interferons. *Herba Polonica* 40, no. 1–2:42–49.

Ueda, S., Y. Kuwabara, N. Hirai, H. Sasaki, and T. Sugahara. 1991. Antimutagenic capacities of different kinds of vegetables and mushrooms. *Nippon Shokuhin Kogyo Gakkaishi* 38, no. 6:507–514.

***Oxytropis falcate* (Leguminosae) (Fabales)**   **Anti-tumor**

Location: Eurasia, North America, with some species also found in the Arctic region
    Appearance: perennial hairy plant

    Flowers: pink, purple, white, or yellow

Tradition: it has a great reputation in the medicine used in Tibet, where it is referred to as the 'King of the Herbs'. It is traditionally used for detoxification, anti-inflammation, pain relief, and to cure sores.
    Parts used: roots
    Active ingredients: (–)-7-methoxy-dihydroflavone, 2′,4′-dihydroxy-chalcone, dalbergin, beta-sitosterol, tetrahydroflemichapparin-B
    Documented target cancers:
- Essential oil and flavonoids of *O. falcata* are reported to inhibit proliferation of human hepatocellular carcinoma SMMC-7721 cells *in vitro* and significantly inhibit growth of H22 solid tumors transplanted in mice.

References

Wang, D., W. Tang, G. Yang, and B. Cai. 2010. Anti-inflammatory, antioxidant and cytotoxic activities of flavonoids from *Oxytropis falcata* Bunge. *Chinese Journal of Natural Medicines* 8, no. 6:461–465.

Wang, D., H. Yang, and G.M. Yang. 2010. Volatile compositions from a Tibetan medicine: *Oxytropis falcata* Bunge. *Natural Product Research and Development* 22, no. 4:614–619.

Yang, G.M., R. Yan, Z.X. Wang, F.F. Zhang, Y. Pan, and B.C. Cai. 2013. Antitumor effects of two extracts from *Oxytropis falcata* on hepatocellular carcinoma *in vitro* and *in vivo*. *Chinese Journal of Natural Medicines* 11, no. 5:519–524.

Zhang, D., R. Jiang, E.K. Hong, et al. 2017. The pharmacologically active components of *Oxytropis falcata* bunge reduce ischemic-reperfusion injury in the rat heart. *Legume Research: An International Journal* 40, no. 2:264–270.

***Pachysandra terminalis* (Buxaceae) (Buxales)**   **Anti-metastatic**

Location: Japan, Korea, China
    Appearance: evergreen shrub, approximately 15 cm

    Leaves: dark glossy, toothed at the edges
    Flowers: white at the beginning of summer

Parts used: fruit
    Active ingredients: pregnane alkaloids, terminamines H-J
    Documented target cancers:

- Several terminamines and alkaloids from *Pachysandra terminalis* are reported to inhibit the migration of MB-MDA-231 breast cancer cells induced by the chemokine epithelial growth factor.
- Active ingredients of the plant are reported to exhibit significant anti-metastasis activities.

References

Zhai, H.Y., C. Zhao, N. Zhang, et al. 2012. Alkaloids from *Pachysandra terminalis* inhibit breast cancer invasion and have potential for development as antimetastasis therapeutic agents. *Journal of Natural Products* 75, no. 7:1305–1311.

Zhao, C., C.C. Gan, M.N. Jin, S.A. Tang, N. Qin, and H.Q. Duan. 2014. Antitumor metastasis pregnane alkaloids from *Pachysandra terminalis*. *Journal of Asian Natural Products Research* 16, no. 5:440–446.

Paeonia moutan (Paeoniaceae) (Saxifragales)   **Anti-migratory**

Location: Asia, Eastern Europe, Western and North America
  Appearance: decidiuous shrub

  Leaves: rather large, without glands and stipules and with anomocytic stomata
  Flowers: large, bisexual, mostly single at the end of the stem

Tradition: significant as a status symbol in Chinese politics and culture
  Parts used: bark, roots, rhizomes
  Active ingredients: paeonol
  Documented target cancers:
  • The root bark of *Paeonia moutan* is reported to inhibit monolayer and anchorage-independent growth and interrupted coordinated migration of cell lines derived from human oral squamous cell carcinoma (HSC2, 3, 4, SAS).

References

Hu, S., G. Shen, W. Zhao, F. Wang, X. Jiang, and D. Huang. 2010. Paeonol, the main active principles of *Paeonia moutan*, ameliorates alcoholic steatohepatitis in mice. *Journal of Ethnopharmacology* 128, no. 1:100–106.

Li, C., K. Yazawa, S. Kondo, et al. 2012. The root bark of *Paeonia moutan* is a potential anticancer agent in human oral squamous cell carcinoma cells. *Anticancer Research* 32, no. 7:2625–2630.

Paeonia officinalis L. (Paeonia (Paeony) (Ranunculaceae)   **Tumor inhibitor**

Location: only grows wild on an island called the Steep Holmes, in the Severn. Great Britain
  Appearance:

  Stem: green (red when quite young), about 1 m high
  Root: composed of several roundish, thick knobs of tubers, which hang below each other, connected by strings
  Leaves: composed of several unequal lobes, which are cut into many segments
  Flowers: deep purple, fragrant
  In bloom: late spring

Tradition: the genus is supposed to have been named after the physician Paeos, who cured gods of wounds received during the Trojan War with the aid of this plant. In ancient times it was connected with many superstitions. It was used as antispasmodic and tonic.
  Parts used: root
  Active ingredients: LH-RH antagonist and a weak anti-estrogen on the uterine DNA synthesis in immature rats

Figure 9.19 (See color insert.) *Paeonia officinalis.* © 2001 Horticopia, Inc.

Documented target cancers: intestinal metaplasia, atypical hyperplasia of the gastric mucosa (*Paeonia lactiflora*), uterine myomas (*Paeonia lactiflora, Paeonia suffruticosa*)

Further details:

- Shi-Quan-Da-Bu-Tang (Ten Significant Tonic Decoction), or SQT (Juzentaihoto, TJ-48), was formulated by Taiping Hui-Min Ju (Public Welfare Pharmacy Bureau) in Chinese Song Dynasty in AD 1200. It is prepared by extracting a mixture of ten medical herbs (*Rehmannia glutinosa, Paeonia lactiflora, Liqusticum wallichii, Angelica sinesis, Glycyrrhiza uralensis, Poria cocos, Atractylodes macrocephala, Panax ginseng, Astragalus membranaceus*, and *Cinnamomum cassia*) that tone the blood and vital energy and strengthen health and immunity. This potent and popular prescription has traditionally been used against anemia, anorexia, extreme exhaustion, fatigue, kidney and spleen insufficiency, and general weakness, particularly after illness.
- *Paeonia alba* is one of the herbal constituents of Xiao Wei Yan Powder (some of the other constituents are *Smilax glabrae, Hedyotis diffusae, Taraxacum mongolicum, Caesalpinia sappan, Cyperus rotundus, Bletilla striata*, and *Glycyrrhiza uralensis*). This preparation has been used for the treatment of intestinal metaplasia and atypical hyperplasia of the gastric mucosa of chronic gastritis, administered orally at 5–7 g/d. After two to four months of administration, the total remission rate exceeded 90%.
- The root of *Paeonia lactiflora* Pall. and the root bark of *Paeonia suffruticosa* Andr. (Paeoniaceae) are components of Kuei-chih-fu-ling-wan (Keishi-bukuryo-gan), a traditional Chinese herbal remedy which contains another three components: the bark of *Cinnamomum cassia* Bl. (Lauraceae), seeds of *Prunus persica* Batsch. or *P. persiba* Batsch.var.davidiana Maxim. (Rosaceae), and carpophores of *Poria cocos* Wolf. (Polyporaceae). This prescription has been frequently used in the treatment of gynecological disorders such as hypermenorrhea, dysmenorrhea, and sterility. After treatment with the preparation, clinical symptoms of hypermenorrhea and dysmenorrhea were improved in more than 90% of the cases with shrinking of uterine myomas in roughly 60% of the cases (Figure 9.19).

References

Aburada, M., S. Takeda, E. Ito, M. Nakamura, and E. Hosoya. 1983. Protective effects of juzentaihoto, dried decoctum of 10 Chinese herbs mixture, upon the adverse effects of mitomycin C in mice. *Journal of Pharmacobio-Dynamics* 6, no. 12:1000–1004.

Liu, X.R., W.Q. Han, and D.R. Sun. 1992. Treatment of intestinal metaplasia and atypical hyperplasia of gastric mucosa with xiao wei yan powder. *Chung Kuo Chung HSI i Chieh Ho Tsa Chih* 12, no. 10(602–603):580.

Sakamoto, S., H. Yoshino, Y. Shirahata, K. Shimodairo, and R. Okamoto. 1992. Pharmacotherapeutic effects of kuei-chih-fu-ling-wan (keishi-bukuryo-gan) on human uterine myomas. *American Journal of Chinese Medicine* 20, no. 3–4:313–317.

Zee-Cheng, R.K. 1992. Shi-quan-da-bu-tang (ten significant tonic decoction), SQT. A potent Chinese biological response modifier in cancer immunotherapy, potentiation and detoxification of anticancer drugs.
Methods and Findings in Experimental and Clinical Pharmacology 14, no. 9:725–736.

Paeonia suffruticosa (Paeoniaceae) (Saxifragales)   **Anti-proliferative Anti-tumor**

Location: Asia, Eastern Europe, Western and North America
  Appearance: decidiuous shrub

  Leaves: large, compound, without glands and stipules and with anomocytic stomata
  Flowers: large, bisexual, mostly single at the end of the stem

Tradition: widely used in traditional Chinese medicine for the treatment of blood-heat and blood-
stasis syndrome
  Parts used: bark, roots, rhizomes, seeds
  Active ingredients: oligostilbenes, transresveratrol, paeonol
  Documented target cancers:

- Several oligostilbenes from the seeds of *P. suffruticosa* are reported to show potent antiproliferative
  and antimetastasis effects in a panel of human lung, breast, and bone cancer cell lines.
- The plant extract is reported to reduce cell viability and inhibit cell invasion activity in bladder
  cancer cells, with lower cytotoxicity in normal urotheliums.
- *In vivo*, the plant extract is reported to decrease the bladder tumor size without altering the blood
  biochemical parameters in a mouse orthotopic bladder cancer model.

References

Gao, Y., and C. He. 2017. Anti-proliferative and anti-metastasis effects of ten oligostilbenes from the seeds of
Paeonia suffruticosa on human cancer cells. *Oncology Letters* 13, no. 6:4371–4377.
Lin, M.Y., Y.R. Lee, S.Y. Chiang, et al. 2013. Cortex moutan induces bladder cancer cell death via apoptosis
and retards tumor growth in mouse bladders. *Evidence-Based Complementary and Alternative Medicine*
2013:207279.

Panax ginseng (Araliaceae) (Apiales)   **Anti-cancer**

Location: Asia
  Appearance: herbaceous plant (40–60 cm)

  Leaves: palmoid, complex, with elliptical toothed tips
  Flowers: white with yellow and green colorations, umbrella formations

Tradition: used in traditional medicine in China for thousands of years. It is used as a general tonic
or adaptogen with chronically ill patients and is frequently featured in traditional medicine pre-
scriptions from China, Japan, and Korea.
  Parts used: roots, leafs
  Active ingredients: saponins, ginzenozites a, b, c, panaxoxides
  Precautions: in large quantities it can cause nervousness, insomnia, overstimulation, headaches,
and skin and gastrointestinal problems.
  Documented target cancers:

- The anticancer properties of *Panax ginseng* have been studied since the 1980s, and formulations of the plant are reported to be effective against liver, ovarian, skin, stomach, lung, utery, kidney, and mammary gland cancers.
- Animal data suggest that treatment with *Panax ginseng* extracts decreases the incidence of cancer in various animal models.

References

Chang, Y.S., E.K. Seo, C. Gyllenhaal, and K.I. Block. 2003. *Panax* ginseng: A role in cancer therapy? *Integrative Cancer Therapies* 2, no. 1:13–33.

Shin, H.R., J.Y. Kim, T.K. Yun, G. Morgan, and H. Vainio. 2000. The cancer-preventive potential of *Panax ginseng*: A review of human and experimental evidence. *Cancer Causes and Control* 11, no. 6:565–576.

Wang, C.Z., S. Anderson, W. Du, T.C. He, and C.S. Yuan. 2016. Red ginseng and cancer treatment. *Chinese Journal of Natural Medicines* 14, no. 1:7–16.

Panax notoginseng (Araliaceae) (Apiales)　　　　　　　　　　　　　　　　　　　　　**Anti-cancer**

Location: Asia

　Appearance: herbaceous plant

　Leaves: palmoid, complex, with elliptical toothed tips
　Flowers: white

Tradition: it is traditionally used as a hemostatic medicine to control internal and external bleeding in China for thousands of years.

　Parts used: roots

　Active ingredients: notoginseng saponins, protoparaxotriol saponins, panasadiol saponins, ginsenoside Rg1, ginsenoside Rb1, ginsenoside Re, and notoginsenoside R1

　Documented target cancers:

- *Panax notoginseng* saponins are reported to exhibit a dose-dependent effect on impairing viability, *in vitro* migration and invasion *in vivo* of 4T1 cell, a highly metastatic mouse breast carcinoma cell line.
- *Panax notoginseng* ethanol extract is reported to affect human colorectal cancer cells by inhibiting cell migration, invasion, and adhesion and regulating the expression of metastasis-associated signaling molecules.
- *Panax notoginseng* saponins are reported to selectively impair the survival of Lewis lung carcinoma cells.

References

Hsieh, S.L., S. Hsieh, Y.H. Kuo, J.J. Wang, J.C. Wang, and C.C. Wu. 2016. Effects of *Panax notoginseng* on the metastasis of human colorectal cancer cells. *American Journal of Chinese Medicine* 44, no. 4:851–870.

Su, P., L. Wang, S.J. Du, W.F. Xin, and W.S. Zhang. 2014. Advance in studies of *Panax notoginseng* saponins on pharmacological mechanism of nervous system disease. *Zhongguo Zhong Yao za Zhi = Zhongguo Zhongyao Zazhi = China Journal of Chinese Materia Medica* 39, no. 23:4516–4521.

Wang, P., J. Cui, X. Du, et al. 2014. *Panax notoginseng* saponins (PNS) inhibit breast cancer metastasis. *Journal of Ethnopharmacology* 154, no. 3:663–671.

Yang, Q., P. Wang, J. Cui, W. Wang, Y. Chen, and T. Zhang. 2016. *Panax notoginseng* saponins attenuate lung cancer growth in part through modulating the level of Met/miR-222 axis. *Journal of Ethnopharmacology* 193:255–265.

Yang, X., X. Xiong, H. Wang, and J. Wang. 2014. Protective effects of *Panax notoginseng* saponins on cardio-vascular diseases: A comprehensive overview of experimental studies. *Evidence-Based Complementary and Alternative Medicine* 2014:204840.

Panax quinquefolium (Linn.) (Ginseng) (Araliaceae)	Immunomodulator

Location: in most of the forests of the countries of Southeast Asia but also in USA and Canada

Origin: Manchuria, China, and other parts of eastern Asia

Appearance:

Stem: simple, erect about 30.5 cm high
Root: It is 10–25 cm long and 1–2 cm diameter.
Leaves: each divided into five finely toothed leaflets
Flowers: single terminal umbel, with a few small, yellowish flowers

Degree of rarity: low, as it is cultivated

Tradition: the root has been used for centuries in traditional Chinese medicine. They believe that it makes those who use it stronger and younger.

Parts used: roots (Ginseng radix)

Active ingredients: Ginsenosides (saponins), ginsan (acidic polysaccharide), panaxytriol, and panaxydol (polyacetylenic alcohols)

Particular value: it is used particularly for dyspepsia, vomiting, and nervous disorders.

Documented target cancers:

- Ginsenosides appear to have an antitumor promoting activity and antimetastatic action in several cancers such as ovarian cancer, breast cancer, stomach cancer, and melanoma.
- Ginsan has antineoplastic activity. It is proved to induce Th1 cell and macrophage cytokines.
- Antineoplastic activity, cancer chemoprevention, effects on cytochemical components of SGC-823 gastriccarcinoma (in cell culture), Ehrlich ascites tumor cells (mouse), inhibition of autochthonous tumor, effects on adenocarcinoma of the human ovarian, stomach cancer, melanoma cells.
- Human breast cancer MCF-7 cells
- HCT116 human colon carcinoma cells

Further details:

- Panaxytriol (possible action) is cytotoxic. It is responsible for inhibition of mitochondrial respiration. Panaxydol has antiproliferative activity and its affinity for target cell membrane.
- Red ginseng is a traditional Chinese medicine. Its extract A and B are the active components of *Panax ginseng*. As it is considered as a tonic many studies have taken on ginseng and immune function of the human body. Some studies refer to the effects of red ginseng extracts on transplantable tumors, proliferation of lymphocyte. It is proven that in a two-stage model, red ginseng extracts had a significant cancer chemoprevention. At the dose of 50–400 mg/kg, the extracts could inhibit DMBA/Croton oil-induced skin papilloma in mice and decrease the incidence of papilloma. The red ginseng extract B seems to have a stronger antioxidative effect than that of extract A. Those doses (50 approximately 400 mg/kg) could significantly inhibit the growth of transplantable mouse sarcoma S180 and melanoma B16. In lower doses (extract A 0.5 mg/ml and B 0.1 and 0.25 mg/ml) might effectively promote the transformation of T lymphocyte.
- Another study took place in Korea with Korean red ginseng, evaluating the effects of ginseng in inhibition or prevention of carcinogenesis. It was administered orally to ICR newborn mice. Tumors

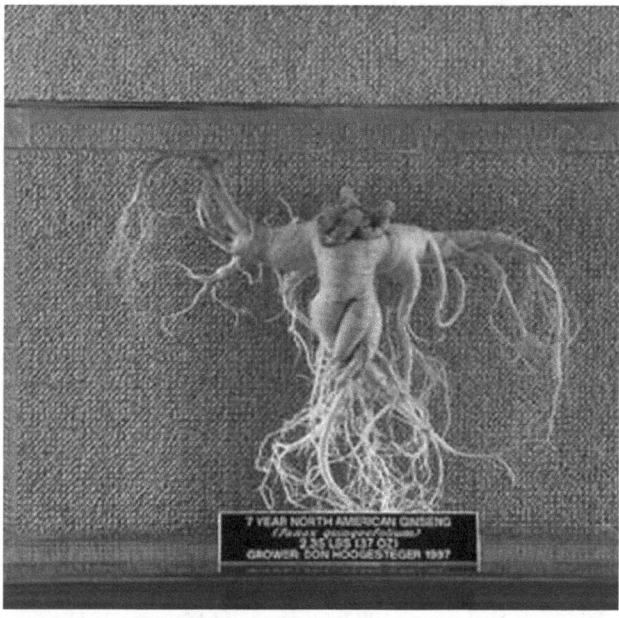

Figure 9.20   (See color insert.) *Ginseng/Panax.*

were induced by various chemical carcinogens within 24 h after birth. The newborn mice were injected in the ubscapular region with 9, 10-Dimethyl-1, 2-benzanthracene (DMBA), urethane, and aflatoxin B1. There was autopsy of the mice immediately following sacrifice and examination of all their organs (histopathological examinations, weight, etc.). The decrease in the average diameter and in the weight of lung adenomas was over 23%, while the incidence of diffuse pulmonary infiltration decreased by 63%. The results of the study indicate that Korean red ginseng extract inhibited the incidence and also the proliferation of tumors induced by DMBA, urethane, and aflatoxin B1.

- *Panax ginseng*: most of the compounds come from the methanolic extract of the root. Only the roots are used in medicine. It contains ginsenosides, ginsan, panaxytriol, and panaxydol. A new chloride is also produced that is cytotoxic. It is used against various human cancers such as stomach, breast, ovarian, lung, leukemia, hepatoma, and adenocarcinomas. Oral administration or by injection (shenmai injection).
- *Panax vietnamensis*: the root contains the ginsenosides: majonoside -R2, ginsenoside -R2, and ginsenoside -Rg1. It is used for its inhibitory effects on tumor growth (human ovarian cancer cells) and for its antitumor promoting activity. *Ginsenoside* -Rg1 seems to down-regulate glycocorticoid receptors and displays synergistic effects with CAMP.
- *Panax quinquefolius* L.: it is the American ginseng. In studies, the extract of the root was used, and the administration was oral. Ginsenosides were contained, also, and the effects showed a decrease of serum gamma globulin and IgG1 isotype (in mice) and ps2 expression in MCF-7 breast cancer cells.
- *Panax ginseng (red)*: it is the Korean *Panax ginseng*. In medicine the extract of the root is used: A and B which contain: ginsenoside -Rg3, -Rb2, -Rh2, -Rh4, -20(R), -20(S). Inhibits the tumor metastasis, tumor angiogenesis, improves the cell immune system. In studies related to stomach cancer the shenmai injection is used, produced by red ginseng extract (Figure 9.20).

References

Bernart, M.W., J.H. Cardellina, M.S. Balaschak, M.R. Alexander, R.H. Shoemaker, and M.R. Boyd. 1996. Cytotoxic falcarinol oxylipins from Dendropanax arboreus. *Journal of Natural Products* 59, no. 8:748–753.

Kim, K.H., Y.S. Lee, I.S. Jung, S.Y. Park, H.Y. Chung, I.R. Lee, and Y.S. Yun. 1998. Acidic polysaccharide from *Panax ginseng*, ginsan, induces Th1 cell and macrophage cytokines and generates LAK cells in synergy with rIL-2. *Planta Medica* 64, no. 2:110–115.

Kim, Y.W., D.K. Song, W.H. Kim, et al. 1997. Long-term oral administration of ginseng extract decreases serum gamma-globulin and IgG1 isotype in mice. *Journal of Ethnopharmacology* 58, no. 1:55–58.

King, M.L., and L.L. Murphy. 2010. Role of cyclin inhibitor protein p21 in the inhibition of HCT116 human colon cancer cell proliferation by American ginseng (*Panax quinquefolius*) and its constituents. *Phytomedicine* 17, no. 3–4:261–268.

Lee, Y.J., E. Chung, K.Y. Lee, Y.H. Lee, B. Huh, and S.K. Lee. 1997. Ginsenoside-Rg1, one of the major active molecules from panax ginseng, is a functional ligand of glucocorticoid receptor. *Molecular and Cellular Endocrinology* 133, no. 2:135–140.

Qiu, Y.K., D.Q. Dou, L.P. Cai, et al. 2009. Dammarane-type saponins from *Panax quinquefolium* and their inhibition activity on human breast cancer MCF-7 cells. *Fitoterapia* 80, no. 4:219–222.

Wakabayashi, C., H. Hasegawa, J. Murata, and I. Saiki. 1997. *In vivo* antimetastatic action of ginseng pro-topanaxadiol saponins is based on their intestinal bacterial metabolites after oral administration. *Oncology Research* 9, no. 8:411–417.

Wakabayashi, C., K. Murakami, H. Hasegawa, J. Murata, and I. Saiki. 1998. An intestinal bacterial metabolite of ginseng protopanaxadiol saponins has the ability to induce apoptosis in tumor cells. *Biochemical and Biophysical Research Communications* 246, no. 3:725–730.

Xiaoguang, C., L. Hongyan, L. Xiaohong, F. Zhaodi, L. Yan, T. Lihua, and H. Rui. 1998. Cancer chemopreventive and therapeutic activities of red ginseng. *Journal of Ethnopharmacology* 60, no. 1:71–78.

Yamamoto, M., A. Kumagai, and Y. Yamamura. 1983. Plasma lipid-lowering and lipogenesis-stimulating actions of ginseng saponins in tumor-bearing rats. *American Journal of Chinese Medicine* 11, no. 1–4:88–95.

Yun, T.K., Y.S. Yun, and I.W. Han. 1983. Anticarcinogenic effect of long-term oral administration of red ginseng on newborn mice exposed to various chemical carcinogens. *Cancer Detection and Prevention* 6, no. 6:515–525.

***Pandanus odoratissimus* (Pandanaceae) (Pandanales)**   **Anti-cancer**

Location: India, Asia, North Australia, Philippines, and along the coasts of the Pacific ocean
  Appearance: small tree

  Leaves: sword shaped, blue-green, aromatic

Tradition: traditionally recommended by the Indian Ayurvedic medicines for treatment of headache, rheumatism, spasm, cold/flu, epilepsy, wounds, boils, scabies, leucoderma, ulcers, colic, hepatitis, smallpox, leprosy, syphilis, and cancer and as a cardiotonic, antioxidant, dysuric, and aphrodisiac.

  Parts used: roots, leaves, fruit
  Active ingredients: lignans, isoflavonoids, cumestrol, alkaloids, glucosides
  Documented target cancers:

- Aqueous extract of roots and leaves of *P. odoratissimus* is reported to possess *in vivo* anti-cancer activity in Ehrlich ascites carcinoma induced liquid tumors in Swiss albino mice.
- Aqueous extract of *P. odoratissimus* is reported effective on calu-6 (non-small cell lung cancer cell lines), PBMC (peripheral blood mononuclear cells), and WI (lung fibroblast cell lines).

References

Adkar, P.P., and V.H. Bhaskar. 2014. Pandanus odoratissimus (Kewda): A review on ethnopharmacology, phytochemistry, and nutritional aspects. *Advances in Pharmacological Sciences* 2014:120895.

Gowtham Raj, G., H. Sara Varghese, S. Kotagiri, and V. Swamy B.M. 2014. Evaluation of anti-cancer potential of aqueous extract of Pandanus odoratissimus (Y. Kimura) Hatus. Forma ferreus, by *in vivo* ascitic tumor model in Swiss albino mice. *Pharmacognosy Journal* 6, no. 1:57–62.

Raj, G.G., H.S. Varghese, S. Kotagiri, B.M. Vrushabendra Swamy, A. Swamy, and R.K. Pathan. 2014. Anticancer studies of aqueous extract of roots and leaves of Pandanus odoratissimus f. Ferreus (Y. Kimura) Hatus: An *in vitro* approach. *Journal of Traditional and Complementary Medicine* 4, no. 4:279–284.

Pao pereira (Apocynaceae) (Gentianales)   **Cytotoxic**

Location: Amazon rain forest

Appearance: tree

Tradition: in Brazil it was used as antipyretic and antiperiodic while native Indians used it to support the immune system.

Parts used: crust of the trunk

Active ingredients: b –carboline, flavopereirine (PB 100), indole

Documented target cancers:

- *Pao pereira* is reported to selectively inhibit ovarian cancer cell growth in different ovarian cancer cell lines, compared to an normal immortalized epithelial cell line.
- The plant extract is reported to suppress metastatic castration-resistant prostate cancer PC3 cell growth in a dose- and time-dependent manner.
- A β-carboline alkaloid-enriched extract from the plant is reported to suppress prostate cancer cells.

References

Bemis, D.L., J.L. Capodice, M. Desai, A.E. Katz, and R. Buttyan. 2009. β-Carboline alkaloid–enriched extract from the Amazonian rain forest tree *Pao Pereira* suppresses prostate cancer cells. *Journal of the Society for Integrative Oncology* 7, no. 2:59–65.

Chang, C., W. Zhao, B. Xie, et al. 2014. *Pao Pereira* extract suppresses castration-resistant prostate cancer cell growth, survival, and invasion through inhibition of NFκB signaling. *Integrative Cancer Therapies* 13, no. 3:249–258.

Yu, J., and Q. Chen. 2014. The plant extract of *Pao Pereira* potentiates carboplatin effects against ovarian cancer. *Pharmaceutical Biology* 52, no. 1:36–43.

Paris polyphylla (Melanthiaceae) (Liliales)   **Cytotoxic Anti-tumor**

Location: China, Korea, Japan, Indochina

Appearance: perennial, flowering plant

Leaves: thin spiral, pointy, green with smooth edges

Tradition: in Nepal it is used to neutralize poison in domesticated animals. A root decoction is used in the treatment of bites from poisonous snakes, ulcers, Japanese encephalitis B, and diphtheria.

Parts used: roots, rhizomes, stems, leaves, fruits, seeds

Active ingredients: saponines

Documented target cancers:

- *Paris polyphylla* ethanol extract is reported to inhibit prostate cancer growth *in vitro* and *in vivo*, induce apoptosis, and cause cell cycle arrest.

- The plant extract has been reported to inhibit the growth and proliferation on esophageal cancer ECA109 cells.
- The aqueous extract of *Paris polyphylla* is reported to inhibit human ovarian carcinoma cell line (OVCAR-3 cells).

References

Li, F.R., P. Jiao, S.T. Yao, et al. 2012. *Paris polyphylla* Smith extract induces apoptosis and activates cancer suppressor gene connexin26 expression. *Asian Pacific Journal of Cancer Prevention* 13, no. 1:205–209.

Wang, C.W., C.J. Tai, C.Y. Choong, Y.C. Lin, B.H. Lee, Y.C. Shi, and C.J. Tai. 2016. Aqueous extract of *Paris polyphylla* (AEPP) inhibits ovarian cancer via suppression of peroxisome proliferator-activated receptor-gamma coactivator (PGC)-1alpha. *Molecules* 21, no. 6:727.

Zhang, D., K. Li, C. Sun, et al. 2018. Anti-cancer effects of *Paris Polyphylla* ethanol extract by inducing cancer cell apoptosis and cycle arrest in prostate cancer cells. *Current Urology* 11, no. 3:144–150.

Parmelia sulcata (Parmeliaceae) (Lecanorales)   **Anti-tumor**

Location: forests of Arizona, coastal oaks in California, and stones in the high mountains of northern Baja California

Appearance: foliose lichen consisting of a tholos formed by planes, foliage, blue gray lobes of 4–20 cm. The pods have a network of sharp edges and pits, giving the lichen a hammer face.

Tradition: used for the treatment of cranial and pimmonic diseases

Parts used: dry thalli

Active ingredients: salazinic acid, lecanoric acid, protocetraric acid

Documented target cancers:

- Methanol extracts of the *P. sulcata* have shown antitumor activity in intestinal adenocarcinoma.
- *P. sulcata* has caused cell growth arrest in human breast cancer cells (MCF-7 and MDA-MB-231).
- Antitumor activity in FemX (human melanoma) and LS174 (human intestinal carcinoma) cancer cell lines.

Further details:
- *Parmelia sulcata* is very sensitive to gaseous pollutants and has been proposed as a bio-indicator of low-medium air pollution.

References

Ari, F., E. Ulukaya, S. Oran, S. Celikler, S. Ozturk, and M.Z. Ozel. 2015. Promising anticancer activity of a lichen, *Parmelia sulcata* Taylor, against breast cancer cell lines and genotoxic effect on human lymphocytes. *Cytotechnology* 67, no. 3:531–543.

Kosanic, M., B. Rankovic, T. Stanojkovic, P. Vasiljevic, and N. Manojlovic. 2014. Biological activities and chemical composition of lichens from Serbia. *Excli Journal* 13:1226–1238.

Manojlović, N., B. Ranković, M. Kosanić, P. Vasiljević, and T. Stanojković. 2012. Chemical composition of three Parmelia lichens and antioxidant, antimicrobial and cytotoxic activities of some their major metabolites. *Phytomedicine* 19, no. 13:1166–1172.

Zambare, V.P., and L.P. Christopher. 2012. Biopharmaceutical potential of lichens. *Pharmaceutical Biology* 50, no. 6:778–798.

Passiflora tetrandra (Passifloraceae)	Cytotoxic

Parts used: leaves

Active ingredients: 4-Hydroxy-2-cyclopentenone

Documented target cancers: 4-Hydroxy-2-cyclopentenone is cytotoxic to P388 murine leukemia cells (IC_{50} of less than 1 microgram/ml).

Further details:

- 4-Hydroxy-2-cyclopentenone is also responsible for the anti-bacterial activity of an extract of leaves from *Passiflora tetrandra* with minimum inhibitory doses (MID) of ca. 10 micrograms/disk against *Escherichia coli*, *Bacillus subtilis*, and *Pseudomonas aeruginosa*.

References

Perry, N.B., G.D. Albertson, J.W. Blunt, A.L.J. Cole, M.H.G. Munro, and J.R.L. Walker. 1991. JR4-Hydroxy-2-cyclopentenone: An anti-Pseudomonas and cytotoxic component from *Passiflora tetrandra*. *Planta Medica* 57, no. 2:129–131.

Suffredini, I.B., M.L.B. Paciencia, A.D. Varella, and R.N. Younes. 2006. *In vitro* prostate cancer cell growth inhibition by Brazilian plant extracts. *Die Pharmazie* 61, no. 8:722–724.

Petroselinum sativum (Apiaceae) (Apiales)	Anti-tumor Cytotoxic

Location: areas with Mediterranean climate

Appearance: biennial plant of low growth

Leaves: small, feathered
Flowers: small, greenish yellow and are carried in shades of inflorescence

Tradition:

- Used in traditional medicine as a diuretic, it is effective in relieving the body of stones and treating jaundice, dysentery, cystitis, cough, asthma, shortness of breath, and dyspnoea.
- An infusion of roots and seeds is taken after delivery to promote lactation and help shrink the uterus.
- It has been used to treat eye infections, while a cotton-impregnated cloth is known to relieve toothache or ear pain. It is also used to prevent hair loss.

Parts used: leaves

Active ingredients: apigenin and myristicin

Precautions: regular use of parsley requires dietary compensation with potassium, given parsley's potassium-depleting diuretic effect.

Documented target cancers:

- Myristicin, isolated from *Petroselinum sativum*, is reported to inhibit the *in vivo* formation of tumors in benzo[a]pyrene-induced cancer on female mice.
- *Petroselinum sativum* extracts have been shown cytotoxic against human breast cancer MCF-7 and human hepatocellular carcinoma HepG2 cells.
- Apigenin is shown to enhance the activities of anti-metastatic CD26 protein in human fatty tissue carcinoma cells. Futhermore it is reported to inhibit cell growth, sensitize cancer cells in apoptosis, and block the growth of blood vessels that serve the developing tumor.
- Apigenin has exhibited anti-gastric cancer progression effects in *Helicobacter pylori*-infected Mongolian gerbils.

References

Farshori, N.N., E.S. Al-Sheddi, M.M. Al-Oqail, J. Musarrat, A.A. Al-Khedhairy, and M.A. Siddiqui. 2013. Anticancer activity of *Petroselinum sativum* seed extracts on MCF-7 human breast cancer cells. *Asian Pacific Journal of Cancer Prevention* 14, no. 10:5719–5723.

Farshori, N.N., E.S. Al-Sheddi, M.M. Al-Oqail, J. Musarrat, A.A. Al-Khedhairy, and M.A. Siddiqui. 2014. Cytotoxicity assessments of *Portulaca oleracea* and *Petroselinum sativum* Seed extracts on human hepatocellular carcinoma cells (HepG2). *Asian Pacific Journal of Cancer Prevention* 15, no. 16:6633–6638.

Kuo, C.H., B.C. Weng, C.C. Wu, S.F. Yang, D.C. Wu, and Y.C. Wang. 2014. Apigenin has anti-atrophic gastritis and anti-gastric cancer progression effects in *Helicobacter pylori*-infected Mongolian gerbils. *Journal of Ethnopharmacology* 151, no. 3:1031–1039.

Lefort, E.C., and J. Blay. 2011. The dietary flavonoid apigenin enhances the activities of the anti-metastatic protein CD26 on human colon carcinoma cells. *Clinical and Experimental Metastasis* 28, no. 4:337–349.

Lefort, É.C., and J. Blay. 2013. Apigenin and its impact on gastrointestinal cancers. *Molecular Nutrition and Food Research* 57, no. 1:126–144.

Zheng, G.Q., P.M. Kenney, J. Zhang, and L.K. Lam. 1992. Inhibition of benzo [a] pyrene-induced tumorigenesis by myristicin, a volatile aroma constituent of parsley leaf oil. *Carcinogenesis* 13, no. 10:1921–1923.

Pharbitis nil (Convolvulaceae) (Solanales)   **Cytotoxic Pro-apoptotic**

Location: China

Appearance: perennial herbaceous climbing plant

Leaves: simple, heart-shaped
Flowers: large, funnel-shaped
Fruit: spherical or oval shape, smooth

Tradition: the seeds of the plant are used as contraceptive in Korea. They have intense, sweet, warm, calming, laxative, anthelmintic, anti-inflammatory action. They are useful in scabies, dyspepsia, flatulence, bronchitis, arthralgia, headache, liver disease, splenic disease, and fever. Seeds also act as a anticonvulsant, anticholinergic, and blood cleansers.

Parts used: mainly the seeds

Active ingredients: pharbosides A–F, pharboside G, 7β,16β,17-trihydroxy-ent-kauran-6α,19-olide, 6β,7β,16α,17-tetrahydroxy-ent-kauranoic acid, 6β,7β,16β,17-tetrahydroxy-ent-kauranoic acid

Documented target cancers:

- Various substances isolated from ethanol extract of the *Pharbitis nil* seeds have exhibited moderate cytotoxic activity against A549 (non-small cell lung carcinoma), SK-OV-3 (ovary malignant ascites), SK-MEL-2 (skin melanoma), and HCT (colon adenocarcinoma) cells.
- The plant extract is reported to induce apoptotic cell death *in vitro* in breast cancer MCF-7 cells.
- *Pharbitis nil* has been reported to induce inhibition of growth and apoptosis of human gastric cancer cells of the AGS line.

References

Ju, J.H., M.J. Jeon, W. Yang, K.M. Lee, H.S. Seo, and I. Shin. 2011. Induction of apoptotic cell death by *Pharbitis nil* extract in HER2-overexpressing MCF-7 cells. *Journal of Ethnopharmacology* 133, no. 1:126–131.

Kim, K.H., S.U. Choi, and K.R. Lee. 2009. Diterpene glycosides from the seeds of *Pharbitis nil*. *Journal of Natural Products* 72, no. 6:1121–1127.

Ko, S.G., S.H. Koh, C.Y. Jun, C.G. Nam, H.S. Bae, and M.K. Shin. 2004. Induction of apoptosis by *Saussurea lappa* and *Pharbitis nil* on AGS gastric cancer cells. *Biological and Pharmaceutical Bulletin* 27, no. 10:1604–1610.

Phellinus linteus (Hymenochaetaceae) (Hymenochaetales)	Anti-proliferative Cytotoxic

Location: Korea and neighboring areas in China
    Appearance: thick, hard, woody, cork mushroom

Stem: thick, dark brown to black

Tradition: it has been described in Asian herbal medicine literature to be effective on a diverse range of diseases, including improving blood circulation, enhancing detoxication and hepatoprotection of human body, combating allergy and diabetes, curing oral ulcer, and alleviating gastroenteric disorder or lymphatic disease.

    Parts used: entire mushroom
    Active ingredients: hipsolon, PLS-1, atractylenolide I
    Documented target cancers:

- The PLS-1 polysaccharide isolated from the mycelia of *Phellinus linteus* has potent *in vitro* anti-proliferative activity on S-180 sarcoma cells through apoptosis.
- The ethanolic extract of the plant in combination with low-dose of 5-FU is reported to inhibit the growth rates of breast cancer cells (MDA-MB-231) and can cause autophagy.
- The bioactive compound hispolon has been shown to inhibit the growth of estrogen receptor positive human breast cancer cells. In addition, it has been reported to contribute to the activation of caspases and causes apoptosis in human nasopharyngeal carcinomas. Finally, it induces apoptosis and sensitizes human cancer cells by modulating necrosis receptors.
- Atractylenolide I, isolated from *Phellinus linteus* ethyl acetate extract cultivated on brown rice, is reported to inhibit the growth of HT-29 human colon cancer cells.

References

Chen, Y.C., H.Y. Chang, J.S. Deng, et al. 2013. Hispolon from *Phellinus linteus* induces G0/G1 cell cycle arrest and apoptosis in NB4 human leukaemia cells. *American Journal of Chinese Medicine* 41, no. 6:1439–1457.

Hsieh, M.J., S.Y. Chien, Y.E. Chou, C.J. Chen, J. Chen, and M.K. Chen. 2014. Hispolon from *Phellinus linteus* possesses mediate caspases activation and induces human nasopharyngeal carcinomas cells apoptosis through ERK1/2, JNK1/2 and p38 MAPK pathway. *Phytomedicine* 21, no. 12:1746–1752.

Jang, E.H., S.Y. Jang, I.H. Cho, D. Hong, B. Jung, M.J. Park, and J.H. Kim. 2015. Hispolon inhibits the growth of estrogen receptor positive human breast cancer cells through modulation of estrogen receptor alpha. *Biochemical and Biophysical Research Communications* 463, no. 4:917–922.

Jeon, T.I., C.H. Jung, J.Y. Cho, D.K. Park, and J.H. Moon. 2013. Identification of an anticancer compound against HT-29 cells from Phellinus linteus grown on germinated brown rice. *Asian Pacific Journal of Tropical Biomedicine* 3, no. 10:785–789.

Kim, J.H., Y.C. Kim, and B. Park. 2016. Hispolon from *Phellinus linteus* induces apoptosis and sensitizes human cancer cells to the tumor necrosis factor-related apoptosis-inducing ligand through upregulation of death receptors. *Oncology Reports* 35, no. 2:1020–1026.

Lee, W.Y., K.F. Hsu, T.A. Chiang, and C.J. Chen. 2015. *Phellinus linteus* extract induces autophagy and synergizes with 5-fluorouracil to inhibit breast cancer cell growth. *Nutrition and Cancer* 67, no. 2:275–284.

Mei, Y., H. Zhu, Q. Hu, Y. Liu, S. Zhao, N. Peng, and Y. Liang. 2015. A novel polysaccharide from myce-lia of cultured *Phellinus linteus* displays antitumor activity through apoptosis. *Carbohydrate Polymers* 124:90–97.

Zhu, T., S.H. Kim, and C.Y. Chen. 2008. A medicinal mushroom: *Phellinus linteus. Current Medicinal Chemistry* 15, no. 13:1330–1335.

Phyllanthus emblica (Phyllanthaceae) (Malpighiales)   **Pro-apoptotic**

Location: tropical Asian countries (mainly India)

Appearance: medium-sized tree with many branches, usually 10–30m tall

Leaves: purple-green

Tradition: in traditional Indian medicine, dried and fresh fruits of the plant are used. All parts of the plant are used in various Ayurvedic/Unani medicine herbal preparations, including the fruit, seed, leaves, root, bark, and flowers.

Parts used: almost all of the plant, except the root system

Active ingredients: ascorbic acid, galic acid, progestin A, progalin A, galic acid

Precautions: high doses may cause burning sensation during urination. It should also be avoided in case of acute diarrhea or dysentery.

Documented target cancers:

- Progalin A alond with gallic acid, isolated from the leaves of *Phyllanthus emblica*, are reported to exhibit immune toxicity *in vitro* and contribute to the human liver carcinoma BEL-7404 apoptosis.
- Polyphenol extracts from the plant appear to induce apoptosis of cancerous cervical cells by cell cycle arrest in G2/M phase.
- Galic acid from the leaves of *Phyllanthus emblica* is shown to induce apoptosis in human hepatoma BEL-7404 cells.

References

Dhir, H., K. Agarwal, A. Sharma, and G. Talukder. 1991. Modifying role of *Phyllanthus emblica* and ascorbic acid against nickel clastogenicity in mice. *Cancer Letters* 59, no. 1:9–18.

Dhir, H., A.K. Roy, A. Sharma, and G. Talukder. 1990. Modification of clastogenicity of lead and aluminium in mouse bone marrow cells by dietary ingestion of Phyllanthus emblica fruit extract. *Mutation Research* 241, no. 3:305–312.

Huang, J.L., and Z.G. Zhong. 2011. Study of galic acid extracted from the leaves of *Phyllanthus emblica* on apoptotic mechanism of human hepatocellular carcinoma cells BEL-7404. *Zhong Yao Cai = Zhongyaocai = Journal of Chinese Medicinal Materials* 34, no. 2:246–249.

Suresh, K., and D.M. Vasudevan. 1994. Augmentation of murine natural killer cell and antibody depen-dent cellular cytotoxicity activities by *Phyllanthus emblica*, a new immunomodulator. *Journal of Ethnopharmacology* 44, no. 1:55–60.

Zhong, Z.G., D.P., Wu, J.L. Huang, H. Liang, Z.H. Pan, W.Y. Zhang, and H.M. Lu. 2011. Progallin A isolated from the acetic ether part of the leaves *of Phyllanthus emblica* L. induces apoptosis of human hepato-cellular carcinoma BEL-7404 cells by up-regulation of Bax expression and down-regulation of Bcl-2 expression. *Journal of Ethnopharmacology* 133, no. 2:765–772.

Phyllantus niruri (Phyllanthus) (Euphorbiaceae)	**Antitumor Cytotoxic**

Location: Northern Asia
Appearance:

Stem: 0.5 m high, erect, red
Leaves: small, green, oblonged, and feathered
Flowers: greenish white

Parts used: root, fruit, leaf
Active ingredients: glycosides: phyllanthoside, phyllanthostatin (*Phyllanthus acuminatus*)
Particular value: it is used as antitumor, antileukemic (*Phyllanthus acuminatus*), antiviral, cytotoxic, chemopreventive.
Indicative dosage and application: against the growth of the murine P-388 lymphocytic leukemia cell line in a dose of 0.35 micrograms/ml
Documented target cancers:

- Treatment of acute and chronic hepatitis B and healthy carriers of HBV. Hepatocellular carcinoma (liver cancer) (*Phyllanthus urinaria*, *Phyllanthus amarus*)
- Dalton's lymphoma ascites (DLA) tumor (mice) (*Phyllanthus emblica*)
- Phyllanthostatin inhibits the growth of the murine P-388 lymphocytic leukemia cell line.
- A dry extract of *Phyllanthus niruri* is reported to induce apoptosis in human liver carcinoma cells.

Further details:

- The aqueous extract of *Phyllanthus amarus* contains some components that are able to inhibit *in vitro* HBsAg secretion in a dose-dependent manner. Various hepatoma cell lines, such as the Alexander cell line, a human derived cell line which has the property of secreting HBsAg in the supernatant.
- Extracts of *Phyllanthus amarus* have also been shown to inhibit the DNA polymerase of HBV and woodchuck hepatitis virus (WHV) *in vitro*.
- *Phyllanthus niruri* L.: the MeOH extract of the dried leaf contains niruriside, a potent antiviral compound.
- *Phyllanthus emblica*: is an excellent source of vitamin C (ascorbate) and, when administered orally, has been found to enhance natural killer (NK) cell activity and antibody dependent cellular cytotoxicity. Enhanced activity was highly significant on days three, five, seven, and nine after tumor inoculation with respect to the untreated tumor-bearing control. The following has been documented: (a) an absolute requirement for a functional NK cell or K cell population in order that *P. emblica* can exert its effect on tumor-bearing animals, and (b) the antitumor activity of *P. emblica* is mediated primarily through the ability of the drug to augment natural killer cell mediated cytotoxicity.
- Aqueous extracts of edible dried fruits of *Phyllanthus emblica* prevented the incidence of carcinogenesis in mice treated with nickel chloride. Ascorbic acid, a major constituent of the fruit, fed to the mice for seven consecutive days in equivalent concentration as that present in the fruit, however, could only alleviate the cytotoxic effects induced by low doses of nickel; at the higher doses it was ineffective. The greater efficacy of the fruit extract could be due to the interaction of its various natural components rather than to any single constituent.
- The roots of *Phyllanthus acuminatus* contain the glycosides, phyllanthoside (a major antineoplastic constituent), and phyllanthostatin which inhibits (ED50 = 0.35 micrograms/ml) the growth of the murine P-388 lymphocytic leukemia cell line. This species contains also didesacetylphyllanthostatin and descinnamoylphyllanthocindiol.

References

Blumberg, B.S., I. Millman, P.S. Venkateswaran, and S.P. Thyagarajan. 1989. Hepatitis B virus and hepatocellular carcinoma-treatment of HBV carriers with *Phyllanthus amarus*. *Cancer Detection and Prevention* 14, no. 2:195–201.

Blumberg, B.S., I. Millman, P.S. Venkateswaran, and S.P. Thyagarajan. 1990. Hepatitis B virus and primary hepatocellular carcinoma: Treatment of HBV carriers with *Phyllanthus amarus*. *Vaccine* 8 Suppl:S86–S92.

de Araújo Júnior, R.F., T.P. de Souza, J.G.L. Pires, et al. 2012. A dry extract of *Phyllanthus niruri* protects normal cells and induces apoptosis in human liver carcinoma cells. *Experimental Biology and Medicine* 237, no. 11:1281–1288.

Dhir, H., K. Agarwal, A. Sharma, and G. Talukder. 1991. Modifying role of *Phyllanthus emblica* and ascorbic acid against nickel clastogenicity in mice. *Cancer Letters* 59, no. 1:9–18.

Jayaram, S., and S.P. Thyagarajan. 1996. Inhibition of HBsAg secretion from Alexander cell line by *Phyllanthus amarus*. *Indian Journal of Pathology and Microbiology* 39, no. 3:211–215.

Ji, X.H., Y.Z. Qin, W.Y. Wang, J.Y. Zhu, and X.T. Liu. 1993. Effects of extracts from *Phyllanthus urinaria* L. on HBsAg production in PLC/PRF/5 cell line. *Chung Kuo Chung Yao Tsa Chih* 18, no. 8(496–498):511.

Pettit, G.R., D.E. Schaufelberger, R.A. Nieman, C. Dufresne, and J.A. Saenz-Renauld. 1990. Antineoplastic agents, 177. Isolation and structure of phyllanthostatin 6. *Journal of Natural Products* 53, no. 6:1406–1413.

Qian-Cutrone, J., S. Huang, J. Trimble, et al. 1996. Niruriside, a new HIV REV/RRE binding inhibitor from *Phyllanthus niruri*. *Journal of Natural Products* 59, no. 2:196–199.

Suresh, K., and D.M. Vasudevan. 1994. Augmentation of murine natural killer cell and antibody dependent cellular cytotoxicity activities by *Phyllanthus emblica*, a new immunomodulator. *Journal of Ethnopharmacology* 44, no. 1:55–60.

Yeh, S.F., C.Y. Hong, Y.L. Huang, T.Y. Liu, K.B. Choo, and C.K. Chou. 1993. Effect of an extract from *Phyllanthus amarus* on hepatitis B surface antigen gene expression in human hepatoma cells. *Antiviral Research* 20, no. 3:185–192.

Phyllanthus watsonii (Phyllanthaceae) (Malpighiales)   **Cytotoxic**

Location: Endau Rompin area of Malaysia

Appearance: small shrub

In bloom: bottled, thin bunch of up to 15 centimeters

Tradition: phyllanthus plants were used in folk medicine to treat diabetes, anemia, bronchitis, and hepatitis.

Parts used: aerial parts

Active ingredients: friedelin, lupeol, gluchidone, gluchidonol

Documented target cancers:

- *P. watsonii* extracts and fractions are reported to selectively inhibit the growth of the human gynecological cancer cells SKOV-3 and Ca Ski and the colon cancerous cells HT-29 by inducing apoptosis and cell cycle differentiation.
- Methanol, hexane, and ethylacetate isolates from leaves of the plant were cytotoxic and selectively inhibited growth and abrupt growth of MCF-7 cells.

References

Ramasamy, S., N. Abdul Wahab, N. Zainal Abidin, and S. Manickam. 2013. Effect of extracts from Phyllanthus watsonii Airy Shaw on cell apoptosis in cultured human breast cancer MCF-7 cells. *Experimental and Toxicologic Pathology* 65, no. 3:341–349.

Ramasamy, S., N. Abdul Wahab, N. Zainal Abidin, S. Manickam, and Z. Zakaria. 2012. Growth inhibition of human gynecologic and colon cancer cells by Phyllanthus watsonii through apoptosis induction. *PLOS ONE* 7, no. 4:e34793.

Tang, Y.Q., I. Jaganath, R. Manikam, and S.D. Sekaran. 2013. Phyllanthus suppresses prostate cancer cell, PC-3, proliferation and induces apoptosis through multiple signalling pathways (MAPKs, PI3K/Akt, NFB, and Hypoxia). *Evidence-Based Complementary and Alternative Medicine* 2013:1–13.

Phyllosticta cirsii (Botryosphaeraceae) (Botryosphaeriales)   **Cytotoxic**

Appearance: fungal pathogen isolated from diseased *Cirsium arvense* leaves. It forms irregular white spots with brown outline in the foliage.

Active ingredients: phyllostoxin and phyllostin

Documented target cancers:

- Phyllostictine A, a novel oxazatricycloalkenone isolated from *Phyllosticta cirsii*, is reported to display *in vitro* growth-inhibitory activity both in normal and cancer cells without actual bioselectivity, while proliferating cells appear significantly more sensitive to phyllostictine A than nonproliferating ones.

References

Evidente, A., A. Cimmino, A. Andolfi, M. Vurro, M.C. Zonno, and A. Motta. 2008. Phyllostoxin and phyllostin, bioactive metabolites produced by *Phyllosticta cirsii*, a potential mycoherbicide for Cirsium arvense biocontrol. *Journal of Agricultural and Food Chemistry* 56, no. 3:884–888.

Le Calvé, B., B. Lallemand, C. Perrone, et al. 2011. *In vitro* anticancer activity, toxicity and structure–activity relationships of phyllostictine A, a natural oxazatricycloalkenone produced by the fungus *Phyllosticta cirsii*. *Toxicology and Applied Pharmacology* 254, no. 1:8–17.

Physalis minima (Solanaceae) (Solanales)   **Cytotoxic**

Location: Central and South America

Appearance: annual plant that reaches about 50 cm high

Leaves: hairy

Flowers: brittle yellow to orange blossoms

Tradition: the fruit is said to be appetizing, diuretic, laxative, and tonic. The juice from the leaves mixed with mustard oil and water has been used as a remedy for ear pain.

Parts used: fruit, seeds, roots, leaves

Active ingredients: fusalin F

Precautions: the calyx is toxic and should not be eaten. Although no such reference has been made to the particular plant, it belongs to a genus where many of the species have poisonous leaves and strains, although full mature fruits are usually edible.

Documented target cancers:

- Chloroform extract from *Physalis minima* is reported to significantly reduce the growth of human T-47D breast cancer cells relative to other extracts.
- The active ingredient fusalin F induces apoptosis-based cytotoxicity in T-47D breast cancer cells.
- Anti-cancer properties and cytotoxic activities of the chloroform extract have been observed in NCI-H23 (human pulmonary adenocarcinoma).

References

Leong, O.K., T.S.T. Muhammad, and S.F. Sulaiman. 2011. Cytotoxic activities of Physalis minima L. chloroform extract on human lung adenocarcinoma NCI-H23 cell lines by induction of apoptosis. *Evidence-Based Complementary and Alternative Medicine* 2011:1–10.

Navdeep, S., B. Anisha, S.D. Harcharan, and S. Vivek. 2015. Perspectives and possibilities of Indian species of genus Physalis (L.) – A comprehensive review. *European Journal of Pharmaceutical and Medical Research* 2, no. 2:326–353.

Ooi, K.L., T.S.T. Muhammad, and S.F. Sulaiman. 2010. Growth arrest and induction of apoptotic and non-apoptotic programmed cell death by, *Physalis minima* L. chloroform extract in human ovarian carcinoma Caov-3 cells. *Journal of Ethnopharmacology* 128, no. 1:92–99.

Ooi, K.L., T.S.T. Muhammad, and S.F. Sulaiman. 2013. Physalin F from *Physalis minima* L. triggers apoptosis-based cytotoxic mechanism in T-47D cells through the activation caspase-3-and c-myc-dependent pathways. *Journal of Ethnopharmacology* 150, no. 1:382–388.

Ooi, K.L., T.S.T. Tengku Muhammad, C.H. Lim, and S.F. Sulaiman. 2010. Apoptotic effects of Physalis minima L. chloroform extract in human breast carcinoma T-47D cells mediated by c-myc-, p53-, and caspase-3-dependent pathways. *Integrative Cancer Therapies* 9, no. 1:73–83.

Phytolacca americana (Phytolaccaceae) (Caryophyllales)	**Cytotoxic**
	Anti-tumor

Location: Central America

  Appearance: perennial, downy, herbaceous plant

  Leaves: rough textured, alternating
  Flowers: five sepals and no petal
  Fruit: glowing purple color and layered beads

Tradition: the plant is traditionally used in the treatment of diseases associated with a weakened immune system. The fresh root is used as a poultice in bruises, rheumatic pains, etc., while washing from the roots is applied to edema and sprains. A tea made from fruit is used in the treatment of rheumatism, dysentery, etc. The leaves are laxatives, emesis, and expectorants.

  Parts used: fruit, roots, leaves

  Active ingredients: americanin A

  Documented target cancers:

- The neolignan, americanin A, isolated the from the seeds of *Phytolacca americana* has shown anticancer activity (antiproliferative and tumor suppressive properties) in human colon cancer cells. Americanin A inhibited proliferation of human colon cancer HCT116 cells both *in vitro* and *in vivo*.
- Nano-triterpenoids of *Phytolacca decandra* (syn. *P. americana*) have shown potent antitumor potential against A549 adenocarcinoma. The nanotriperpenoids were isolated from an ethanolic extract of the *Phytolacca decandra* plant and coated with biodegradable, non-toxic poly (lactide-co-glycolide) polymers for better bioavailability targeted delivery.
- Oleanolic acid isolated from the ethanolic extract of the *Phytolacca decandra* appears to induce apoptosis in A375 melanoma cells.

References

Das, J., S. Das, A. Paul, A. Samadder, and A.R. Khuda-Bukhsh. 2014. Strong anticancer potential of nano-triterpenoid from Phytolacca decandra against A549 adenocarcinoma via a Ca2+-dependent mitochondrial apoptotic pathway. *Journal of Acupuncture and Meridian Studies* 7, no. 3:140–150.

Ghosh, S., K. Bishayee, and A.R. Khuda-Bukhsh. 2014. Oleanolic acid isolated from ethanolic extract of Phytolacca decandra induces apoptosis in A375 skin melanoma cells: Drug-DNA interaction and signaling cascade. *Journal of Integrative Medicine* 12, no. 2:102–114.

Jung, C., J.Y. Hong, S.Y. Bae, S.S. Kang, H.J. Park, and S.K. Lee. 2015. Antitumor activity of Americanin A isolated from the seeds of *Phytolacca americana* by regulating the ATM/ATR signaling pathway and the Skp2–p27 axis in human colon cancer cells. *Journal of Natural Products* 78, no. 12:2983–2993.

Ravikiran, G., A.B. Raju, and Y. Venugopal. 2011. *Phytolacca americana*: A review. *International Journal Resource Pharmacologia Biomed Science* 2, no. 1:942–946.

Pimenta dioica (Myrtaceae) (Myrtales)   **Cytotoxic Anti-tumor**

Location: Caribbean, West Indies, South Mexico, Central America

Appearance: evergreen bush tree, reaching a height of 10–18 m. It has grayish-white bark that peels in thin sheets.

Leaves: large, glossy, leathery, aromatic, opposite and elliptical in shape, about 8 inches long and 2 inches wide. At the leaf axils, pyramidal cymes of small white flowers develop.

Flowers: male and female; the female flowers are fruiting and develop into clusters of pea-sized, brownish green, spicy berries with one or two seeds.

Tradition: the dried fruit is used as the main ingredient of gastronomy in many cuisines (e.g. Caribbean, Palestine), deodorant (combines aromas of carnation, black pepper, nutmeg, and cinnamon on leaves and fruits).

Parts used: the immature fruit (berries), especially the rind and shell. Other parts of the plant have medicinal uses as well.

Active ingredients: eugenol, phenolic acids, catechins, phenylpropanoids, diterpenoids, and lupeol

Documented target cancers:

- Human hepatocellular and breast carcinoma cells Hep-G2 and MCF-7: polyphenols isolated from the methanol extract of *P. dioica* leaves remarkably inhibit the cell growth of Hep-G2 and HCT-116 cells with less effect on MCF-7 cells.
- Human breast cancer (BrCa) cells *in vitro* and *in vivo*

References

Marzouk, M.S., F.A. Moharram, M.A. Mohamed, A.M. Gamal-Eldeen, and E.A. Aboutabl. 2007. Anticancer and antioxidant tannins from Pimenta dioica leaves. *Zeitschrift Fur Naturforschung. C, Journal of Biosciences* 62, no. 7–8:526–536.

Zhang, L., and B.L. Lokeshwar. 2012. Medicinal properties of the Jamaican pepper plant *Pimenta dioica* and allspice. *Current Drug Targets* 13, no. 14:1900–1906.

Zhang, L., N. Shamaladevi, G.K. Jayaprakasha, B.S. Patil, and B.L. Lokeshwar. 2015. Polyphenol-rich extract of *Pimenta dioica* berries (allspice) kills breast cancer cells by autophagy and delays growth of triple negative breast cancer in athymic mice. *Oncotarget* 6, no. 18:16379–16395.

Pinus massoniana (Pinaceae) (Pinales)   **Cytotoxic Anti-migratory**

Location: Taiwan, southern China, Hong Kong, and northern Vietnam

Appearance: evergreen tree reaching 25–45 m in height, with a broad, rounded crown of long branches. Its bark is thick, grayish-brown, scaly plated at the base of the trunk, orange-red, thin, and flaking higher on the trunk.

Leaves: needle-like, dark green, with two per fascicle, 12–20 cm long and 0.8–1 mm wide, the persistent fascicle sheath 1.5–2 cm long

Flowers: The cones are ovoid, 4–7 cm long, chestnut-brown, opening when mature in late winter to 4–6 cm broad.

Tradition:

- The leaves give a characteristic smoke flavor to tea. The needle leaves are used for producing a green dye.
- Cut or boiled leaves are used in rheumatism and intestinal parasites, pollen for cardiac diseases, turpentine is antiseptic and diuretic. Steam baths with the plant are used for rheumatism and respiratory system treatment.

Parts used: nodes, vine, stem, resin, root, pollen, leaves

    Active ingredients: diterpenoids in petroleum ether extract

    Documented target cancers:

- The less polar diterpenoids isolated from the petroleum ether extract of *Pinus massoniana* resin had strong cytotoxicity against A431 and A549 cancer cells (epithelial and lung cell cancer).
- *P. massoniana* bark extract has been shown to inhibit migration of the lung cancer A549 cells.
- *P. massoniana* bark extract significantly reduced the growth of ovarian cancer cells (ovarian cancer cell line A2780) and induced dose-dependent apoptosis.

References

Liu, J., J. Bai, G. Jiang, et al. 2015. Anti-tumor effect of *Pinus massoniana* bark proanthocyanidins on ovarian cancer through induction of cell apoptosis and inhibition of cell migration. *PLOS ONE* 10, no. 11:e0142157.

Mao, P., E. Zhang, Y. Chen, L. Liu, D. Rong, Q. Liu, and W. Li. 2017. *Pinus massoniana* bark extract inhibits migration of the lung cancer A549 cell line. *Oncology Letters* 13, no. 2:1019–1023.

Yang, N.Y., L. Liu, W.W. Tao, J.A. Duan, and L.J. Tian. 2010. Diterpenoids from *Pinus massoniana* resin and their cytotoxicity against A431 and A549 cells. *Phytochemistry* 71, no. 13:1528–1533.

Pinus sylvestris (Pinaceae) (Pinales)   **Cytotoxic**

Location: forests in Europe, Asia, northeastern America

    Appearance: evergreen coniferous tree growing up to 35 m in height

    Leaves: needle-shaped, glaucous blue-green, often darker green to dark yellow-green in winter (2.5–5 cm long and 1–2 mm broad), produced in fascicles of two with a persistent gray 5–10 mm basal sheath. On vigorous young trees the leaves can be twice as long and occasionally occur in fascicles of three or four on the tips of strong shoots.

Tradition: the resin is used as antiseptic, anti-rheumatic, diuretic, for respiratory infections, kidneys', bladder's, and rheumatisms' treatment. The leaves are used for lassitude, skin diseases, and somnolence. The seeds are used for bronchitis, asthma, and turbeculosis.

    Parts used: bark

    Active ingredients: turpentine, terpenoids, and carboxyl acids from needles, monotrupic hydrocarbons from wood, stump phenols, flavonoids (polydelflinidin, quercetin, taxifolin)

    Precautions: pine pollen can cause allergic reactions, even in people who test negatively to pine skin tests.

Documented target cancers:
- Breast cancer cells: the needle extract was found to suppress the viability of several human cancer cell lines showing some selectivity to estrogen receptor negative breast cancer cells, MDA-MB-231(half maximal inhibitory concentration [IC_{50}] 35 µg/ml) in comparison with estrogen receptor-positive breast cancer cells, MCF-7 (IC_{50} 86 µg/ml).

Reference

Hoai, N.T., H.V. Duc, do T. Thao, A. Orav, and A. Raal. 2015. Selectivity of *Pinus sylvestris* extract and essential oil to estrogen-insensitive breast cancer cells *Pinus sylvestris* against cancer cells. *Pharmacognosy Magazine* 11 Suppl 2:S290–S295.

Piper umbellatum (Piperaceae) (Piperales)	Anti-tumor

Location: tropical America, tropical woods in Japan, Africa, islands of Indian Ocean
Appearance: succulent, perennial plant (syn. *Pothomorphe umbellata*)

Stem: up to 1.5–2 (rarely 4) m
Leaves: alternate, simple, and entire, stipules absent, petiole 6.5–30 cm long
Flowers: minute, bisexual, sessile

Tradition: used in ceremonies (religious, purifying, etc.), as a bait for fish in Ghana, and as an ingredient in many cuisines (spice or raw as a vegetable). Used in traditional medicine as a diuretic and for kidney related diseases.

Parts used: stem-pith, (young) leaves, inflorescence, bark, and basal part of the stem

Active ingredients: terpenes, alkaloids, flavonoids, sterols, bicyclogermacrene, cadinene. The catechol 4-nerolidylcatechol (4-NC) is the major compound.

Documented target cancers: Ehrlich ascitic tumor-bearing animals

References

Iwamoto, L.H., D.B. Vendramini-Costa, P.A. Monteiro, et al. 2015. Anticancer and anti-inflammatory activities of a standardized dichloromethane extract from *Piper umbellatum* L. leaves. *Evidence-Based Complementary and Alternative Medicine* 2015:1–8.

Longato, G.B., L.Y. Rizzo, I.M.D. Sousa, et al. 2011. *In vitro* and *in vivo* anticancer activity of extracts, fractions, and eupomatenoid-5 obtained from *Piper regnellii* leaves. *Planta Medica* 77, no. 13:1482–1488.

Sacoman, J.L., K.M. Monteiro, A. Possenti, G.M. Figueira, M.A. Foglio, and J.E. Carvalho. 2008. Cytotoxicity and antitumoral activity of dichloromethane extract and its fractions from *Pothomorphe umbellata*. *Brazilian Journal of Medical and Biological Research = Revista Brasileira de Pesquisas Medicas e Biologicas* 41, no. 5:411–415.

Pistacia integerrima (Anacardiaceae) (Sapindales)	Cytotoxic

Location: India, America, China, Japan, England
Appearance: dioecious deciduous tree, reaching 25 meters high, fertilized by the wind

Leaves: 20–25 cm in length, with or without terminal leaflet; leaflets four to five pairs, lanceolate, coriaceous, and base oblique. The odor of this shrub is peculiar. Leaves are dark green in color which turn bright red in autumn.

Flowers: in lateral panicles, male compact, pubescent, female lax, and elongate. Plants wear the flowers and fruits in spring and have large clusters of tawny colored fruit in winter.

Tradition:

- Wood, firewood and charcoal, leaves as animal feed, leaf and peel pigments
- Edible fruits
- Uses as antimicrobial, antioxidant, and analgesic. Traditional medication for animal diseases
- Used against cough, asthma, diarrhea, fever, vomiting, and scorpion and snake bites. Regulates digestive and respiratory systems.

Parts used: galls, fruit

Active ingredients: alkaloids, steroids, terpenoids, flavonoids, anthraquinones, and saponin compounds

Documented target cancers:
- The crude extract inhibited breast cancer cells (MCF-7) cell viability in a dose-dependent manner, with the IC_{50} value calculated at 90.9 µg/ml.

References

Bibi, Y., S. Nisa, M. Zia, A. Waheed, S. Ahmed, and M.F. Chaudhary. 2012. The study of anticancer and anti-fungal activities of *Pistacia integerrima* extract *in vitro*. *Indian Journal of Pharmaceutical Sciences* 74, no. 4:375–379.

Bibi, Y., M. Zia, and A. Qayyum. 2015. Review – An overview of *Pistacia integerrima* a medicinal plant species: Ethnobotany, biological activities and phytochemistry. *Pakistan Journal of Pharmaceutical Sciences* 28, no. 3:1009–1013.

Plicosepalus curviflorus (Loranthaceae)   **Cytotoxic**

Location: Africa, Arabian Peninsula, and the Middle East

Appearance: aerial hemiparasitic plant

Tradition: the stems of *P. curviflorus* are used for the treatment of cancer in Yemen

Parts used: dried powdered leaves

Active ingredients: esters of flavonoic gallic acid (1,2), with compounds such as quercetin and isomers of pentahydroxyflavan-5-O-gallic acid ester

Documented target cancers:
- Human cell lines: MCF-7 cells (breast cancer cell line), HepG-2 (liver cancer cell line), HCT-116 (colon cancer cell line), Hep-2 (laryngeal cancer cell line), HeLa (cervical cancer cell line)

Reference

Fawzy, G.A., A.M. Al-Taweel, and S. Perveen. 2014. Anticancer activity of flavane gallates isolated from *Plicosepalus curviflorus*. *Pharmacognosy Magazine* 10 Suppl 3:S519–S523.

Plumeria sp. (Plumeria)   **Cytotoxic**

Location: warm tropical areas of the Pacific Islands, Caribbean, South America, and Mexico (Plumeria rubra: Indonesia, Thailand)

Figure 9.21 (See color insert.) *Plumeria.*

Appearance:

Stem: 10–12 m, widely spaced thick succulent branches
Leaves: round or pointed, long leather, fleshy leaves in clusters near the branch tips
Flowers: large, waxy, red, white, yellow, pink, and multiple pastels, fragrant
In bloom: early summer through the early fall months

Tradition: traditional medicinal plant of Thailand
    Parts used: bark
    Active ingredients:

- Petroleum-ether- and $CHCl_3$-soluble extracts: (1) iridoids: fulvoplumierin, allamcin and allamandin, (2) 2,5-dimethoxy-p-benzoquinone
- H_2O-soluble extract: (1) iridoids: plumericin, isoplumericin, (2) lignan: liriodendrin

Documented target cancers:

- Murine lymphocytic leukemia (P-388) and a number of human cancer cell-types (breast, colon, fibrosarcoma, lung, melanoma, KB)
- Dalton lymphoma ascites in mice

Further details:

- The iridoids: plumericin, isoplumericin except their cytotoxic activity, they also have antibacterial activity.
- Five additional iridoids, 15-demethylplumieride, plumieride, alpha-allamcidin], beta-allamcidin, and 13-O-trans-p-coumaroylplumieride, were obtained as inactive constituents. Compound 15-demethylplumieride was found to be a novel natural product, and its structure was determined by spectroscopic methods and by conversion to plumieride (Figure 9.21).

References

Borchert, R., and G. Rivera. 2001. Photoperiodic control of seasonal development and dormancy in tropical stem-succulent trees. *Tree Physiology* 21, no. 4:213–221.

França, O.O., R.T. Brown, and C.A. Santos. 2000. Uleine and demethoxyaspidospermine from the bark of *Plumeria lancifolia*. *Fitoterapia* 71, no. 2:208–210.

Guevara, A.P., E. Amor, and G. Russell. 1996. Antimutagens from *Plumeria acuminata* Ait. *Mutation Research* 12, no. 361(2–3):67–72.

Hamburger, M.O., G.A. Cordell, and N. Ruangrungsi. 1991. Traditional medicinal plants of Thailand. XVII. Biologically active constituents of *Plumeria rubra*. *Journal of Ethnopharmacology* 33, no. 3:289–292.

Kardono, L.B., S. Tsauri, K. Padmawinata, J.M. Pezzuto, and A.D. Kinghorn. 1990. Cytotoxic constituents of the bark of *Plumeria rubra* collected in Indonesia. *Journal of Natural Products* 53, no. 6:1447–1455.

Muir, C.K., and K.F. Hoe. 1982. Pharmacological action of leaves of *Plumeria acuminata*. *Planta Medica* 44, no. 1:61–63.

Radford, D.J., A.D. Gillies, J.A. Hinds, and P. Duffy. 1986. Naturally occurring cardiac glycosides. *Medical Journal of Australia* 144, no. 10:540–544.

Radha, R., S. Kavimani, and V. Ravichandran. 2008. Antitumour activity of methanolic extract of Plumeria alba L. leaves against Dalton lymphoma ascites in mice. *International Journal of Health Research* 1, no. 2:79–85.

Rekha, J.B., and B. Jayakar. 2011. Anti cancer activity of ethanolic extract of leaves of Plumeria rubra (Linn). *Current Pharmacologia Research* 1, no. 2:175.

Sharma, G., M.K. Chahar, S. Dobhal, et al. 2011. Phytochemical constituents, traditional uses, and pharmacological properties of the genus Plumeria. *Chemistry and Biodiversity* 8, no. 8:1357–1369.

Tan, G.T., J.M. Pezzuto, A.D. Kinghorn, and S.H. Hughes. 1991. Evaluation of natural products as inhibitors of human immunodeficiency virus type 1 (HIV-1) reverse transcriptase. *Journal of Natural Products* 54, no. 1:143–154.

Pogostemon benghalensis **(Lamiaceae) (Lamiales)**   **Anti-tumor**

Location: Bangladesh, India, China, Thailand, Sri Lanka
  Appearance: aromatic herb

Stem: strong, solid, angular
Leaves: opposite; petiole 2.5 cm long; blade ovate, 13 × 6 cm, base cuneate, margin double dentate, apex acuminate
Flowers: inflorescence is a verticillaster, arranged in a terminal false spike, about 7 cm long, at base branched into more than two lateral spikes.

Tradition:

- The leaves produce an oil with a characteristic aroma (similar to cedar) that contains a bitter resin and is used as a tonic. The leaf oil is used as antifungal, while the acetone leaf extract acts as insecticide, leech repellent, antidepressant, aphrodisiac, and as a dye component. The ashes of the stem are used as a grain fertilizer.
- All parts of the plant and especially the fresh leaves are used to treat wounds; the leaves are used in the treatment of kidney stones and hemorrhoids.

Parts used: whole plant
  Active ingredients: transcaryophyllene, germanacrine B, δ-caduene, β-octane, γ-oleem, and caryophyllene oxide
  Documented target cancers:
- Both hydroethanolic and aqueous extract of the plant have been shown to increase median survival time, reduce solid tumor volume, and normalize hematological parameters in Ehrlich ascites carcinoma (EAC) tumor-bearing mice.

Reference

Patel, M.S., B.V. Antala, E. Dowerah, R. Senthilkumar, and M. Lahkar. 2014. Antitumor activity of *Pogostemon benghalensis* Linn. on Ehrlich ascites carcinoma tumor bearing mice. *Journal of Cancer Research and Therapeutics* 10, no. 4:1071–1075.

Polanisia dodecandra L. (Polanisia) (Capparaceae)　　　　　　　　　　　**Cytotoxic**

Location: plants are found from Quebec and Maryland to southern Saskatchewan and Manitoba south to Arkansas and northern Mexico at elevations under 6,000 feet.

　Appearance:

Stem: 0.3–1 m, simple, strong dark odor
Leaves: 5 cm long and bear three leaflets about an inch long
Flowers: 20 flowers are clustered at the top of the plant. 1 cm long, white with purple basis
In bloom: May to October

Active ingredients: *Flavonols*: 5,3′-dihydroxy-3,6,7,8,4′-pentamethoxyflavone [1], 5,4′-dihydroxy-3,6,7,8,3′-pentamethoxyflavone [2]

　Documented target cancers:

- It is used in: central nervous system cancer (SF-268, SF-539, SNB-75, U-251)
- Non-small cell lung cancer (HOP-62, NCI-H266, NCI-H460, NCI-H522)
- Small cell lung cancer (DMS-114)
- Ovarian cancer (OVCAR-3, SK-OV-3)
- Colon cancer (HCT-116)
- Renal cancer (UO-31)
- Melanoma cell line (SK-MEL-5)
- Leukemia cell lines (HL-60 [TB], SR)
- Medulloblastoma (TE-671) tumor cells

Further details:

- 5,3′-dihydroxy-3,6,7,8,4′-pentamethoxyflavone inhibited tubulin polymerization ($IC_{50} = 0.83$ +/- 0.2 μM) and the binding of radiolabeled colchicine to tubulin with 59% inhibition when present in equimolar concentrations with colchicine. It is the first example of a flavonol that exhibits potent inhibition of tubulin polymerization and, therefore, warrants further investigation as an antimitotic agent.

References

Shi, Q., K. Chen, L. Li, et al. 1995. Antitumor agents, 154. Cytotoxic and antimitotic flavonols from *Polanisia dodecandra*. *Journal of Natural Products* 58, no. 4:475–482.

Wang, H.K., Y. Xia, Z.Y. Yang, S.L. Natschke, and K.H. Lee. 1998. Recent advances in the discovery and development of flavonoids and their analogues as antitumor and anti-HIV agents. *Advances in Experimental Medicine and Biology* 439:191–225.

Polyalthia barnesii (Polyalthia) (Annonaceae)　　　　　　　　　　　**Cytotoxic**

Parts used: stem bark
　Active ingredients:

- Clerodane diterpenes (cytotoxic): 16 alpha-hydroxycleroda-3,13(14)Z-dien-15,16-olide
- 3 beta, 16 alpha-dihydroxycleroda-4(18),13(14)Z-dien-15,16-olide and 4 beta, 16 alpha-dihydroxy clerod-13(14)Z-en-15,16-olide

Documented target cancers: the above compounds are found to exhibit broad cytotoxicity against a panel of human cancer cell lines.

　Further details:

• The (three) cytotoxic clerodane diterpenes were purified from an ethyl acetate-soluble extract of the stem bark of *Polyalthia barnesii*, namely, 16 alpha-hydroxycleroda-3,13(14)Z-dien-15,16-olide.

Reference

Ma, X., I.S. Lee, H.B. Chai, et al. 1994. Cytotoxic clerodane diterpenes from *Polyalthia barnesii*. *Phytochemistry* 37, no. 6:1659–1662.

Polyalthia evecta (Annonaceae) (Magnoliales)   **Cytotoxic**

Location: Indo-China, northeastern Thailand, Vietnam
  Appearance: small shrub

  Stem: hairy
  Fruit: aggregate, an umbel-like arrangement
  Flowers: on long pedicel with small bracteole

Tradition: used as folk medicine by China Li ethnic minority for prevention of fever, hypertension, and inhibition of cancer.
  Parts used: roots, leaves
  Active ingredients: evectic acid and furan (roots), stannic acid in leaf extracts
  Documented target cancers:

• SPC-A-1 (human lung cancer cell Line), BEL-7402 (human hepatocellular carcinoma cell line), SGC-7901 (human gastric cancer cell line), and K562 (human myelogenous leukemia cell line)
• NCI-H460 cells: large cell lung cancer
• Selective cytotoxic against HepG2 cells (human liver cancer cell line) compared to Vero cells (normal)

References

Machana, S., N. Weerapreeyakul, S. Barusrux, K. Thumanu, and W. Tanthanuch. 2012. Synergistic anticancer effect of the extracts from Polyalthia evecta caused apoptosis in human hepatoma (HepG2) cells. *Asian Pacific Journal of Tropical Biomedicine* 2, no. 8:589–596.

Prayong, P., S. Barusrux, and N. Weerapreeyakul. 2008. Cytotoxic activity screening of some indigenous Thai plants. *Fitoterapia* 79, no. 7–8:598–601.

Sashidhara, K.V., S.P. Singh, R. Kant, P.R. Maulik, J. Sarkar, S. Kanojiya, and K. Ravi Kumar. 2010. Cytotoxic cycloartane triterpene and rare isomeric bisclerodane diterpenes from the leaves of *Polyalthia longifolia* var. pendula. *Bioorganic and Medicinal Chemistry Letters* 20, no. 19:5767–5771.

Suedee, A., I.O. Mondranondra, A. Kijjoa, et al. 2007. Constituents of *Polyalthia jucunda* and their cytotoxic effect on human cancer cell lines. *Pharmaceutical Biology* 45, no. 7:575–579.

Yuan, Y., G.J. Huang, T.S. Wang, and G.Y. Chen. 2011. *In vitro* screening of five Hainan plants of Polyalthia (Annonaceae) against human cancer cell lines with MTT assay. *Journal of Medicinal Plants Research* 5, no. 5:837–841.

Polyalthia longifolia (Annonaceae) (Magnoliales)   **Anti-leukemic**
                                                                                          Cytotoxic

---

Location: native to India, subtropical and tropical climates

Appearance: evergreen tree up to 20 m high

Stem: straight stem of columnar form and smooth, thick, gray-brown bark
Leaves: smooth green, oval-oblong or oval-lanceolate with length 11–31 cm and width 2–8 cm
Flowers: greenish yellow, crowded in axillary bundles or shades.
In bloom: February–March

Tradition: it is used as antipyretic agent in indigenous systems of medicine. The bark is bitter, acrid, cooling, febrifuge, and anthelminthic. It is useful in skin disease, diabetes, hypertension, and helminthiasis.

Part of the Ayurvedic treatment for uterine disorders.

Parts used: Almost all of the plant

Active ingredients: liriodenine, new halimane diterpene: 3β,5β,16α-trihydroxyhalima-13 (14)-en-15,16-olide, and a new oxoprotoberberine alkaloid: (–)-8-oxopolyalthiaine. Polyalthurea, (+)-rumphiin, 3,4,5-trimethoxy benzoic acid, (–)-seselinone, cannabisin D, allantoin, oxostephanine, and a mixture of beta-sitosterol and stigmasterol

Precautions: methanol extract of *Polyalthia longifolia* leaf, up to the dose level 3,240 mg/kg body weight, does not produce any toxic effects or deaths in Wistar albino rats.

Indicative dosage and application:

- Decoction of bark (50–100 mL) for treating fever

Documented target cancers:

- Leukemic HL-60 cells (Chloroform fraction), A549 (lung cancer) and MCF-7 (breast cancer) cancer cells , cervical cancer, colorectal cancer cells SW-620 colon (IC_{50} value 6.1 µg/mL)

Related species: *Polyalthia rumphii*: isolated compounds from the stem showed significant anticancer activity against SPC-A-1 and BEL-7402 cell lines with IC_{50} values of 1.47 and 1.73 mg/mL. *Polyalthia evecta*: the extracts from *Polyalthia evecta* caused apoptosis in human hepatoma (HepG2) cells.

References

Chanda, S., R. Dave, M. Kaneria, and V. Shukla. 2012. Acute oral toxicity of *Polyalthia longifolia* var. pendula leaf extract in wistar albino rats. *Pharmaceutical Biology* 50, no. 11:1408–1415.

Katkar, K.V., A.C. Suthar, and V.S. Chauhan. 2010. The chemistry, pharmacologic, and therapeutic applications of *Polyalthia longifolia*. *Pharmacognosy Reviews* 4, no. 7:62–68.

Machana, S., N. Weerapreeyakul, S. Barusrux, K. Thumanu, and W. Tanthanuch. 2012. Synergistic anticancer effect of the extracts from *Polyalthia evecta* caused apoptosis in human hepatoma (HepG2) cells. *Asian Pacific Journal of Tropical Biomedicine* 2, no. 8:589–596.

Wang, T.S., Y.P. Luo, J. Wang, M.X. He, M.G. Zhong, Y. Li, and X.P. Song. 2013. (+)-Rumphiin and polyalthurea, new compounds from the stems of *Polyalthia rumphii*. *Natural Product Communications* 8, no. 10:1427–1429.

Polyalthia rumphii (Annonaceae) (Magnoliales)   Cytotoxic

Location: tropical climate

Appearance: small to medium sized tree, up to 15 m high and up to 60 cm in diameter

Stem: oblong
Leaves: elongated lanceolates, length of 10–17cm and width of 3–7 cm, hairless, shiny dark green with secondary prominent neurons seven to ten on each side

Flowers: flowers 4–7 cm in diameter, greenish-yellowish, axillary
In bloom: May–October

Tradition: used in traditional Chinese medicine to fight fever and hypertension
   Parts used: stem
   Active ingredients: bis (2-ethylheptyl) phthalate, dibutyl phthalate, diisobutyl phthalate, epiyangambin, diayangambin, oxostephanine, N-trans-feruloyltyramine
   Documented target cancers:

- Anticancer action against the SPC-A-1 (lung cancer) and BEL-7402 (hepatocellular carcinoma)

Related species: *Polyalthia evecta*: synergistic anticancer effect of the extracts from *Polyalthia evecta* caused apoptosis in human hepatoma (HepG2) cell. *Polyalthia longifolia*: the bark of *P. longifolia* has significant anticancer activity.

Reference

Wang, T., Y. Yuan, J. Wang, C. Han, and G. Chen. 2012. Anticancer activities of constituents from the stem of *Polyalthia rumphii*. *Pakistan Journal of Pharmaceutical Sciences* 25, no. 2:353–356.

Polygala tenuifolia (Polygalaceae) (Fabales)   **Anti-tumor**

Location: East Asia
   Appearance: perennial herb growing to 0.2 m by 0.2 m

Stem: 10–50 cm high
Leaves: linearly lanceolate with length 10–30 mm and width 0.5–1 mm
Flowers: hermaphrodite, purple.
In bloom: May–September

Tradition:

- It is used primarily as an expectorant. It is one of the 50 fundamental herbs used in traditional Chinese medicine, where it is called yuan zhi.
- Used as expectorant and tonic for the treatment of bronchial asthma, chronic bronchitis, and pertussis. The root is antibacterial, cardiotonic, expectorant, hemolytic, and acts as hypotensive, soothing, and tonic. It also enhances the memory.

Parts used: leaves and root
   Active ingredients: triterpenic saponins, xanthanoids, phenols, polysaccharides
   Precautions: the estimated maximum safe human dose is estimated to be 360 mg for a 60 kg human.
   Indicative dosage and application:

- The only human studies on *Polygala tenuifolia* have used 100 mg of an ethanolic extract called BT-11, three times a day, for a total daily dose of 300 mg.

Documented target cancers:

- A purified polysaccharide isolated from the roots of *Polygala tenuifolia* inhibited cellular proliferation of OVCAR-3 ovarian cancer cells and suppressed SKOV3 ovarian xenograft tumor growth in BALB/c mice.

- Two acidic polysaccharide fractions from the roots of *Polygala tenuifolia* significantly inhibited the growth of A549 cells (lung cancer) *in vitro* and exhibited significantly higher antitumor activity against solid tumor A549.

Related species: *Polygala sibirica*: haemolytic, sedative, analgesic, diuretic, expectorant, and nervine. *Polygala japonica*: antiphlogistic, antitussive, carminative, depurative, expectorant, and tonic. *Polygala vulgaris*: contains triterpenoid saponins which promote the clearing of phlegm from the bronchial tubes; valuable herb for the treatment of respiratory problems such as chronic bronchitis, bronchial asthma, and convulsive coughs.

Related compounds: research suggests that *Polygala tenuifolia* is similar in mechanism and effectiveness to ketamine, which interacts with N-methyl-D-aspartic acid and its receptors. Rodent studies assessing *Polygala tenuifolia*'s anti-amnesiac and adaptogen-like properties suggest that the herb can increase the level of compounds responsible for brain growth, like nerve growth factor (NGF) and brain-derived neurotrophic factor (BDNF).

References

Ikeya, Y., S. Takeda, M. Tunakawa, H. Karakida, K. Toda, T. Yamaguchi, and M. Aburada. 2004. Cognitive improving and cerebral protective effects of acylated oligosaccharides in *Polygala tenuifolia*. *Biological and Pharmaceutical* Bulletin 27, no. 7:1081–1085.

Xin, T., F. Zhang, Q. Jiang, et al. 2012. Purification and antitumor activity of two acidic polysaccharides from the roots of *Polygala tenuifolia*. *Carbohydrate Polymers* 90, no. 4:1671–1676.

Yao, H., P. Cui, D. Xu, Y. Liu, Q. Tian, and F. Zhang. 2018. A water-soluble polysaccharide from the roots of *Polygala tenuifolia* suppresses ovarian tumor growth and angiogenesis *in vivo*. *International Journal of Biological Macromolecules* 107, no. A:713–718.

Zhang, F., X. Song, L. Li, et al. 2015. *Polygala tenuifolia* polysaccharide PTP induced apoptosis in ovarian cancer cells via a mitochondrial pathway. *Tumour Biology* 36, no. 4:2913–2919.

Polygonum bistorta/Persicaria bistorta (Polygonaceae) (Caryophyllales)	Cytotoxic Anti-tumor

Location: Europe and north and west Asia
Appearance: perenial herbaceous plant growing to 0.5 m at a fast rate

Stem: straight, plain, smooth, linear, reaches 1 m
Leaves: large, lanceolate with corrugated and long stems
Flowers: pink, rarely white in sparse, cylindrical ears, hermaphrodite, and pollinated by insects
In bloom: June–September.

Tradition:

- The roots and leaves are used as a remedy for wounds. The plant has been cultivated both for its medicinal uses as well as a vegetable.
- Use as astringent, antidiarrhoeal, anticatarrhal, antihaemorrhagic, demulcent, anti-inflammatory, styptic, tonic, alterative, diuretic action.

Parts used: all parts of the plant, especially the root
Active ingredients: tannins
Precautions: some members of this genus are reported to cause photosensitivity in susceptible people. Many species also contain oxalic acid. People suffering from kidney stone should be skeptical about enriching their diet with plants of this genus.

Indicative dosage and application: a gargle, to treat mouth and throat infections, can be made by using one or two grams of the dried rhizomes to one cup of warm water. For commercial tincture, the usual recommended dosage is 1 to 3 ml three times daily.

Documented target cancers:

- Chloroform and hexane of the plant have shown moderate to very good activity against P388 (murine lymphocytic leukemia), HL60 (human leukemia), and L2 (Lewis lung carcinoma) cell lines
- *P. bistorta* aqueous extract oral administration delayed tumor (hepatocellular cancer cell line HCCLM3) growth in a xenograft model.

References

Liu, Y.H., Y.P. Weng, H.Y. Lin, et al. 2017. Aqueous extract of *Polygonum bistorta* modulates proteostasis by ROS-induced ER stress in human hepatoma cells. *Scientific Reports* 7:41437.
Manoharan, K.P., D. Yang, A. Hsu, and B.T. Huat. 2007. Evaluation of *Polygonum bistorta* for anticancer potential using selected cancer cell lines. *Medicinal Chemistry* 3, no. 2:121–126.

Polygonum tinctorium/Persicaria tinctorial (Polygonaceae)	(Caryophyllales) Anti-tumor, Anti-leukemic

Location: Eastern Europe and Asia
  Appearance: frost tender annual, growing to 0.8 m

Stem: erect, 50–80 cm tall, usually branched
Leaves: oval or elliptical 3–8 × 2–4 cm, green or dark blue-green when dried
Flowers: pink, hermaphrodite (have both male and female organs), pollinated by insects
In bloom: July–August

Tradition: the indigo pigment in the leaves is used in the treatment of freckles, pimples, erysipelas, mumps, thrush, epidemic parotitis, infantile convulsions, and high febrile conditions of children. The stems and the leaves are used as anti-inflammatory, antiphlogistic, antipyretic, and depurative.

  Parts used: leaves
  Active ingredients: tryptanthrin
  Documented target cancers:

- Tryptanthrin, a bioactive ingredient of *Polygonum tinctorium* Lour., has potent cytocidal effects on various human leukemia cells *in vitro*.
- Tryptanthrin has been shown to induce remarkable necrotic and apoptotic changes in malignant tumor cells (murine glioblastoma, colon cancer, and malignant melanoma).

Further details:

- A blue dye is obtained from the leaves of this plant. The leaves produce about 4–5% indigo by hydrolysis and acidification.

Related species: *Polygonum hydropiper*: anti-inflammatory, astringent, carminative, diaphoretic, diuretic, emmenagogue, stimulant, stomachic, styptic. Used in the treatment of a wide range of ailments including diarrhoea, dyspepsia, itching skin, excessive menstrual bleeding, and haemorrhoids.

References

Kimoto, T. 1999. Cytotoxic effects of substances in indigo plant (Polygonum tinctorium Lour.) on malignant tumor cells. *Nature Medicine* 53:72–79.

Kimoto, T., K. Hino, S. Koya-Miyata, et al. 2001. Cell differentiation and apoptosis of monocytic and promyelocytic leukemia cells (U-937 and HL-60) by tryptanthrin, an active ingredient of Polygonum tinctorium Lour. *Pathology International* 51, no. 5:315–325.

Portulaca oleracea (Portulacaceae) (Caryophyllales)	Anti-proliferative
	Anti-tumor

Location: North Africa and Southern Europe, Middle East, and the Indian Subcontinent, Malesia, and Australia.

Appearance: annual, growing to 0.3 m (1 ft) by 0.3 m (1 ft) at a fast rate

Stem: sometimes flushed red or purple, not articulated, prostrate or decumbent, diffuse, much branched
Leaves: oblong oval 0.6–4cm long, dark green
Flowers: yellow, two to six in acute or axillary clusters
In bloom: June–September

Tradition: known as Ma Chi Xian in traditional Chinese medicine. Its leaves are used for insect or snake bites, boils, sores, pain from bee stings, bacillary dysentery, diarrhea, hemorrhoids, postpartum bleeding, and intestinal bleeding. Used for the treatment of thrush and dandruff and to treat stomach pains and headaches.

Parts used: leaves and seed

Active ingredients: portulacerebroside A (PCA) (ceresherosides)

Particular value: rich source of omega-3 fatty acids, believed to be important for preventing heart attacks and strengthening the immune system

Documented target cancers:

- Exposure to seed oil of *Portulaca oleracea* is reported to result in significant cytotoxicity and inhibition of growth of the human liver cancer (HepG2) and human lung cancer (A-549) cell lines.
- *P. oleracea* extract inhibits the growth of colon cancer stem cells (HT-29) in a dose-dependent manner.
- Seed oil of *Portulaca oleracea* significantly inhibited tumor cell growth in cervical cancer HeLa cells, esophageal cancer Eca-109 cells, and breast cancer MCF-7 cells.

References

Al-Sheddi, E.S., N.N. Farshori, M.M. Al-Oqail, J. Musarrat, A.A. Al-Khedhairy, and M.A. Siddiqui. 2015. *Portulaca oleracea* seed oil exerts cytotoxic effects on human liver cancer (HepG2) and human lung cancer (A-549) cell lines. *Asian Pacific Journal of Cancer Prevention* 16, no. 8:3383–3387.

Guo, G., L. Yue, S. Fan, S. Jing, and L.J. Yan. 2016. Antioxidant and antiproliferative activities of purslane seed oil. *Journal of Hypertension* 5, no. 2:218.

Jin, H., L. Chen, S. Wang, and D. Chao. 2017. *Portulaca oleracea* extract can inhibit nodule formation of colon cancer stem cells by regulating gene expression of the Notch signal transduction pathway. *Tumour Biology* 39, no. 7:1010428317708699.

Zhou, Y.X., H.L. Xin, K. Rahman, S.J. Wang, C. Peng, and H. Zhang. 2015. *Portulaca oleracea* L.: A review of phytochemistry and pharmacological effects. *BioMed Research International* 2015:925631.

Premna herbacea (Laminaceae) (Laminales)   **Cytotoxic**

Location: Himalayas, Assam, West Bengal, Bihar, Orissa, and the Deccan peninsula
  Appearance: herbaceous plant of low growth and a creeping woody rhizome

  Leaves: simple, anatomically elongated above 13 cm, 6.5 cm wide, with full and irregular toothed
    margins

Tradition: the root bark is used for toothache, for fever, cough, rheumatism, also used for bronchitis,
hypertension, tumors, inflammation.
  Parts used: roots and fruit
  Active ingredients: diterpenoid quinonemethides, isobharangin, bharangin, diterpenoid
quinonemethide
  Documented target cancers:

- Bharangin, isolated from the hexane extract of the root nodules is reported to exhibit cytotoxic
  properties against the P-338 cell line.
- Various compounds of the plant are reported cytotoxic against human cancer cell lines, including
  breast (MCF-7), lung (A549), uterine cervix (HELA), liver (HEPG2), and prostate (PC-3).

References

Awasthee, N., V. Rai, S.S. Verma, K.S. Francis, M.S. Nair, and S.C. Gupta. 2018. Anti-cancer activities of
  Bharangin against breast cancer: Evidence for the role of NF-κB and lncRNAs. *Biochimica et Biophysica
  Acta (BBA) – General Subjects* 1862, no. 12:2738–2749.
Dhamija, I., N. Kumar, S.N. Manjula, V. Parihar, M.M. Setty, and K.S.R. Pai. 2013. Preliminary evaluation of *in
  vitro* cytotoxicity and *in vivo* antitumor activity of *Premna herbacea* Roxb. in Ehrlich ascites carcinoma
  model and Dalton's lymphoma ascites model. *Experimental and Toxicologic Pathology* 65, no. 3:235–242.
Gupta, S.C., R. Kannappan, J. Kim, et al. 2011. Bharangin, a diterpenoid quinonemethide, abolishes constitu-
  tive and inducible nuclear factor-κB (NF-κB) activation by modifying p65 on cysteine 38 residue and
  reducing inhibitor of nuclear factor-κB α kinase activation, leading to suppression of NF-κB-regulated
  gene expression and sensitization of tumor cells to chemotherapeutic agents. *Molecular Pharmacology*
  80, no. 5:769–781.

Prunella vulgaris (Lamiaceae) (Laminales)   **Anti-metastatic Pro-apoptotic**

Location: Europe, Asia, North America, mostly found in Ireland
  Appearance: herbaceous plant

  Leaves: lanceolate, toothed spear, reddish on the edge
  Flowers: two lips, tubular in shape, upper lip is purple and lower lip is often white

Tradition: used commonly as dietary supplements in world. Traditional Chinese medicine for use in
treating sore throat, swelling of the thyroid gland, jaundice, fever, infectious hepatitis, dermatosis,
skin allergies, and for accelerating wound healing.
  Active ingredients: rutin, rosmarinic acid, quercetin, caffeic acid
  Documented target cancers:

- An aqueous extract isolated from *Prunella vulgaris* is reported to significantly inhibit lung metastasis and tumor cell growth in mice.

- A *P. vulgaris* extract of 60% ethanol was found to have the ability to regulate cell cycle and induce apoptosis in lung cancer SPC-A-1 cells.
- *Prunella vulgaris* was found to promote apoptosis in human well-differentiated thyroid carcinoma cells.

References

Choi, J.H., E.H. Han, Y.P. Hwang, et al. 2010. Suppression of PMA-induced tumor cell invasion and metastasis by aqueous extract isolated from *Prunella vulgaris* via the inhibition of NF-κB-dependent MMP-9 expression. *Food and Chemical Toxicology* 48, no. 2:564–571.

Feng, L., X. Jia, M. Zhu, Y. Chen, and F. Shi. 2010. Chemoprevention by *Prunella vulgaris* L. extract of non-small cell lung cancer via promoting apoptosis and regulating the cell cycle. *Asian Pacific Journal of Cancer Prevention* 11, no. 5:1355–1358.

Kim, S.H., C.Y. Huang, C.Y. Tsai, S.Y. Lu, C.C. Chiu, and K. Fang. 2012. The aqueous extract of *Prunella vulgaris* suppresses cell invasion and migration in human liver cancer cells by attenuating matrix metalloproteinases. *American Journal of Chinese Medicine* 40, no. 3:643–656.

Pseudolarix kaempferi (Pseudoradix) (Pinaceae)	Cytotoxic

Location: East China
  Appearance: evergreen tree with height reaching up to 40 m

  Leaves: small like pins

Tradition:

- It is mainly used in Chinese herbal medicine where it is considered one of the fundamental herbs.
- The stem and bark are used in the treatment of ringworm; the bark has fungicidal activity.

Parts used: trunk and seeds
  Active ingredients:

- Triterpene lactones pseudolarolides A, B, C, and D and
- Diterpene acids pseudolaric acid-A and -B

Documented target cancers: against:

- Human cancer cell lines: KB (nasopharyngeal), A-549 (lung), and HCT-8 (colon) (pseudolarolide B, pseudolaric acid-A and -B)
- Murine leukemia cell line (P-388) (pseudolarolide B, pseudolaric acid-A and -B)
- Potent cytotoxicity against three human cancer cell lines: KB, A-549, and HCT-8

Further details:
- The seeds contain the triterpene lactones pseudolarolides A, B, C, and D and the diterpene acids pseudolaric acid-A and -B.

References

Chen, G.F., Z.L. Li, D.J. Pan, et al. 1993. The isolation and structural elucidation of four novel triterpene lactones, pseudolarolides A, B, C, and D, from *Pseudolarix kaempferi*. *Journal of Natural Products* 56, no. 7:1114–1122.

Liu, P., H. Guo, W. Wang, et al. 2007. Cytotoxic diterpenoids from the bark of *Pseudolarix kaempferi* and their structure–activity relationships. *Journal of Natural Products* 70, no. 4:533–537.

Pan, D.J., Z.L. Li, C.Q. Hu, K. Chen, J.J. Chang, and K.H. Lee. 1990. The cytotoxic principles of *Pseudolarix kaempferi*: Pseudolaric acid-A and-B and related derivatives1. *Planta Medica* 56, no. 4:383–385.

Wu, X.D., J. He, Y. Shen, et al. 2012. Pseudoferic acids A–C, three novel triterpenoids from the root bark of *Pseudolarix kaempferi*. *Tetrahedron Letters* 53, no. 7:800–803.

Psidium guajava (Myrtaceae) (Myrtales)   **Cytotoxic**

Location: Caribbean, Central America, South America
   Appearance: small tree up to 33 feet high, with stretched branches with a smooth, thin, copper-colored bark

   Leaves: fragrant when crushed, oval or oblong-elliptical, skinny, with obvious parallel veins
   Flowers: slightly fragrant, white, individually or in small groups, in the armpits of the leaves

Tradition: infusion of leaves has been prescribed in India for brain diseases, nephritis, and cachexia, for the treatment of irritated bowel, treatment of open wounds, and rashes; the fruit is also suitable for stomach disorders, fighting parasites, is anti-diarrhea, anti-inflammatory, and inhibits the growth of bacteria.
   Parts used: fruit
   Active ingredients: ascorbic acid, apigenin, lycopene
   Precautions: it has a cardiac suppressive effect and is contraindicated in certain heart conditions.
   Documented target cancers:

* Human colon cancer cell lines: SW480, HT-29, and Caco-2
* Human oral epidermal carcinoma
* Pancreatic cancer: AsPC-1 (human Caucasian adenocarcinoma), CD18 (human pancreatic cancer), MIA PaCa2 (human Caucasian pancreatic carcinoma), and S2-013
* PC-3 prostate cancer cells

References

Correa, M.G., J.S. Couto, and A.J. Teodoro. 2016. Anticancer properties of *Psidium guajava* – A mini-review. *Asian Pacific Journal of Cancer Prevention* 17, no. 9:4199–4204.

Díaz-de-Cerio, E., V. Verardo, A.M. Gómez-Caravaca, A. Fernández-Gutiérrez, and A. Segura-Carretero. 2017. Health effects of *Psidium guajava* L. leaves: An overview of the last decade. *International Journal of Molecular Sciences* 18, no. 4:897.

Ryu, N.H., K.R., Park, S.M. Kim, et al. 2012. A hexane fraction of guava leaves (*Psidium guajava* L.) induces anticancer activity by suppressing AKT/mammalian target of rapamycin/ribosomal p70 S6 kinase in human prostate cancer cells. *Journal of Medicinal Food* 15, no. 3:231–241.

Psoralea corylifolia (Fabaceae) (Fabales)   **Cytotoxic Anti-tumor**

Location: Guizhou, Sichuan, Yunnan, China
   Appearance: shrub, 4 m high

   Leaves: complex, with several pairs of leaflets and a leaflet terminal
   Flowers: pea-shaped, blue, purple, and white

Tradition: used externally for the treatment of various skin conditions including leprosy, whites, and hair loss, used to treat Vitiligo and psoriasis

Parts used: fruit, root

Active ingredients: psoralidin, psoralen, isopsoralen

Documented target cancers:

- Psoralidin has been shown to induce cytotoxicity against gastric cancer cells (SNU-1, SNU-16), colon (HT-29), and breast (MCF-7). Apoptosis in androgen-dependent (LNCaP, C4-2B) and androgen-independent (DU-145, PC-3) prostate cancer cells and inhibited the growth of PC-3 tumor xenografts in nude mice.

References

Bronikowska, J., E. Szliszka, D. Jaworska, Z.P. Czuba, and W. Krol. 2012. The coumarin psoralidin enhances anticancer effect of tumor necrosis factor-related apoptosis-inducing ligand (TRAIL). *Molecules* 17, no. 6:6449–6464.

Panno, M.L., and F. Giordano. 2014. Effects of psoralens as anti-tumoral agents in breast cancer cells. *World Journal of Clinical Oncology* 5, no. 3:348–358.

Psychotria sp. (Psychotria) (Psychotrieae)   **Cytotoxic**

Location: Pacific Islands

Appearance:

Stem: slender, grows partly underground
Root: fibrous rootlets
In bloom: January–February

Parts used: aerial parts and stem bark

Active ingredients: alkaloids

Documented target cancers:

- All members of the series exhibited readily detected cytotoxic activity against proliferating and non-proliferating Vero (African green monkey kidney) cells in culture.
- Hodgkinsine A exhibited substantial antiviral activity against a DNA virus, herpes simplex type 1, and an RNA virus, vesicular stomatitis virus.
- Three cyclotides (psyle A, C, and E) from *Psychotria leptothyrsa* were reported to exhibit potent cytotoxicity against the breast cancer line, MCF-7, and its drug-resistant subline MCF-7/ADR.

Further details:

- *Calycodendron milnei*, a species endemic to the Vate Islands (New Hebrides), synthesizes a series of Nb-methyltryptamine-derived alkaloids made by linking together two to eight pyrrolidinoindoline units. Nine alkaloids of this class have been isolated from the aerial parts and stem bark of *Calycodendron milnei* and examined for potential application as anti-cancer and anti-infective agents. All members of the series showed readily detected anti-bacterial, anti-fungal, and anti-candidal activities using both tube dilution and disc diffusion assay methods. The most potent antimicrobial alkaloids were hodgkinsine A and quadrigemine C, which exhibited minimum inhibitory concentration (MIC) values as low as 5 micrograms/ml.

References

Fragoso, V., N.C. do Nascimento, D.J. Moura, A.C. e Silva, M.F. Richter, J. Saffi, and A.G. Fett-Neto. 2008. Antioxidant and antimutagenic properties of the monoterpene indole alkaloid psychollatine and the crude foliar extract of *Psychotria umbellata* Vell. *Toxicology in vitro* 22, no. 3:559–566.

Gerlach, S.L., R. Rathinakumar, G. Chakravarty, U. Göransson, W.C. Wimley, S.P. Darwin, and D. Mondal. 2010. Anticancer and chemosensitizing abilities of cycloviolacin O2 from Viola odorata and psyle cyclotides from *Psychotria leptothyrsa*. *Biopolymers* 94, no. 5:617–625.
Saad, H.E., S.H. El-Sharkawy, and W.T. Shier. 1995. Biological activities of pyrrolidinoindoline alkaloids from *Calycodendron milnei*. *Planta Medica* 61, no. 4:313–316.

Pterodon pubescens (Fabaceae) (Fabales)   **Cytotoxic**

Location: Central and South America

Appearance: deciduous tree with ellipsoid crown, cylindrical trunk that can reach 30–60 cm in diameter

Tradition: for the sore throat and respiratory disorders (bronchitis and tonsillitis), analgesic action, cleansing action; tonic also has hypoglycemic activities.

Parts used: fruit and other parts

Active ingredients: 6α-acetoxi 7β-hydroxy-vouacapan, 6α,7β-diacetoxyvouacapan, 7β-diacetoxyvouacapan

Documented target cancers:

UACC- (ovarian), PCV3 (prostate), HT-29 (colon), 786-0 (renal) K562 (leukemia), MCF-7 (breast), and NOI-ADR/RES (ovaries with multi-drug resistance phenotype)

References

Hoscheid, J., and M.L.C. Cardoso. 2015. Sucupira as a potential plant for arthritis treatment and other diseases. *Arthritis* 2015:379459.
Spindola, H.M., J.Ed Carvalho, A.L.T.G. Ruiz, et al. 2009. Furanoditerpenes from *Pterodon pubescens* Benth with selective *in vitro* anticancer activity for prostate cell line. *Journal of the Brazilian Chemical Society* 20, no. 3:569–575.

Pulsatilla koreana (Ranunculaceae) (Ranunculales)   **Anti-tumor Cytotoxic**

Location: Korea, China, borders between China and North Korea

Appearance: small tree up to 33 feet high, with stretched branches with a smooth, thin, copper-colored bark

Leaves: feathered in a basic rosette, with five white-haired leaflets
Flowers: bell-shaped, red in purple, with six petals and a silky surface

Tradition: for amoebic dysentery, malaria, epistaxis, leucorrhea, internal hemorrhoids, but also as a contraceptive. The root is anti-inflammatory and antiparasitic.

Parts used: roots and other parts

Active ingredients: lupane, oleanane, hederagenin

Documented target cancers:

- *P. koreana* extract is reported to suppress the growth of anaplastic thyroid cancer cells in a dose-dependent manner and inhibit tumor growth and weight in a mouse xenograft model.
- Several saponins isolated from the root of *Pulsatilla koreana* have shown potent cytotoxic activity against A-549 (lung cancer), SK-OV-3 (ovarian cancer), SK-MEL-2 (skin cancer), and HCT-15 (colon cancer) cell lines.
- The petroleum ether fraction of the plant is reported to exhibit inhibitory effects against various cancer cell lines.

References

Bang, S.C., J.H. Lee, G.Y. Song, D.H. Kim, M.Y. Yoon, and B.Z. Ahn. 2005. Antitumor activity of *Pulsatilla koreana* saponins and their structure–activity relationship. *Chemical and Pharmaceutical Bulletin* 53, no. 11:1451–1454.

Kim, Y., S.B. Kim, Y.J. You, and B.Z. Ahn. 2002. Deoxypodophyllotoxin; the cytotoxic and antiangiogenic component from *Pulsatilla koreana*. *Planta Medica* 68, no. 3:271–274.

Park, B.H., K.H. Jung, M.K. Son, J.H. Seo, H.S. Lee, J.H. Lee, and S.S. Hong. 2013. Antitumor activity of *Pulsatilla koreana* extract in anaplastic thyroid cancer via apoptosis and anti-angiogenesis. *Molecular Medicine Reports* 7, no. 1:26–30.

Punica granatum (Lythraceae) (Myrtales)   **Anti-proliferative**

Location: Middle East and the Caucasus, North Africa and tropical Africa, Indian subcontinent, Central Asia, drier parts of Southeast Asia, the Mediterranean basin, California, and Arizona

Appearance: shrub and grows up to 6–10 meters high

Leaves: opposite, glossy, narrowly elongated, entire, 3–7 cm long and 2 cm wide
Flowers: bright red, 3 cm in diameter

Parts used: leaves, fruit, seed

Active ingredients: corilagin, quercetin, pseudopelletierine

Tradition: used as an antiparasitic agent, a 'blood tonic', and to heal aphthae, diarrhea, and ulcers. Also serves as a remedy for diabetes in the Unani system of medicine practiced in the Middle East and India.

Documented target cancers:

- Fermented juice and pericarp polyphenols of the plant are reported to show anti-proliferative effect in breast cancer cells.
- Research in mice has shown that the fruit extract inhibits tumorigenesis in lung cancer and skin cancer models (Figure 9.22).

Figure 9.22   (See color insert.) *Punica granatum.*

References

Jurenka, J.S. 2008. Therapeutic applications of pomegranate (*Punica granatum* L.): A review. *Alternative Medicine Review* 13, no. 2:128–144.

Kim, N.D., R. Mehta, W. Yu, et al. 2002. Chemopreventive and adjuvant therapeutic potential of pomegranate (*Punica granatum*) for human breast cancer. *Breast Cancer Research and Treatment* 71, no. 3:203–217.

Rabdosia rubescens (Rabdosia) (Lamiaceae)   **Cytotoxic**

Location: west China
  Tradition: it is a traditional medicinal herb of China.
  Active ingredients:

- Unsaturated lactone: 10-epi-olguine (*Rabdosia ternifolia* (D. Don) Hara)
- Oridonin (*Rabdosia rubescens*) (cytotoxic + cisplatin, inducing DNA damage)
- Diterpenoids: enmein-, oridonin-, and trichorabdal-type (*Rabdosia trichocarpa*)
- Antitumor constituent: rabdophyllin G (*Rabdosia macrophylla*)
- Ponicidin

Particular value: important use in medicine for fighting cancer
  Precautions: careful use because of its toxicity
  Documented target cancers:

- Human cancer cell lines, Ehrlich ascites carcinoma (mice), antileukemic
- Lung cancer cells *in vitro*
- MCF-7 and MDA-MB-231 breast cancer cells

Further details:

- Trichorabdal-type diterpenoids showed the highest antitumor activity against Ehrlich ascites carcinoma in mice. *In vitro* activity against HeLa cells and *in vivo* activity against P388 lymphocytic leukemia were also determined but no synergistic increase in activity due to plural active sites was observed in those cases.
- From August 1974 to January 1987, 650 cases of moderately and advanced esophageal carcinoma were treated with a combination of chemotherapy and *Rabdosia rubescens* and/or traditional Chinese medicinal prescription. After treatment, 40 patients survived for over five years (five-year survival rate 6.15%), 32 for over six years, 23 for more than ten years, five for more than 15 years, and 20 died of tumors (16 cases) or other diseases (four cases). There were 20 patients living for more than 18 years. Analyzing the data, it is believed that the age, the state of activity, the length of illness, the effectiveness of primary treatment, the multi-course extensive therapy, long-term maintenance treatment, etc., are all important factors affecting the results of drug treatment.
- One hundred and fifteen patients with inoperable esophageal carcinoma were treated by either chemotherapy alone or chemotherapy plus *Rabdosia rubescens*. In group A, out of 31 patients treated with pingyangmycin (P) and nitrocaphane (N), ten (32.3%) responded to the treatment. Among them, two showed partial response (greater than 50% tumor regression) and eight minimal response (greater than 50% tumor regression). In group B, out of 84 patients treated with PN plus *Rabdosia rubescens*, 59 (70.2%) responded. Of them, ten showed complete response (100% tumor regression), 16 partial response, and 33 minimal response. The one-year survival rates of group A and B were 13.6% and 41.3%. Statistical significance was present in these two groups both in the response rate and one-year survival rate. As regards the drug toxicity, there was no significant difference between these two groups. Alopecia, anorexia, nausea, and hyperpyrexia occurred in more than 30% of patients. Mild leukopenia and thrombocytopenia and interstitial pneumonia were noted in some patients, and two patients died of toxicity in the lungs.

References

Bai, N., K. He, Z. Zhou, et al. 2010. Ent-kaurane diterpenoids from Rabdosia rubescens and their cytotoxic effects on human cancer cell lines. *Planta Medica* 76, no. 2:140–145.

Cheng, P.Y., M.J. Xu, Y.L. Lin, and J.C. Shi. 1982. The structure of rabdophyllin G, an antitumor constituent of *Rabdosia macrophylla. Yao Hsueh Hsueh Pao* 17, no. 12:917–921.

Fuji, K., M. Node, M. Sai, E. Fujita, S. Takeda, and N. Unemi. 1989. Terpenoids. LIII. Antitumor activity of trichorabdals and related compounds. *Chemical and Pharmaceutical Bulletin* 37, no. 6:1472–1476.

Gao, Z.G., Q.X. Ye, and T.M. Zhang. 1993. Synergistic effect of oridonin and cisplatin on cytotoxicity and DNA cross-link against mouse sarcoma S180 cells in culture. *Zhongguo Yao Li Xue Bao = Acta Pharmacologica Sinica* 14, no. 6:561–564.

Hsieh, T.C., E.K. Wijeratne, J.Y. Liang, A.L. Gunatilaka, and J.M. Wu. 2005. Differential control of growth, cell cycle progression, and expression of NF-κB in human breast cancer cells MCF-7, MCF-10A, and MDA-MB-231 by ponicidin and oridonin, diterpenoids from the Chinese herb Rabdosia rubescens. *Biochemical and Biophysical Research Communications* 337, no. 1:224–231.

Ikezoe, T., Y. Yang, K. Bandobashi, et al. 2005. Oridonin, a diterpenoid purified from Rabdosia rubescens, inhibits the proliferation of cells from lymphoid malignancies in association with blockade of the NF-κB signal pathways. *Molecular Cancer Therapeutics* 4, no. 4:578–586.

Liu, J.J., R.W. Huang, D.J. Lin, et al. 2006. Ponicidin, an ent-kaurane diterpenoid derived from a constituent of the herbal supplement PC-SPES, Rabdosia rubescens, induces apoptosis by activation of caspase-3 and mitochondrial events in lung cancer cells *in vitro. Cancer Investigation* 24, no. 2:136–148.

Lu, G.H., F.P. Wang, J.M. Pezzuto, T.C. Tam, I.D. Williams, and C.T. Che. 1997. 10-Epi-olguine from *Rabdosia ternifolia. Journal of Natural Products* 60, no. 4:425–427.

Nagao, Y., N. Ito, T. Kohno, H. Kuroda, and E. Fujita. 1982. Antitumor activity of *Rabdosia* and *Teucrium* diterpenoids against P 388 lymphocytic leukemia in mice. *Chemical and Pharmaceutical Bulletin (Tokyo)* 30, no. 2:727–729.

Wang, R.L. 1993. A report of 40 cases of esophageal carcinoma surviving for more than 5 years after treatment with drugs. Chung Hua Chung Liu Tsa Chih 15, no. 4:300–302.

Wang, R.L., B.L. Gao, M.L. Xiong, et al. 1986. Potentiation by Rabdosia rubescens on chemotherapy of advanced esophageal carcinoma. *Chung Hua Chung Liu Tsa Chih* 8, no. 4:297–299.

Rhus succedanea (Sumach) (Anacardiaceae)　　　　　　　　　　　　　　　**Tumor inhibitor Cytotoxic**

Location: Japan
　Appearance:

　Stem: 1.2 m high
　Leaves: pinnate

Tradition: as the bark is rich in tannin, it is used in candle-making, for adulterating white beeswax, and in making pomades. Japan Wax is obtained in Japan by expression and heat or by the action of solvents from the fruit of sumach.

　Parts used: bark, root, fruit
　Active ingredients:

- Tyrosinase inhibitor: 2-hydroxy-4-methoxybenzaldehyde
- Hinokiflavone (cytotoxic)

Particular value: the root-bark is astringent and diuretic. Used in diabetes

　Documented target cancers: compounds from the EtOH extract of the sap of *Rhus succedanea* have been reported cytotxic against the cancer cell lines HeLa, Huh7, HCT116, and LoVo, and rat C6 glioma.

Further details:

- The root of *Rhus vulgaris* contains 2-hydroxy-4-methoxybenzaldehyde, which is also found in two other East African medicinal plants: the root of *Mondia whitei* (Hook) Skeels (Asclepiaceae) and the bark of *Sclerocarya caffra* Sond (Anacardiaceae).
- The fruit of *Rhus succedanea* consists almost entirely of palmitin and free palmitic acid, and is not a true wax.

References

Kubo, I., and I. Kinst-Hori. 1999. 2-Hydroxy–4-methoxybenzaldehyde: A potent tyrosinase inhibitor from African medicinal plants. *Planta Medica* 65, no. 1:19–22.

Wang, H.K., Y. Xia, Z.Y. Yang, S.L. Natschke, and K.H. Lee. 1998. Recent advances in the discovery and development of flavonoids and their analogues as antitumor and anti-HIV agents. *Advances in Experimental Medicine and Biology* 439:191–225.

Wu, P.L., S.B. Lin, C.P. Huang, and R.Y.Y. Chiou. 2002. Antioxidative and cytotoxic compounds extracted from the sap of *Rhus succedanea*. *Journal of Natural Products* 65, no. 11:1719–1721.

Rubia cordifolia L. (Rubia (Bengal madder)) (Rubiaceae)	Antitumor

Location: India

Appearance:

Stem: 3 m high, stalks are very weak so that they often lie along the ground preventing the plant from rising.

Root: main and side roots, the side roots run under the surface of the ground for some distance sending up shoots.

Leaves: have spines along the midrib on the underside.

Flowers: the flower-shoots spring from the joints in pairs, the loose spikes of yellow.

In bloom: June

Parts used: root

Active ingredients:

- The root of *R. cordifolia*: naphthohydroquinones, naphthohydroquinone dimers, naphthohydroquinone, naphthoquinone, anthraquinones, naphthohydroquinone dimer, bicyclic hexapeptides: RA-XI, -XII, XIII, -XIV, -XV, and -XVI (P388)
- *Rubia akane, R. cordifolia*: cyclic hexapeptide: RA-700

Indicative dosage and application:

- RA-700 was given from 0.2 to 1.4 mg/m^2 in single i.v. dose study, from 0.4 to 2.0 mg/m^2 in five-day i.v.

Documented target cancers:

- Various tumors *in vivo* and *in vitro* (such as: P388, L1210, L5178Y, B16 melanoma, Lewis lung carcinoma, and sarcoma-180)
- The methanol fraction of *Rubia cordifolia* extract is reported to exhibit potent inhibition of human cervical cancer cell lines and larynx carcinoma cells.

Further details:

- RA-700 has been tested in a phase I clinical study conducted by the RA-700 clinical study group consisting of six institutions. A single dose administration and five-day schedule administration were evaluated with 14 patients, respectively. RA-700 was given from 0.2 to 1.4 mg/m^2 in single i.v. dose study, from 0.4 to 2.0 mg/m^2 in five-day i.v. schedule study. Nausea and vomiting, fever, stomach ache, mild hypotension, and slight abnormality of electric-cardiogram were observed as the toxicities. In a pharmacokinetic study, the elimination half-lives ($t_{1/2}$) of RA-700 in plasma were 55 min of alpha-phase and 3.9 h of beta-phase by single dose study, and 23–25 min of alpha-phase and 6–14 h of beta-phase by five-day schedule study. Accumulation was not found by five-day schedule administration, and metabolites were not observed in plasma and urine. It seems that RA-700 is metabolized by the liver and excreted in the feces. In conclusion, the maximum tolerated dose was 1.4 mg/m^2 for five-day schedule administration.
- Further studies have shown that: (1) changes in cardiac function were noted in both groups. (2) Changes in blood pressure, sigma QRS, ejection fraction, and fractional shortening of the second group tended to be more extreme than those of the first group. Care for continuity is a concern with long-term and high doses of RA-700. (3) Because of the small sample, we could find no relationship between the changes in cardiac function and the injection doses of RA-700. (4) Therefore, the cardiac function must be checked by giving anti-neoplastic drugs to neoplastic patients.
- The antitumor activity of RA-700 was evaluated in comparison with deoxy-bouvardin and vincristine (VCR). As regards the proliferation of L1210 cultured cells, the cytotoxicity of RA-700 was similar to that of VCR but superior to that of deoxy-bouvardin. The IC$_{50}$ value of RA-700 was 0.05 mcg/ml under our experimental conditions. RA-700 inhibited the incorporation of ^{14}C-leucine at a concentration at which no effects were observed on the incorporation of 3H-thymidine and 3H-uridine in L1210 culture cells *in vitro*. The antitumor activity of RA-700 was similar to that of deoxy-bouvardin and VCR against P388 leukemia. Daily treatment with RA-700 at an optimal dose resulted in 118% ILS. As with deoxy-bouvardin and VCR, the therapeutic efficacy of RA-700 depends on the time schedule. RA-700 showed marginal activity against L1210 leukemia (50% ILS), similar to that of deoxy-bouvardin but inferior to that of VCR. RA-700 inhibited Lewis tumor growth in the early stage after tumor implantation, whereas deoxy-bouvardin and VCR did not. As regards toxicity, a slight reduction of peripheral WBC counts was observed with the drug but no reduction of RBC and platelet counts. BUN, creatinine, GPT, and GOT levels in plasma did not change with the administration of the drug.
- Another anticancer principle isolated from *Rubia cordifolia* is RC-18, which has been used against a spectrum of experimental murine tumors, viz. P388, L1210, L5178Y, B16 melanoma, Lewis lung carcinoma, and sarcoma-180. RC-18 exhibited significant increase in life span of ascites leukemia P388, L1210, L5178Y, and a solid tumor B16 melanoma. However, it failed to show any inhibitory effect on solid tumors, Lewis lung carcinoma, and sarcoma 180. Promising results against a spectrum of experimental tumors suggest that RC-18 may lead to the development of a potential anticancer agent.
- Madder root, *Rubia tinctorum* L., is a traditional herbal medicine used against kidney stones. This species contains lucidin, a hydroxyanthraquinone derivative present in this plant, which is mutagenic in bacteria and mammalian cells. In these respects, the use of madder root for medicinal purposes is associated with a carcinogenic risk.

References

Alegre, A., J. Díaz-Mediavilla, J. San-Miguel, et al. 1998. Autologous peripheral blood stem cell transplantation for multiple myeloma: A report of 259 cases from the Spanish Registry. Spanish Registry for Transplant in MM (Grupo Español de Trasplante Hematopoyético-GETH) and PETHEMA. *Bone Marrow Transplantation* 21, no. 2:133–140.

Brunet, S., A., Urbano-Ispizua, C. Solano, et al. 1997. Allogenic transplant of non-manipulated hematopoietic progenitor peripheral blood cells. Spanish experience of 79 cases. *Sangre* 42 Suppl 1:42–43.

García Laraña, J., J. Díaz Mediavilla, R. Martínez, et al. 1997. Maintenance treatment with interferon-alfa in multiple myeloma after autotransplantation of peripheral blood progenitor cells. Spanish Register of Transplantation in myeloma. *Sangre* 42 Suppl 1:38–41.

Ghosh, S., M. Das Sarma, A. Patra, and B. Hazra. 2010. Anti-inflammatory and anticancer compounds isolated from Ventilago madraspatana Gaertn., *Rubia cordifolia* Linn. and Lantana camara Linn. *Journal of Pharmacy and Pharmacology* 62, no. 9:1158–1166.

López, A., J. de la Rubia, F. Arriaga, C. Jiménez, G.F. Sanz, N. Carpio, and M.L. Marty. 1998. Severe hemolytic anemia due to multiple red cell alloantibodies after an ABO-incompatible allogeneic bone marrow transplant. *Transfusion* 38, no. 3:247–251.

Lopez, F., I. Jarque, G. Martin, et al. 1998. Invasive fungal infections in patients with blood disorders. *Medicina Clínica (Barc)* 110, no. 11:401–405.

Patel, P.R., A.A. Nagar, R.C. Patel, D.K. Rathod, and V.R. Patel. 2010. In-vitro anticancer activity of Rubia cordifolia against Hela and Hep-2 cell lines. *Phytomedicine* 2:44–46.

Sanz, M.A., J. de la Rubia, S. Bonanad, et al. 1998. Prolonged molecular remission after PML/RAR alpha-positive autologous peripheral blood stem cell transplantation in acute promyelocytic leukemia: Is relevant pretransplant minimal residual disease in the graft? *Leukemia* 12, no. 6:992–995.

Son, J.K., S.J. Jung, J.H. Jung, et al. 2008. Anticancer constituents from the roots of *Rubia cordifolia* L. *Chemical and Pharmaceutical Bulletin* 56, no. 2:213–216.

Urbano-Ispizua, A., C. Solano, S. Brunet, et al. 1997. Allogeneic transplant of CD34+ peripheral blood cells from HLA-identical donors: Spanish experience of 40 cases. *Sangre* 42 Suppl 1:44–45.

Urbano-Ispizua, A., C. Solano, S. Brunet, et al. 1998. Allogeneic transplantation of selected CD34+ cells from peripheral blood: Experience of 62 cases using immunoadsorption or immunomagnetic technique. Spanish Group of Allo-PBT. *Bone Marrow Transplantation* 22, no. 6:519–525.

Westendorf, J., W. Pfau, and A. Schulte. 1998. Carcinogenicity and DNA adduct formation observed in ACI rats after long-term treatment with madder root, Rubia tinctorum L. *Carcinogenesis* 19, no. 12:2163–2168.

***Salvia sclarea* (Salvia (Clarry)) (Lamiaceae)**   **Anti-leukemic**

Location: Middle Europe

Appearance:

Stem: it's a biennial plant with square brownish stems 0.5–1 m high, hairy, with few new branches.

Leaves: are arranged in pairs, almost stalkless, almost as large as the hand, heart shaped, and covered with velvety hairs.

Flowers: are interspersed with large colored, membraneous bracts, longer than the spiny calyx. Blue or white

In bloom: summer

Tradition: this herb was first brought into use by the wine merchants of Germany and later was employed as a substitute for sophisticating beer, communicating considerable bitterness, and intoxicating property. In ancient times and in middle times it was used for its curative properties.

Parts used: seeds

Active ingredients: a specific lectin from the seeds of *Salvia sclarea* and the diterpenoids ferruginol, salvipisone, aethiopinone, and 1-oxoaethiopinone from the roots

Documented target cancers:

• Inhibitory activity against human erythroleukemic cell line K562 and leukemia cells Jurkat, HL-60, and NALM-6

Further details:

• From the seeds of *Salvia sclarea* (SSA) was isolated a lectin specific for GalNac-Ser/Thr studied in human erythroleukemic cell line K562. Another study proved that glycoproteins from the human T leukemia cells Jurkat were found to bind to the GalNac-Ser/Thr specific lectin from SSA. Studies show that this specific lectin has an inhibitory activity against the human erythroleukemic cell line K562 and T leukemia cells.

- Some strong natural antioxidants like carnosol were proved to exhibit anti-inflammatory and inhibitory effects with regard to tumor-initiation activities in mice test systems. Also some sage compounds (ursolic and/or oleanolic acid) that show no antioxidant may turn out to be promising in future research of inflammation and of cancer prevention. A squalene derived triterpenoid ursolic acid (structure 29) and its isomer oleanolic acid (structure 30) (up to 4% in sage leaves, dry weight basis) act anti-inflammatory and inhibit tumorigenesis in mouse skin. Recent data on the anti-inflammatory activity of sage (*S. officinalis* L.) extracts when applied topically (ID_{50} = 2040 µg/cm^2) and evaluated as oedema inhibition after Croton oil-induced dermatitis in mouse ear, confirm/suggest ursolic acid to be the main active ingredient, responsible for sage anti-inflammatory effect. The data on the pharmacological effects of these metabolites promise new therapeutic possibilities of sage extracts.

- Ursolic acid showed significant cytotoxicity in lymphatic leukemia cells P-388 (ED_{50} = 3.15 µg/ml) and L-1210 (ED_{50} = 4.00 µg/ml) as well as human lung carcinoma cell A-549 (ED_{50} = 4.00 µg/ml) (Lee et al. 1987, Fang and Mc Laughlin 1989). Both carnosol and ursolic acid are referred to as being strong inhibitors of 12-*O*-tetradecanoylphorbol-13-acetate (TPA)-induced ornithine decarboxylase activity and of TPA-induced tumor promotion in mouse skin. The tumorigenesis-prevention potential of ursolic acid was comparable to that of retinoic acid (RA)—a known inhibitor of tumor promotion. Both ursolic acid- and oleanolic acid-treatment (41 nmol of each), when applied continuously before each TPA-treatment (4.1 nmol), delayed the formation of papillomas in mouse skin, significantly reduced the rate of papilloma-bearing mice and reduced the number of papillomas per mouse, when compared with the control group (only TPA treatment). Ursolic acid acted more effectively in a single application before initial TPA-treatment when compared to the effect of RA and/or oleanolic acid. So, the mechanism of the inhibitory action of ursolic acid (inhibition of the first critical cellular event in tumor promotion step caused by TPA) may differ slightly from those of RA and/or oleanolic acid, which block a critical second stage process in tumor promotion by TPA (induction of ornithine decarboxylase and polyamine levels).

- A possible tumorigenesis preventing effect can be predicted for abietane diterpene galdosol (structure 1), isolated from *S. canariensis* L., which showed significant cytostatic activity (ID_{50} = 0.50 µg/ml) when inhibition of development of single-layer culture of HeLA 229 cells was measured in *in vitro* experiment.

- One of the most dangerous environmental sources of cytogenetic damage is ionizing radiation, which acts either directly or by secondary reactions and induces ionization in tissues. Interaction of ionizing radiation with water and other protoplasmatic constituents in oxidative metabolism causes formation of harmful oxygen radicals. DNA lesions, caused by reactive oxygen species in mammalian cells, are the initial event which may lead to possible mutagenesis and/or carcinogenesis and form the basis of spontaneous cancer incidence. Free radicals play an important role in preventing deleterious alterations in cellular DNA and genotoxic effects caused by ionizing radiation in mammalian tissues. Many drugs and chemicals (for example sulfhydryl compounds) are known to increase the survival rate in animals. Based on animal models studies, *S. miltiorrhiza* and its extracts were shown to have a potential to prevent X-radiation-induced pulmonary injuries and high dosage gamma-irradiation-induced platelet aggregation lesions.

- The antiproliferative activity of tanshinones against five human tumor cells, i.e. A-549 (lung), SK-OV-3 (ovary), SK-MEL-2 (melanoma), XF-498 (central nerve system), and HCT-15 (colon), was evaluated by sulforhodamine-B method. Eighteen isolated tanshinones exhibited significant but presumably nonspecific cytotoxicity against all tested tumor cells, which might be attributed to common naphtoquinone skeleton rather than to substituents attached to it. Methylenetanshiquinone (structure 32) and tanshindiol C (structure 33) exhibited most powerful cytotoxic effects against tested tumor cells, with IC_{50} ranging from 0.4 µg/ml in A-549 cells to 2.2 µg/ml in SK-MEL-2 cells and IC_{50} from 0.3 µg/ml in SK-MEL-2 cells to 0.9 µg/ml in SK-OV-3 cancer cell lines, respectively.

- From *S. przewalskii* Maxim. var. *mandarinorum* Stib., a strong bacteriostatic compound, przewaquinone A (structure 34) was isolated. Przewaquinone A was reported (Xiao and Fu 1987) to possess potential for inhibiting Lewis lung carcinoma and melanoma B-16 (Figure 9.23).

Figure 9.23   (See color insert.) *Salvia sclarea.*

References

Darias, V., L. Bravo, R. Rabanal, C.C. Sánchez-Mateo, and D.A. Martín-Herrera. 1990. Cytostatic and antibacterial activity of some compounds isolated from several Lamiaceae species from the Canary Islands. *Planta Medica* 56, no. 1:70–72.

Du, H., Z. Qian, and Z. Wang. 1990. Prevention of radiation injury of the lungs by *Salvia miltiorrhiza* in mice. *Zhong Xi Yi Jie He za Zhi = Chinese Journal of Modern Developments in Traditional Medicine* 10, no. 4:230–231.

Fang, X.P., and J.L. McLaughlin. 1989. Ursolic acid, a cytotoxic component of the berries of Ilex verticillata. *Fitoterapia* 61, no. 1:176–177.

Fiore, G., C. Nencini, F. Cavallo, et al. 2006. *In vitro* antiproliferative effect of six Salvia species on human tumor cell lines. *Phytotherapy Research* 20, no. 8:701–703.

Hanawalt, P.C. 1998. Genomic instability: Environmental invasion and the enemies within. *Mutation Research* 400, no. 1–2:117–125.

Huang, M.T., C.T. Ho, Z.Y. Wang, et al. 1994. Inhibition of skin tumorigenesis by rosemary and its constituents carnosol and ursolic acid. *Cancer Research* 54, no. 3:701–708.

Lee, A.R., W.L. Wu, W.L. Chang, H.C. Lin, and M.L. King. 1987. Isolation and bioactivity of new tanshinones. *Journal of Natural Products* 50, no. 2:157–160.

Lutz, W.K. 1998. Dose-response relationships in chemical carcinogenesis: Superposition of different mechanisms of action, resulting in linear-nonlinear curves, practical treshholds, J-shapes. *Mutation Research* 405, no. 2:117–124.

Medeiros, A., S. Bianchi, J.J. Calvete, et al. 2000. Biochemical and functional characterization of the Tn-specific lectin from *Salvia sclarea* seeds. *European Journal of Biochemistry* 267, no. 5:1434–1440.

Różalski, M., Ł. Kuźma, H. Wysokińska, and U. Krajewska. 2006. Cytotoxic and proapoptotic activity of diterpenoids from *in vitro* cultivated Salvia sclarea roots. Studies on the leukemia cell lines. *Zeitschrift für Naturforschung: Teil C* 61, no. 7–8:483–488.

Ryu, S.Y., C.O. Lee, and S.U. Choi. 1997. *In vitro* cytotoxicity of Tanshinones from *Salvia miltiorrhiza*. *Planta Medica* 63, no. 4:339–342.

Tokuda, H., H. Ohigashi, K. Koshimizu, and Y. Ito. 1986. Inhibitory effects of ursolic and oleanolic acid on skin tumor promotion by 12-o-tetradecanoilphorbol-13-acetate. *Cancer Letters* 33, no. 3:279–285.

Wang, H.F., X.D. Li, Y.M. Chen, L.B. Yuan, and W.O. Foye. 1991. Radiation-protective and platelet aggregation inhibitory effects of five traditional Chinese drugs and acetylsalicylic acid following high-dose-
gamma-irradiation. *Journal of Ethnopharmacology* 34, no. 2–3:215–219.

Xiao, P.G. 1989. Excerpts of the Chinese pharmacopoeia. In *Herbs, Spices and Medicinal Plants: Recent
advances in Botany, Horticulture and Pharmacology*. L.E. Craker, J.E. Simon (Eds.)., Vol. 4. Oryx
Press:42–114.

Sargassum bacciferum (Sargassum) (Fucaceae)	Antimetastatic

Location: North Atlantic Ocean
  Appearance:

> *Thallus*: coarse, light yellow or brownish-green, erect, 0.5–1 m in height. Attaches itself to the rocks
> by branched, rootlike, woody extremities, developed from the base of the stalk. The front is almost
> funnel shaped, narrow and trap shaped at the base, the rest is flat and leaf-like in form, wavy, many
> times divided into two, with erect divisions having a very strong, broad, compressed midrib running
> to the apex.

Parts used: dried mass of root, stem and leaves
  Active ingredients:

- Aqueous extract: Fucoidan polysaccharides
- Methanolic extract: dihydroxysargaquinone

Documented target cancers:

- Antimetastatic: lung cancer, Ehrlich carcinoma in mice
- Antileukemic (dihydroxysargaquinone)
- Immunostimulatory
- Cytotoxic (dihydroxysargaquinone)

Further details:

- *Sargassum thunbergii*, the brown seaweed umitoranoo contains neutral and acidic polysaccharides.
  Antitumor activity has been attributed to two fractions, GIV-A ([alpha] 25D -127 degrees and mol.
  wt., 19,000) and GIV-B ([alpha]25D -110 degrees and mol. wt., 13,500). These compounds were
  found to be a fucoidan or L-fucan containing approx. 30% sulfate ester groups per fucose residue,
  about 10% uronic acid, and less than 2% protein.
- GIV-A markedly inhibited the growth of Ehrlich ascites carcinoma at the dose of 20 mg/kg per
  day ×10 with no sign of toxicity in mice. It acts as a so-called activator of the reticuloendothelial system. Fucoidan enhanced the phagocytosis and chemiluminescence of macrophages. By the
  immunofluorescent method, binding of the third component of complement (C3) cleavage product
  to macrophages and the proportion of C3 positive cells were increased. These results suggest that
  the antitumor activity of fucoidan is related to the enhancement of immune responses. The present
  results indicate that fucoidan may open new perspectives in cancer chemotherapy.
- *Sargassum fulvellum* contains a polysaccharide fraction (either a sulphated peptidoglycuronogly-
  can or a sulphated glycuronoglycan) with remarkable tumor-inhibiting effect against sarcoma-180
  implanted subcutaneously in mice.
- *Sargassum tortile*: the CCl4 partition fractions from methanolic extracts of this species contain
  dihydroxysargaquinone, which is cytotoxic against cultured P-388 lymphocytic leukemia cells.
- *Sargassum kjellmanianum* is also effective in the *in vivo* growth inhibition of the implanted sar-
  coma-180 cells (Figure 9.24).

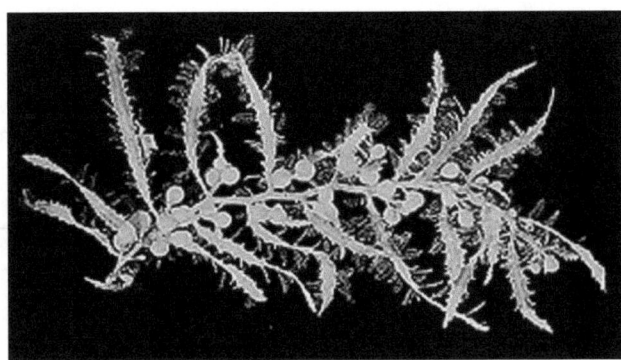

Figure 9.24   (See color insert.) *Sargassum.*

References

Amagata, T., K. Minoura, and A. Numata. 1998. Cytotoxic metabolites produced by a fungal strain from a *Sargassum alga. Journal of Antibiotics* 51, no. 4:432–434.

Iizima-Mizui, N., M. Fujihara, J. Himeno, K. Komiyama, I. Umezawa, and T. Nagumo. 1985. Antitumor activity of polysaccharide fractions from the brown seaweed *Sargassum kjellmanianum. Kitasato Archives of Experimental Medicine* 58, no. 3:59–71.

Itoh, H., H. Noda, H. Amano, and H. Ito. 1995. Immunological analysis of inhibition of lung metastases by fucoidan (GIV-A) prepared from brown seaweed *Sargassum thunbergii. Anticancer Research* 15, no. 5(B):1937–1947.

Itoh, H., H. Noda, H. Amano, C. Zhuaug, T. Mizuno, and H. Ito. 1993. Antitumor activity and immunological properties of marine algal polysaccharides, especially fucoidan, prepared from *Sargassum thunbergii* of Phaeophyceae. *Anticancer Research* 13, no. 6 (A):2045–2052.

Numata, A., S. Kanbara, C. Takahashi, R. Fujiki, M. Yoneda, E. Fujita, and Y. Nabeshima. 1991. Cytotoxic activity of marine algae and a cytotoxic principle of the brown alga *Sargassum tortile. Chemical and Pharmaceutical Bulletin* 39, no. 8:2129–2131.

Yamamoto, I., T. Nagumo, M. Takahashi, M. Fujihara, Y. Suzuki, and N. Iizima. 1981. Antitumor effect of seaweeds. III. Antitumor effect of an extract from *Sargassum kjellmanianum. Japanese Journal of Experimental Medicine* 51, no. 3:187–189.

Yamamoto, I., M. Takahashi, T. Suzuki, H. Seino, and H. Mori. 1984. Antitumor effect of seaweeds. IV. Enhancement of antitumor activity by sulfation of a crude fucoidan fraction from *Sargassum kjellmanianum. Japanese Journal of Experimental Medicine* 54, no. 4:143–151.

Zhuang, C., H. Itoh, T. Mizuno, and H. Ito. 1995. Antitumor active fucoidan from the brown seaweed, umitoranoo (*Sargassum thunbergii*). *Bioscience, Biotechnology, and Biochemistry* 59, no. 4:563–567.

***Scutellaria baicalensis* Georgii (Scutellaria (Scullcap)) (Labiatae)**   **Antitumor**

Location: US, Great Britain
   Appearance:

Stem: square, 15–45 cm high, somewhat slender, either paniculately branched or in small specimens
Root: perennial and creeping root-stock
Leaves: opposite downy leaves, oblong and tapering, heart-shaped at the base, 1–5 cm long, notched and shortly petioles
Flowers: in pairs, each growing from the axils of the upper, leaf-like bracts, bright blue with white inside
In bloom: July–September

Parts used: the whole herb
  Active ingredients:

- Flavonoids: baicalin, baicalein, and wogonin
- Flavones: 5,7,2'-trihydroxy- and 5,7,2',3'-tetrahydroxyflavone

Documented target cancers:

- Hepatoma cell lines, Pliss' lymphosarcoma, Epstein-Barr virus, skin cancer (mice)
- *Scutellaria baicalensis* is reported to strongly inhibit cell growth of squamous cell carcinoma (SCC-25, KB), breast cancer (MCF-7), hepatocellular carcinoma (HepG2), prostate carcinoma (PC-3 and LNCaP), and colon cancer (KM-12 and HCT-15).
- Aqueous extracts of *S. baicalensis* roots are reported to induce apoptosis and therefore suppress growth of lymphoma and myeloma cell lines.
- Human malignant brain tumor cells
- Human prostate cancer cell lines (LNCaP, androgen dependent, and PC-3, androgen independent)

Further details:

- *Scutellaria baicalensis* Georgi (methanol extract) contains the flavonoids baicalin, baicalein, and wogonin which induce the quinone reductase in the Hepa 1c1c7 murine hepatoma cell line. Baicalin may be the major active principle of QR induction mediated by scutellaria radix extract. In addition, the flavones 5,7,2'-trihydroxy- and 5,7,2',3'-tetrahydroxyflavone exhibit remarkable inhibitory effects on mouse skin tumor promotion in an *in vivo* two-stage carcinogenesis test and on the Epstein–Barr virus early antigen activation.
- The advancement of Pliss' lymphosarcoma in rats was shown to be associated with disorders of platelet-mediated hemostasis presenting with either lowered or increased aggregation activity of platelets. In the latter case, a direct correlation was observed between functional activity of thrombocytes, on the one hand, and degree of tumor advancement and its metastatic activity, on the other. The extract of *Scutellaria baicalensis* Georgi was shown to produce a normalizing effect on platelet-mediated hemostasis whatever the pattern of alteration that points to the adaptogenic activity of the drug. This activity is thought to be responsible for the drug's antitumor and, particularly, metastasis-preventing effect.
- In experiments with murine and rat transplantable tumors, *Scutellaria baicalensis* Georgi extract treatment was shown to ameliorate cyclophosphamide and 5-fluorouracil-induced myelotoxicity and to decrease tumor cell viability. This was partly attributed to a pronounced antistressor action of the extract and its normalizing effect on some homeostatic parameters.
- As a supplement to conventional chemotherapy: cytostatic therapy of patients with lung cancer is attended with decrease in the relative number of T-lymphocytes and their theophylline-resistant population. Patients who were given SB showed a tendency towards increase of these parameters during antitumor chemotherapy. The immunoregulation index (IRI) in this case was approximately twice the background values during the whole period of investigation. The inclusion of SB in the therapeutic complex promotes increase in the number of immunoglobulins A at a stable level of immunoglobulin G.
- Glial cells have a role in maintaining the function of neural cells. A study was undertaken to clarify the effects of baicalin and baicalein, flavonoids isolated from an important medicinal plant *Scutellariae Radix* (the root of *Scutellaria baicalensis* Georgi), on glial cell function using C6 rat glioma cells. Baicalin and baicalein caused concentration-dependent inhibition of a histamine-induced increase in intracellular Ca^{2+} concentrations ($[Ca^{2+}]i$). The potency of baicalein was significantly greater than that of baicalin. The noradrenaline- and carbachol-induced increase in $[Ca^{2+}]i$ was also inhibited by baicalein, and both drugs inhibited histamine-induced accumulation of total [3H]inositol phosphates, consistent with their inhibition of the increase in $[Ca^{2+}]i$. These results suggest that baicalin and baicalein inhibit $[Ca^{2+}]i$ elevation by reducing phospholipase C activity. The inhibitory effects of baicalin and baicalein on $[Ca^{2+}]i$ elevation might be important in the

interpretation of their pharmacological action on glial cells, such as inhibition of Ca^{2+}-required enzyme phospholipase A2.

- Hemopoiesis was studied in 88 patients with lung cancer during antitumor chemotherapy and its combination with a dry SB extract. Administration of the plant preparation was accompanied with hemopoiesis stimulation, intensification of bone-marrow erythro- and granulocytopoiesis, and increase in the content of circulating precursors of the type of erythroid and granulomonocytic colony-forming units.

- Isolation of E-1-(4'-Hydroxyphenyl)-but-1-en-3-one from *Scutellaria barbata*.

- Ten known glycosidic compounds, betulalbuside A (1), 8-hydroxylinaloyl,3-O-beta-D-glucopyrano side (2) (monoterpen glycosides), ipolamide (3) (iridoid glycoside), acteoside (verbascoside) (4), leucosceptoside A (5), martynoside (6), forsythoside B (7), phlinoside B (8), phlinoside C (9), and teuerioside (10) (phenylpropanoid glycosides), were isolated from methanolic extracts of *Phlomis armeniaca* and *Scutellaria salviifolia* (Labiatae). Structure elucidations were carried out using 1H-, 13C-NMR, and FAB-MS spectra, as well as chemical evidence. The cytotoxic and cytostatic activities of isolated compounds were investigated by the 3-[4,5-dimethylthiazol-2-yl]-2,5-diphenyltetr azolium bromide (MTT) method. Among the glycosides obtained here, caffeic acid-containing phenylpropanoid (or phenethyl alcohol or phenylethanoid) glycosides were found to show activity against several kinds of cancer cells. However, they didn't affect the growth and viability of primary-cultured rat hepatocytes. Study of the structure–activity relationship indicated that ortho-dihydroxy aromatic systems of phenylpropanoid glycosides are necessary for their cytotoxic and cytostatic activities.

- *Oldenlandia diffusa* (OD) and *Scutellaria barbata* (SB) have been used in traditional Chinese medicine for treating liver, lung, and rectal tumors while *Astragalus membranaceus* (AM) and *Ligustrum lucidum* (LL) are often used as adjuncts in cancer therapy. The effects of aqueous extracts of these four herbs on aflatoxin B1 (AFB1)-induced mutagenesis were investigated, *Salmonella typhimurium* TA100 as the bacterial tester strain and rat liver 9,000 × g supernatant as the activation system. The effects of these herbs on [3H]AFB1 binding to calf-thymus DNA were assessed. Organosoluble and water-soluble metabolites of AFB1 were extracted and analyzed by high-performance liquid chromatography (HPLC). Mutagenesis assays revealed that all of these herbs produced a concentration-dependent inhibition of histidine-independent revertant (His+) colonies induced by AFB1. At a concentration of 1.5 mg/plate, SB and OD in combination exhibited an additive effect. The trend of inhibition of these four herbs on AFB1-induced mutagenesis was: SB greater than LL greater than AM. LL, OD, and SB significantly inhibited AFB1 binding to DNA, reduced AFB1-DNA adduct formation, and also significantly decreased the formation of organosoluble metabolites of AFB1. These data suggest that these Chinese medicinal herbs possess cancer chemopreventive properties.

References

Ducki, S., J.A. Hadfield, N.J. Lawrence, C.Y. Liu, A.T. McGown, and X. Zhang. 1996. Isolation of E-1-(4'-hydroxyphenyl)-but-1-en-3-one from *Scutellaria barbata*. *Planta Medica* 62, no. 2:185–186.

Goldberg, V.E., V.M. Ryzhakov, M.G. Matiash, et al. 1997. Dry extract of *Scutellaria baicalensis* as a hemostimulant in antineoplastic chemotherapy in patents with lung cancer. *Eksperimental'Naya i Klinicheskayafarmakologiya* 60, no. 6:28–30.

Kumagai, T., C.I. Müller, J.C. Desmond, Y. Imai, D. Heber, and H.P. Koeffler. 2007. Scutellaria baicalensis, a herbal medicine: Anti-proliferative and apoptotic activity against acute lymphocytic leukemia, lymphoma and myeloma cell lines. *Leukemia Research* 31, no. 4:523–530.

Kyo, R., N. Nakahata, I. Sakakibara, M. Kubo, and Y. Ohizumi. 1998a. Effects of Sho-saiko-to, San'o-shashin-to and Scutellariae Radix on intracellular Ca2+ mobilization in C6 rat glioma cells. *Biological and Pharmaceutical Bulletin* 21, no. 10:1067–1071.

Kyo, R., N. Nakahata, I. Sakakibara, M. Kubo, and Y. Ohizumi. 1998b. Baicalin and baicalein, constituents of an important medicinal plant, inhibit intracellular Ca2+ elevation by reducing phospholipase C activity in C6 rat glioma cells. *Journal of Pharmacy and Pharmacology* 50, no. 10:1179–1182.

Park, H.J., Y.W. Lee, H.H. Park, Y.S. Lee, I.B. Kwon, and J.H. Yu. 1998. Induction of quinone reductase by a methanol extract of *Scutellaria baicalensis* and its flavonoids in murine Hepa 1c1c7 cells. *European Journal of Cancer Prevention* 7, no. 6:465–471.

Razina, T.G., E.P. Zueva, V.I. Litvinenko, and I.P. Kovalev. 1998. A semisynthetic flavonoid from the Baikal skullcap (*Scutellaria baicalensis*) as an agent to enhance the efficacy of chemotherapy in experimental tumors. *Eksperimental'Naya i Klinicheskaya Farmakologiya* 61, no. 2:54–56.

Saracoglu, I., M. Inoue, I. Calis, and Y. Ogihara. 1995. Studies on constituents with cytotoxic and cytostatic activity of two Turkish medicinal plants *Phlomis armeniaca* and *Scutellaria salviifolia*. *Biological and Pharmaceutical Bulletin* 18, no. 10:1396–1400.

Scheck, A.C., K. Perry, N.C. Hank, and W.D. Clark. 2006. Anticancer activity of extracts derived from the mature roots of *Scutellaria baicalensis* on human malignant brain tumor cells. *BMC Complementary and Alternative Medicine* 6, no. 1:27.

Smolianinov, E.S., V.E. Goldberg, M.G. Matiash, et al. 1997. Effect of *Scutellaria baicalensis* extract on the immunologic status of patients with lung cancer receiving antineoplastic chemotherapy. *Eksperimental'Naya i Klinicheskaya Farmakologiya* 60, no. 6:49–51.

Yamashiki, M., A. Nishimura, H. Suzuki, S. Sakaguchi, and Y. Kosaka. 1997. Effects of the Japanese herbal medicine 'Sho-saiko-to' (TJ-9) on *in vitro* interleukin-10 production by peripheral blood mononuclear cells of patients with chronic hepatitis C. *Hepatology* 25, no. 6:1390–1397.

Ye, F., S. Jiang, H. Volshonok, J. Wu, and D.Y. Zhang. 2007. Molecular mechanism of anti-prostate cancer activity of *Scutellaria baicalensis* extract. *Nutrition and Cancer* 57, no. 1:100–110.

Ye, F., L. Xui, J. Yi, W. Zhang, and D.Y. Zhang. 2002. Anticancer activity of Scutellaria baicalensis and its potential mechanism. *Journal of Alternative and Complementary Medicine* 8, no. 5:567–572.

Zhao, Q., X.Y. Chen, and C. Martin. 2016. *Scutellaria baicalensis*, the golden herb from the garden of Chinese medicinal plants. *Science Bulletin* 61, no. 18:1391–1398.

Scutellaria barbata (Lamiaceae) (Lamiales)	Cytotoxic

Location: Korea and southern China
  Appearance: perennial herb

  Leaves: slightly toothed, bevelled or triangular in shape
  Flowers: cyan-purple with hairy crown

Active ingredients: phytol, wogonin, luteolin, and hispidulin
  Tradition: it is used with other herbs in traditional Chinese medicine to treat bacterial infections, hepatitis, and tumors.
  Documented target cancers:

- Ethanol extracts of *S. barbata* are reported to greatly inhibit lung cancer A549 cell growth.
- The chloroform fraction of *Scutellaria barbata* has been shown to be cytotoxic against human breast cancer cell lines (MCF-7 and MDA-MB-435S), human cervix carcinoma cell lines (HeLa), human liver carcinoma cell lines (Bel-7402 and HepG2), human renal adenocarcinoma cell line (ACHN), murine sarcoma (S180), and Ehrlich's ascites carcinoma (EAC) cell lines.
- The methylene chloride fraction of SB has been demonstrated to induce apoptosis in human U937 leukemia cell line.

References

Cha, Y.Y., E.O. Lee, H.J. Lee, et al. 2004. Methylene chloride fraction of *Scutellaria barbata* induces apoptosis in human U937 leukemia cells via the mitochondrial signaling pathway. *Clinica Chimica Acta: International Journal of Clinical Chemistry* 348, no. 1–2:41–48.

Yin, X., J. Zhou, C. Jie, D. Xing, and Y. Zhang. 2004. Anticancer activity and mechanism of *Scutellaria barbata* extract on human lung cancer cell line A549. *Life Sciences* 75, no. 18:2233–2244.

Yu, J., H. Liu, J. Lei, W. Tan, X. Hu, and G. Zou. 2007. Antitumor activity of chloroform fraction of *Scutellaria barbata* and its active constituents. *Phytotherapy Research* 21, no. 9:817–822.

Selaginella tamariscina (Selaginellaceae) (Selaginellales)   **Anti-tumor**

Location: Russia, China, India, Japan, Philippines, Korea, Taiwan, Thailand
  Appearance: evergreen perennial plant that reaches a height of up to 45 cm
  Tradition: traditional Chinese herb for the therapy of chronic trachitis
  Active ingredients: sumaflavone and amentoflavone
  Documented target cancers:

- Ethanol extract of *Selaginella tamariscina* is reported to provide strong inhibition of tumor growth in Lewis lung carcinoma inoculated mice.
- Fractions of *Selaginella tamariscina* extracts showed significant tumoricidal effects against cultured human leukemia cells whereas these fractions did not affect normal human lymphocytes.
- *Selaginella tamariscina* extract and amentoflavone synergistically increased the doxorubicin antiproliferative effect in A549 and NCI-H460 human lung cancer cells *in vitro*. Additionally, they significantly inhibited A549 tumor growth in animal experiments.
- Antimetastatic activities of *Selaginella tamariscina* on lung cancer cells *in vitro* and *in vivo* have also been reported.

References

Jung, Y.J., E.H. Lee, C.G. Lee, et al. 2017. AKR1B10-inhibitory *Selaginella tamariscina* extract and amentoflavone decrease the growth of A549 human lung cancer cells *in vitro* and *in vivo*. *Journal of Ethnopharmacology* 202:78–84.

Le, M.H., T.T. Do, T.H. Hoang, V.M. Chau, and T.D. Nguyen. 2012. Toxicity and anticancer effects of an extract from *Selaginella tamariscina* on a mice model. *Natural Product Research* 26, no. 12:1130–1134.

Lee, C.W., H.J. Choi, H.S. Kim, et al. 2008. Biflavonoids isolated from *Selaginella tamariscina* regulate the expression of matrix metalloproteinase in human skin fibroblasts. *Bioorganic and Medicinal Chemistry* 16, no. 2:732–738.

Lee, I.S., A. Nishikawa, F. Furukawa, K.I. Kasahara, and S.U. Kim. 1999. Effects of *Selaginella tamariscina* on *in vitro* tumor cell growth, p53 expression, G1 arrest and *in vivo* gastric cell proliferation. *Cancer Letters* 144, no. 1:93–99.

Yang, S.F., S.C. Chu, S.J. Liu, Y.C. Chen, Y.Z. Chang, and Y.S. Hsieh. 2007. Antimetastatic activities of *Selaginella tamariscina* (Beauv.) on lung cancer cells *in vitro* and *in vivo*. *Journal of Ethnopharmacology* 110, no. 3:483–489.

Seseli mairei (Apiaceae)   **Antitumor**

Location: China
  Tradition: leaves are used for making salads
  Parts used: roots
  Active ingredients: Cytotoxic *polyacetylene: seselidiol*
  Documented target cancers:

Figure 9.25   (See color insert.) *Seseli.*

- Cytotoxicity against KB, P-388, and L-1210 tumor cells

Further details:
- Seselidiol is a new polyacetylene, that has been isolated from the roots of *Seseli mairei.* On the basis of chemical and spectroscopic evidence, its structure has been established as heptadeca-1,8(Z)-diene-4,6-diyne-3,10-diol. Seselidiol and its acetate have been demonstrated to show moderate cytotoxicity against KB, P-388, and L-1210 tumor cells (Figure 9.25).

Reference

Hu, C.Q., J.J. Chang, and K.H. Lee. 1990. Antitumor agents, 115. Seselidiol, a new cytotoxic polyacetylene from *Seseli mairei. Journal of Natural Products* 53, no. 4:932–935.

Siegesbeckia glabrescens **(Asteraceae) (Asterales)**   **Cytotoxic**

Location: Asia
   Appearance: annual herb

   Leaves: opposite, blade-like, with nerves covered with white fluff on their lower surface

Tradition: used in ancient China to treat rheumatism
   Parts used: various dried parts of the plant
   Active ingredients: siegenolides A and B
   Precautions: beverage overdose of the plant can lead to gastrointestinal discomfort, loose stools, and nausea.
   Documented target cancers:

- *S. glabrescens* extract is reported to inhibit the proliferation, attachment, and migration of SKOV-3 human ovarian cancer cells.
- *Siegesbeckia glabrescens* extract has been shown to induce apoptosis in breast cancer estrogen-receptor (ER)-positive (MCF-7) and ER-negative (MDA-MB-231) cell lines.

References

Cho, Y.R., S.W. Choi, and D.W. Seo. 2013. The *in vitro* antitumor activity of *Siegesbeckia glabrescens* against ovarian cancer through suppression of receptor tyrosine kinase expression and the signaling pathways. *Oncology Reports* 30, no. 1:221–226.

Jeon, C.M., I.S. Shin, N.R. Shin, et al. 2014. *Siegesbeckia glabrescens* attenuates allergic airway inflammation in LPS-stimulated RAW 264.7 cells and OVA induced asthma murine model. *International Immunopharmacology* 22, no. 2:414–419.

Jun, S.Y., Y.H. Choi, and H.M. Shin. 2006. *Siegesbeckia glabrescens* induces apoptosis with different pathways in human MCF-7 and MDA-MB-231 breast carcinoma cells. *Oncology Reports* 15, no. 6:1461–1467.

Kim, Y.S., H. Kim, E. Jung, et al. 2012. A novel antibacterial compound from *Siegesbeckia glabrescens*. *Molecules* 17, no. 11:12469–12477.

Wu, Q., H. Li, S.Y. Lee, H.J. Lee, and J.H. Ryu. 2015. New cytotoxic sesquiterpenoids from *Siegesbeckia glabrescens*. *Molecules* 20, no. 2:2850–2856.

***Sinojackia sarcocarpa* (Styracaceae) (Ericales)**   **Cytotoxic Anti-tumor**

Location: China
   Appearance: trees or shrub, 7 to 10 m high

   Leaves: oval blade with a round base
   Flowers: white, 4–6 cm in size

Tradition: *Sinojackia sarcocarpa* has been used to treat many diseases such as obstructive thromboangitis, arthritis variants, and angina pectoris.
   Parts used: aerial parts
   Active ingredients: pentacyclic triterpenoids
   Documented target cancers:

- A novel triterpenoid from the leaves of *Sinojackia sarcocarpa*, 2α, 3α, 19β, 23β-tetrahydroxyurs-12-en-28-oic acid, is reported to exhibit significantly higher cytotoxicity to the cancer cell lines A2780 and HepG2 than to the noncancerous IOSE144 and QSG7701.
- Two pentacyclic triterpenoids from the leaves of *Sinojackia sarcocarpa* exhibited inhibition on murine sarcoma S(180) tumors.

References

Liu, S., M. Cao, D. Li, et al. 2011. Purification and anticancer activity investigation of pentacyclic triterpenoids from the leaves of *Sinojackia sarcocarpa* LQ. *Natural Product Research* 25, no. 17:1600–1606.

Wang, O., S. Liu, J. Zou, et al. 2011. Anticancer activity of 2α, 3α, 19β, 23β-tetrahydroxyurs-12-en-28-oic acid (THA), a novel triterpenoid isolated from Sinojackia sarcocarpa. *PLOS ONE* 6, no. 6:e21130.

***Solanum lycopersicum* (Solanaceae) (Solanales)**   **Cytotoxic**

Location: endemic plant of South and Central America
   Appearance: erect or spreading annual plant

   Leaves: 10–25 cm long, composed of five to nine leaflets each
   Flowers: yellow, with a diameter of 1–2 cm and five pointed lobes
   Fruit: rail, around 100 g

Tradition: used in folk medicine for acne scar treatment, peeling, eczema, nausea, wounds, and relief of the symptoms of anemia.

Parts used: fruit

Active ingredients: lycopene

Documented target cancers:

- *S. lycopersicum* extract is reported to exert protective potential against DMBA induced skin papillomas and B6 F10 melanomas in skin tumor induced mouse models.

References

Agrawal, R.C., R. Jain, W. Raja, and M. Ovais. 2009. Anticarcinogenic effects of *Solanum lycopersicum* fruit extract on Swiss albino and C57 Bl mice. *Asian Pacific Journal of Cancer Prevention* 10, no. 3:379–382.

Bhuvaneswari, V., and S. Nagini. 2005. Lycopene: A review of its potential as an anticancer agent. *Current Medicinal Chemistry: Anti-Cancer Agents* 5, no. 6:627–635.

Solanum lyratum (Solanaceae) (Solanales)   Cytotoxic

Location: Japan, Cambodia, Korea, Laos, Thailand, Vietnam

Appearance: climbing plant

Leaves: elliptical

Flowers: blue, purple, or white, 3 cm in size

Tradition: used in folk medicine for its detoxifying and antipyretic action. The decoction of the plant is used for the treatment of leucorrhea, swollen thyroid gland, etc. In addition, the leaves of the plant boiled in breast milk are used to treat nausea in babies.

Parts used: whole plant

Active ingredients: diosgenin

Documented target cancers:

- The plant extract has been used to treat esophagus and stomach cancer.
- The plant extract induces apoptosis in human colon adenocarcinoma COLO 205 cells.
- It has been reported as cytotoxic against LLC lung cancer cell line.

References

Hsu, S.C., J.H. Lu, C.L. Kuo, et al. 2008. Crude extracts of *Solanum lyratum* induced cytotoxicity and apoptosis in a human colon adenocarcinoma cell line (colo 205). *Anticancer Research* 28, no. 2 (A):1045–1054.

Lee, J.H., Y.H. Lee, H.J. Lee, et al. 2009. Caspase and mitogen activated protein kinase pathways are involved in *Solanum lyratum* herba induced apoptosis. *Journal of Ethnopharmacology* 123, no. 1:121–127.

Yang, J.S., C.C. Wu, C.L. Kuo, et al. 2012. *Solanum lyratum* extracts induce extrinsic and intrinsic pathways of apoptosis in WEHI-3 murine leukemia cells and inhibit allograft tumor. *Evidence-Based Complementary and Alternative Medicine* 2012:254960.

Solanum nigrum (Solanaceae) (Solanales)   Anti-cancer

Location: endemic in Eurasia

Appearance: annual bush

Leaves: ovate-wavy, heart-shaped
Flowers: off-white, with bright yellow anthers

Tradition: the plant is used as an analgesic, anti-inflammatory, anticonvulsant, and vasodilator. It is applied in the form of ointment or as a juice to combat toothache. It was also traditionally used to treat testicular swelling.

Active ingredients: polysaccharides (SNLWP-1, SNLAP-1, SNLAP-2, SNLP, SNL-AE)
Documented target cancers:
- The plant extract has been reported active agianst prostate cancer and uterine cancer *in vitro*.

References

Li, J., Q. Li, T. Feng, et al. 2007. Antitumor activity of crude polysaccharides isolated from *Solanum nigrum* Linne on U14 cervical carcinoma bearing mice. *Phytotherapy Research* 21, no. 9:832–840.

Li, J., Q. Li, T. Feng, and K. Li. 2008. Aqueous extract of *Solanum nigrum* inhibit growth of cervical carcinoma (U14) via modulating immune response of tumor bearing mice and inducing apoptosis of tumor cells. *Fitoterapia* 79, no. 7–8:548–556.

Mogla, E.H., O.M. Abdalla, W.S. Koko, and A.M. Saadabi. 2014. *In vitro* anticancer activity and cytotoxicity of *Solanum nigrum* on cancer and normal cell lines. *International Journal of Cancer Research* 10, no. 2:74–80.

Solanum tuberosum (Solanaceae) (Solanales)   **Anti-cancer**

Location: Andes (indigenous), cultivated worldwide
Appearance: herbaceous perennial

Flowers: white, pink, red, blue, or purple

Tradition: carrying a raw potato was an old-fashioned remedy against rheumatism, and ladies in former times had special bags or pockets made in their dresses in which to carry one or more small raw potatoes for the purpose of avoiding rheumatism.

Parts used: underground part (tubers), leaves, peel
Active ingredients: solanine, very small amounts of atropine, chlorogenic acid derivatives, and anthocyanin
Documented target cancers:

- Extracts from diverse genetic breeding clones have been reported to inhibit the growth of *in vitro* HT-29 colon cancer cell cultures.
- *S. tuberosum* cultivars are reported to inhibit carcinogenesis in carcinogen injected rats.

References

Thompson, M.D., H.J. Thompson, J.N. McGinley, E.S. Neil, D.K. Rush, D.G. Holm, and C. Stushnoff. 2009. Functional food characteristics of potato cultivars (*Solanum tuberosum* L.): Phytochemical composition and inhibition of 1-methyl-1-nitrosourea induced breast cancer in rats. *Journal of Food Composition and Analysis* 22, no. 6:571–576.

Umadevi, M., P. Sampath, D. Bhowmik, and S. Duraivel. 2013. Health benefits and cons of *Solanum tuberosum*. *Journal of Medicinal Plant Studies* 1, no. 1:16–25.

Zuber, T., D. Holm, P. Byrne, L. Ducreux, M. Taylor, M. Kaiser, and C. Stushnoff. 2015. Optimization of *in vitro* inhibition of HT-29 colon cancer cell cultures by *Solanum tuberosum* L. extracts. *Food and Function* 6, no. 1:72–83.

Sophora alopecuroides (Fabaceae) (Fabales)	Anti-tumor

Location: Southwest and East Asia, Greece, Turkey, and southern Russia
    Appearance: shrub

    Leaves: long, 7 to 12 pairs
    Flowers: creamy

Tradition: used in Chinese medicine for the treatment of some skin and gynacological diseases such as eczema, dermatitis, and colpitis, as well as fever, sore throat, and inflammation.
    Parts used: aerial parts and seeds
    Active ingredients: quinolizidine alkaloids, lupine, sophoridine, matrine, and sophocarpine
    Documented target cancers:
- Total alkaloids of *S. alopecuroides* are reported to induce apoptosis in cervical tumor HeLa cells, with obvious inhibitory effects on cell growth.

References

Atta-ur-Rahman, M.I. Choudhary, K. Parvez, A. Ahmed, F. Akhtar, M. Nur-e-Alam, and N.M. Hassan. 2000. Quinolizidine alkaloids from *Sophora alopecuroides*. *Journal of Natural Products* 63, no. 2:190–192.

Kucukbaci, N., S. Özkan, N. Adiguzel, and F. Tosun. 2011. Characterisation and antimicrobial activity of *Sophora alopecuroides* L. var. alopecuroides alkaloid extracts. *Turkish Journal of Biology* 35, no. 3:379–385.

Li, J.G., X.Y. Yang, and W. Huang. 2016. Total alkaloids of *Sophora alopecuroides* inhibit growth and induce apoptosis in human cervical tumor hela cells *in vitro*. *Pharmacognosy Magazine* 12 (Suppl 2):S253–S256.

Sophora flavescens (Fabaceae) (Fabales)	Anti-cancer

Location: Japan
    Appearance: evergreen shrub of 150 cm high
    Tradition: the root is tonic, helps in the stomach, is diuretic, antipyretic. It is used to treat jaundice, dysentery, diarrhea, and to treat urinary tract infections. Also, internal and external use helps in the treatment of vaginitis, eczema, itching, syphilis, mumps, and allergic reactions. *Sophora flavescens* alkaloid gels, known as Kushen, a compound in traditional Chinese medicine, has been clinically used in China for many years, and there are many clinical reports demonstrating its anti-cancer effect.
    Precautions: the plant is poisonous when used in large quantities.
    Parts used: root
    Active ingredients: cytosine, matrine
    Documented target cancers:
- *Sophora flavescens* alkaloid gels are reported to restrain cervical cancer cell proliferation, inhibit metastasis, induce cell cycle arrest in G2/M phase, and induce cellular apoptosis in cervical cancer cell lines SiHa and C33A.

References

He, X., J. Fang, L. Huang, J. Wang, and X. Huang. 2015. *Sophora flavescens* Ait.: Traditional usage, phytochemistry and pharmacology of an important traditional Chinese medicine. *Journal of Ethnopharmacology* 172:10–29.

Zhou, Y.J., Y.J. Guo, X.L. Yang, and Z.L. Ou. 2018. Anti-cervical cancer role of Matrine, oxymatrine and *Sophora flavescens* alkaloid gels and its mechanism. *Journal of Cancer* 9, no. 8:1357–1364.

***Sophora interrupta* (Fabaceae) (Fabales)**   **Anti-cancer Anti-angiogenic**

Location: India
   Appearance: perennial tropical shrub

   Flowers: golden-yellow, bell-shaped

Tradition: important shrub in the Ayurveda. Various parts of *S. interrupta* are traditionally used to treat different diseases and have antibacterial, anti-inflammatory, anti-allergic, hepatitis, cardiac arrhythmia, and anti-tumor properties of human health as folk medicine.
   Parts used: roots
   Active ingredients: piseatannol, biochanin-A, camphorol
   Documented target cancers:

- *S. interrupta* root ethylacetate extract is reported to exhibit anticancer activity in breast adenocarcinoma cancer (MCF-7) and rostate adenocarcinoma (PC-3) cell lines.
- The plant extract has potential anti-angiogenic compounds that may interfere with VEGF-induced cancer malignancy.

References

Mathi, P., K. Nikhil, N. Ambatipudi, P. Roy, V.R. Bokka, and M. Botlagunta. 2014. In-vitro and in-silico characterization of *Sophora interrupta* plant extract as an anticancer activity. *Bioinformation* 10, no. 3:144–151.
Mathi, P., G.K. Veeramachaneni, K.K. Raj, V.R. Talluri, V.R. Bokka, and M. Botlagunta. 2016. *In vitro* and in silico characterization of angiogenic inhibitors from *Sophora interrupta*. *Journal of Molecular Modeling* 22, no. 10:247.

***Sphaeranthus amaranthoides* (Asteraceae) (Asterales)**   **Cytotoxic**

Location: Sri Lanka and southern India
   Appearance: low annuals shrub with spreading branches

   Leaves: 2 to 4 inches, linear–oblong, narrowed at the base

Tradition: used in folk medicine for the treatment of chronic skin diseases and helps to purify the blood. It works against skin wounds and has mild analgesic action.
   Parts used: leaves
   Active ingredients: chrysosplenol D
   Documented target cancers:

- Breast adenocarcinoma MCF-7 cell line
- A549 human lung carcinoma cell line

References

Gayatri, S., R. Suresh, C.U.M. Reddy, and K. Chitra. 2016. Isolation and Characterization of chemopreventive agent from *Sphaeranthus amaranthoides* Burm F. *Pharmacognosy Research* 8, no. 1:61–65.

Geethalakshmi, R., C. Sakravarthi, T. Kritika, M.A. Arul Kirubakaran, and D.V.L. Sarada. 2013. Evaluation of antioxidant and wound healing potentials of *Sphaeranthus amaranthoides* Burm.f. *BioMed Research International* http://dx.doi.org/10.1155/2013/607109.

Thanigavelan, V., M.P. Kumar, and G.V. Rajamanickam. 2012. Pharmacological study of a siddha holistic herb sivakaranthai-*Sphaeranthus amaranthoides* burm for analgesic and anti-inflammatory activities. *Journal of Applied Pharmaceutical Science* 2, no. 1:95.

Sphaeranthus bullatus (Asteraceae) (Asterales)	Cytotoxic

Location: East Africa

    Appearance: annual herb

    Tradition: the leaves of the plant are used against malaria.

    Parts used: leaves and root

    Active ingredients: carbotacetone derivatives

    Documented target cancers:

- Several carvotacetone compounds of the plant are reported to exhibit strong anticancer activity against SK-MEL, KB, BT-549, and SK-OV-3 solid tumor cells.

Reference

Machumi, F., A. Yenesew, J.O. Midiwo, et al. 2012. Antiparasitic and anticancer carvotacetone derivatives from Sphaeranthus bullatus.Natural product communications 7(9):1123-6

Spirulina platensis (Spirulinaceae) (Spirulinales)	Anti-tumor

Location: Central and South America

    Appearance: cyanobacteria (photosynthetic prokaryote) consisting of many cells

    Parts used: whole organism

    Active ingredients: phycocyanobilin and chlorophyllin

    Documented target cancers:

- A sulfated polysaccharide derived from *Spirulina platensis* has been shown to significantly inhibit the invasion of tumor cells: B16-BL6 melanoma, colon 26 M3.1 carcinoma, and HT-1080 fibrosarcoma.
- It is reported to significantly decrease the proliferation of human pancreatic cancer cell lines *in vitro* and *in vivo* in xenotransplanted nude mice.

References

Koníčková, R., K. Vanková, J. Vaníková, et al. 2014. Anti-cancer effects of blue-green alga *Spirulina platensis*, a natural source of bilirubin-like tetrapyrrolic compounds. *Annals of Hepatology* 13, no. 2:273–283.

Mishima, T., J. Murata, M. Toyoshima, et al. 1998. Inhibition of tumor invasion and metastasis by calcium-spirulan (Ca-SP), a novel sulfated polysaccharide derived from a blue-green alga, *Spirulina platensis*. *Clinical and Experimental Metastasis* 16, no. 6:541–550.

Stellera chamaejasme (Stellera) (Thymelaceae)   **Cytotoxic**

Location: China, rocky slopes and plains and isolated areas of altitude 2,700–4,300 meters
   Appearance: perennial herb

   Flowers: white, red, and dark red

Parts used: root
   Active ingredients: diterpene: gnidimacrin, cycloheptides
   Indicative dosage and application: gnidimacrin has been used at the dosages of 0.02–0.03 mg/
kg ip against mouse leukemia P-388 and L-1210 *in vivo* and showed significant antitumor activities.
   Documented target cancers:

- Human leukemias, stomach cancers, and non-small cell lung cancers *in vitro*
- The extract of *Stellera chamaejasme* is reported to inhibit the growth and induce the apoptosis of human lung cancer NCI-H157 cells.
- The daphnane-type diterpene gnidimacrin, isolated from the root of *Stellera chamaejasme*, was found to strongly inhibit cell growth of human leukemias, stomach cancers, and non-small cell lung cancers *in vitro*.

Further details:

- *Stellera chamaejasme* L.: the root (methanolic extract) contains the daphnane-type diterpene *gnidimacrin*. Gnidimacrin acts as a protein kinase C activator for tumor cells.
- Gnidimacrin was found to strongly inhibit cell growth of human leukemias, stomach cancers, and non-small cell lung cancers *in vitro* at concentrations of $10^{(-9)}$ to $10^{(-10)}$ M. On the other hand, even at $10^{(-6)}$ to $10^{(-5)}$ M, the small cell lung cancer cell line H69 and the hepatoma cell line HLE were refractory to gnidimacrin. The agent showed significant antitumor activity against murine leukemias and solid tumors in an *in vivo* system. In K562, a sensitive human leukemia cell line, *gnidimacrin* induced blebbing of the cell surface, which was completely inhibited by staurosporine at concentrations above $10^{(-8)}$ M, and arrested the cell cycle transiently to G2 and finally the G1 phase at growth-inhibitory concentrations. It inhibited phorbol-12,13-dibutyrate(PDBu) binding to K562 cells and directly stimulated protein kinase C (PKC) activity in the cells in a dose-dependent manner (3–100 nM). Although activation of PKC isolated from refractory H69 cells was observed only with 100 nM gnidimacrin, the degree of activation was lower than that produced by 3 nM in K562 cells.
- Gnidimacrin showed significant antitumor activities against mouse leukemia P-388 and L-1210 *in vivo*. At the dosages of 0.02–0.03 mg/kg ip, the increase in life span (ILS) was 70 and 80%, respectively. Gnidimacrin was also active against murine solid tumors *in vivo*, such as Lewis lung carcinoma, B-16 melanoma, and colon cancer 26. It showed ILSs of 40, 49, and 41% at the dosages of 0.01–0.02 mg/kg ip, respectively. Gnidimacrin strongly inhibited cell proliferation of human cancer cell lines such as leukemia K562, stomach cancers Kato-III, MKN-28, MKN-45, and mouse L-1210 by the MTT assay and colony forming assay *in vitro*. The IC_{50} of gnidimacrin was 0.007–0.00012 microgram/ml.
- Inhibitory effects of *Stellera chamaejasme* on the growth of a transplantable tumor in mice.

References

Feng, W., I. Tetsuro, and Y. Mitsuzi. 1995. The antitumor activities of gnidimacrin isolated from *Stellera chamaejasme* L. *Chung Hua Chung Liu Tsa Chih. Zhonghua Zhong Liu za Zhi* [*Chinese Journal of Oncology*] 17, no. 1:24–26.

Liu, X., and X. Zhu. 2012. *Stellera chamaejasme* L. extract induces apoptosis of human lung cancer cells via activation of the death receptor-dependent pathway. *Experimental and Therapeutic Medicine* 4, no. 4:605–610.

Qiao, L., L. Yang, D. Zhang, J. Zou, and J. Dai. 2011. Studies on chemical constitutes from callus cultures of *Stellera chamaejasme*. *Zhongguo Zhong Yao za Zhi = Zhongguo Zhongyao Zazhi = China Journal of Chinese Materia Medica* 36, no. 24:3457–3462.

Yang, B.Y. 1986. Inhibitory effects of *Stellera chamaejasme* on the growth of a transplantable tumor in mice. Chung Yao Tung Pao 11, no. 1:58–59.

Yoshida, M., W. Feng, N. Saijo, and T. Ikekawa. 1996. Antitumor activity of daphnane-type diterpene gnidimacrin isolated from *Stellera chamaejasme* L. *International Journal of Cancer* 66, no. 2:268–273.

Strychnos Nux-vomica (Strychnos) (Loganiaceae)   **Cytotoxic Poisonous**

Location: India, in the Malay Archipelago
   Appearance:

   Stem: medium-sized tree with short, thick trunk
   Root: very bitter
   Leaves: opposite
   Flowers: small, greeny-white

Tradition: the powdered seeds are employed in atonic dyspepsia. The tincture of *Nux Vomica* is often used in mixtures, for its stimulant action on the gastro-intestinal tract.

   Parts used: seeds

   Active ingredients: strychnopentamine (a dimeric indole alkaloid) from *Strychnos usambarensis*

   Particular value: strychnine is the chief alkaloid constituent of the seeds and acts as a bitter. It improves the pulse and raises blood pressure, acts as a tonic to the circulatory system in cardiac failure, but in small doses, because it can be poisonous.

   Precautions: application of the drug can cause partial haemolysis and liver damage.

   Indicative dosage and application:

- Four subcutaneous injections of 1.5 mg strychnopentamine (one per day) induce a significant decrease of the number of Ehrlich ascites tumor cells.
- Strychnopentamine at a relatively low concentration (less than 1 microgram) after 72 h of treatment on B16 melanoma cells and on non-cancer human fibroblasts cultured *in vitro*.

Documented target cancers:

- Against Ehrlich ascites tumor cells with a significant increase of the survival of the treated mice.
- Strychnopentamine applied on B16 melanoma cells and on non-cancer human fibroblasts cultured *in vitro* strongly inhibits cell proliferation and induces cell death.
- Induced apoptosis and repressed migration of human hepatocellular carcinoma cell lines HepG2 and SMMC-7721.

Further details:

- Strychnopentamine (SP) is an alkaloid isolated from *Strychnos usambarensis* Gilg and is a potential anticancer agent, which strongly inhibits cell proliferation and induces cell death on B16 melanoma cells and on non-cancer human fibroblasts cultured *in vitro* and induces a significant decrease of the number of Ehrlich ascites tumor cells.
- Strychnopentamine, in a low concentration (less than 1 microgram), after 72 h showed that incorporation of thymidine and leucine by B16 cells significantly decreases after only 1 h of treatment. SP

induces the formation of dense lamellar bodies and vacuolization in the cytoplasm, intense blebbing at the cell surface, and various cytological alterations leading to cell death.

- Three more alkaloids isolated from *Strychnos usambarensis* on cancer cells in culture.

References

Bassleer, R., M.C. Depauw-Gillet, B. Massart, J.M. Marnette, P. Wiliquet, M. Caprasse, and L. Angenot. 1982. Effects of three alkaloids isolated from *Strychnos usambarensis* on cancer cells in culture. *Planta Medica* 45, no. 2:123–126.

Deng, X., F. Yin, X. Lu, B. Cai, and W. Yin. 2006. The apoptotic effect of brucine from the seed of *Strychnos nux-vomica* on human hepatoma cells is mediated via Bcl-2 and Ca2+ involved mitochondrial pathway. *Toxicological Sciences* 91, no. 1:59–69.

Deng, X.K., W. Yin, W.D. Li, et al. 2006. The anti-tumor effects of alkaloids from the seeds of *Strychnos nux-vomica* on HepG2 cells and its possible mechanism. *Journal of Ethnopharmacology* 106, no. 2:179–186.

Quetin-Leclercq, J., B. Bouzahzah, A. Pons, et al. 1993. Strychnopentamine, a potential anticancer agent. *Planta Medica* 59, no. 1:59–62.

Quetin-Leclercq, J., M.C. De Pauw-Gillet, L. Angenot, and R. Bassleer. 1991. Effects of strychnopentamine on cells cultured *in vitro*. *Chemico-Biological Interactions* 80, no. 2:203–216.

Shu, G., X. Mi, J. Cai, et al. 2013. Brucine, an alkaloid from seeds of Strychnos nux-vomica Linn., represses hepatocellular carcinoma cell migration and metastasis: The role of hypoxia inducible factor 1 pathway. *Toxicology Letters* 222, no. 2:91–101.

Symphytum officinale L. (Comfrey) (Boraginaceae)   Antimitotic Carcinogenic

---

Probably nature's most famous wound-healing species, comfrey has been often referred to as a cancer-fighting drug. Quite ironically, its use may actually increase the possibility of contracting the disease.

Location: Europe and temperate Asia. Usually in watery areas

Appearance:

Stem: leafy, angular, covered with bristly hairs, 60–90 cm high
Root: fibrous, fleshy, and spindle-shaped
Leaves: radical leaves are very large (they decrease in size), shape ovate, covered with rough hairs.
Flowers: yellow or purple, growing on short stalks, scorpoid in form
In bloom: May–July

Degree of rarity: low

Tradition: a green vegetable (roots and leaves). Decoction used as an herbal tea.

Parts used: Leaves, root

Active ingredients: pyrrolizidine alkaloid-N-oxides: 7-acetyl intermedine, 7-acetyl lycopsamine, lycopsamine, intermedine, symphytine

Documented carcinogenic properties:

- Its crude watery extract and its protein fraction stimulate the *in vivo* proliferation of neoplastic cells and exert an antimitotic effect on human T lymphocytes.
- When digested, it may cause hepatocellular adenomas (at least in rats!).
- Contains hepatotoxic pyrrolizidine alkaloids.
- Alkaloid fractions obtained from the roots demonstrate antimitotic and mutagenic activities against both animal and plant cells.
- Mutagenic in rat liver with reported tumorigenesis in the livers of Big Blue transgenic rats.

Further details:

- The crude watery extract of *Symphytum officinale* and certain protein and carbohydrate components had remarkable effects on the respiratory burst of human PMN granulocytes stimulated via Fc receptors.
- Pyrrolizidine alkaloids have been linked to liver and lung cancers and a range of other deleterious effects. Some comfrey-containing products were found to contain measurable quantities of one or more of the hepatotoxic pyrrolizidine alkaloids, in ranges from 0.1 to 400.0 ppm. Products containing comfrey leaf in combination with one or more other ingredients were found to contain the lowest alkaloid levels. Highest levels were found in bulk comfrey root, followed by bulk comfrey leaf.
- The carcinogenicity of *Symphytum officinale* L. was studied in inbred ACI rats. Three groups of 19–28 rats each were fed comfrey leaves for 480–600 days; four additional groups of 15–24 rats were fed comfrey roots for varying lengths of time. A control group was given a normal diet. Mutations were induced in all experimental groups that received the diets containing comfrey roots and leaves. Hemangioendothelial sarcoma of the liver was infrequently induced.
- Mutagenic and antimitotic effects have been attributed to aqueous solutions of alkaloid fractions obtained from infusions of *Symphytum officinale* L.

References

Aftab, K., F. Shaheen, F.V. Mohammad, M. Noorwala, and V.U. Ahmad. 1996. Phyto-pharmacology of saponins from *Symphytum officinale* L. *Advances in Experimental Medicine and Biology* 404:429–442.

Ahmad, V.U., M. Noorwala, F.V. Mohammad, and B. Sener. 1993. A new triterpene glycoside from the roots of *Symphytum officinale*. *Journal of Natural Products* 56, no. 3:329–334.

Ahmad, V.U., M. Noorwala, F.V. Mohammad, B. Sener, A.H. Gilani, and K. Aftab. 1993. Symphytoxide A, a triterpenoid saponin from the roots of *Symphytum officinale*. *Phytochemistry* 32, no. 4:1003–1006.

Barbakadze, V.V., E.P. Kemertelidze, I.L. Targamadze, A.S. Shashkov, and A.I. Usov. 2002. Novel biologically active polymer of 3-(3,4-dihydroxyphenyl)glyceric acid from two types of the comphrey *Symphytum asperum* and *S. caucasicvum* (*Boraginoceae*). *Bioorganicheskaya Khimiya* 28, no. 4:362–366.

Barthomeuf, C.M., E. Debiton, V.V. Barbakadze, and E.P. Kemertelidze. 2001. Evaluation of the dietetic and therapeutic potential of a high molecular weight hydroxycinnamate-derived polymer from *Symphytum asperum Lepech*. Regarding its antioxidant, antilipoperoxidant, antiinflammatory, and cytotoxic properties. *Journal of Agricultural and Food Chemistry* 49, no. 8:3942–3946.

Behninger, C., G. Abel, E. Röder, V. Neuberger, and W. Göggelmann. 1989. Studies on the effect of an alkaloid extract of *Symphytum officinale* on human lymphocyte cultures. *Planta Medica* 55, no. 6:518–522.

Betz, J.M., R.M. Eppley, W.C. Taylor, and D. Andrzejewski. 1994. Determination of pyrrolizidine alkaloids in commercial comfrey products (*Symphytum* sp.). *Journal of Pharmaceutical Sciences* 83, no. 5:649–653.

Couet, C.E., C. Crews, and A.B. Hanley. 1996. Analysis, separation, and bioassay of pyrrolizidine alkaloids from comfrey (*Symphytum officinale*). *Natural Toxins* 4, no. 4:163–167.

Furmanowa, M., J. Guzewska, and B. Bełdowska. 1983. Mutagenic effects of aqueous extracts of *Symphytum officinale L.* and of its alkaloidal fractions. *Journal of Applied Toxicology* 3, no. 3:127–130.

Hirono, I., H. Mori, and M. Haga. 1978. Carcinogenic activity of *Symphytum officinale*. *Journal of the National Cancer Institute* 61, no. 3:865–869.

Johnson, B.M., J.L. Bolton, and R.B. Van Breemen. 2001. Screening botanical extracts or quinoid metabolites. *Chemical Research in Toxicology* 14, no. 11:1546–1551.

Kim, N.C., N.H. Oberlies, D.R. Brine, R.W. Handy, M.C. Wani, and M.E. Wall. 2001. Isolation of symlandine from the roots of common comfrey (*Symphytum officinale*) using countercurrent chromatography. *Journal of Natural Products* 64, no. 2:251–253.

Lenghel, V., D.L. Radu, P. Chirilă, and A. Olinescu. 1995. The influence of some vegetable extracts on the *in vitro* adherence of mouse and human lymphocytes to nylon fibers. *Roumanian Archives of Microbiology and Immunology* 54, no. 1–2:15–30.

Mei, N., L. Guo, P.P. Fu, R.H. Heflich, and T. Chen. 2005. Mutagenicity of comfrey (*Symphytum officinale*) in rat liver. *British Journal of Cancer* 92, no. 5:873–875.

Mei, N., L. Guo, L. Zhang, et al. 2006. Analysis of gene expression changes in relation to toxicity and tumorigenesis in the livers of Big Blue transgenic rats fed comfrey (*Symphytum officinale*). *BMC Bioinformatics* 7, no. 2:S16.

Mohammad, F.V., M. Noorwala, V.U. Ahmad, and B. Sener. 1995a. A bidesmosidic hederagenin hexasaccharide from the roots of *Symphytum officinale*. *Phytochemistry* 40, no. 1:213–218.

Mohammad, F.V., M. Noorwala, V.U. Ahmad, and B. Sener. 1995b. Bidesmosidic triterpenoidal saponins from the roots of *Symphytum officinale*. *Planta Medica* 61, no. 1:94.

Mroczek, T., K. Glowniak, and A. Wlaszczyk. 2002. Simultaneous determination of N-oxides and free bases of pyrrolizidine alkaloids by cation-exchange solid-phase extraction and ion-pair high-performance liquid chromatography. *Journal of Chromatography A* 949, no. 1–2:249–262.

Noorwala, M., F.V. Mohammad, V.U. Ahmad, and B. Sener. 1994. A bidesmosidic triterpene glycoside from the roots of *Symphytum officinale*. *Phytochemistry* 36, no. 2:439–443.

Olinescu, A., G. Manda, M. Neagu, S. Hristescu, and C. Daşanu. 1993. Action of some proteic and carbohydrate components of *Symphytum officinale* upon normal and neoplastic cells. *Roumanian Archives of Microbiology and Immunology* 52, no. 2:73–80.

Stickel, F., and H.K. Seitz. 2000. The efficacy and safety of comfrey. *Public Health Nutricao* 3, no. 4(A):501–508.

Tabebuia avellanedae (Bignoniaceae) (Lamiales)   **Cytotoxic Pro-apoptotic**

Location: Latin America, Amazon, US
  Appearance: evergreen, gigantic tropical tree

  Leaves: broad
  Flowers: pink to violet

Tradition: known as a useful medicinal plant since the Incan Era
  Parts used: bark, heartwood
  Active ingredients: β-lapachone, naphthoquinones
  Documented target cancers:

- β-lapachone, a quinone obtained from the bark of *Tabebuia avellanedae*, is reported to inhibit the viability of the human hepatoma cell line HepG2 by inducing apoptosis.
- A stereoselective synthesized compound based on the biologically active naphthoquinones from *Tabebuia avellanedae* is reported to exhibit potent cytotoxicity against PC-3 prostate, A549 lung, and MCF-7 breast cancer cell lines.

References

Byeon, S.E., J.Y. Chung, Y.G. Lee, B.H. Kim, K.H. Kim, and J.Y. Cho. 2008. *In vitro* and *in vivo* anti-inflammatory effects of taheebo, a water extract from the inner bark of *Tabebuia avellanedae*. *Journal of Ethnopharmacology* 119, no. 1:145–152.

Woo, H.J., K.Y. Park, C.H. Rhu, et al. 2006. β-Lapachone, a quinone isolated from *Tabebuia avellanedae*, induces apoptosis in HepG2 hepatoma cell line through induction of Bax and activation of caspase. *Journal of Medicinal Food* 9, no. 2:161–168.

Yamashita, M., M. Kaneko, A. Iida, H. Tokuda, and K. Nishimura. 2007. Stereoselective synthesis and cytotoxicity of a cancer chemopreventive naphthoquinone from *Tabebuia avellanedae*. *Bioorganic and Medicinal Chemistry Letters* 17, no. 23:6417–6420.

Tamarindus indica **(Tamarinds) (Leguminosae)**   **Immunomodulator**

Synonyms: Implee. *Tamarinus officinalis* (Hook)
   Location: India, tropical Africa
   Appearance:

   Stem: large handsome tree with spreading branches and a thick straight trunk, 12 m high
   Leaves: alternate, abruptly pinnated
   Flowers: fragrant, yellow-veined, red and purple filaments

Degree of rarity: low, as it is cultivated in West Indies
   Tradition: in Mauritius the Creoles mix salt with the pulp and use it as a liniment for rheumatism and make a decoction of the bark for asthma. The Bengalese employ tamarind pulp in dysentery and in times of scarcity use it as food. The natives of India consider that it is unsafe to sleep under the tree owing to the acid they exhale during the moisture of the night.
   Parts used: fruits freed from brittle outer part of pericarp
   Active ingredients: polysaccharide
   Particular value: it is used as a cathartic, astrigent, febrifuge, antiseptic, refrigerant. It is useful in correcting bilious disorders. A Tamarind pulp is made which is considered as a useful drink in febrile conditions and a good diet in convalescense to maintain a slightly laxative action of the bowels. The pulp is said to weaken the action of resinous cathartics but is frequently prescribed with them as a vehicle for jalap.
   Documented target cancers:

- Immunomodulatory activities such as phagocytic enhancement, leukocyte migration inhibition, and inhibition of cell proliferation.
- Murine cancer cell lines: DLA (Daltons Lymphoma Ascites) and EAC (Ehrlich Ascites Carcinoma)
- A549, human lung adenocarcinoma, KB oral cancer, and MCF-7 breast cancer cell lines

Further details:
- A polysaccharide isolated and purified from *Tamarindus indica* shows immunomodulatory activities such as phagocytic enhancement, leukocyte migration inhibition ,and inhibition of cell proliferation. These properties suggest that this polysaccharide from *T. indica* may have some biological applications.

References

Aravind, S.R., M.M. Joseph, S. Varghese, P. Balaram, and T.T. Sreelekha. 2012. Antitumor and immunopotentiating activity of polysaccharide PST001 isolated from the seed kernel of Tamarindus indica: An *in vivo* study in mice. *Scientific World Journal* 2012:1–14.

Aravind, S.R., M.J. Manu, S.H.E.E.J.A. Varghese, P.R.A.B.H.A.B. Balaram, and T.T. Sreelekha. 2012. Polysaccharide PST001 isolated from the seed kernel of Tamarindus indica induces apoptosis in murine cancer cells. *International Journal Life Science Pharmazeuten Resource* 2:159–172.

Sreelekha, T.T., T. Vijayakumar, R. Ankanthil, K.K. Vijayan, and M.K. Nair. 1993. Immunomodulatory effects of a polysaccharide from *Tamarindus indica*. *Anti-Cancer Drugs* 4, no. 2:209–212.

Terminalia arjuna **(Combretaceae)**   **Anticancer**

Location: indigenous to India, Mauritius medicinal plant
   Appearance: large, evergreen tree, reaches a height of up to 25–30 m

Leaves: elongated conical, green on the top surface of the leaf and brown on the bottom
Flowers: soft yellow flowers
In bloom: March to June

Tradition: commonly known as 'Arjuna', which has been used as a cardiotonic in heart failure, ischemic, cardiomyopathy, atherosclerosis, and myocardium necrosis and has been used for the treatment of different human diseases like blood diseases, anemia, and venereal and viral disease.

Parts used: bark, stem, and leaves

Active ingredients: ellagitanninarjunin along with gallic acid, ethyl gallate, the flavoneluteolin, and tannins

Documented target cancers:

- *T. arjuna* is reported to inhibit the proliferation of human liver cancer cell line HepG2 in a concentration-dependent manner.
- OVCAR-3 (ovarian), SF295 (central nervous system), A498 (renal), NCI-H460 (lung), KM20L2 (colon), and SKMEL-5 (melanoma)
- The antioxidant action of aqueous extract of *T. arjuna* is found to play a role in the anti-carcinogenic activity in lymphoma-bearing AKR mice.

Further details:
- Luteolin has a well established record of inhibiting various cancer cell lines and may account for most of the rationale underlying the use of *T. arjuna* in traditional cancer treatments. Luteolin was also found to exhibit specific activity against the pathogenic bacterium *Neisseria gonorrhoeae*.

References

Amalraj, A., and S. Gopi. 2017. Medicinal properties of *Terminalia arjuna* (Roxb.) Wight & Arn.: A review. *Journal of Traditional and Complementary Medicine* 7, no. 1:65–78.

Kandil, F.E., and M.I. Nassar. 1998. A tannin anti-cancer promotor from *Terminalia arjuna*. *Phytochemistry* 47, no. 8:1567–1568.

Pettit, G.R., M.S. Hoard, D.L. Doubek, J.M. Schmidt, R.K. Pettit, L.P. Tackett, and J.C. Chapuis. 1996. Antineoplastic agents 338. The cancer cell growth inhibitory. Constituents of *Terminalia arjuna* (Combretaceae). *Journal of Ethnopharmacology* 53, no. 2:57–63.

Sivalokanathan, S., M.R. Vijayababu, and M.P. Balasubramanian. 2006. Effects of *Terminalia arjuna* bark extract on apoptosis of human hepatoma cell line HepG2. *World Journal of Gastroenterology* 12, no. 7:1018–1024.

Verma, N., and M. Vinayak. 2009. Effect of *Terminalia arjuna* on antioxidant defense system in cancer. *Molecular Biology Reports* 36, no. 1:159–164.

Trichosanthes kirilowii (Cucurbitaceae) (Cucurbitales)   **Cytotoxic**

Location: East Asia

Appearance: flowering herb

Tradition: in traditional Chinese medicine it is used to reduce fevers, swelling, and coughing. A starch extracted from the root is used for abscesses, amenorrhea, jaundice, and polyuria. Modern Chinese medicinal uses include the management of diabetes and use as an abortifacient. The plant has been used for centuries in the treatment of tumors.

Parts used: seed, fruit, root, leaves

Active ingredients: trichosanthin, karasurin, curcubitacin D, dihydrocucurbitacin D

Documented target cancers:

- The active ingredients curcubitacin D and dihydrocucurbitacin D ghave been reported to induce apoptosis by caspase-3 by phosphorylation in hepatocellular carcinoma cells.
- The active ingredient trichosanthin exerts specific cytotoxic activity towards trophoblastic tumor cells.
- A serine protease extracted from *Trichosanthes kirilowii* is reported to reduce the viability of human colorectal adenocarcinoma cells.

References

Song, L., J. Chang, and Z. Li. 2016. A serine protease extracted from *Trichosanthes kirilowii* induces apoptosis via the PI3K/AKT-mediated mitochondrial pathway in human colorectal adenocarcinoma cells. *Food and Function* 7, no. 2:843–854.

Takahashi, N., Y. Yoshida, T. Sugiura, K. Matsuno, A. Fujino, and U. Yamashita. 2009. Cucurbitacin D isolated from *Trichosanthes kirilowii* induces apoptosis in human hepatocellular carcinoma cells *in vitro*. *International Immunopharmacology* 9, no. 4:508–513.

Tsao, S.W., K.T. Yan, and H.W. Yeung. 1986. Selective killing of choriocarcinoma cells *in vitro* by trichosanthin, a plant protein purified from root tubers of the Chinese medicinal herb *Trichosanthes kirilowii*. *Toxicon* 24, no. 8:831–840.

Trifolium pratense L. (Clover, Red) (Leguminosae)	Chemopreventive

Synonyms: trefoil, purple clover

Location: throughout Europe, central and northern Asia from the Mediterranean to the Arctic Circle and high up into the mountains

Appearance:

Stem: several stems 0.3 to 0.6 m high
Root: one root, slightly hairy
Leaves: ternate, leaflets ovate, nearly smooth

Tradition: fomentations and poultices of the herb have been used as local.

Parts used: leaves, flowers

Active ingredients: isoflavone biochanin A

Particular value: the fluid extract is used as an expectorant and antispasmodic.

Documented target cancers: the ability of the isoflavone biochanin A to inhibit carcinogen activation in cells in culture suggests that *in vivo* studies of this compound as a potential chemopreventive agent are warranted. Additionally the extracts exhibit antiangiogenic effects.

Further details:

- Based on the epidemiological evidence for a relationship between consumption of certain foods and decreased cancer incidence in humans, an assay was developed to screen and fractionate plant extracts for chemopreventive potential. This assay measures effects on the metabolism of [3H] benzo(a)pyrene [B(a)P] in hamster embryo cell cultures. Screening of several plant extracts has generated a number of activity leads. The 95% ethyl alcohol extract of one of these actives, *Trifolium pratense* L. Leguminosae, red clover, significantly inhibited the metabolism of B(a)P and decreased the level of binding of B(a)P to DNA by 30 to 40%. Using activity-directed fractionation by solvent partitioning and then silica gel chromatography, a major active compound was isolated and identified as the isoflavone, biochanin A. The pure compound decreased the metabolism of B(a)P by 54% in comparison to control cultures and decreased B(a)P-DNA binding by 37 to 50% at a dose of 25 micrograms/ml. These studies demonstrate that the hydrocarbon metabolism assay can detect and guide the fractionation of potential anticarcinogens from plants.

- The tannins, delphinidin and procyanidin, were isolated from flowers of white clover (*Trifolium repens*) and the leaves of Arnot Bristly Locust (*Robina fertilis*), respectively, and tested for mutagenic properties in a range of systems. There was no evidence for either compound causing significant levels of frameshift or base-pair mutagenesis in bacterial mutagenicity assays, although both were weakly positive in a bacterial DNA-repair test. Both compounds very slightly increased the frequency of petite mutagenesis in Saccharomyces cerevisiae strain D5. In V79 Chinese hamster cells, both were efficient inducers of micronuclei. In each of these test systems, increasing the potential of the compound for metabolic activation by addition of 'S9' mix had little effect on toxicity or mutagenicity of either tannin. It would seem that potential chromosome-breaking activity of condensed tannins could represent a carcinogenic hazard for animals grazing on pastures of white clover in flower. It may also have wider implications for human carcinogenesis by some, if not all, condensed tannins.

References

Booth, N.L., C.R. Overk, P. Yao, et al. 2006. The chemical and biologic profile of a red clover (*Trifolium pratense* L.) phase II clinical extract. *Journal of Alternative and Complementary Medicine* 12, no. 2:133–139.

Cassady, J.M., T.M. Zennie, Y.H. Chae, M.A. Ferin, N.E. Portuondo, and W.M. Baird. 1988. Use of a mammalian cell culture benzo(a)pyrene metabolism assay for the detection of potential anticarcinogens from natural products: Inhibition of metabolism by biochanin A, an isoflavone from *Trifolium pratense* L. *Cancer Research* 48, no. 22:6257–6261.

Ferguson, L.R., P. van Zijl, W.D. Holloway, and W.T. Jones. 1985. Condensed tannins induce micronuclei in cultured V79 Chinese hamster cells. *Mutation Research* 158, no. 1–2:89–95.

Krenn, L., and D.H. Paper. 2009. Inhibition of angiogenesis and inflammation by an extract of red clover (*Trifolium pratense* L.). *Phytomedicine* 16, no. 12:1083–1088.

Tripterygium hypoglaucum (Celastraceae) (Celastrales)	Anti-proliferative Anti-tumor

---

Location: native in Korea, Japan
  Appearance: deciduous, coniferous shrub

  Leaves: blade, usually oval or oval-rounded, sometimes oblong or elliptical oval
  Flowers: yellowish or white
  Fruit: green-white with three wings, autumn-red

Tradition: in China, *T. wilfordii* has historically been used as a treatment of rheumatoid arthritis.
  Parts used: leaves, root, flowers
  Active ingredients: triepoxide,triplatin (celastrol), triptogelin A-1
  Documented target cancers:

- Celastrol isolated from the plant is reported to inhibit the proliferation of human breast cells by induction of apoptosis mediated by the caspase-dependent mitochondrial pathway.
- Triptogelin A-1, one of the active sesquiterpenes, showed remarkable inhibitory effects for the promotion of murine skin tumor in an *in vivo* two-stage carcinogenicity assay.
- Total alkaloids of *Tripterygium hypoglaucum* have been reported to inhibit the growth of colon cancer cells *in vitro* in a significant dose-dependent manner and to significantly reduce both tumor weight and volume in xenograft mice.

Further details:

- *Tripterigium wilfordii* has been used non-medicinally as an insecticide against maggots or larvae and as a rat and bird poison.

Related compounds: triptolide isolated from the *T. wilfordii* plant can limit pancreatic tumors in mice.

References

Jiang, X., X.C. Huang, L. Ao, et al. 2014. Total alkaloids of *Tripterygium hypoglaucum* (levl.) Hutch inhibits tumor growth both *in vitro* and *in vivo*. *Journal of Ethnopharmacology* 151, no. 1:292–298.

Ujita, K., Y. Takaishi, H. Tokssuda, H. Nishino, A. Iwashima, and T. Fujita. 1993. Inhibitory effects of triptogelin A-1 on 12-O-tetradecanoylphorbol-13-acetate-induced skin tumor promotion. *Cancer Letters* 68, no. 2–3:129–133.

Yang, H.S., J.Y. Kim, J.H. Lee, et al. 2011. Celastrol isolated from *Tripterygium regelii* induces apoptosis through both caspase-dependent and-independent pathways in human breast cancer cells. *Food and Chemical Toxicology* 49, no. 2:527–532.

***Tropaeolum majus* (Nasturtium) (Tropaeolaceae)**   **Antitumor**

Synonyms: garden nasturtium, Indian cress
    Location: South American Andes from Bolivia to Columbia
    Appearance:

Leaves: rounded or kidney shaped, with wavy-margins. Are pale green, about 0.5–1.25 cm across, and are borne on long petioles like an umbrella.

Flowers: bright and happy little flowers. They typically have five petals, although there are double and semi-double varieties. The flowers are about 0.25–0.5 cm in diameter and come in a kaleidoscope of colors including russet, pink, yellow, orange, scarlet, and crimson.

Degree of rarity: low
    Parts used: flowers, leaves, and immature seed
    Active ingredients: benzyl glucosinolate which, through enzymatic hydrolysis, results in the production of benzyl isothiocyanate (BITC).
    Particular value: the dwarf, bushy nasturtiums add rainbows of cheerful color in annual beds and borders. Use the trailing forms on low fences or trellises, on a gravelly or sandy slope, or in a hanging container. Many gardeners include nasturtiums in the salad garden.
    Indicative dosage and application:

- Shows promising cytotoxicity in the low µMolar range (0.86 to 9.4 µM).
- Toxic effects at a dose of 200 mg/kg (within 24 h of drug administration) but no reduction in tumor mass.

Documented target cancers:

- BITC has shown *in vitro* anticancer properties against a variety of human and murine tumor cell lines: human ovarian carcinoma cell lines (SKOV-3, 41-M, CH1, CH1cisR), a human lung tumor (H-69), a murine leukemia (L-1210), and a murine plasmacytoma (PC6/sens).

Figure 9.26 (See color insert.) *Tropaeolum.*

Further details:

- Cultured cells of *Tropaeolum majus* produce significant amounts of benzyl glucosinolate. The *in vitro* anticancer properties of BITC against a variety of human and murine tumor cell lines have been studied by four independent methods: SRB, MTT, cell counting, and clonogenic assays. Regardless of the assay used, BITC showed promising cytotoxicity in the low µMolar range (0.86 to 9.4 µM) against four human ovarian carcinoma cell lines (SKOV-3, 41-M, CH1, CH1cisR), a human lung tumor (H-69), a murine leukemia (L-1210), and a murine plasmacytoma (PC6/sens). The L1210 cells were most sensitive. BITC administered to mice bearing the ADJ/PC6 plasmacytoma subcutaneous tumor showed toxic effects at a dose of 200 mg/kg (within 24 h of drug administration) but no reduction in tumor mass. However, the growth inhibitory properties of BITC against a range of tumor cell types warrant further *in vivo* anti-tumor evaluation as well as its biotechnological production (Figure 9.26).

Reference

Pintão, A.M., M.S. Pais, H. Coley, L.R. Kelland, and I.R. Judson. 1995. *In vitro* and *in vivo* antitumor activity of benzyl isothiocyanate: A natural product from *Tropaeolum majus*. *Planta Medica* 61, no. 3:233–236.

***Tylophora indica* (Apocynaceae) (Gentianales)**   **Cytotoxic**

Location: native to India.

Appearance: climbing plant

Flowers: thin, greenish-yellow or greenish-purple, with oblique pointed petals

Leaves: oval-oblong to elliptic oblong with narrow nose in the base, thick-plush on the bottom surface and smooth on top

Fruit: follicle

Tradition: Tylophora has been used traditionally in the Ayurvedic system for diarrhea probably due to its anti-inflammatory and antimicrobial actions, although human studies have not confirmed this use. Moreover, it has been traditionally used as a folk remedy in certain regions of India for the treatment of bronchial asthma, bronchitis, rheumatism, and dermatitis. In the latter half of the nineteenth century, it was called Indian ipecacuahna, as the roots of the plant have often been employed as an effective substitute for ipecac. Its use for inducing vomiting led to tylophora's inclusion in the Pharmacy of Bengal in 1884.

Parts used: roots, leaves

Active ingredients: tylophorine, tylophorinine, kaempferol

Documented target cancers:

- Alkaloids isolated from tylophora are reported to cause apoptosis in erythroleukemic cells.
- Ethanolic extracts of *Tylophora indica* have been reported cytotoxic against colon cancer HCT-15 cell line.

References

Ganguly, T., and A. Khar. 2002. Induction of apoptosis in a human erythroleukemic cell line K562 by Tylophora alkaloids involves release of cytochrome c and activation of caspase 3. *Phytomedicine* 9, no. 4:288–295.

Khanna, C., S. Singh, and M. Vyas. 2018. Diverse pharmacological potentials of an indigenous climber *Tylophora indica* (Burm. F.) Merr.: A review. *Journal of Pharmacy Research* 12, no. 3:389.

Pratheesh, K.V., V.J. Shine, J. Emima, G.L. Renju, and R. Rajesh. 2014. Study on the anti-cancer activity of *Tylophora indica* Leaf extracts on human colorectal cancer cells. *International Journal of Pharmacognosy and Phytochemical Research* 6, no. 2:355–361.

Rani, A.S., S. Patnaik, G. Sulakshanaand, and B. Saidulu. 2012. Review of Tylophora indica-an antiasthmatic plant. *FS Jornal Resource Basic Appliations Science* 1, no. 3:20–21.

Rao, K.V., R.A. Wilson, and B. Cummings. 1971. Alkaloids of Tylophora III: New alkaloids of Tylophora indica (burm) Merrill and Tylophora dalzellii hook. *Journal of Pharmaceutical Sciences* 60, no. 11:1725–1726.

***Uncaria macrophylla* (Rubiaceae) (Gentianales)**   **Cytotoxic**

Location: South China and Indo-China

Appearance: flowering, climbing plant

Leaves: blade, oval-elliptic, hairless or heart-shaped

In bloom: axillary

Tradition: used in traditional Chinese herbal medicine as spasmolytic, analgesic, and sedative treatments for many symptoms associated with hypertension and cerebrovascular disorders.

Parts used: branches with the hooks

Active ingredients: triterpene urosol, oxindole, macrophyllines A and B, corynantheidine

Documented target cancers:

- A new oleanolic triterpene, $3\beta,6\beta,19\alpha$-trihydroxy-12-oleanen-28-oic acid, is reported with weak cytotoxicity against breast and hepatocellular carcinoma, MCF-7 and HepG2 cell lines.
- Corynantheidine isolated from the the aerial parts of *Uncaria macrophylla* has shown moderate cytotoxicity against SW480 cells (adenocarcinoma of the large intestine).

References

Sun, G., X. Zhang, X. Xu, J. Yang, M. Zhong, and J. Yuan. 2012. A new triterpene from *Uncaria macrophylla* and its antitumor activity. *Molecules* 17, no. 2:1883–1889.

Wang, K., X.Y. Zhou, Y.Y. Wang, et al. 2011. Macrophyllionium and macrophyllines A and B, oxindole alkaloids from *Uncaria macrophylla*. *Journal of Natural Products* 74, no. 1:12–15.

Zhang, Q., C. Lin, W. Duan, X. Wang, and A. Luo. 2013. Preparative separation of six Rhynchophylla alkaloids from *Uncaria macrophylla* wall by pH-zone refining counter-current chromatography. *Molecules* 18, no. 12:15490–15500.

Uncaria tomentosa (Rubiaceae) (Gentianales)   **Cytotoxic Pro-apoptotic**

Location: native to tropical rainforests of the Amazon
    Appearance: woody oak

Leaves: oval-oblong or elliptical with a smooth edge
Flowers: yellow-white

Tradition: the plant has been reportedly used by indigenous people in the Andes to treat inflammation, rheumatism, gastric ulcers, tumors, dysenterysss, and as birth control. Also, it is is popular in South American folk medicine for treating intestinal complaints, gastric ulcers, arthritis, and to promote wound healing.
    Parts used: young branches with hooks, root
    Active ingredients: catechin, proanthocyanides, β-sitosterol, oxindole, oleanolic acid
    Documented target cancers:

*   Breast cancer (*in vitro* research)
*   Colorectal cancer (oxindole isolated from *U. tomentosa* is reported to cause apoptosis in colon adenocarcinomas).

References

de Oliveira, L.Z., I.L. Farias, M.L. Rigo, et al. 2014. Effect of *Uncaria tomentosa* extract on apoptosis triggered by oxaliplatin exposure on HT29 cells. *Evidence-Based Complementary and Alternative Medicine* 2014:274786.

Núñez, C., I. Lozada-Requena, T. Ysmodes, D. Zegarra, F. Saldaña, and J. Aguilar. 2015. Immunomodulation of *Uncaria tomentosa* over dendritic cells, IL-12 and profile TH1/TH2/TH17 in breast cancer. *Revista Peruana de Medicina Experimental y Salud Publica* 32, no. 4:643–651.

Valeriana officinalis (Valerian) (Valerianaceae)   **Cytotoxic**

Synonyms: Amantilla, Setwall, All-Heal
    Location: throughout, mainly in Europe and Northern Asia, in meadows, borders of rivers, and open woods on moist soil
    Appearance:

Stem: erect, up to 1.5–2 m high
Root: conical root-stock or rizhome.
Leaves: opposite, pinnate, up to 20 cm long

Flowers: pink and small, in umbel-like clusters, 5–6 mm long, with a stinking odor (as the whole plant)
Fruit: capsule
In bloom: May–September

Degree of rarity: low.

Tradition: the term *Phu*, a synonym of the root of *valerian*, indicates its stinking scent. The species has probably derived its name from *Valerius*, who first used it in medicine or the Latin word *valere* ('to be in health'). *Valerian* is referred to as a calminative in medical texts of the Middle Age.

Biology: the rhizome develops underground for several years before a flowering stem emerges (only one shoot per root). The plant can be propagated either by runners or by seed. For cultivation, adequate fertilization is recommended.

Parts used: root

Active ingredients: valerianic acid, borneol, a-pirene, camphene, valtrate, choline, valerianates (valerianic acid combines with various bases), chatarine, and valerianine (alkaloids from the root)

Particular value: *valerian* is a powerful nervine, stimulant, carminative, and antispasmodic. It allays pain and promotes sleep. Oil of valerian is used as a remedy for cholera (in a form of cholera drops). The juice of the fresh root (Energetene of *valerian*) has been recommended as a narcotic in insomnia and as anti-convulsant in epilepsy.

Precautions: toxic in high doses. It can cause central paralysis, giddiness, headache, agitation, decrease sensibility, motility, and reflex excitability, nausea.

Indicative dosage and application: still testing. A proposal dose is 300 and 500 mg/kg/24 h (in rats) but not yet confirmed.

Documented target cancers:

- Weak cell growth inhibition in 549 (human lung adenocarcinoma)
- HCT116 (human colon carcinoma)
- SK-BR-3 (human breast carcinoma) and
- HepG2 (human hepatoma) cell lines

Further details:

- Reiterated administration of *Valeriana officinalis* to laboratory animals has been associated with toxic effects. Rats received 300 and 600 mg/kg/24 h of the drug for 30 days. During the period of the treatment, animals' weight and blood pressure have been measured. On the end of the treatment the animals have been sacrificed. The principal organs have been weighed, and in blood samples collected hematological and biochemical parameters have been determined. This work is concerned with pharmacological properties which are related to the two plants. The influence of the drugs on the behavior, the pain, the intestinal peristalsis, and strychnine convulsions are reported.
- Colchicine-treated suspension cultures of *Valeriana wallichii* produce higher amounts of valepotriates than did the respective untreated cultures. The ability to produce valepotriates in the treated culture remains in the absence of colchicine even if the chromosome status returns to normal. When the colchicine treatment is repeated, a further increase in valepotriate production can be obtained. Besides known valepotriates, a series of 14 new compounds, hitherto not described for the parent plant, were isolated from the cell suspension culture. Eight of them are also found in plant parts in minor amounts, but six seem to be present only in tissue cultures of *V. wallichii*.
- Different *in vitro* cultures of Valerianaceae were analysed for valepotriate content [(iso)valtrate, acevaltrate, didrovaltrate] in a study on properties of production *in vitro* (plant species, growth conditions, differentiation level, valepotriate content of the medium after growth). The *in vitro* cultures were: callus cultures of *Valeriana officinalis* L., *Valerianella locusta* L., and *Centranthus ruber* L.DC.; a suspension culture of *Valeriana officinalis* L. and a root organ culture of *Centranthus ruber* L.DC. All of the cultures produced valepotriates *in vitro* in different amounts. None of the media that had served for growth contained any valepotriates. In order to characterize the *in vitro*

Figure 9.27   (See color insert.) *Valeriana.*

growth more precisely, different parameters (such as fresh and dry weight, lipid and nitrogen content, and *(iso)valtrate* content) were analysed at different time intervals during a growth period in one of the cultures (callus culture of *Valeriana officinalis* L.).

- It is possible directly to separate and analyse, quantitatively and qualitatively, the valepotriates from *Valeriana* crude extracts or from commercial *Valeriana* preparations by high-performance liquid chromatography. The separations are achieved on 4 or 8 mm I.D. columns packed with silica gel (particle size 10 micron) with n-hexane-ethyl acetate mixtures as eluent. A refractive index detection system is necessary for determining all of the valepotriates. If the concentration differences between didrovaltratum and valtratum are very great, an ultraviolet (UV) detector must be used, and the determination must be conducted in two steps. For valtratum drugs UV detection alone will suffice. As internal standards p-dimethylaminobenzaldehyde should be used for extracts and preparations from valtratum races and benzaldehyde in the presence of didrovaltratum races. This determination is superior to the combined thin-layer chromatographic-hydroxamic acid method used hitherto with respect to time consumption, precision, and sensitivity (Figure 9.27).

References

Albrecht, M., and W. Berger. 1995. Psychopharmaceuticals and safety in traffic. Zeits Allegmeinmed 71:1215–1221.

Becker, H., and S. Chavadej. 1985. Valepotriate production of normal and colchicine-treated cell suspension cultures of *Valeriana wallichii*. *Journal of Natural Products* 48, no. 1:17–21.

Becker, H., R. Schrall, and W. Hartmann. 1977. Callus cultures of a valerian species. 1. Installation of a callus culture of *Valeriana wallichii* DC and 1st analytical studies. *Archiv der Pharmazie (Weinheim)* 310, no. 6:481–484.

Brown, D.J. 1996. *Herbal Prescriptions for Better Health.* Prima Publishing, Rocklin, CA.

Bucková, A., K. Grznár, M. Haladová, and E. Eisenreichová. 1977. Active substances in *Valeriana officinalis* L. Cesk Farm 26, No. 7. *Ceskoslovenska Farmacie* 26, no. 7:308–309.

Cavadas, C., I. Araújo, M.D. Cotrim, T. Amaral, A.P. Cunha, T. Macedo, and C.F. Ribeiro. 1995. *In vitro* study on the interaction of Valeriana officinalis L. extracts and their amino acids on GABAA receptor in rat brain. *Arzneimittel-Forschung* 45, no. 7:753–755.

Czabajska, W., M. Jaruzelski, and D. Ubysz. 1976. New methods in the cultivation of *Valeriana officinalis*. *Planta Medica* 30, no. 1:9–13.

Della Loggia, R., A. Tubaro, and C. Redaelli. 1981. Evaluation of the activity on the mouse CNS of several plant extracts and a combination of them. *Rivista di Neurologia* 51, no. 5:297–310.

Dressing, H., S. Köhler, and W.E. Müller. 1996. Improvement of sleep quality with a high-dose valerian/lemon balm preparation: A bo-controlled double-blind study. *Psychopharmakotherapie* 6:32–40.

Fehri, B., J.M. Aiache, K. Boukef, A. Memmi, and B. Hizaoui. 1991. *Valeriana officinalis* and *Crataegus oxyacantha*: Toxicity from repeated administration and pharmacologic investigations. *Journal de Pharmacie de Belgique* 46, no. 3:165–176.

Fursa, M.S. 1980. Composition of the flavonoids of *Valeriana officinalis* from the Asiatic part of the USSR. Farm Zhurnal 3:72–73.

Han, Z.Z., Z.H. Yan, Q.X. Liu, X.Q. Hu, J. Ye, H.L. Li, and W.D. Zhang. 2012. Acylated iridoids from the roots of Valeriana officinalis var. latifolia. *Planta Medica* 78, no. 15:1645–1650.

Hendriks, H., R. Bos, D.P. Allersma, T.M. Malingré, and A.S. Koster. 1981. Pharmacological screening of valerenal and some other components of essential oil of *Valeriana officinalis*. *Planta Medica* 42, no. 1:62–68.

Hromádková, Z., A. Ebringerová, and P. Valachovic. 2002. Ultrasound-assisted extraction of water-soluble polysaccharides from the roots of valerian (*Valeriana officinalis* L.). *Ultrasonics Sonochemistry* 9, no. 1:37–44.

Janot, M.M., J. Guilhem, O. Contz, G. Venera, and E. Cionga. 1979. Contribution to the study of valerian alcaloids (*Valeriana officinalis*, L.): Actinidine and naphthyridylmethylketone, a new alkaloid. *Annales Pharmaceutiques Francaises* 37, no. 9–10:413–420

Kohnen, R., and W.D. Oswald. 1988. The effects of valerian, propranolol and their combination on activation performance and mood of healthy volunteers under social stress conditions. *Pharmacopsychiatry* 21, no. 6:447–448.

Kornilievs' kyĭ, I., M.S. Fursa, A.S. Rybal'chenko, and K. Koreshchuk. 1979. Flavonoid makeup of Valeriana officinalis from the southern and central provinces of the Ukraine. Farm Zh. 4:71–72.

Leathwood, P.D., and F. Chauffard. 1982–1983. Quantifying the effects of mild sedatives. *Journal of Psychiatric Research* 17, no. 2:115–122.

Leathwood, P.D., and F. Chauffard. 1985. Aqueous extract of valerian reduces latency to fall asleep in man. *Planta Medica* 51, no. 2:144–148.

Leathwood, P.D., F. Chauffard, E. Heck, and R. Munoz-Box. 1982. Aqueous extract of valerian root (*Valeriana officinalis* L.) improves sleep quality in man. *Pharmacology, Biochemistry, and Behavior* 17, no. 1:65–71.

Mennini, T., and P. Bernasconi. 1993. *In vitro* study on the interaction of extracts and pure compounds from *Valeriana officinalis* roots with GABA, benzodiazepine and barbiturate receptors. *Fitoterapia* 64:291–300.

Nikul'shina, N.I., V.A. Talan, V.G. Bukharov, and V.M. Ivanova. 1969. Valeroside A – A glycoside from valerian (*Valeriana officinalis* L.). *Farmatsiia* 18, no. 6:44–47.

Pank, F., H.J. Hannig, J. Hauschild, and B. Zygmunt. 1980. Chemical weed control in the cropping of medicinal plants. Part 1: Valerian (*Valeriana officinalis* L.). *Die Pharmazie* 35, no. 2:115–119.

Paris, R., P. Besson, and A. Hérisset. 1966. Tests of 'industrial lyophilization' of medicinal plants. 3. *Valeriana officinalis* L. influence of lyophilization on the quality of the drug. *Annales Pharmaceutiques Francaises* 24, no. 11:669–674.

Perebeĭnos, V.S. 1974. Permissible content of stalk residues in crude *Valeriana officinalis*. *Farmatsiia* 23, no. 3:72–76.

Santos, M.S., F. Ferreira, A.P. Cunha, A.P. Carvalho, C.F. Ribeiro, and T. Macedo. 1994. Synaptosomal GABA release as influenced by valerian root extract – Involvement of the GABA carrier. *Archives Internationales de Pharmacodynamie et de Therapie* 327, no. 2:220–231.

Suomi, J., S.K. Wiedmer, M. Jussila, and M.L. Riekkola. 2001. Determination of iridoid glycosides by micellar electrokinetic capillary chromatography-mass spectrometry with use of the partial filling technique. *Electrophoresis* 22, no. 12:2580–2587.

Tamamura, K., M. Kakimoto, M. Kawaguchi, and T. Iwasaki. 1973. Pharmacological studies on the constituents of crude drugs and plants. 1. Pharmacological actions of *Valeriana officinalis* Linne. var. latifolia Miquel. *Yakugaku Zasshi* 93, no. 5:599–606.

Tittel, G., and H. Wagner. 1978. High-performance liquid chromatographic separation and quantitative determination of valepotriates in valeriana drugs and preparations. *Journal of Chromatography* 1, no. 148(2):459–468.

Torssell, K., and K. Wahlberg. 1967. Isolation, structure and synthesis of alkaloids from *Valeriana officinalis* L. *Acta Chem. Scandinavica* 21, no. 1:53–62.

Tucakov, J. 1965. Comparative ethnomedical study of Valeriana officinalis L. *Glas – Srpska akademija nauka i umetnosti* 18:131–150.

Tufik, S., K. Fujita, Mde L. Seabra, and L.L. Lobo. 1994. Effects of a prolonged administration of valepotriates in rats on the mothers and their offspring. *Journal of Ethnopharmacology* 41, no. 1–2:39–44.

Verzarne Petri, G. 1974. Biosynthesis of alkaloids, valtrates and volatile oils in the roots of *Valeriana officinalis* L. from radioactive precursors. *Acta Pharmaceutica Hungarica* 0, (0 Suppl 1):54–65.

Violon, C., N. Van Cauwenbergh, and A. Vercruysse. 1983. Valepotriate content in different *in vitro* cultures of Valerianaceae and characterization of *Valeriana officinalis* L. callus during a growth period. *Pharmaceutisch Weekblad. Scientific Edition* 5, no. 5:205–209.

Wagner, H., R. Schaette, L. Hörhammer, and J. Hölzl. 1972. Dependence of the valepotriate and essential oil content in *Valeriana officinalis* L.s.l. *Arzneimittel-Forschung* 22, no. 7:1204–1209.

Yang, G.Y., and W. Wang. 1994. Clinical studies on the treatment of coronary heart disease with *Valeriana officinalis var latifolia. Zhongguo Zhong Xi Yi Jie He Za Zhi Zhongguo Zhongxiyi Jiehe Zazhi = Chinese Journal of Integrated Traditional and Western Medicine* 14, no. 9:540–542.

Zhang, B.H., H.P. Meng, T. Wang, et al. 1982. Effects of *Valeriana officinalis* L. extract on cardiovascular system. *Yao Xue Xue Bao = Acta Pharmaceutica Sinica* 17, no. 5:382–384. 192.

Ventilago maderaspatana (Rhamnaceae) (Rosales)   **Cytotoxic**

Location: Indian-Malaysian peninsula
  Appearance: large, woody, evergreen climbing shrub

  Leaves: smooth or thin velvety
  Flowers: small, yellow-green
  Fruit: oak, yellowish

Tradition: *V. maderaspatana* is traditionally used to treat disorders like skin problems, fever, and diabetes. The root bark is used as a carminative, stomachic, stimulant, thermogenic, alexeteric, flatulence, and tonic. Bark and leaves are used to cure malarial fever, while tender branches are also used to treat vertigo. Latex of this plant is used to cure edema.

  Parts used: bark, root, leaf, seed

  Active ingredients: ventinone-A & B, pseudozerbin, physicion, emodin, insulinicin, xanthorin, xanthorin-5-methyl, freedelin, oleanolic acid, and 1-hydroxy naphthoquinone

  Documented target cancers:

- The compounds oleanolic acid and 1-hydroxy naphthoquinone exhibited promising cytotoxicity against A375 (malignant skin melanoma) cells.
- The ethanolic extract of the plant showed moderate cytotoxicity towards breast cancer MCF-7 cell line.

References

Chittethu, A.B., S. Sathianarayanan, A. Nair, E. Varghese, R.V. Gopal, and K.S. Sreelakshmi. 2011. Prelimimary phytochemical screening and cytotoxic activity of ethanolic extract of *Ventilago madraspatana* against human breast cancer. *International Journal of Pharmacology and Biological Sciences* 5, no. 2:75–78.

Ghosh, S., M. Das Sarma, A. Patra, and B. Hazra. 2010. Anti-inflammatory and anticancer compounds isolated from *Ventilago madraspatana* Gaertn., *Rubia cordifolia* Linn. and *Lantana camara* Linn. *Journal of Pharmacy and Pharmacology* 62, no. 9:1158–1166.

Periyasamy, K., and K. Saravanan. 2016. Ethnobotanical, phytochemical and pharmceutical studies of medicinal plant, *Ventilago maderaspatana* gaertn (red creeper): A review. *International Journal of Current Pharmaceutical Research* 8, no. 1:16–18.

Veratrum dahuricum (Liliaceae) (Liliales)	**Anti-tumor**	**Cytotoxic**

Location: South Korea, China, Siberia
  Appearance: herbaceous, tuberous rhizome

  Leaves: very large, dense, ovoid, and wrap the shoot alternately
  Flowers: colorless or greenish, the form of straw-shaped bunch at the end of the stem

Tradition:
  Native Americans used the juice from the roots to poison arrows before combat. The dried powdered root was also used as an insecticide.
  Western American Indians combined amounts of the winter-harvested root of this plant with *Salvia dorii* to potentiate its effects and reduce the toxicity of the herb. The plant's teratogenic properties and ability to induce severe birth defects were well known to Native Americans.
  Parts used: rhizomes, roots
  Active ingredients: verramin, veramitaline, cyclopamine, germerin, zervin, zerbic acid, pseudozerbin, veratrosin, veratrine A, B, resveratrol
  Documented target cancers:

- Cyclopamine is reported to exert an inhibitory activity against the growth of PANC-1 tumors in mice (pancreatic tumor).
- *In vitro*, the cytotoxic activity of six alkaloids (veratramine, veramitaline, veratrosine, cyclopamine, zerbin, and germerin) was investigated and germerin was found to have selective cytotoxicity against two cell lines: SW1990 and NCI-H249 (pancreatic cancer and lung cancer).
- Inhibitory effect of resveratol on T24 cells (bladder carcinoma)

Precautions:

- The rhizome dust can irritate the mucous membrane of the eye, nose, and mouth.
- Its alkaloids are powerful poisons.

Related species: *V. album* var. Dahuricum (White Helleborough): the therapeutic use of plant tincture greatly reduces blood pressure. It is used for severe cases of hypertension, myasthenia gravis, and progressive muscular dystrophy.

References

Tang, J., H.L. Li, Y.H. Shen, et al. 2010. Antitumor and antiplatelet activity of alkaloids from *Veratrum dahuricum*. *Phytotherapy Research* 24, no. 6:821–826.

Tang, J., H.L. Li, Y.H. Shen, H.Z. Jin, S.K. Yan, R.H. Liu, and W.D. Zhang. 2008. Antitumor activity of extracts and compounds from the rhizomes of *Veratrum dahuricum*. *Phytotherapy Research* 22, no. 8:1093–1096.

Xuetao, M.A., M. Liu, Z. Wu, and Y. Li. 2015. Anti-proliferative activity of active constituents in *Veratrum dahuricum* on human bladder cancer T24 cells. *Biomedical Research* 26, no. 4:672–676.

***Vernonia scorpiodes* (Asteraceae) (Asterales)**   **Anti-tumor Anti-leukemic**

Location: South America, Brazil
   Appearance: lianous herb, climbing with many branches

Leaves: simple, blade, elliptic-lanceolate
Flowers: intense purple color

Tradition: used in Brazil folk medicine, popularly known as piracá. The topical application of the alcoholic extract of its fresh leaves is widely used to treat a variety of skin disorders, including chronic wounds such as ulcers of the lower limbs.
   Parts used: leaves, root
   Active ingredients: triterpenes, steroids, flavonoids, sesquiterpenes, polyacetylenic γ-lactone
   Documented target cancers:

- Polyacetylene may be very promising in cancer research, since some of its derivatives exhibit interesting *in vitro* inhibitory activity against tumor, leukemic and lung cancer cell lines.
- The dichloromethane fraction of *V. scorpioides* leaf extract is reported to totally inhibit tumor development when in direct contact with tumor cells and also ascitic tumor development with *in vitro* treatment on Ehrlich ascitic and solid tumor-bearing mice.

References

Machado, A.L., F.M. Aragão, P.N. Bandeira, et al. 2013. Chemical constituents of *Vernonia scorpioides* (Lam) Pers. (Asteraceae). *Química Nova* 36, no. 4:540–543.
Pagno, T., L.Z. Blind, M.W. Biavatti, and M.R.O. Kreuger. 2006. Cytotoxic activity of the dichloromethane fraction from *Vernonia scorpioides* (Lam.) Pers. (Asteraceae) against Ehrlich's tumor cells in mice. *Brazilian Journal of Medical and Biological Research = Revista Brasileira de Pesquisas Medicas e Biologicas* 39, no. 11:1483–1491.
Pollo, L.A., C.F. Bosi, A.S. Leite, et al. 2013. Polyacetylenes from the leaves of *Vernonia scorpioides* (Asteraceae) and their antiproliferative and antiherpetic activities. *Phytochemistry* 95:375–383.
Pollo, L.A.E., M.J. Frederico, A.J. Bortoluzzi, F.R.M.B. Silva, and M.W. Biavatti. 2018. A new polyacetylene glucoside from *Vernonia scorpioides* and its potential antihyperglycemic effect. *Chemico-Biological Interactions* 279:95–101.
Rauh, L.K., C.D. Horinouchi, A.M. Loddi, et al. 2011. Effectiveness of *Vernonia scorpioides* ethanolic extract against skin inflammatory processes. *Journal of Ethnopharmacology* 138, no. 2:390–397.

***Viola odorata* (Violet sweet) (Violaceae)**   **Cytotoxic**

Other names: sweet-scented violet
   Location: in tropical and temperate regions of the world, in deciduous woods and hedges
   Appearance:

Stem: slightly hairy, up to 10 cm high
Root: stolon, up to 20 cm long
Leaves: rounded, sagittate to heart-shaped, slightly hairy, alternate, up to 6 cm long. The two halves of the young leaves are rolled in two coils.
Flowers: deep purple (occasionally white or pink), fragrant, with yellow stamens, 0.5–1.5 cm
Fruit: three-valved capsule

In bloom: February–April. Flowers produced in autumn are very small, with no apparent flower-like structure and not fragrant (*cleistogamous*) but are highly seed-setting.

Biology: a perennial plant, violet is propagated either by seed or cuttings (scions). The flowers are great attractors of bees and other insects, due to their high honey content. It is recommended to avoid cultivation near air-polluted areas, because the hairy parts can become accumulating points for smog.

Degree of rarity: very low

Tradition: the species is supposed to have derived its name from *Viola*, the Latin form of the Greek name *Ione* or *Io*, who was turned into a plant by her beloved Jupiter, the flowers emerging right above the earth so that she could use them as food. Another Greek myth claims that violet emerged on the spot where a resting Orpheus laid his lyre. Homer and Virgil have mentioned the calming and sedative properties of the plant. Exactly the same properties made the species associated with death, as referred to by Shakespeare in *Hamlet*.

Parts used: whole plant fresh, flowers, and leaves dried, rhizomes

Active ingredients: Cyclopentenyl cytosine, Cycloviolacin O2

Particular value: violet flowers possess slightly laxative properties, well known in the form of syrup. It is also used in ague, epilepsy inflammatation of the eyes, sleeplessness.

Precautions: rhizomes are strongly emetic and purgative.

Indicative dosage and application: it has not yet standardized a dose, for example on human glioblastoma cells the levels of the drug range from 0.01 to 1 µM.

Documented target cancers: Cyclopentenyl cytosine (CPEC) and Cycloviolacin O2 exert antiproliferative effects against a wide variety of human and murine tumor lines.

Further details:

- Cyclopentenyl cytosine (CPEC) inhibits the proliferation of tumor cell lines, including a panel of human gliosarcoma and astrocytoma lines. This effect is produced primarily by the 5′-triphosphate metabolite CPEC-TP, an inhibitor of cytidine-5′-triphosphate (CTP) synthase (EC 6.3.4.2). This has been demonstrated, for example, on human glioblastoma cells obtained at surgery and exposed to the drug at levels ranging from 0.01 to 1 µM for 24 h. Dose-dependent accumulation of CPEC-TP was accompanied by a concomitant decrease in CTP pools, with 50% depletion of the latter being achieved at a CPEC level of ca. 0.1 µM. Human glioma cell proliferation was inhibited 50% by 24-h exposure to 0.07 µM CPEC. Post-exposure decay of CPEC-TP was slow, with a half time of 30 h. DNA cytometry showed a dose-dependent shift in cell cycle distribution, with an accumulation of cells in S-phase. The pharmacological effects of CPEC on freshly excised glioblastoma cells are quantitatively similar to those seen in a range of established tissue culture lines, including human glioma, colon carcinoma, and MOLT-4 lymphoblasts, supporting the recommendation that the drug may be advantageous for the treatment of human glioblastoma.

References

Agbaria, R., J.A. Kelley, J. Jackman, J.J. Viola, Z. Ram, and E.J. Oldfield. 1997. Antiproliferative effects of cyclopentenyl cytosine (NSC 75575) in human glioblastoma cells. *DG Oncoles* 9, no. 3:111–118.

Crucitti, F., G. Doglietto, D. Frontera, G. Viola, and M. Buononato. 1995. Carcinoma of the pancreatic head area. Therapy: Resectability and surgical management of resectable tumors. *Rays* 20, no. 3:304–315.

De Berardis, B., G. Torresini, V. Viola, G. Imondi, S. Marinelli, and F. Di Pietrantonio. 2000. Recurrent giant retroperitoneal leiomyosarcoma. Report of a clinical case. *Il Giornale di Chirurgia* 21, no. 5:239–241.

Fazio, V., V. Messina, A. Marino, F. Di Trapani, and V. Viola. 2002. Treatment with self-expanding metallic enteral stents in occlusion caused by neoplastic stenosis of the sigmoid and rectum. *Chirurgia Italiana* 54, no. 2:233–239.

Ferrara, F., M. Annunziata, E.M. Schiavone, et al. 2001. High-dose idarubicin and busulphan as conditioning for autologous stem cell transplantation in acute myeloid leukemia: A feasibility study. *Hematology Journal* 2, no. 4:214–219.

Ferrara, F., S. Palmieri, B. Pocali, et al. 2002. De novo acute myeloid leukemia with multilineage dysplasia: Treatment results and prognostic evaluation from a series of 44 patients treated with fludarabine, cytarabine and G-CSF (FLAG). *European Journal of Haematology* 68, no. 4:203–209.

Frontera, D., G. Doglietto, G. Viola, and F. Crucitti. 1995. Carcinoma of the pancreatic head area. Epidemiology, natural history and clinical findings. *Rays* 20, no. 3:226–236.

Gerlach, S.L., R. Rathinakumar, G. Chakravarty, U. Göransson, W.C. Wimley, S.P. Darwin, and D. Mondal. 2010. Anticancer and chemosensitizing abilities of cycloviolacin O2 from Viola odorata and psyle cyclotides from Psychotria Leptothyrsa. *Biopolymers* 94, no. 5:617–625.

Krygier, G., K. Lombardo, C. Vargas, et al. 2001. Familial uveal melanoma: Report on three sibling cases. *British Journal of Ophthalmology* 85, no. 8:1007–1008.

Longo, V.D., K.L. Viola, W.L. Klein, and C.E. Finch. 2000. Reversible inactivation of superoxidesensitive aconitase in Abeta1–42-treated neuronal cell lines. *Journal of Neurochemistry* 75, no. 5:1977–1985.

Madajewicz, S., P. Hentschel, P. Burns, et al. 2000. Phase I chemotherapy study of biochemical modulation of folinic acid and fluorouracil by gemcitabine in patients with solid tumor malignancies. *Journal of Clinical Oncology* 18, no. 20:3553–3557.

Morini, A., V. Manera, C. Boninsegna, L. Viola, and D. Orrico. 2000. Mandibular drop resulting from bilateral metastatic trigeminal neuropathy as the presenting symptom of lung cancer. *Journal of Neurology* 247, no. 8:647–649.

Palmieri, S., L. Sebastio, G. Mele, et al. 2002. High-dose cytarabine as consolidation treatment for patients with acute myeloid leukemia with t(8;21). *Leukemia Research* 26, no. 6:539–543.

Villani, F., G. Viola, C. Vismara, A. Laffranchi, A. Di Russo, S. Viviani, and V. Bonfante. 2002. Lung function and serum concentrations of different cytokines in patients submitted to radiotherapy and intermediate/high dose chemotherapy for Hodgkin's disease. *Anticancer Research* 22, no. 4:2403–2408.

Wikstroemia indica (Wikstroemia) (Thymelaeaceae)	Anti-leukemic

Location: Guam and Micronesia
  Appearance:

Stem: shrub with smooth, reddish bark
Leaves: opposite, light green that is rounded at both ends
Flowers: small, yellowish green, grow in racemes from the leaf axils

Active ingredients: Daphnoretin, tricin, kaempferol-3-O-beta-D-glucopyranoside, and (+)-nortrachelogenin, wikstroelides
  Documented target cancers:

- It is used against Ehrlich ascites carcinoma (mice) (daphnoretin), as antileukemic (tricin, kaempferol-3-O-beta-D-glucopyranoside, and nortrachelogenin), and against P-388 lymphocytic leukemia.
- Daphnoretin from the bark of *Wikstroemia indica* has been reported to exhibit significant inhibition on the proliferation of cancer CNE and HeLa cells.

Further details:

- *Wikstroemia indica* (Thymelaeaceae): the bark contains kaempferol-3-O-beta-D-glucopyranoside, huratoxin, pimelea factor P2, wikstroelides A-G, daphnane-type diterpenoids (wikstroelides H-O), tricin, kaempferol-3-O-beta-D-glucopyranoside, (+)-nortrachelogenin, daphnoretin, tricin, kaempferol-3-O-beta-D-glucopyranoside, and (+)-nortrachelogenin.
- The ethanol extracts of *Wikstroemia foetida* var. oahuensis and *Wikstroemia uva-ursi* showed antitumor activity against the P-388 lymphocytic leukemia (3PS) test system. One PS-active constituent of both plants was the lignan wikstromol (Figure 9.28).

Figure 9.28 (See color insert.) *Wikstroemia indica.*

References

Abe, F., Y. Iwase, T. Yamauchi, K. Kinjo, S. Yaga, M. Ishii, and M. Iwahana. 1998. Minor daphnane-type diterpenoids from *Wikstroemia retusa. Phytochemistry* 47, no. 5:833–837.

Li, Y.M., L. Zhu, J.G. Jiang, L. Yang, and D.Y. Wang. 2009. Bioactive components and pharmacological action of *Wikstroemia indica* (L.) CA Mey and its clinical application. *Current Pharmaceutical Biotechnology* 10, no. 8:743–752.

Lu, C.L., Y.M. Li, G.Q. Fu, et al. 2011. Extraction optimisation of daphnoretin from root bark of *Wikstroemia indica* (L.) CA and its anti-tumour activity tests. *Food Chemistry* 124, no. 4:1500–1506.

Wang, H.K., Y. Xia, Z.Y. Yang, S.L. Natschke, and K.H. Lee. 1998. Recent advances in the discovery and development of flavonoids and their analogues as antitumor and anti-HIV agents. *Advances in Experimental Medicine and Biology* 439:191–225.

Withania somnifera (Solanaceae) (Solanales)	Cytotoxic

Location: Mediterranean region through tropical Africa to South Africa, Canary and Cape Verde Islands to the Middle East and Arabia, India, Sri Lanka, and southern China

Appearance: small shrub up to 2 m high

Leaves: alternate, simple, broadly ovate, obovate, or oblong
Fruit: red fruit covered by the brownish, papery, inflated calyx

Tradition: in traditional medicine in southern Africa the leaves are used to heal open as well as septic, inflamed wounds, abscesses, inflammation, haemorrhoids, rheumatism, and syphilis.

Parts used: leaves and roots

Active ingredients: alkaloids and steroids (withanolides)

Documented target cancers:

- *W. somnifera* has been reported to exhibit inherent metastatic and selective inhibitory potential against PC3 prostate cancer cells.
- An aqueous extract of the root of *W. somnifera* is reported to decrease pro-inflammatory cytokine levels in leukemic THP-1 cells, which may alleviate cancer cachexia and excessive leukemic cell growth.

References

Naidoo, D.B., A.A. Chuturgoon, A. Phulukdaree, K.P. Guruprasad, K. Satyamoorthy, and V. Sewram. 2018. *Withania somnifera* modulates cancer cachexia associated inflammatory cytokines and cell death in leukaemic THP-1 cells and peripheral blood mononuclear cells (PBMC's). *BMC Complementary and Alternative Medicine* 18, no. 1:126.

Sehrawat, A., S.K. Samanta, S.H. Kim, E.R. Hahm, and S.V. Singh 2017. Scientific evidence for anticancer effects of withania somnifera and its primary bioactive component withaferin A. In: *Science of Ashwagandha: Preventive and Therapeutic Potentials*. Kaul, S., and R. Wadhwa (Eds.). Springer, Cham,175–196.

Setty Balakrishnan, A., A.A. Nathan, M. Kumar, S. Ramamoorthy, and S.K. Ramia Mothilal. 2017. *Withania somnifera* targets interleukin-8 and cyclooxygenase-2 in human prostate cancer progression. *Prostate International* 5, no. 2:75–83.

Xanthium strumarium (Cocklebur) (Compositae)   **Cytotoxic**

Location: South Europe, in America near sea coast, central Asia northwards to the Baltic
   Appearance:

Stem: coarse, erect, annual, 0.3–0.6 m high
Leaves: on long stalks, large, broadly heart-shaped, coarsely toothed or angular in both sides
Flowers: heads, greenish yellow, terminal clusters on short racemes, upper ones male, lower female

Parts used: the whole plant
   Active ingredients: xanthatin
   Particular value: a valuable and sure specific in the treatment of hydrophobia
   Precautions: intoxication
   Indicative dosage and application: under investigation
   Documented target cancers:

* Serofibrinous ascites, edema of the gallbladder wall, and lobular accentuation of the liver
* Cultured human tumor cells: A549 (non-small cell lung), SK-OV-3 (ovary), SK-MEL-2 (melanoma), XF498 (central nervous system), and HCT-15 (colon) *in vitro*
* Active fractions of the methanolic extract of *Xanthium strumarium* are reported to increase the life-span of tumor-bearing animals.

Further details:
* Cocklebur (*Xanthium strumarium*) fed to feeder pigs was associated with acute to subacute hepatotoxicosis. Cotyledonary seedings fed at 0.75 to 3% of body weight or ground bur fed at 20 to 30% of the ration caused acute depression, convulsions, and death. Principle gross lesions were marked serofibrinous ascites, edema of the gallbladder wall, and lobular accentuation of the liver. Acute to subacute centrilobular hepatic necrosis was present microscopically. The previously reported toxic principle, hydroquinone, was not recovered from the plant or bur of *X. strumarium*. Authentic hydroquinone administered orally failed to produce lesions typical of cocklebur intoxication but did produce marked hyperglycemia. Carboxyatractyloside recovered from the aqueous extract of *X. strumarium* and authentic carboxyatractyloside, when fed to pigs, caused signs and lesions typical of cocklebur intoxication. Marked hypoglycemia and elevated serum glutamic oxaloacetic transaminase and serum isocitric dehydrogenase concentrations occurred in pigs with acute hepatic necrosis that had received either cocklebur seedlings, ground bur, or carboxyatractyloside

References

Aranjani, J.M., A. Manuel, C. Mallikarjuna Rao, et al. 2013. Preliminary evaluation of *in vitro* cytotoxicity and *in vivo* antitumor activity of *Xanthium strumarium* in transplantable tumors in mice. *American Journal of Chinese Medicine* 41, no. 1:145–162.

Battle, R.W., J.K. Gaunt, and D.L. Laidman. 1976. The effect of photoperiod on endogenous gamma-tocopherol and plastochromanol in leaves of *Xanthium strumarium* L. (cocklebur). *Biochemical Society Transactions* 4, no. 3:484–486.

Chu, T.R., and Y.C. Wei. 1965. Studies on the principal unsaturated fatty acids of the seed oil of *Xanthium strumarium* L. *Yao Xue Xue Bao. Yao Xue Xue Bao = Acta Pharmaceutica Sinica* 12, no. 11:709–712.

Cole, R.J., B.P. Stuart, J.A. Lansden, and R.H. Cox. 1980. Isolation and redefinition of the toxic agent from cocklebur (*Xanthium strumarium*). *Journal of Agricultural and Food Chemistry* 28, no. 6:1330–1332.

Hatch, R.C., A.V. Jain, R. Weiss, and J.D. Clark. 1982. Toxicologic study of carboxyatractyloside (active principle in cocklebur – *Xanthium strumarium*) in rats treated with enzyme inducers and inhibitors and glutathione precursor and depletor. *American Journal of Veterinary Research* 43, no. 1:111–116.

Jain, S.R. 1968. Investigations on antileucodermic activity of *Xanthium strumarium*. *Planta Medica* 16, no. 4:467–468.

Kapoor, V.K., A.S. Chawla, A.K. Gupta, and K.L. Bedi. 1976. Studies on the oil of *Xanthium strumarium*. *Journal of the American Oil Chemists' Society* 53, no. 8:524.

Khafagy, S.M., N.N. Sabry, A.M. Metwally, and S.F. el-Naggar. 1974. Phytochemical investigation of *Xanthium strumarium*. *Planta Medica* 26, no. 1:75–78.

Kim, Y.S., J.S. Kim, S.H. Park, et al. 2003. Two cytotoxic sesquiterpene lactones from the leaves of Xanthium strumarium and their *in vitro* inhibitory activity on farnesyltransferase. *Planta Medica* 69, no. 4:375–377.

Kuo, Y.C., C.M. Sun, W.J. Tsai, J.C. Ou, W.P. Chen, and C.Y. Lin. 1998. Chinese herbs as modulators of human mesangial cell proliferation: Preliminary studies. *Journal of Laboratory and Clinical Medicine* 132, no. 1:76–85.

Kupiecki, F.P., C.D. Ogzewalla, and F.M. Schell. 1974. Isolation and characterization of a hypoglycemic agent from *Xanthium strumarium*. *Journal of Pharmaceutical Sciences* 63, no. 7:1166–1167.

McMillan, C. 1973. Partial fertility of artificial hybrids between Asiatic and American cockleburs (*Xanthium strumarium* L.). *Nature: New Biology* 246, no. 153:151–153.

Pashchenko, M.M., and G.P. Pivnenko. 1970. Polyphenol substances in *Xanthium riparium* and *Xanthium strumarium*. Farm Zhurnal 25, no. 6:41–43.

Ramírez-Erosa, I., Y. Huang, R.A. Hickie, R.G. Sutherland, and B. Barl. 2007. Xanthatin and xanthinosin from the burs of *Xanthium strumarium* L. as potential anticancer agents. *Canadian Journal of Physiology and Pharmacology* 85, no. 11:1160–1172.

Roussakis, C., I. Chinou, C. Vayas, C. Harvala, and J.F. Verbist. 1994. Cytotoxic activity of xanthatin and the crude extracts of *Xanthium strumarium*. *Planta Medica* 60, no. 5:473–474.

Sila, V.I., and L.V. Lisenko. 1971. A pharmacological study of the sum of *Xanthium strumarium* alkaloids. Farm Zhurnal 26, no. 2:71–73.

Stuart, B.P., R.J. Cole, and H.S. Gosser. 1981. Cocklebur (*Xanthium strumarium*, L. var. *strumarium*) intoxication in swine: Review and redefinition of the toxic principle. *Veterinary Pathology* 18, no. 3:368–383.

Xylopia aromatica (Annonaceae)   **Cytotoxic**

Parts used: bark

Active ingredients: Annonaceous acetogenins: asimicin, venezenin, xylopien, xylomaterin, xylopianin, xylopiacin, xylomaticin, annomontacin, gigantetronenin, gigantetrocin A, and annonacin

Documented target cancers:

- Acetogenins showed cytotoxicity, comparable or superior to adriamycin, against three human solid tumor cell lines.
- The hexane extracts of *X. aromatica* root wood are reported with IC_{50} values ranging from 5 to 20 g/mL against HCT-8 (human colon carcinoma), HL-60 (leukemia), SF-295 (brain), and MDA-MB-435 (melanoma) tumor cell lines.

Further details:

- *Xylopia aromatica*: the bark (EtOH extract) contains the acetogenins we have already mentioned. These acetogenins showed reduction of the 10-keto of one to the racemic OH-10 derivative enhanced the bioactivity, as did the conversion of one to six and seven. Venezenin, like other Annonaceous acetogenins, showed inhibition of oxygen uptake by rat liver mitochondria and demonstrated that the THF ring may not be essential to this mode of action.
- The ethyl acetate extract of *Periconia atropurpurea*, an endophytic fungus obtained from the leaves of *Xylopia aromatica*, is reported to exhibit cytotoxicity against human cervix carcinoma (HeLa) and Chinese hamster ovary (CHO) cells.

References

Colman-Saizarbitoria, T., J. Zambrano, N.R. Ferrigni, Z.M. Gu, J.H. Ng, D.L. Smith, and J.L. McLaughlin. 1994. Bioactive annonaceous acetogenins from the bark of *Xylopia aromatica*. *Journal of Natural Products* 57, no. 4:486–493.
de Mesquita, M.L., J.E. de Paula, C. Pessoa, et al. 2009. Cytotoxic activity of Brazilian Cerrado plants used in traditional medicine against cancer cell lines. *Journal of Ethnopharmacology* 123, no. 3:439–445.
Teles, H.L., R. Sordi, G.H. Silva, et al. 2006. Aromatic compounds produced by Periconia atropurpurea, an endophytic fungus associated with *Xylopia aromatica*. *Phytochemistry* 67, no. 24:2686–2690.

Zanthoxylum heitzii (Rutaceae) (Sapindales)   **Cytotoxic**

Location: warm temperate and subtropical areas worldwide
   Appearance: spiny tree with a large crown of ascending branches

   Leaves: alternate

Tradition: used in folk medicine for bark-gonorrhea, dexterity, joint pain, malaria, rheumatism, toothache, and heart problems.
   Parts used: stem, fruit
   Active ingredients: syringic acid, juglon, luteolin, benzophenanthridines, furoquinolines, aporphines, canthinones, acridones, sesamin
   Documented target cancers:

- The methanol extract of fruits and barks are reported to significantly inhibit cell proliferation of human promyelocytic leukemia HL-60 cells.
- Extract from the bark is reported to have significant cytotoxic activity against human leukemia THP-1 and prostate cancer PC-3 cell lines.

References

Dzoyem, J.P., S.K. Guru, C.A. Pieme, et al. 2013. Cytotoxic and antimicrobial activity of selected Cameroonian edible plants. *BMC Complementary and Alternative Medicine* 13, no. 1:78.
Pieme, C.A., G.K. Santosh, E.M. Tekwu, et al. 2014. Fruits and barks extracts of *Zanthozyllum heitzii* a spice from Cameroon induce mitochondrial dependent apoptosis and Go/G1 phase arrest in human leukemia HL-60 cells. *Biological Research* 47, no. 1:54.

| *Zea mays* (Poaceae) (Poales) | **Anti-cancer Anti-metastatic** |

Location: native to Mexico and Central America, cultivated worldwide
    Appearance: annual cereal crop

Leaves: simple and interchangeable

Tradition: decoction of maize silk, roots, leaves, and cob are used in folk medicine for bladder problems, nausea, vomiting, and stomach complaints.
    Parts used: all parts
    Active ingredients: polysaccharides, peptides
    Documented target cancers:

- Corn pectic polysaccharide from *Zea mays* is reported to effectively inhibit invasion and metastasis in experimental animals.
- *Zea mays* leaf extracts are reported to exhibit anticancer activity on oxidative stress-induced Hep2 (laryngeal carcinoma) cells.
- *Zea mays* corn peptides are reported as apoptosis inducers in the human hepatoma HepG2 cell line and as effective inhibitors of tumor growth in H22 hepatocellular carcinoma-bearing mice.

References

Balasubramanian, K., and P.R. Padma. 2013. Anticancer activity of *Zea mays* leaf extracts on oxidative stress-induced HepG2 Cells. *Journal of Acupuncture and Meridian Studies* 6, no. 3:149–158.

Díaz-Gómez, J.L., F. Castorena-Torres, R.E. Preciado-Ortiz, and S. García-Lara. 2017. Anti-cancer activity of maize bioactive peptides. *Frontiers in Chemistry* 5:44.

Jayaram, S., S. Kapoor, and S.M. Dharmesh. 2015. Pectic polysaccharide from corn (*Zea mays* L.) effectively inhibited multi-step mediated cancer cell growth and metastasis. *Chemico-Biological Interactions* 235:63–75.

Li, J.T., J.L. Zhang, H. He, Z.L. Ma, Z.K. Nie, Z.Z. Wang, and X.G. Xu. 2013. Apoptosis in human hepatoma HepG2 cells induced by corn peptides and its anti-tumor efficacy in H22 tumor bearing mice. *Food and Chemical Toxicology* 51:297–305.

Rouf Shah, T., K. Prasad, and P. Kumar. 2016. Maize—A potential source of human nutrition and health: A review. *Cogent Food and Agriculture* 2, no. 1:1-9.

| *Zieridium pseudobtusifolium* (Rutaceae) | **Tumor inhibitor cytotoxic** |

Active ingredients: flavonols: 5,3'-dihydroxy-3,6,7,8,4'-pentamethoxyflavone, digicitrin, 5-hydroxy-3,6,7,8,3',4'-hexamethoxyflavone, 3-O-demethyldigicitrin, 3,5,3'-trihydroxy-6,7,8,4'-tetramethoxyflavone, and 3,5-dihydroxy-6,7,8,3',4'-pentamethoxyflavone
    Indicative dosage and application:

- IC_{50} 0.04 micrograms/ml against (KB) human nasopharyngeal carcinoma cells
- IC_{50} 12 μM inhibited tubulin

Documented target cancers:

- Cytotoxic activity against KB cells
- Human nasopharyngeal carcinoma cells
- Inhibits tubulin assembly into microtubules

Further details:

- Bioassay-guided fractionation of the extracts of *Zieridium pseudobtusifolium* and *Acronychia porteri* led to the isolation of 5,3'-dihydroxy-3,6,7,8,4'-pentamethoxyflavone which showed activity against (KB) human nasopharyngeal carcinoma cells (IC_{50} 0.04 micrograms/ml) and inhibited tubulin assembly into microtubules (IC_{50} 12 μM). All of the mentioned (in the active ingredients) flavonols showed cytotoxic activity against KB cells.

Reference

Lichius, J.J., O. Thoison, A. Montagnac, et al. 1994. Antimitotic and cytotoxic flavonols from Zieridium pseudobtusifolium and Acronychia porteri. *Journal of Natural Products* 57, no. 7:1012–1016.

Zingiber officinale (Zingiberaceae) (Zingiberales)	Cytotoxic

Location: tropical regions, jungles
  Appearance: tender herbaceous perennial plant with aromatic rhizome

  Leaves: narrowly lanceolate
  Flowers: white or yellowish-green

Tradition: the root is used in folk medicine to help relieve dizziness, sweating, nausea, and vomiting caused by sickness or nausea, sore throat, headache, ulcerative colitis, menstrual disorders, arthritis pain, fever, flu, and depressive symptoms.
  Parts used: root
  Active ingredients: 6-gingerol
  Documented target cancers:

- Ginger extract is reported to significantly reduce the elevated expression of pro-inflammatory cytokine transcriptional factors in rat models of liver cancer.
- Exposure of ginger leaf extract to human colorectal cancer cells (HCT116, SW480, and LoVo cells) is reported to reduce the cell viability and induce apoptosis in a dose-dependent manner.
- 6-gingerol is reported cytotoxic against human colon cancer (HCT15), mouse leukemic monocyte macrophage (Raw 264.7), and murine fibro sarcoma (L929) cells.

References

Habib, S.H.M., S. Makpol, N.A.A. Abdul Hamid, S. Das, W.Z.W. Ngah, and Y.A.M. Yusof. 2008. Ginger extract (*Zingiber officinale*) has anti-cancer and anti-inflammatory effects on ethionine-induced hepatoma rats. *Clinics* 63, no. 6:807–813.

Kumara, M., M.R. Shylajab, P.A. Nazeemc, and T. Babu. 2017. 6-Gingerol is the most potent anticancerous compound in ginger (*Zingiber officinale* Rosc). *Journal of Developing Drugs* 6:167.

Park, G.H., J.H. Park, H.M. Song, et al. 2014. Anti-cancer activity of ginger (*Zingiber officinale*) leaf through the expression of activating transcription factor 3 in human colorectal cancer cells. *BMC Complementary and Alternative Medicine* 14, no. 1:408.

Zornia brasiliensis (Fabaceae) (Dalbergieae)	Cytotoxic

Location: Brazilian territory
  Appearance: small perennial shrub

  Flowers: yellow

Tradition: commonly known as 'urinária' in Brazil, it is typically used as a diuretic and for the treatment of venereal diseases.

Parts used: whole plant

Active ingredients: zornioside, D-pinitol, roseoside, trans-nerolidol, trans-karyophyllene, α-humulene, farnesene

Documented target cancers:

- The leaf essential oil of *Z. brasiliensis* is reported to present promising cytotoxicity against B16-F10 (mouse melanoma), HepG2 (human hepatocellular carcinoma), K562 (human chronic myelocytic leukemia), and HL-60 (human promyelocytic leukemia) cell lines. Furthermore, the oil showed *in vivo* inhibition of tumor growth in mice inoculated with B16-F10 mouse melanoma.
- The compound zornioside is reported to be selectively cytotoxic for HL60 leukemia cells.

References

Costa, E.V., L.R.A. Menezes, S.L. Rocha, et al. 2015. Antitumor properties of the leaf essential oil of Zornia brasiliensis. *Planta Medica* 81, no. 7:563–567.

Nascimento, Y.M., L.S. Abreu, R.L. Lima, et al. 2018. Zornioside, a dihydrochalcone C-glycoside, and other compounds from *Zornia brasiliensis*. *Revista Brasileira de Farmacognosia* 28, no. 2:192–197.

***Zornia diphylla* (Fabaceae) (Dalbergieae)** **Anti-tumor**

Location: tropical regions in New Mexico, Brazil, West Indies, Australia

Appearance: perennial non-climbing herb

Flowers: yellow, hidden in the bracts

Active ingredients: isoflavones

Documented target cancers:

- The n-hexane extract of the plant is reported to show promising *in vitro* and *in vivo* antitumor activity against Dalton's lymphoma ascites (DLA) cells.
- Several isoflavones, isolated from *Zornia diphylla*, showed anti-tumor activities against the esophageal squamous cell carcinoma TE13 and the murine melanoma B16 cancer cell lines.

References

Arunkumar, R., S.A. Nair, K.B. Rameshkumar, and A. Subramoniam. 2014. The essential oil constituents of *Zornia diphylla* (L.) Pers, and anti-inflammatory and antimicrobial activities of the oil. *Records of Natural Products* 8, no. 4.

Arunkumar, R., S.A. Nair, and A. Subramoniam. 2012. Induction of cell-specific apoptosis and protection of mice from cancer challenge by a steroid positive compound from *Zornia diphylla* (L.) Pers. *Journal of Pharmacology and Pharmacotherapeutics* 3, no. 3:233–241.

Ren, F.Z., Y.Q. Gao, X.X. Cheng, L.H. Li, S.H. Chen, and Y.I. Zhang. 2012. Study on chemical constituents of *Zornia diphylla*. *Chinese Pharmaceutical Journal* 47:179–181.

Appendix

Chemical names: 1,4-Naphthoquinone (Naphthalene-1,4-dione; 1,4-Naphthalenedione; P-Naphthoquinone)
PubChem CID: 8530

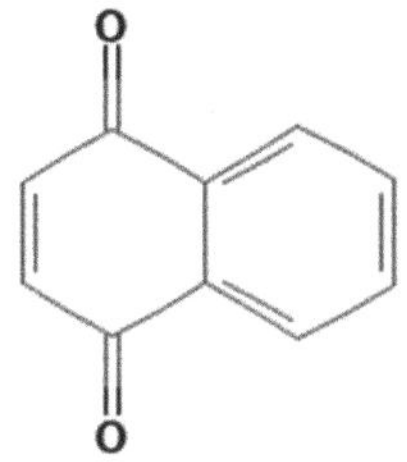

1,4-Naphthoquinone

Chemical names: 2-Hydroxy-4-methoxybenzaldehyde (4-Methoxysalicylaldehyde; Benzaldehyde, 2-hydroxy-4-methoxy-)
PubChem CID: 69600

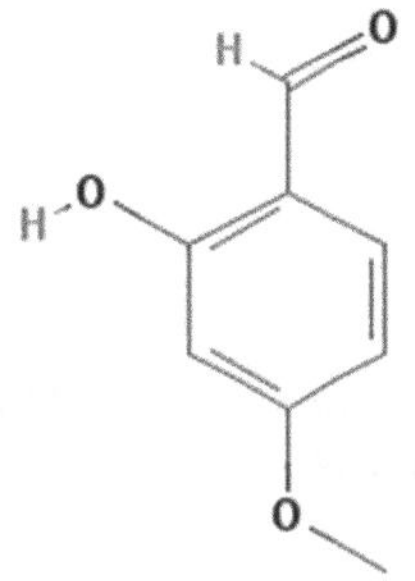

2-Hydroxy-4-methoxybenzaldehyde

Chemical names: 4,5,7-trihydroxyflavanone (4,5,7-Trihydroxy-2-phenyl-4H-chromen-3-one)
PubChem CID: 17873337

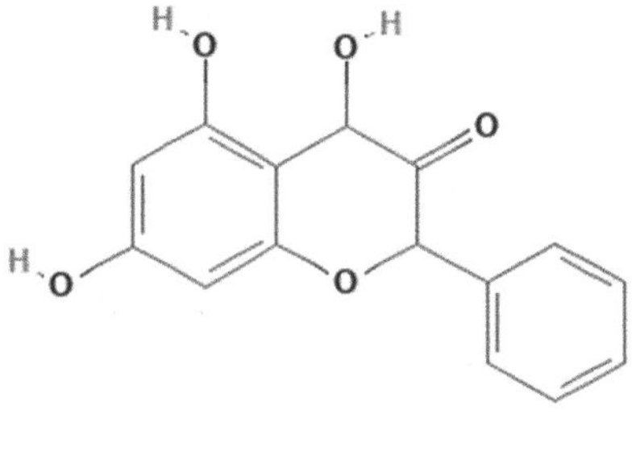

4,5,7-trihydroxyflavanone

Chemical names: 5-Fluorouracil (Fluorouracil; 51-21-8; 5-FU; Fluoroplex)
    PubChem CID: 3385

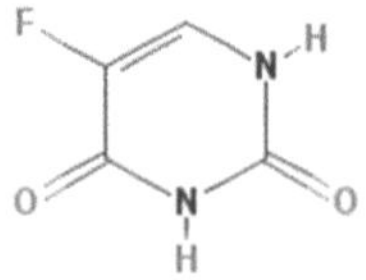

5-Fluorouracil

Chemical names: 6-Gingerol (Gingerol; 23513-14-6; [6]-Gingerol; (6)-Gingerol)
    PubChem CID: 442793

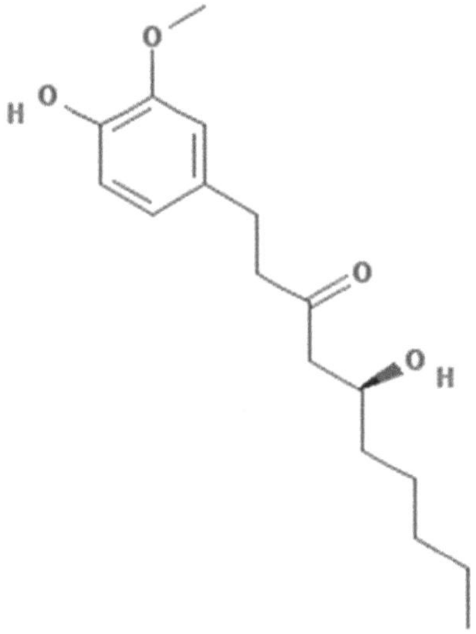

6-Gingerol

Chemical names: 7,12-Dimethylbenzanthracene (DMBA; 7,12-Dimethylbenzanthracene; 7,12-DIMETHYLBENZ(A)ANTHRACENE)
    PubChem CID: 6001

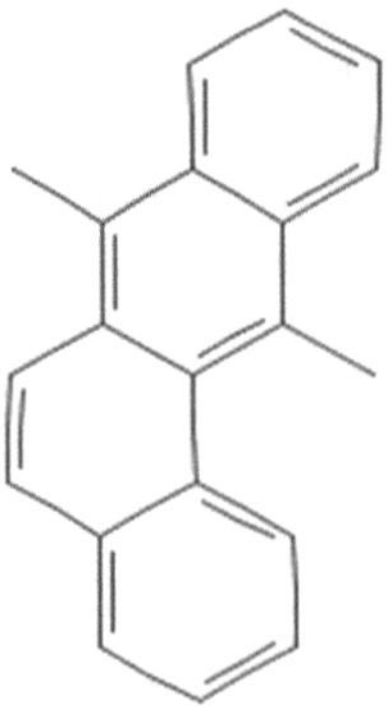

7,12-Dimethylbenzanthracene

Chemical names: 9-Ethoxyaristololactam (122739-09-7; Benzo[f]-1,3-dioxolo[4,5]benz[1,2,3-cd]indol-5(6H)-one,7-ethoxy-8-methoxy-; ACMC-20dchx; AC1L4ZWG)
PubChem CID: 195368

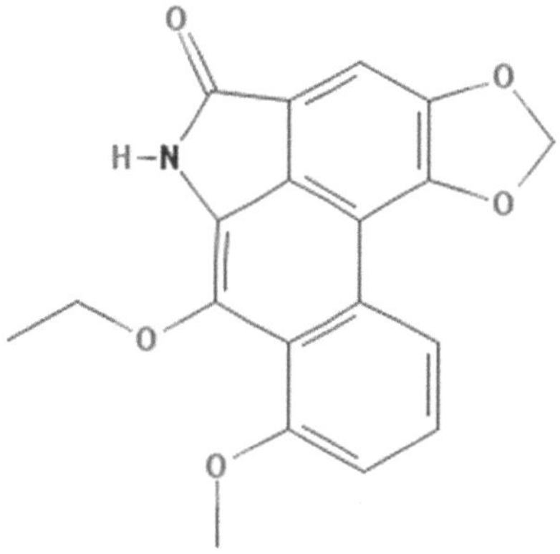

9-Ethoxyaristololactam

Chemical names: 9-Methoxycanthin-6-one (8-Methoxycanthin-6-one; 74991-91-6; CHEBI:66699)
PubChem CID: 9881423

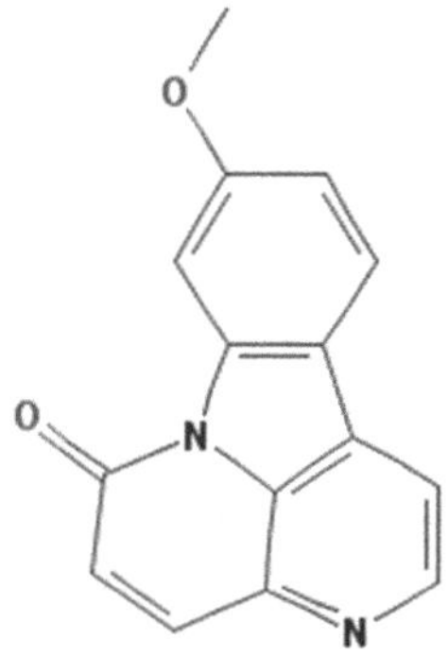

9-Methoxycanthin-6-one

Chemical names: 22-Hydroxytingenone (Tingenin B; 22-Hydroxytingenone; 50656-68-3)
PubChem CID: 73147

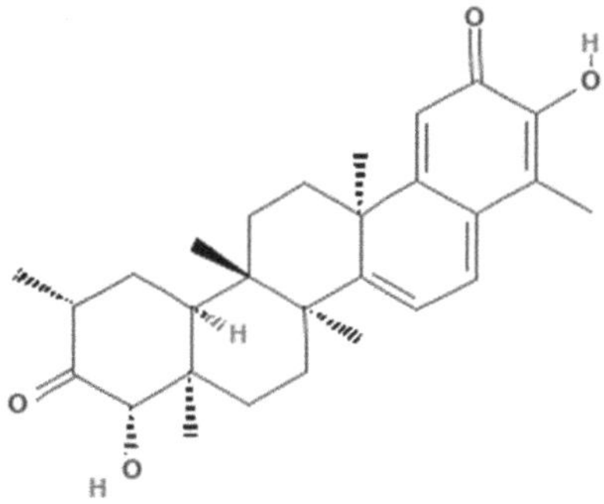

22-Hydroxytingenone

Chemical names: Acronine (ACRONYCINE; Compound 42339; 7008-42-6; Acromycine)
    PubChem CID: 345512

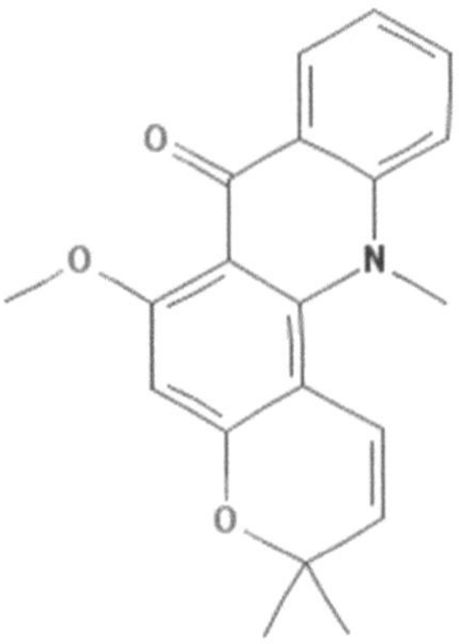

Acronine

Chemical names: Adenosine diphosphate (Adenosine 5'-diphosphate; 58-64-0; ADP; Adenosine-5'-diphosphate)
    PubChem CID: 6022

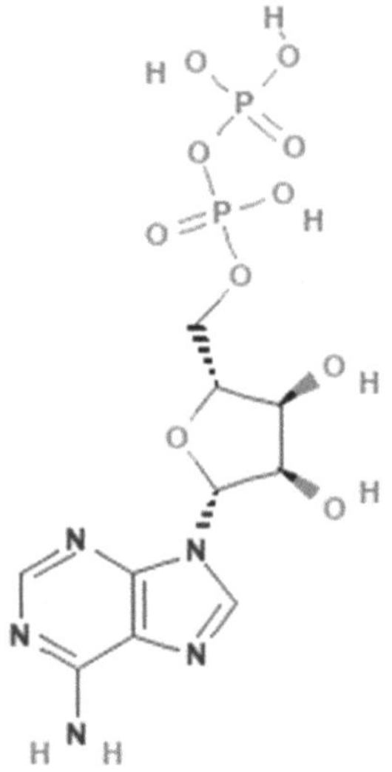

Adenosine diphosphate

Chemical names: Aflatoxin B1 (1162-65-8; Aflatoxin B; NSC 529592; HSDB 3453)
   PubChem CID: 186907

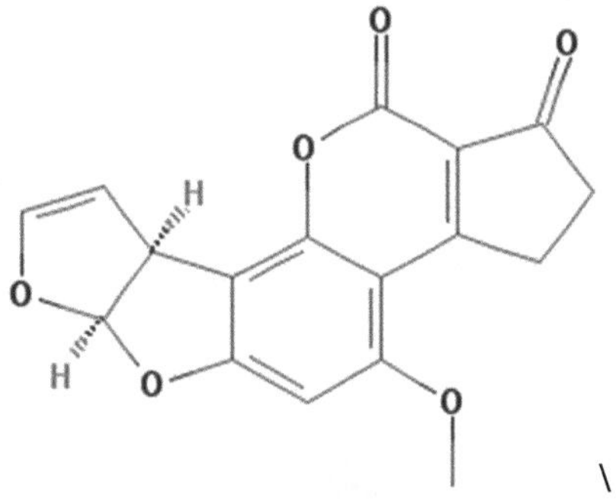

Aflatoxin B1

Chemical names: Alantolactone (Helenine; 546-43-0; Eupatal; Alant camphor)
   PubChem CID: 72724

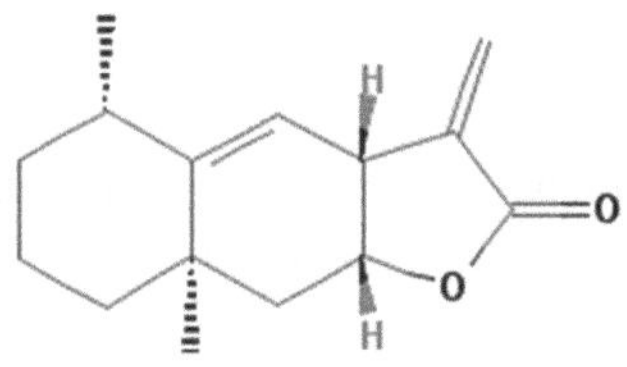

Alantolactone

Allamandin (C09766; 51820-82-7; CHEBI:2592)
   PubChem CID: 5281540

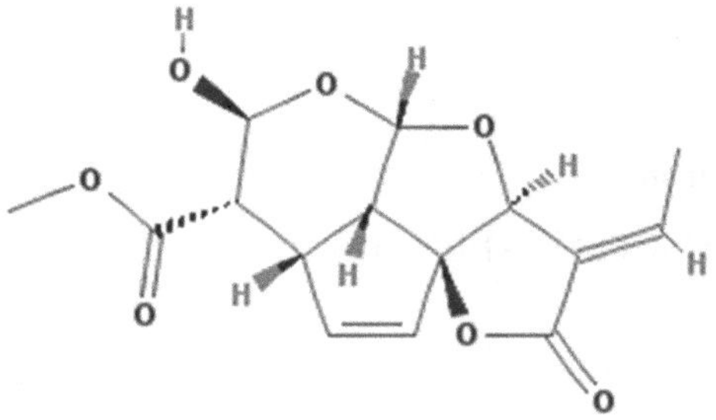

Allamandin

Chemical names: Allamcin (93452-23-4)
   PubChem CID: 5477870

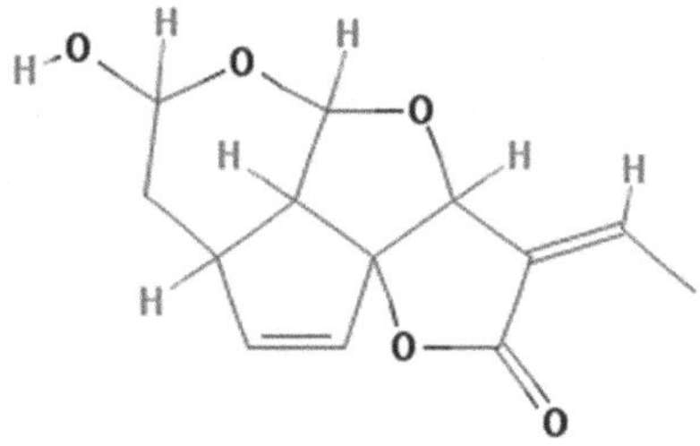

Allamcin

Chemical names: Aluminum isopropoxide (Aluminium isopropoxide; 555-31-7; Aluminum isopropylate; Aluminum triisopropoxide)
PubChem CID: 11143

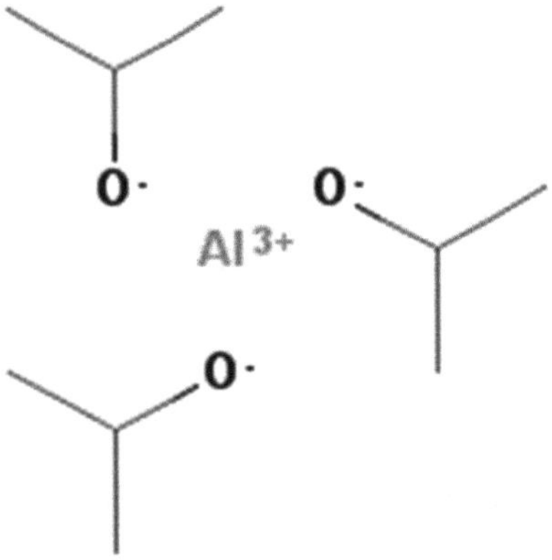

Aluminum isopropoxide

Chemical names: Ammonium phosphate monobasic (Ammonium dihydrogen phosphate; 7722-76-1; Ammonium biphosphate; Ammonium acid phosphate)
PubChem CID: 24402

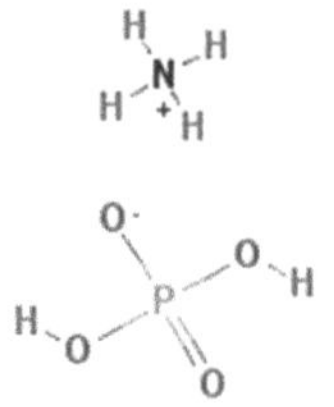

Ammonium phosphate monobasic

Chemical names: Angelicin (ISOPSORALEN; 523-50-2)
PubChem CID: 10658

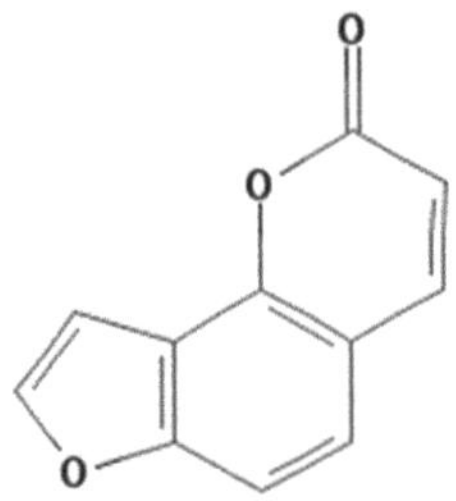

Angelicin

Chemical names: Anhydrovinblastine; 3′,4′-Anhydrovinblastine; AVLB; CHEBI:1322
  PubChem CID: 443324

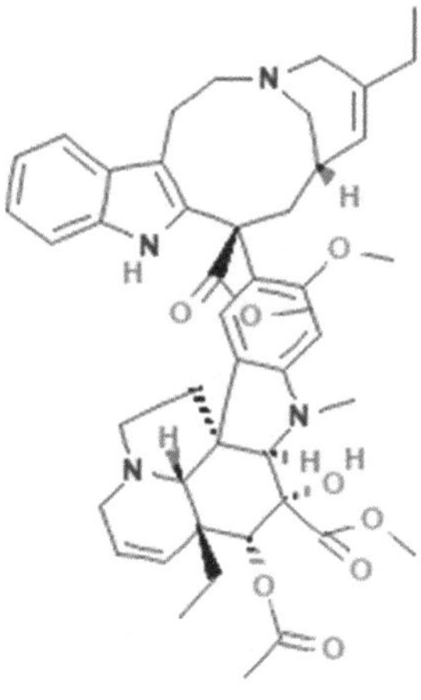

Anhydrovinblastine

Chemical names: Arachidonic Acid (506-32-1; Arachidonate; Immunocytophyte; (all-Z)-5,8,11,14-Eicosatetraenoic acid)
  PubChem CID: 444899

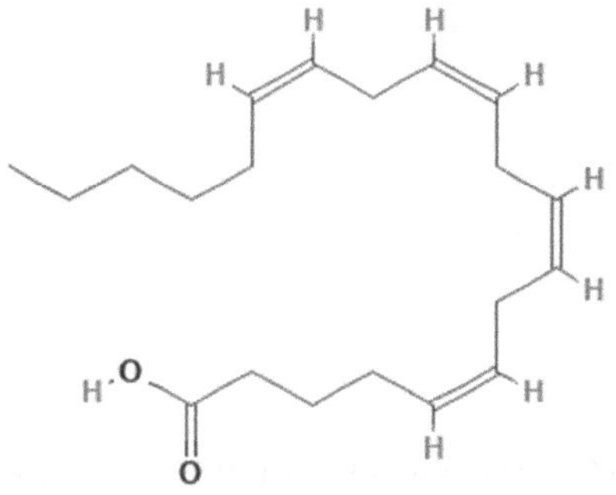

Arachidonic Acid

Chemical names: Ascorbic Acid (L-ascorbic acid; Vitamin C)
  PubChem CID: 54670067

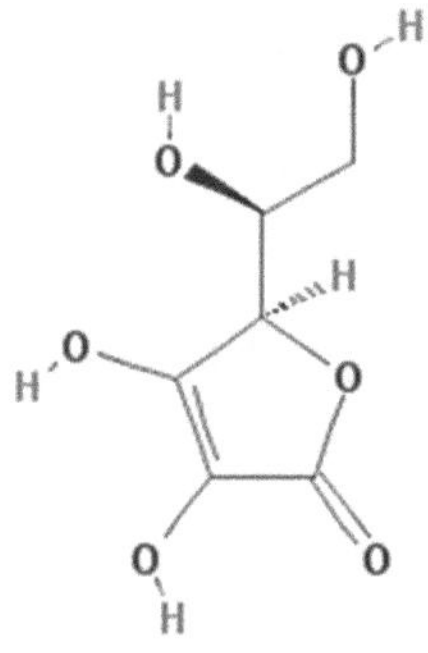

Ascorbic Acid

Chemical names: Baccatin III; 27548-93-2; BACCATINE III; UNII-40K5PZ0K67
   PubChem CID: 65366

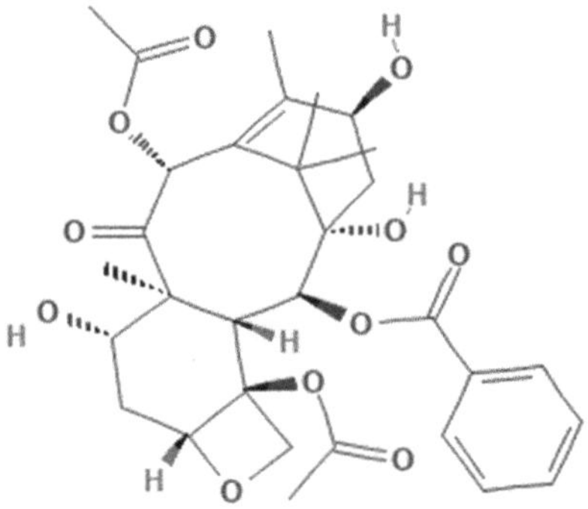

Baccatin III

Chemical names: Baccatin IV; 57672-77-2; MolPort-005-945-407
   PubChem CID: 15275710

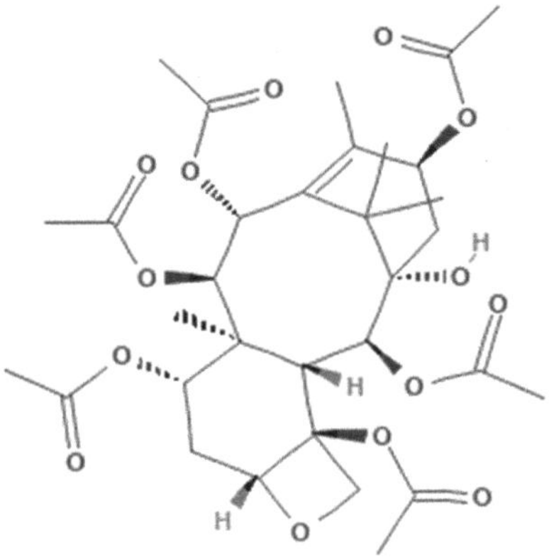

Baccatin IV

Chemical names: Baccatin (FMQSPIDOGLAJKQ-GPFIYXOFSA-N; 66107-60-6)
   PubChem CID: 101316887

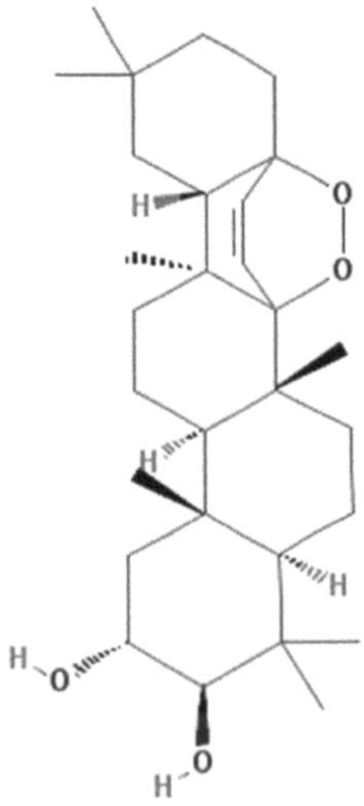

Baccatin

Chemical names: Baicalein (491-67-8; 5,6,7-Trihydroxyflavone; 5,6,7-trihydroxy-2-phenyl-4H-c hromen-4-one; Noroxylin)
  PubChem CID: 5281605

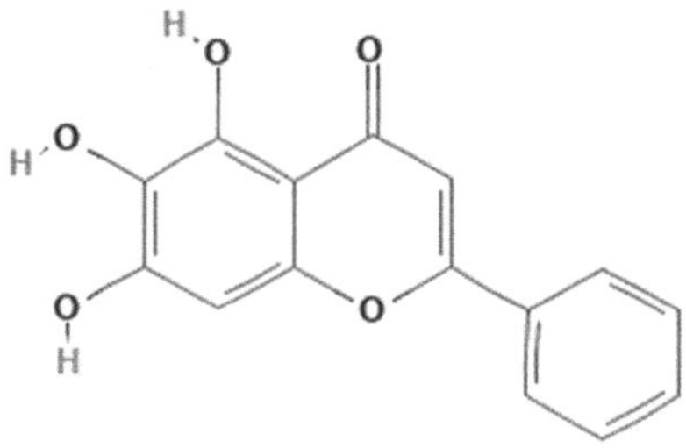

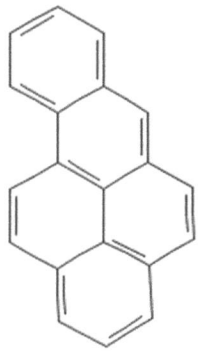

Baicalein

Chemical names: Benzo(A)Pyrene (50-32-8; 3,4-Benzopyrene; BENZO(A)PYRENE; Benzo[pqr] tetraphene)
  PubChem CID: 2336

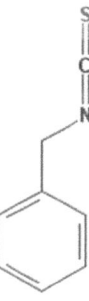

Benzo(A)Pyrene

Chemical names: Benzyl Isothiocyanate (622-78-6; (Isothiocyanatomethyl)benzene; Benzylisothiocyanate; Benzyl mustard oil)
  PubChem CID: 2346

Benzyl Isothiocyanate

Chemical names: beta-Carotene (7235-40-7; Beta Carotene; Beta,beta-Carotene; Betacarotene)
   PubChem CID: 5280489

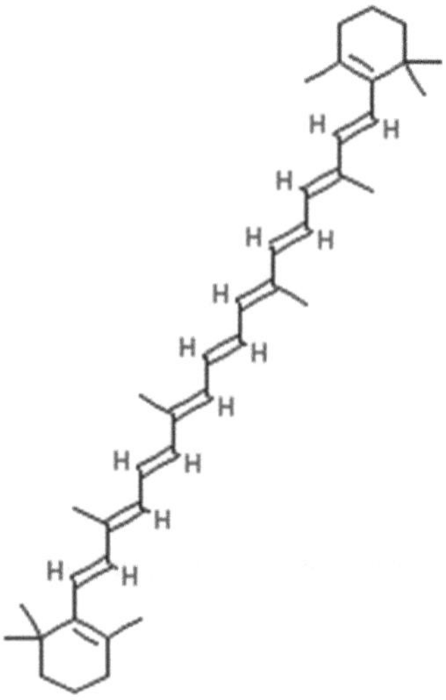

beta-Carotene

Chemical names: Biochanin A (491-80-5; Biochanin; 4′-Methylgenistein; 5,7-Dihydroxy-4′-methoxyisoflavone)
   PubChem CID: 5280373

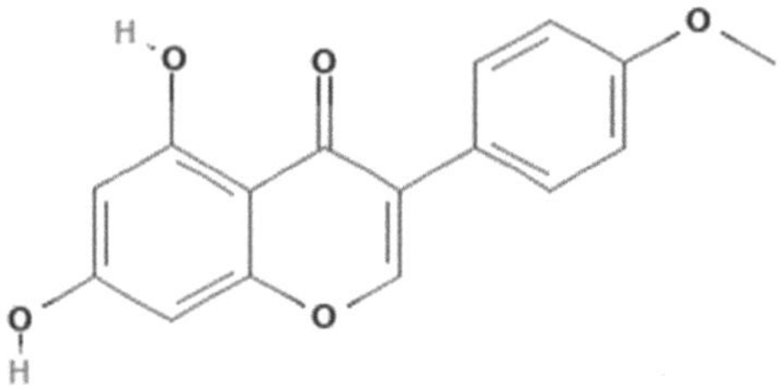

Biochanin A

Chemical names: CABAZITAXEL; 183133-96-2; Jevtana; Taxoid XRP6258; Cabazitaxelum
   PubChem CID: 9854073

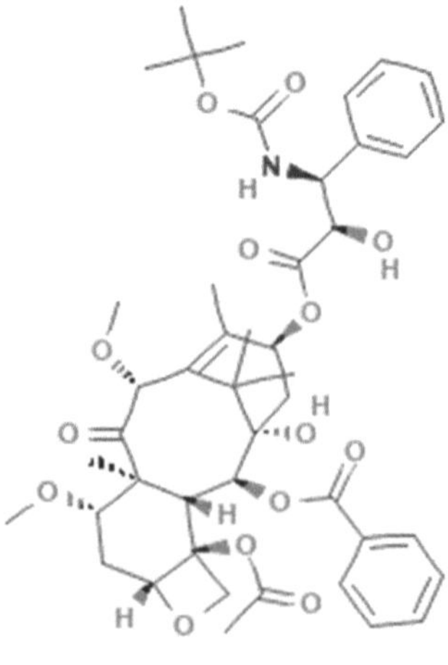

Cabazitaxel

Chemical names: Caffeine (58-08-2; Guaranine; 1,3,7-Trimethylxanthine; Thein)
    PubChem CID: 2519

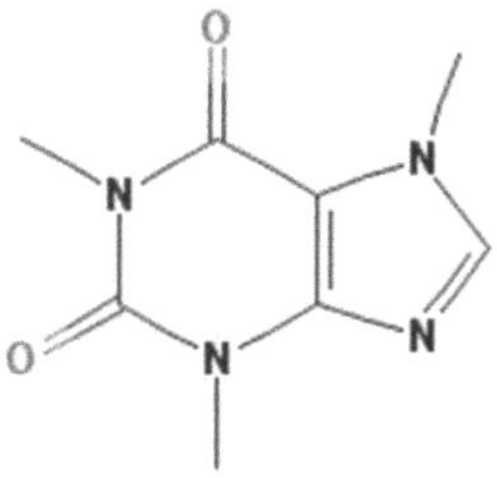

Caffeine

Chemical names: Camptothecin; Camptothecine; (S)-(+)-Camptothecin; 7689-03-4; Campathecin; (+)-Camptothecine
    PubChem CID: 24360

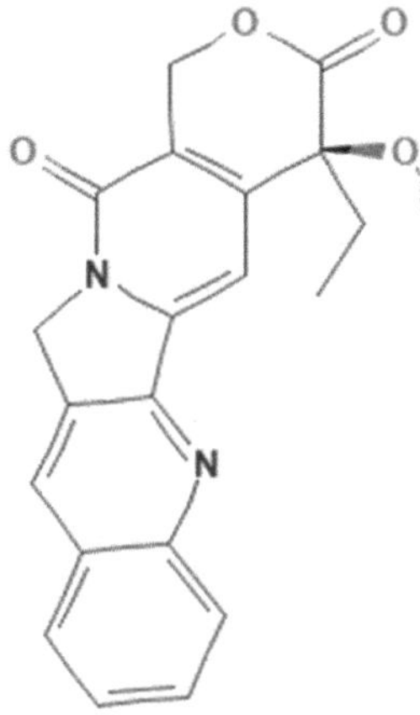

Camptothecin

Chemical names: Catharanthine; 2468-21-5; UNII-WT0YJV846J
    PubChem CID: 5458190

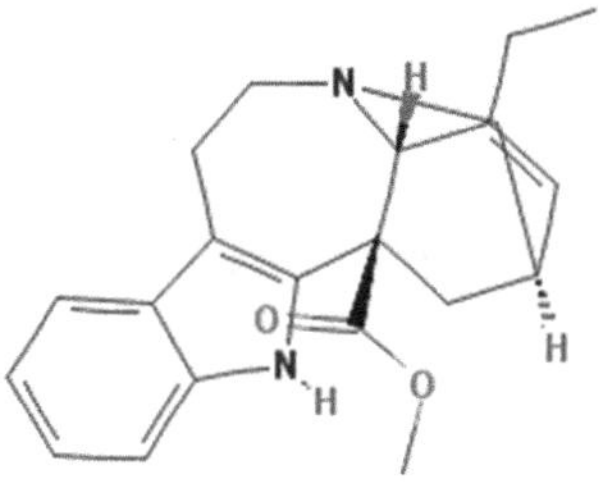

Catharanthine

Chemical names: Canthin-6-one (479-43-6; 6H-Indolo[3,2,1-de][1,5]naphthyridin-6-one; UNII-3FK17S759N; 6H-Indolo(3,2,1-de)(1,5)naphthyridin-6-one)
    PubChem CID: 97176

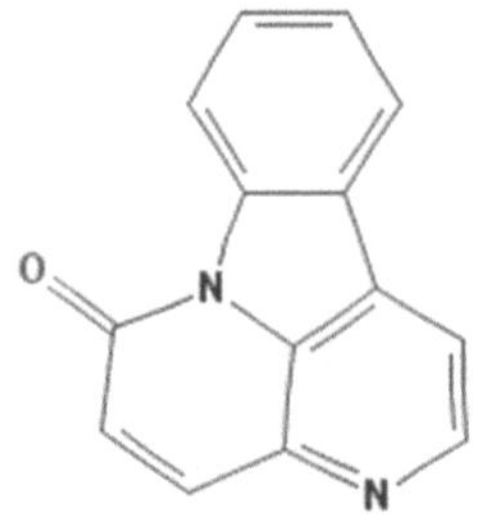

Canthin-6-one

Chemical names: Carbamoylcholine chloride (Carbachol; Carbamoylcholine chloride; 51-83-2; Carbamylcholine chloride)
    PubChem CID: 5831

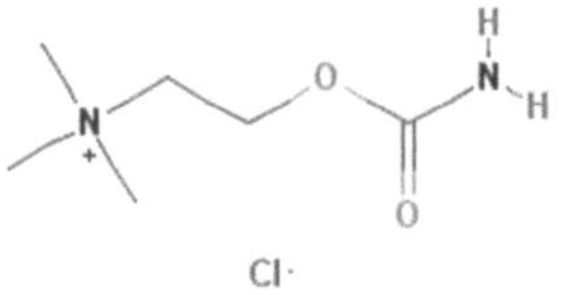

Carbamoylcholine chloride

Chemical names: Carvacrol (5-Isopropyl-2-methylphenol; 499-75-2; Isopropyl-o-cresol; O-Thymol)
    PubChem CID: 10364

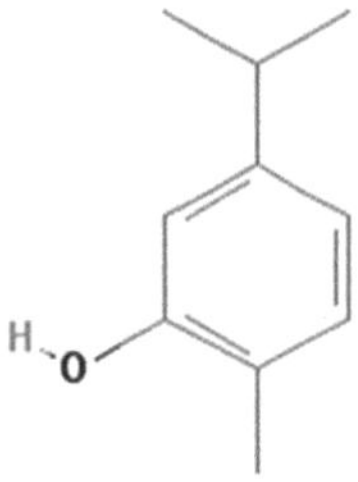

Carvacrol

Chemical names: Z-DEVD-FMK; 210344-95-9; Caspase-3 Inhibitor
    PubChem CID: 16760394

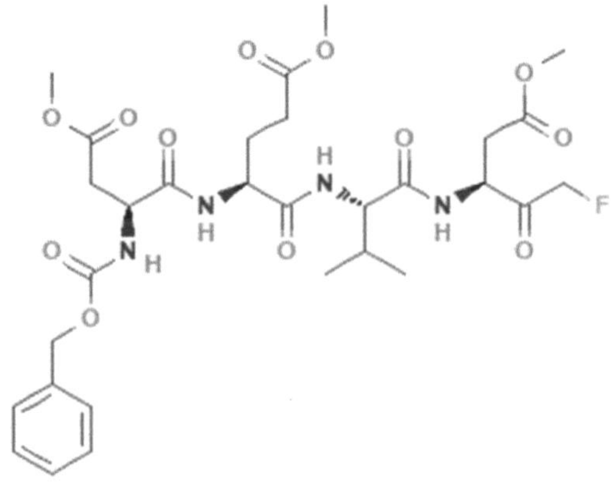

Caspase-3 Inhibitor

Chemical names: Catechin ((+)-catechin; CATECHIN; Cianidanol; 154-23-4; Catechuic acid)
   PubChem CID: 9064

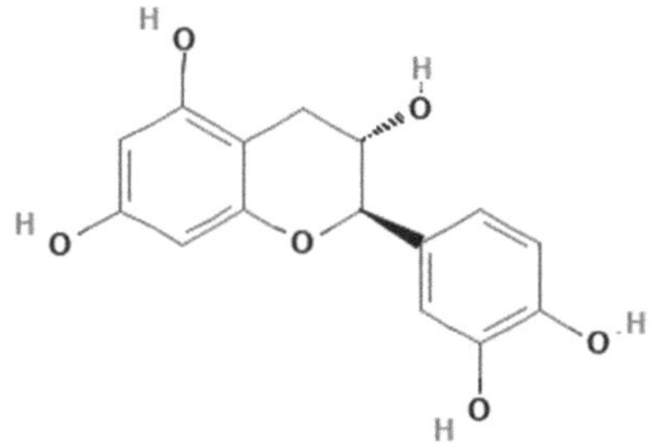

Catechin

Chemical names: Cisplatin (14913-33-8; Trans-Dichlorodiamineplatinum(II); Azane; dichloroplatinum; H6Cl2N2Pt)
   PubChem CID: 5702198

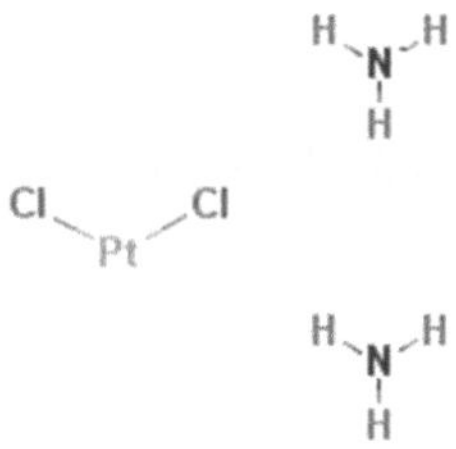

Cisplatin

Chemical names: Colchicine (Colchicina; Colchicin; Colchisol)
   PubChem CID: 6167

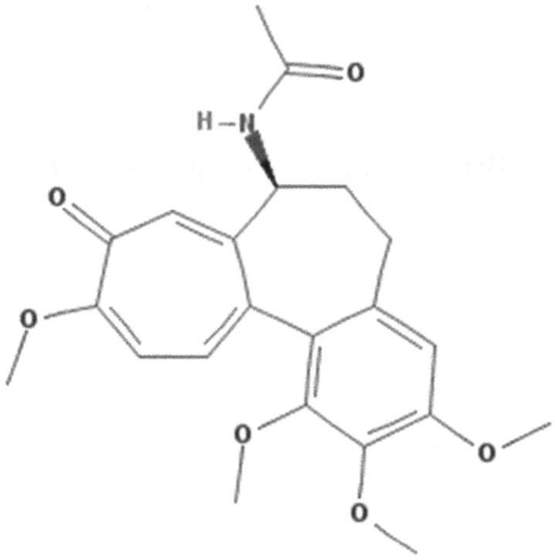

Colchicine

Chemical names: Curcumin (Diferuloylmethane; Natural yellow 3; Turmeric yellow)
    PubChem CID: 969516

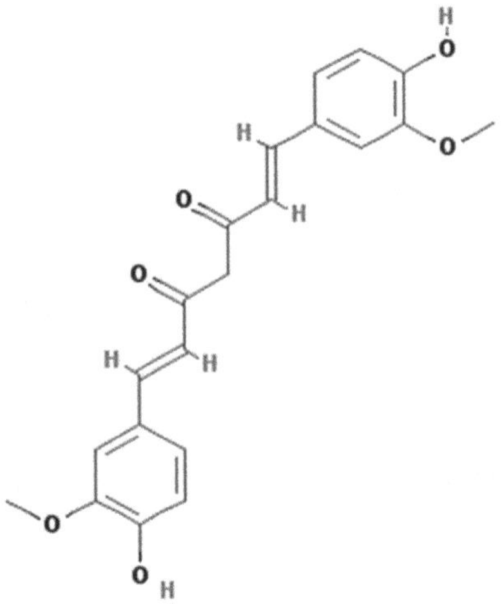

Curcumin

Chemical names: Cyclophosphamide (Cyclophosphamid; Cyclophosphane; Cytophosphan)
    PubChem CID: 2907

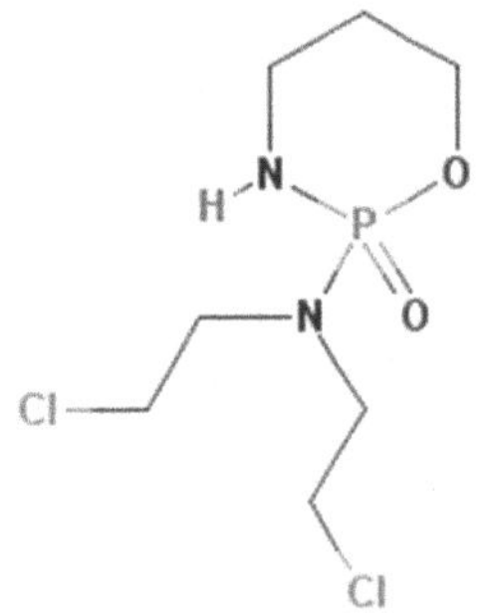

Cyclophosphamide

Chemical names: D-(+)-Lactose (Beta-Lactose; Beta-D-Lactose; Lactose; 5965-66-2; Fast-flo Lactose)
    PubChem CID: 6134

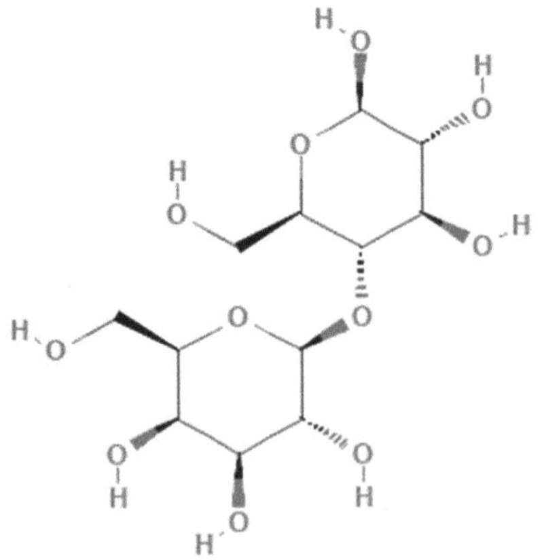

D-(+)-Lactose

Chemical names: Deoxypodophyllotoxin (Anthricin; (–)-Deoxypodophyllotoxin; Desoxypodophyl-lotoxin)

PubChem CID: 345501

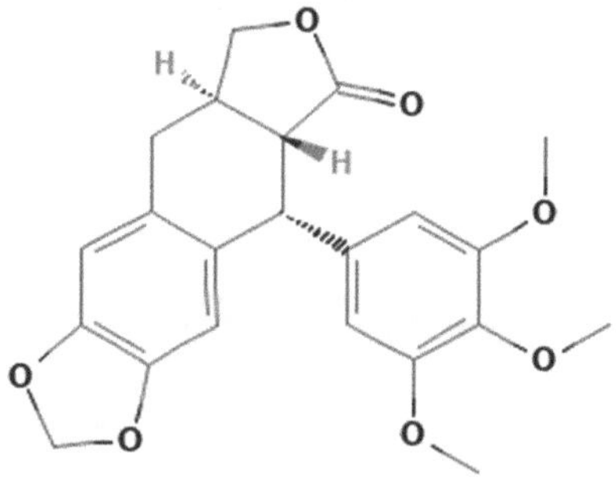

Deoxypodophyllotoxin

Chemical names: D-Galactose (D-Galactopyranose; Galactose; Galactopyranose)

PubChem CID: 6036

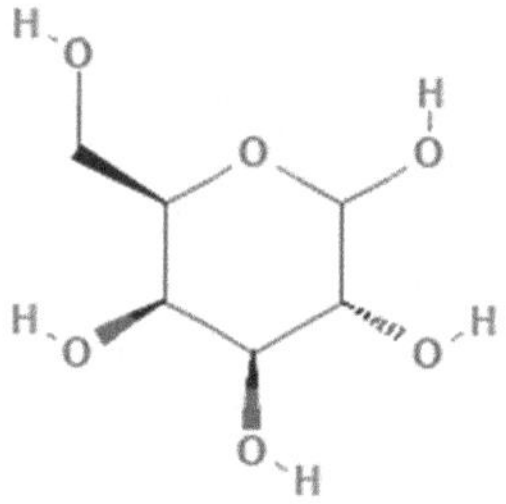

D-Galactose

Chemical names: Dichloromethane (Methylene chloride; Methylene dichloride; 75-09-2; Methane, dichloro-)

PubChem CID: 6344

Dichloromethane

Chemical names: Dihydrofolate (Dihydrofolic acid; Dihydrofolate; 7,8-Dihydrofolic acid)

PubChem CID: 98792

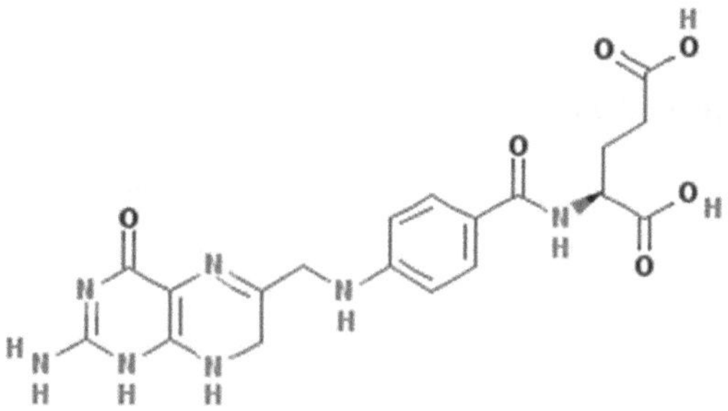

Dihydrofolate

Chemical names: Docetaxel (Taxotere; 114977-28-5; Docetaxel Anhydrous; Docetaxol; RP-56976; EmDOC; Docetaxel, Trihydrate)
　　PubChem CID: 148124

Docetaxel

Chemical names: Dodecylbenzenesulfonic acid, sodium salt (Sodium O-dodecylbenzenesulfonate; SDBS; Sodium o-dodecylbenzenesulfonate)
　　PubChem CID: 23662403

Dodecylbenzenesulfonic acid, sodium salt

Chemical names: Ellagic acid (Benzoaric acid; Lagistase; Eleagic acid)
PubChem CID: 5281855

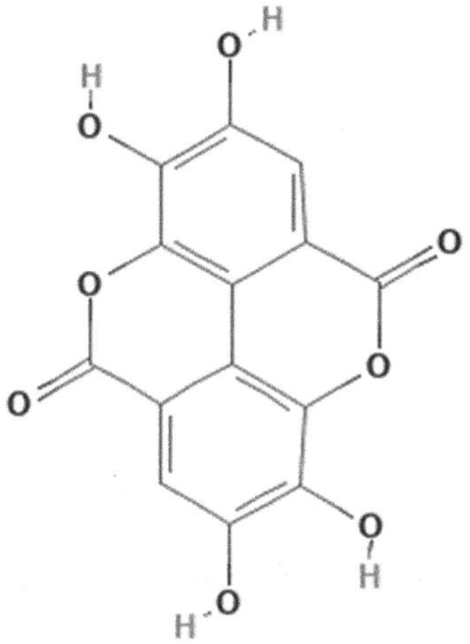

Ellagic acid

Chemical names: Ethyl Arachidonate (Arachidonic acid ethyl ester; 1808-26-0; UNII-3NU6034AW3; 5,8,11,14-Eicosatetraenoic acid, ethyl ester)
PubChem CID: 5367369

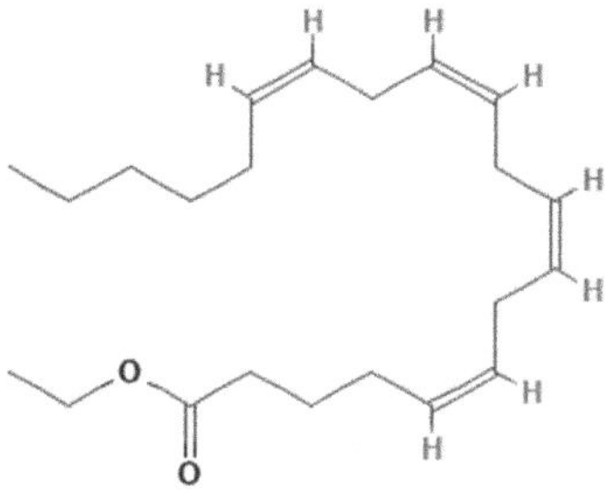

Ethyl Arachidonate

Chemical names: Etoposide (33419-42-0; VePesid; Toposar; Trans-Etoposide)
PubChem CID: 36462

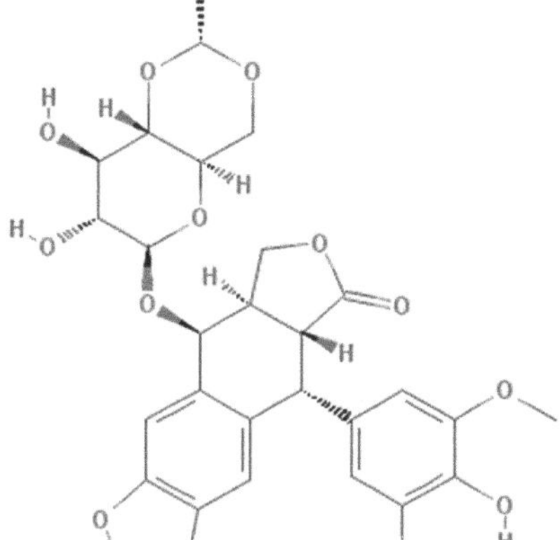

Etoposide

Chemical names: Eupatorin (3′,5-Dihydroxy-4′,6,7-trimethoxyflavone; Eupatorine)
   PubChem CID: 97214

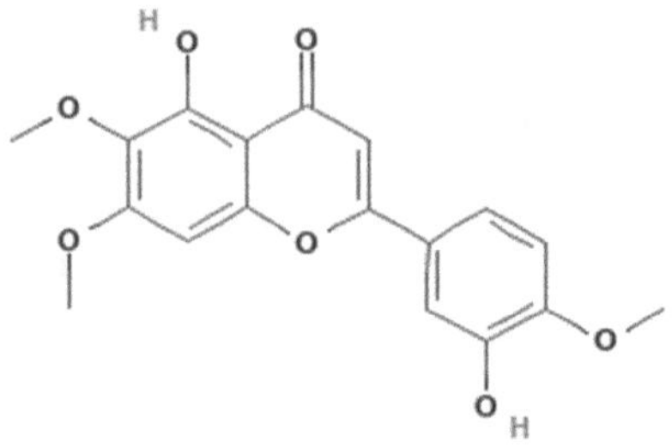

Eupatorin

Chemical names: Fagaronine (2-Hydroxy-3,8,9-trimethoxy-5-methylbenzo(c)phenanthridine)
   PubChem CID: 40305

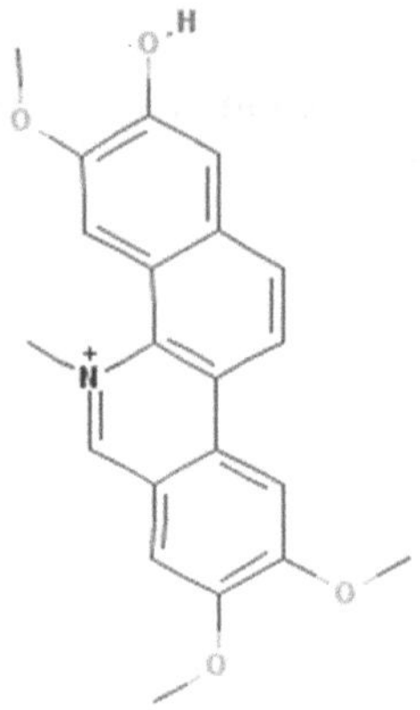

Fagaronine

Chemical names: Falcarinol (Panaxynol; CHEBI:66722; 1,9-Heptadecadiene-4,6-diyn-3-ol; (3R,9 Z)-heptadeca-1,9-dien-4,6-diyn-3-ol)
   PubChem CID: 5281149

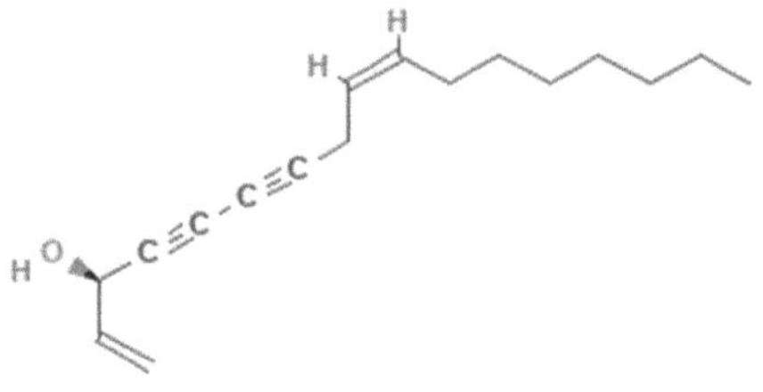

Falcarinol

Chemical names: Genistein (Prunetol; Genisteol; 4',5,7-Trihydroxyisoflavone)
    PubChem CID: 5280961

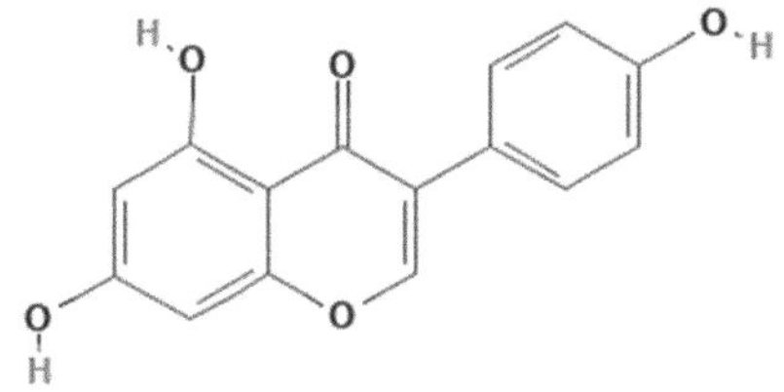

Genistein

Chemical names: Glycyrrhetinic acid (Enoxolone; 471-53-4; Uralenic acid; Glycyrrhetic acid)
    PubChem CID: 10114

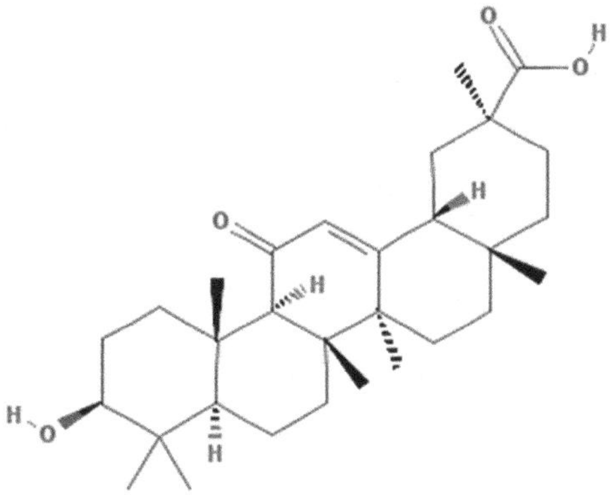

Glycyrrhetinic acid

Chemical names: Glycyrrhizic acid (Glycyrrhizin; Glycyrrhizinic acid; 1405-86-3; Glycyron)
    PubChem CID: 14982

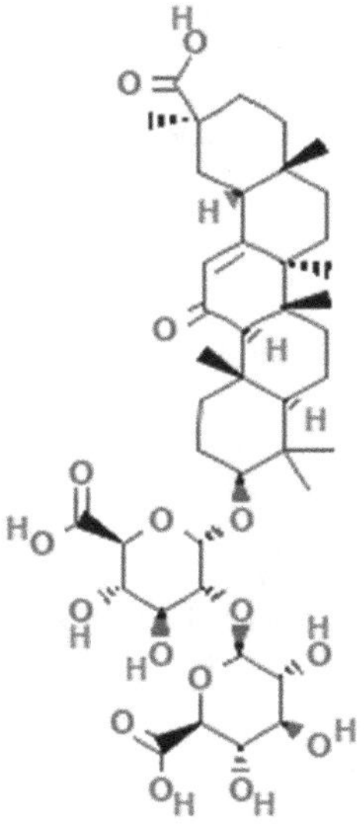

Glycyrrhizic acid

Chemical names: Goniothalamicin (2(5H)-Furanone)
   PubChem CID: 44593503

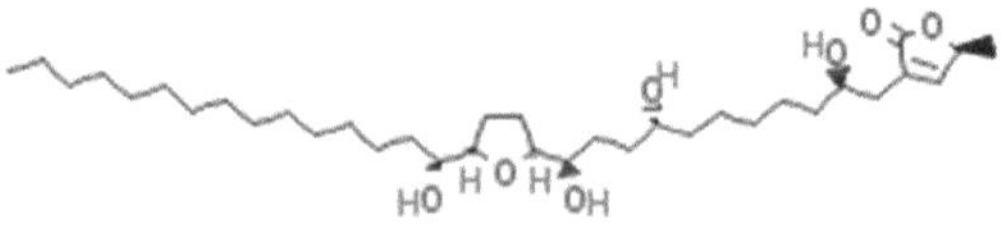

Goniothalamicin

Chemical names: Gyrophoric acid (548-89-0; UNII-BAQ44A6C6H; C24H20O10; NSC646006)
   PubChem CID: 135728

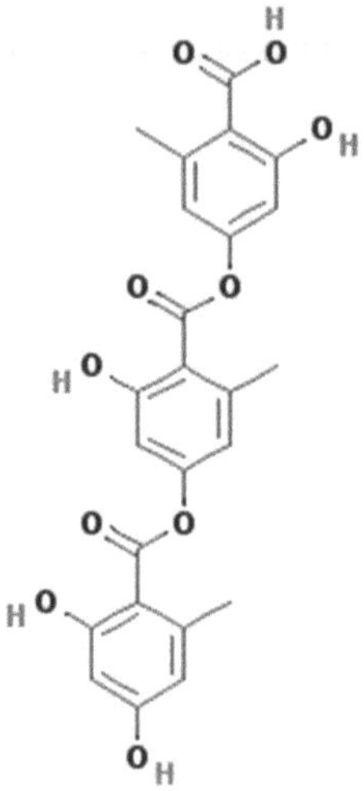

Gyrophoric acid

Chemical names: Helenalin (6754-13-8; PF 56; Helenalin A; HSDB 3490)
   PubChem CID: 23205

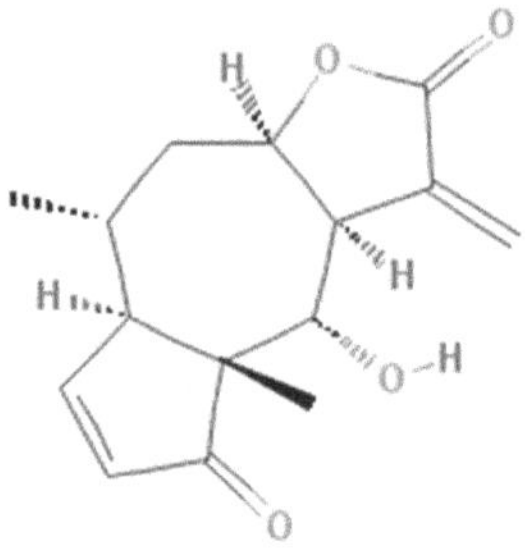

Helenalin

Chemical names: Hexane (N-Hexane; 110-54-3; Esani; Skellysolve B)
   PubChem CID: 8058

Hexane

Chemical names: Hydroquinone (1,4-benzenediol; Benzene-1,4-diol; 123-31-9; Quinol)
    PubChem CID: 785

Hydroquinone

Chemical names: Hypericin (548-04-9; Hypericum red; Cyclosan; Hipericina)
    PubChem CID: 5281051

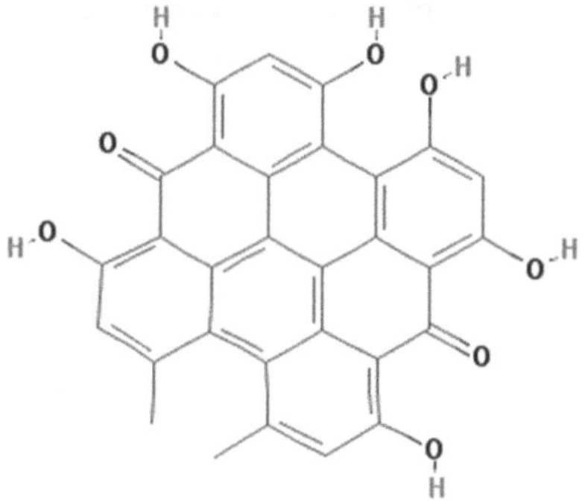

Hypericin

Chemical names: Indole (1H-Indole; 120-72-9; Indol; 2,3-Benzopyrrole)
    PubChem CID: 798

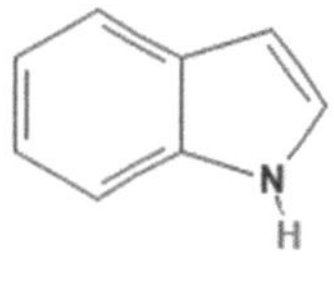

Indole

Chemical names: Irinotecan; 97682-44-5; Camptosar; (+)Irinotecan; Irinotecanum; Irinotecanum
    PubChem CID: 60838

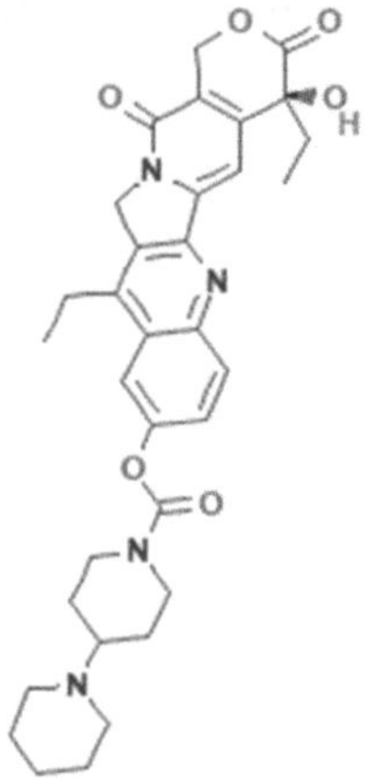

Chemical names: Isoflavone (574-12-9; 3-phenyl-4H-chromen-4-one; 3-phenylchromen-4-one; Isoflavon)
    PubChem CID: 72304

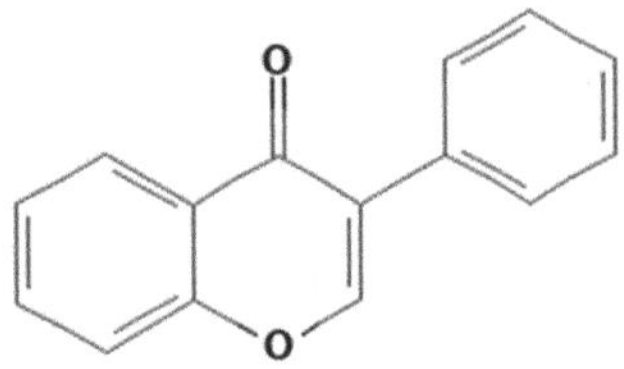

Isoflavone

Chemical names: Isotretinoin (3-cis-Retinoic acid; Accutane; Roaccutane; Claravis)
    PubChem CID: 5282379

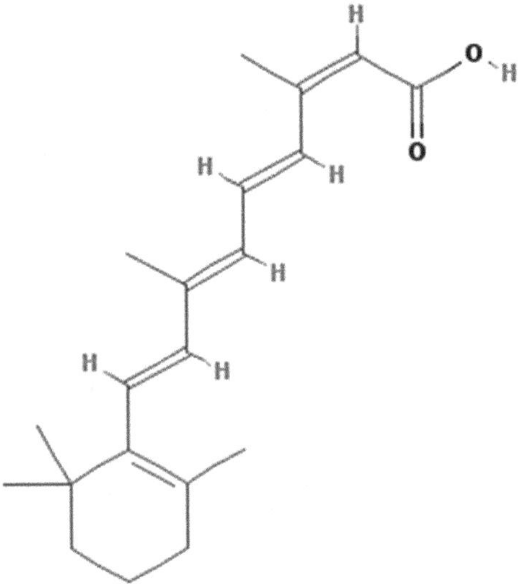

Isotretinoin

Chemical names: L-(+)-Arabinose (DL-Arabinose; (2R,3S,4S)-2,3,4,5-tetrahydroxypentanal; Aldehydo-L-arabinose)
    PubChem CID: 5460291

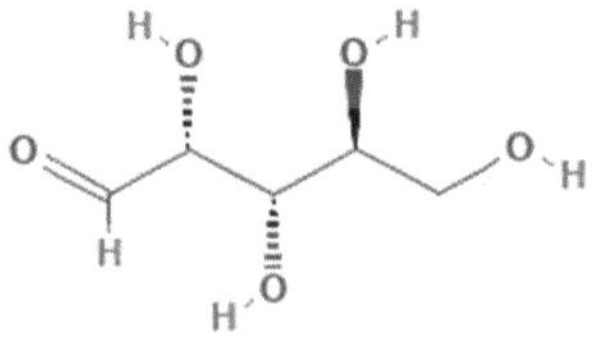

L-(+)-Arabinose

Chemical names: Lapachol (84-79-7; Tecomin; Bethabarra wood; Greenhartin)
PubChem CID: 3884

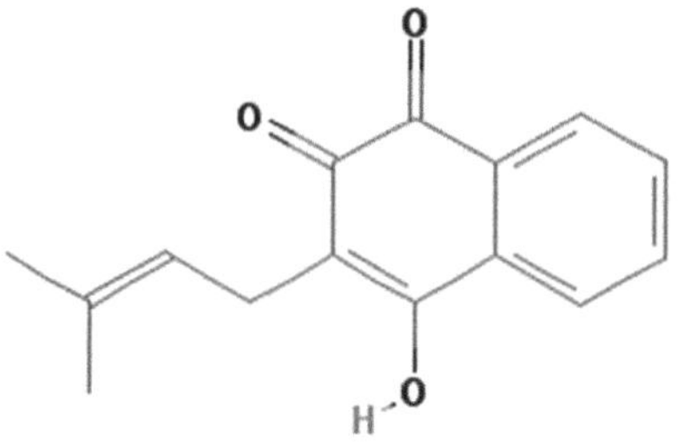

Lapachol

Chemical names: Levodopa (L-dopa; 59-92-7; Dopar; 3,4-dihydroxy-L-phenylalanine)
PubChem CID: 6047

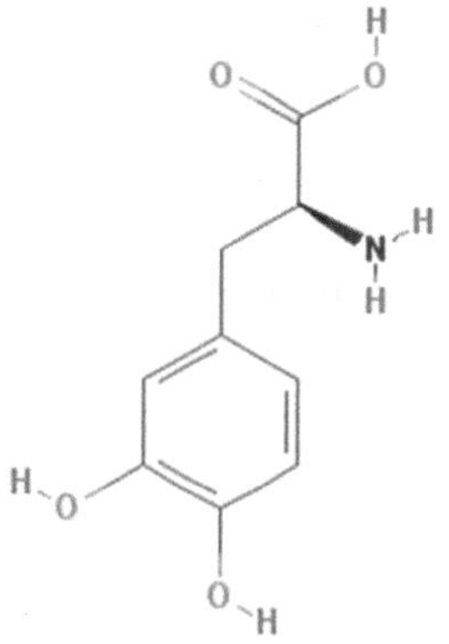

Levodopa

Chemical names: Linolenic acid (Alpha-Linolenic acid; 463-40-1; Linolenate; (9Z,12Z,15Z)-octadeca-9,12,15-trienoic acid)
PubChem CID: 5280934

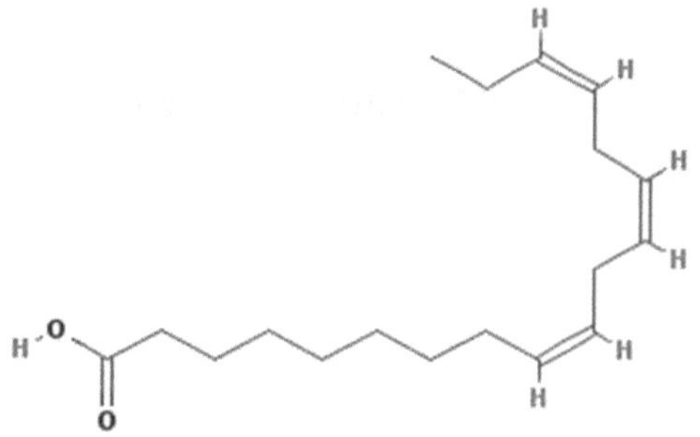

Linolenic Acid

Chemical names: L-Malic acid, sodium salt (Sodium L-malate; Disodium L-malate; Disodium (S)-malate)
    PubChem CID: 6097189

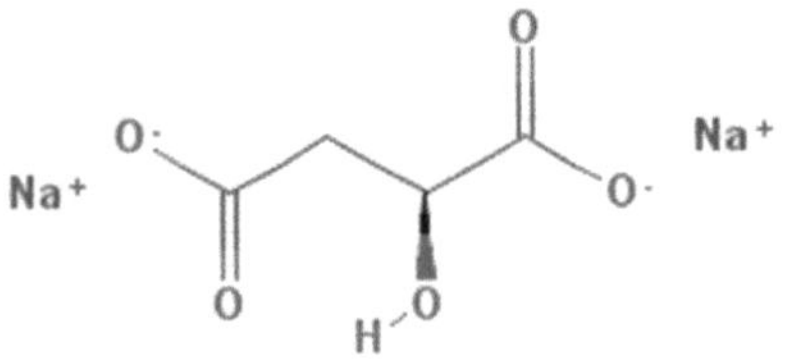

L-Malic acid, sodium salt

Chemical names: L-Rhamnose (Rhamnose; L-Rhamnopyranose; 6-Deoxy-L-Mannopyranose; 6-Deoxy-L-Mannose)
    PubChem CID: 25310

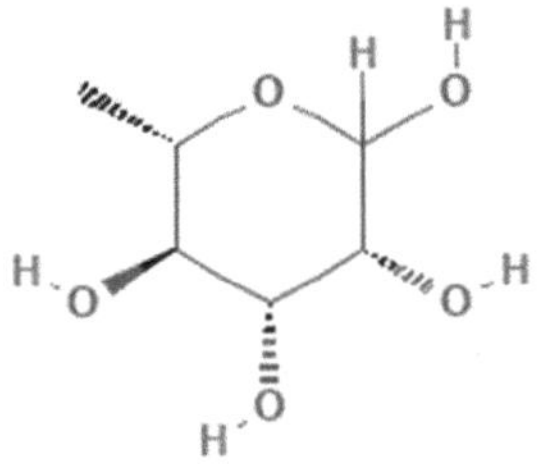

L-Rhamnose

Chemical names: L-tyrosine (Tyrosine; 60-18-4; (S)-Tyrosine; P-Tyrosine)
    PubChem CID: 6057

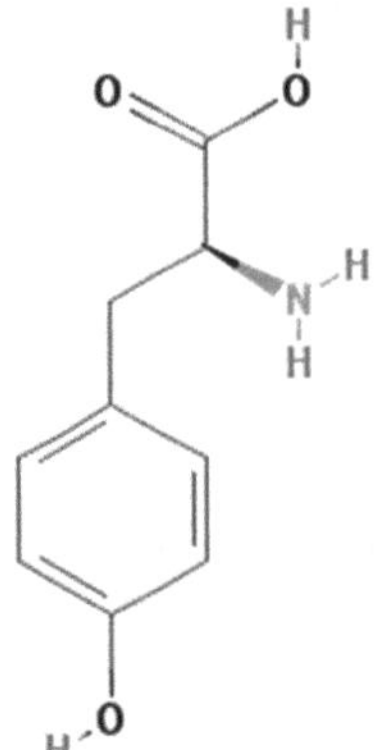

L-tyrosine

Chemical names: Luteolin; 491-70-3; 3',4',5,7-Tetrahydroxyflavone; Digitoflavone; Luteolol
    PubChem CID: 5280445

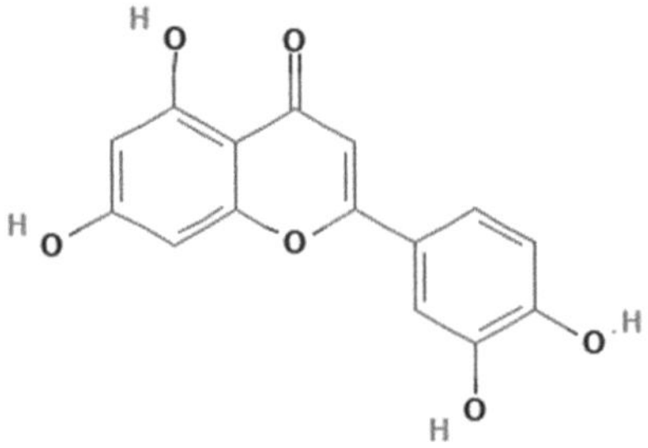

Luteolin

Chemical names: Maytansine (Maitansine; Maitansinum; Maitansina; Maytansine [USAN])
   PubChem CID: 5281828

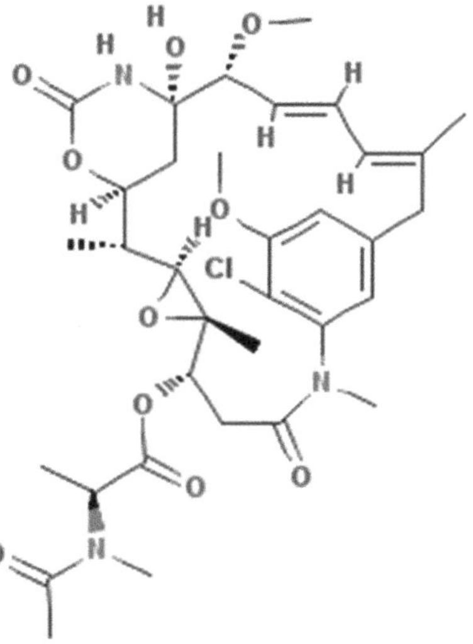

Maytansine

Chemical names: Methotrexate (59-05-2; Rheumatrex; Abitrexate; Amethopterin)
   PubChem CID: 126941

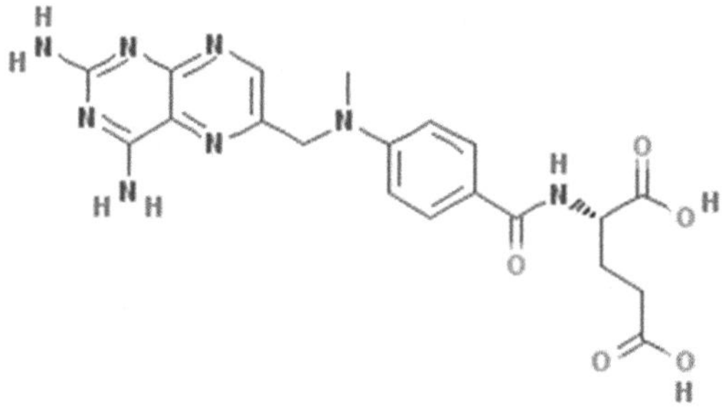

Methotrexate

Chemical names: Methyl methanesulfonate (Methyl mesylate; Methanesulfonic acid methyl ester; Methanesulfonic acid)
   PubChem CID: 4156

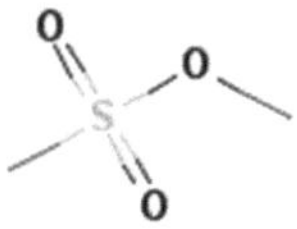

Methyl methanesulfonate

Chemical names: Mitomycin C (Mitomycin; 50-07-7; Ametycine; Mutamycin)
    PubChem CID: 5746

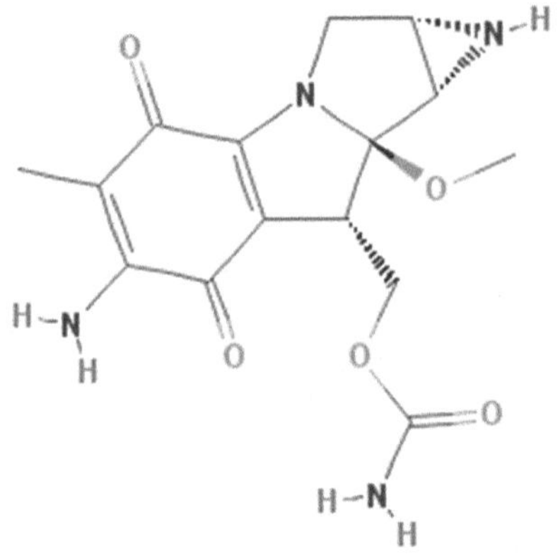

Mitomycin C

Chemical names: Mitopodozide (Proreside; Proresid; Proresipar; Podophyllic Acid Ethylhydrazide)
    PubChem CID: 251643

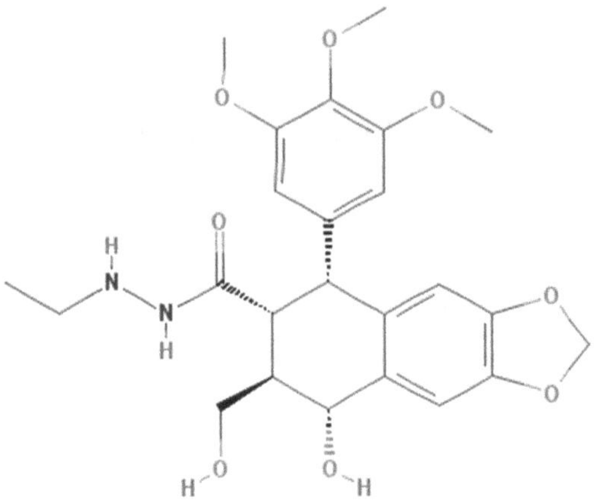

Mitopodozide

Chemical names: N-Acetyl-D-Galactosamine (2-Acetamido-2-Deoxy-D-Galactopyranose; GalNAc; 14215-68-0; D-GalNAc)
    PubChem CID: 35717

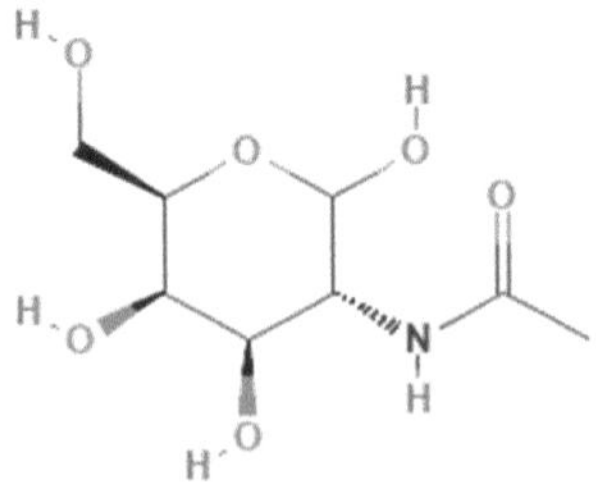

N-Acetyl-D-Galactosamine

Chemical names: N-Acetylgalactosamine (Alpha-N-Acetyl-D-galactosamine; Alpha-GalNAc; TN saccharide; Alpha-GalpNAc; GalNAc-alpha; 14215-68-0)
PubChem CID: 84265

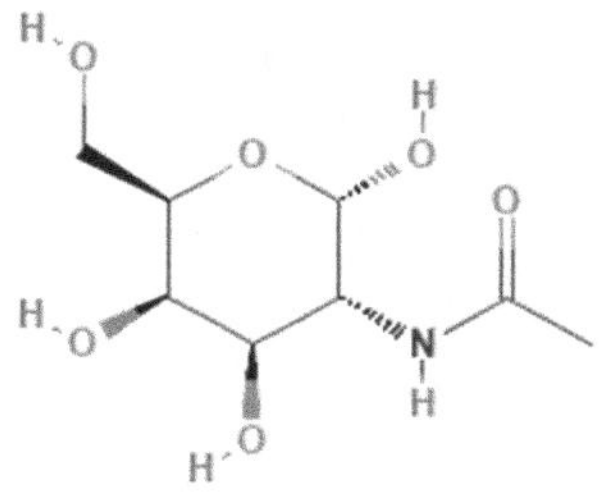

N-Acetylgalactosamine

Chemical names: Naringenin (5,7-Dihydroxy-2-(4-hydroxyphenyl)chroman-4-one; 67604-48-2; 4',5,7-Trihydroxyflavanone; Naringenine)
PubChem CID: 932

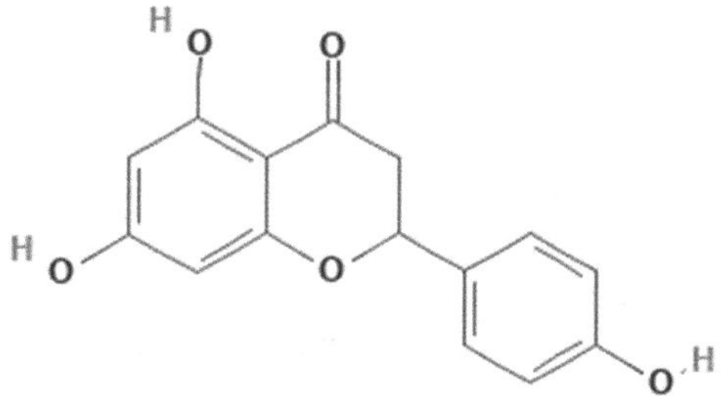

Naringenin

Chemical names: Neurolenin B (Neurolenin B; CHEBI:66621; (3aS,4S,5R,6S,8Z,10R,11aR)-5-(acetyloxy)-6-hydroxy-6,10-dimethyl-3-methylidene-2,7-dioxo-2,3,3a,4,5,6,7,10,11,11a-decahydrocyclodeca[b]furan-4-yl 3-methylbutanoate)
PubChem CID: 49799795

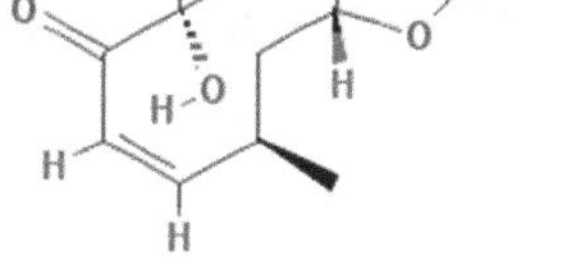

Neurolenin B

Chemical names: Nickel chloride (Nickel(II) chloride; 7718-54-9; Nickel dichloride; Nickelous chloride)
PubChem CID: 24385

Nickel Chloride

Chemical names: N-Methyl-N′-nitro-N-nitrosoguanidine (1-Methyl-3-nitro-1-nitrosoguanidine; 70-25-7; Methylnitronitrosoguanidine)
   PubChem CID: 9562060

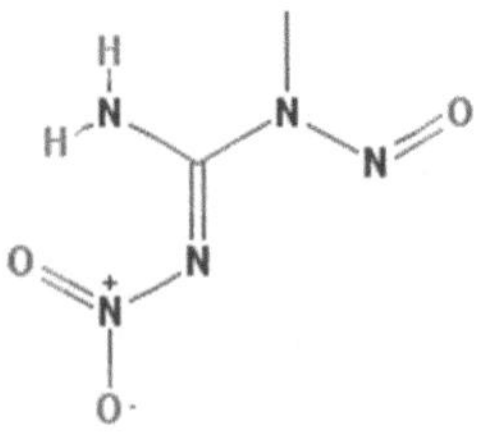

N-Methyl-N′-nitro-N-nitrosoguanidine

Chemical names: N-Nitrosopyrrolidine (1-Nitrosopyrrolidine; 930-55-2; Pyrrolidine, 1-nitroso-; Nitrosopyrrolidine)
   PubChem CID: 13591

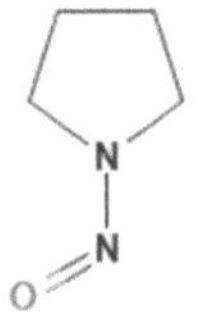

N-Nitrosopyrrolidine

Chemical names: Norepinephrine (L-Noradrenaline; Levarterenol; Arterenol; Noradrenaline)
   PubChem CID: 439260

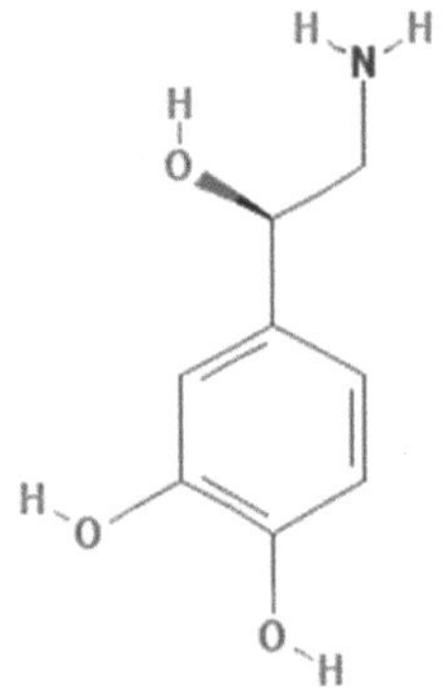

Norepinephrine

Chemical names: Oleic acid (112-80-1; Cis-9-Octadecenoic acid; Cis-Oleic acid; Elaidoic acid)
   PubChem CID: 445639

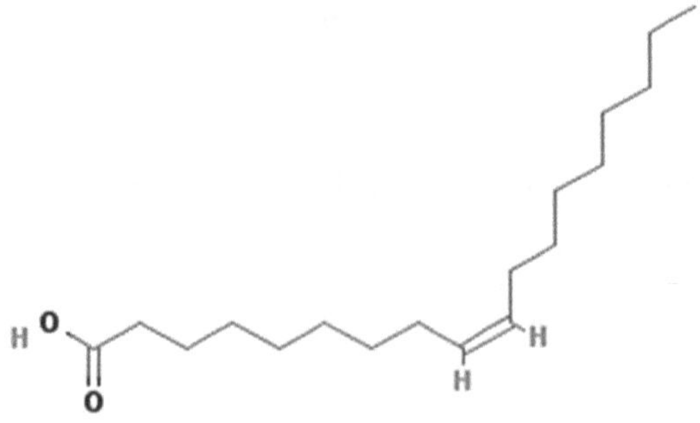

Oleic acid

Chemical names: Paclitaxel (TAXOL; 33069-62-4; Taxol A; Abraxane)
  PubChem CID: 36314

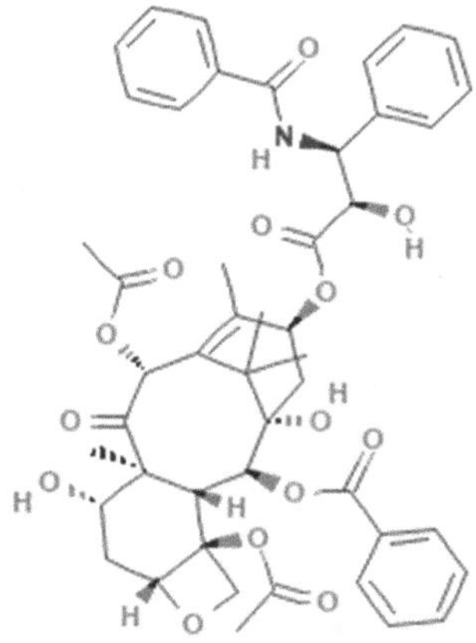

Paclitaxel

Chemical names: Parthenolide (20554-84-1; (–)-Parthenolide; CHEBI:7939; Parthenolide, Tanacetum parthenium)
  PubChem CID: 7251185

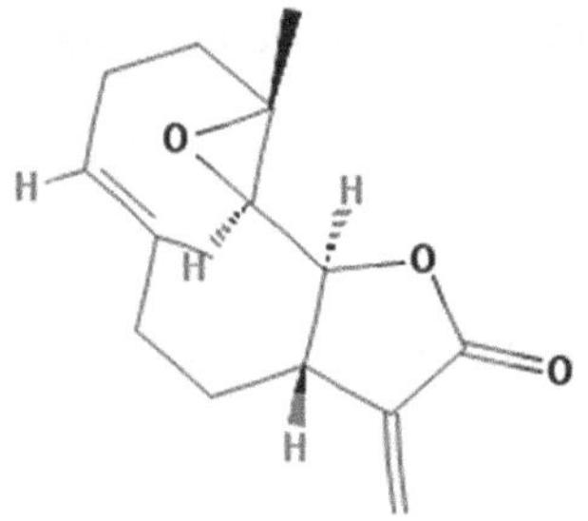

Parthenolide

Chemical names: Phloroglucinol (108-73-6; Benzene-1,3,5-triol; 1,3,5-trihydroxybenzene; 1,3,5-benzenetriol)
  PubChem CID: 359

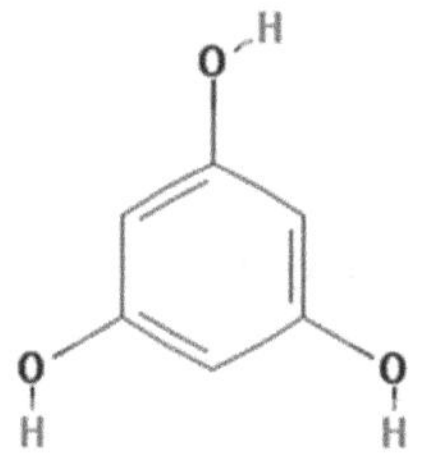

Phloroglucinol

Chemical names: Phyllanthoside (Phylanthoside; CCRIS 670; NSC 328426; NSC-328426)
PubChem CID: 6436218

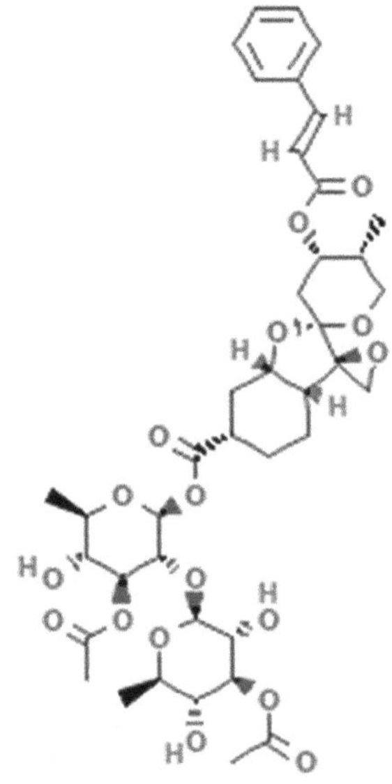

Phyllanthoside

Chemical names: Picrolonic Acid (550-74-3; 3-Methyl-4-nitro-1-(4-nitrophenyl)-1H-pyrazol-5(4H)-one; UNII-AW40R29U5B; NSC 5049)
PubChem CID: 68369

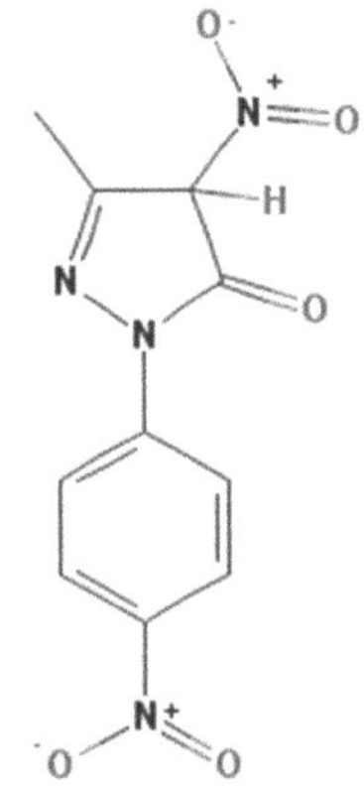

Picrolonic Acid

Chemical names: Piperidine (10-89-4; Hexahydropyridine; Cyclopentimine; Azacyclohexane)
PubChem CID: 8082

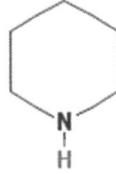

Piperidine

Chemical names: Plumbagin (481-42-5; Plumbagine; 5-Hydroxy-2-methyl-1,4-naphthoquinone; Plumbaein)
PubChem CID: 10205

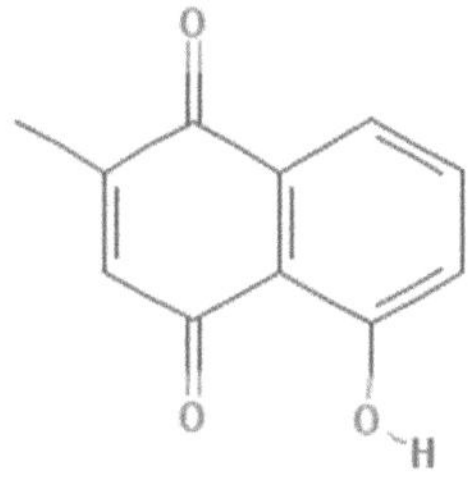

Plumbagin

Chemical names: Plumericin (77-16-7; UNII-L0126U506Z; CHEBI:8274; PLUMERICINE)
PubChem CID: 5281545

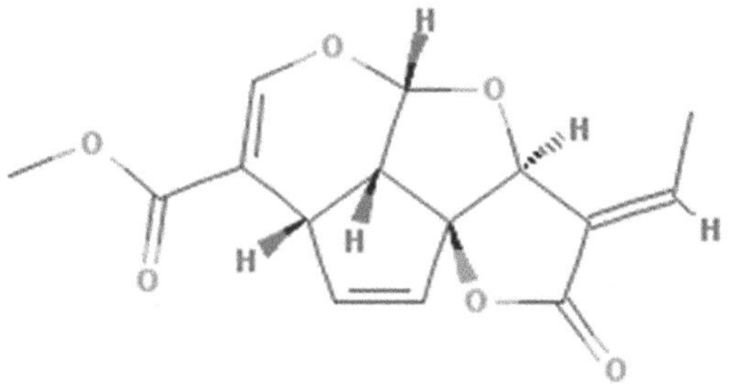

Plumericin

Chemical names: Podophyllotoxin (Podofilox; 518-28-5; Condylox; Condylinej)
PubChem CID: 10607

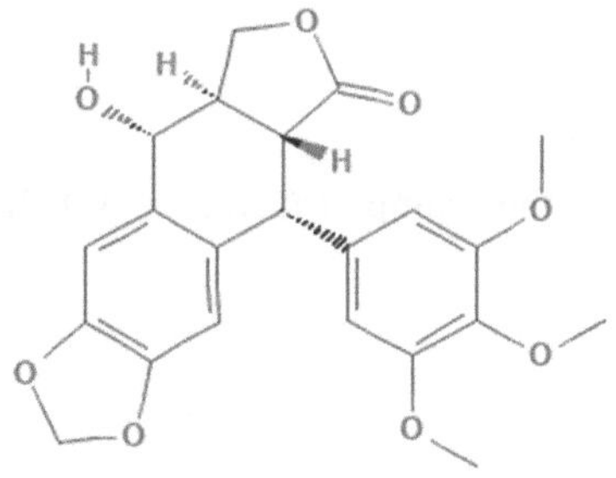

Podophyllotoxin

Chemical names: Psoralen (66-97-7; 7H-Furo[3,2-g]chromen-7-one; Ficusin; Furocoumarin)
PubChem CID: 6199

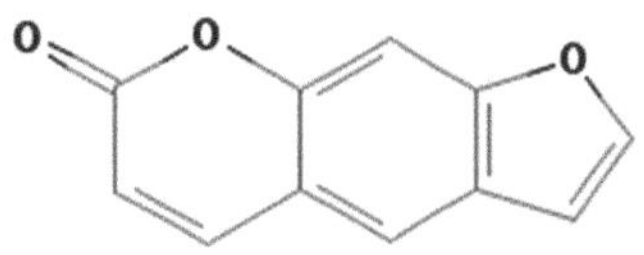

Psoralen

Chemical names: Quercetin (117-39-5; Meletin; Sophoretin; Quercetine)
    PubChem CID: 5280343

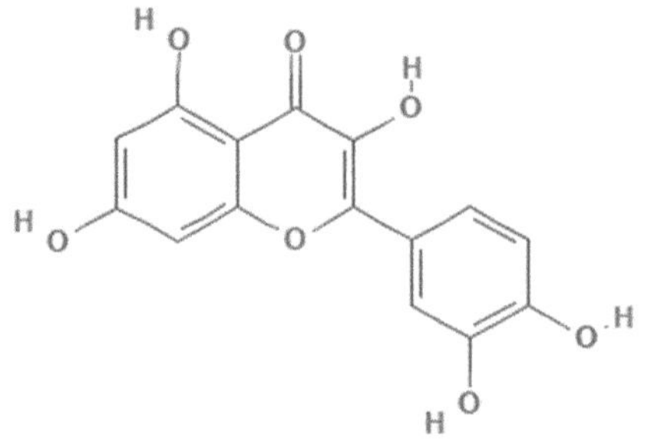

Quercetin

Chemical names: Resveratrol; 501-36-0; Trans-Resveratrol; 3,4′,5-Trihydroxystilbene; 3,5,4′-Trihydroxystilbene; (E)-Resveratrol; 3,4′,5-Stilbenetriol; (E)-5-(4-Hydroxystyryl)Benzene-1,3-Diol.
    PubChem CID: 445154

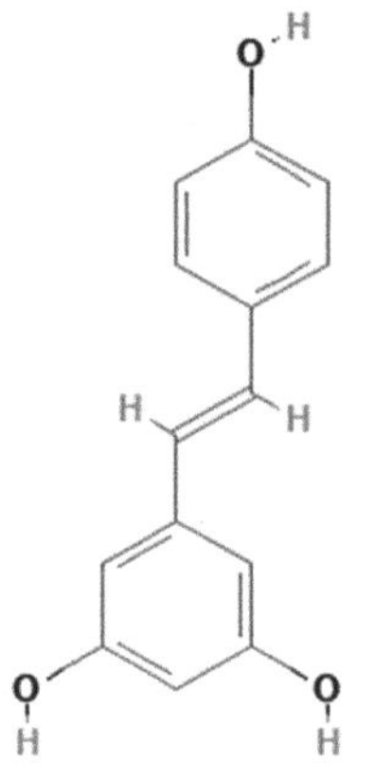

Resveratrol

Chemical names: Rubitecan; 9-Nitrocamptothecin; 91421-42-0; Orathecin; 9-Nitro-20(S)-Camptothecin; Camptogen
    PubChem CID: 472335

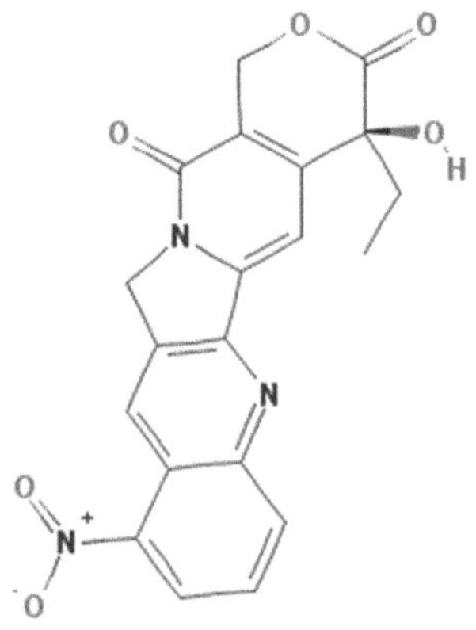

Chemical names: SELENIUM DIOXIDE; Selenium Oxide; Selenium (IV) Oxide
  PubChem CID: 24007

Selenium Dioxide

Chemical names: Silibinin; Silybin; 22888-70-6; SILYMARIN; Flavobin; Silymarin I; Silybin A;
Silybine
  PubChem CID: 31553

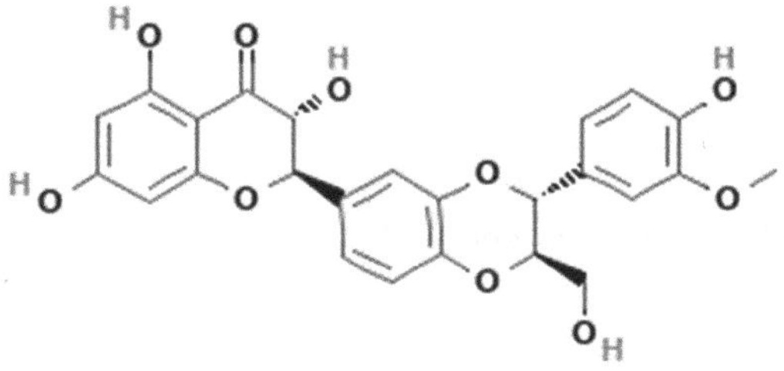

Silibinin

Chemical names: Sophocarpine (6483-15-4; 13,14-Didehydromatridin-15-one; 145572-44-7;
(–)-Sophocarpine)
  PubChem CID: 115269

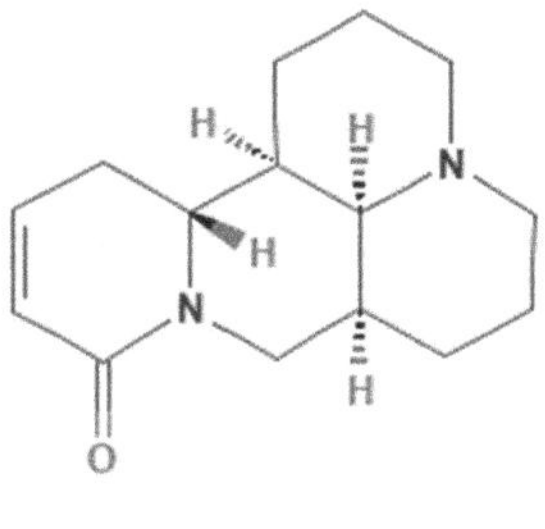

Sophocarpine

Chemical names: Tabun (Taboon A; Gelan I; Trilon 83; Ethyl dimethylphosphoramidocyanidate)
  PubChem CID: 6500

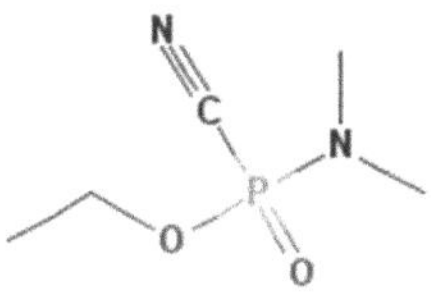

Tabun

Chemical names: Taraxeryl acetate (Taraxerol acetate; 2189-80-2; Taraxerylacetate; AC1L3RRW)
    PubChem CID: 94225

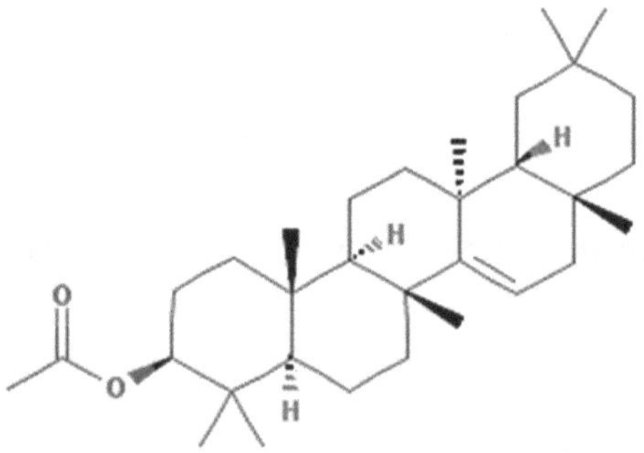

Taraxeryl acetate

Chemical names: Paclitaxel (TAXOL; 33069-62-4; Taxol A; Abraxane; Paxene; Plaxicel; Paxceed)
    PubChem CID: 36314

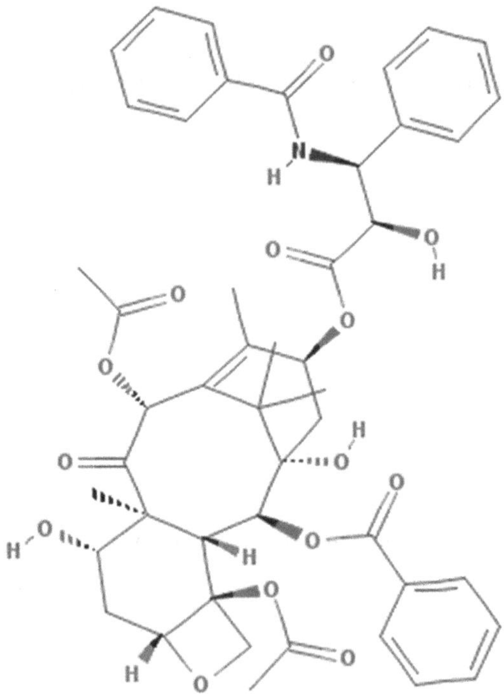

Taxol/Paclitaxel

Chemical names: Taxotere (Docetaxel; Docetaxel anhydrous; Docetaxol)
     PubChem CID: 148124

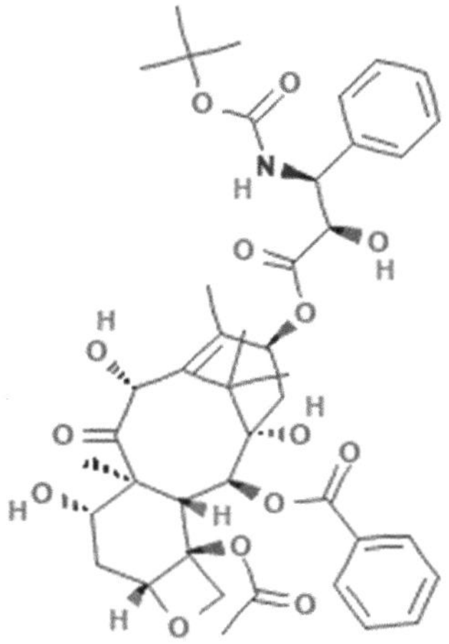

Taxotere

Chemical names: Teniposide (29767-20-2; Vumon; Vehem; VM-26; Teniposido; Teniposidum; Teniposidum)
     PubChem CID: 452548

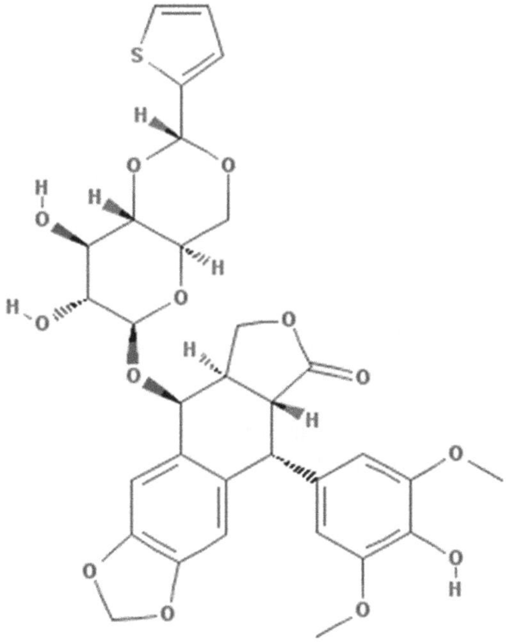

Tenoposide

Chemical names: Tetrahydrofuran (Oxolane; 109-99-9; Furan, tetrahydro-; Furanidine)
     PubChem CID: 8028

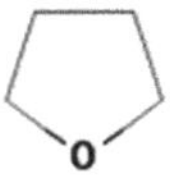

Tetrahydrofuran

Chemical names: Thymol (89-83-8; 2-Isopropyl-5-methylphenol; 5-Methyl-2-isopropylphenol; Thyme camphor)
   PubChem CID: 6989

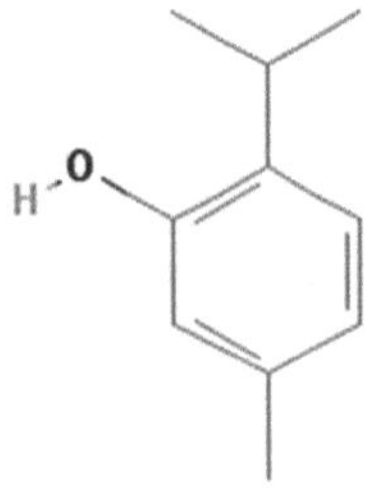

Thymol

Topotecan; 123948-87-8; Hycamtin; Topotecan lactone; Hycamptamine; Topotecane
   PubChem CID: 60700

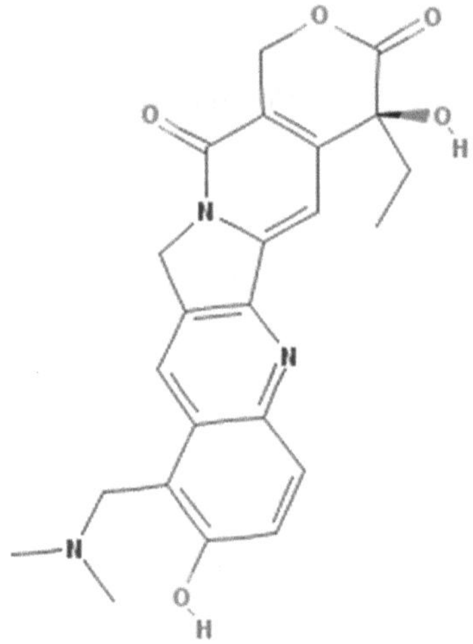

Topotecan

Chemical names: Tricine (5704-04-1; N-Tris(hydroxymethyl)methylglycine; N-(Tri(hydroxymethyl) methyl)glycine; N-[Tris(hydroxymethyl)methyl]glycine)
   PubChem CID: 79784

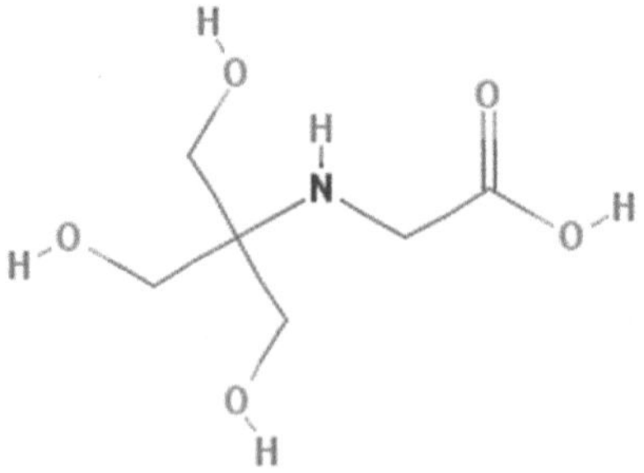

Tricine

Chemical names: Tubulosine (Marckine; Tubulosan-8'-ol, 10,11-dimethoxy-; 2632-29-3; 10,11-Dimethoxytubulosan-8'-ol)
PubChem CID: 72341

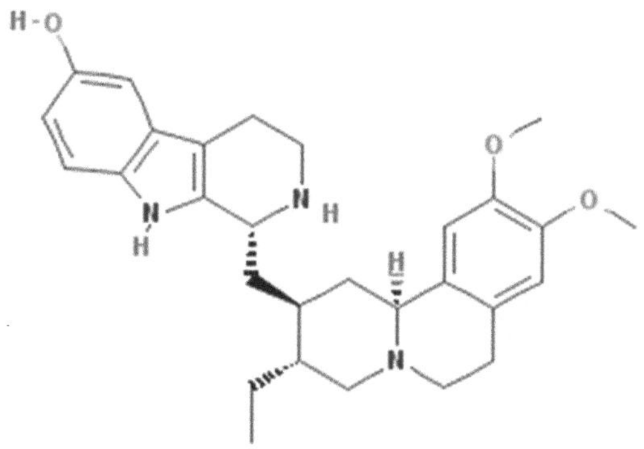

Tubulosine

Chemical names: Urethane (ETHYL CARBAMATE; 51-79-6; Urethan; Ethylurethane)
PubChem CID: 5641

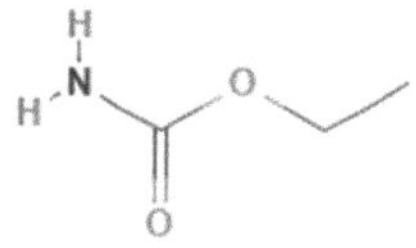

Urethane

Chemical names: Ursolic acid (Malol; Prunol; Urson; 77-52-1)
PubChem CID: 64945

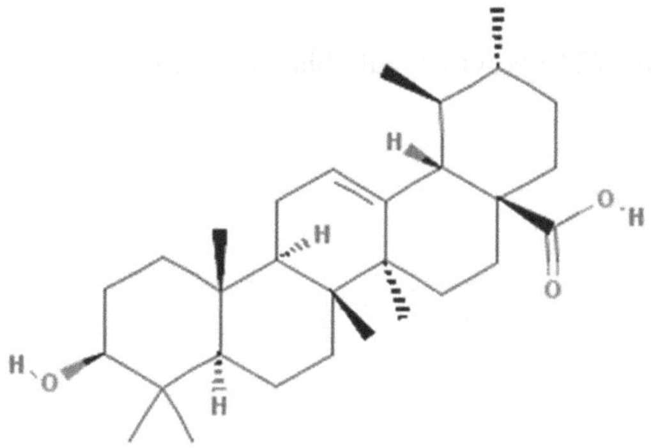

Ursolic acid

Chemical names: Valtrate (18296-44-1; Valepotriate; Valtratum; UNII-L3JQ035X9B)
    PubChem CID: 442436

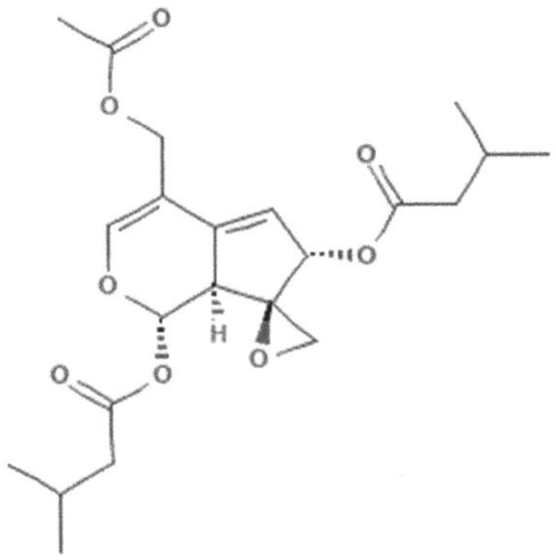

Valtrate

Chemical names: Vinblastine (Vinblastin; Vincaleucoblastin; Vincaleukoblastine; Vinblastinum)
    PubChem CID: 13342

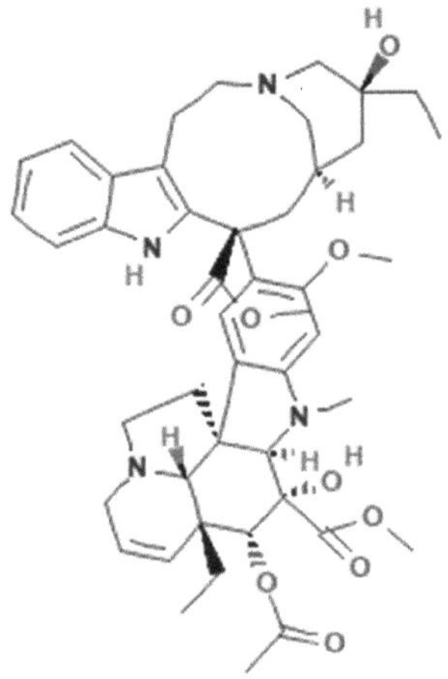

Vinblastine

Chemical names: Vincristine (22-Oxovincaleukoblastine; Leurocristine; 57-22-7; Vincrystine)
    PubChem CID: 5978

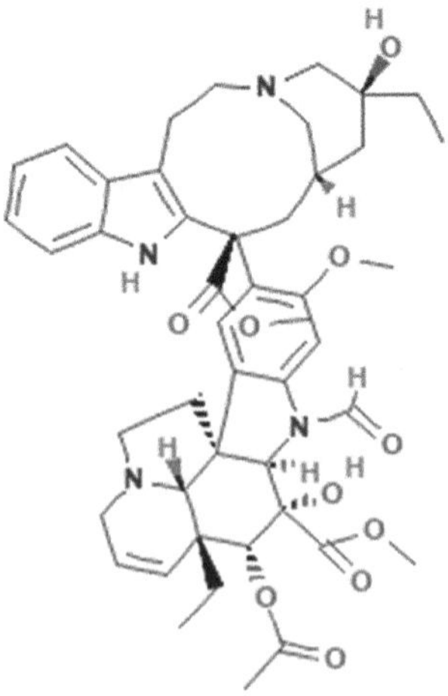

Vincristine

Chemical names: Vinflunine (162652-95-1; Javlor (TN); C45H54F2N4O8; Javlor; L-0070; D0OT9S; SCHEMBL8403)

PubChem CID: 6918295

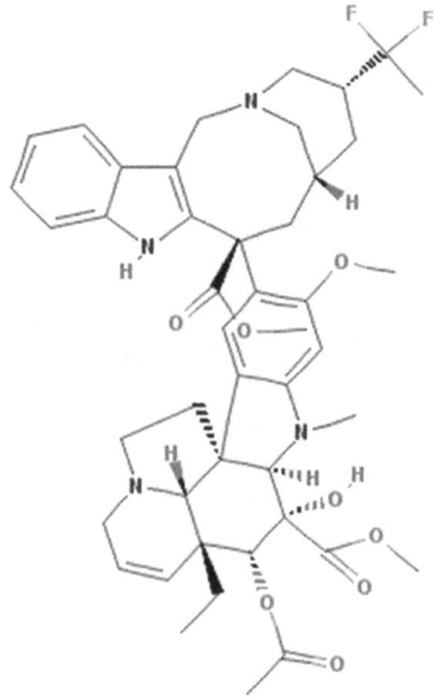

Vinflunine

Chemical names: Vinorelbine (Navelbine; Vinorelbinum; Vinorelbina; Exelbine; Navelbine Base; 71486-22-1; Nor-5′-Anhydrovinblastine)

PubChem CID: 5311497

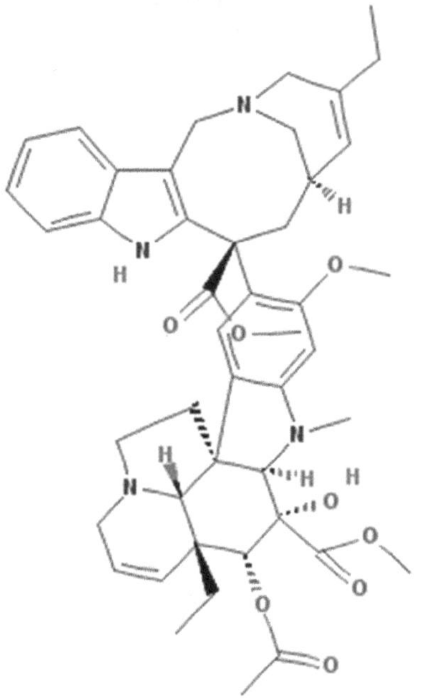

Vinorelbine

Chemical names: Viscotoxin A3 (VT A3)
   PubChem CID: 135304435

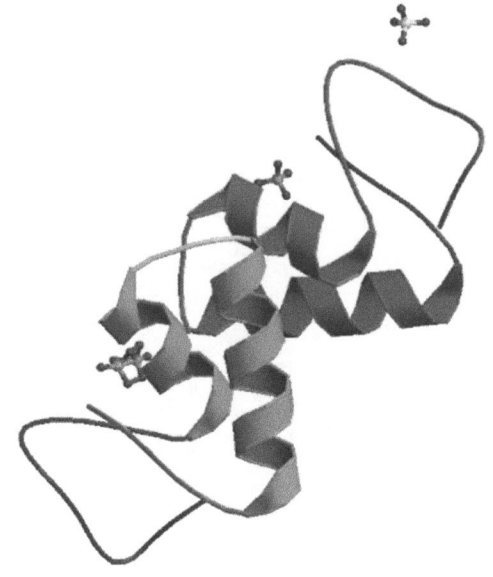

Viscotoxin A3

   A NOTE FOR CHEMICAL COMPOUNDS:
   All Chemical Compounds have the following reference:

REFERENCE

https://pubchem.ncbi.nlm.nih.gov

Index